The Main Functional Groups (*continued*)

Structure	Class of Compound	Specific Example	Name	Use
C. Containing nitrogen				
$—NH_2$	primary amine	$CH_3CH_2NH_2$	ethylamine	intermediate for dyes, medicinals
—NHR	secondary amine	$(CH_3CH_2)_2NH$	diethylamine	pharmaceuticals
$—NR_2$	tertiary amine	$(CH_3)_3N$	trimethylamine	insect attractant
$—C{\equiv}N$	nitrile	$CH_2{=}CH—C{\equiv}N$	acrylonitrile	orlon manufacture
D. Containing oxygen and nitrogen				
$—\overset{+}{N}(=O)O^-$	nitro compounds	CH_3NO_2	nitromethane	rocket fuel
$—C(=O)—NH_2$	primary amide	$HC(=O)NH_2$	formamide	softener for paper
E. Containing halogen				
—X	alkyl or aryl halide	CH_3Cl	methyl chloride	refrigerant, local anesthetic
$—C(=O)—X$	acid (acyl) halide	$CH_3C(=O)Cl$	acetyl chloride	acetylating agent
F. Containing sulfur				
—SH	thiol	CH_3CH_2SH	ethanethiol	odorant to detect gas leaks
—S—	thioether	$(CH_2{=}CHCH_2)_2S$	allyl sulfide	odor of garlic
$—S(=O)_2—OH$	sulfonic acid	$CH_3—C_6H_4—SO_3H$	*para*-toluenesulfonic acid	strong organic acid

Get a Better Grade in Chemistry!

Log in now to the #1 online homework and tutorial system for chemistry. Score better on exams, get homework help, and more!

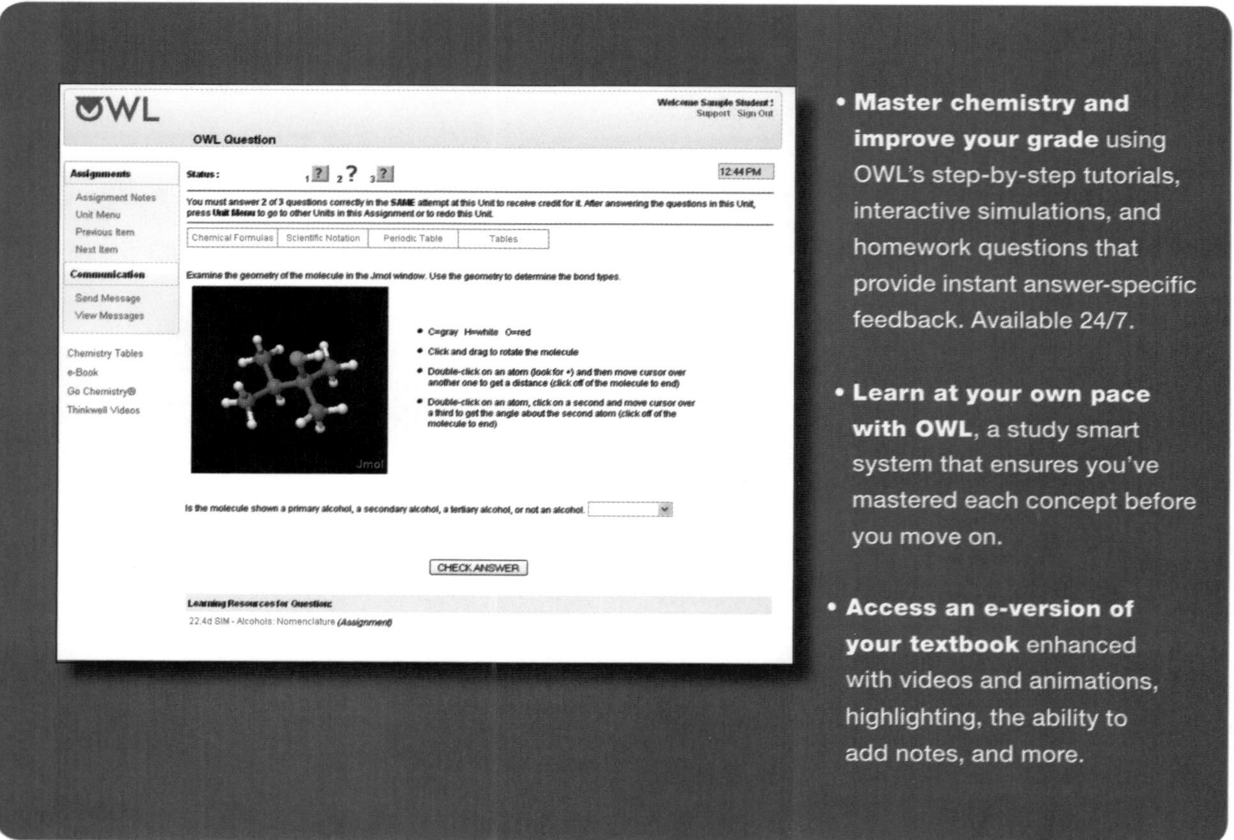

To get started, use the access code that may have been packaged with your text or purchase access online. Check with your instructor to verify that OWL is required for your course before purchasing.

www.cengage.com/OWL

Organic Chemistry

A SHORT COURSE

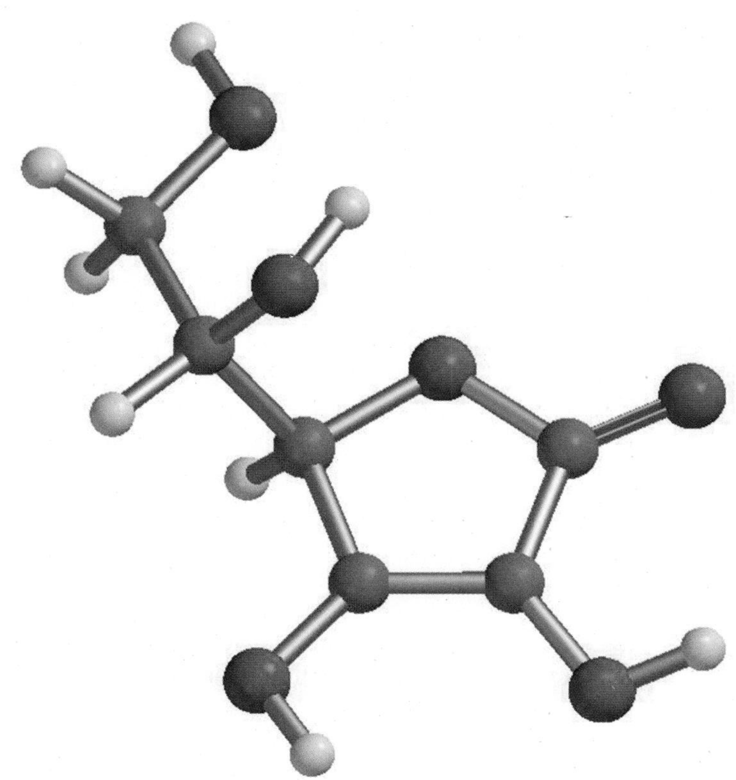

13th EDITION

Organic Chemistry

A SHORT COURSE

David J. Hart
The Ohio State University

Christopher M. Hadad
The Ohio State University

Leslie E. Craine
Central Connecticut State University

Harold Hart
Michigan State University

BROOKS/COLE
CENGAGE Learning™

Australia • Brazil • Japan • Korea • Mexico • Singapore • Spain • United Kingdom • United States

Organic Chemistry: A Short Course, 13th Edition
David J. Hart, Christopher M. Hadad, Leslie E. Craine, and Harold Hart

Acquiring Sponsoring Editor: Christopher D. Simpson

Developmental Editor: Naomi Friedman

Assistant Editor: Jon Olafsson

Senior Media Editor: Lisa Weber

Media Editor: Stephanie Van Camp

Marketing Manager: Barb Bartoszek

Marketing Assistant: Kevin Carroll

Marketing Communications Manager: Linda Yip

Content Project Manager: Teresa L. Trego

Art Director: John Walker

Print Buyer: Linda Hsu

Rights Acquisitions Specialist: Don Schlotman

Production Service: PreMediaGlobal

Text Designer: tani hasegawa

Photo Researcher: Bill Smith Group

Copy Editing: PreMediaGlobal

Illustration: PreMediaGlobal

OWL producers: Stephen Battisti, Cindy Stein, David Hart (Center for Educational Software Development, University of Massachusetts, Amherst)

Cover Designer: Lee Friedman

Cover Image: © Spencer Jones/FoodPix

Compositor: PreMediaGlobal

Library of Congress Control Number: 2010939442

Student Edition:

ISBN-13: 978-1-111-42556-2

ISBN-10: 1-111-42556-6

Brooks/Cole
20 Davis Drive
Belmont, CA 94002-3098
USA

Printed in the United States of America
1 2 3 4 5 6 7 14 13 12 11 10

Brief Contents

Contents

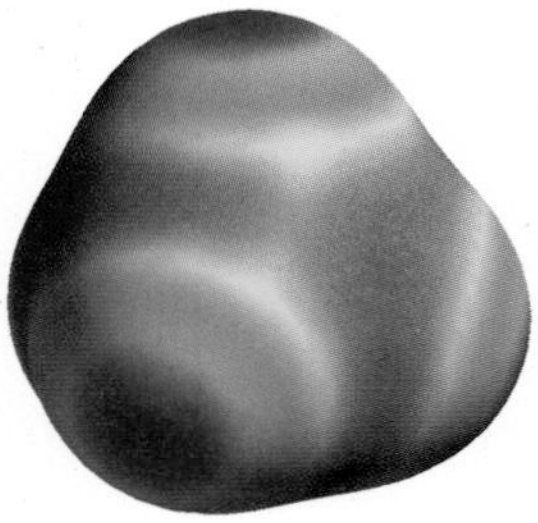

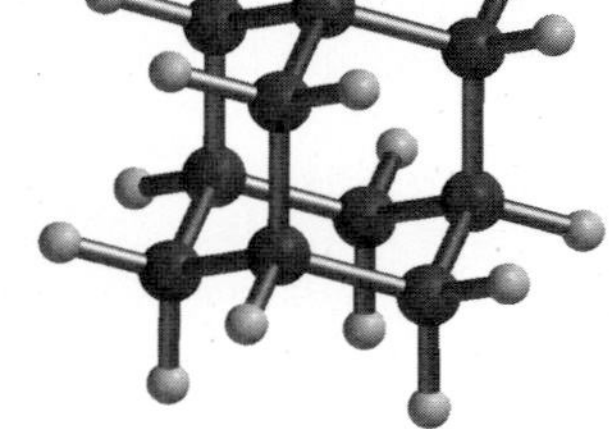

5 Stereoisomerism 147

6 Organic Halogen Compounds; Substitution and Elimination Reactions 182

7 Alcohols, Phenols, and Thiols 206

8 Ethers and Epoxides 234

9 Aldehydes and Ketones 253

10 Carboxylic Acids and Their Derivatives 287

13 Heterocyclic Compounds 390

14 Synthetic Polymers 411

15 Lipids and Detergents 438

Preface

Purpose

Over fifty years have passed since the first edition of this text was published. Although the content and appearance of the book have changed over time, our purpose in writing *Organic Chemistry: A Short Course* remains constant: to present a brief introduction to modern organic chemistry in a clear and engaging manner.

This book was written for students who, for the most part, will not major in chemistry, but whose main interest requires some knowledge of organic chemistry, such as agriculture, biology, human or veterinary medicine, pharmacy, nursing, medical technology, health sciences, engineering, nutrition, and forestry. To encourage these students to enjoy the subject as we do, we have made a special effort to relate the practical applications of organic chemistry to biological processes and everyday life. The success of this approach is demonstrated by the widespread use of this textbook by hundreds of thousands of students in the United States and around the world, via its numerous translations.

Organic Chemistry: A Short Course is designed for a one-semester introductory course, but it can be readily adapted to other course types. Often, it is used in a one- or two-quarter course. In some countries (France and Japan, for example), it serves as an introductory text for chemistry majors, followed by a longer and more detailed full-year text. It has even been used in the United States for a one-year science majors course (with suitable supplementation by the instructor). In many high schools, it is used as the text for a second-year course, following the usual introductory general chemistry course.

New to the 13th Edition

The text was critically revised to clarify difficult content and to improve the presentation. In addition to many small changes, major changes to this edition have focused on improving graphics throughout the text in a pedagogically useful manner. For example, (1) some new ball-and-stick structures have been added to help students visualize molecules in three dimensions; (2) many additional problems have been written, and many of these problems require students to develop their three-dimensional visualization skills; (3) in some locations, new graphics and some electrostatic potential maps have been added in order to help in discussions of acid–base chemistry; and (4) several energy diagrams are used to illustrate the structural changes that occur as reactions proceed from reactants to products.

Other changes include increased use of the arrow-pushing formalism to facilitate teaching and understanding of reaction mechanisms. Users of the last edition who enjoyed the "A Word About . . ." essays will enjoy the addition of five new activities, many of which cover topics related to green chemistry. Each essay references one or more related problem that will allow students to apply the knowledge they have gained from these features.

We are very conscious of the need to keep the book to a manageable size for the one-semester course. Outdated information has been deleted and, in some cases, replaced with new material. In the end, users will find this edition practically identical in length to the previous one.

Organization

The organization is fairly classical, with some exceptions. After an introductory chapter on bonding, isomerism, and an overview of the subject (Chapter 1), the next three chapters treat saturated, unsaturated, and aromatic hydrocarbons in sequence. The concept of reaction mechanism is presented early, and examples are included in virtually all subsequent chapters. Stereoisomerism is also introduced early, briefly in Chapters 2 and 3, and then given separate attention in a full chapter (Chapter 5). Halogenated compounds are used in Chapter 6 as a vehicle for introducing aliphatic substitution and elimination mechanisms and dynamic stereochemistry.

Chapters 7 through 10 cover oxygen functionality in order of the increasing oxidation state of carbon—alcohols and phenols, ethers and epoxides, aldehydes and ketones, and acids and their derivatives. Brief mention of sulfur analogs is made in these chapters. Chapter 11 deals with amines. Chapters 2 through 11 treat every main functional group and constitute the heart of the course. Chapter 12 then takes up spectroscopy, with an emphasis on nuclear magnetic resonance (NMR) and applications to structure determination. This chapter handles the student's question: How do you know that those molecules really have the structures you say they have?

Next come two chapters on topics not always treated in introductory texts but that are especially important in practical organic chemistry—Chapter 13 on heterocyclic compounds and Chapter 14 on polymers. The book ends with four chapters on biologically important substances—lipids; carbohydrates; amino acids, peptides, and proteins; and nucleic acids.

"A Word About . . ." Essays

Although relevant applications of organic chemistry are stressed throughout the text, short sections under the general rubric "A Word About . . ." emphasize applications to other branches of science and to human life. These sections, which have been a popular feature, appear at appropriate places within the text rather than as isolated essays. They stand out from the text so that instructors can easily refer to these sections, as desired. There are forty-one of these essays on topics ranging from polycyclic aromatic hydrocarbons and cancer to ether and anesthesia, water treatment and the chemistry of enols and enolates, green chemistry and ibuprofen, sweetness and sweeteners, and DNA and crime.

"A Closer Look At . . ." Online Activities

Many resources on topics relevant to organic chemistry have become readily available over the Internet. Each "A Closer Look At . . ." section features a guided tour of selected topics in organic chemistry through directed activities based on selected Web sites. Instructors may assign these activities as desired, using them as a basis for class discussion or as a springboard for projects. Eleven of these activities appear in this edition, covering topics such as mass spectrometry and carbon dating (Chapter 12), Nobel laureates and protein chemistry (Chapter 17), and the polymerase chain reaction (Chapter 18).

Examples and Problems

Problem solving is essential to learning organic chemistry. Examples (worked-out problems) appear at appropriate places within each chapter to help students develop these skills. These examples and their solutions are clearly marked. Unsolved problems that provide immediate learning reinforcement are included in each chapter and are supplemented with an abundance of end-of-chapter problems. The combined number of examples and problems is over 1,000—an average of almost 60 per chapter.

OWL

OWL for Organic Chemistry

Instant Access OWL with eBook for text (6 months) ISBN-10: 1-111-47131-2, ISBN-13: 978-1-111-47131-6

By Steve Hixson and Peter Lillya of the University of Massachusetts, Amherst, and William Vining of the State University of New York at Oneonta. End-of chapter questions by David W. Brown, Florida Gulf Coast University. **OWL** Online Web Learning offers more assignable, gradable content (including end-of chapter questions specific to this textbook) and more reliability and flexibility than any other system. OWL's powerful course management tools allow instructors to control due dates, number of attempts, and whether students see answers or receive feedback on how to solve problems. OWL includes the **YouBook**, a Flash-based eBook that is interactive and customizable. It features a text edit tool that allows instructors to modify the textbook narrative as needed. With YouBook, instructors can quickly re-order entire sections and chapters or hide any content they don't teach to create an eBook that perfectly matches their syllabus. Instructors can further customize the YouBook by publishing web links. The YouBook also includes animated figures, video clips, highlighting, notes, and more.

Developed by chemistry instructors for teaching chemistry, OWL is the only system specifically designed to support **mastery learning**, where students work as long as they need to master each chemical concept and skill. OWL has already helped hundreds of thousands of students master chemistry through a wide range of assignment types, including tutorials, interactive simulations, and algorithmically generated homework questions that provide instant, answer-specific feedback.

OWL is continually enhanced with online learning tools to address the various learning styles of today's students such as:

- **Quick Prep** review courses that help students learn essential skills to succeed in General and Organic Chemistry
- **Jmol** molecular visualization program for rotating molecules and measuring bond distances and angles

In addition, when you become an OWL user, you can expect service that goes far beyond the ordinary. For more information or to see a demo, please contact your Cengage Learning representative or visit us at **www.cengage.com/owl.**

Student Ancillaries

Study Guide and Solutions Manual Written by the authors of the main text, this guide contains chapter summaries and learning objectives, reaction summaries, mechanism summaries, answers to all text problems, and sample test questions. Download a sample chapter from the Student Companion Website, which is accessible from www.cengage.com ISBN-10: 1-111-42585-X, ISBN-13: 978-1-111-42585-2.

Laboratory Manual Written by Leslie Craine and T. K. Vinod, this manual contains thirty experiments that have been tested with thousands of students. Most of the preparative experiments contain procedures on both macroscale and microscale, thus adding considerable flexibility for the instructor and the opportunity for both types of laboratory experience for the student. Experiments involving molecular modeling now contain computer-modeling activities in addition to activities based on traditional modeling kits. The experiments, capable of being completed in a two- or three-hour lab period, are a good mix of techniques, preparations, tests, and applications. Hazardous chemicals on the OSHA list have been avoided, care has been taken to minimize contact with solvents, and updated caution notes and waste disposal instructions are included. ISBN-10: 1-111-42584-1, ISBN-13: 978-1-111-42584-8

Student Companion Website The Student Companion Website includes a glossary, flashcards, and an interactive periodic table, which are accessible from www.cengagebrain.com.

CengageBrain.com App Now, students can prepare for class anytime and anywhere using the CengageBrain.com application developed specifically for the Apple iPhone® and iPod touch®, which allows students to access free study materials—book-specific quizzes, flash cards, related Cengage Learning materials and more—so they can study the way they want, when they want to . . . even on the go. For more information about this complimentary application, please visit www.cengagebrain.com.

Visit CengageBrain.com To access these and additional course materials, please visit www.cengagebrain.com. At the CengageBrain.com home page, search for the ISBN (from the back cover of your book) using the search box at the top of the page. This will take you to the product page where these resources can be found. (Instructors can log in at login.cengage.com.)

Instructor Ancillaries

A complete suite of customizable teaching tools accompanies *Organic Chemistry: A Short Course*. Whether available in print, online, or on CD, these integrated resources are designed to save you time and help make class preparation, presentation, assessment, and course management more efficient and effective.

PowerLecture With ExamView® Instructor's CD/DVD Package
ISBN-10: 1-111-42587-6, ISBN-13: 978-1-111-42587-6
PowerLecture is a one-stop digital library and presentation tool that includes:

- Prepared **Microsoft® PowerPoint® Lecture Slides** that cover all key points from the text in a convenient format that you can enhance with your own materials or with the supplied interactive video and animations for personalized, media-enhanced lectures.
- Image libraries in PowerPoint and JPEG formats that contain **digital files for all text art, all photographs, and all numbered tables** in the text. These files can be used to create your own transparencies or PowerPoint lectures.
- ***Instructor's Resource Manual*** written by Christopher M. Hadad that offers a transition guide, tables suggesting the approximate number of lectures to devote to each

Apple, iPhone, iPod touch, and iTunes are trademarks of Apple Inc., registered in the U.S. and other countries.

chapter, summaries of the worked examples and problems, a chapter-by-chapter outline listing those sections that are most important, and answers to the review problems on synthesis that are featured in the Study Guide and Solutions Manual.

- ***Instructor's Resource Guide* for the *Laboratory Manual*** written by Christopher M. Hadad of The Ohio State University, that contains detailed discussions of experiments and answers to all of the prelab exercise questions and most of the questions in the report sheets contained in the Laboratory Manual.
- **Sample chapters from the *Study Guide and Solutions Manual.*** We provide sample chapters of this in Adobe Acrobat PDF format as a courtesy to instructors who may wish to recommend it to students.
- **ExamView Computerized Testing** that enables you to create customized tests of up to 250 items in print or online using more than 700 questions carefully matched to the corresponding text sections. Tests can be taken electronically or printed for class distribution.

Faculty Companion Website

Accessible from www.cengagebrain.com. this Web site provides downloadable files for the Instructor's Resource Guide for the Lab Manual, Instructor's Resource Manual, as well as WebCT and Blackboard versions of Exam*View*® Computerized Testing.

Cengage Learning Custom Solutions develops personalized solutions to meet your course needs. Match your learning materials to your syllabus and create the perfect learning solution—your customized text will contain the same thought-provoking, scientifically sound content, superior authorship, and stunning art that you've come to expect from Cengage Learning Brooks/Cole texts, yet in a more flexible format. Visit www.cengage.com/custom to start building your book today.

Signature Labs . . . for the customized laboratory. Signature Labs is Cengage Learning's digital library of tried-and-true labs that help you take the guesswork out of running your chemistry laboratory. Select just the experiments you want from hundreds of options and approaches. Provide your students with only the experiments they will conduct so that you will get the results you seek. Visit www.signaturelabs.com to begin building your manual today.

Acknowledgments

We would like to thank the following reviewers for diligently contributing their insights to this edition of *Organic Chemistry:*

Scott W. Cowley, *Colorado School of Mines*; Sarah A. Cummings, *University of Nevada, Reno*; J. Brent Friesen, *Dominican University*; Michael Harmata, *University of Missouri-Columbia*; Marjorie J. Hummel, *Governors State University*; and Barbara Oviedo Mejia, *California State University, Chico.*

We have incorporated many of their recommendations, and the book is much improved as a result.

One pleasure of authorship is receiving letters from students (and their teachers) who have benefited from the book. We thank all who have written to us, from all parts of the world, since the last edition; many of the suggestions have been incorporated into this revision. We are happy to hear from users and nonusers, faculty and students, who have suggestions for further improvement.

David J. Hart
Department of Chemistry, The Ohio State University

Christopher M. Hadad
Department of Chemistry, The Ohio State University

Leslie E. Craine
Department of Chemistry, Central Connecticut State University

Harold Hart
Emeritus Professor of Chemistry, Michigan State University

A WORD ABOUT... Oranges and the Cover of This Book

Recent editions of this book featured apples, limes, and onions. Continuing with the theme of edible fruits and organic chemistry, the cover of this edition features a photograph of an orange peel and the chemical structure of the ester, octyl acetate. Many esters (Chapter 10) are rather pleasant-smelling substances and are responsible for the flavor and fragrance of many fruits. Among the more common are octyl acetate (oranges), pentyl acetate (bananas), ethyl butanoate (pineapples), and pentyl butanoate (apricots). Natural flavors can be exceedingly complex. For example, no fewer than 53 esters have been identified among the volatile constituents of Bartlett pears! In the chemical industry, mixtures of esters are used as perfumes and artificial flavorings.

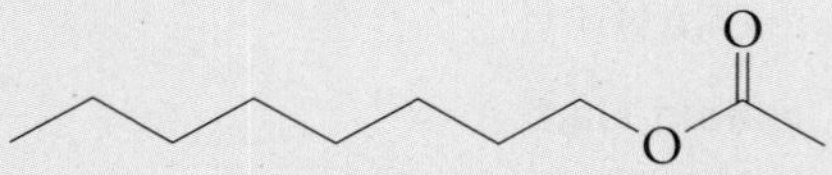

octyl acetate (oranges)

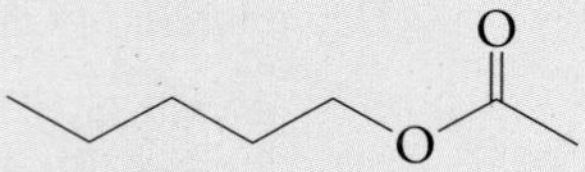

pentyl acetate (bananas)

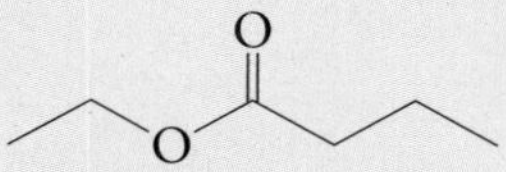

ethyl butanoate (pineapples)

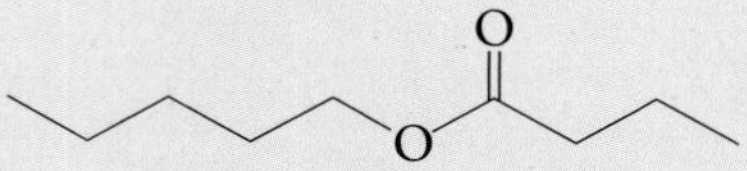

pentyl butanoate (apricots)

The therapeutic properties of oranges are legendary, due to the fact that oranges, like other citrus fruit, contain vitamin C, L-ascorbic acid. Humans, monkeys, guinea pigs, and a few other vertebrates lack an enzyme that is essential for the biosynthesis of L-ascorbic acid from D-glucose (Chapter 16). L-Ascorbic acid is needed for collagen synthesis (collagen is the structural protein of skin, connective tissue, tendon, cartilage, and bone; Chapter 17). Hence, ascorbic acid must be included in the diet of humans and these other species. The lack of L-ascorbic acid in one's diet leads to diseases that plagued sailors for centuries. These included scurvy, a disease that results in weak blood vessels, hemorrhaging, loosening of teeth, lack of ability to heal wounds, and eventual death.

L-Ascorbic acid resembles a monosaccharide (Chapter 16), but its structure has several unusual features. The compound has a five-membered unsaturated lactone ring (Chapter 10) with two hydroxyl groups (Chapter 7) attached to a carbon-carbon double bond (Chapter 3), an enediol structure.

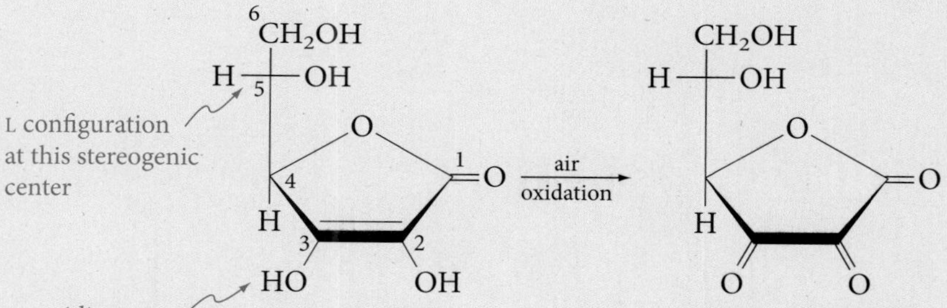

L configuration at this stereogenic center

acidic proton

air oxidation

L-ascorbic acid (vitamin C)
[mp 192°C (decomposes)]
pleasant, sharp-acid taste

dehydroascorbic acid

As a consequence of this structural feature, L-ascorbic acid is easily oxidized to dehydroascorbic acid. Both forms are biologically effective as a vitamin. There is no carboxyl group in L-ascorbic acid; nevertheless, vitamin C is an acid with a pK_a of 4.17.

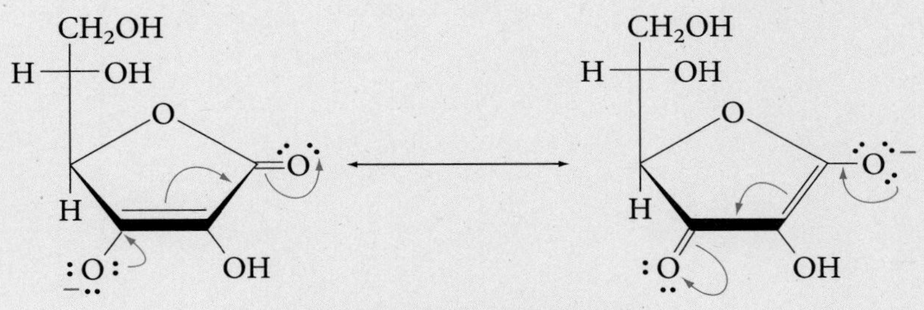

resonance stabilized ascorbate anion

An orange peel is a rich source of organic compounds, including lipids and flavonoids. In order to prevent water loss, the orange peel wax is composed of esters of unsaturated fatty acids (Chapter 15) and long-chain alcohols. The unsaturated fatty acids are dominated by linoleic, oleic, linolenic, arachidic and erucic acids (Chapter 15), and the long-chain alcohols (Chapter 7) contain typically thirty-two or thirty-four carbons. The large hydrocarbon units in the orange peel aid in the retention of water inside the fruit. Orange peel wax also contains other fatty acid conjugates, including glycolipids and phospholipids (Chapter 15). The orange peel is also a source of flavonoids, including hesperidin. Many of these flavonoid compounds have been studied for their biological properties, including potential effectiveness as anti-cancer, anti-inflammatory, and anti-allergy effects.

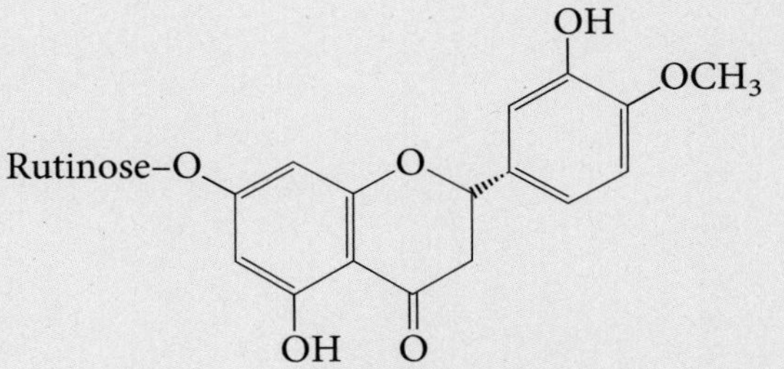

hesperidin

As you read this book, you will come across numerous "A Word About . . ." sections that relate the chemistry you are studying to our world. We encourage you to return to this "A Word About . . ." with greater understanding after you have completed your course in organic chemistry.

Natural and synthetic organic compounds are everywhere in the environment and in our material culture.

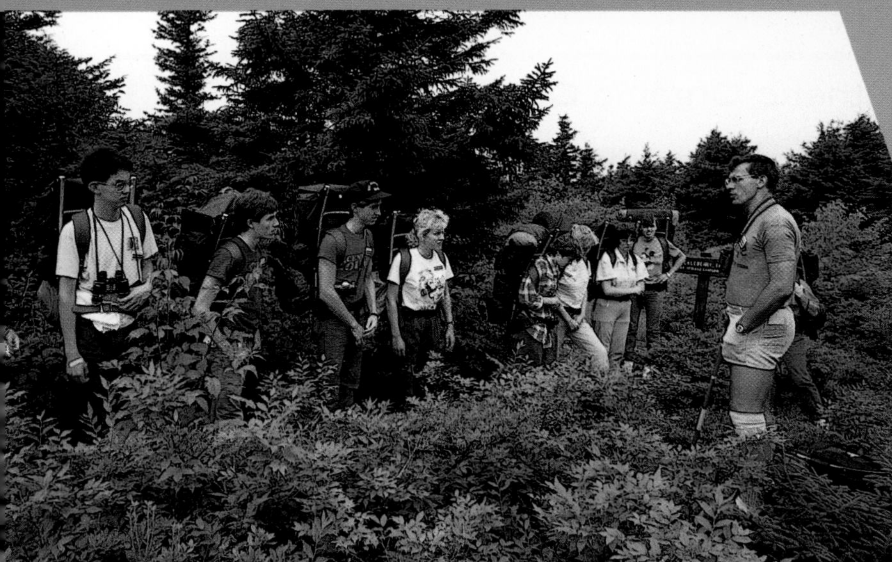

To the Student

In this introduction, we will briefly discuss organic chemistry and its importance in a technological society. We will also explain how this course is organized and give you a few hints that may help you to study more effectively.

What Is Organic Chemistry About?

The term *organic* suggests that this branch of chemistry has something to do with *organisms,* or living things. Originally, organic chemistry did deal only with substances obtained from living matter. Years ago, chemists spent much of their time extracting, purifying, and analyzing substances from animals and plants. They were motivated by a natural curiosity about living matter and also by the desire to obtain from nature ingredients for medicines, dyes, and other useful products.

It gradually became clear that most compounds in plants and animals differ in several respects from those that occur in nonliving matter, such as minerals. In particular, most compounds in living matter are made up of the same few elements: carbon, hydrogen, oxygen, nitrogen, and sometimes sulfur, phosphorus, and a few others. Carbon is virtually always present. This fact led to our present definition: **Organic chemistry** is the chemistry of carbon compounds. This definition broadens the scope of the subject to include not only compounds from nature but also synthetic compounds—compounds invented by organic chemists and prepared in their laboratories.

Organic chemistry is the chemistry of carbon compounds.

Synthetic Organic Compounds

Scientists used to believe that compounds occurring in living matter were different from other substances and that they contained some sort of intangible vital force that imbued them with life. This idea discouraged chemists from trying to make organic compounds in the laboratory. But in 1828, the German chemist Friedrich Wöhler, then 28 years old, accidentally prepared urea, a well-known constituent of urine, by heating the inorganic (or mineral) substance ammonium cyanate. He was quite excited about this result, and in a letter to his former teacher, the Swedish chemist Jöns Jacob Berzelius, he wrote, "I can make urea without the necessity of a kidney, or even of an animal, whether man or dog." This experiment and others like it gradually discredited the vital-force theory and opened the way for modern synthetic organic chemistry.

Synthesis consists of piecing together small simple molecules to make larger, more complex molecules.

Synthesis usually consists of piecing together small, relatively simple molecules to make larger, more complex ones. To make a molecule that contains many atoms from molecules that contain fewer atoms, one must know how to link atoms to each other—that is, how to make and break chemical bonds. Wöhler's preparation of urea was accidental, but synthesis is much more effective when it is carried out in a controlled and rational way so that when all the atoms are assembled, they will be connected to one another in the correct manner to give the desired product.

Chemical bonds are made or broken during chemical reactions. In this course, you will learn about quite a few reactions that can be used to make new bonds and that are therefore useful in the synthesis of pharmaceuticals and industrial chemicals.

Why Synthesis?

At present, the number of organic compounds that have been synthesized in research laboratories is far greater than the number isolated from nature. Why is it important to know how to synthesize molecules? There are several reasons. For one, it might be important to synthesize a natural product in the laboratory to make the substance more widely available at lower cost than it would be if the compound had to be extracted from its natural source. Some examples of compounds first isolated from nature but now produced synthetically for commercial use are vitamins, amino acids, dyes for clothing, fragrances, and the moth-repellent camphor. Although the term *synthetic* is sometimes frowned upon as implying something artificial or unnatural, these synthetic natural products are in fact identical to the same compounds extracted from nature.

Another reason for synthesis is to create new substances that may have new and useful properties. Synthetic fibers such as nylon and Orlon, for example, have properties that make them superior for some uses to natural fibers such as silk, cotton, and hemp. Most pharmaceutical drugs used in medicine are synthetic (including aspirin, ether, Novocain, and ibuprofen). The list of synthetic products that we take for granted is long indeed—plastics, detergents, insecticides, and oral contraceptives are just a few. All of these are compounds of carbon; all are organic compounds.

Finally, organic chemists sometimes synthesize new compounds to test chemical theories—and sometimes they synthesize compounds just for the fun of it. Certain geometric structures, for example, are aesthetically pleasing, and it can be a challenge to make a molecule in which the carbon atoms are arranged in some regular way. One example is the hydrocarbon cubane, C_8H_8. First synthesized in 1964, its molecules have eight carbons at the corners of a cube, each carbon with one hydrogen and three other carbons connected to it. Cubane is more than just aesthetically pleasing. The bond angles in cubane are distorted from normal because of its geometry. Studying the chemistry of cubane therefore gives chemists information about how the distortion of carbon–carbon and carbon–hydrogen bonds affects their chemical behavior.

Although initially of only theoretical interest, the special properties of cubane may eventually lead to its practical use in medicine and in explosives.

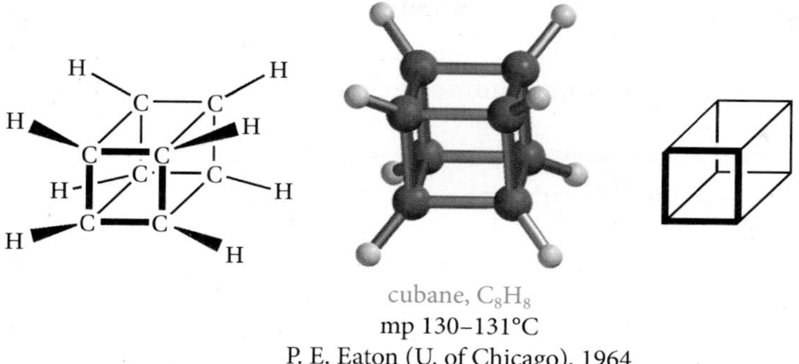

cubane, C_8H_8
mp 130–131°C
P. E. Eaton (U. of Chicago), 1964

Organic Chemistry in Everyday Life

Organic chemistry touches our daily lives. We are made of and surrounded by organic compounds. Almost all of the reactions in living matter involve organic compounds, and it is impossible to understand life, at least from the physical point of view, without knowing some organic chemistry. The major constituents of living matter—proteins, carbohydrates, lipids (fats), nucleic acids (DNA and RNA), cell membranes, enzymes, hormones—are organic, and later in the book, we will describe their chemical structures. These structures are quite complex. To understand them, we will first have to discuss simpler molecules.

Other organic substances include the gasoline, oil, and tires for our cars; the clothing we wear; the wood for our furniture; the paper for our books; the medicines we take; and plastic containers, camera film, perfume, carpeting, and fabrics. Name it, and the chances are good that it is organic. Daily, in the paper, on the Internet, or on television, we encounter references to polyethylene, epoxys, Styrofoam, nicotine, polyunsaturated fats, and cholesterol. All of these terms refer to organic substances; we will study them and many more like them in this book.

In short, organic chemistry is more than just a branch of science for the professional chemist or for the student preparing to become a physician, dentist, veterinarian, pharmacist, nurse, or agriculturist. It is part of our technological culture.

Organization

Organic chemistry is a vast subject. Some molecules and reactions are simple; others are quite complex. We will proceed from the simple to the complex by beginning with a chapter on bonding, with special emphasis on bonds to carbon. Next, there are three chapters on organic compounds containing only two elements: carbon and hydrogen (called hydrocarbons). The second of these chapters (Chapter 3) contains an introduction to organic reaction mechanisms and a discussion of reaction equilibria and rates. These are followed by a chapter that deals with the three-dimensionality of organic compounds. Next, we add other elements to the carbon and hydrogen framework, halogens in Chapter 6, oxygen and sulfur in Chapters 7 through 10, and nitrogen in Chapter 11. At that point, we will have completed an introduction to all the main classes of organic compounds.

Spectroscopy is a valuable tool for determining organic structures—that is, the details of how atoms and groups are arranged in organic molecules. We take up this topic in Chapter 12. Next comes a chapter on heterocyclic compounds, many of which are important in medicine and in natural products. It is followed by a chapter on polymers, which highlights one of the most important industrial uses of organic chemistry. The last four chapters deal with the organic chemistry of four major classes of biologically important molecules: the lipids, carbohydrates, proteins, and nucleic acids. Because the structures of these molecules of nature are rather complex, we leave them for last. But with the background knowledge of simpler molecules that you will have acquired by then, these compounds and their chemistry will be clearer and more understandable.

To help you organize and review new material, we have placed a *Reaction Summary* and a *Mechanism Summary* at the end of each chapter in which new reactions and new reaction mechanisms are introduced.

"A Word About . . ."

In each chapter after the first, you will find special sections under the general heading "A Word About . . ." These are short, self-contained articles that expand on the main subject of the chapter. They may deal with intellectual curiosities (the first one, on impossible organic structures); industrial applications (petroleum, gasoline, and octane number in Chapter 3 and industrial alcohols in Chapter 7); organic chemistry in biology or medicine (polycyclic aromatic hydrocarbons and cancer in Chapter 4 and morphine and other nitrogen-containing drugs in Chapter 13); or just fun topics (sweetness and sweeteners in Chapter 16). They provide a convenient break at various points in each chapter, and we hope that you will enjoy them.

"A Closer Look At . . ."

Throughout the text, there are several sections titled "A Closer Look At . . ." These sections feature various online activities on selected topics of organic chemistry. *A Closer Look At* activities provide a wonderful basis for classroom discussion, as well as offer a springboard for potential projects.

The Importance of Problem Solving

One key to success in studying organic chemistry is problem solving. Each chapter in this book contains numerous facts that must be digested. Also, the subject matter builds continuously so that to understand each new topic, it is essential to have the preceding information clear in your mind and available for recall. To learn all these materials, careful study of the text is necessary, but it is *not sufficient*. Practical knowledge of how to use the facts is required, and such skill can be obtained only through the solving of problems and mastery of the concepts.

This book contains several types of problems. Some, called *Examples*, contain a *Solution*, so you can see how to work such problems. Throughout a chapter, examples are usually followed by similar *Problems*, designed to reinforce your learning immediately by allowing you to be sure that you understand the new material just presented. These *Problems* will be of most value if you work them when you come across them as you read the book. At the end of each chapter, *Additional Problems* enable you to practice your problem-solving skills and evaluate your retention of material. The end-of-chapter problems are grouped by topics. In general, problems that simply test your knowledge come first and more challenging problems follow.

Try to work as many problems as you can. If you have trouble, two sources of help we suggest are your instructor and the *Study Guide and Solutions Manual* that accompanies this text. If you visit your instructors with your questions, you are likely to find that they are thrilled to be asked to help, and they may provide you with the insight you need to better understand a concept or problem. The study guide provides answers to the problems and explains how to solve them. It also provides you with review materials and additional problems that do not appear in the textbook. Problem solving is time-consuming, but it will pay off in an understanding of the subject.

And now, let us begin.

Methyl butyrate and propyl acetate, organic flavor and fragrance molecules found in apples and pears, respectively, are structural isomers (Sec. 1.8).

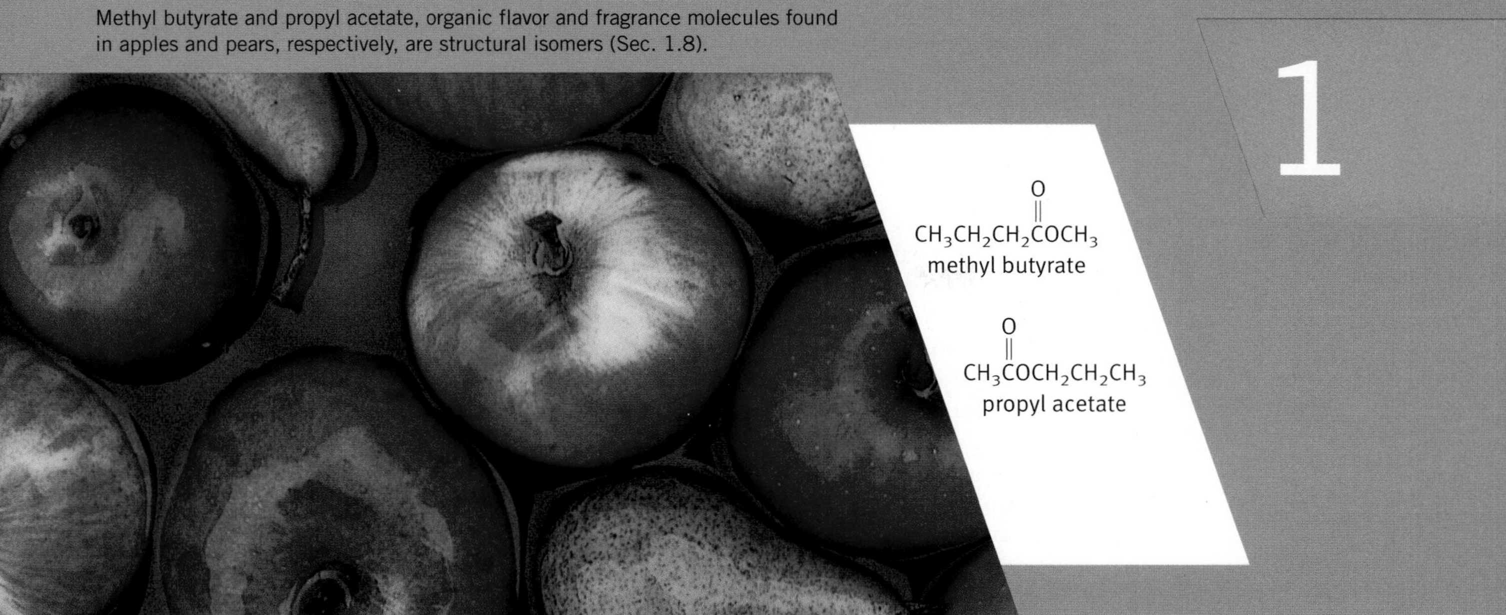

Jerry Howard/Positive Images

1 Bonding and Isomerism

Why does sucrose (table sugar) melt at 185°C, while sodium chloride (table salt)—melts at a much higher temperature, 801°C? Why do both of these substances dissolve in water, while olive oil does not? Why does the molecule methyl butyrate smell like apples, while the molecule propyl acetate, which contains the same number and kind of atoms, smells like pears? To answer questions such as these, you must understand how atoms bond with one another and how molecules interact with one another. Bonding is the key to the structure, physical properties, and chemical behavior of different kinds of matter.

Perhaps you have already studied bonding and related concepts in a beginning chemistry course. Browse through each section of this chapter to see whether it is familiar, and try to work the problems. If you can work the problems, you can safely skip that section. But if you have difficulty with any of the problems within or at the end of this chapter, study the entire chapter carefully because we will use the ideas developed here throughout the rest of the book.

Online homework for this chapter can be assigned in OWL, an online homework assessment tool.

1.1 How Electrons Are Arranged in Atoms

An **atom** consists of a small, dense **nucleus** containing positively charged **protons** and neutral **neutrons** and surrounded by negatively charged **electrons**. The **atomic number** of an element equals the number of protons in its nucleus; its **atomic weight** is the sum of the number of protons and neutrons in its nucleus.

Atoms contain a small, dense **nucleus** surrounded by **electrons**. The nucleus is positively charged and contains most of the mass of the atom. The nucleus consists of **protons**, which are positively charged, and **neutrons**, which are neutral. (The only exception is hydrogen, whose nucleus consists of only a single proton.) In a neutral atom, the positive charge of the nucleus is exactly balanced by the negative charge of the electrons that surround it. The **atomic number** of an element is equal to the number of protons in its nucleus (and to the number of electrons around the nucleus in a neutral atom). The **atomic weight** is approximately equal to the sum of the number of protons and the number of neutrons in the nucleus; the electrons are not counted because they are very light by comparison. The periodic table on the inside back cover of this book shows all the elements with their atomic numbers and weights.

We are concerned here mainly with the atom's electrons because their number and arrangement provide the key to how a particular atom reacts with other atoms to form molecules. Also, we will deal only with electron arrangements in the lighter elements because these elements are the most important in organic molecules.

Electrons are located in **orbitals**. Orbitals are grouped in **shells**. An orbital can hold a maximum of two electrons.

Electrons are concentrated in certain regions of space around the nucleus called **orbitals**. Each orbital can contain a maximum of two electrons. The orbitals, which differ in shape, are designated by the letters *s, p,* and *d.* In addition, orbitals are grouped in **shells** designated by the numbers 1, 2, 3, and so on. Each shell contains different types and numbers of orbitals, corresponding to the shell number. For example, shell 1 contains only one type of orbital, designated the 1*s* orbital. Shell 2 contains two types of orbitals, 2*s* and 2*p,* and shell 3 contains three types, 3*s,* 3*p,* and 3*d.* Within a particular shell, the number of *s, p,* and *d* orbitals is 1, 3, and 5, respectively (Table 1.1). These rules permit us to count how many electrons each shell will contain when it is filled (last column in Table 1.1). Table 1.2 shows how the electrons of the first 18 elements are arranged.

Valence **electrons** are located in the outermost shell. The **kernel** of the atom contains the nucleus and the inner electrons.

The first shell is filled for helium (He) and all elements beyond, and the second shell is filled for neon (Ne) and all elements beyond. Filled shells play almost no role in chemical bonding. Rather, the outer electrons, or **valence electrons**, are mainly involved in chemical bonding, and we will focus our attention on them.

Table 1.3 shows the valence electrons, the electrons in the outermost shell, for the first 18 elements. The element's symbol stands for the **kernel** of the element (the nucleus plus the filled electron shells), and the dots represent the valence electrons. The elements are arranged in groups according to the periodic table, and (except for helium) these group numbers correspond to the number of valence electrons.

Armed with this information about atomic structure, we are now ready to tackle the problem of how elements combine to form chemical bonds.

Table 1.1 ■ Numbers of Orbitals and Electrons in the First Three Shells

	Number of orbitals of each type			
Shell number	***s***	***p***	***d***	**Total number of electrons when shell is filled**
1	1	0	0	2
2	1	3	0	8
3	1	3	5	18

Table 1.2 ■ Electron Arrangements of the First 18 Elements

Atomic number	Element	Number of electrons in each orbital				
		$1s$	$2s$	$2p$	$3s$	$3p$
1	H	1				
2	He	2				
3	Li	2	1			
4	Be	2	2			
5	B	2	2	1		
6	C	2	2	2		
7	N	2	2	3		
8	O	2	2	4		
9	F	2	2	5		
10	Ne	2	2	6		
11	Na	2	2	6	1	
12	Mg	2	2	6	2	
13	Al	2	2	6	2	1
14	Si	2	2	6	2	2
15	P	2	2	6	2	3
16	S	2	2	6	2	4
17	Cl	2	2	6	2	5
18	Ar	2	2	6	2	6

Table 1.3 ■ Valence Electrons of the First 18 Elements

Group	I	II	III	IV	V	VI	VII	VIII
	$\mathrm{H}\cdot$							$\mathrm{He}:$
	$\mathrm{Li}\cdot$	$\mathrm{Be}\cdot$	$\cdot\overset{\cdot}{\mathrm{B}}\cdot$	$\cdot\overset{\cdot}{\underset{\cdot}{\mathrm{C}}}\cdot$	$\cdot\overset{\cdot}{\underset{\cdot}{\mathrm{N}}}:$	$\cdot\overset{\cdot\cdot}{\underset{\cdot}{\mathrm{O}}}:$	$:\overset{\cdot\cdot}{\underset{\cdot}{\mathrm{F}}}:$	$:\overset{\cdot\cdot}{\underset{\cdot\cdot}{\mathrm{Ne}}}:$
	$\mathrm{Na}\cdot$	$\overset{\cdot}{\mathrm{Mg}}\cdot$	$\cdot\overset{\cdot}{\mathrm{Al}}\cdot$	$\cdot\overset{\cdot}{\underset{\cdot}{\mathrm{Si}}}\cdot$	$\cdot\overset{\cdot}{\underset{\cdot}{\mathrm{P}}}:$	$\cdot\overset{\cdot\cdot}{\underset{\cdot}{\mathrm{S}}}:$	$:\overset{\cdot\cdot}{\underset{\cdot}{\mathrm{Cl}}}:$	$:\overset{\cdot\cdot}{\underset{\cdot\cdot}{\mathrm{Ar}}}:$

1.2 Ionic and Covalent Bonding

An early, but still useful, theory of chemical bonding was proposed in 1916 by Gilbert Newton Lewis, then a professor at the University of California, Berkeley. Lewis noticed that the **inert gas** helium had only two electrons surrounding its nucleus and that the next inert gas, neon, had 10 such electrons (2 + 8; see Table 1.2). He concluded that atoms of these gases must have very stable electron arrangements *because these elements do not combine with other atoms.* He further suggested that other atoms might react in such a way in order to achieve these stable arrangements. This stability could be achieved in one of two ways: by complete transfer of electrons from one atom to another or by sharing of electrons between atoms.

An **inert gas** has a stable electron configuration.

1.2.a Ionic Compounds

Ionic compounds are composed of positively charged **cations** and negatively charged **anions**.

Ionic bonds are formed by the transfer of one or more valence electrons from one atom to another. Because electrons are negatively charged, the atom that gives up electrons becomes positively charged, a **cation**. The atom that receives electrons becomes negatively charged, an **anion**. The reaction between sodium and chlorine atoms to form sodium chloride (ordinary table salt) is a typical electron-transfer reaction.*

$$\underset{\text{sodium atom}}{\text{Na}\cdot} + \underset{\text{chlorine atom}}{\cdot\ddot{\underset{\cdot\cdot}{\text{Cl}}}:} \longrightarrow \underset{\text{sodium cation}}{\text{Na}^{+}} + \underset{\text{chloride anion}}{:\ddot{\underset{\cdot\cdot}{\text{Cl}}}:^{-}} \qquad (1.1)$$

The sodium atom has only one valence electron (it is in the third shell; see Table 1.2). By giving up that electron, sodium achieves the electron arrangement of neon. At the same time, it becomes positively charged, a sodium cation. The chlorine atom has seven valence electrons. By accepting an additional electron, chlorine achieves the electron arrangement of argon and becomes negatively charged, a chloride anion. Atoms, such as sodium, that tend to give up electrons are said to be **electropositive**. Often such atoms are metals. Atoms, such as chlorine, that tend to accept electrons are said to be **electronegative**. Often such atoms are nonmetals.

Electropositive atoms give up electrons and form cations.

Electronegative atoms accept electrons and form anions.

EXAMPLE 1.1

Write an equation for the reaction of magnesium (Mg) with fluorine (F) atoms.

$$\dot{\text{Mg}}\cdot + \cdot\ddot{\underset{\cdot\cdot}{\text{F}}}: + \cdot\ddot{\underset{\cdot\cdot}{\text{F}}}: \longrightarrow \text{Mg}^{2+} + 2\,:\ddot{\underset{\cdot\cdot}{\text{F}}}:^{-}$$

Solution Magnesium has two valence electrons. Since each fluorine atom can accept only one electron (from the magnesium) to complete its valence shell, two fluorine atoms are needed to react with one magnesium atom.

PROBLEM 1.1 Write an equation for the reaction of sodium atoms (Na) with chlorine atoms (Cl).

The product of eq. 1.1 is sodium chloride, an ionic compound made up of equal numbers of sodium and chloride ions. In general, ionic compounds form when strongly electropositive atoms and strongly electronegative atoms interact. The ions in a crystal of an ionic substance are held together by the attractive force between their opposite charges, as shown in Figure 1.1 for a sodium chloride crystal.

In a sense, the ionic bond is not really a bond at all. Being oppositely charged, the ions attract one another like the opposite poles of a magnet. In the crystal, the ions are packed in a definite arrangement, but we cannot say that any particular ion is bonded or connected to any other particular ion. And, of course, when the substance is dissolved, the ions separate and are able to move about in solution relatively freely.

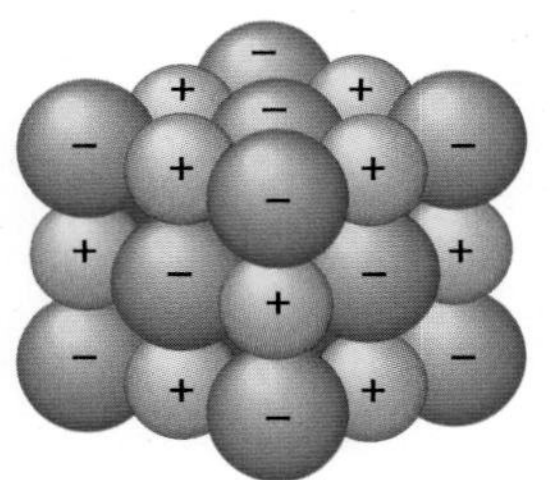

Figure 1.1
Sodium chloride, NaCl, is an ionic crystal. The purple spheres represent sodium ions, Na^+, and the green spheres are chloride ions, Cl^-. Each ion is surrounded by six oppositely charged ions, except for those ions that are at the surface of the crystal.

*The curved arrow in eq. 1.1 shows the movement of one electron from the valence shell of the sodium atom to the valence shell of the chlorine atom. The use of curved arrows to show the movement of electrons is explained in greater detail in Section 1.13.

EXAMPLE 1.2

What charge will a beryllium ion carry?

Solution As seen in Table 1.3, beryllium (Be) has two valence electrons. To achieve the filled-shell electron arrangement of helium, it must lose both of its valence electrons. Thus, the beryllium cation will carry two positive charges and is represented by Be^{2+}.

PROBLEM 1.2 Using Table 1.3, determine what charge the ion will carry when each of the following elements reacts to form an ionic compound: Al, Li, S, and O.

Generally speaking, within a given horizontal row in the periodic table, the more electropositive elements are those farthest to the left, and the more electronegative elements are those farthest to the right. Within a given vertical column, the more electropositive elements are those toward the bottom, and the more electronegative elements are those toward the top.

EXAMPLE 1.3

Which atom is more electropositive?

a. lithium or beryllium

b. lithium or sodium

Solution

a. The lithium nucleus has less positive charge (+3) to attract electrons than the beryllium nucleus (+4). It takes less energy, therefore, to remove an electron from lithium than it does to remove one from beryllium. Since lithium loses an electron more easily than beryllium, lithium is the more electropositive atom.
b. The valence electron in the sodium atom is shielded from the positive charge of the nucleus by two inner shells of electrons, whereas the valence electron of lithium is shielded by only one inner shell. It takes less energy, therefore, to remove an electron from sodium; so, sodium is the more electropositive element.

PROBLEM 1.3 Using Table 1.3, determine which is the more electropositive element: sodium or aluminum, carbon or nitrogen, carbon or silicon.

PROBLEM 1.4 Using Table 1.3, determine which is the more electronegative element: fluorine or chlorine, oxygen or fluorine, nitrogen or phosphorus.

PROBLEM 1.5 Judging from its position in Table 1.3, do you expect carbon to be electropositive or electronegative?

1.2.b The Covalent Bond

Elements that are neither strongly electronegative nor strongly electropositive, or that have similar electronegativities, tend to form bonds by sharing electron pairs rather than completely transferring electrons. A **covalent bond** involves the mutual sharing of one or more electron pairs between atoms. Two (or more) atoms joined by covalent bonds constitute a **molecule**. When the two atoms are identical or have

A **covalent bond** is formed when two atoms share one or more electron pairs. A **molecule** consists of two or more atoms joined by covalent bonds.

equal electronegativities, the electron pairs are shared equally. The hydrogen molecule is an example.

$$\underset{\text{hydrogen atoms}}{H\cdot + H\cdot} \longrightarrow \underset{\text{hydrogen molecule}}{H:H} + \text{heat} \quad (1.2)$$

Each hydrogen atom can be considered to have filled its first electron shell by the sharing process. That is, each atom is considered to "own" all of the electrons it shares with the other atom, as shown by the loops in these structures.

(H:)H H(:H)

EXAMPLE 1.4

Write an equation similar to eq. 1.2 for the formation of a chlorine molecule from two chlorine atoms.

$$:\ddot{\underset{..}{Cl}}\cdot + \cdot\ddot{\underset{..}{Cl}}: \longrightarrow :\ddot{\underset{..}{Cl}}:\ddot{\underset{..}{Cl}}: + \text{heat}$$

Solution One electron pair is shared by the two chlorine atoms. In that way, each chlorine completes its valence shell with eight electrons (three unshared pairs and one shared pair).

PROBLEM 1.6 Write an equation similar to eq. 1.2 for the formation of a fluorine molecule from two fluorine atoms.

Bond energy (BE) is the energy necessary to break a mole of covalent bonds. The amount of energy depends on the type of bond broken.

When two hydrogen atoms combine to form a molecule, heat is liberated. Conversely, this same amount of heat (energy) has to be supplied to a hydrogen molecule to break it apart into atoms. To break apart 1 mole (2 g) of hydrogen molecules into atoms requires 104 kcal (or 435 kJ*) of heat, quite a lot of energy. This energy is called the **bond energy**, or **BE**, and is different for bonds between different atoms (see Table A in the Appendix).

The **bond length** is the average distance between two covalently bonded atoms.

The H—H bond is a very strong bond. The main reason for this is that the shared electron pair is attracted to *both* hydrogen nuclei, whereas in a hydrogen atom, the valence electron is associated with only one nucleus. But other forces in the hydrogen molecule tend to counterbalance the attraction between the electron pair and the nuclei. These forces are the repulsion between the two like-charged nuclei and the repulsion between the two like-charged electrons. A balance is struck between the attractive and the repulsive forces. The hydrogen atoms neither fly apart nor do they fuse together. Instead, they remain connected, or bonded, and vibrate about some equilibrium distance, which we call the **bond length**. For a hydrogen molecule, the bond length (that is, the average distance between the two hydrogen nuclei) is 0.74 Å.** The length of a covalent bond depends on the atoms that are bonded and the number of electron pairs shared between the atoms. Bond lengths for some typical covalent bonds are given in Table B in the Appendix.

---

*Although most organic chemists use the kilocalorie as the unit of heat energy, the currently used international unit is the kilojoule; 1 kcal = 4.184 kJ. In this text, the kilocalorie will be used. If your instructor prefers to use kJ, multiply kcal × 4.184 (or × 4 for a rough estimate) to convert to kJ.

**Å, or angstrom unit, is 10^{-8} cm, so the H—H bond length is 0.74×10^{-8} cm. Although the angstrom is commonly used by organic chemists, another unit often used for bond lengths is the picometer (pm; 1 Å = 100 pm). To convert the H—H bond length from Å to pm, multiply 0.74 × 100. The H—H bond length is 74 pm. In this text, the angstrom will be used as the unit for bond lengths.

1.3 Carbon and the Covalent Bond

Now let us look at carbon and its bonding. We represent atomic carbon by the symbol $\cdot\dot{\underset{\cdot}{C}}\cdot$ where the letter C stands for the kernel (the nucleus plus the two 1*s* electrons) and the dots represent the valence electrons.

With four valence electrons, the valence shell of carbon is half filled (or half empty). Carbon atoms have neither a strong tendency to lose all their electrons (and become C^{4+}) nor a strong tendency to gain four electrons (and become C^{4-}). Being in the middle of the periodic table, *carbon is neither strongly electropositive nor strongly electronegative.* Instead, it usually forms covalent bonds with other atoms by sharing electrons. For example, carbon combines with four hydrogen atoms (each of which supplies one valence electron) by sharing four electron pairs.* The substance formed is known as methane. Carbon can also share electron pairs with four chlorine atoms, forming tetrachloromethane.**

```
     H                      H
     •x                     |
  H x C x H      or     H — C — H
     x•                     |
     H                      H
              methane

     ••
    :Cl:                    Cl
  ••  •x  ••                |
 :Cl x C x Cl:   or    Cl — C — Cl
  ••  x•  ••                |
    :Cl:                    Cl
     ••
        tetrachloromethane
       (carbon tetrachloride)
```

By sharing electron pairs, the atoms complete their valence shells. In both examples, carbon has eight valence electrons around it. In methane, each hydrogen atom completes its valence shell with two electrons, and in tetrachloromethane, each chlorine atom fills its valence shell with eight electrons. In this way, all valence shells are filled and the compounds are quite stable.

The shared electron pair is called a covalent bond because it bonds or links the atoms by its attraction to both nuclei. The single bond is usually represented by a dash, or a single line, as shown in the structures above for methane and tetrachloromethane.

EXAMPLE 1.5

Draw the structure for chloromethane (also called methyl chloride), CH_3Cl.

Solution

```
     H                      H
     ••  ••                 |
  H : C : Cl:    or     H — C — Cl
     ••  ••                 |
     H                      H
```

PROBLEM 1.7 Draw the structures for dichloromethane (also called methylene chloride), CH_2Cl_2, and trichloromethane (chloroform), $CHCl_3$.

*To designate electrons from different atoms, the symbols • and x are often used. But the electrons are, of course, identical.

**Tetrachloromethane is the systematic name, and carbon tetrachloride is the common name. We discuss how to name organic compounds later.

1.4 Carbon–Carbon Single Bonds

The unique property of carbon atoms—that is, the property that makes it possible for millions of organic compounds to exist—is their ability to share electrons not only with different elements but also with other carbon atoms. For example, two carbon atoms may be bonded to one another, and each of these carbon atoms may be linked to other atoms. In ethane and hexachloroethane, each carbon is connected to the other carbon *and* to three hydrogen atoms or three chlorine atoms. Although they have two carbon atoms instead of one, these compounds have chemical properties similar to those of methane and tetrachloromethane, respectively.

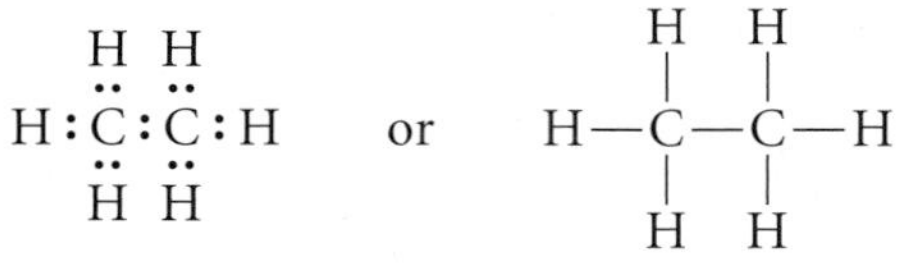

$$\begin{array}{c} \text{H}\ \ \text{H} \\ \text{H}:\ddot{\text{C}}:\ddot{\text{C}}:\text{H} \\ \text{H}\ \ \text{H} \end{array} \quad \text{or} \quad \begin{array}{c} \text{H}\quad\text{H} \\ | \quad\ | \\ \text{H}-\text{C}-\text{C}-\text{H} \\ | \quad\ | \\ \text{H}\quad\text{H} \end{array} \qquad\qquad \begin{array}{c} :\ddot{\text{Cl}}:\ :\ddot{\text{Cl}}: \\ :\ddot{\text{Cl}}:\ddot{\text{C}}\ :\ \ddot{\text{C}}:\ddot{\text{Cl}}: \\ :\underset{\cdot\cdot}{\text{Cl}}:\ :\underset{\cdot\cdot}{\text{Cl}}: \end{array} \quad \text{or} \quad \begin{array}{c} \text{Cl}\quad\text{Cl} \\ | \quad\ | \\ \text{Cl}-\text{C}-\text{C}-\text{Cl} \\ | \quad\ | \\ \text{Cl}\quad\text{Cl} \end{array}$$

ethane 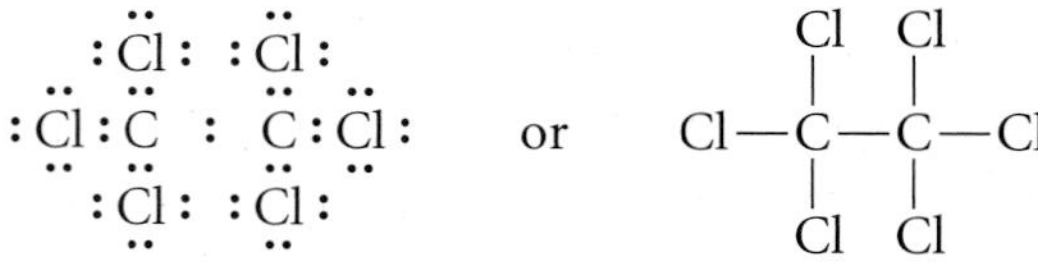 hexachloroethane

The carbon–carbon bond in ethane, like the hydrogen–hydrogen bond in a hydrogen molecule, is a purely covalent bond, with the electrons being shared *equally* between the two identical carbon atoms. As with the hydrogen molecule, heat is required to break the carbon–carbon bond of ethane to give two CH_3 fragments (called methyl radicals). A **radical** is a molecular fragment with an odd number of unshared electrons.

A **radical** is a molecular fragment with an odd number of unshared electrons.

$$\begin{array}{c} \text{H}\ \ \text{H} \\ | \quad | \\ \text{H}-\text{C}:\text{C}-\text{H} \\ | \quad | \\ \text{H}\ \ \text{H} \end{array} \xrightarrow{\text{heat}} \begin{array}{c} \text{H}\qquad\quad\text{H} \\ | \qquad\quad\ | \\ \text{H}-\text{C}\cdot + \cdot\text{C}-\text{H} \\ | \qquad\quad\ | \\ \text{H}\qquad\quad\text{H} \end{array} \tag{1.3}$$

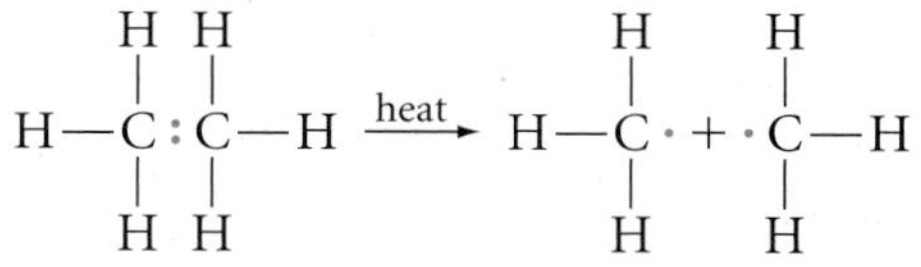

ethane two methyl radicals

However, less heat is required to break the carbon–carbon bond in ethane than is required to break the hydrogen–hydrogen bond in a hydrogen molecule. The actual amount is 88 kcal (or 368 kJ) per mole of ethane. The carbon–carbon bond in ethane is longer (1.54 Å) than the hydrogen–hydrogen bond (0.74 Å) and also somewhat weaker. Breaking carbon–carbon bonds by heat, as represented in eq. 1.3, is the first step in the *cracking* of petroleum, an important process in the manufacture of gasoline (see "A Word About Petroleum, Gasoline, and Octane Number" on pages 102–103).

EXAMPLE 1.6

What do you expect the length of a C—H bond (as in methane or ethane) to be?

Solution It should measure somewhere between the H—H bond length in a hydrogen molecule (0.74 Å) and the C—C bond length in ethane (1.54 Å). The actual value is about 1.09 Å, close to the average of the H—H and C—C bond lengths.

PROBLEM 1.8 The Cl—Cl bond length is 1.98 Å. Which bond will be longer, the C—C bond in ethane or the C—Cl bond in chloromethane?

There is almost no limit to the number of carbon atoms that can be linked, and some molecules contain as many as 100 or more carbon–carbon bonds. This ability of an element to form chains as a result of bonding between the same atoms is called **catenation**.

Catenation is the ability of an element to form chains of its own atoms through covalent bonding.

PROBLEM 1.9 Using the structure of ethane as a guide, draw the structure for propane, C_3H_8.

1.5 Polar Covalent Bonds

As we have seen, covalent bonds can be formed not only between identical atoms (H—H, C—C) but also between different atoms (C—H, C—Cl), provided that the atoms do not differ too greatly in electronegativity. However, if the atoms are different from one another, the electron pair may not be shared equally between them. Such a bond is sometimes called a **polar covalent bond** because the atoms that are linked carry a partial negative and a partial positive charge.

A **polar covalent bond** is a covalent bond in which the electron pair is not shared equally between the two atoms.

The hydrogen chloride molecule provides an example of a polar covalent bond. Chlorine atoms are more electronegative than hydrogen atoms, but even so, the bond that they form is covalent rather than ionic. However, the shared electron pair is attracted more toward the chlorine, which therefore is slightly negative with respect to the hydrogen. This bond polarization is indicated by an arrow whose head is negative and whose tail is marked with a plus sign. Alternatively, a partial charge, written as $\delta+$ or $\delta-$ (read as "delta plus" or "delta minus"), may be shown:

$$\overset{\longmapsto}{\mathrm{H} : \ddot{\underset{..}{\mathrm{Cl}}}:} \quad \text{or} \quad \overset{\delta+}{\mathrm{H}} : \overset{\delta-}{\ddot{\underset{..}{\mathrm{Cl}}}}: \quad \text{or} \quad \overset{\delta+}{\mathrm{H}}-\overset{\delta-}{\ddot{\underset{..}{\mathrm{Cl}}}}:$$

The bonding electron pair, which is shared *unequally*, is displaced toward the chlorine.

You can usually rely on the periodic table to determine which end of a polar covalent bond is more negative and which end is more positive. As we proceed from left to right across the table within a given period, the elements become *more* electronegative, owing to increasing atomic number or charge on the nucleus. The increasing nuclear charge attracts valence electrons more strongly. As we proceed from the top to the bottom of the table within a given group (down a column), the elements become *less* electronegative because the valence electrons are shielded from the nucleus by an increasing number of inner-shell electrons. From these generalizations, we can safely predict that the atom on the right in each of the following bonds will be negative with respect to the atom on the left:

$$\overset{\longmapsto}{\begin{matrix}\text{C—N}\\\text{C—O}\end{matrix}} \qquad \overset{\longmapsto}{\begin{matrix}\text{C—Cl}\\\text{C—Br}\end{matrix}} \qquad \overset{\longmapsto}{\begin{matrix}\text{H—O}\\\text{H—S}\end{matrix}} \qquad \overset{\longmapsto}{\begin{matrix}\text{Br—Cl}\\\text{Si—C}\end{matrix}}$$

The carbon–hydrogen bond, which is so common in organic compounds, requires special mention. Carbon and hydrogen have nearly identical electronegativities, so the C—H bond is almost purely covalent. The electronegativities of some common elements are listed in Table 1.4.

Table 1.4 ■ Electronegativities of Some Common Elements

Group

I	II	III	IV	V	VI	VII
H 2.2						
Li 1.0	Be 1.6	B 2.0	C 2.5	N 3.0	O 3.4	F 4.0
Na 0.9	Mg 1.3	Al 1.6	Si 1.9	P 2.2	S 2.6	Cl 3.2
K 0.8	Ca 1.0					Br 3.0
						I 2.7

Key: < 1.0; 1.0–1.4; 1.5–1.9; 2.0–2.4; 2.5–2.9; 3.0–3.4

EXAMPLE 1.7

Indicate any bond polarization in the structure of tetrachloromethane.

Solution

$$\begin{array}{c} Cl^{\delta-} \\ | \\ {}^{\delta-}Cl-C^{\delta+}-Cl^{\delta-} \\ | \\ Cl^{\delta-} \end{array}$$

Chlorine is more electronegative than carbon. The electrons in each C—Cl bond are therefore displaced toward the chlorine.

PROBLEM 1.10 Predict the polarity of the N—Cl bond and of the S—O bond.

PROBLEM 1.11 Draw the structure of the refrigerant dichlorodifluoromethane, CCl_2F_2 (CFC-12), and indicate the polarity of the bonds. (The C atom is the central atom.)

PROBLEM 1.12 Draw the formula for methanol, CH_3OH, and (where appropriate) indicate the bond polarity with an arrow, $\leftarrow\!\!\!+$. (The C atom is bonded to three H atoms and the O atom.)

1.6 Multiple Covalent Bonds

To complete their valence shells, atoms may sometimes share more than one electron pair. Carbon dioxide, CO_2, is an example. The carbon atom has four valence electrons, and each oxygen has six valence electrons. A structure that allows each atom to complete its valence shell with eight electrons is

$$\overset{\times\times}{\underset{\times\times}{O}} \overset{\times}{\underset{\times}{:}} C \overset{\times}{\underset{\times}{:}} \overset{\times\times}{\underset{\times\times}{O}} \quad \text{or} \quad \overset{\times\times}{\underset{\times\times}{O}}{=}C{=}\overset{\times\times}{\underset{\times\times}{O}} \quad \text{or} \quad O{=}C{=}O$$

A B C

In structure A, the dots represent the electrons from carbon, and the x's are the electrons from the oxygens. Structure B shows the bonds' and oxygens' unshared electrons, and structure C shows only the covalent bonds. Two electron pairs are shared between carbon and oxygen. Consequently, the bond is called a **double bond.** Each oxygen atom also has two pairs of **nonbonding electrons**, or **unshared electron pairs**. The loops in the following structures show that each atom in carbon dioxide has a complete valence shell of eight electrons:

In a **double bond**, two electron pairs are shared between two atoms.

Nonbonding electrons, or **unshared electron pairs**, reside on one atom.

$$\overset{\times\times}{\underset{\times\times}{O}} \overset{\times}{\underset{\times}{:}} C \overset{\times}{\underset{\times}{:}} \overset{\times\times}{\underset{\times\times}{O}} \qquad \overset{\times\times}{\underset{\times\times}{O}} \overset{\times}{\underset{\times}{:}} C \overset{\times}{\underset{\times}{:}} \overset{\times\times}{\underset{\times\times}{O}} \qquad \overset{\times\times}{\underset{\times\times}{O}} \overset{\times}{\underset{\times}{:}} C \overset{\times}{\underset{\times}{:}} \overset{\times\times}{\underset{\times\times}{O}}$$

Hydrogen cyanide, HCN, is an example of a simple compound with a **triple bond,** a bond in which three electron pairs are shared.

In a **triple bond**, three electron pairs are shared between two atoms.

$$H \overset{\times}{\cdot} C \overset{\times\times\times}{\vdots} N \overset{\times}{\times} \quad \text{or} \quad H{-}C{\equiv}N\overset{\times}{\times} \quad \text{or} \quad H{-}C{\equiv}N$$

hydrogen cyanide

PROBLEM 1.13 Show with loops how each atom in hydrogen cyanide completes its valence shell.

EXAMPLE 1.8

Determine what, if anything, is wrong with the following electron arrangement for carbon dioxide:

$$:O:::C::\ddot{O}:$$

Solution The formula contains the correct total number of valence electrons (16), and each oxygen is surrounded by 8 valence electrons, which is correct. However, what is wrong is that the carbon atom has 10 valence electrons, 2 more than is allowable.

PROBLEM 1.14 Show what is wrong with each of the following electron arrangements for carbon dioxide:

a. $:O:::C:::O:$ b. $:\underset{\cdot\cdot}{\ddot{O}}:\underset{\cdot\cdot}{\ddot{C}}:\underset{\cdot\cdot}{\ddot{O}}:$ c. $:\underset{\cdot\cdot}{\ddot{O}}:C:::O:$

PROBLEM 1.15 Methanal (formaldehyde) has the formula H_2CO. Draw a structure that shows how the valence electrons are arranged.

PROBLEM 1.16 Draw an electron-dot structure for carbon monoxide, CO.

Hydrocarbons are compounds composed of just hydrogen and carbon atoms.

Carbon atoms can be connected to one another by double bonds or triple bonds, as well as by single bonds. Thus, there are three **hydrocarbons** (compounds with just carbon and hydrogen atoms) that have two carbon atoms per molecule: ethane, ethene, and ethyne.

```
    H  H         H         H
    |  |          \       /
 H—C—C—H           C=C         H—C≡C—H
    |  |          /       \
    H  H         H         H
  ethane          ethene          ethyne
               (ethylene)      (acetylene)
```

They differ in that the carbon–carbon bond is single, double, or triple, respectively. They also differ in the number of hydrogens. As we will see later, these compounds have different chemical reactivities because of the different types of bonds between the carbon atoms.

EXAMPLE 1.9

Draw the structure for C_3H_6 having one carbon–carbon double bond.

Solution First, draw the three carbons with one double bond.

C=C—C

Then add the hydrogens in such a way that each carbon has eight electrons around it (or in such a way that each carbon has four bonds).

```
    H  H  H
    |  |  |
 H—C=C—C—H
          |
          H
```

PROBLEM 1.17 Draw three different structures that have the formula C_4H_8 and have one carbon–carbon double bond.

1.7 Valence

The **valence** of an element is the number of bonds that an atom of the element can form.

The **valence** of an element is simply the number of bonds that an atom of the element can form. The number is usually equal to the *number of electrons needed to fill the valence shell.* Table 1.5 gives the common valences of several elements. Notice the difference between the number of valence electrons and the valence. Oxygen, for example, has six valence electrons but a valence of only 2. The *sum* of the two numbers is equal to the number of electrons in the filled shell.

The valences in Table 1.5 apply whether the bonds are single, double, or triple. For example, carbon has four bonds in each of the structures we have written so far: methane, tetrachloromethane, ethane, ethene, ethyne, carbon dioxide, and so on. These common valences are worth remembering, because they will help you to write correct structures.

Table 1.5 ■ Valences of Common Elements

Element	H·	·C· (4 dots)	·N: (5 dots)	·O: (6 dots)	:F: (7 dots)	:Cl: (7 dots)
Valence	1	4	3	2	1	1

EXAMPLE 1.10

Using dashes for bonds, draw a structure for C_3H_4 that has the proper valence of 1 for each hydrogen and 4 for each carbon.

Solution There are three possibilities:

$$H-\underset{\displaystyle H}{\overset{\displaystyle H}{C}}-C\equiv C-H \qquad \overset{H}{\underset{H}{}}\!\!>C=C=C<\!\!\overset{H}{\underset{H}{}} \qquad \overset{H}{}\!\!\setminus C=C/\!\!\overset{H}{}\ \text{(ring closed by }\underset{H\ \ H}{C}\text{)}$$

A compound that corresponds to each of these three different arrangements of the atoms is known.

PROBLEM 1.18 Use dashes for bonds and use the valences given in Table 1.5 to write a structure for each of the following:

a. CH_3F b. CH_5N

PROBLEM 1.19 Does C_2H_5 represent a stable molecule?

In Example 1.10, we saw that three carbon atoms and four hydrogen atoms can be connected to one another in three different ways, each of which satisfies the valences of both kinds of atoms. Let us take a closer look at this phenomenon.

1.8 Isomerism

The **molecular formula** of a substance tells us the numbers of different atoms present, but a **structural formula** tells us how those atoms are arranged. For example, H_2O is the molecular formula for water. It tells us that each water molecule contains two hydrogen atoms and one oxygen atom. But the structural formula H—O—H tells us

The **molecular formula** of a substance gives the number of different atoms present; the **structural formula** indicates how those atoms are arranged.

A CLOSER LOOK AT... Green Chemistry

Conduct research on the Internet to find more information on green chemistry and to answer the following questions.

Green Chemistry

1. What is green chemistry?
2. What are the 12 principles of green chemistry?
3. What is the Pollution Prevention Act of 1990?
4. What are some organizations involved in promoting green chemistry? How are these organizations promoting green chemistry?

Hangers Cleaners

1. How does Hangers Cleaners make use of carbon dioxide?
2. What are the advantages of using CO_2 over traditional dry cleaning solvents? Explain why you think so.

Ionic Liquids and Their Uses

1. What are some of the advantages of using ionic liquids as solvents in chemical reactions?
2. What environmental problems posed by traditional solvents are avoided by using ionic liquids?

more than that. The structural formula gives us the connectivity between atoms and tells us that the hydrogens are connected to the oxygen (and not to each other).

Isomers are molecules with the same number and kinds of atoms but different arrangements of the atoms. **Structural** (or **constitutional**) **isomers** have the same molecular formula but different structural formulas.

It is sometimes possible to arrange the same atoms in more than one way and still satisfy their valences. Molecules that have the same kinds and numbers of atoms but different arrangements are called **isomers**, a term that comes from the Greek (*isos*, equal, and *meros*, part). **Structural** (or **constitutional**) **isomers** are compounds that have the same molecular formula, but different structural formulas. Let us look at a particular pair of isomers.

Two very different chemical substances are known, each with the molecular formula C_2H_6O. One of these substances is a colorless liquid that boils at 78.5°C, whereas the other is a colorless gas at ordinary temperatures (boiling point (bp) −23.6°C). The only possible explanation is that the atoms must be arranged differently in the molecules of each substance and that these arrangements are somehow responsible for the fact that one substance is a liquid and the other is a gas.

For the molecular formula C_2H_6O, two (and only two) structural formulas are possible that satisfy the valence requirement of 4 for carbon, 2 for oxygen, and 1 for hydrogen. They are:

```
   H  H                    H       H
   |  |                    |       |
H—C—C—O—H      and     H—C—O—C—H
   |  |                    |       |
   H  H                    H       H
```

ethanol (ethyl alcohol) bp 78.5°C

methoxymethane (dimethyl ether) bp −23.6°C

In one formula, the two carbons are connected to one another by a single covalent bond; in the other formula, each carbon is connected to the oxygen. When we complete the valences by adding hydrogens, each arrangement requires six hydrogens. Many kinds of experimental evidence verify these structural assignments. We leave for later chapters (Chapters 7 and 8) an explanation of why these arrangements of atoms produce substances that are so different from one another.

Ethanol and methoxymethane are structural isomers. They have the same molecular formula but different structural formulas. Ethanol and methoxymethane differ in physical and chemical properties as a consequence of their different molecular structures. In general, structural isomers are different compounds. They differ in physical and chemical properties as a consequence of their different molecular structures.

PROBLEM 1.20 Draw structural formulas for the three possible isomers of C_3H_8O.

1.9 Writing Structural Formulas

In a **continuous chain**, atoms are bonded one after another.

You will be writing structural formulas throughout this course. Perhaps a few hints about how to do so will be helpful. Let's look at another case of isomerism. Suppose we want to write out all possible structural formulas that correspond to the molecular formula C_5H_{12}. We begin by writing all five carbons in a **continuous chain**.

C—C—C—C—C

a continuous chain

This chain uses up one valence for each of the end carbons and two valences for the carbons in the middle of the chain. Each end carbon therefore has three valences left for bonds to hydrogens. Each middle carbon has only two valences for bonds to hydrogens. As a consequence, the structural formula in this case is written as:

```
    H   H   H   H   H
    |   |   |   |   |
H—C—C—C—C—C—H
    |   |   |   |   |
    H   H   H   H   H
```

pentane, bp 36°C

To find structural formulas for the other isomers, we must consider **branched chains**. For example, we can reduce the longest chain to only four carbons and connect the fifth carbon to one of the middle carbons, as in the following structural formula:

In a **branched chain**, some atoms form branches from the longest continuous chain.

```
C—C—C—C
  |
  C
```

a branched chain

If we add the remaining bonds so that each carbon has a valence of 4, we see that three of the carbons have three hydrogens attached, but the other carbons have only one or two hydrogens. The molecular formula, however, is still C_5H_{12}.

```
    H   H   H   H
    |   |   |   |
H—C—C—C—C—H
    |   |   |   |
    H   |   H   H
        |
    H—C—H
        |
        H
```

2-methylbutane, bp 28°C
(isopentane)

Suppose we keep the chain of four carbons and try to connect the fifth carbon somewhere else. Consider the following chains:

```
C—C—C—C     C—C—C—C     C—C—C—C
|                 |         |
C                 C         C
```

Do we have anything new here? *No!* The first two structures have five-carbon chains, exactly as in the formula for pentane, and the third structure is identical to the branched chain we have already drawn for 2-methylbutane—a four-carbon chain with a one-carbon branch attached to the second carbon in the chain (counting now from the right instead of from the left). Notice that for every drawing of pentane, you can draw a line through all five carbon atoms without lifting your pencil from the paper. For every drawing of 2-methylbutane, a continuous line can be drawn through exactly four carbon atoms.*

But there is a third isomer of C_5H_{12}. We can find it by reducing the longest chain to only three carbons and connecting two one-carbon branches to the middle carbon.

```
    C
    |
  C—C—C
    |
    C
```

*Using a molecular model kit (see note on p. 37) to construct the carbon chains as drawn will help you to see which representations are identical and which are different.

If we fill in the hydrogens, we see that the middle carbon has no hydrogens attached to it.

```
            H
            |
         H—C—H
   H        |       H
   |        |       |
H—C————C————C—H
   |        |       |
   H        |       H
         H—C—H
            |
            H
```

2,2-dimethylpropane, bp 10°C
(neopentane)

So we can draw three (and only three) different structural formulas that correspond to the molecular formula C_5H_{12}, and in fact, we find that only three different chemical substances with this formula exist. They are commonly called *n*-pentane (*n* for normal, with an unbranched carbon chain), isopentane, and neopentane.

PROBLEM 1.21 To which isomer of C_5H_{12} does each of the following structural formulas correspond?

```
   H          H
   |          |
H—C—H     H—C—H
   |    H     |
   |    |     |
H—C————C————C—H
   |    |     |
   H    H     H
```

```
        H
        |
     H—C—H
  H     |     H
  |     |     |
H—C————C————C—H
  |     |     |
  H     H     |
           H—C—H
              |
              H
```

```
   H      H      H
   |      |      |
H—C————C————C—H
   |      |      |
   H      |      H
          |      H
          |      |
       H—C————C—H
          |      |
          H      H
```

1.10 Abbreviated Structural Formulas

Structural formulas like the ones we have written so far are useful, but they are also somewhat cumbersome. They take up a lot of space and are tiresome to write out. Consequently, we often take some shortcuts that still convey the meaning of structural formulas. For example, we may abbreviate the structural formula of ethanol (ethyl alcohol) from:

```
   H  H
   |  |
H—C—C—O—H     to     CH3—CH2—OH     or     CH3CH2OH
   |  |
   H  H
```

Each formula clearly represents ethanol rather than its isomer methoxymethane (dimethyl ether), which can be represented by any of the following structures:

```
   H     H
   |     |
H—C—O—C—H     to     CH3—O—CH3     or     CH3OCH3
   |     |
   H     H
```

The structural formulas for the three pentanes can be abbreviated in a similar fashion.

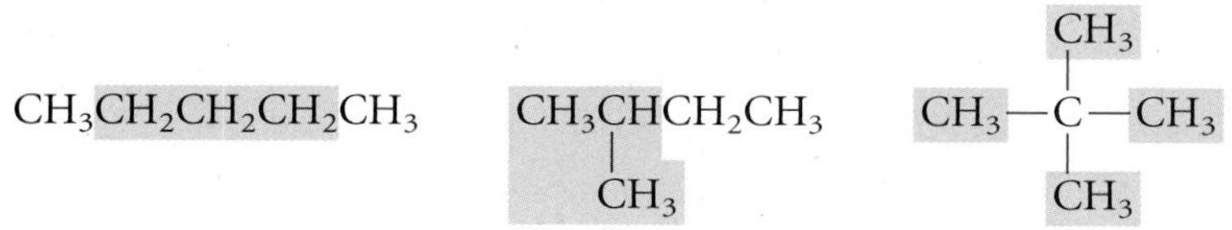

```
                                                   CH3
                                                   |
CH3CH2CH2CH2CH3        CH3CHCH2CH3          CH3—C—CH3
                          |                        |
                          CH3                      CH3
```

n-pentane isopentane neopentane

Sometimes these formulas are abbreviated even further. For example, they can be printed on a single line in the following ways:

$CH_3(CH_2)_3CH_3$ — *n*-pentane

$(CH_3)_2CHCH_2CH_3$ — isopentane

$(CH_3)_4C$ — neopentane

EXAMPLE 1.11

Write a structural formula that shows all bonds for each of the following:

a. $CH_3CCl_2CH_3$ b. $(CH_3)_2C(CH_2CH_3)_2$

Solution

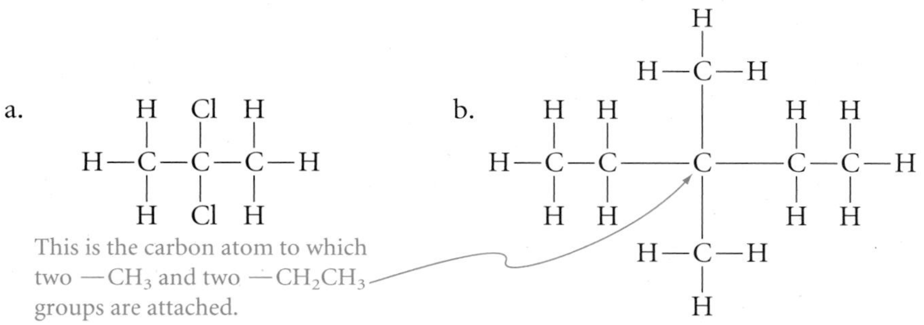

```
                                                  H
                                                  |
                                               H—C—H
a.     H  Cl  H              b.      H  H         |          H  H
       |  |   |                      |  |         |          |  |
    H—C—C—C—H                  H—C—C———————C———————C—C—H
       |  |   |                      |  |         |          |  |
       H  Cl  H                      H  H         |          H  H
                                               H—C—H
                                                  |
                                                  H
```

This is the carbon atom to which two $—CH_3$ and two $—CH_2CH_3$ groups are attached.

PROBLEM 1.22 Write a structural formula that shows all bonds for each of the following:

a. $(CH_3)_3CCH_2CH_2OH$ b. $Cl_2C{=}CCl_2$

Perhaps the ultimate abbreviation of structures is the use of lines to represent the carbon framework:

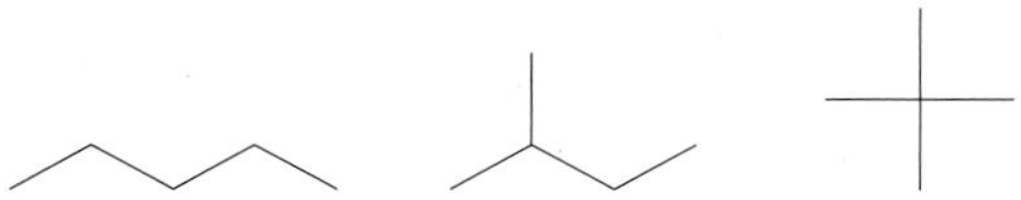

n-pentane isopentane neopentane

In these formulas, *each line segment is understood to have a carbon atom at each end.* The hydrogens are omitted, but we can quickly find the number of hydrogens on each carbon by subtracting from four (the valence of carbon) the number of line segments that emanates from any point. Multiple bonds are represented by multiple line segments. For example, the hydrocarbon with a chain of five carbon atoms and a double bond between the second and third carbon atoms (that is, $CH_3CH{=}CHCH_2CH_3$) is represented as follows:

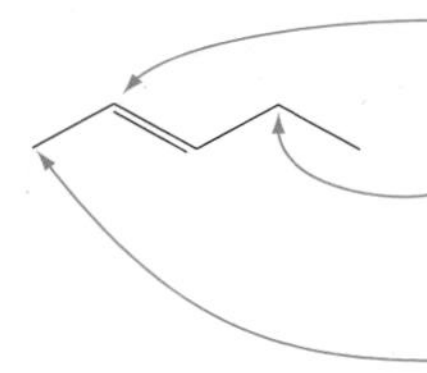

Three line segments emanate from this point; therefore, this carbon has one hydrogen (4 − 3 = 1) attached to it.

Two line segments emanate from this point; therefore, this carbon has two hydrogens (4 − 2 = 2) attached to it.

One line segment emanates from this point; therefore, this carbon has three hydrogens (4 − 1 = 3) attached to it.

EXAMPLE 1.12

Write a more detailed structural formula for .

Solution

$$CH_3{-}\underset{}{\overset{CH_2}{\overset{\|}{C}}}{-}CH_2{-}CH_3$$

or

$$H{-}\underset{H}{\overset{H}{C}}{-}\overset{CH_2}{\overset{\|}{C}}{-}\underset{H}{\overset{H}{C}}{-}\underset{H}{\overset{H}{C}}{-}H$$

PROBLEM 1.23 Write a more detailed structural formula for .

EXAMPLE 1.13

Write a line-segment formula for $CH_3CH_2CH{=}CHCH_2CH(CH_3)_2$.

Solution

PROBLEM 1.24 Write a line-segment formula for $(CH_3)_2C{=}CHCH(CH_3)_2$.

1.11 Formal Charge

So far, we have considered only molecules whose atoms are neutral. But in some molecules, one or more atoms may be charged, either positively or negatively. Because such charges usually affect the chemical reactions of such molecules, it is important to know how to tell where the charge is located.

Consider the formula for hydronium ion, H_3O^+, the product of the reaction of a water molecule with a proton.

$$H{-}\ddot{\underset{\cdot\cdot}{O}}{-}H + H^+ \longrightarrow \left[H{-}\underset{\cdot\cdot}{\overset{H}{\overset{|}{O}}}{-}H \right]^+ \tag{1.4}$$

hydronium ion

The structure has eight electrons around the oxygen and two electrons around each hydrogen, so that all valence shells are complete. Note that there are eight valence electrons altogether. Oxygen contributes six, and each hydrogen contributes one, for a total of nine, but the ion has a single positive charge, so one electron must have been given away, leaving eight. Six of these eight electrons are used to form three O—H single bonds, leaving one unshared electron pair on the oxygen.

Although the entire hydronium ion carries a positive charge, we can ask, "Which atom, in a formal sense, bears the charge?" To determine **formal charge**, we consider each atom to "own" *all* of its unshared electrons plus only *half* of its shared electrons (one electron from each covalent bond). We then subtract this total from the number of valence electrons in the neutral atom to get the formal charge. This definition can be expressed in equation form as follows:

The **formal charge** on an atom in a covalently bonded molecule or ion is the number of valence electrons in the neutral atom minus the number of covalent bonds to the atom and the number of unshared electrons on the atom.

$$\text{Formal charge} = \begin{array}{l}\text{number of valence electrons}\\ \text{in the neutral atom}\end{array} - \left(\begin{array}{c}\text{unshared}\\ \text{electrons}\end{array} + \begin{array}{c}\text{half the shared}\\ \text{electrons}\end{array}\right) \quad (1.5)$$

or, in a simplified form,

$$\text{Formal charge} = \begin{array}{l}\text{number of valence electrons}\\ \text{in the neutral atom}\end{array} - (\text{dots} + \text{bonds})$$

Let us apply this definition to the hydronium ion.

For each hydrogen atom:
Number of valence electrons in the neutral atom $= 1$
Number of unshared electrons $= 0$
Half the number of the shared electrons $= 1$
Therefore, the formal charge $= 1 - (0 + 1) = 0$

For the oxygen atom:
Number of valence electrons in the neutral atom $= 6$
Number of unshared electrons $= 2$
Half the number of the shared electrons $= 3$
Therefore, the formal charge $= 6 - (2 + 3) = +1$

Thus, it is the oxygen atom that formally carries the $+1$ charge in the hydronium ion.

EXAMPLE 1.14

On which atom is the formal charge in the hydroxide ion, OH^-?

Solution The electron-dot formula is

$$\left[:\ddot{\underset{\cdot\cdot}{O}}:H\right]^-$$

Oxygen contributes six electrons, hydrogen contributes one, and there is one more for the negative charge, for a total of eight electrons. The formal charge on oxygen is $6 - (6 + 1) = -1$, so the oxygen carries the negative charge. (So instead, you might see hydroxide written as HO^- to reflect the negative charge on oxygen.) The hydrogen is neutral.

PROBLEM 1.25 Calculate the formal charge on the nitrogen atom in ammonia, NH_3; in the ammonium ion, NH_4^+; and in the amide ion, NH_2^-.

Now let us look at a slightly more complex situation involving electron-dot structures and formal charge.

1.12 Resonance

In electron-dot structures, a pair of dots or a dash represents a bond between just two atoms. But sometimes, an electron pair is involved with more than two atoms in the process of forming bonds. Molecules and ions in which this occurs cannot be adequately represented by a single electron-dot structure. As an example, consider the structure of the carbonate ion, CO_3^{2-}.

The total number of valence electrons in the carbonate ion is 24 (4 from the carbon, $3 \times 6 = 18$ from the three oxygens, *plus* 2 more electrons that give the ion its negative charge; these 2 electrons presumably have been donated by some metal, perhaps one each from two sodium atoms). An electron-dot structure that completes the valence shell of eight electrons around the carbon and each oxygen is

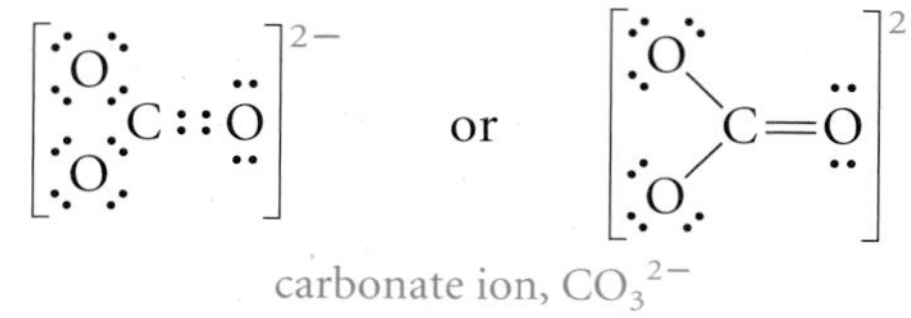

or

carbonate ion, CO_3^{2-}

The structure contains two carbon–oxygen *single* bonds and one carbon–oxygen *double* bond. Application of the definition for formal charge shows that the carbon is formally neutral, each singly bonded oxygen has a formal charge of -1, and the doubly bonded oxygen is formally neutral.

PROBLEM 1.26 Show that the last sentence of the preceding paragraph is correct.

When we wrote the electron-dot structure for the carbonate ion, our choice of which oxygen atom would be doubly bonded to the carbon atom was purely arbitrary. There are in fact *three exactly equivalent* structures that we might write.

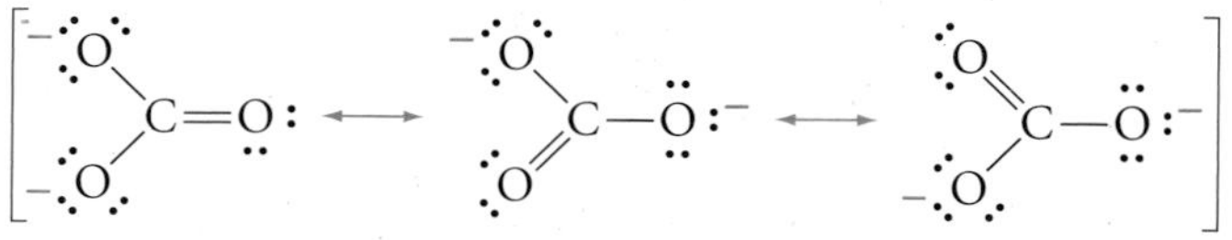

three equivalent structures for the carbonate ion

In each structure there is one C=O bond and there are two C—O bonds. These structures have the same arrangement of the atoms. They differ from one another *only* in the arrangement of the electrons.

The three structures for the carbonate ion are redrawn below, with curved arrows to show how electron pairs can be moved to convert one structure to another:

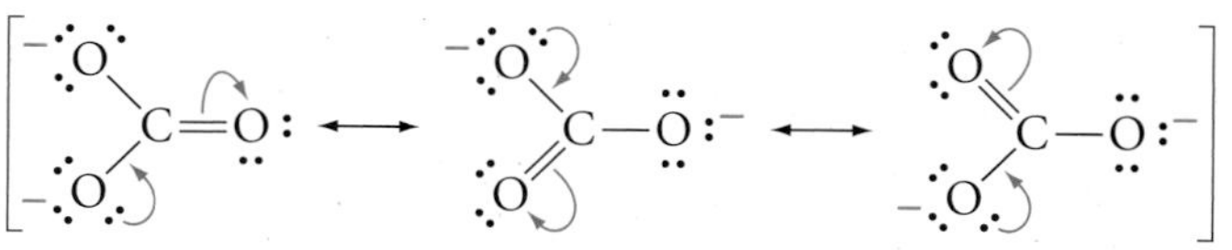

Chemists use curved arrows to keep track of a change in the location of electrons. A detailed explanation of the use of curved arrows is given in Section 1.13.

Resonance structures of a molecule or ion are two or more structures with identical arrangements of the atoms but different arrangements of the electrons. If resonance structures can be written, the true structure of the molecule or ion is a **resonance hybrid** of the contributing resonance structures.

Physical measurements tell us that *none of the foregoing structures accurately describes the real carbonate ion.* For example, although each structure shows two different types of bonds between carbon and oxygen, we find experimentally that *all three carbon–oxygen bond lengths are identical: 1.31 Å.* This distance is intermediate between the normal C=O (1.20 Å) and C—O (1.41 Å) bond lengths. To explain this fact, we usually say that the real carbonate ion has a structure that is a **resonance hybrid** of the three contributing **resonance structures**. It is as if we could take an average of the three structures. In the real carbonate ion, the two formal negative charges are spread *equally* over the three oxygen atoms, so that each oxygen atom carries two-thirds of a negative charge. It is important to note that the carbonate ion does not physically alternate among three resonance structures but has in fact one structure—a *hybrid* of the three resonance structures.

Whenever we can write two or more structures for a molecule with different arrangements of the electrons but identical arrangements of the atoms, we call these structures *resonance structures.* Resonance is very different from isomerism, for which the atoms themselves are arranged differently. When resonance is possible, the substance is said to have a structure that is a resonance hybrid of the various contributing structures. We use a double-headed arrow ($\longleftrightarrow$) between contributing structures to distinguish resonance from an equilibrium between different compounds, for which we use $\rightleftharpoons$.

Each carbon–oxygen bond in the carbonate ion is neither single nor double, but something in between—perhaps a one-and-one-third bond (any particular carbon–oxygen bond is single in two contributing structures and double in one). Sometimes we represent a resonance hybrid with one formula by writing a solid line for each full bond and a dotted line for each partial bond (in the carbonate ion, the dots represent one-third of a bond).

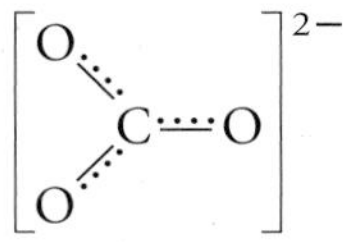

carbonate ion
resonance hybrid

PROBLEM 1.27 Draw the three equivalent contributing resonance structures for the nitrate ion, NO_3^-. What is the formal charge on the nitrogen atom and on each oxygen atom in the individual structures? What is the charge on the oxygens and on the nitrogen in the resonance hybrid structure? Show with curved arrows how the structures can be interconverted.

1.13 Arrow Formalism

Arrows in chemical drawings have specific meanings. For example, in Section 1.12 we used curved arrows to move electrons to show the relatedness of the three resonance structures of the carbonate ion. Just as it is important to learn the structural representations and names of molecules, it is important to learn the language of arrow formalism in organic chemistry.

Curved arrows show how electrons are moved in resonance structures and in reactions.

1. **Curved arrows** are used to show how electrons are moved in resonance structures and in reactions. Therefore, curved arrows always start at the initial position of electrons and end at their final position. In the example given below, the arrow that points from the C═O bond to the oxygen atom in the structure on the left indicates that the two electrons in one of the covalent bonds between carbon and oxygen are moved onto the oxygen atom:

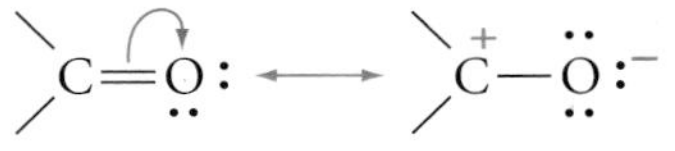

   Note that the carbon atom in the structure on the right now has a formal positive charge, and the oxygen has a formal negative charge. Notice also that when a pair of electrons in a polar covalent bond is moved to one of the bonded atoms, *it is moved to the more electronegative atom,* in this case oxygen. In the following example, the arrow that points from the unshared pair of electrons on the oxygen atom to a point between the carbon and oxygen atoms in the structure on the left indicates that the unshared pair of electrons on the oxygen atom moves between the oxygen and carbon atoms to form a covalent bond:

$$>\overset{+}{C}-\ddot{\underset{..}{O}}:^{-} \longleftrightarrow >C=\ddot{\underset{..}{O}}:$$

Note that both carbon and oxygen have formal charges of 0 in the structure on the right.

Fishhook arrows indicate the movement of only a single electron.

A curved arrow with half a head is called a **fishhook**. This kind of arrow is used to indicate the movement of a single electron. In eq. 1.6, two fishhooks are used to show the movement of each of the two electrons in the C—C bond of ethane to a carbon atom, forming two methyl radicals (see eq. 1.3):

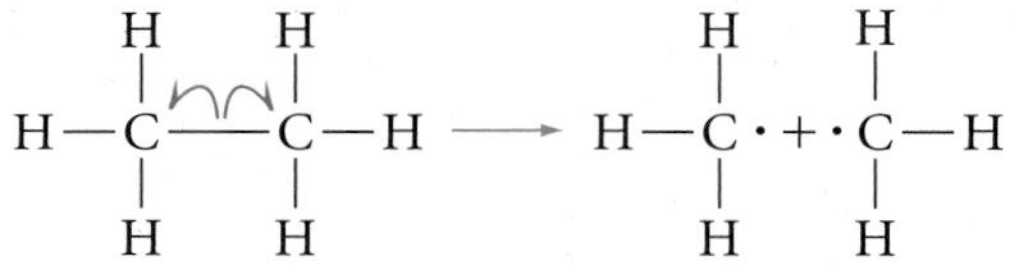

$$H_3C-CH_3 \longrightarrow H_3C\cdot + \cdot CH_3 \quad (1.6)$$

Straight arrows point from reactants to products in chemical reaction equations.

2. **Straight arrows** point from reactants to products in chemical reaction equations. An example is the straight arrow pointing from ethane to the two methyl radicals in eq. 1.6. Straight arrows with half-heads are commonly used in pairs to indicate that a reaction is *reversible.*

$$A + B \rightleftharpoons C + D$$

A double-headed straight arrow between two structures indicates resonance structures.

A **double-headed straight arrow** (⟷) between two structures indicates that they are resonance structures. Such an arrow does not indicate the occurrence of a chemical reaction. The double-headed arrows between resonance structures (Sec. 1.12) for the C=O bond are shown above.

EXAMPLE 1.15

Using correct arrow formalism, write the contributors to the resonance hybrid structure of the acetate ion, $CH_3CO_2^-$. Indicate any formal charges.

Solution There are two equivalent resonance structures for the acetate ion. Each one has a formal negative charge on one of the oxygen atoms.

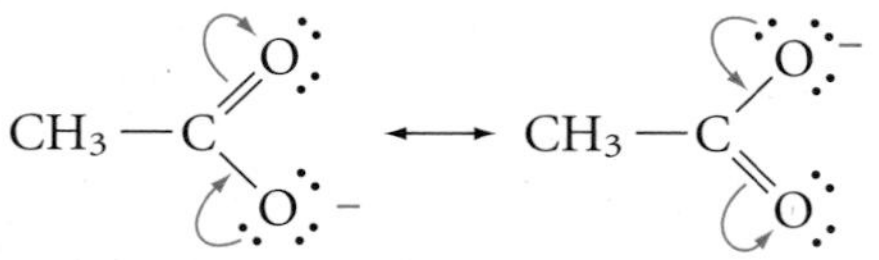

$$CH_3-C(=O)-O^- \longleftrightarrow CH_3-C(-O^-)=O$$

Notice that when one pair of electrons from oxygen is moved to form a covalent bond with carbon, a pair of electrons in a covalent bond between carbon and the other oxygen atom is moved to oxygen. This is necessary to ensure that the carbon atom does not exceed its valence of 4.

PROBLEM 1.28 Using correct arrow formalism, write the contributors to the resonance hybrid of azide ion, a linear ion with three connected nitrogens, N_3^-. Indicate the formal charge on each nitrogen atom.

We will use curved arrows throughout this text as a way of keeping track of electron movement. Several curved-arrow problems are included at the end of this chapter to help you get used to drawing them.

1.14 The Orbital View of Bonding; the Sigma Bond

Although electron-dot structures are often useful, they have some limitations. The Lewis theory of bonding itself has some limitations, especially in explaining the three-dimensional geometries of molecules. For this purpose in particular, we will discuss how another theory of bonding, involving orbitals, is more useful.

The atomic orbitals named in Section 1.1 have definite shapes. The *s* orbitals are spherical. The electrons that fill an *s* orbital confine their movement to a spherical region of space around the nucleus. The three *p* orbitals are dumbbell shaped and mutually perpendicular, oriented along the three coordinate axes, *x*, *y*, and *z*. Figure 1.2 shows the shapes of these orbitals.

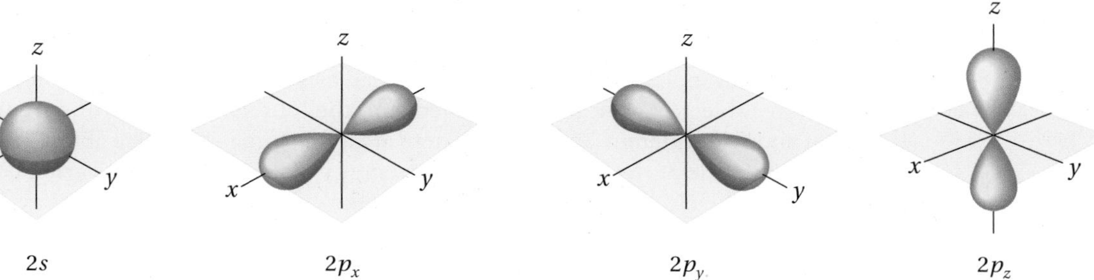

Figure 1.2
The shapes of the *s* and *p* orbitals used by the valence electrons of carbon. The nucleus is at the origin of the three coordinate axes.

In the orbital view of bonding, atoms approach each other in such a way that their atomic orbitals can *overlap* to form a bond. For example, if two hydrogen atoms form a hydrogen molecule, their two spherical 1*s* orbitals combine to form a new orbital that encompasses both of the atoms (see Figure 1.3). This orbital contains both valence electrons (one from each hydrogen). Like atomic orbitals, each **molecular orbital** can contain no more than two electrons. In the hydrogen molecule, these electrons mainly occupy the space between the two nuclei.

A **molecular orbital** is the space occupied by electrons in a molecule.

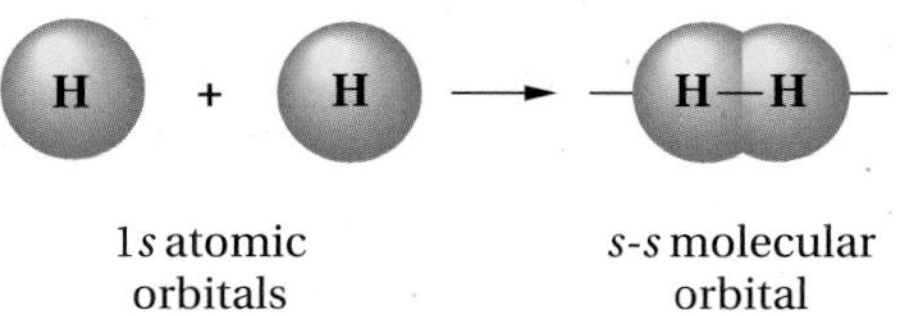

Figure 1.3
The molecular orbital representation of covalent bond formation between two hydrogen atoms.

The orbital in the hydrogen molecule is cylindrically symmetric along the H—H internuclear axis. Such orbitals are called **sigma (σ) orbitals**, and the bond is referred to as a **sigma bond**. Sigma bonds may also be formed by the overlap of an *s* and a *p* orbital or of two *p* orbitals, as shown in Figure 1.4.*

A **sigma (σ) orbital** lies along the axis between two bonded atoms; a pair of electrons in a sigma orbital is called a **sigma bond**.

*Two properly aligned *p* orbitals can also overlap to form another type of bond, called a π (pi) bond. We discuss this type of bond in Chapter 3.

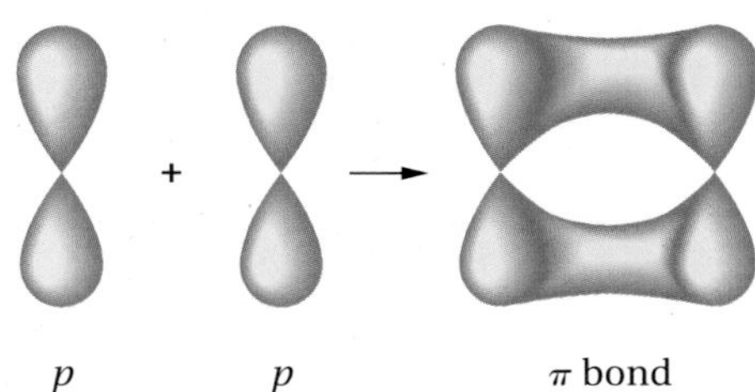

▪ **Figure 1.4**
Orbital overlap to form σ bonds.

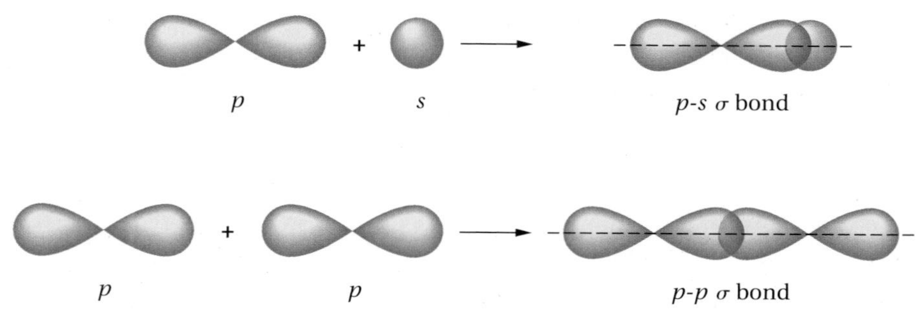

Let us see how these ideas apply to bonding in carbon compounds.

1.15 Carbon sp^3 Hybrid Orbitals

▪ **Figure 1.5**
Distribution of the six electrons in a carbon atom. Each dot stands for an electron.

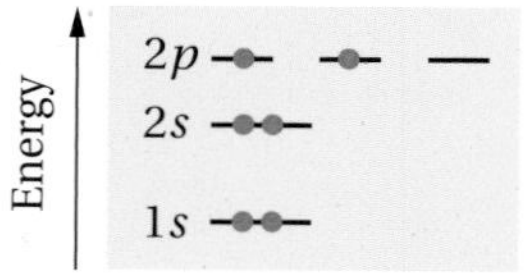

In a carbon atom, the six electrons are arranged as shown in Figure 1.5 (compare with carbon in Table 1.2). The 1*s* shell is filled, and the four valence electrons are in the 2*s* orbital and two different 2*p* orbitals. There are a few things to notice about Figure 1.5. The energy scale at the left represents the energy of electrons in the various orbitals. The farther the electron is from the nucleus, the greater its potential energy, because it takes energy to keep the electron (negatively charged) and the nucleus (positively charged) apart. The 2*s* orbital has a slightly lower energy than the three 2*p* orbitals, which have equal energies (they differ from one another only in orientation around the nucleus, as shown in Figure 1.2). The two highest energy electrons are placed in different 2*p* orbitals rather than in the same orbital, because this keeps them farther apart and thus reduces the repulsion between these like-charged particles. One *p* orbital is vacant.

We might get a misleading idea about the bonding of carbon from Figure 1.5. For example, we might think that carbon should form only two bonds (to complete the partially filled 2*p* orbitals) or perhaps three bonds (if some atom donated two electrons to the empty 2*p* orbital). But we know from experience that this picture is wrong. Carbon usually forms *four* single bonds, and often these bonds are all equivalent, as in CH_4 or CCl_4. How can this discrepancy between theory and fact be resolved?

An sp^3 **hybrid orbital** is a *p*-shaped orbital that is one part *s* and three parts *p* in character.

One solution, illustrated in Figure 1.6, is to mix or combine the four atomic orbitals of the valence shell to form four identical hybrid orbitals, each containing one valence electron. In this model, the hybrid orbitals are called sp^3 **hybrid orbitals** because each one has one part *s* character and three parts *p* character. As shown in Figure 1.6, each sp^3 orbital has the same energy: less than that of the 2*p* orbitals but greater than that of the 2*s* orbital. The shape of sp^3 orbitals resembles the shape of *p* orbitals, except that the dumbbell is lopsided, and the electrons are more likely to be found in the lobe that extends out the greater distance from the nucleus, as shown in Figure 1.7. The four sp^3 hybrid orbitals of a single carbon atom are directed toward the corners of a regular tetrahedron, also shown in Figure 1.7. This particular geometry puts each orbital as far from the other three orbitals as it can be and thus minimizes repulsion when the orbitals are filled with electron pairs. The angle between any two of the four bonds formed from sp^3 orbitals is approximately 109.5°, the angle made by lines drawn from the center to the corners of a regular tetrahedron.

▪ **Figure 1.6**
Unhybridized vs. sp^3 hybridized orbitals on carbon. The dots stand for electrons. (Only the electrons in the valence shell are shown; the electrons in the 1*s* orbital are omitted because they are not involved in bonding.)

Energy
2p
2s
sp^3
Atomic orbitals of carbon
Four equivalent sp^3 hybrid orbitals

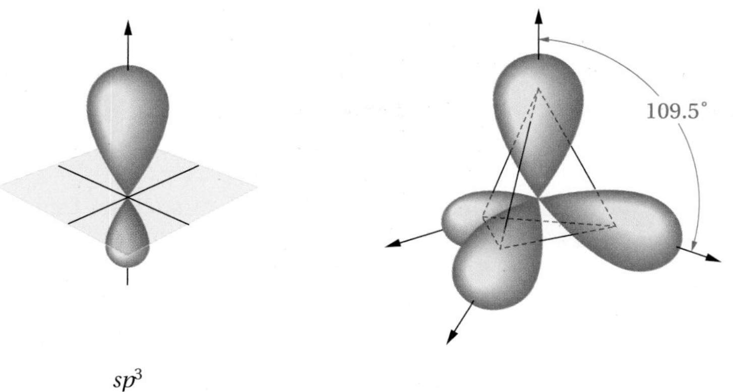

▰ Figure 1.7
An sp^3 orbital extends mainly in one direction from the nucleus and forms bonds with other atoms in that direction. The four sp^3 orbitals of any particular carbon atom are directed toward the corners of a regular tetrahedron, as shown in the right-hand part of the figure (in this part of the drawing, the small "back" lobes of the orbitals have been omitted for simplification, although they can be important in chemical reactions).

Hybrid orbitals can form sigma bonds by overlap with other hybrid orbitals or with nonhybridized atomic orbitals. Figure 1.8 shows some examples.

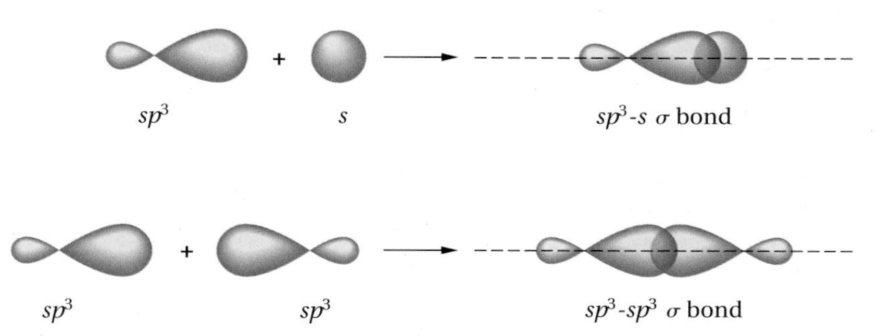

▰ Figure 1.8
Examples of sigma (σ) bonds formed from sp^3 hybrid orbitals.

1.16 Tetrahedral Carbon; the Bonding in Methane

We can now describe how a carbon atom combines with four hydrogen atoms to form methane. This process is pictured in Figure 1.9. The carbon atom is joined to each hydrogen atom by a sigma bond, which is formed by the overlap of a carbon sp^3 orbital with a hydrogen 1*s* orbital. The four sigma bonds are directed from the carbon nucleus to the corners of a regular tetrahedron. In this way, the electron pair in any one bond experiences minimum repulsion from the electrons in the other bonds. Each H—C—H **bond angle** is the same, 109.5°. To summarize, in methane, there are four sp^3–*s* C—H sigma bonds, each directed from the carbon atom to one of the four corners of a regular tetrahedron.

A **bond angle** is the angle made by two covalent bonds to the same atom.

H H H H

methane, CH_4

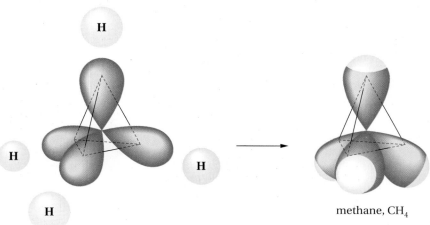

▰ Figure 1.9
A molecule of methane, CH_4, is formed by the overlap of the four sp^3 carbon orbitals with the 1*s* orbitals of four hydrogen atoms. The resulting molecule has the geometry of a regular tetrahedron and contains four sigma bonds of the sp^3–*s* type.

PROBLEM 1.29 Considering the repulsion that exists between electrons in different bonds, give a reason why a planar geometry for methane would be less stable than the tetrahedral geometry.

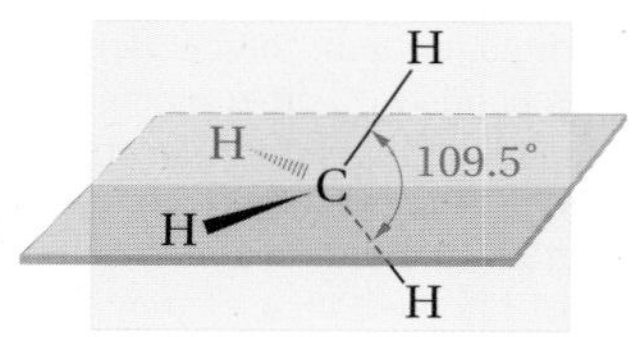

■ **Figure 1.10**
The carbon and two of the hydrogens in methane form a plane that perpendicularly bisects the plane formed by the carbon and the other two hydrogens.

Because the tetrahedral geometry of carbon plays such an important role in organic chemistry, it is a good idea to become familiar with the features of a regular tetrahedron. One feature is that *the center and any two corners of a tetrahedron form a plane that is the perpendicular bisector of a similar plane formed by the center and the other two corners.* In methane, for example, any two hydrogens and the carbon form a plane that perpendicularly bisects the plane formed by the carbon and the other two hydrogens. These planes are illustrated in Figure 1.10.

The geometry of carbon with four single bonds, as in methane, is commonly represented as shown in Figure 1.11a, in which the **solid lines** lie in the plane of the page, the **dashed wedge** goes behind the plane of the paper, and the **solid wedge** extends out of the plane of the paper toward you. Structures drawn in this way are sometimes called **3D** (that is, three-dimensional) **structures**.

Two other 3D representations commonly used are the ball-and-stick model (Figure 1.11b) and the space-filling model (Figure 1.11c). The ball-and-stick model emphasizes the bonds that connect the atoms, while the space-filling model emphasizes the space occupied by the atoms.

In addition, a 3D representation, called an **electrostatic potential map**, is sometimes used to show the distribution of electrons in a molecule (Figure 1.11d). Red indicates partial negative charge (greater electron density), and blue indicates partial positive

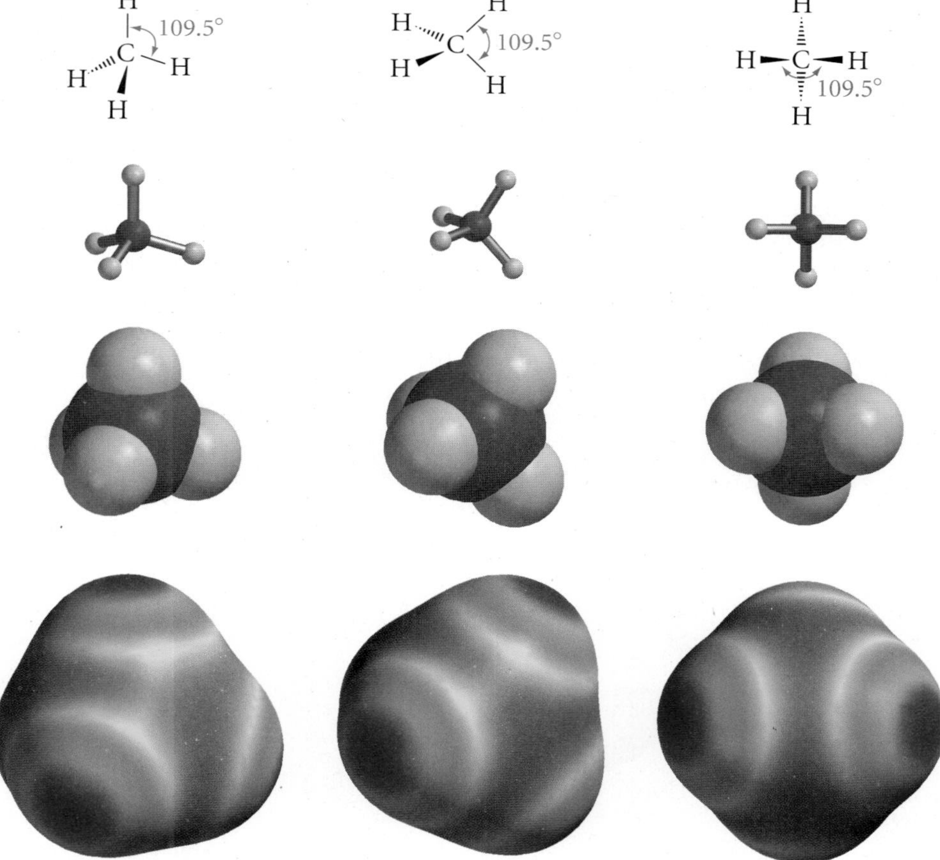

(a) In a **3D structure, solid lines** lie in the plane of the page (C and H in C—H lie in the plane). **Dashed wedges** extend behind the plane (H in C⋯H lies behind the plane). **Solid wedges** project out toward you (H in C◄H is in front of the plane).

(b) A **ball-and-stick model** of a molecule emphasizes the bonds that connect atoms.

(c) A **space-filling model** emphasizes the space occupied by the atoms.

(d) An **electrostatic potential map** shows the distribution of electrons in a molecule. Red indicates partial negative charge, and blue indicates partial positive charge.

■ **Figure 1.11**
Four representations of methane.

charge (less electron density). This representation is useful for showing whether a molecule is polar or nonpolar (see Sec. 2.7) and can give us insight into the chemical behavior of a molecule, as we will see when we look at the chemistry of functional groups in later chapters.

Now that we have described single covalent bonds and their geometry, we are ready to tackle, in the next chapter, the structure and chemistry of saturated hydrocarbons. But before we do that, we present a brief overview of organic chemistry, so that you can see how the subject will be organized for study.

Because carbon atoms can be linked to one another or to other atoms in so many different ways, the number of possible organic compounds is almost limitless. Literally millions of organic compounds have been characterized, and the number grows daily. How can we hope to study this vast subject systematically? Fortunately, organic compounds can be classified according to their structures into a relatively small number of groups. Structures can be classified both according to the molecular framework (sometimes called the carbon *skeleton*) and according to the groups that are attached to that framework.

1.17 Classification According to Molecular Framework

The three main classes of molecular frameworks for organic structures are acyclic, carbocyclic, and heterocyclic compounds.

1.17.a Acyclic Compounds

By **acyclic** (pronounced a′-cyclic), we mean *not cyclic.* Acyclic organic molecules have chains of carbon atoms but no rings. As we have seen, the chains may be unbranched or branched.

Acyclic compounds contain no rings. **Carbocyclic compounds** contain rings of carbon atoms. **Heterocyclic compounds** have rings containing at least one atom that is *not* carbon.

unbranched chain of eight carbon atoms

branched chain of eight carbon atoms

Pentane is an example of an acyclic compound with an unbranched carbon chain, whereas isopentane and neopentane are also acyclic but have branched carbon frameworks (Sec. 1.9). Figure 1.12 shows the structures of a few acyclic compounds that occur in nature.

geraniol	heptane	2-heptanone
(structure with OH)	$CH_3(CH_2)_5CH_3$	$CH_3C(=O)(CH_2)_4CH_3$
(oil of roses)	(petroleum)	(oil of cloves)
bp 229–230°C	bp 98.4°C	bp 151.5°C
A branched chain compound used in perfumes	A hydrocarbon present in petroleum, used as a standard in testing the octane rating of gasoline	A colorless liquid with a fruity odor, in part responsible for the "peppery" odor of blue cheese

■ **Figure 1.12**
Examples of natural acyclic compounds, their sources (in parentheses), and selected characteristics.

Foodcollection/Getty Images

2-Heptanone contributes to the "peppery" odor of blue cheese.

1.17.b Carbocyclic Compounds

Carbocyclic compounds contain rings of carbon atoms. The smallest possible carbocyclic ring has three carbon atoms, but carbon rings come in many sizes and shapes. The rings may have chains of carbon atoms attached to them and may contain multiple bonds. Many compounds with more than one carbocyclic ring are known. Figure 1.13 shows the structures of a few carbocyclic compounds that occur in nature. Five- and six-membered rings are most common, but smaller and larger rings are also found.

Figure 1.13
Examples of natural carbocyclic compounds with rings of various sizes and shapes. The source and special features of each structure are indicated below it.

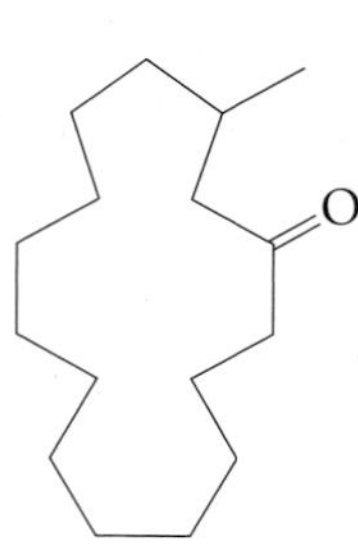

muscone
(musk deer)
bp 327–330°C

A 15-membered ring ketone, used in perfumes

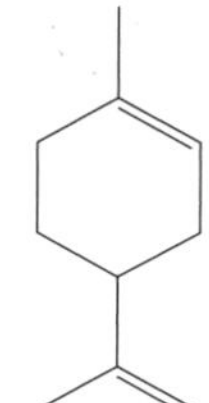

limonene
(citrus fruit oils)
bp 178°C

A ring with two side chains, one of which is branched

benzene
(petroleum)
mp 5.5°C, bp 80.1°C

A very common ring

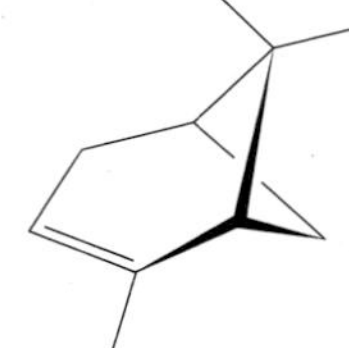

α-pinene
(turpentine)
bp 156.2°C

A bicyclic molecule; one would have to break *two* bonds to make it acyclic

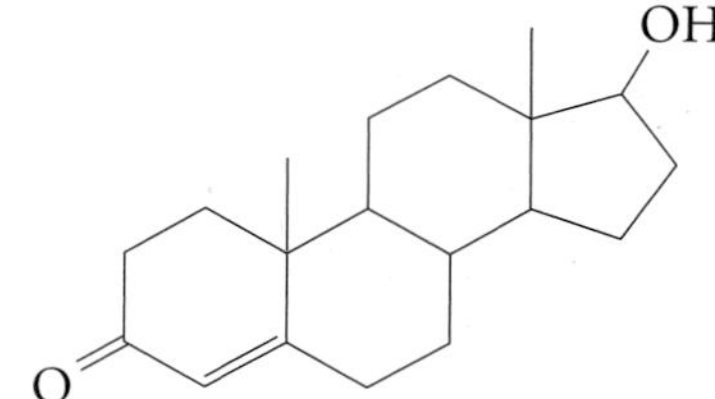

testosterone
(testes)
mp 155°C

A male sex hormone in which several rings of common sizes are *fused* together; that is, they share two adjacent carbon atoms

Musk deer, source of muscone.

1.17.c Heterocyclic Compounds

Heterocyclic compounds make up the third and largest class of molecular frameworks for organic compounds. In heterocyclic compounds, at least one atom in the ring must be a heteroatom, an atom that is *not* carbon. The most common heteroatoms are oxygen, nitrogen, and sulfur, but heterocyclics with other elements are also known. More than one heteroatom may be present and, if so, the heteroatoms may be alike or different. Heterocyclic rings come in many sizes, may contain multiple bonds, may have carbon chains or rings attached to them, and in short may exhibit a great variety of structures. Figure 1.14 shows the structures of a few natural products that contain heterocyclic rings. In these abbreviated structural formulas, the symbols for the heteroatoms are shown, but the carbons are indicated using lines only.

The structures in Figures 1.12 through 1.14 show not only the molecular frameworks, but also various groups of atoms that may be part of or attached to the frameworks. Fortunately, these groups can also be classified in a way that helps simplify the study of organic chemistry.

Clover, a source of coumarin.

nicotine
bp 246°C

Present in tobacco, nicotine has two heterocyclic rings of different sizes, each containing one nitrogen.

adenine
mp 360–365°C
(decomposes)

One of the four heterocyclic bases of DNA, adenine contains two fused heterocyclic rings, each of which contains two heteroatoms (nitrogen).

penicillin-G
(amorphous solid)

One of the most widely used antibiotics, penicillin has two heterocyclic rings, the smaller of which is crucial to biological activity.

coumarin
mp 71°C

Found in clover and grasses, coumarin produces the pleasant odor of new-mown hay.

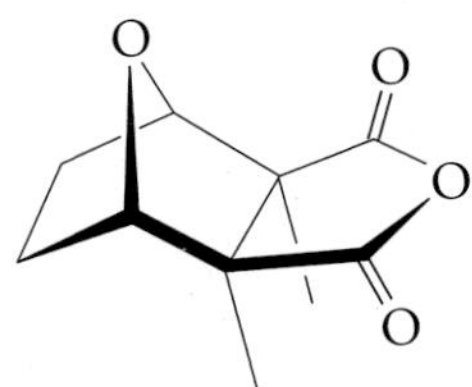

α-terthienyl
mp 92–93°C

This compound, with three linked sulfur-containing rings, is present in certain marigold species.

cantharidin
mp 218°C

This compound, an oxygen heterocycle, is the active principle in cantharis (also known as Spanish fly), a material isolated from certain dried beetles of the species *Cantharis vesicatoria* and incorrectly thought by some to increase sexual desire.

■ Figure 1.14
Examples of natural heterocyclic compounds having a variety of heteroatoms and ring sizes.

1.18 Classification According to Functional Group

Functional groups are groups of atoms that have characteristic chemical properties regardless of the molecular framework to which they are attached.

Certain groups of atoms have chemical properties that depend only moderately on the molecular framework to which they are attached. These groups of atoms are called **functional groups.** The hydroxyl group, —OH, is an example of a functional group, and compounds with this group attached to a carbon framework are called alcohols. In most organic reactions, some chemical change occurs at the functional group, but the rest of the molecule keeps its original structure. This maintenance of most of the structural formula throughout a chemical reaction greatly simplifies our study of organic chemistry. It allows us to focus attention on the chemistry of the various functional groups. We can study classes of compounds instead of having to learn the chemistry of each individual compound.

Some of the main functional groups that we will study are listed in Table 1.6, together with a typical compound of each type. Although we will describe these classes of compounds in greater detail in later chapters, it would be a good idea for you to become familiar with their names and structures now. If a particular functional group is mentioned before its chemistry is discussed in detail, and you forget what it is, you can refer to Table 1.6 or to the inside front cover of this book.

PROBLEM 1.30 What functional groups can you find in the following natural products? (Their formulas are given in Figures 1.12, 1.13, and 1.14.)

a. testosterone b. penicillin-G c. muscone d. α-pinene

Table 1.6 ▪ The Main Functional Groups

	Structure	Class of compound	Specific example	Common name of the specific example
A. Functional groups that are a part of the molecular framework	$-\overset{\vert}{\underset{\vert}{C}}-\overset{\vert}{\underset{\vert}{C}}-$	alkane	$CH_3—CH_3$	ethane, a component of natural gas
	$>C=C<$	alkene	$CH_2═CH_2$	ethylene, used to make polyethylene
	$—C≡C—$	alkyne	$HC≡CH$	acetylene, used in welding
	(benzene ring)	arene	(benzene ring)	benzene, raw material for polystyrene and phenol
B. Functional groups containing oxygen				
1. With carbon–oxygen single bonds	$-\overset{\vert}{\underset{\vert}{C}}-OH$	alcohol	CH_3CH_2OH	ethyl alcohol, found in beer, wines, and liquors
	$-\overset{\vert}{\underset{\vert}{C}}-O-\overset{\vert}{\underset{\vert}{C}}-$	ether	$CH_3CH_2OCH_2CH_3$	diethyl ether, once a common anesthetic

(continued)

Table 1.6 ■ continued

	Structure	Class of compound	Specific example	Common name of the specific example
*2. With carbon–oxygen double bonds**	$-\overset{O}{\overset{\Vert}{C}}-H$	aldehyde	$CH_2{=}O$	formaldehyde, used to preserve biological specimens
	$-\overset{\vert}{\underset{\vert}{C}}-\overset{O}{\overset{\Vert}{C}}-\overset{\vert}{\underset{\vert}{C}}-$	ketone	$CH_3\overset{O}{\overset{\Vert}{C}}CH_3$	acetone, a solvent for varnish and rubber cement
3. With single and double carbon–oxygen bonds	$-\overset{O}{\overset{\Vert}{C}}-OH$	carboxylic acid	$CH_3\overset{O}{\overset{\Vert}{C}}-OH$	acetic acid, a component of vinegar
	$-\overset{O}{\overset{\Vert}{C}}-O-\overset{\vert}{\underset{\vert}{C}}-$	ester	$CH_3\overset{O}{\overset{\Vert}{C}}-OCH_2CH_3$	ethyl acetate, a solvent for nail polish and model airplane glue
*C. Functional groups containing nitrogen***	$-\overset{\vert}{\underset{\vert}{C}}-NH_2$	primary amine	$CH_3CH_2NH_2$	ethylamine, smells like ammonia
	$-C{\equiv}N$	nitrile	$CH_2{=}CH-C{\equiv}N$	acrylonitrile, raw material for making Orlon
D. Functional group with oxygen and nitrogen	$-\overset{O}{\overset{\Vert}{C}}-NH_2$	primary amide	$H-\overset{O}{\overset{\Vert}{C}}-NH_2$	formamide, a softener for paper
E. Functional group with halogen	$-X$	alkyl or aryl halide	CH_3Cl	methyl chloride, refrigerant and local anesthetic
F. Functional groups containing sulfur[†]	$-\overset{\vert}{\underset{\vert}{C}}-SH$	thiol (also called mercaptan)	CH_3SH	methanethiol, has the odor of rotten cabbage
	$-\overset{\vert}{\underset{\vert}{C}}-S-\overset{\vert}{\underset{\vert}{C}}-$	thioether (also called sulfide)	$(CH_2{=}CHCH_2)_2S$	diallyl sulfide, has the odor of garlic

*The $\rangle C{=}O$ group, present in several functional groups, is called a **carbonyl group**. The $-\overset{O}{\overset{\Vert}{C}}-OH$ group of acids is called a **carboxyl group** (a contraction of *carb*onyl and hydr*oxyl*).

The $-NH_2$ group is called an **amino group.

[†]Thiols and thioethers are the sulfur analogs of alcohols and ethers.

ADDITIONAL PROBLEMS

OWL Interactive versions of these problems are assignable in OWL.

Valence, Bonding, and Lewis Structures

1.31 Show the number of valence electrons in each of the following atoms. Let the element's symbol represent its kernel, and use dots for the valence electrons.

a. chlorine b. nitrogen c. sulfur
d. carbon e. oxygen f. potassium

1.32 Use the relative positions of the elements in the periodic table (Table 1.3 or inside back cover) to classify the following substances as ionic or covalent:

a. F_2 b. SiF_4 c. CH_4
d. O_2 e. $CaCl_2$ f. NaI
g. NaBr h. ClF

1.33 When a solution of salt (sodium chloride) in water is treated with a silver nitrate solution, a white precipitate forms immediately. When tetrachloromethane is shaken with aqueous silver nitrate, no such precipitate is produced. Explain these facts in terms of the types of bonds present in the two chlorides.

1.34 For each of the following elements, determine (1) how many valence electrons it has and (2) what its common valence is:

a. N b. C c. F
d. O e. P f. S

1.35 Write a structural formula for each of the following compounds, using a line to represent each single bond and dots for any unshared electron pairs:

a. CH_3OH b. CH_3CH_2Cl c. C_3H_8
d. $CH_3CH_2NH_2$ e. C_2H_5F f. CH_2O

1.36 Draw a structural formula for each of the following covalent molecules. Which bonds are polar? Indicate the polarity by proper placement of the symbols $\delta+$ and $\delta-$.

a. BF_3 b. CH_3F c. CO_2
d. Cl_2 e. SF_6 f. CH_4
g. SO_2 h. CH_3OH

1.37 Consider the X—H bond, in which X is an atom other than H. The H in a polar bond is more acidic (more easily removed) than the H in a nonpolar bond. Considering bond polarity, which hydrogen in acetic acid, $CH_3C(=O)—OH$, do you expect to be most acidic? Write an equation for the reaction between acetic acid and potassium carbonate.

Structural Isomers

1.38 Draw structural formulas for all possible isomers having the following molecular formulas:

a. C_3H_8 b. C_3H_7F c. $C_2H_2Br_2$
d. C_3H_6 e. C_4H_9Cl f. $C_3H_6Cl_2$
g. C_3H_8S h. $C_2H_4F_2$

1.39 Draw structural formulas for the five isomers of C_6H_{14}. As you write them out, try to be systematic, starting with a consecutive chain of six carbon atoms.

Structural Formulas

1.40 For each of the following abbreviated structural formulas, write a structural formula that shows all of the bonds:

a. $CH_3(CH_2)_4CH_3$
b. $(CH_3)_3CCH_2CH_2CH_3$
c. $(CH_3CH_2)_2NH$
d. $CH_3CH_2SCH_2CH_3$
e. $ClCH_2CH_2OH$
f. $(CH_3CH_2)_2CHOH$

1.41 Write structural formulas that correspond to the following abbreviated structures, and show the correct number of hydrogens on each carbon:

a. 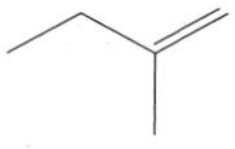
b. 
c. 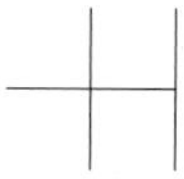
d. 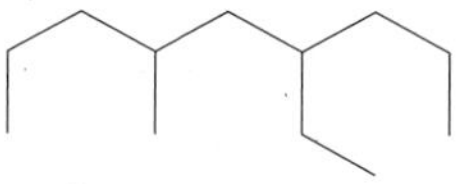
e.  O
f. 
g. 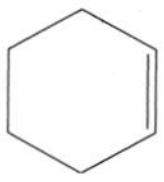
h.  O

1.42 For each of the following abbreviated structural formulas, write a line-segment formula (like those in Problem 1.41).

a. $CH_3(CH_2)_4CH_3$

b. $(CH_3)_2CHCH_2CH_2\overset{\overset{\displaystyle O}{\|}}{C}CH_3$

c. $CH_3\underset{\underset{\displaystyle OH}{|}}{C}HCH_2C(CH_3)_3$

d. 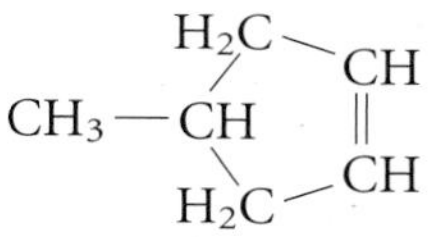 $CH_3—CH$ bonded to H_2C and H_2C, which are bonded to $CH=CH$ (ring: $H_2C—CH=CH—H_2C$)

1.43 An abbreviated formula of 2-heptanone is shown in Figure 1.12.

a. How many carbons does 2-heptanone have?
b. What is its molecular formula?
c. Write a more detailed structural formula for it.

1.44 What is the *molecular formula* for each of the following compounds? Consult Figures 1.13 and 1.14 for the abbreviated structural formulas.

a. nicotine
b. adenine
c. limonene
d. coumarin
e. benzene

Formal Charge, Resonance, and Curved-Arrow Formalism

1.45 Write electron-dot formulas for the following species. Show where the formal charges, if any, are located.

a. cyanide ion, CN^-
b. nitric acid, $HONO_2$
c. dimethyl ether, CH_3OCH_3
d. ammonium ion, NH_4^+
e. nitrous acid, HONO
f. carbon monoxide, CO
g. boron trifluoride, BF_3
h. hydrogen peroxide, H_2O_2
i. bicarbonate ion, HCO_3^-

1.46 Draw electron-dot formulas for the two contributors to the resonance hybrid structure of the nitrite ion, NO_2^-. (Each oxygen is connected to the nitrogen.) What is the charge on each oxygen in each contributor and in the hybrid structure? Show by curved arrows how the electron pairs can relocate to interconvert the two structures.

1.47 Write the structure obtained when electrons move as indicated by the curved arrows in the following structure:

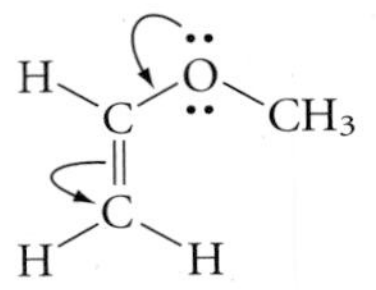

Does each atom in the resulting structure have a complete valence shell of electrons? Locate any formal charges in each structure.

1.48 Consider each of the following highly reactive carbon species. What is the formal charge on carbon in each of these structures?

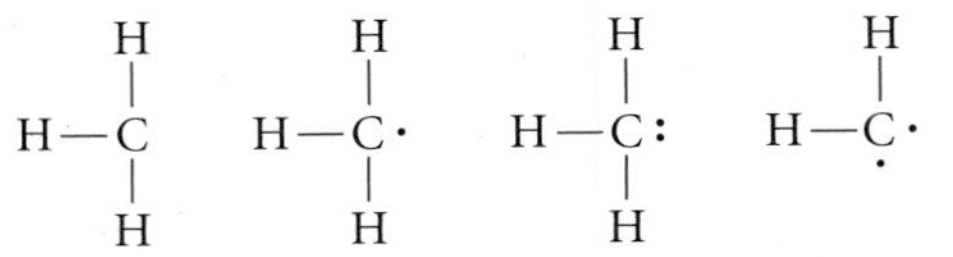

1.49 Add curved arrows to the following structures to show how electron pairs must be moved to interconvert the structures, and locate any formal charges.

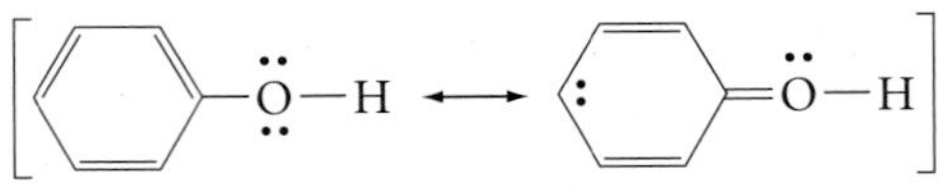

1.50 Add curved arrows to show how electrons must move to form the product from the reactants in the following equation, and locate any formal charges.

$$CH_3-\ddot{O}H + CH_3-\overset{\overset{\cdot\cdot}{:O}}{\overset{\|}{C}}-OCH_3 \longrightarrow CH_3-\overset{:\ddot{O}:}{\overset{|}{C}}-OCH_3 \text{ (with } H\ddot{O}-CH_3 \text{ on C)}$$

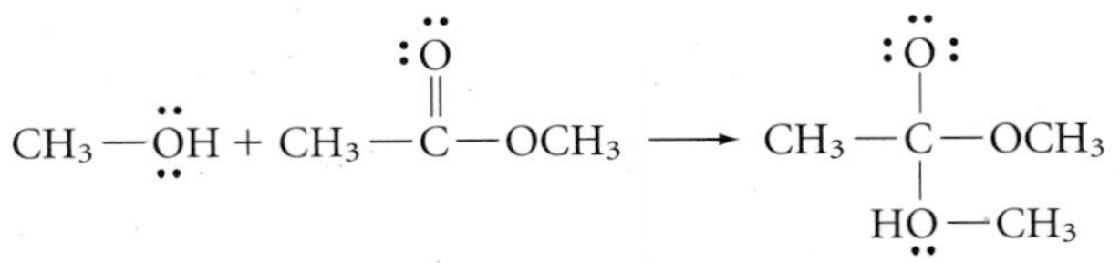

Electronic Structure and Molecular Geometry

1.51 Fill in any unshared electron pairs that are missing from the following formulas:

a. $(CH_3CH_2)_2NH$

b. $CH_3\overset{O}{\overset{\|}{C}}-OH$

c. $CH_3CH_2SCH_2CH_3$

d. $CH_3OCH_2CH_2OH$

1.52 Make a drawing (similar to the right-hand part of Figure 1.6) of the electron distribution that will be expected in nitrogen atoms if the *s* and *p* orbitals are hybridized to sp^3. Based on this model, predict the geometry of the ammonia molecule, NH_3.

1.53 The ammonium ion, NH_4^+, has a tetrahedral geometry analogous to that of methane. Explain this structure in terms of atomic and molecular orbitals.

1.54 Use lines, dashed wedges, and solid wedges to show the geometry of CF_4 and CH_3SH.

1.55 Silicon is just below carbon in the periodic table. Predict the geometry of silicon tetrafluoride, SiF_4.

1.56 Examine the three ball-and-stick models shown below:

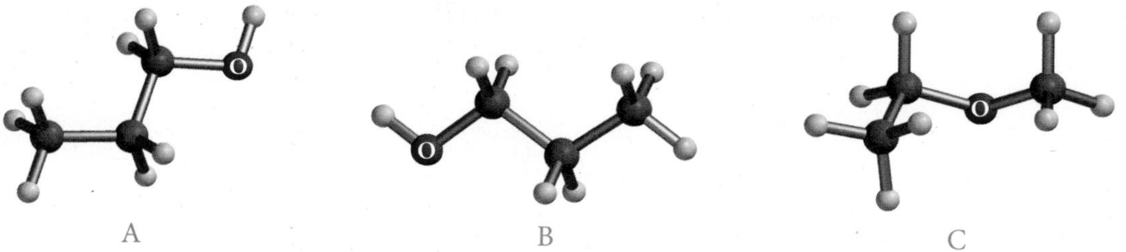

a. Redraw the three structures using solid lines, dashed wedges, and solid wedges (see Figure 1.11).
b. What is the relationship, identical or isomers, between structures A and B? Between structures A and C?

Classification of Organic Compounds

1.57 Write a structural formula that corresponds to the molecular formula C_3H_6O and is

a. acyclic
b. carbocyclic
c. heterocyclic

1.58 Draw a chemical structure for diacetyl (2,3-butanedione), a diketone (see Table 1.6) with the $C_4H_6O_2$ chemical formula. Diacetyl is used as a flavoring for microwave popcorn, but has been under scrutiny of late as a possible causative agent for bronchiolitis obliterans, often referred to as popcorn lung disease.

1.59 Divide the following compounds into groups that might be expected to exhibit similar chemical behavior:

a. C_4H_{10}
b. CH_3OCH_3
c. C_3H_7OH
d. C_8H_{18}
e. $HOCH_2CH_2CH_2OH$
f. CH_3NH_2
g. $CH_3CH_2CH_3$
h. CH_3OH
i. $(CH_3)_2CHNH_2$
j. C_3H_7OH
k. $CH_3CH_2OCH_3$
l. $H_2NCH_2CH_2NH_2$

1.60 Using Table 1.6, write a structural formula for each of the following:

a. an alcohol, C_3H_8O
b. an ether, $C_4H_{10}O$
c. an aldehyde, C_3H_6O
d. a ketone, C_3H_6O
e. a carboxylic acid, $C_3H_6O_2$
f. an ester, $C_5H_{10}O_2$

1.61 Many organic compounds contain more than one functional group. An example is phenylalanine (shown below), one of the simple building blocks of proteins (Chapter 17).

O
‖
HO—C—$CHCH_2$—(benzene ring)
|
NH_2

phenylalanine

a. What functional groups are present in phenylalanine?
b. Redraw the structure, adding all unshared electron pairs.
c. What is the molecular formula of phenylalanine?
d. Draw another structural isomer that has this formula. What functional groups does this isomer have?

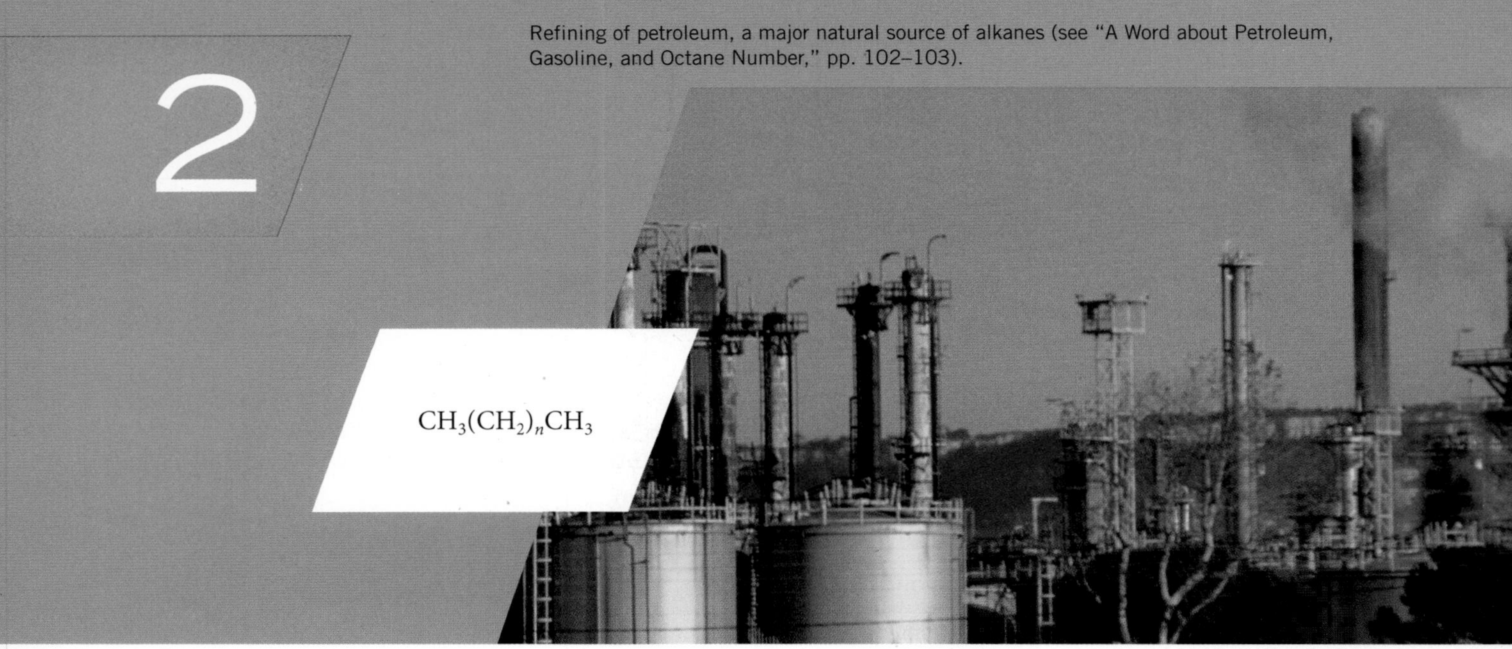

Refining of petroleum, a major natural source of alkanes (see "A Word about Petroleum, Gasoline, and Octane Number," pp. 102–103).

Sami Sarkis/Photodisc/Getty images

Alkanes and Cycloalkanes; Conformational and Geometric Isomerism

The main components of petroleum and natural gas, resources that now supply most of our fuel for energy, are **hydrocarbons**, compounds that contain only carbon and hydrogen. There are three main classes of hydrocarbons, based on the types of carbon–carbon bonds present. **Saturated hydrocarbons** contain only carbon–carbon *single* bonds. **Unsaturated hydrocarbons** contain carbon–carbon *multiple* bonds—double bonds, triple bonds, or both. **Aromatic hydrocarbons** are a special class of cyclic compounds related in structure to benzene.*

Saturated hydrocarbons are known as **alkanes** if they are acyclic, or as **cycloalkanes** if they are cyclic. Let us look at their structures and properties.

Online homework for this chapter can be assigned in OWL, an online homework assessment tool.

*Unsaturated and aromatic hydrocarbons are discussed in Chapters 3 and 4, respectively.

2.1 The Structures of Alkanes

The simplest alkane is methane. Its tetrahedral three-dimensional structure was described in the previous chapter (see Figure 1.9). Additional alkanes are constructed by lengthening the carbon chain and adding an appropriate number of hydrogens to complete the carbon valences (for examples, see Figure 2.1* and Table 2.1).

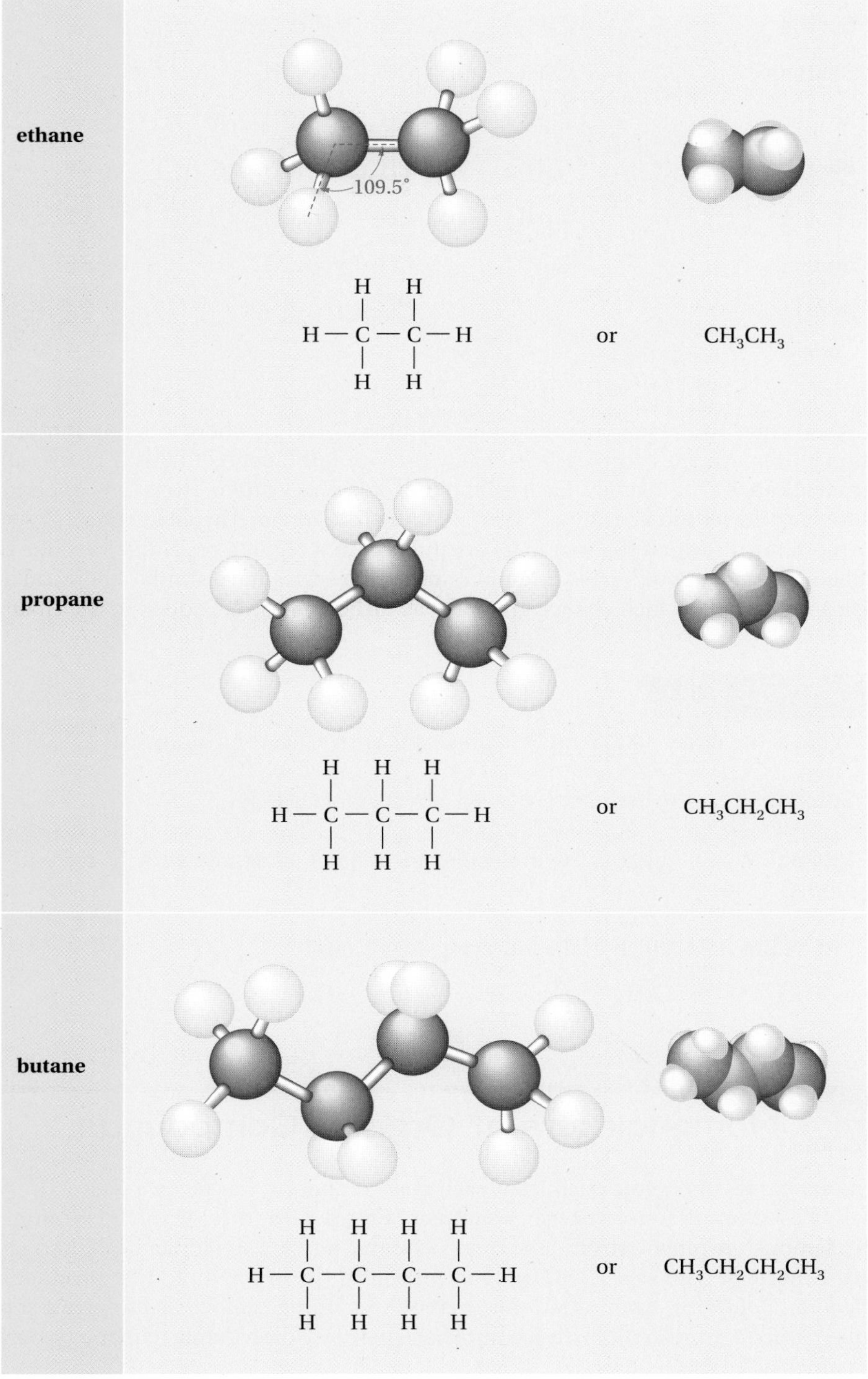

Figure 2.1 Three-dimensional models of ethane, propane, and butane. The ball-and-stick models at the left show the way in which the atoms are connected and depict the correct bond angles. The space-filling models at the right are constructed to scale and give a better idea of the molecular shape, though some of the hydrogens may appear hidden.

*Molecular models can help you visualize organic structures in three dimensions. They will be extremely useful to you throughout this course, especially when we consider various types of isomerism. Relatively inexpensive sets are usually available at stores that sell textbooks, and your instructor can suggest which kind to buy. If you cannot locate or afford a set, you can create models that are adequate for most purposes from toothpicks (for bonds) and marshmallows, gum drops, or jelly beans (for atoms).

Alkanes are **saturated hydrocarbons**, containing only carbon–carbon single bonds. **Cycloalkanes** contain rings. **Unsaturated hydrocarbons** contain carbon–carbon double or triple bonds. **Aromatic hydrocarbons** are cyclic compounds structurally related to benzene.

Table 2.1 ■ Names and Formulas of the First Ten Unbranched Alkanes

Name	Number of carbons	Molecular formula	Structural formula	Number of structural isomers
methane	1	CH_4	CH_4	1
ethane	2	C_2H_6	CH_3CH_3	1
propane	3	C_3H_8	$CH_3CH_2CH_3$	1
butane	4	C_4H_{10}	$CH_3CH_2CH_2CH_3$	2
pentane	5	C_5H_{12}	$CH_3(CH_2)_3CH_3$	3
hexane	6	C_6H_{14}	$CH_3(CH_2)_4CH_3$	5
heptane	7	C_7H_{16}	$CH_3(CH_2)_5CH_3$	9
octane	8	C_8H_{18}	$CH_3(CH_2)_6CH_3$	18
nonane	9	C_9H_{20}	$CH_3(CH_2)_7CH_3$	35
decane	10	$C_{10}H_{22}$	$CH_3(CH_2)_8CH_3$	75

Unbranched alkanes are called **normal alkanes**, or *n*-alkanes.

A —CH_2— group is called a **methylene group**.

Compounds of a **homologous series** differ by a regular unit of structure and share similar properties.

All alkanes fit the general molecular formula C_nH_{2n+2}, where n is the number of carbon atoms. Alkanes with carbon chains that are unbranched (Table 2.1) are called **normal alkanes** or *n*-alkanes. Each member of this series differs from the next higher and the next lower member by a —CH_2— group (called a **methylene group**). A series of compounds in which the members are built up in a regular, repetitive way like this is called a **homologous series**. Members of such a series have similar chemical and physical properties, which change gradually as carbon atoms are added to the chain.

EXAMPLE 2.1

What is the molecular formula of an alkane with six carbon atoms?

Solution If $n = 6$, then $2n + 2 = 14$. The formula is C_6H_{14}.

PROBLEM 2.1 What is the molecular formula of an alkane with 12 carbon atoms?

PROBLEM 2.2 Which of the following are alkanes?

a. C_7H_{16} b. C_7H_{12} c. C_8H_{16} d. $C_{29}H_{60}$

2.2 Nomenclature of Organic Compounds

In the early days of organic chemistry, each new compound was given a name that was usually based on its source or use. Examples (Figs. 1.13 and 1.14) include limonene (from lemons), α-pinene (from pine trees), coumarin (from the tonka bean, known to South American natives as *cumaru*), and penicillin (from the mold that produces it, *Penicillium notatum*). Even today, this method of naming can be used to give a short, simple name to a molecule with a complex structure. For example, cubane (p. xxvii) was named after its shape.

It became clear many years ago, however, that one could not rely only on common or trivial names and that a systematic method for naming compounds was needed. Ideally, the rules of the system should result in a unique name for each compound. Knowing the rules and seeing a structure, one should be able to write the systematic name. Seeing the systematic name, one should be able to write the correct structure.

Eventually, internationally recognized systems of nomenclature were devised by a commission of the International Union of Pure and Applied Chemistry; they are known as the IUPAC (pronounced "eye-you-pack") systems. In this book, we will use mainly IUPAC names. However, in some cases, the common name is so widely used that we will ask you to learn it (for example, formaldehyde [common] is used in preference to methanal [systematic], and cubane is much easier to remember than its systematic name, pentacyclo[4.2.0.$0^{2,5}$.$0^{3,8}$.$0^{4,7}$]octane).

2.3 IUPAC Rules for Naming Alkanes

1. The general name for acyclic saturated hydrocarbons is *alkanes*. The *-ane* ending is used for all saturated hydrocarbons. This is important to remember because later other endings will be used for other functional groups.
2. Alkanes without branches are named according to the *number of carbon atoms*. These names, up to ten carbons, are given in the first column of Table 2.1.
3. For alkanes with branches, the **root name** is that of the longest continuous chain of carbon atoms. For example, in the structure

The **root name** of an alkane is that of the longest continuous chain of carbon atoms.

$$\begin{array}{c} \quad\ CH_3\ \ CH_3 \\ \quad\ |\qquad | \\ CH_3-CH-CH-CH_2-CH_3 \end{array} \quad \text{or} \quad \begin{array}{c} \qquad\quad CH_3\ \ CH_3 \\ \qquad\quad |\qquad | \\ CH_3-CH-CH-CH_2-CH_3 \end{array}$$

the longest continuous chain (in color) has five carbon atoms. The compound is therefore named as a substituted *pent*ane, even though there are seven carbon atoms altogether.

4. Groups attached to the main chain are called **substituents**. Saturated substituents that contain only carbon and hydrogen are called **alkyl groups**. An alkyl group is named by taking the name of the alkane with the same number of carbon atoms and changing the *-ane* ending to *-yl*.

   In the previous example, each substituent has only one carbon. Derived from methane by removing one of the hydrogens, a one-carbon substituent is called a **methyl group**.

Substituents are groups attached to the main chain of a molecule. Saturated substituents containing only C and H are called **alkyl groups**.

The one-carbon alkyl group derived from methane is called a **methyl group**.

$$\begin{array}{c} H \\ | \\ H-C-H \\ | \\ H \end{array} \qquad \begin{array}{c} H \\ | \\ H-C- \\ | \\ H \end{array} \quad \text{or} \quad CH_3- \quad \text{or} \quad Me-$$

methane — methyl group

The names of substituents with more than one carbon atom will be described in Section 2.4.

5. The main chain is numbered in such a way that the first substituent encountered along the chain receives the lowest possible number. Each substituent is then located by its name and by the number of the carbon atom to which it is attached. When two or more identical groups are attached to the main chain, prefixes such

as *di-*, *tri-*, and *tetra-* are used. *Every substituent must be named and numbered,* even if two identical substituents are attached to the same carbon of the main chain. The compound

```
          CH3  CH3
 1       2|   3|     4      5
CH3 — CH — CH — CH2 — CH3
```

is correctly named 2,3-dimethylpentane. The name tells us that there are two methyl substituents, one attached to carbon-2 and one attached to carbon-3 of a five-carbon saturated chain.

6. If two or more different types of substituents are present, they are listed alphabetically, except that prefixes such as *di-* and *tri-* are not considered when alphabetizing.
7. Punctuation is important when writing IUPAC names. IUPAC names for hydrocarbons are written as one word. Numbers are separated from each other by commas and are separated from letters by hyphens. There is no space between the last named substituent and the name of the parent alkane that follows it.

To summarize and amplify these rules, we take the following steps to find an acceptable IUPAC name for an alkane:

1. Locate the longest continuous carbon chain. This gives the name of the parent hydrocarbon. For example,

```
        C—C                    C—C
        |                      |
C—C—C—C        not      C—C—C—C
```

2. Number the longest chain beginning at the end nearest the first branch point. For example,

```
    C     C                      C     C
    |     |                      |     |
C—C—C—C—C—C      not      C—C—C—C—C—C
6 5 4 3 2 1                1 2 3 4 5 6
```

If there are two equally long continuous chains, select the one with the most branches. For example,

```
  C                          C
1 2|  3  4  5  6             |  3  4  5  6
C—C—C—C—C—C       not      C—C—C—C—C—C
    |                           |
    C—C                         C—C
                                2  1
 two branches                 one branch
```

If there is a branch equidistant from each end of the longest chain, begin numbering nearest to a third branch:

```
  C C       C                  C C       C
  | |       |                  | |       |
C—C—C—C—C—C—C       not      C—C—C—C—C—C—C
1 2 3 4 5 6 7                7 6 5 4 3 2 1
2,3,6-trimethylheptane       2,5,6-trimethylheptane
```

If there is no third branch, begin numbering nearest the substituent whose name has alphabetic priority:

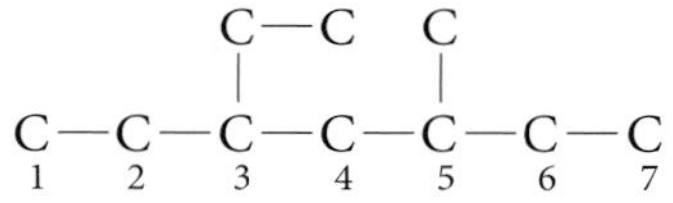

```
    C—C  C                              C—C  C
    |    |                              |    |
C—C—C—C—C—C—C          not          C—C—C—C—C—C—C
1 2 3 4 5 6 7                       7 6 5 4 3 2 1
```

3-ethyl-5-methylheptane — *not* — 5-ethyl-3-methylheptane

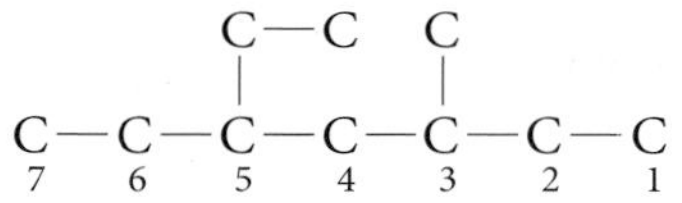

3. Write the name as one word, placing substituents in alphabetic order and using proper punctuation.

EXAMPLE 2.2

Give an IUPAC name for $CH_3—C(CH_3)_2—CH_2CH_2CH_3$.

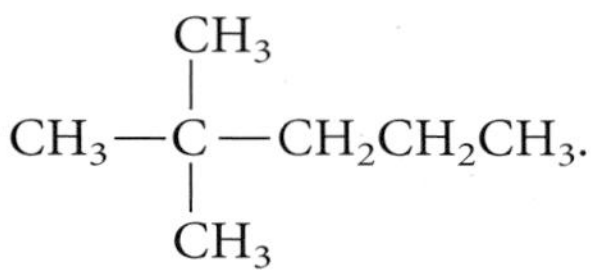

Solution $\overset{1}{C}H_3—\overset{2}{C}(CH_3)_2—\overset{3}{C}H_2\overset{4}{C}H_2\overset{5}{C}H_3$ 2,2-dimethylpentane

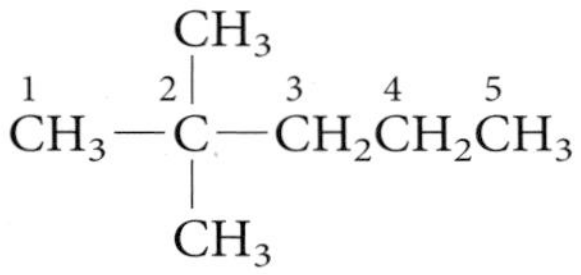

PROBLEM 2.3 Give an IUPAC name for the following compounds:

a. 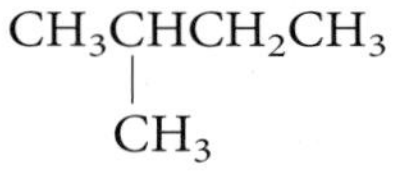 $CH_3CH(CH_3)CH_2CH_3$ b. 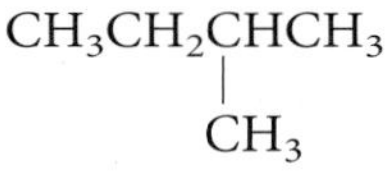 $CH_3CH_2CH(CH_3)CH_3$ c. 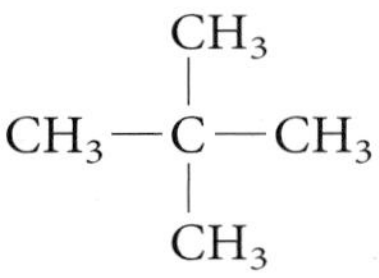 $CH_3—C(CH_3)_2—CH_3$

2.4 Alkyl and Halogen Substituents

As illustrated for the methyl group, alkyl substituents are named by changing the *-ane* ending of alkanes to *-yl*. Thus the two-carbon alkyl group is called the **ethyl group**, from ethane.

The two-carbon alkyl group is the **ethyl group**. The **propyl group** and the **isopropyl group** are three-carbon groups attached to the main chain by the first and second carbons, respectively.

CH_3CH_3 (ethane) $CH_3CH_2—$ or $C_2H_5—$ or Et— (ethyl group)

When we come to propane, there are two possible alkyl groups, depending on which type of hydrogen is removed. If a *terminal* hydrogen is removed, the group is called a **propyl group**.

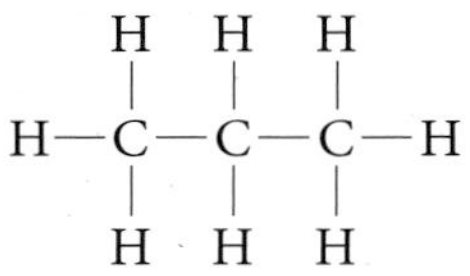

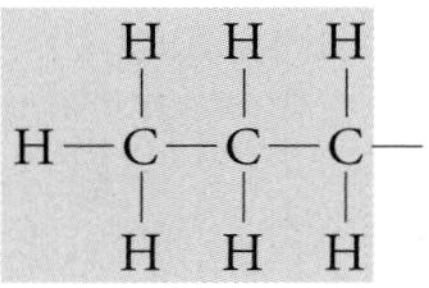

```
  H H H          H H H
  | | |          | | |
H—C—C—C—H      H—C—C—C—     or   CH3CH2CH2—   or   Pr—
  | | |          | | |
  H H H          H H H
```

propyl group

But if a hydrogen is removed from the *central* carbon atom, we get a different isomeric propyl group, called the **isopropyl** (or 1-methylethyl)* **group**.

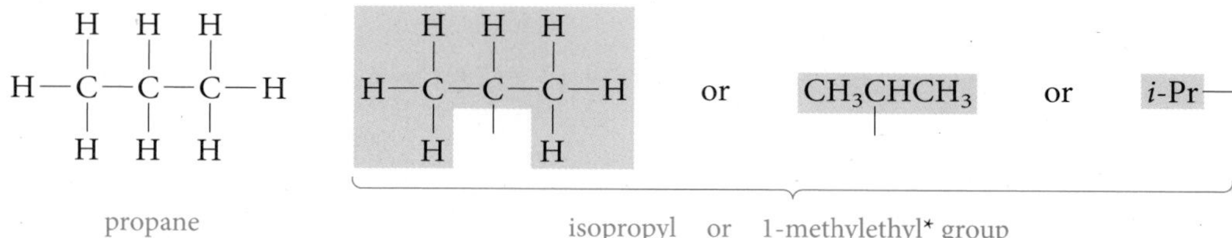

There are four different butyl groups. The butyl and *sec*-butyl groups are based on *n*-butane, while the isobutyl and *tert*-butyl groups come from isobutane.

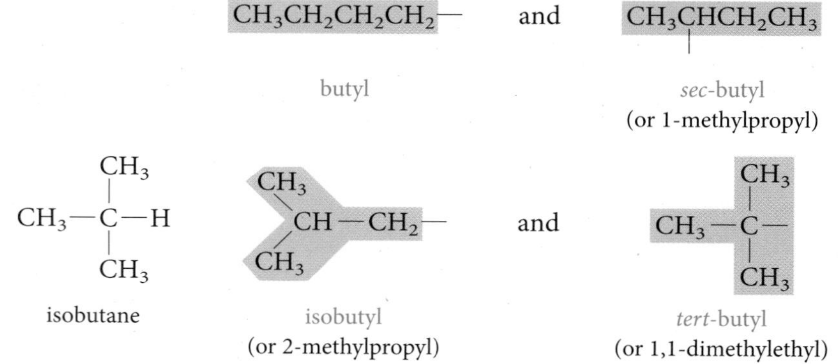

These names for the alkyl groups with up to four carbon atoms are very commonly used, so you should memorize them.

R is the general symbol for an alkyl group.

The letter **R** is used as a general symbol for an alkyl group. The formula R—H therefore represents any alkane, and the formula R—Cl stands for any alkyl chloride (methyl chloride, ethyl chloride, and so on).

Halogen substituents are named by changing the *-ine* ending of the element to *-o*.

F—	Cl—	Br—	I—
fluoro-	chloro-	bromo-	iodo-

EXAMPLE 2.3

Give the common and IUPAC names for $CH_3CH_2CH_2Br$.

Solution The common name is propyl bromide (the common name of the alkyl group is followed by the name of the halide). The IUPAC name is 1-bromopropane, the halogen being named as a substituent on the three-carbon chain.

PROBLEM 2.4 Give an IUPAC name for CH_2ClF.

PROBLEM 2.5 Write the formula for each of the following compounds:

a. isobutyl chloride
b. isopropyl bromide
c. 2-chlorobutane
d. *tert*-butyl iodide
e. propyl fluoride
f. general formula for an alkyl bromide

*The name 1-methylethyl for this group comes about by regarding it as a substituted ethyl group.

$\overset{2}{C}H_3\overset{1}{C}H_2$— (ethyl)

$\overset{2}{C}H_3\overset{1}{C}H$— with CH_3 on C-1 (1-methylethyl)

2.5 Use of the IUPAC Rules

The examples given in Table 2.2 illustrate how the IUPAC rules are applied for particular structures. Study each example to see how a correct name is obtained and how to avoid certain pitfalls.

It is important not only to be able to write a correct IUPAC name for a given structure, but also to do the converse: Write the structure given the IUPAC name. In this case, first write the longest carbon chain and number it, then add the substituents to the correct carbon atoms, and finally fill in the formula with the correct number of hydrogens at each carbon. For example, to write the formula for 2,2,4-trimethylpentane, we go through the following steps:

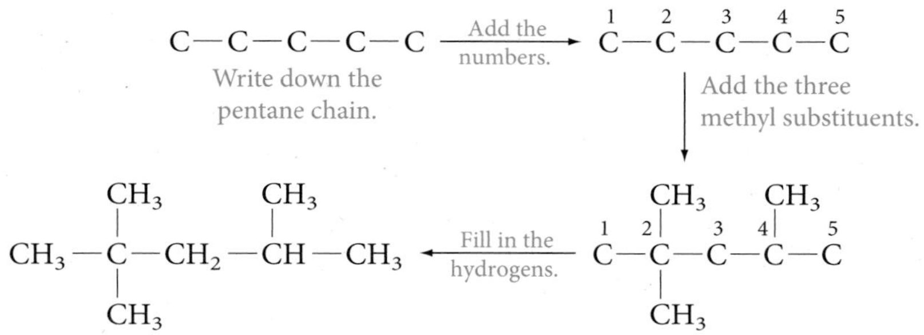

C—C—C—C—C (Write down the pentane chain.) —Add the numbers.→ $\overset{1}{C}-\overset{2}{C}-\overset{3}{C}-\overset{4}{C}-\overset{5}{C}$

↓ Add the three methyl substituents.

$\overset{1}{C}-\underset{CH_3}{\overset{CH_3}{\overset{|}{C}}}\,{}^2-\overset{3}{C}-\overset{CH_3}{\overset{|}{C}}\,{}^4-\overset{5}{C}$

←Fill in the hydrogens.

$CH_3-\underset{CH_3}{\overset{CH_3}{C}}-CH_2-\overset{CH_3}{CH}-CH_3$

2,2,4-trimethylpentane

Table 2.2 ■ Examples of Use of the IUPAC Rules

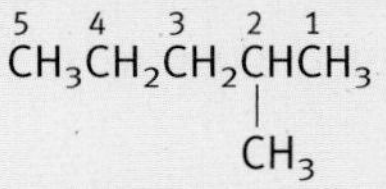 $\overset{5}{C}H_3\overset{4}{C}H_2\overset{3}{C}H_2\overset{2}{C}H\overset{1}{C}H_3$ with CH_3 on C2 2-methylpentane (*not* 4-methylpentane)	The ending *-ane* tells us that all the carbon–carbon bonds are single; *pent-* indicates five carbons in the longest chain. We number them from right to left, starting closest to the branch point.
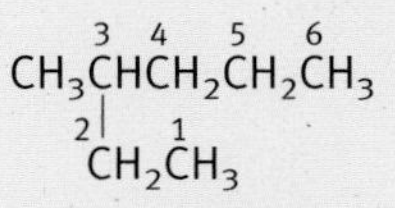 $CH_3\overset{3}{C}H\overset{4}{C}H_2\overset{5}{C}H_2\overset{6}{C}H_3$ with $\overset{2}{C}H_2\overset{1}{C}H_3$ on C3 3-methylhexane (*not* 2-ethylpentane or 4-methylhexane)	This is a six-carbon saturated chain with a methyl group on the third carbon. We would usually write the structure as $CH_3CH_2\underset{CH_3}{CH}CH_2CH_2CH_3$.
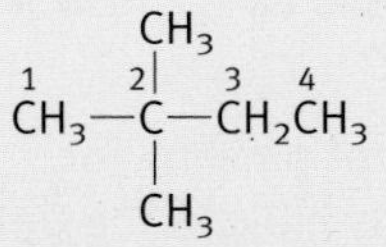 $\overset{1}{C}H_3-\overset{2}{C}(CH_3)_2-\overset{3}{C}H_2\overset{4}{C}H_3$ 2,2-dimethylbutane (*not* 2,2-methylbutane or 2-dimethylbutane)	There must be a number for each substituent, and the prefix *di-* says that there are two methyl substituents.
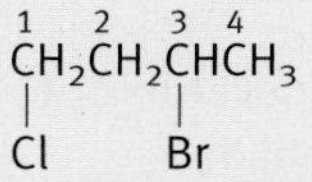 $\overset{1}{C}H_2\overset{2}{C}H_2\overset{3}{C}H\overset{4}{C}H_3$ with Cl on C1 and Br on C3 3-bromo-1-chlorobutane (*not* 1-chloro-3-bromobutane or 2-bromo-4-chlorobutane)	First, we number the butane chain from the end closest to the first substituent. Then we name the substituents in alphabetical order, regardless of position number.

PROBLEM 2.6 Name the following compounds by the IUPAC system:

a. $CH_3CHFCH_2CH_3$

b.
```
              Br
              |
(CH3)3CCHCH(CH3)2
```

PROBLEM 2.7 Write the structure for 3,3-dimethyloctane.

PROBLEM 2.8 Explain why 1,3-difluorobutane is a correct IUPAC name, but 1,3-dimethylpentane is *not* a correct IUPAC name.

2.6 Sources of Alkanes

Petroleum and **natural gas** are the two most important natural sources of alkanes.

The two most important natural sources of alkanes are **petroleum** and **natural gas.** Petroleum is a complex liquid mixture of organic compounds, many of which are alkanes or cycloalkanes. For more details about how petroleum is refined to obtain gasoline, fuel oil, and other useful substances, read "A Word about Petroleum, Gasoline, and Octane Number" on pages 102–103.

Natural gas, often found associated with petroleum deposits, consists mainly of methane (about 80%) and ethane (5% to 10%), with lesser amounts of some higher alkanes. Propane is the major constituent of liquefied petroleum gas (LPG), a domestic fuel used mainly in rural areas and mobile homes. Butane is the gas of choice in some areas. Natural gas is becoming an energy source that can compete with and possibly surpass oil. In the United States, there are about a million miles of natural gas pipelines distributing this energy source to all parts of the country. Natural gas is also distributed worldwide via huge tankers. To conserve space, the gas is liquefied (−160°C), because 1 cubic meter (m^3) of liquefied gas is equivalent to about 600 m^3 of gas at atmospheric pressure. Large tankers can carry more than 100,000 m^3 of liquefied gas.

A CLOSER LOOK AT... Natural Gas

Conduct research on the Internet to find more information and answer questions about natural gas.

Natural Gas as a Fossil Fuel

1. What are the physical properties of natural gas?
2. Why does natural gas used for home heating (or from the Bunsen burner in your chemistry lab) have an odor?
3. Why do the physical properties of natural gas make it safer to use than most other fossil fuels?
4. How is natural gas produced in nature? What keeps it from escaping to the surface of the earth? How is it obtained from its natural source?
5. How many miles (or kilometers) of pipeline for transporting natural gas currently exist in the United States?
6. What percent of U.S. energy needs are supplied by natural gas?
7. What are the major geographic locations of natural gas in the United States and Canada?

2.7 Physical Properties of Alkanes and Nonbonding Intermolecular Interactions

Alkanes are insoluble in water. This is because water molecules are *polar,* whereas alkanes are *nonpolar* (all the C—C and C—H bonds are nearly purely covalent). The O—H bond in a water molecule is strongly polarized by the high electronegativity of oxygen (Sec. 1.5). This polarization places a partial positive charge on the hydrogen atom and a partial negative charge on the oxygen atom (Figure 2.2(a)). As a result, the hydrogen atoms in one water molecule are strongly attracted to the oxygen atoms in other water molecules, and the small size of the H atoms allows the molecules to approach each other very closely. This special attraction is called **hydrogen bonding** (Figure 2.2(b)).* To intersperse alkane and water molecules, we would have to break up the hydrogen bonding interactions between water molecules, which would require considerable energy. Alkanes, with their nonpolar C—H bonds, cannot replace hydrogen bonding among water molecules with attractive alkane–water interactions that are comparable in strength, so mixing alkane molecules and water molecules is *not* an energetically favored process.

The mutual insolubility of alkanes and water is used to advantage by many plants. Alkanes often constitute part of the protective coating on leaves and fruits. If you have ever polished an apple, you know that the skin, or cuticle, contains waxes. Constituents of these waxes include the normal alkanes $C_{27}H_{56}$ and $C_{29}H_{60}$. The leaf wax of cabbage and broccoli is mainly n-$C_{29}H_{60}$, and the main alkane of tobacco leaves is n-$C_{31}H_{64}$. Similar hydrocarbons are found in beeswax. The major function of plant waxes is to prevent water loss from the leaves or fruit.

Alkanes have lower boiling points for a given molecular weight than most other organic compounds. This is because they are nonpolar molecules. They are constantly

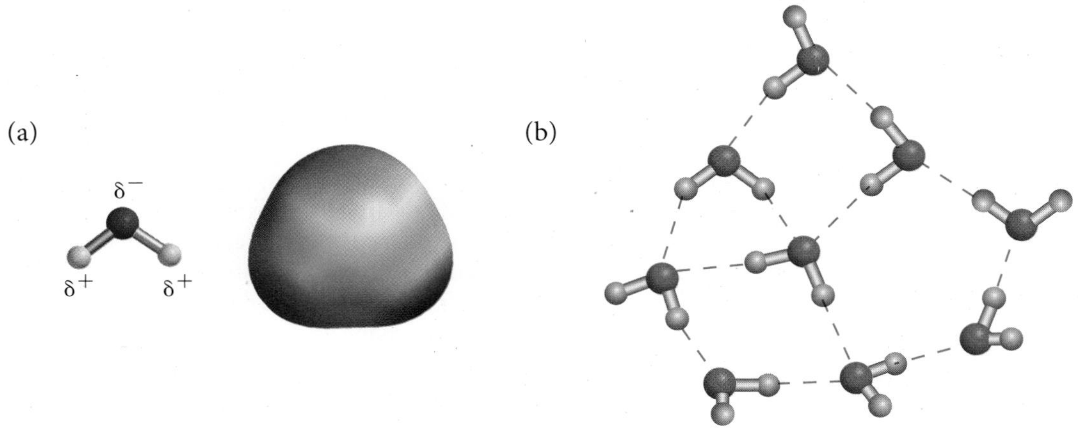

■ **Figure 2.2**
(a) Electrostatic potential map for a water molecule, showing the partial negative charge (red) for the oxygen atom and partial positive charge (blue) for the hydrogens. (b) Interaction of partial positive and negative charges in a network of water molecules via bridging hydrogen bonds.

*Molecules that contain N—H and F—H covalent bonds also have hydrogen bonding interactions with molecules containing N, O, or F atoms.

A CLOSER LOOK AT... Hydrogen Bonding

Conduct research on the Internet to find more information about hydrogen bonding and to answer the following questions.

Water and Ice

1. What is the partial charge on the O atom? On each H atom?
2. If the water molecule were linear (H—O—H) instead of bent O , would it still be polar?
   H H
3. What is the average length (in angstroms) of a hydrogen bond between two water molecules?
4. How is a hydrogen bond different from a covalent bond?

Water, Ice, and Hydrogen Bonding

1. Search online for information on water, ice, and hydrogen bonds.
   a. How many hydrogen bonds does a water molecule make in liquid water? In ice?
   b. Why is ice less dense than liquid water? How is this related to hydrogen bonding?
2. One property of alkanes is that they are not soluble in water. This is related to interactions between molecules. Let us explore the connection to hydrogen bonding.
   a. What is required for hydrogen bonding between molecules?
   b. Can alkane molecules form hydrogen bonds to each other or to water molecules?
   c. When the oil tanker Exxon *Valdez* broke apart near the Alaskan coast, the oil (composed mostly of alkanes) that spilled did not dissolve in the water. It floated on top. Explain these facts in terms of intermolecular interactions and the properties of alkanes and water molecules.

moving, and the electrons in a nonpolar molecule can become unevenly distributed within the molecule, causing the molecule to have partially positive and partially negative ends. The *temporarily* polarized molecule causes its neighbor to become temporarily polarized as well, and these molecules are weakly attracted to each other. Such interactions between molecules are called **van der Waals attractions**.

δ+ δ− δ+ δ−
δ+ δ− δ+ δ−

Because they are weak attractions, the process of separating molecules from one another (which is what we do when we convert a liquid to a gas) requires relatively little energy, and the boiling points of these compounds are relatively low. Figure 2.3 shows the boiling points of some alkanes. Since these attractive forces can only operate over short distances between the surfaces of molecules, *the boiling points of alkanes rise as the chain length increases and fall as the chains become branched and more nearly spherical in shape.* Figure 2.4 illustrates the effect of molecular shape on van der Waals attractions.

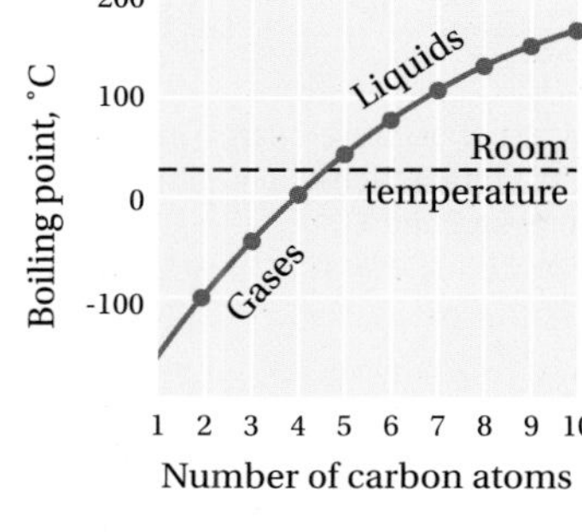

Name	Formula	Boiling point, °C
pentane	$CH_3CH_2CH_2CH_2CH_3$	36
2-methylbutane (isopentane)	$CH_3CHCH_2CH_3$ with CH_3 on C-2	28
2,2-dimethylpropane (neopentane)	$CH_3—C(CH_3)_2—CH_3$	10

■ **Figure 2.3**
As shown by the curve, the boiling points of the normal alkanes rise smoothly as the length of the carbon chain increases. Note from the table, however, that chain branching causes a decrease in boiling point (each compound in the table has the same number of carbons and hydrogens, C_5H_{12}).

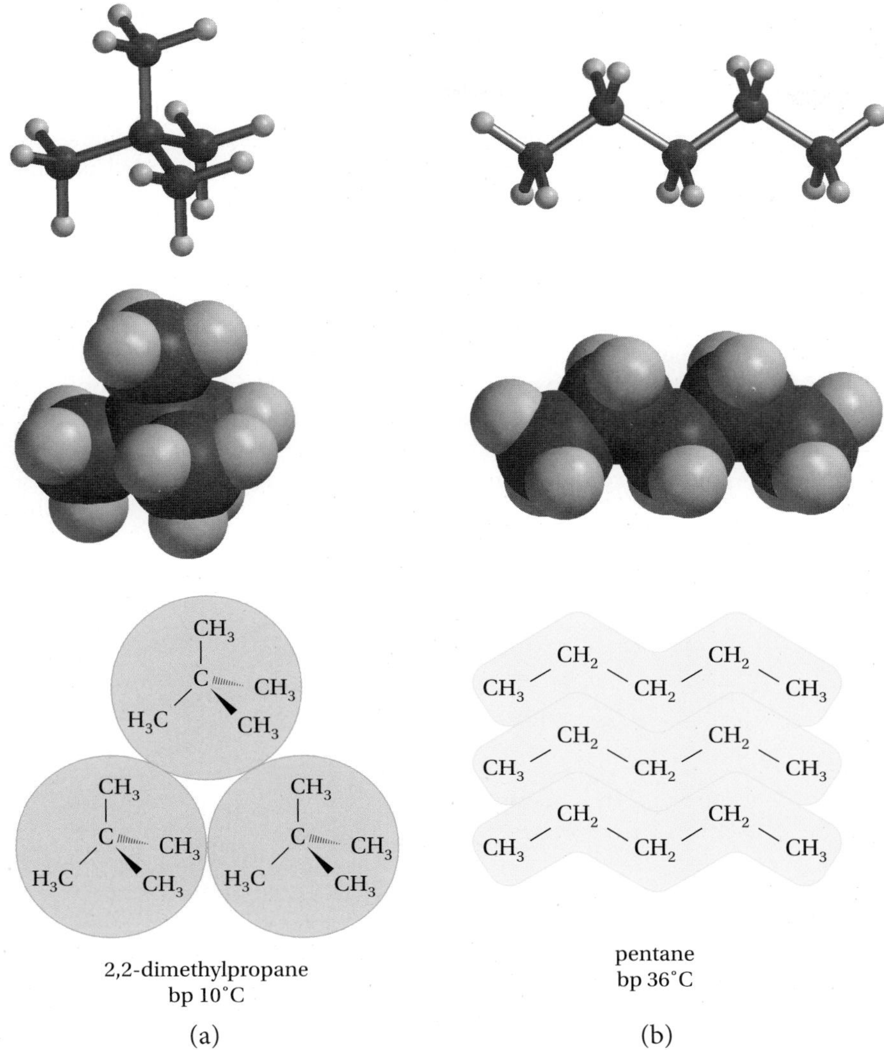

Figure 2.4
Ball-and-stick (top) and space-filling (middle) representations for 2,2-dimethylpropane (a) and pentane (b) showing the relatively spherical nature of 2,2-dimethylpropane as compared to the rod-shaped nature of pentane (bottom).

Despite having the same molecular weight, the rod-shaped pentane molecules have more surface area available for contact between them than the spherical 2,2-dimethylpropane molecules. Pentane molecules, therefore, experience more van der Waals attractions (hence, higher boiling point) than do 2,2-dimethylpropane molecules.

Hydrogen bonding and van der Waals attractions are examples of **nonbonding intermolecular interactions**. These kinds of interactions have important consequences for the properties and behavior of molecules, and we will encounter more examples as we continue to explore the chemistry of different classes of organic compounds.

Hydrogen bonding and **van der Waals attractions** are **nonbonding intermolecular interactions.**

2.8 Conformations of Alkanes

The shapes of molecules often affect their properties. A simple molecule like ethane, for example, can have an infinite number of shapes as a consequence of rotating one carbon atom (and its attached hydrogens) with respect to the other carbon atom. These arrangements are called **conformations** or **conformers**. Conformers are **stereoisomers**, isomers in which the atoms are connected in the same order but are arranged differently in space. Two possible conformers for ethane are shown in Figure 2.5.*

Different **conformations** (shapes) of the same molecule that are interconvertible by rotation around a single bond are called **conformers** or **rotamers**. Conformers are **stereoisomers**, isomers with the same atom connectivity but different spatial arrangements of atoms.

*Build a 3D model of ethane (using a molecular model kit) and use it as you read this section to model the staggered and eclipsed conformations shown in Figure 2.5.

In the staggered conformation of ethane, each C—H bond on one carbon bisects an H—C—H angle on the other carbon. In the eclipsed conformation, C—H bonds on the front and back carbons are aligned. By rotating one carbon 60° with respect to the other, we can interconvert staggered and eclipsed conformations. Between these two extremes are an infinite number of intermediate conformations of ethane.

The staggered and eclipsed conformations of ethane can be regarded as **rotamers** because each is convertible to the other by rotation about the carbon–carbon bond. Such rotation about a single bond occurs easily because the amount of overlap of the sp^3 orbitals on the two carbon atoms is unaffected by rotation about the sigma bond (see Figure 1.8). Indeed, there is enough energy available at room temperature for the staggered and eclipsed conformers of ethane to interconvert rapidly. Consequently, the conformers cannot be separated from one another. We know from various types of physical evidence, however, that both forms are not equally stable. The staggered conformation is the most stable (has the lowest potential energy) of all ethane conformations, while the eclipsed conformation is the least stable (has the highest potential energy). At room temperature, the staggered conformation is practically the only conformation present.

staggered   eclipsed   (2.1)

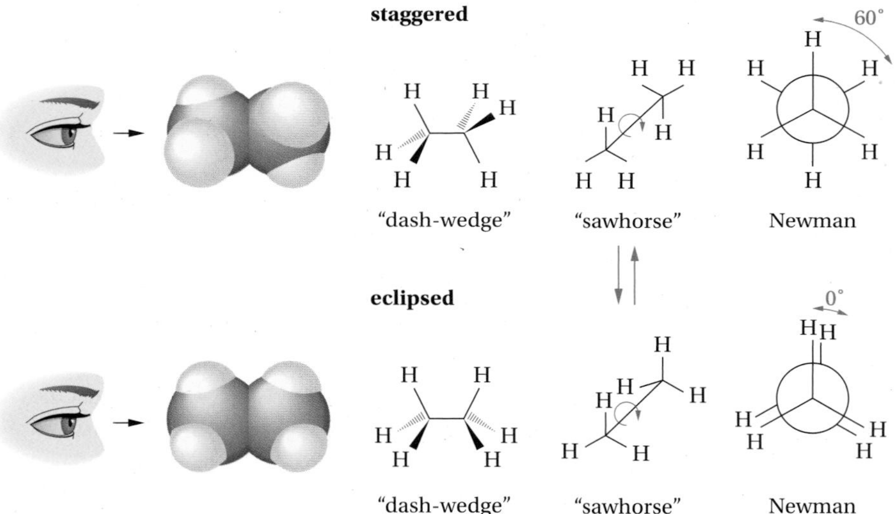

▪ **Figure 2.5**
Two of the possible conformations of ethane: staggered and eclipsed. Interconversion is easy via a 60° rotation about the C—C bond, as shown by the curved arrows. The structures at the left are space-filling models. In each case, the next structure is a "dash-wedge" structure, which, if viewed as shown by the eyes, converts to the "sawhorse" drawing, or the Newman projection at the right, an end-on view down the C—C axis. In the Newman projection, the circle represents two connected carbon atoms. Bonds on the "front" carbon go to the center of the circle, and bonds on the "rear" carbon go only to the edge of the circle.

EXAMPLE 2.4

Draw the Newman projections for the staggered and eclipsed conformations of propane.

Solution

CH_3
staggered

The projection formula is similar to that of ethane, except for the replacement of one hydrogen with methyl.

eclipsed

Rotation of the "rear" carbon of the staggered conformation by 60° gives the eclipsed conformation shown.

We are looking down the C_1—C_2 bond.

PROBLEM 2.9 Draw Newman projections for two different *staggered* conformations of butane (looking end-on at the bond between carbon-2 and carbon-3), and predict which of the two conformations is more stable. (If you have a model kit, build a model of butane to help you visualize the Newman projections.)

The most important thing to remember about conformers is that they are just different forms of a single molecule that can be interconverted by rotational motions about single (sigma) bonds. More often than not, there is sufficient thermal energy for this rotation at room temperature. Consequently, at room temperature, it is usually not possible to separate conformers from one another.

Now let us look at the structures of cycloalkanes and their conformations.

2.9 Cycloalkane Nomenclature and Conformation

Cycloalkanes are saturated hydrocarbons that have at least one ring of carbon atoms. A common example is cyclohexane.

or

Structural and abbreviated structural formulas for cyclohexane

Cycloalkanes are named by placing the prefix *cyclo-* before the alkane name that corresponds to the number of carbon atoms in the ring. The structures and names of the first six unsubstituted cycloalkanes are as follows:

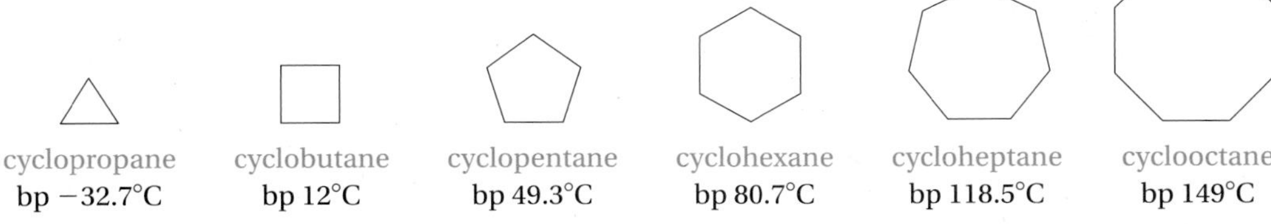

Alkyl or halogen substituents attached to the rings are named in the usual way. If only one substituent is present, no number is needed to locate it. If there are several substituents, numbers are required. One substituent is always located at ring carbon number 1, and the remaining ring carbons are then numbered consecutively in a way that gives the other substituents the lowest possible numbers. With different substituents, the one with highest alphabetic priority is located at carbon 1. The following examples illustrate the system:

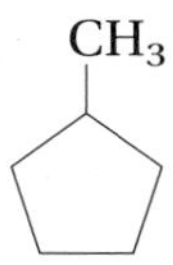

methylcyclopentane (*not* 1-methylcyclopentane)

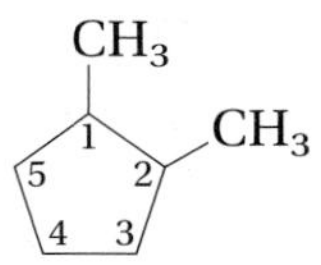

1,2-dimethylcyclopentane (*not* 1,5-dimethylcyclopentane)

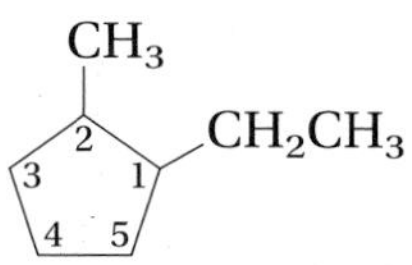

1-ethyl-2-methylcyclopentane (*not* 2-ethyl-1-methylcyclopentane)

PROBLEM 2.10 The general formula for an alkane is C_nH_{2n+2}. What is the corresponding formula for a cycloalkane with one ring?

PROBLEM 2.11 Draw the structural formulas for

a. 1,3-dimethylcyclohexane
b. 1,2,3-trichlorocyclopropane
c. 1,2-diethylcyclobutane
d. 1-ethyl-3-fluorocyclopentane

PROBLEM 2.12 Give IUPAC names for

a. 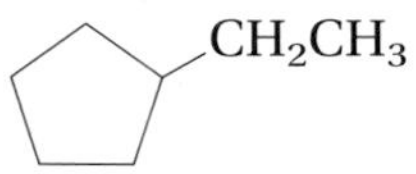 CH_2CH_3 b. 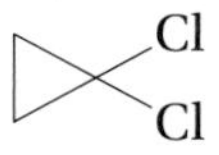 Cl Cl c. 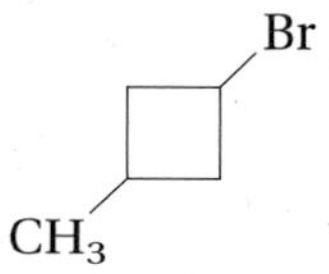 Br CH_3

What are the conformations of cycloalkanes? Cyclopropane, with only three carbon atoms, is necessarily planar (because three points determine a plane). The C—C—C angle is only 60° (the carbons form an equilateral triangle), much less than the usual sp^3 tetrahedral angle of 109.5°. The hydrogens lie above and below the carbon plane, and hydrogens on adjacent carbons are eclipsed.

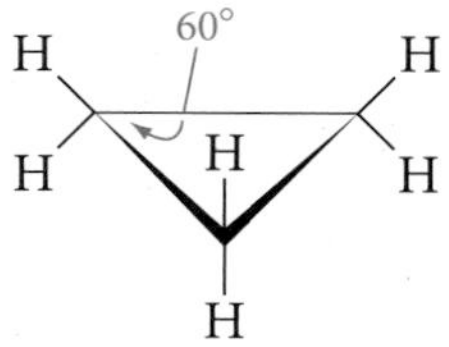

cyclopropane

EXAMPLE 2.5

Explain why the hydrogens in cyclopropane lie above and below the carbon plane.

Solution Refer to Figure 1.10. The carbons in cyclopropane have a geometry similar to that shown there, except that the C—C—C angle is "squeezed" and is smaller than tetrahedral. In compensation, the H—C—H angle is expanded and is larger than tetrahedral, approximately 120°.

The H—C—H plane perpendicularly bisects the C—C—C plane, which, as drawn here, lies in the plane of the paper.

Cycloalkanes with more than three carbon atoms are nonplanar and have "puckered" conformations. In cyclobutane and cyclopentane, puckering allows the molecule to adopt the most stable conformation (with the least strain energy). Puckering introduces strain by making the C—C—C angles a little smaller than they would be if the molecules were planar; however, less eclipsing of the adjacent hydrogens compensates for this.

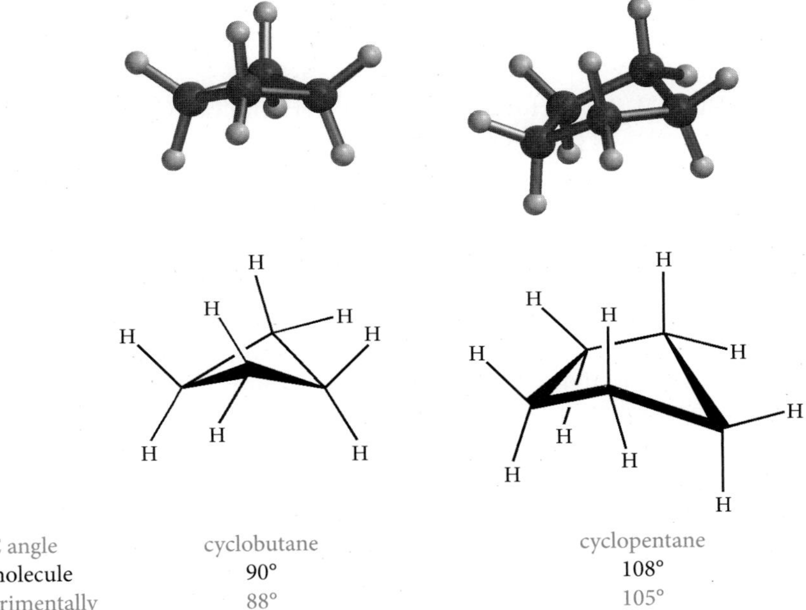

Six-membered rings are rather special and have been studied in great detail because they are very common in nature. If cyclohexane were planar, the internal C—C—C angles would be those of a regular hexagon, 120°—quite a bit larger than the normal tetrahedral angle (109.5°). The resulting strain prevents cyclohexane from being planar (flat). Its most favored conformation is the **chair conformation**, an arrangement in which all of the C—C—C angles are 109.5° and all of the hydrogens on adjacent carbon atoms are perfectly staggered. Figure 2.6 shows models of the cyclohexane chair conformation.* (If a set of molecular models is available, it would be a good idea

In the **chair conformation** of cyclohexane, the six **axial** hydrogen atoms lie above and below the mean plane of the ring, while the six **equatorial** hydrogens lie in the plane.

---

*__Diamond__ is one naturally occurring form of carbon. In the diamond crystal, the carbon atoms are connected to one another in a structure similar to the chair form of cyclohexane, except that all of the hydrogens are replaced by carbon atoms, resulting in a continuous network of carbon atoms. The hydrocarbons **adamantane** and **diamantane** show the beginnings of the diamond structure in their fusing of chair cyclohexanes. For a fascinating article on diamond structure, see "Diamond Cleavage" by M. F. Ansell in *Chemistry in Britain,* **1984,** 1017–1021.

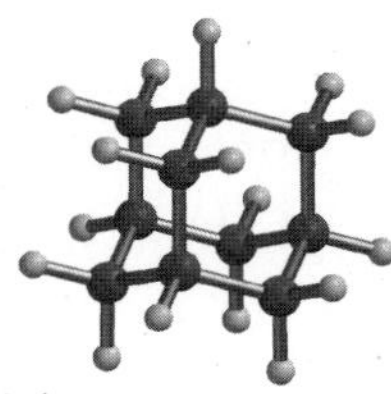

adamantane ($C_{10}H_{16}$) mp 268–269°C

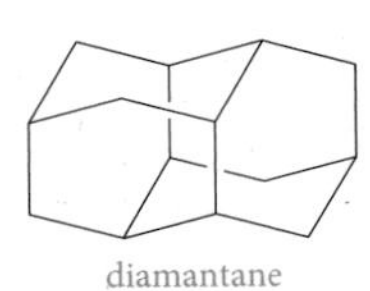
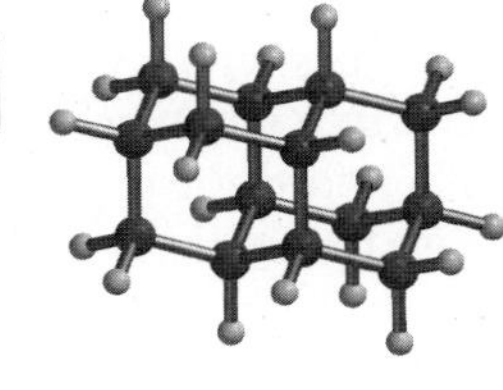

diamantane ($C_{14}H_{20}$) mp 236–237°C

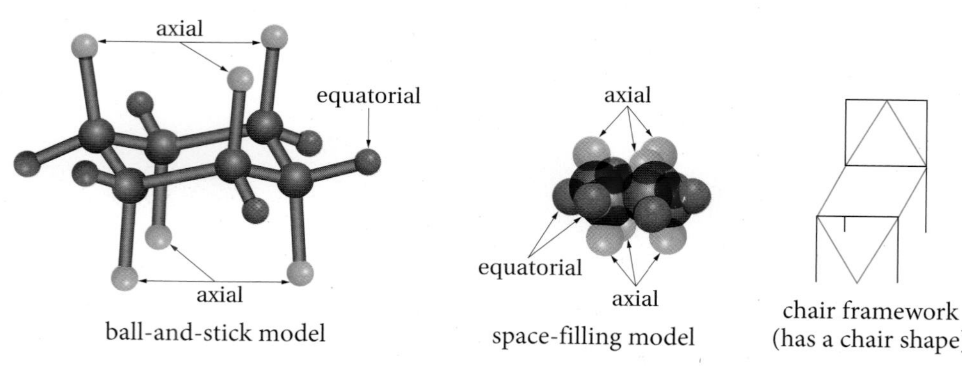

■ **Figure 2.6**
The chair conformation of cyclohexane, shown in ball-and-stick (left) and space-filling (center) models. The axial hydrogens, depicted in green, lie above or below the mean plane of the carbons, and the six equatorial hydrogens, depicted in red, lie approximately in that mean plane. The carbon atoms are depicted in gray. The origin of the chair terminology is illustrated at the right.

for you to construct a cyclohexane model to better visualize the concepts discussed in this and the next two sections.)

PROBLEM 2.13 How are the H—C—H and C—C—C planes at any one carbon atom in cyclohexane related? (Refer, if necessary, to Example 2.5.)

In the chair conformation, the hydrogens in cyclohexane fall into two sets, called **axial** and **equatorial**. Three axial hydrogens lie above and three lie below the average plane of the carbon atoms; the six equatorial hydrogens lie approximately in that plane. By a motion in which alternate ring carbons (say, 1, 3, and 5) move in one direction (down) and the other three ring carbons move in the opposite direction (up), one chair conformation can be converted into another chair conformation in which all axial hydrogens have become equatorial, and all equatorial hydrogens have become axial.

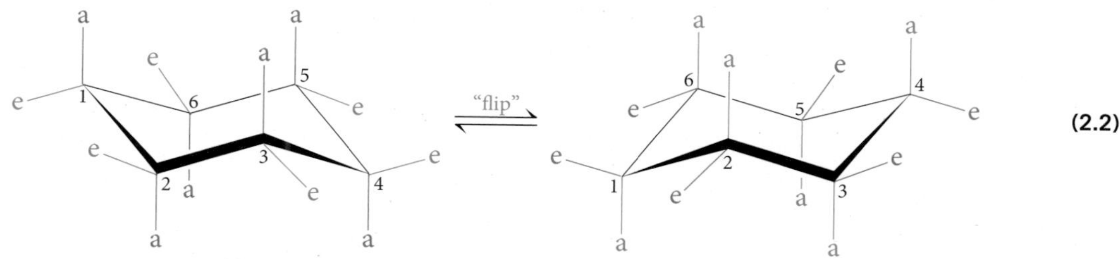

(2.2)

Axial bonds (red) in the left structure become equatorial bonds (red) in the right structure when the ring "flips."

At room temperature, this flipping process is rapid, but at low temperatures (say, −90°C), it slows down enough that the two different types of hydrogens can actually be detected by proton nuclear magnetic resonance (NMR) spectroscopy (see Chapter 12).

Cyclohexane conformations have another important feature. If you look carefully at the space-filling model of cyclohexane (Figure 2.6), you will notice that *the three axial hydrogens on the same face of the ring are close to each other.* If an axial hydrogen is replaced by a larger substituent (such as a methyl group), the axial crowding is even worse. Therefore, the preferred conformation is the one in which the larger substituent, in this case the methyl group, is equatorial (Figure 2.7).

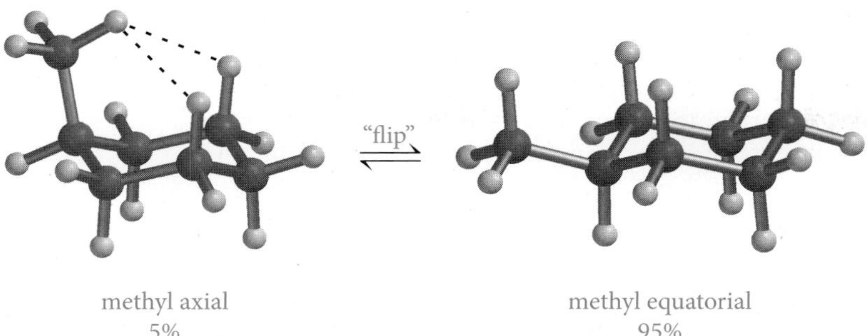

▪ Figure 2.7
Conformational equilibrium between the axial (left) and equatorial (right) isomers of methylcyclohexane. The steric interactions between the axial methyl and the 1,3-diaxial hydrogens are evident in the left ball-and-stick structure.

PROBLEM 2.14 Another puckered conformation for cyclohexane, one in which all C—C—C angles are the normal 109.5°, is the boat conformation.

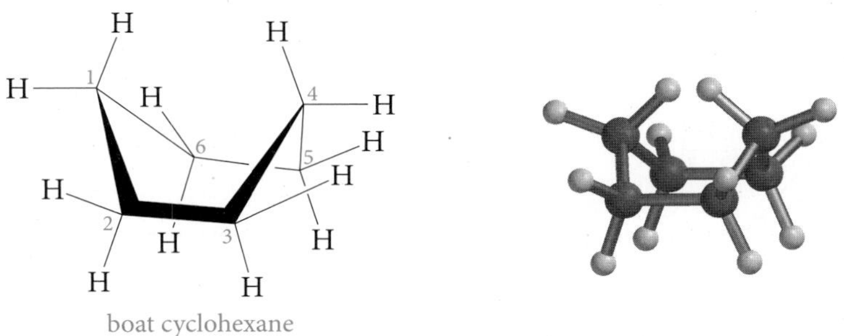

Explain why this conformation is very much less stable than the chair conformation. (*Hint:* Note the arrangement of hydrogens as you sight along the bond between carbon-2 and carbon-3; a molecular model will help you answer this problem.)

PROBLEM 2.15 For *tert*-butylcyclohexane, only one conformation, with the *tert*-butyl group equatorial, is detected experimentally. Explain why this conformational preference is greater than that for methylcyclohexane (see Figure 2.7).

The six-membered ring in the chair conformation is a common structural feature of many organic molecules, including sugar molecules (Sec. 16.7) like glucose, where one ring carbon is replaced by an oxygen atom.

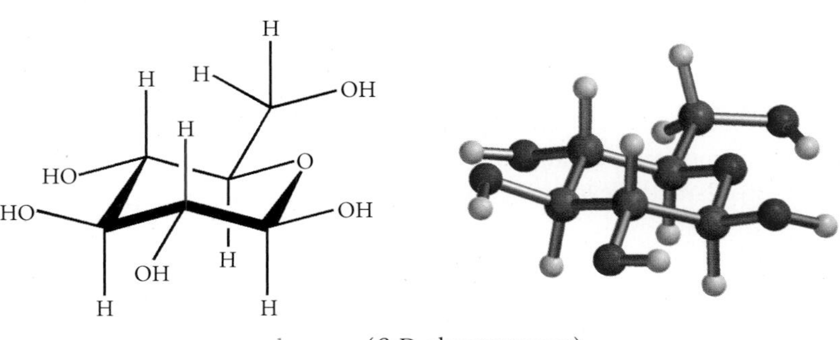

glucose (β-D-glucopyranose)

Notice that the bulkier group on each carbon is in the equatorial position. The conformations of sugars will be studied in greater detail in Chapter 16.

Before we proceed to reactions of alkanes and cycloalkanes, we need to consider a type of isomerism that may arise when two or more carbon atoms in a cycloalkane have substituents.

2.10 *Cis–Trans* Isomerism in Cycloalkanes

Cis–trans isomers of cycloalkanes are a type of stereoisomer, also called **geometric** stereoisomers, in which substituents are on the same side (*cis*) or on the opposite sides (*trans*) of the ring.

Stereoisomerism deals with molecules that have the same order of attachment of the atoms, but different arrangements of the atoms in space. ***Cis–trans* isomerism** (sometimes called **geometric isomerism**) is one kind of stereoisomerism, and it is most easily understood with a specific case. Consider, for example, the possible structures of 1,2-dimethylcyclopentane. For simplicity, let us neglect the slight puckering of the ring and draw it as if it were planar. The two methyl groups may be on the same side of the ring plane or they may be on the opposite sides.

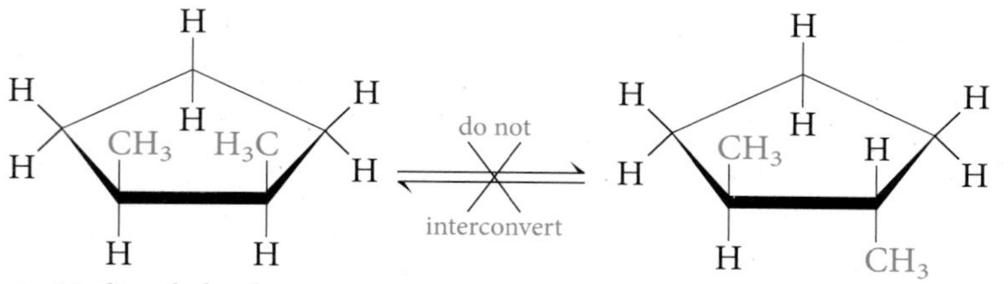

do not interconvert

cis-1,2-dimethylcyclopentane
bp 99°C

trans-1,2-dimethylcyclopentane
bp 92°C

The methyl groups are said to be *cis* (Latin, on the same side) or *trans* (Latin, across) to each other.

Cis–trans isomers differ from one another only in the way that the atoms or groups are positioned in space. Yet this difference is sufficient to give them different physical and chemical properties. (Note, for example, the boiling points for the two 1,2-dimethylcyclopentane structures.) Therefore, *cis–trans* isomers are unique compounds. Unlike conformers, they are not readily interconverted by rotation around carbon–carbon bonds. In this example, the cyclic structure limits rotation about the ring bonds. To interconvert these dimethylcyclopentanes, one would have to break open, rotate, and re-form the ring, or carry out some other bond-breaking process.

Cis–trans isomers can be separated from each other and kept separate, usually without interconversion at room temperature. *Cis–trans* isomerism can be important in determining the biological properties of molecules. For example, a molecule in which two reactive groups are *cis* will interact differently with an enzyme or biological receptor site than will its isomer with the same two groups *trans*.

PROBLEM 2.16 Draw the structure for the *cis* and *trans* isomers of

a. 1,3-dibromocyclopentane
b. 1-chloro-2-methylcyclopropane

A WORD ABOUT... Isomers—Possible and Impossible

Table 2.1 shows that there are 75 structural isomers of the alkane $C_{10}H_{22}$. How many such isomers do you think there might be if we double the number of carbons ($C_{20}H_{42}$)? The answer is 366,319! And if we double the number of carbons again ($C_{40}H_{82}$)? Exactly 62,481,801,147,341. Of course, no one sits down with pencil and paper or molecular models and determines these numbers by constructing all the possibilities; it could take a lifetime. Complex mathematical formulas have been developed to compute these numbers.

Although we can write some isomers' formulas on paper, they are structurally impossible and cannot be synthesized. Consider, for example, the series of alkanes obtained by replacing the hydrogens of methane with methyl groups and then repeating that process on the product indefinitely. You can see from the following drawings that we build up molecules in this way with a central core of carbon atoms and a surface of hydrogen atoms.

$$CH_4 \longrightarrow C(CH_3)_4 \longrightarrow C[C(CH_3)_3]_4 \longrightarrow C\{C[C(CH_3)_3]_3\}_4$$

$$CH_4 \longrightarrow C_5H_{12} \longrightarrow C_{17}H_{36} \longrightarrow C_{53}H_{108}$$

In three dimensions, the molecules are nearly spherical. Of these compounds, only the first two are known (methane and 2,2-dimethylpropane). The $C_{17}H_{36}$ hydrocarbon (tetra-*t*-butylmethane or, more accurately, 3,3-di-*t*-butyl-2,2,4,4-tetramethylpentane) has not yet been synthesized, and if it ever is, it will be an exceptionally strained molecule. The reason is simply that there is not enough room for all the methyl groups on the surface of the molecule.

Repetitive branching in compounds can lead to molecular shapes that resemble this tree.

Therefore, if this $C_{17}H_{36}$ isomer is ever prepared, its bond angles and bond lengths are likely to be severely distorted from the normal. There is almost no possibility of synthesizing the $C_{53}H_{108}$ isomer in this series; its structure is too strained. It is interesting to note that, like these hydrocarbons, the growth of trees, sponges, and other biological structures is similarly limited by the ratio of surface area to volume. (For more on this subject, see the article by R. E. Davies and P. J. Freyd, *J. Chem. Educ.* **1989**, *66,* 278–281.)

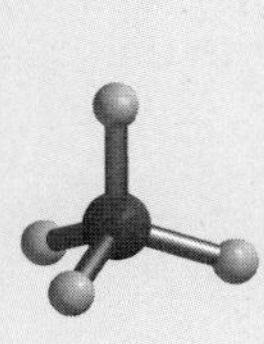

methane

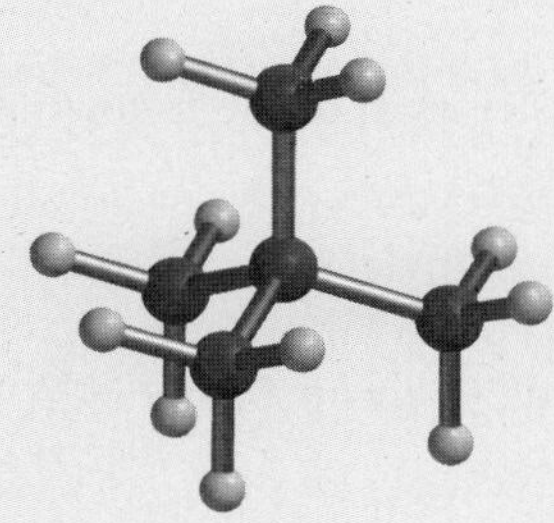

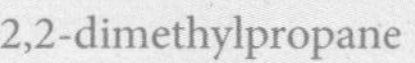

2,2-dimethylpropane

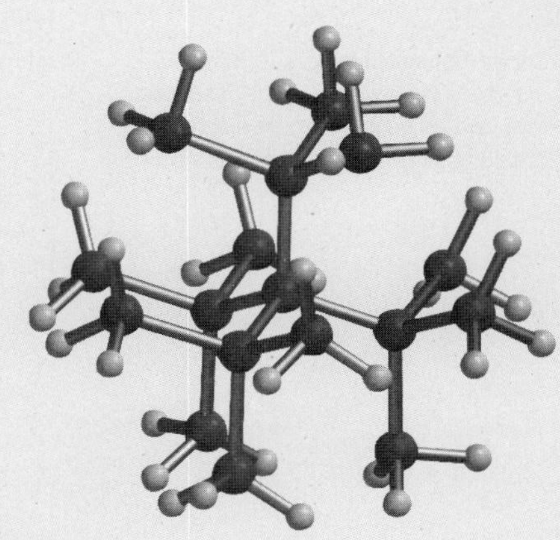

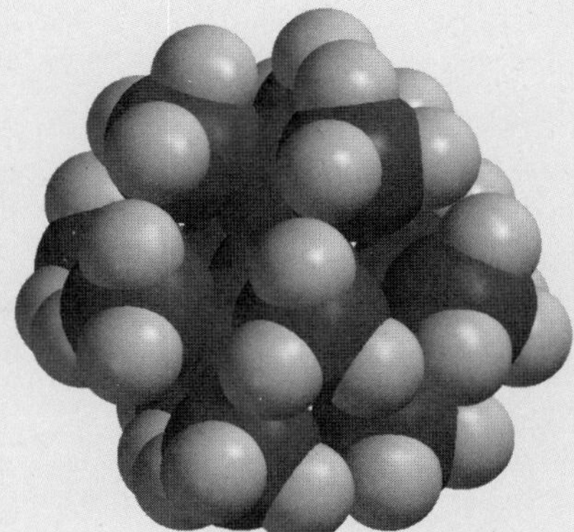

tetra-*t*-butylmethane ($C_{17}H_{36}$)

2.11 Summary of Isomerism

At this point, it may be useful to summarize the relationships of the several types of isomers we have discussed so far. These relationships are outlined in Figure 2.8.

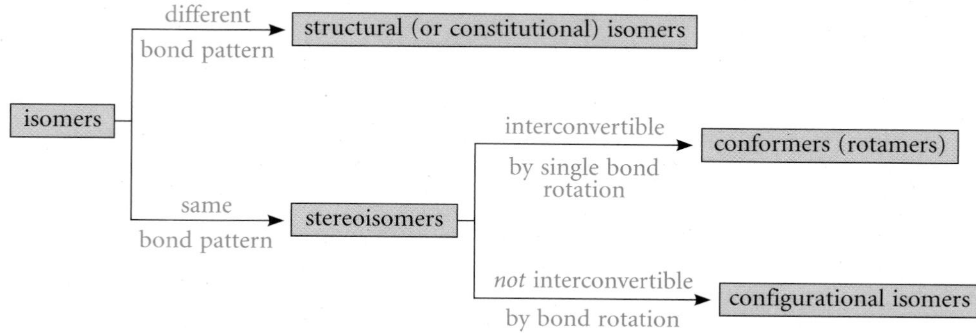

Figure 2.8
The relationships of the various types of isomers.

The first thing to look at in a pair of isomers is their bonding patterns (or atom connectivities). If the bonding patterns are *different,* the compounds are structural (or constitutional) isomers. But if the bonding patterns are the *same,* the compounds are stereoisomers. Examples of structural isomers are ethanol and methoxymethane (page 14) or the three isomeric pentanes (page 15). Examples of stereoisomers are the staggered and eclipsed forms of ethane (page 48) or the *cis* and *trans* isomers of 1,2-dimethylcyclopentane (page 54).

If compounds are stereoisomers, we can make a further distinction as to isomer type. If *single-bond rotation* easily interconverts the two stereoisomers (as with staggered and eclipsed ethane), we call them conformers. If the two stereoisomers can be interconverted only by breaking and remaking bonds (as with *cis*- and *trans*-1,2-dimethylcyclopentane), we call them **configurational isomers.***

Configurational isomers (such as *cis–trans* isomers) are stereoisomers that can only be interconverted by breaking and remaking bonds.

PROBLEM 2.17 Classify each of the following isomer pairs according to the scheme in Figure 2.8.

a. *cis*- and *trans*-1,2-dimethylcyclohexane
b. chair and boat forms of cyclohexane
c. 1-fluoropropane and 2-fluoropropane

2.12 Reactions of Alkanes

All of the bonds in alkanes are single, covalent, and nonpolar. Hence alkanes are relatively inert. Alkanes ordinarily do not react with most common acids, bases, or oxidizing and reducing agents. Because of this inertness, alkanes can be used as solvents for extraction or crystallization as well as for carrying out chemical reactions of other substances. However, alkanes do react with some reagents, such as molecular oxygen and the halogens. We will discuss those reactions here.

*Remember that conformers are different conformations of the same molecule, whereas configurational isomers are different molecules. Geometric isomers (*cis–trans* isomers) are one type of configurational isomer. As we will see in Chapter 3, geometric isomerism also occurs in alkenes. Also, we will see other types of configurational isomers in Chapter 5.

2.12.a Oxidation and Combustion; Alkanes as Fuels

The most important use of alkanes is as fuel. With excess oxygen, alkanes burn to form carbon dioxide and water. Most important, the reactions evolve considerable heat (that is, the reactions are **exothermic**).

Exothermic reactions evolve heat.

$$\underset{\text{methane}}{CH_4} + 2\,O_2 \longrightarrow CO_2 + 2\,H_2O + \text{heat } (212.8 \text{ kcal/mol}) \tag{2.3}$$

$$\underset{\text{butane}}{C_4H_{10}} + \tfrac{13}{2}\,O_2 \longrightarrow 4\,CO_2 + 5\,H_2O + \text{heat } (688.0 \text{ kcal/mol}) \tag{2.4}$$

These combustion reactions are the basis for the use of hydrocarbons for heat (natural gas and heating oil) and for power (gasoline). An initiation step is required—usually ignition by a spark or flame. Once initiated, the reaction proceeds spontaneously and exothermically.

In methane, all four bonds to the carbon atom are C—H bonds. In carbon dioxide, its combustion product, all four bonds to the carbon are C—O bonds. **Combustion** is an **oxidation reaction**, the replacement of C—H bonds by C—O bonds. In methane, carbon is in its most reduced form, and in carbon dioxide, it is in its most oxidized form. Intermediate oxidation states of carbon are also known, in which only one, two, or three of the C—H bonds are converted to C—O bonds. It is not surprising, then, that if insufficient oxygen is available for complete combustion of a hydrocarbon, *partial* oxidation may occur, as illustrated in eqs. 2.5 through 2.8.

Combustion of hydrocarbons is an **oxidation reaction** in which C—H bonds are replaced with C—O bonds.

$$2\,CH_4 + 3\,O_2 \longrightarrow \underset{\text{carbon monoxide}}{2\,CO} + 4\,H_2O \tag{2.5}$$

$$CH_4 + O_2 \longrightarrow \underset{\text{carbon}}{C} + 2\,H_2O \tag{2.6}$$

$$CH_4 + O_2 \longrightarrow \underset{\text{formaldehyde}}{CH_2O} + H_2O \tag{2.7}$$

$$2\,C_2H_6 + 3\,O_2 \longrightarrow \underset{\text{acetic acid}}{2\,CH_3CO_2H} + 2\,H_2O \tag{2.8}$$

Hisham Ibrahim/PhotoDisc/Getty Images

Auto tailpipe exhaust fumes contain condensed water.

Toxic carbon monoxide in exhaust fumes (eq. 2.5), soot emitted copiously from trucks with diesel engines (eq. 2.6), smog resulting in part from aldehydes (eq. 2.7), and acid buildup in lubricating oils (eq. 2.8) are all prices we pay for being a motorized society.* However, incomplete hydrocarbon combustion is occasionally useful, as in the manufacture of carbon blacks (eq. 2.6) used for automobile tires, and lampblack, a pigment used in ink.

EXAMPLE 2.6

In which compound is carbon more oxidized, formaldehyde (CH_2O) or formic acid (HCO_2H)?

Solution Draw the structures:

H₂C=O (formaldehyde) H—C(=O)—OH (formic acid)

Formic acid is the more oxidized form (three C—O and one C—H bond, compared to two C—O and two C—H bonds in formaldehyde).

---

*You may have noticed white exhaust fumes coming from car tailpipes in cold weather. Combustion of hydrocarbons produces water (eqs. 2.3–2.8), so what you see is condensed water from the combustion of gasoline.

PROBLEM 2.18 Which of the following represents the more oxidized form of carbon?

a. methanol (CH_3OH) or formaldehyde (CH_2O)
b. methanol (CH_3OH) or dimethyl ether (CH_3OCH_3)
c. formaldehyde (CH_2O) or methyl formate (HCO_2CH_3)*

A WORD ABOUT... Alternative Energy: The Benefits of Hydrogen

Our world's current dependence on petroleum and fossil fuels is significant. Besides burning these fossil fuels for energy, we use crude oil to prepare chemicals that are then used to make almost everything—the clothing we wear, the medicines we take, and the plastics, carpeting, and fabrics we use. Concerns about global warming, especially for carbon dioxide (CO_2) emissions, have further exacerbated worries about crude oil consumption and its impact on the environment and the world's population. Recently, a greater push has begun for finding alternative sources that can meet our energy demand and can provide for our chemical needs.

Alternative and renewable energy sources are being investigated, and these sources include solar energy, nuclear, biomass, wind, hydroelectric, and wave technology. Biomass, material obtained from living or recently lived organisms, provides an excellent source of carbohydrates (Chapter 16), lignins (Chapter 7), and lipids (Chapter 15). Much research has been expended into the conversion of carbohydrates, especially from corn and sugar beets, to value-added chemicals, including ethanol (CH_3CH_2OH, Chapter 7). Ethanol is used as a chemical to produce other products, as a fuel additive for gasoline-combustion engines, and for a myriad of other applications.

However, carbohydrates as precursors to chemical feedstocks present a conundrum with regard to our food supply. Should we take a food source and convert it to something that we burn in an engine instead of feeding the world's people? Moreover, due to its use in generating ethanol as a fuel additive (Chapter 8), the demand for corn caused the price of many food products to rise dramatically as corn is used to feed cattle as well as to make a host of consumed products.

Alternatively, lipids provide tremendous opportunities as they can be "harvested" from various sources, such as microbial, algal, vegetable, and animal origins, and there is significant current interest in preparing biofuels, such as biodiesel, from lipids (Chapter 15).

Another source of energy that is under investigation is hydrogen (H_2), particularly for mobile sources. Hydrogen offers many advantages over many alternative fuel sources. First, hydrogen (H_2) is a clean fuel that does not emit CO_2 when burned. A significant portion of the global CO_2 emissions comes from mobile sources, specifically cars and other transportation. When a carbon source, such as methane, is burned, CO_2 is generated (eq. 2.3). To minimize CO_2 emissions into the atmosphere, the fuel must not contain carbon. Thus, H_2 is a better choice. Second, the combustion of H_2 is very exothermic, releasing 68.4 kcal/mol of energy when one mole of H_2 is burned.

$$H_2 + \tfrac{1}{2}\,O_2 \rightarrow H_2O + \text{heat (68.4 kcal/mol)}$$

Methane, on the other hand, releases 212.8 kcal/mol of energy. Most importantly, for H_2 as a fuel, the emissions are pretty simple, just water. No CO_2, NO, or NO_2 is generated—the latter two are related to formation of tropospheric ozone, which is deleterious to human health.

Our current combustion engines have to be modified for using H_2 as this fuel ignites rapidly, but this issue has been resolved by vehicle manufacturers. Transporting hydrogen, though, is a challenge. Gaseous H_2 takes up a lot of space. Tankers can liquefy H_2 to very low temperatures to reduce its volume for transport, but this is not so easy for a family sedan.

Hence, an emerging area of chemical research is how to make H_2 from easily transportable sources. For example, three units of H_2 could be generated from ethanol (CH_3CH_2OH). A person could fill up a vehicle at the "ethanol pump," and, as needed, some on-board catalyst could convert it into H_2 for combustion in the engine. This work is at the forefront of basic and applied research.

See Problems 2.49 and 2.53.

*See Table 1.6 for the generic structure of an ester.

2.12.b Halogenation of Alkanes

When a mixture of an alkane and chlorine gas is stored at low temperatures in the dark, no reaction occurs. In sunlight or at high temperatures, however, an exothermic reaction occurs. One or more hydrogen atoms of the alkane are replaced by chlorine atoms. This reaction can be represented by the general equation

$$R{-}H + Cl{-}Cl \xrightarrow[\text{heat}]{\text{light or}} R{-}Cl + H{-}Cl \quad (2.9)$$

or, specifically for methane:

$$\underset{\text{methane}}{CH_4} + Cl{-}Cl \xrightarrow[\text{or heat}]{\text{sunlight}} \underset{\substack{\text{chloromethane}\\ \text{(methyl chloride)}\\ \text{bp } -24.2^\circ\text{C}}}{CH_3Cl} + HCl \quad (2.10)$$

The reaction is called **chlorination**. This process is a **substitution reaction**, as a chlorine is substituted for a hydrogen.

An analogous reaction, called **bromination**, occurs when the halogen source is bromine.

$$R{-}H + Br{-}Br \xrightarrow[\text{heat}]{\text{light or}} R{-}Br + HBr \quad (2.11)$$

Chlorination of hydrocarbons is a **substitution reaction** in which a chlorine atom is substituted for a hydrogen atom. Likewise in **bromination** reactions, a bromine atom is substituted for a hydrogen atom.

If excess halogen is present, the reaction can continue further to give polyhalogenated products. Thus, methane and excess chlorine can give products with two, three, or four chlorines.*

$$CH_3Cl \xrightarrow{Cl_2} \underset{\substack{\text{dichloromethane}\\ \text{(methylene chloride)}\\ \text{bp } 40^\circ\text{C}}}{CH_2Cl_2} \xrightarrow{Cl_2} \underset{\substack{\text{trichloromethane}\\ \text{(chloroform)}\\ \text{bp } 61.7^\circ\text{C}}}{CHCl_3} \xrightarrow{Cl_2} \underset{\substack{\text{tetrachloromethane}\\ \text{(carbon tetrachloride)}\\ \text{bp } 76.5^\circ\text{C}}}{CCl_4} \quad (2.12)$$

By controlling the reaction conditions and the ratio of chlorine to methane, we can favor formation of one or another of the possible products.

PROBLEM 2.19 Write the names and structures of all possible products for the bromination of methane.

With longer chain alkanes, mixtures of products may be obtained even at the first step.** For example, with propane,

$$\underset{\text{propane}}{CH_3CH_2CH_3} + Cl_2 \xrightarrow[\text{or heat}]{\text{light}} \underset{\substack{\text{1-chloropropane}\\ (n\text{-propyl chloride})}}{CH_3CH_2CH_2Cl} + \underset{\substack{\text{2-chloropropane}\\ \text{(isopropyl chloride)}}}{CH_3\underset{\substack{|\\ Cl}}{C}HCH_3} + HCl \quad (2.13)$$

When larger alkanes are halogenated, the mixture of products becomes even more complex; individual isomers become difficult to separate and obtain pure, so halogenation

*Note that we sometimes write the formula of one of the reactants (in this case Cl_2) over the arrow for convenience, as in eq. 2.12. We also sometimes omit obvious inorganic products (in this case, HCl).

**Note that we often do not write a balanced equation, especially when more than one product is formed from a single organic reactant. Instead, we show, on the right side of the equation, the structures of *all* of the important organic products, as in eq. 2.13.

A WORD ABOUT... Methane, Marsh Gas, and Miller's Experiment

Methane is commonly found in nature wherever bacteria decompose organic matter in the absence of oxygen, as in marshes, swamps, or the muddy sediment of lakes—hence, its common name, *marsh gas.* In China, methane has been collected from the mud at the bottom of swamps for use in domestic cooking and lighting. Methane is similarly formed from bacteria in the digestive tracts of certain ruminant animals, such as cows.

The scale of methane production by bacteria is considerable. The earth's atmosphere contains an average of 1 part per million of methane. Because our planet is small and because methane is light compared to most other air constituents (O_2, N_2), one would expect most of the methane to have escaped from our atmosphere, and it has been calculated that the equilibrium concentration should be very much less than is observed. The reason, then, for the relatively high observed concentration is that at the same time that methane escapes from the atmosphere, it is constantly replenished by bacterial decay of plant matter.

In cities, the amount of methane in the atmosphere reaches much higher levels, up to several parts per million. The peak concentrations come in the early morning and late afternoon, directly correlated with the peaks of automobile traffic. Fortunately, methane, which constitutes about 50% of urban atmospheric hydrocarbon pollutants, seems to have no direct harmful effect on human health.

Methane can accumulate in coal mines, where it is a hazard because, when mixed with 5% to 14% of air, it is explosive. Also, miners can be asphyxiated by it (due to lack of sufficient oxygen). Dangerous concentrations of methane can be detected readily by a variety of safety devices.

Hydrogen is the most common element in the solar system (it constitutes about 87% of the sun's mass). It therefore seemed reasonable to think that, when the planets were formed, other elements should have been present in reduced (not oxidized) forms: carbon as methane, nitrogen as ammonia, and oxygen as water. Indeed, some of the outer planets (Saturn and Jupiter) still have atmospheres that are rich in methane and ammonia.

A now-famous experiment by Stanley L. Miller (working in the laboratory of H. C. Urey at Columbia University) supports the idea that life could have arisen in a reducing atmosphere. Miller found that when mixtures of methane, ammonia, water, and hydrogen were subjected to electric discharges (to simulate lightning), some organic compounds (amino acids, for example) that are important to biology and necessary for life were formed. Similar results have since been obtained using heat or ultraviolet light in place of electric discharges (it seems likely that the earth's early atmosphere was subjected to much more ultraviolet radiation than it is now). When oxygen was added to these simulated primeval atmospheres, no amino acids were produced—strong evidence that the earth's original atmosphere did *not* contain free oxygen.

Coal mining in Appalachia.

In the years since Miller's experiment, ideas about the chemistry of life's origin have become more precise as a consequence of much experimentation and of exploration in outer space. We now know that the earth's primary atmosphere was formed mainly by degassing the molten interior rather than by accretion from the solar nebula. It seems likely that the main carbon sources in the earth's early atmosphere were CO_2 and CO, *not* methane as assumed by Miller, and that nitrogen was present mainly as N_2 rather than as ammonia. Repetition of Miller-type experiments with these assumed primordial atmospheres again gave biomolecules.

Miller's experiment provided a model for much work in the branch of a science now called **chemical evolution** or **prebiotic chemistry**, the study of chemical events that may have taken place on earth or elsewhere in the universe leading to the appearance of the first living cell. For additional reading, you can consult *Chemical Evolution* by Stephen F. Mason, Clarendon Press, Oxford, 1991.

See Problem 2.54.

tends not to be a useful way to synthesize specific alkyl halides. With unsubstituted *cycloalkanes*, however, where all of the hydrogens are equivalent, a single pure organic product can be obtained:

$$\text{cyclopentane} + Br_2 \xrightarrow{\text{light}} \text{bromocyclopentane (cyclopentyl bromide)} + HBr \qquad (2.14)$$

PROBLEM 2.20 Write the structures of all possible products of *mono*chlorination of pentane. Note the complexity of the product mixture, compared to that from the corresponding reaction with *cyclo*pentane (eq. 2.14).

PROBLEM 2.21 How many organic products can be obtained from the monobromination of heptane? Of cycloheptane?

PROBLEM 2.22 Do you think that the bromination of 2,2-dimethylpropane might be synthetically useful?

2.13 The Free-Radical Chain Mechanism of Halogenation

One may well ask how halogenation occurs. Why is light or heat necessary? Equations 2.9 and 2.10 express the *overall* reaction for halogenation. They describe the structures of the reactants and the products, and they show necessary reaction conditions or catalysts over the arrow. But they do *not* tell us exactly how the products are formed from the reactants.

A **reaction mechanism** is a step-by-step description of the bond-breaking and bond-making processes that occur when reagents react to form products. In the case of halogenation, various experiments show that this reaction occurs in several steps, and not in one magical step. Indeed, halogenation occurs via a **free-radical chain** of reactions.

A **reaction mechanism** is a step-by-step description of the bond-breaking and bond-making processes that occur when reagents react to form products.

A **free-radical chain reaction** includes a **chain-initiating step**, **chain-propagating steps**, and **chain-terminating steps**.

The **chain-initiating step** is the breaking of the halogen molecule into two halogen atoms.

$$\textit{initiation} \quad :\ddot{Cl}:\ddot{Cl}: \xrightarrow[\text{or heat}]{\text{light}} :\ddot{Cl}\cdot + :\ddot{Cl}\cdot \qquad (2.15)^*$$

chlorine molecule → chlorine atoms

The Cl—Cl bond is weaker than either the C—H bond or the C—C bond (compare the bond energies, Table A in the Appendix), and is therefore the easiest bond to break by supplying heat energy. When light is the energy source, molecular chlorine (Cl_2) absorbs visible light but alkanes do not. Thus, once again, the Cl—Cl bond would break first.

*Recall from Section 1.13 that we use a "fishhook," or half-headed arrow, ⤻, to show the movement of only *one* electron, whereas we use a complete (double-headed) arrow, ⤻, to describe the movement of an electron *pair*.

The **chain-propagating steps** are

propagation

$$\mathrm{R{-}H + \cdot\ddot{\underset{..}{Cl}}: \longrightarrow R\cdot + H{-}Cl} \quad \text{(2.16)}$$

alkyl radical

$$\mathrm{R\cdot + Cl{-}Cl \longrightarrow R{-}Cl + \cdot\ddot{\underset{..}{Cl}}:} \quad \text{(2.17)}$$

alkyl chloride

Chlorine atoms are very reactive, because they have an incomplete valence shell (seven electrons instead of the required eight). They may either recombine to form chlorine molecules (the reverse of eq. 2.15) or, if they collide with an alkane molecule, abstract a hydrogen atom to form hydrogen chloride and an alkyl radical R•. Recall from Section 1.4 that a radical is a fragment with an odd number of unshared electrons. The space-filling models in Figure 2.1 show that alkanes seem to have an exposed surface of hydrogens covering the carbon skeleton. So it is most likely that, if a halogen atom collides with an alkane molecule, it will hit the hydrogen end of a C—H bond.

Like a chlorine atom, the alkyl radical formed in the first step of the chain (eq. 2.16) is very reactive (note that the alkyl radical, like the halogen radical, has an incomplete octet). If the alkyl radical was to collide with a chlorine molecule (Cl_2), it could form an alkyl chloride molecule and a chlorine atom (eq. 2.17). The chlorine atom formed in this step can then react to repeat the sequence. When you add eq. 2.16 and eq. 2.17, you get the overall equation for chlorination (eq. 2.9). In each chain-propagating step, a radical (or atom) is consumed, but another radical (or atom) is formed and can continue the chain. Almost all of the reactants are consumed, and almost all of the products are formed in these steps.

Were it not for **chain-terminating steps**, all of the reactants could, in principle, be consumed by initiating a single reaction chain. However, because many chlorine molecules react to form chlorine atoms in the chain-initiating step, many chains are started simultaneously. Quite a few radicals are present as the reaction proceeds. If any two radicals combine, the chain will be terminated. Three possible chain-terminating steps are

termination

$$\mathrm{:\ddot{\underset{..}{Cl}}\cdot + \cdot\ddot{\underset{..}{Cl}}: \longrightarrow Cl{-}Cl} \quad \text{(2.18)}$$

$$\mathrm{R\cdot + \cdot R \longrightarrow R{-}R} \quad \text{(2.19)}$$

$$\mathrm{R\cdot + \cdot\ddot{\underset{..}{Cl}}\cdot \longrightarrow R{-}Cl} \quad \text{(2.20)}$$

No new radicals are formed in these reactions, so the chain is broken or, as we say, terminated. Note that eq. 2.20 is a useful reaction as it leads to a desired product, but the other termination reactions lead to regeneration of the starting halogen (Cl_2, eq. 2.18) or a byproduct (eq. 2.19).

PROBLEM 2.23 Show that when eq. 2.16 and eq. 2.17 are added, the overall equation for chlorination (eq. 2.9) results.

PROBLEM 2.24 Write equations for all of the steps (initiation, propagation, and termination) in the free-radical chlorination of methane to form methyl chloride.

PROBLEM 2.25 Account for the experimental observation that small amounts of ethane and chloroethane are produced during the monochlorination of methane. (*Hint:* Consider the possible chain-terminating steps.)

REACTION SUMMARY

1. Reactions of Alkanes and Cycloalkanes

a. Combustion (Sec. 2.12a)

$$C_nH_{2n+2} + \left(\frac{3n+1}{2}\right)O_2 \longrightarrow nCO_2 + (n+1)H_2O$$

b. Halogenation (Sec. 2.12b)

$$R{-}H + X_2 \xrightarrow[\text{or light}]{\text{heat}} R{-}X + H{-}X \quad (X = Cl, Br)$$

MECHANISM SUMMARY

1. Halogenation (Sec. 2.13)

Initiation $:\ddot{X}:\ddot{X}: \xrightarrow[\text{or heat}]{\text{light}} :\ddot{X}\cdot + :\ddot{X}\cdot$

chlorine molecule — chlorine atoms

Propagation $R{-}H + \cdot\ddot{X}: \longrightarrow R\cdot + H{-}X$

alkyl radical

$R\cdot + X{-}X \longrightarrow R{-}X + \cdot\ddot{X}:$

alkyl chloride

Termination $:\ddot{X}\cdot + \cdot\ddot{X}: \longrightarrow X{-}X$

$R\cdot + \cdot R \longrightarrow R{-}R$

$R\cdot + \cdot\ddot{X}: \longrightarrow R{-}X$

(X = Cl, Br)

ADDITIONAL PROBLEMS

OWL Interactive versions of these problems are assignable in OWL.

Alkane Nomenclature and Structural Formulas

2.26 Write structural formulas for the following compounds:

a. 2-methylhexane
b. 2,3-dimethylbutane
c. 4-ethyl-2,2-dimethylheptane
d. 2-bromo-4-methyloctane
e. 1,1-diiodocyclobutane
f. 2-chlorobutane
g. 1-isopropyl-1,3-dimethylcyclohexane
h. 1,1,2-trifluoropropane
i. 1,1,3,3-tetrachlorocyclopropane

2.27 Write expanded formulas for the following compounds and name them using the IUPAC system:

a. $(CH_3)_3CCH_2CH_2CH_3$
b. $CH_3(CH_2)_2CH_3$
c. $(CH_3)_2CHCH_2CH_2CH_3$
d. $CH_3CCl_2CF_3$
e. $(CH_2)_4$
f. $CH_3CH_2CHFCH_3$
g. EtBr
h. $ClCH_2CH_2Cl$
i. *i*-PrCl

2.28 Give both common and IUPAC names for the following compounds:

a. CH_3Br
b. CH_3CH_2Cl
c. CH_2Cl_2
d. $(CH_3)_2CHBr$
e. CHI_3
f. $(CH_3)_3CCl$
g. $CH_3CH_2CH_2CH_2F$

2.29 Write a structure for each of the compounds listed. Explain why the name given here is incorrect, and give a correct name in each case.

a. 2,3-fluoropropane
b. 1-methylbutane
c. 2-ethylbutane
d. 1-methyl-2-ethylcyclopropane
e. 1,1,3-trimethylhexane
f. 4-bromo-3-methylbutane
g. 1,3-dimethylcyclopropane

2.30 Chemical substances used for communication in nature are called *pheromones.* The pheromone used by the female tiger moth to attract the male is the 18-carbon-atom alkane 2-methylheptadecane. Write its structural formula.

2.31 Write the structural formulas for all isomers of each of the following compounds, and name each isomer by the IUPAC system. (The number of isomers is indicated in parentheses.)

a. C_4H_{10} (2)
b. C_4H_9Cl (4)
c. C_3H_6FCl (5)
d. $C_3H_6Br_2$ (4)
e. $C_2H_2ClBr_3$ (3)
f. C_5H_{12} (3)

2.32 Write structural formulas and names for all possible cycloalkanes having each of the following molecular formulas. Be sure to include *cis–trans* isomers when appropriate. Name each compound by the IUPAC system.

a. C_5H_{10} (there are 6)
b. C_6H_{12} (there are 16)

2.33 Write a correct IUPAC name for each of the following structures:

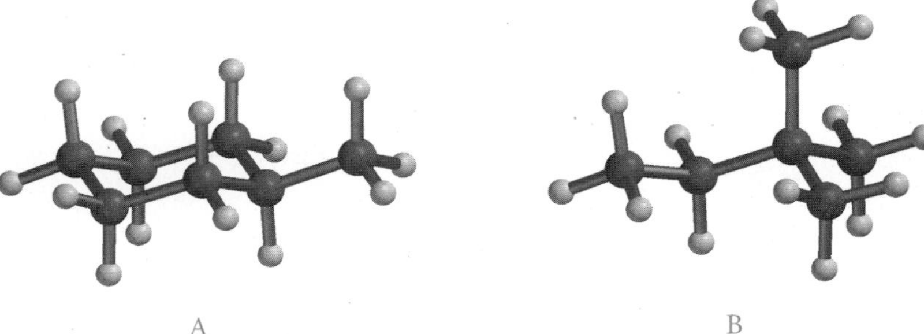

2.34 Draw the tetra-ethyl analog of 2,2-dimethylpropane as discussed in "A Word About... Isomers—Possible and Impossible" on page 55. What is the correct IUPAC name for this compound? Make models of this compound and 2,2-dimethylpropane and compare how sterically congested they are.

Alkane Properties and Intermolecular Interactions

2.35 Without referring to tables, arrange the following five hydrocarbons in order of increasing boiling point. (*Hint:* Draw structures or make models of the five hydrocarbons to see their shapes and sizes.)

a. 2-methylhexane
b. heptane
c. 3,3-dimethylpentane
d. hexane
e. 2-methylpentane

Explain your answer in terms of intermolecular interactions.

2.36 Arrange the following liquids in order from least soluble in hexane to most soluble in hexane:

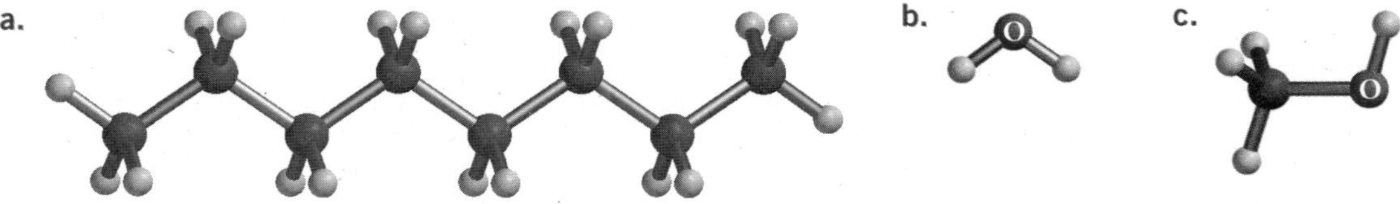

Explain your answer in terms of intermolecular interactions.

Conformations of Alkanes

2.37 In Problem 2.9, you drew two staggered conformations of butane (looking end-on down the bond between carbon-2 and carbon-3). There are also two eclipsed conformations around this bond. Draw Newman projections for them. Arrange all four conformations in order of decreasing stability.

2.38 Draw the Newman projections for the unique staggered and eclipsed conformers of 2-methylbutane.

2.39 Draw all possible staggered and eclipsed conformations of 1-bromo-2-chloroethane (see the following ball-and-stick model), using Newman projections. Underneath each structure, draw the corresponding "dash-wedge" and "sawhorse" structures. Rank the conformations in order of decreasing stability.

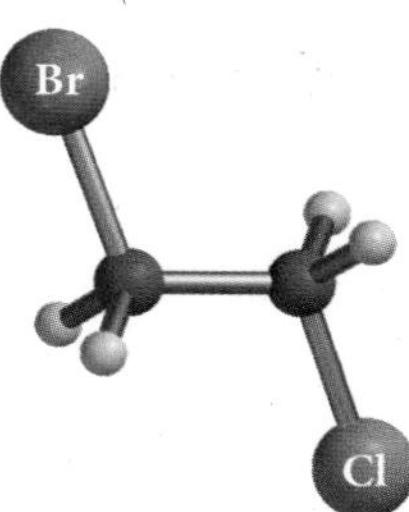

Conformations of Cycloalkanes; *Cis–Trans* Isomerism

2.40 Draw the formula for the preferred conformation of

a. *cis*-1,4-dimethylcyclohexane
b. *trans*-1-isopropyl-3-methylcyclohexane
c. 1,1-diethylcyclopentane
d. ethylcyclohexane

2.41 Name the following *cis–trans* pairs:

a.

H H
Cl Cl

H Cl
Cl H

b.

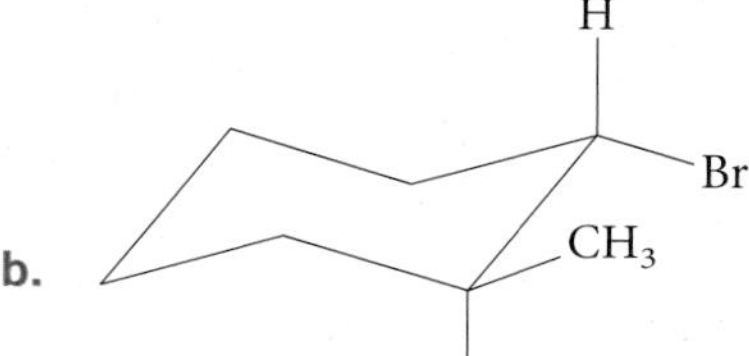

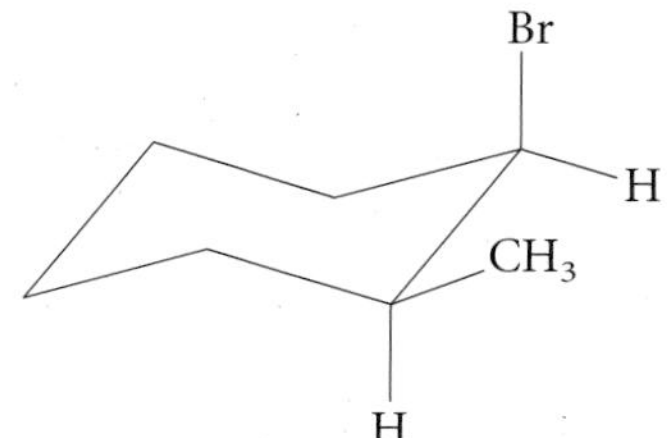

2.42 Explain with the aid of conformational structures why *cis*-1,3-dimethylcyclohexane is more stable than *trans*-1,3-dimethylcyclohexane, whereas the reverse order of stability is observed for the 1,2 and 1,4 isomers. (Constructing models will help you with this problem.)

2.43 Draw structural formulas for all possible dimethylcyclobutanes. Include *cis–trans* isomers.

2.44 Which will be more stable, *cis*- or *trans*-1,4-di-*tert*-butylcyclohexane? Explain your answer by drawing conformational structures for each compound.

2.45 Examine the relationships of isomers as described in Figure 2.8 (p. 56). Then classify the following pairs of structures as structural isomers, conformers, configurational (*cis–trans*) isomers, or identical.

a. the pairs of compounds in Problem 2.41

b. 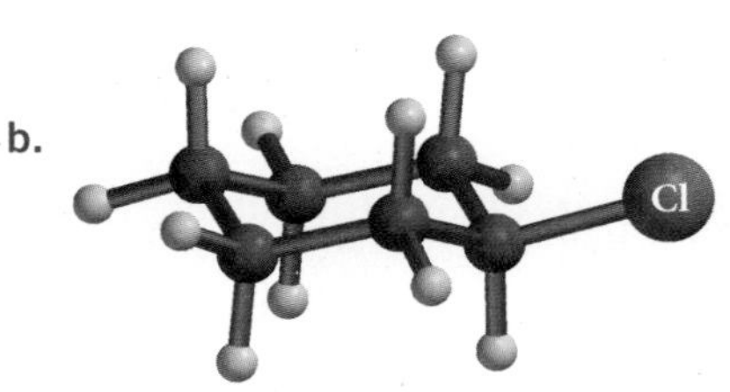

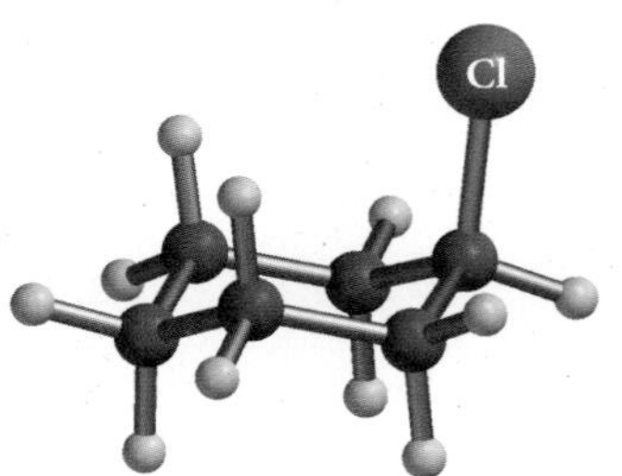

c. Cl, H, CH_3, H, Cl, H and Cl, H, H, H, Cl, CH_3

d. Cl, H, CH_3, H, Cl, H and Cl, H, Cl, H, CH_3, H

e. $CH_3CHCH_2CH_2CH_3$ (with CH_3 on C-2) and $CH_3CH_2CH_2CHCH_3$ (with CH_3 on C-4) (careful!)

2.46 Draw structural formulas for all possible difluorocyclohexanes. Include *cis–trans* isomers.

Reactions of Alkanes: Combustion and Halogenation

2.47 How many monobromination products can be obtained from each of the following polycyclic alkanes?

a. 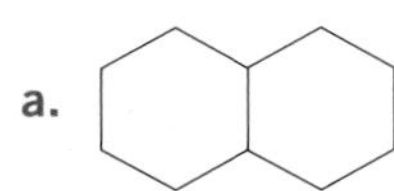

b. 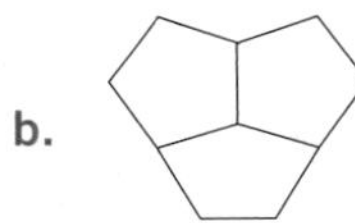

c. 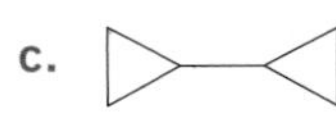

d. $CH_3CH_2CH_2CH_3$

e. CH_3

2.48 Using structural formulas, write equations for each of the following combustion reactions (see Reaction Summary 1.a, p. 63):

a. the complete combustion of propane
b. the complete combustion of pentane
c. the complete combustion of butane

2.49 Petroleum diesel is a complex fuel mixture, typically composed of alkanes with carbon chains containing 8 to 21 carbon atoms. Using a 10-carbon alkane as an example, write a balanced equation for the complete combustion of this fuel to CO_2 and H_2O.

2.50 Using structural formulas, write equations for the following halogenation reactions (see Reaction Summary 1.b, p. 63), and name each organic product:

a. the monochlorination of propane
b. the monobromination of cyclopentane
c. the complete chlorination of butane
d. the monobromination of methylcyclohexane

2.51 From the dichlorination of propane, four isomeric products with the formula $C_3H_6Cl_2$ were isolated and designated A, B, C, and D. Each was separated and further chlorinated to give one or more trichloropropanes, $C_3H_5Cl_3$. A and B gave three trichloro compounds, C gave one, and D gave two. Deduce the structures of C and D. One of the products from A was identical to the product from C. Deduce structures for A and B. (*Hint:* Start by drawing the structures of all four dichlorinated propane isomers.)

2.52 Write all of the steps in the free-radical chain mechanism for the monochlorination of ethane (see Mechanism Summary, p. 63).

$$CH_3CH_3 + Cl_2 \longrightarrow CH_3CH_2Cl + HCl$$

What trace by-products would you expect to be formed as a consequence of the chain-terminating steps?

2.53 Using methanol (CH_3OH) as a fuel source, write a balanced reaction for the formation of H_2. Do the same for $C_{10}H_{22}$ (a component of diesel fuel). For an H_2 fuel cell, why would it be better to use $C_{10}H_{22}$ rather than methanol? What would be a possible disadvantage?

2.54 As noted in the "A Word about…Methane, Marsh Gas, and Miller's Experiment" on page 60, methane can be formed in muddy sediments because of the reducing environment (i.e., lack of oxygen). Write a balanced reaction for the conversion of glucose ($C_6H_{12}O_6$, Sec. 2.9) with molecular hydrogen (H_2) to form methane and water.

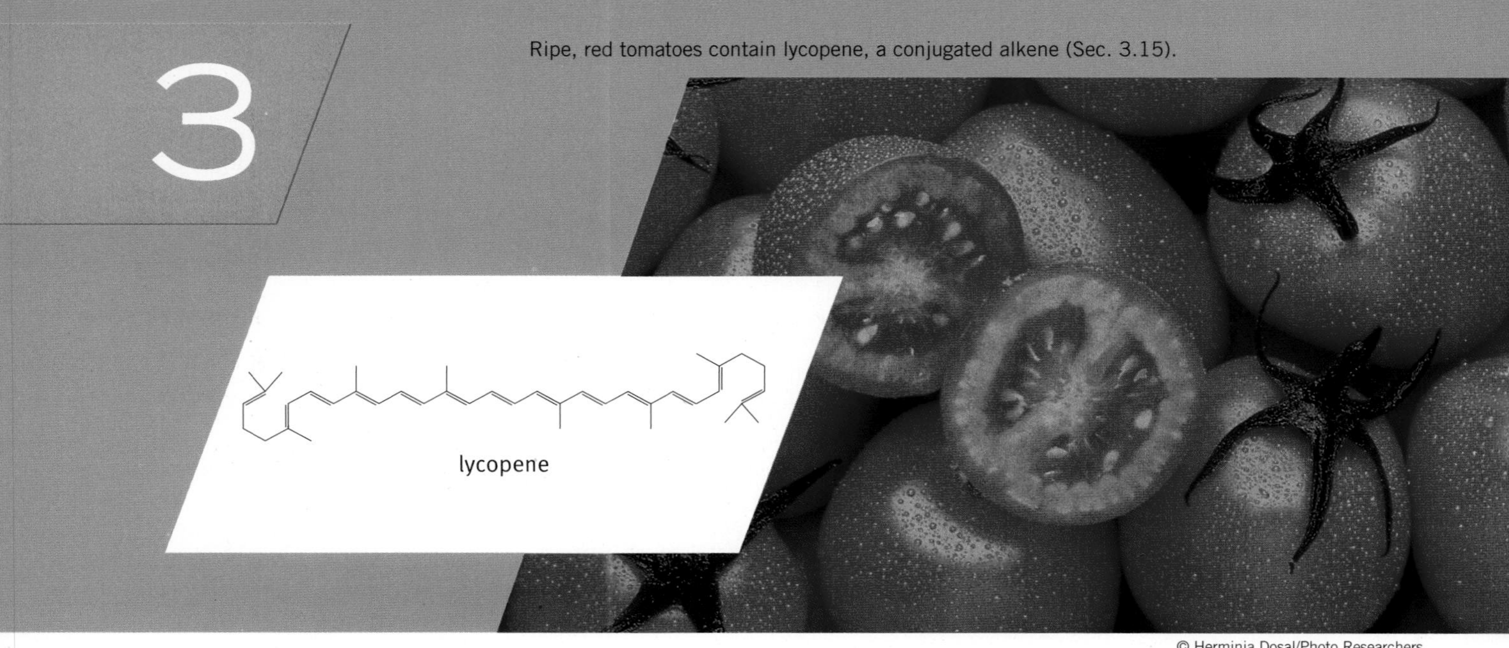

Ripe, red tomatoes contain lycopene, a conjugated alkene (Sec. 3.15).

© Herminia Dosal/Photo Researchers

Alkenes and Alkynes

Alkenes are compounds containing carbon–carbon double bonds. The simplest alkene, ethene, is a plant hormone (see "A Word About . . . Ethylene," pp. 98–99) and an important starting material for the manufacture of other organic compounds (see Figure 3.13, p. 98). The alkene functional group is found in sources as varied as citrus fruits (limonene, Fig. 1.12), steroids (cholesterol, Sec. 15.9), and insect pheromones (muscalure; see "A Word about the Gypsy Moth's Epoxide," p. 245). Alkenes have physical properties similar to those of alkanes (Sec. 2.7). They are less dense than water and, being nonpolar, are not very soluble in water. As with alkanes, compounds with four or fewer carbons are colorless gases, whereas higher homologs are volatile liquids.

Alkynes, compounds containing carbon–carbon triple bonds, are similar to alkenes in their physical properties and chemical behavior. In this chapter, we will examine the structure and chemical reactions of these two classes of compounds. We will also examine briefly the relationship between chemical reactions and energy.

Online homework for this chapter can be assigned in OWL, an online homework assessment tool.

3.1 Definition and Classification

Hydrocarbons that contain a carbon–carbon double bond are called **alkenes**; those with a carbon–carbon triple bond are **alkynes**.* Their general formulas are

Alkenes and **alkynes** are **unsaturated** hydrocarbons containing carbon–carbon double bonds and carbon–carbon triple bonds, respectively.

$$\underset{\text{alkenes}}{C_nH_{2n}} \qquad \underset{\text{alkynes}}{C_nH_{2n-2}}$$

Both of these classes of hydrocarbons are **unsaturated**, because they contain fewer hydrogens per carbon than alkanes (C_nH_{2n+2}). Alkanes can be obtained from alkenes or alkynes by adding 1 or 2 moles of hydrogen.

$$\left.\begin{array}{l} \underset{\text{alkene}}{RCH{=}CHR} \xrightarrow[\text{catalyst}]{H_2} \\ \underset{\text{alkyne}}{RC{\equiv}CR} \xrightarrow[\text{catalyst}]{2H_2} \end{array}\right\} \longrightarrow \underset{\text{alkane}}{RCH_2CH_2R} \tag{3.1}$$

Compounds with more than one double or triple bond exist. If two double bonds are present, the compounds are called **alkadienes** or, more commonly, **dienes**. There are also trienes, tetraenes, and even polyenes (compounds with *many* double bonds, from the Greek *poly,* many). Polyenes are responsible for the color of carrots (β-carotene, p. 76) and tomatoes (lycopene, p. 68). Compounds with more than one triple bond, or with double and triple bonds, are also known.

Alkadienes, or **dienes,** contain two C—C double bonds that can be **cumulated** (next to each other), **conjugated** (separated by one C—C single bond), or **nonconjugated** (separated by more than one C—C single bond).

EXAMPLE 3.1

What are all of the structural possibilities for the compound C_3H_4?

Solution The formula C_3H_4 corresponds to the general formula C_nH_{2n-2}; thus, a C_3H_4 compound is four hydrogens less than the corresponding alkane (C_nH_{2n+2}), C_3H_8. The C_3H_4 compound could have one triple bond, two double bonds, or one ring and one double bond. For their structures, see the solution to Example 1.10 on page 13.

PROBLEM 3.1 What are all of the structural possibilities for C_4H_6? (Nine compounds, four acyclic and five cyclic, are known.)

When two or more multiple bonds are present in a molecule, it is useful to classify the structure further, depending on the relative positions of the multiple bonds. Double bonds are said to be **cumulated** when they are right next to one another. When multiple bonds *alternate* with single bonds, they are called **conjugated**. When more than one single bond comes between multiple bonds, the latter are isolated or **nonconjugated**.

cumulated	conjugated	nonconjugated (isolated)
C=C=C	C=C—C=C	C=C—C—C=C
C=C=C=C	C=C—C≡C	C≡C—C—C—C≡C

*An old but still used synonym for alkenes is *olefins.* Alkynes are also called *acetylenes,* after the first member of the series.

PROBLEM 3.2 Which of the following compounds have conjugated multiple bonds?

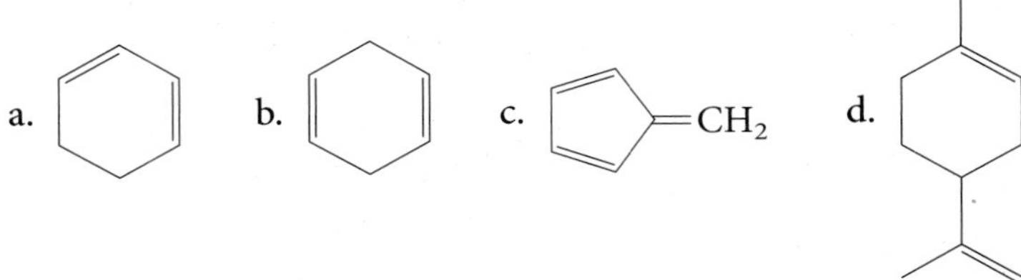

3.2 Nomenclature

The IUPAC rules for naming alkenes and alkynes are similar to those for alkanes (Sec. 2.3), but a few rules must be added for naming and locating the multiple bonds.

1. The ending *-ene* is used to designate a carbon–carbon double bond. When more than one double bond is present, the ending is *-diene, -triene,* and so on. The ending *-yne* (rhymes with wine) is used for a triple bond (*-diyne* for two triple bonds and so on). Compounds with a double *and* a triple bond are *-enynes.*
2. Select the longest chain that includes *both* carbons of the double or triple bond. For example,

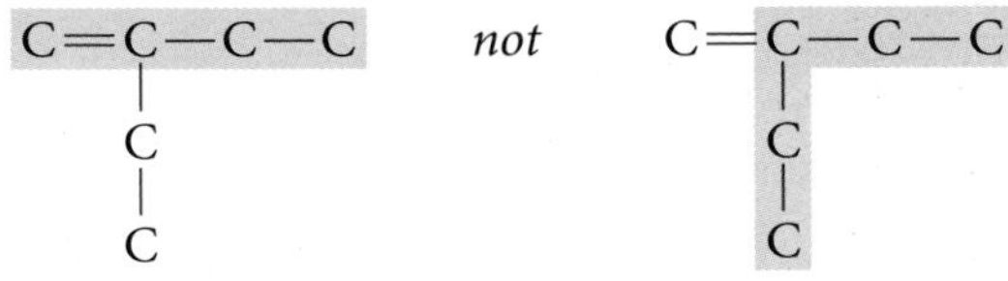

named as a butene, not as a pentene

3. Number the chain from the end nearest the multiple bond so that the carbon atoms in that bond have the lowest possible numbers.

$$\overset{1}{C}-\overset{2}{C}=\overset{3}{C}-\overset{4}{C}-\overset{5}{C} \quad \textit{not} \quad \overset{5}{C}-\overset{4}{C}=\overset{3}{C}-\overset{2}{C}-\overset{1}{C}$$

If the multiple bond is equidistant from both ends of the chain, number the chain from the end nearest the first branch point.

$$\overset{1}{C}-\underset{\textstyle C}{\overset{2}{C}}=\overset{3}{C}-\overset{4}{C} \quad \textit{not} \quad \overset{4}{C}-\underset{\textstyle C}{\overset{3}{C}}=\overset{2}{C}-\overset{1}{C}$$

4. Indicate the position of the multiple bond using the *lower numbered carbon atom* of that bond. For example,

$$\overset{1}{C}H_2=\overset{2}{C}H\overset{3}{C}H_2\overset{4}{C}H_3 \qquad \text{1-butene, } \textit{not} \text{ 2-butene}$$

5. If more than one multiple bond is present, number the chain from the end nearest the first multiple bond.

$$\overset{1}{C}=\overset{2}{C}-\overset{3}{C}=\overset{4}{C}-\overset{5}{C} \quad \textit{not} \quad \overset{5}{C}=\overset{4}{C}-\overset{3}{C}=\overset{2}{C}-\overset{1}{C}$$

If a double and a triple bond are equidistant from the end of the chain, the *double* bond receives the lowest numbers. For example,

$$\overset{1}{C}=\overset{2}{C}-\overset{3}{C}\equiv\overset{4}{C} \quad \textit{not} \quad \overset{4}{C}=\overset{3}{C}-\overset{2}{C}\equiv\overset{1}{C}$$

Let us see how these rules are applied. The first two members of each series are

CH_3CH_3 ethane	$CH_2{=}CH_2$ ethene	$HC{\equiv}CH$ ethyne
$CH_3CH_2CH_3$ propane	$CH_2{=}CHCH_3$ propene	$HC{\equiv}CCH_3$ propyne

The root of the name (*eth-* or *prop-*) tells us the number of carbons, and the ending (*-ane, -ene,* or *-yne*) tells us whether the bonds are single, double, or triple. No number is necessary in these cases, because in each instance, only one structure is possible.

With four carbons, a number is necessary to locate the double or triple bond.

$\overset{1}{C}H_2{=}\overset{2}{C}H\overset{3}{C}H_2\overset{4}{C}H_3$ 1-butene

$\overset{1}{C}H_3\overset{2}{C}H{=}\overset{3}{C}H\overset{4}{C}H_3$ 2-butene

$H\overset{1}{C}{\equiv}\overset{2}{C}\overset{3}{C}H_2\overset{4}{C}H_3$ 1-butyne

$\overset{1}{C}H_3\overset{2}{C}{\equiv}\overset{3}{C}\overset{4}{C}H_3$ 2-butyne

Branches are named in the usual way.

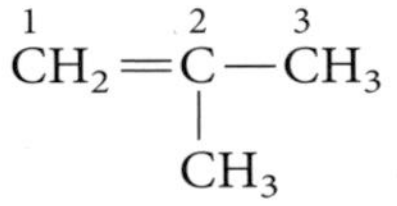

$\overset{1}{C}H_2{=}\overset{2}{C}(CH_3){-}\overset{3}{C}H_3$ methylpropene (isobutylene)

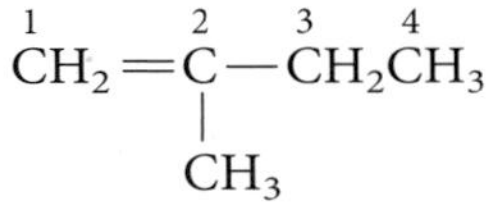

$\overset{1}{C}H_2{=}\overset{2}{C}(CH_3){-}\overset{3}{C}H_2\overset{4}{C}H_3$ 2-methyl-1-butene

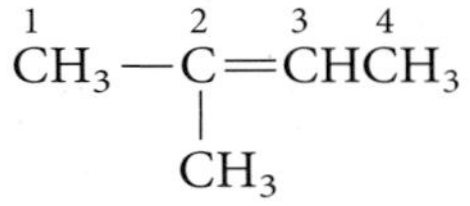

$\overset{1}{C}H_3{-}\overset{2}{C}(CH_3){=}\overset{3}{C}H\overset{4}{C}H_3$ 2-methyl-2-butene

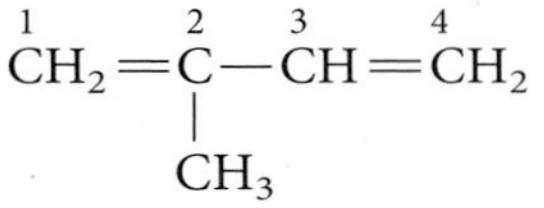

$\overset{1}{C}H_2{=}\overset{2}{C}(CH_3){-}\overset{3}{C}H{=}\overset{4}{C}H_2$ 2-methyl-1,3-butadiene (isoprene)

Note how the rules are applied in the following examples:

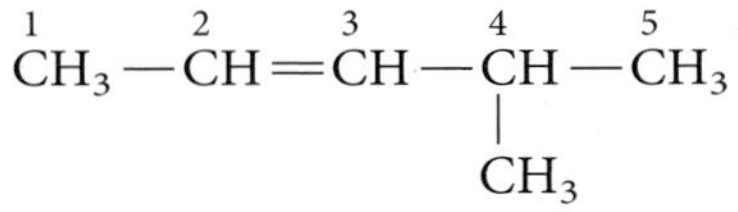

$\overset{1}{C}H_3{-}\overset{2}{C}H{=}\overset{3}{C}H{-}\overset{4}{C}H(CH_3){-}\overset{5}{C}H_3$

4-methyl-2-pentene
(*Not* 2-methyl-3-pentene; the chain is numbered so that the double bond gets the lower number.)

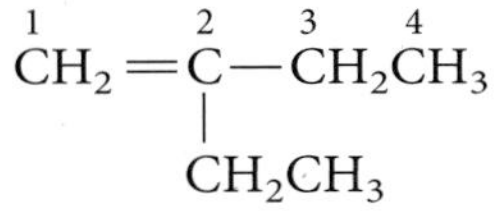

$\overset{1}{C}H_2{=}\overset{2}{C}(CH_2CH_3){-}\overset{3}{C}H_2\overset{4}{C}H_3$

2-ethyl-1-butene
(Named this way, even though there is a five-carbon chain present, because that chain does not include both carbons of the double bond.)

$\overset{1}{C}H_2{=}\overset{2}{C}H{-}\overset{3}{C}H{=}\overset{4}{C}H_2$

1,3-butadiene
(Note the *a* inserted in the name, to help in pronunciation.)

With cyclic hydrocarbons, we start numbering the ring with the carbons of the multiple bond.

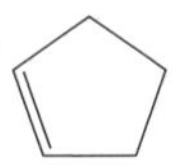

cyclopentene
(No number is necessary, because there is only one possible structure.)

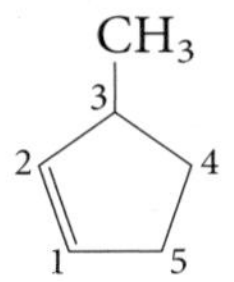

3-methylcyclopentene
(Start numbering at, and number through the double bond; 5-methylcyclopentene and 1-methyl-2-cyclopentene are incorrect names.)

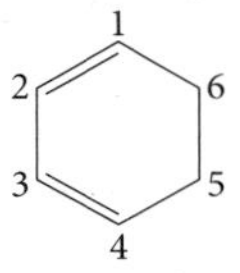

1,3-cyclohexadiene

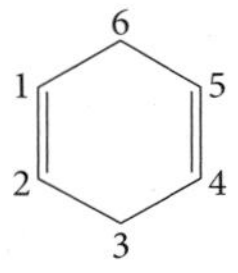

1,4-cyclohexadiene

PROBLEM 3.3 Name each of the following structures by the IUPAC system:

a. $CH_2{=}C(Cl)CH_3$ b. $(CH_3)_2C{=}C(CH_2CH_3)_2$ c. $FCH{=}CHCH_2CH_3$

d. (cyclohexene ring with CH_3 on a double-bond carbon) e. $CH_2{=}C(Br)CH{=}CH_2$ f. $CH_3(CH_2)_3C{\equiv}CCH_3$

EXAMPLE 3.2

Write the structural formula for 3-methyl-2-pentene.

Solution To get the structural formula from the IUPAC name, first write the longest chain or ring, number it, and then locate the multiple bond. In this case, note that the chain has five carbons and that the double bond is located between carbon-2 and carbon-3:

$$\overset{1}{C}{-}\overset{2}{C}{=}\overset{3}{C}{-}\overset{4}{C}{-}\overset{5}{C}$$

Next, add the substituent:

$$\overset{1}{C}{-}\overset{2}{C}{=}\underset{\displaystyle CH_3}{\overset{3}{C}}{-}\overset{4}{C}{-}\overset{5}{C}$$

Finally, fill in the hydrogens:

$$CH_3{-}CH{=}\underset{\displaystyle CH_3}{C}{-}CH_2{-}CH_3$$

PROBLEM 3.4 Write structural formulas for the following:

a. 1,4-dichloro-2-pentene
b. 3-hexyne
c. 1,2-diethylcyclobutene
d. 2-bromo-1,3-pentadiene

In addition to the IUPAC rules, it is important to learn a few common names. For example, the simplest members of the alkene and alkyne series are frequently referred to by their older common names, ethylene, acetylene, and propylene.

$CH_2{=}CH_2$	$HC{\equiv}CH$	$CH_3CH{=}CH_2$
ethylene	acetylene	propylene
(ethene)	(ethyne)	(propene)

Two important groups also have common names. They are the vinyl and allyl groups (their IUPAC names are in parentheses below), shown on the left. These groups are used in common names, illustrated in the examples on the right.

$CH_2{=}CH{-}$	$CH_2{=}CHCl$
vinyl	vinyl chloride
(ethenyl)	(chloroethene)
$CH_2{=}CH{-}CH_2{-}$	$CH_2{=}CH{-}CH_2Cl$
allyl	allyl chloride
(2-propenyl)	(3-chloropropene)

PROBLEM 3.5 Write the structural formula for

a. vinylcyclohexane
b. allylcyclobutane

3.3 Some Facts about Double Bonds

Carbon–carbon double bonds have some special features that are different from those of single bonds. For example, each carbon atom of a double bond is connected to only *three* other atoms (instead of four atoms, as with sp^3 tetrahedral carbon). We speak of such a carbon as being **trigonal**. Furthermore, the two carbon atoms of a double bond and the four atoms that are attached to them lie in a single plane. This planarity is shown in Figure 3.1 for ethylene. The H—C—H and H—C═C angles in ethylene are approximately 120°. Although rotation occurs freely around single bonds, *rotation around double bonds is restricted.* Ethylene does not adopt any other conformation except the planar one. The doubly bonded carbons with two attached hydrogens do not rotate with respect to each other. Finally, carbon–carbon double bonds are shorter than carbon–carbon single bonds.

A **trigonal** carbon atom is bonded to only three other atoms.

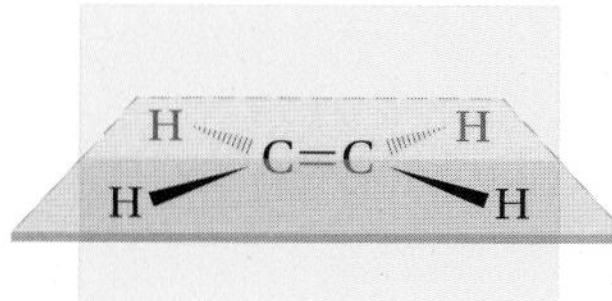

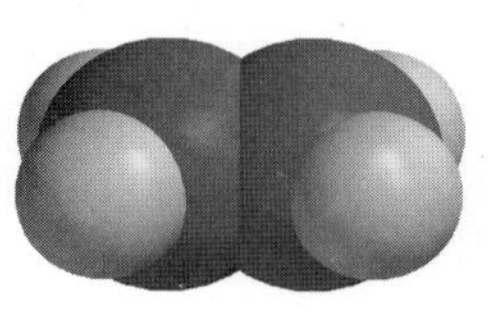
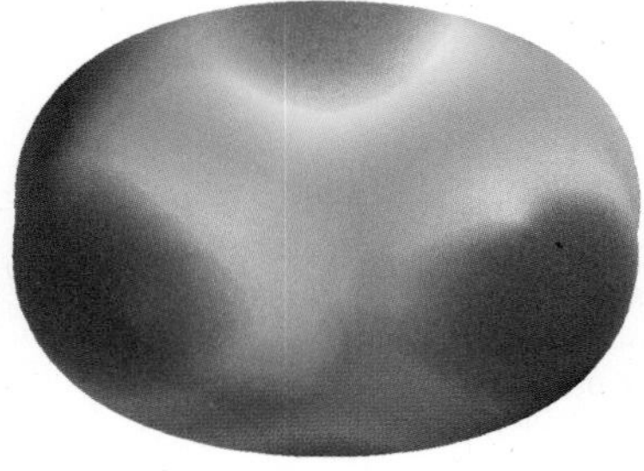

Figure 3.1
Four models of ethylene: The dash-wedge, ball-and-stick, and space-filling models show that the four atoms attached to a carbon–carbon double bond lie in a single plane. The electrostatic potential map shows the electron density (red) above and below the plane that passes through the carbon and hydrogen nuclei.

These differences between single and double bonds are summarized in Table 3.1. Let us see how the orbital model for bonding can explain the structure and properties of double bonds.

TABLE 3.1 ▪ Comparison of C—C and C═C Bonds

Property	C—C	C═C
1. Number of atoms attached to a carbon	4 (tetrahedral)	3 (trigonal)
2. Rotation	relatively free	restricted
3. Geometry	many conformations are possible; staggered is preferred	planar
4. Bond angle	109.5°	120°
5. Bond length	1.54 Å	1.34 Å

3.4 The Orbital Model of a Double Bond; the Pi Bond

sp^2**-Hybridized orbitals** are one part *s* and two parts *p* in character and are directed toward the three vertices of an equilateral triangle. The angle between two sp^2 orbitals is 120°.

Figure 3.2 shows what must happen with the atomic orbitals of carbon to accommodate trigonal bonding, bonding to only three other atoms. The first part of this figure is exactly the same as Figure 1.6 (p. 24). But now we combine only *three* of the orbitals, to make *three equivalent* sp^2**-hybridized orbitals** (called sp^2 because they are formed by combining one *s* and two *p* orbitals). These orbitals lie in a plane and are directed to the corners of an equilateral triangle. The angle between them is 120°. This angle is preferred because repulsion between electrons in each orbital is minimized. Three valence electrons are placed in the three sp^2 orbitals. The fourth valence electron is placed in the remaining 2*p* orbital, whose axis is perpendicular to the plane formed by the three sp^2 hybrid orbitals (see Figure 3.3).

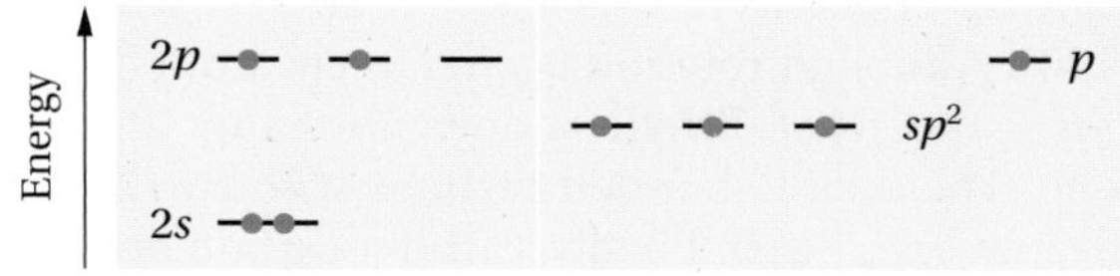

■ **Figure 3.2**
Unhybridized vs. sp^2-hybridized orbitals on carbon.

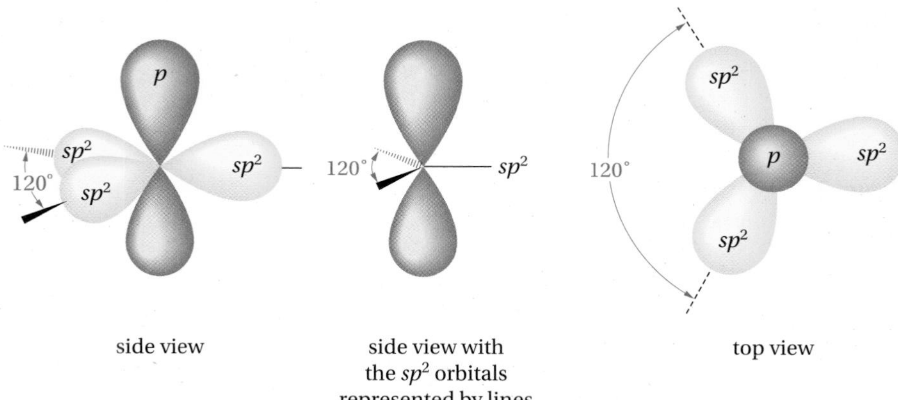

■ **Figure 3.3**
A trigonal carbon showing three sp^2 hybrid orbitals in a plane with a 120° angle between them. The remaining *p* orbital is perpendicular to the sp^2 orbitals. There is a small back lobe to each sp^2 orbital, which has been omitted for ease of representation.

A **sigma (σ) bond** is formed from the *end-on* overlap of two orbitals. A **pi (π) bond** is formed by lateral overlap of *p* orbitals on adjacent atoms.

Now let us see what happens when two sp^2-hybridized carbons are brought together to form a double bond. The process can be imagined as occurring stepwise (Figure 3.4). One of the two bonds, formed by *end-on* overlap of two sp^2 orbitals, is a **sigma (σ) bond** (p. 23). The second bond of the double bond is formed differently. If the two carbons are aligned with the *p* orbitals on each carbon parallel, lateral overlap can occur, as shown at the bottom of Figure 3.4. The bond formed by lateral *p*-orbital overlap is called a **pi (π) bond**. The bonding in ethylene is summarized in Figure 3.5.

The orbital model explains the facts about double bonds listed in Table 3.1. Rotation about a double bond is restricted because, for rotation to occur, we would have to "break" the pi bond, as seen in Figure 3.6. For ethylene, it takes about 62 kcal/mol (259 kJ/mol) to break the pi bond, much more thermal energy than is available at room temperature. With the pi bond intact, the sp^2 orbitals on each carbon lie in a single plane. The 120° angle between those orbitals minimizes repulsion between the electrons in them. Finally, the carbon–carbon double bond is shorter than the carbon–carbon single bond because the two shared electron pairs draw the nuclei closer together than a single pair does.

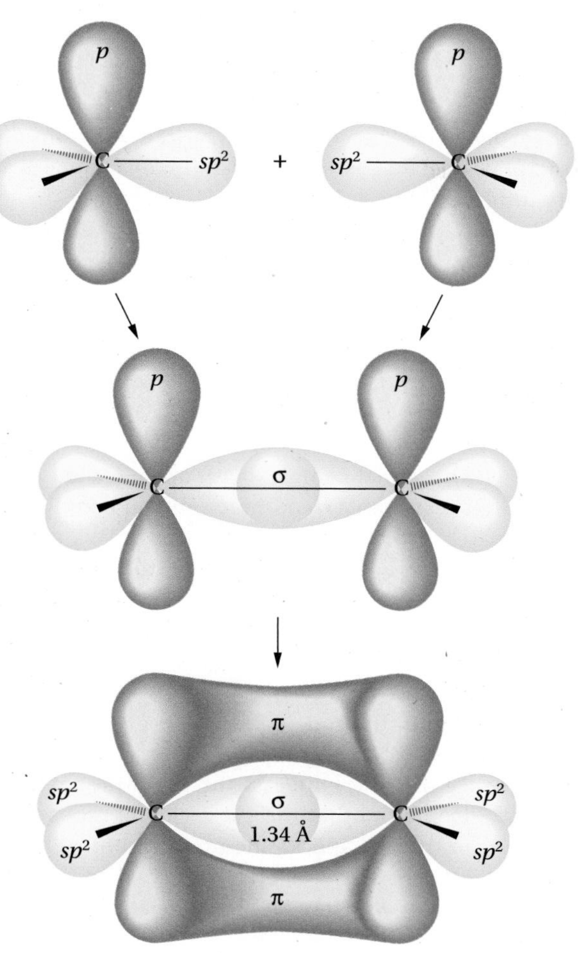

■ Figure 3.4

Schematic formation of a carbon–carbon double bond. Two sp^2 carbons form a sigma (σ) bond (end-on overlap of two sp^2 orbitals) and a pi (π) bond (lateral overlap of two properly aligned p orbitals).

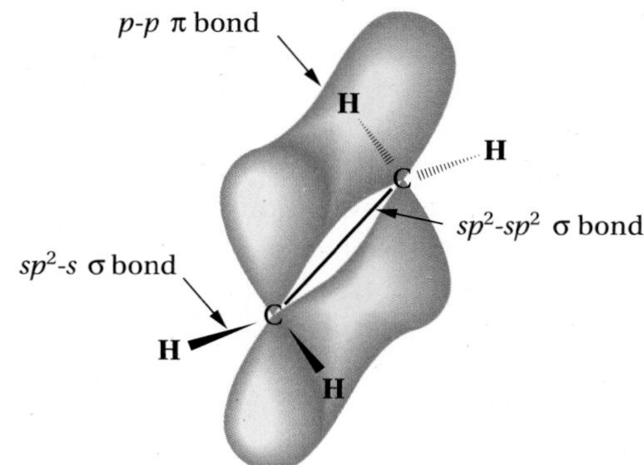

■ Figure 3.5

The bonding in ethylene consists of one sp^2–sp^2 carbon–carbon σ bond, four sp^2–s carbon–hydrogen σ bonds, and one p–p π bond.

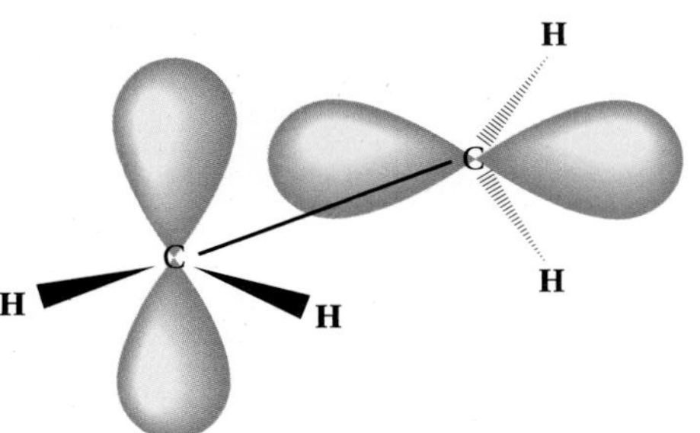

■ Figure 3.6

Rotation of one sp^2 carbon 90° with respect to another orients the p orbitals perpendicular to one another so that no overlap (and therefore no π bond) is possible.

A WORD ABOUT... The Chemistry of Vision

Andrew Ward/Life File/Photodisc/Getty Images

Carrots contain β-carotene.

Color in organic molecules is usually associated with extended conjugated systems of double bonds. A good example is ***β*-carotene**, a yellow-orange pigment found in carrots and many other plants. This $C_{40}H_{56}$ hydrocarbon has 11 carbon–carbon double bonds in conjugation. It is the biological precursor of the C_{20} unsaturated alcohol **vitamin A** (also called retinol), which in turn leads to the key substance involved in vision, **11-*cis*-retinal**. Notice in Figure 3.8 that the conversion of vitamin A to 11-*cis*-retinal involves not only oxidation of the alcohol group ($—CH_2OH$) to an aldehyde ($—CH{=}O$), but also *trans* → *cis* isomerism at the $C_{11}—C_{12}$ double bond.

Cis–trans isomerism plays a key role in the process of vision. The rod cells in the retina of the eye contain a red, light-sensitive pigment called **rhodopsin**. This pigment consists of the protein **opsin** combined at its active site with 11-*cis*-retinal. When visible light with the appropriate energy is absorbed by rhodopsin, the complexed *cis*-retinal is isomerized to the *trans isomer.* This process is fantastically fast, occurring in only picoseconds (10^{-12} seconds). As you can see from their structures, the shapes of the *cis* and *trans* isomers are very different.

11 12 CH=O

trans

trans-retinal

The *trans*-retinal complex with opsin (called metarhodopsin-II) is less stable than the *cis*-retinal complex, and it dissociates into opsin and *trans*-retinal. This change in geometry triggers a response in the rod nerve cells, which is transmitted to the brain and perceived as vision.

If this were all that happened, we would be able to see for only a few moments, because all of the 11-*cis*-retinal present in the rod cells would be quickly consumed. Fortunately, the enzyme *retinal isomerase,* in the presence of light, converts the *trans*-retinal back to the 11-*cis* isomer so that the cycle can be repeated. Calcium ions in the cell and its membrane control how fast the visual system recovers after exposure to light. They also mediate the way in which cells adapt to various

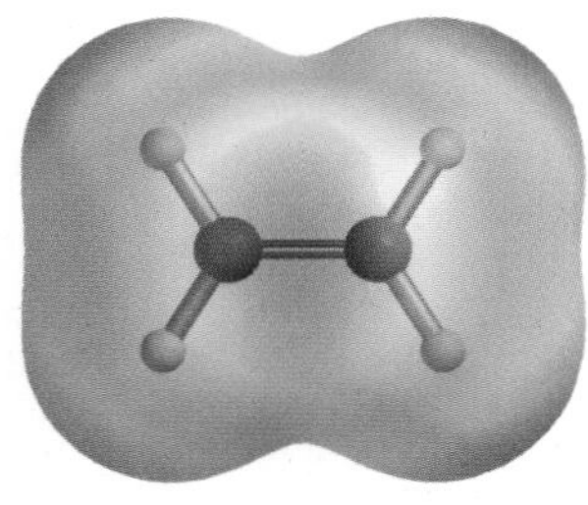

■ Figure 3.7
Electrostatic potential map of ethylene showing accessibility of pi electrons (red) to attack by electron-seeking reagents.

To recap, according to the orbital model, the carbon–carbon double bond consists of one sigma bond and one pi bond. The two electrons in the sigma bond lie along the internuclear axis; the two electrons in the pi bond lie in a region of space above and below the plane formed by the two carbons and the four atoms attached to them. The π electrons are more exposed than the σ electrons and, as we will see, can be attacked by various electron-seeking reagents (see Figure 3.7).

But before we consider reactions at the double bond, let us examine an important result of the restricted rotation around double bonds.

3.5 *Cis–Trans* Isomerism in Alkenes

Because rotation at carbon–carbon double bonds is restricted, *cis–trans* isomerism (geometric isomerism) is possible in appropriately substituted alkenes. For example, 1,2-dichloroethene exists in two different forms:

H, H / C=C / Cl, Cl	Cl, H / C=C / H, Cl
cis-1,2-dichloroethene	*trans*-1,2-dichloroethene
bp 60°C, mp −80°C	bp 47°C, mp −50°C

light levels. The following sequence summarizes the visual cycle:

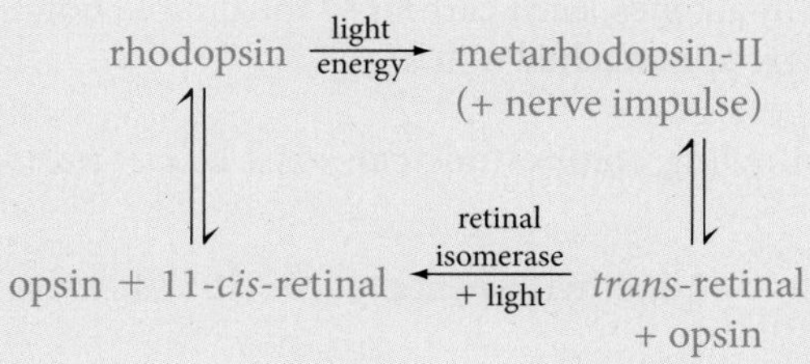

rhodopsin $\xrightarrow[\text{energy}]{\text{light}}$ metarhodopsin-II (+ nerve impulse)

opsin + 11-*cis*-retinal $\xleftarrow[\text{+ light}]{\text{retinal isomerase}}$ *trans*-retinal + opsin

This representation is simplified because there are actually several additional intermediates between rhodopsin and the fully dissociated *trans*-retinal and opsin.

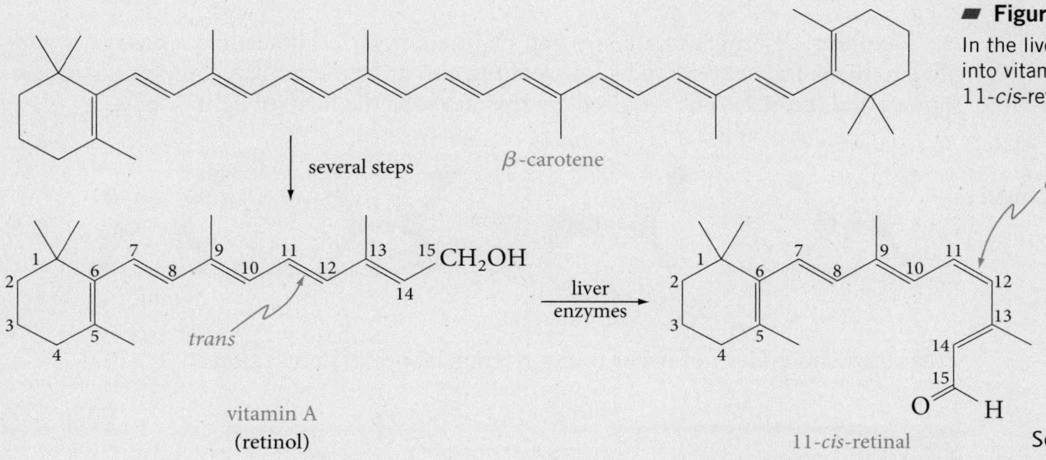

■ **Figure 3.8**
In the liver, β-carotene is converted into vitamin A first and then into 11-*cis*-retinal.

See Problems 3.40 and 3.63.

These stereoisomers are *not* readily interconverted by rotation around the double bond at room temperature. Like *cis–trans* isomers of cycloalkanes, they are configurational stereoisomers and can be separated from one another by distillation, taking advantage of the difference in their boiling points.

EXAMPLE 3.3

Are *cis–trans* isomers possible for 1-butene and 2-butene?

Solution 2-Butene has *cis–trans* isomers, but 1-butene does not.

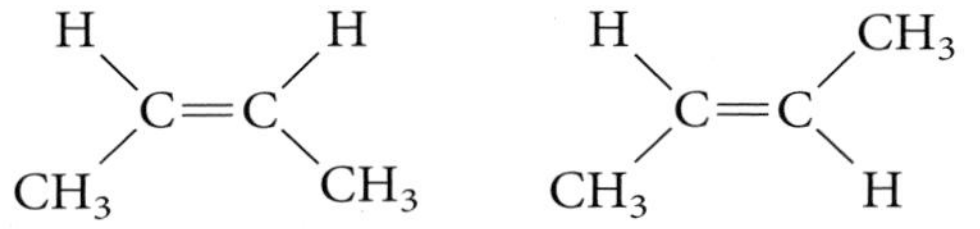

cis-2-butene
bp 3.7°C, mp −139°C

trans-2-butene
bp 0.3°C, mp −106°C

For 1-butene, carbon-1 has two identical hydrogen atoms attached to it; therefore, only one structure is possible.

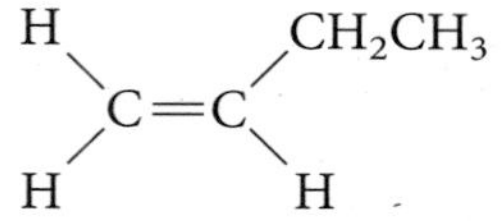

H, H on left carbon; CH_2CH_3, H on right carbon: C=C (1-butene) is identical to H, H on left carbon; H, CH_2CH_3 on right carbon: C=C (1-butene)

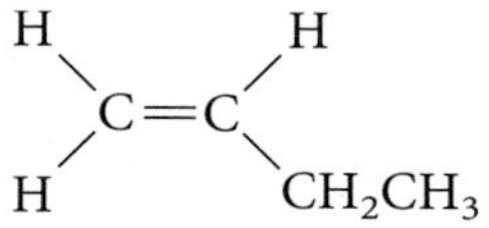

For *cis–trans* isomerism to occur in alkenes, *each* carbon of the double bond must have two different atoms or groups attached to it.

PROBLEM 3.6 Which of the following compounds can exist as *cis–trans* isomers? Draw their structures.

a. pentene b. 3-heptene c. 2-methyl-2-pentene d. 2-hexene

Geometric isomers of alkenes can be interconverted if sufficient energy is supplied to break the pi bond and allow rotation about the remaining, somewhat stronger, sigma bond (eq. 3.2). The required energy may take the form of light or heat.

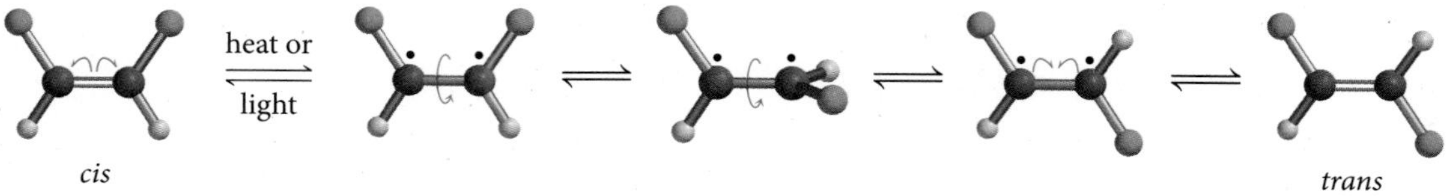

cis $\underset{\text{light}}{\overset{\text{heat or}}{\rightleftharpoons}}$ ⇌ ⇌ ⇌ *trans* **(3.2)**

This conversion does not occur under normal laboratory conditions.

3.6 Addition and Substitution Reactions Compared

We saw in Chapter 2 that, aside from combustion, the most common reaction of alkanes is substitution (for example, halogenation, Sec. 2.12.b). This reaction type can be expressed by a general equation.

$$R{-}H + A{-}B \longrightarrow R{-}A + H{-}B \quad \textbf{(3.3)}$$

where R—H stands for an alkane and A—B may stand for the halogen molecule.

The most common reaction of alkenes is **addition** of a reagent to the carbons of the double bond to give a product with a C—C single bond.

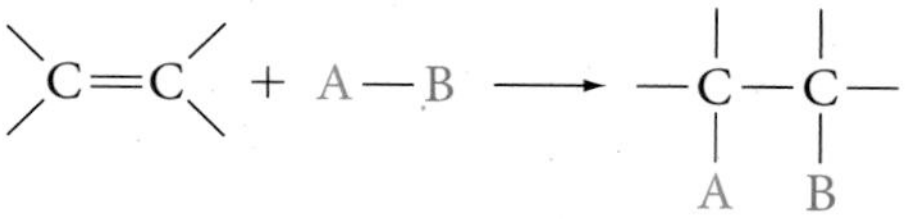

With alkenes, on the other hand, the most common reaction is **addition**:

$$\gt C{=}C\lt + A{-}B \longrightarrow -\underset{A}{\overset{|}{\underset{|}{C}}}-\underset{B}{\overset{|}{\underset{|}{C}}}- \quad \textbf{(3.4)}$$

In an addition reaction, group A of the reagent A—B becomes attached to one carbon atom of the double bond, group B becomes attached to the other carbon atom, and the product has only a single bond between the two carbon atoms.

What bond changes take place in an addition reaction? The pi bond of the alkene is broken, and the sigma bond of the reagent is also broken. Two new sigma bonds are formed. In other words, we break a pi and a sigma bond, and we make two sigma bonds. Because sigma bonds are usually stronger than pi bonds, the net reaction is favorable.

PROBLEM 3.7 Why, in general, is a sigma bond between two atoms stronger than a pi bond between the same two atoms?

3.7 Polar Addition Reactions

Several reagents add to double bonds by a two-step polar process. In this section, we will describe examples of this reaction type, after which we will consider details of the reaction mechanism.

3.7.a Addition of Halogens

Alkenes readily add chlorine or bromine.

$$\underset{\substack{\text{2-butene} \\ \text{bp 1–4°C}}}{CH_3CH{=}CHCH_3} + Cl_2 \longrightarrow \underset{\substack{\text{2,3-dichlorobutane} \\ \text{bp 117–119°C}}}{CH_3\underset{Cl}{\underset{|}{C}}H{-}\underset{Cl}{\underset{|}{C}}HCH_3} \qquad (3.5)$$

$$\underset{\substack{\text{1,4-pentadiene} \\ \text{bp 26.0°C}}}{CH_2{=}CH{-}CH_2{-}CH{=}CH_2} + 2\,Br_2 \longrightarrow \underset{\substack{\text{1,2,4,5-tetrabromopentane} \\ \text{mp 85–86°C}}}{\underset{Br}{\underset{|}{C}}H_2{-}\underset{Br}{\underset{|}{C}}H{-}CH_2{-}\underset{Br}{\underset{|}{C}}H{-}\underset{Br}{\underset{|}{C}}H_2} \qquad (3.6)$$

Usually the halogen is dissolved in some inert solvent such as tri- or tetrachloromethane, and then this solution is added dropwise to the alkene. Reaction is nearly instantaneous, even at room temperature or below. No light or heat is required, as in the case of substitution reactions.

PROBLEM 3.8 Write an equation for the reaction of bromine at room temperature with

a. propene b. 4-methylcyclohexene

The addition of bromine can be used as a chemical test for the presence of unsaturation in an organic compound. Bromine solutions in tetrachloromethane are dark reddish-brown, and both the unsaturated compound and its bromine adduct are usually colorless. As the bromine solution is added to the unsaturated compound, the bromine color disappears. If the compound being tested is saturated, it will not react with bromine under these conditions, and the color will persist.

3.7.b Addition of Water (Hydration)

If an acid catalyst is present, water adds to alkenes. It adds as H—OH, and the products are alcohols.

$$CH_2{=}CH_2 + H{-}OH \xrightarrow{H^+} \underset{\text{ethanol}}{\underset{H}{\underset{|}{C}}H_2{-}\underset{OH}{\underset{|}{C}}H_2} \quad (\text{or } CH_3CH_2OH) \qquad (3.7)$$

$$\underset{\substack{\text{cyclohexene} \\ \text{bp 83.0°C}}}{C_6H_{10}} + H{-}OH \xrightarrow{H^+} \underset{\substack{\text{cyclohexanol} \\ \text{bp 161.1°C}}}{C_6H_{11}OH} \qquad (3.8)$$

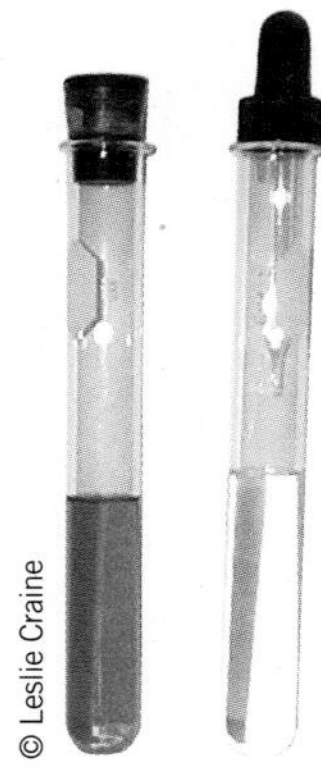

Bromine solution (red-brown) is added to a saturated hydrocarbon (left) and an unsaturated hydrocarbon (right).

An acid catalyst is required in this case because the neutral water molecule is not acidic enough to provide protons to start the reaction. The stepwise mechanism for this reaction is given later in eq. 3.20. Hydration is used industrially and occasionally in the laboratory to synthesize alcohols from alkenes.

PROBLEM 3.9 Write an equation for the acid-catalyzed addition of water to

a. 3-hexene b. 2-butene c. 4-methylcyclopentene

3.7.c Addition of Acids

A variety of acids add to the double bond of alkenes. The hydrogen ion (or proton) adds to one carbon of the double bond, and the remainder of the acid becomes connected to the other carbon.

$$\mathrm{>C{=}C<} + \overset{\delta+}{\mathrm{H}}-\overset{\delta-}{\mathrm{A}} \longrightarrow -\underset{\mathrm{H}}{\overset{|}{\underset{|}{\mathrm{C}}}}-\underset{\mathrm{A}}{\overset{|}{\underset{|}{\mathrm{C}}}}- \quad (3.9)$$

Acids that add in this way are the hydrogen halides (H—F, H—Cl, H—Br, H—I) and sulfuric acid (H—OSO_3H). Here are two typical examples:

$$\underset{\text{ethene}}{CH_2{=}CH_2} + \underset{\text{hydrogen chloride}}{H{-}Cl} \longrightarrow \underset{\text{chloroethane (ethyl chloride)}}{\underset{H\quad\;\; Cl}{CH_2{-}CH_2}} \quad (\text{or } CH_3CH_2Cl) \quad (3.10)$$

$$\text{cyclopentene} + \underset{\text{sulfuric acid}}{H{-}OSO_3H} \longrightarrow \text{cyclopentyl hydrogen sulfate} \quad (3.11)$$

PROBLEM 3.10 Write an equation for each of the following reactions:

a. 2-butene + HCl
b. 3-hexene + HI
c. 4-methylcyclopentene + HBr

Before we discuss the mechanism of these addition reactions, we must introduce a complication that we have carefully avoided in all of the examples given so far.

3.8 Addition of Unsymmetric Reagents to Unsymmetric Alkenes; Markovnikov's Rule

The products of addition of **unsymmetric reagents** to **unsymmetric alkenes** are called **regioisomers. Regiospecific** additions produce only one regioisomer. **Regioselective** additions produce mainly one regioisomer.

Reagents and alkenes can be classified as either **symmetric** or **unsymmetric** with respect to addition reactions. Table 3.2 illustrates what this means. If a reagent and/or an alkene is symmetric, only one addition product is possible. If you check

Table 3.2 ▪ Classification of Reagents and Alkenes by Symmetry with Regard to Addition Reactions

	Symmetric	Unsymmetric
Reagents	Br—Br	H—Br
	Cl—Cl	H—OH
	H—H	H—OSO_3H
Alkenes	$CH_2{=}CH_2$	$CH_3CH{=}CH_2$
	cyclopentene	1-methylcyclopentene (CH_3)
	mirror plane	not a mirror plane

back through all of the equations and problems for addition reactions up to now, you will see that either the alkene or the reagent (or both) was symmetric. But if *both* the reagent *and* the alkene are *unsymmetric,* two products are, in principle, possible.

$$\underset{\text{unsymmetric alkene}}{RCH{=}C} + \underset{\text{unsymmetric reagent}}{X{-}Y} \longrightarrow -\underset{X}{\overset{R}{C}}-\underset{Y}{\overset{H}{C}}- \text{ and/or } -\underset{Y}{\overset{R}{C}}-\underset{X}{\overset{H}{C}}- \quad (3.12)$$

The products of eq. 3.12 are sometimes called **regioisomers**. If a reaction of this type gives *only one* of the two possible regioisomers, it is said to be **regiospecific**. If it gives *mainly one* product, it is said to be **regioselective**.

Let us consider, as a specific example, the acid-catalyzed addition of water to propene. In principle, two products could be formed: 1-propanol or 2-propanol.

$$\underset{\text{propene}}{\overset{3}{C}H_3\overset{2}{C}H{=}\overset{1}{C}H_2} \xrightarrow[H^+]{H-OH} \underset{\text{2-propanol}}{CH_3\underset{OH}{C}HCH_3} \qquad \underset{\text{propene}}{CH_3CH{=}CH_2} \not\xrightarrow[H^+]{H-OH} \underset{\text{1-propanol}}{CH_3CH_2CH_2{-}OH} \quad (3.13)$$

That is, the hydrogen of the water could add to C-1 and the hydroxyl group to C-2 of propene, or vice versa. When the experiment is carried out, *only one product is observed. The addition is regiospecific, and the only product is 2-propanol.*

Most addition reactions of alkenes show a similar preference for the formation of only (or mainly) one of the two possible addition products. Here are some examples.

$$\mathrm{CH_3CH{=}CH_2 + \overset{\delta+}{H}{-}\overset{\delta-}{Cl} \longrightarrow CH_3\underset{\displaystyle Cl}{\underset{|}{C}}HCH_3 \qquad (CH_3CH_2CH_2Cl)\ \text{not observed}} \tag{3.14}$$

$$\mathrm{CH_3\underset{\displaystyle CH_3}{\underset{|}{C}}{=}CH_2 + \overset{\delta+}{H}{-}\overset{\delta-}{OH} \xrightarrow{H^+} CH_3\overset{\displaystyle OH}{\overset{|}{\underset{\displaystyle CH_3}{\underset{|}{C}}}}CH_3 \qquad (CH_3\underset{\displaystyle CH_3}{\underset{|}{C}}HCH_2OH)\ \text{not observed}} \tag{3.15}$$

$$\text{1-methylcyclopentene} + \overset{\delta+}{\mathrm{H}}{-}\overset{\delta-}{\mathrm{I}} \longrightarrow \text{1-iodo-1-methylcyclopentane} \qquad (\text{1-iodo-2-methylcyclopentane})\ \text{not observed} \tag{3.16}$$

Notice that the reagents are all polar, with a positive and a negative end. After studying a number of such addition reactions, the Russian chemist Vladimir Markovnikov formulated the following rule more than 100 years ago: *When an unsymmetric reagent adds to an unsymmetric alkene, the electropositive part of the reagent bonds to the carbon of the double bond that has the greater number of hydrogen atoms attached to it.**

PROBLEM 3.11 Use Markovnikov's Rule to predict which regioisomer predominates in each of the following reactions:

a. 1-pentene + HBr
b. 2-methyl-2-hexene + H_2O (H^+ catalyst)

PROBLEM 3.12 What two products are *possible* from the addition of HCl to 2-octene? Would you expect the reaction to be regiospecific?

Let us now develop a rational explanation for Markovnikov's Rule in terms of modern chemical theory.

3.9 Mechanism of Electrophilic Addition to Alkenes

The pi electrons of a double bond are more exposed to an attacking reagent than are the sigma electrons. The pi bond is also weaker than the σ bond. It is the pi electrons, then, that are involved in additions to alkenes. The double bond can act as a supplier of pi electrons to an electron-seeking reagent (see Figure 3.7, p. 76).

Electrophiles are electron-poor reactants; they seek electrons. **Nucleophiles** are electron-rich reactants; they form bonds by donating electrons to electrophiles.

Polar reactants can be classified as either **electrophiles** or **nucleophiles**. Electrophiles (literally, electron lovers) are electron-poor reagents; in reactions with some other molecule, they seek electrons. They are often positive ions (cations) or otherwise electron-deficient species. Nucleophiles (literally, nucleus lovers), on the other hand, are electron rich; they form bonds by donating electrons to an electrophile.

*Actually, Markovnikov stated the rule a little differently. The form given here is easier to remember and apply. For an interesting historical article on what he actually said, when he said it, and how his name is spelled, see J. Tierney, *J. Chem. Educ.* **1988**, *65*, 1053–54.

$$E^+ + :Nu^- \longrightarrow E:Nu \qquad (3.17)$$

electrophile nucleophile

Let us now consider the mechanism of polar addition to a carbon–carbon double bond, specifically the addition of acids to alkenes. The carbon–carbon double bond, because of its pi electrons, is a nucleophile. The proton (H^+) is the attacking electrophile. As the proton approaches the pi bond, the two pi electrons are used to form a sigma bond between the proton and one of the two carbon atoms. Because this bond uses *both* pi electrons, the other carbon acquires a positive charge, producing a **carbocation**.

A **carbocation** is a positively charged carbon atom bonded to three other atoms.

$$H^+ + C{=}C \longrightarrow H{-}C{-}C^+ \qquad (3.18)$$

carbocation

The resulting carbocations are, however, extremely reactive because there are only six electrons (instead of the usual eight) around the positively charged carbon. The carbocation rapidly combines with some species that can supply it with two electrons, a nucleophile.

$$H{-}C{-}C^+ + Nu:^- \longrightarrow H{-}C{-}C{-}Nu \qquad (3.19)$$

nucleophile

product of addition of H—Nu to an alkene

Examples include the addition of H—Cl, H—OSO_3H, and H—OH to alkenes:

$$H^+ + C{=}C \rightleftharpoons H{-}C{-}C^+ \text{ (carbocation)} \qquad (3.20)$$

with $:\ddot{C}l:^-$ → H—C—C—$:\ddot{C}l:$

with $^-:\ddot{O}SO_3H$ → H—C—C—$:\ddot{O}SO_3H$

with H—$\ddot{O}$—H → H—C—C—O^+(H)H ⇌ H—C—C—$:\ddot{O}$—H + H^+

In these reactions, the electrophile H^+ first adds to the alkene to give a carbocation. Then the carbocation combines with a nucleophile, in these examples, a chloride ion, a bisulfate ion, or a water molecule.

With most alkenes, the first step in this process—the formation of the carbocation—is the slower of the two steps. The resulting carbocation is usually so reactive that combination with the nucleophile is extremely rapid. *Since the first step in these additions is attack by the electrophile, the whole process is called an* **electrophilic addition reaction**.

A reaction in which an electrophile is added to an alkene is called an **electrophilic addition reaction**.

Electrophilic addition of the halogens (Cl_2 and Br_2) to alkenes occurs in a similar manner. Although the mechanism is not identical to that for acids, the end results are the same. For example, when a molecule of Br_2 approaches the pi bond of an alkene, the Br—Br bond becomes polarized: $\overset{\delta+}{Br}—\overset{\delta-}{Br}$. The Br atom closer to the pi bond develops a partial positive charge and thus becomes an electrophile, while the other Br atom develops a partial negative charge and becomes the nucleophile. Although it is impossible to tell from the products, the addition occurs in Markovnikov fashion.*

EXAMPLE 3.4

Because carbocations are involved in the electrophilic addition reactions of alkenes, it is important to understand the bonding in these chemical intermediates. Describe the bonding in carbocations in orbital terms.

Solution The carbon atom is positively charged and therefore has only three valence electrons to use in bonding. Each of these electrons is in an sp^2 orbital. The three sp^2 orbitals lie in one plane with 120° angles between them, an arrangement that minimizes repulsion between the electrons in the three bonds. The remaining *p* orbital is perpendicular to that plane and is vacant.

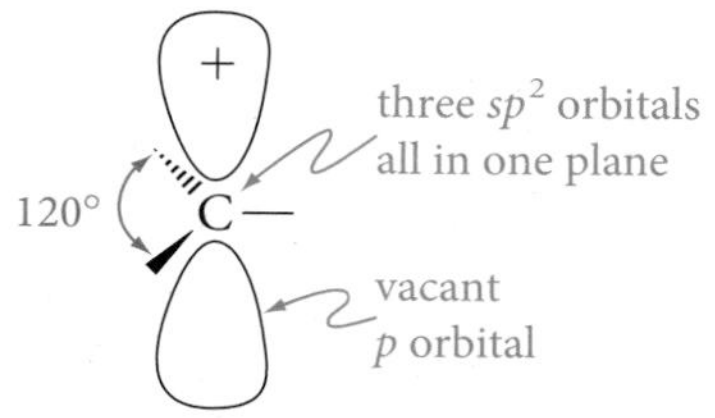

carbocation

3.10 Markovnikov's Rule Explained

To explain Markovnikov's Rule, let us consider a specific example, the addition of H—Cl to propene. The first step is addition of a proton to the double bond. This can occur in two ways, to give either an isopropyl cation or a propyl cation.

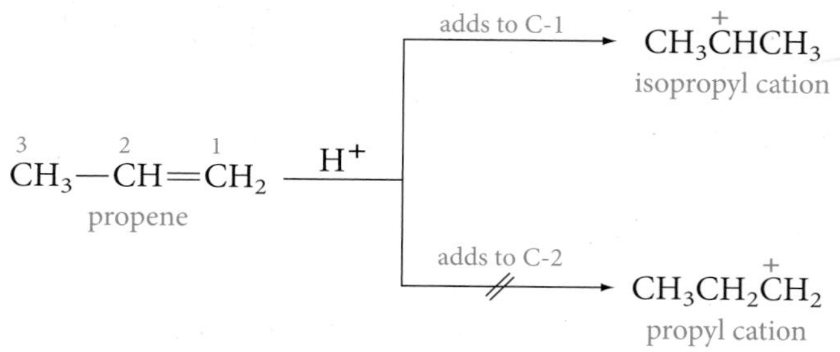

$$\overset{3}{CH_3}—\overset{2}{CH}=\overset{1}{CH_2} \xrightarrow{H^+} \begin{cases} \xrightarrow{\text{adds to C-1}} CH_3\overset{+}{C}HCH_3 \ \text{(isopropyl cation)} \\ \xrightarrow{\text{adds to C-2}}\!\!\!\!\!\!/\!/ \ CH_3CH_2\overset{+}{C}H_2 \ \text{(propyl cation)} \end{cases} \quad \text{propene} \tag{3.21}$$

At this stage of the reaction, the structure of the product is already determined; when combining with chloride ion, the isopropyl cation can give only 2-chloropropane, and the propyl cation can give only 1-chloropropane. The only observed product is 2-chloropropane, so we must conclude that the *proton adds to C-1 to form only the isopropyl cation.* Why?

Carbocations are classified as **primary, secondary,** and **tertiary** when one, two, and three R groups, respectively, are attached to the positively charged carbon atom.

Carbocations can be classified as **tertiary**, **secondary**, or **primary**, depending on whether the positive carbon atom has attached to it three organic groups, two groups,

*Consult your instructor if you are curious about the detailed mechanism.

or only one group. From many studies, it has been established that the stability of carbocations decreases in the following order:

$$\underset{\substack{\text{tertiary (3°)}\\ \text{most stable}}}{R_3C^+} > \underset{\text{secondary (2°)}}{R_2\overset{+}{C}H} >> \underset{\text{primary (1°)}}{R\!-\!\overset{+}{C}H_2} > \underset{\substack{\text{methyl (unique)}\\ \text{least stable}}}{\overset{+}{C}H_3}$$

One reason for this order is the following: A carbocation will be more stable when the positive charge can be spread out, or delocalized, over several atoms in the ion, instead of being concentrated on a single carbon atom. In alkyl cations, this delocalization occurs by drift of electron density to the positive carbon from C—H and C—C sigma bonds that can align themselves with the empty *p* orbital on the positively charged carbon atom (Figure 3.9). If the positively charged carbon center is surrounded by other carbon atoms (alkyl groups), instead of by hydrogen atoms, more C—H and C—C bonds will be available to provide electrons to help delocalize the charge. This is the main reason for the observed stability order of carbocations.

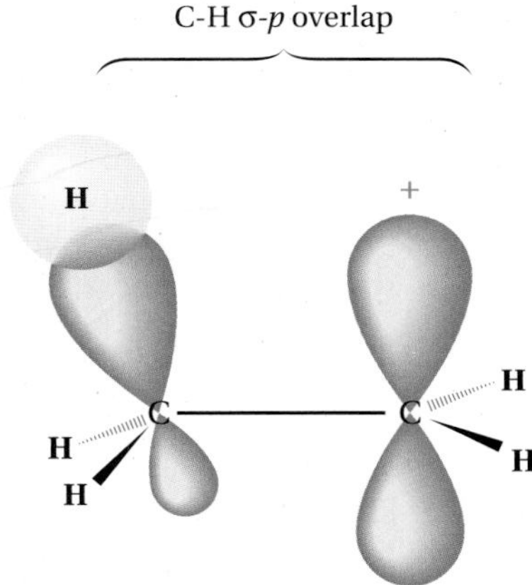

Figure 3.9 Alkyl groups stabilize carbocations by donating electron density from C—H and C—C sigma bonds that can line up with the empty *p* orbital on the positively charged carbon atom.

Markovnikov's Rule can now be restated in modern and more generally useful terms: *The electrophilic addition of an unsymmetric reagent to an unsymmetric double bond proceeds in such a way as to involve the most stable carbocation.*

PROBLEM 3.13 Classify each of the following carbocations as primary, secondary, or tertiary:

a. $CH_3CH_2\overset{+}{C}HCH_3$ b. $(CH_3)_2CH\overset{+}{C}H_2$ c. 1-methylcyclopentyl cation (CH_3 on the ring carbon bearing +)

PROBLEM 3.14 Which carbocation in Problem 3.13 is most stable? Least stable?

PROBLEM 3.15 Write the steps in the electrophilic additions in eqs. 3.15 and 3.16, and in each case, show that reaction occurs via the more stable carbocation.

Discussion of Markovnikov's Rule raises two important general questions about chemical reactions: (1) Under what conditions is a reaction likely to proceed? (2) How rapidly will a reaction occur? We will consider these questions briefly in the next two sections before continuing our survey of the reactions of alkenes.

3.11 Reaction Equilibrium: What Makes a Reaction Go?

A chemical reaction can proceed in two directions. Reactant molecules can form product molecules, and product molecules can react to re-form the reactant molecules. For the reaction*

$$\text{aA} + \text{bB} \rightleftharpoons \text{cC} + \text{dD} \tag{3.22}$$

we describe the chemical equilibrium for the forward and backward reactions by the following equation:

$$K_{eq} = \frac{[\text{C}]^{\text{c}}[\text{D}]^{\text{d}}}{[\text{A}]^{\text{a}}[\text{B}]^{\text{b}}} \tag{3.23}$$

The **equilibrium constant, K_{eq}**, indicates the direction that is favored for a reaction.

In this equation, K_{eq}, the **equilibrium constant**, is equal to the product of the concentrations of the products divided by the product of the concentrations of the reactants. (The small letters a, b, c, and d are the numbers of molecules of reactants and products in the balanced reaction equation.)

The equilibrium constant tells us the direction that is favored for the reaction. If K_{eq} is greater than 1, the formation of products C and D will be favored over the formation of reactants A and B. The preferred direction for the reaction is from left to right. Conversely, if K_{eq} is less than 1, the preferred direction for the reaction is from right to left.

What determines whether a reaction will proceed to the right, toward products? A reaction will occur when the products are lower in energy (more stable) than the reactants. A reaction in which products are higher in energy than reactants will proceed to the left, toward reactants. When products are lower in energy than reactants, heat is given off in the course of the reaction. For example, heat is given off when an acid such as hydrogen bromide (HBr) is added to ethene (eq. 3.24). Such a reaction is **exothermic**.

$$H_2C{=}CH_2 + HBr \rightleftharpoons CH_3CH_2Br \tag{3.24}$$

An **exothermic** reaction evolves heat energy; an **endothermic** reaction takes in heat energy. The chemists' term for heat energy is **enthalpy, *H***.

On the other hand, heat must be added to ethane to produce two methyl radicals (eq. 1.3). This reaction is **endothermic** (takes in heat). The term used by chemists for heat energy is **enthalpy** and is designated by the symbol ***H***. The difference in enthalpy between products and reactants is designated by the symbol ΔH (pronounced "delta H").

For the addition of HBr to ethene, the product (bromoethane) is more stable than the reactants (ethene and HBr), and the reaction proceeds to the right. For this reaction, ΔH is negative (heat is given off), and K_{eq} is much greater than 1 (Figure 3.10a). For the formation of two methyl radicals from ethane, ΔH is positive (heat is absorbed), and K_{eq} is much less than 1 (Figure 3.10b).**

---

*The double arrow indicates that this reaction goes both ways and reaches chemical equilibrium.

Actually, enthalpy is not the only factor that contributes to the energy difference between products and reactants. A factor called **entropy, *S*, also contributes to the total energy difference, which is known as the **Gibbs free-energy difference, ΔG**, according to the equation $\Delta G = \Delta H — T\Delta S$. For most organic reactions, however, the entropy contribution is very small compared to the enthalpy contribution.

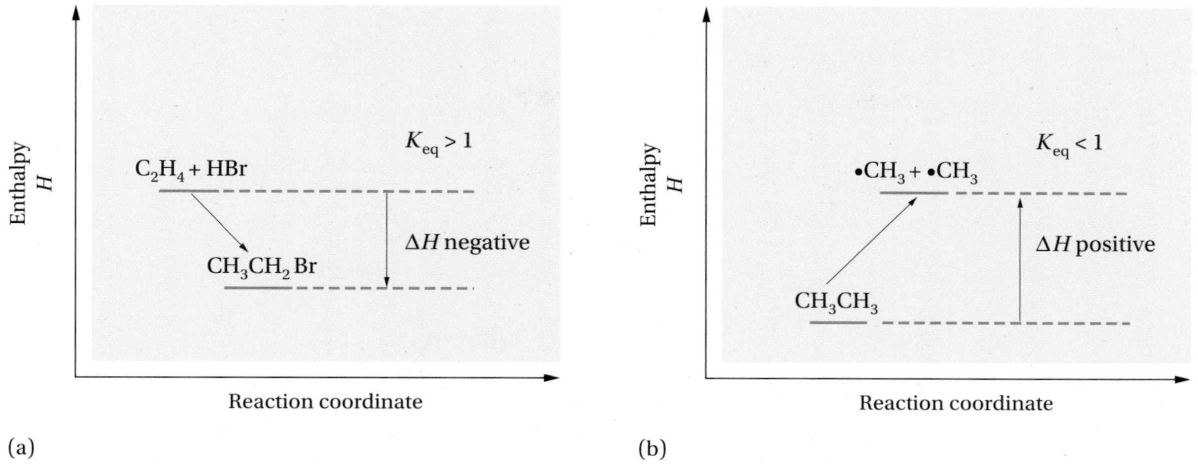

Figure 3.10
(a) The addition of HBr to ethene; the reaction equilibrium lies to the right. (b) The formation of methyl radicals from ethane; the reaction equilibrium lies to the left.

3.12 Reaction Rates: How Fast Does a Reaction Go?

The equilibrium constant for a reaction tells us whether or not products are more stable than reactants. However, *the equilibrium constant does not tell us anything about the rate of a reaction.* For example, the equilibrium constant for the reaction of gasoline with oxygen is very large, but gasoline can be safely handled in air because the reaction is very slow unless a spark is used to initiate it. The rate of addition of HBr to ethene is also very slow, although the reaction is exothermic.

In order to react, molecules must collide with each other with enough energy and with the right orientation so that the breaking and making of bonds can occur. The energy required for this process is a barrier to reaction, and the higher the energy barrier, the slower the reaction.

Chemists use **reaction energy diagrams** to show the changes in energy that occur in the course of a reaction. Figure 3.11 shows the reaction energy diagram for the polar addition of the acid HBr to ethene (eq. 3.24). This reaction occurs in two steps. In the first step, as a proton adds to the double bond, the π bond of the alkene is broken and a C—H σ bond is formed, giving a carbocation intermediate product. The reactants start with the energy shown at the left of the diagram. As the π bond begins to break and the new σ bond begins to form, the structure formed by the reactants reaches a maximum in energy. This structure with maximum energy is called the **transition state** for the first step. This structure cannot be isolated and continues to change until the carbocation product of the first step is fully formed.

A **reaction energy diagram** shows the changes in energy that occur in the course of a reaction. A **transition state** is a structure with maximum energy for a particular reaction step. **Activation energy,** E_a, is the difference in energy between reactants and the transition state, and the activation energy determines the **reaction rate.**

The difference in energy between the transition state and the reactants is called the **activation energy,** E_a. It is this energy that determines the rate of the reaction. If E_a is large, the reaction will be slow. A small E_a means that the reaction will proceed rapidly.

In the second step of the reaction, a new carbon–bromine σ bond is formed. Again, the approach of the bromide ion to the positively charged carbon of the carbocation intermediate causes a rise in energy to a maximum. The structure at this energy maximum is the transition state for the second step. The difference in energy between the carbocation and this transition state is the activation energy E_a for this second step. This structure cannot be isolated and continues to change until the σ bond is fully formed, completing the formation of the product.

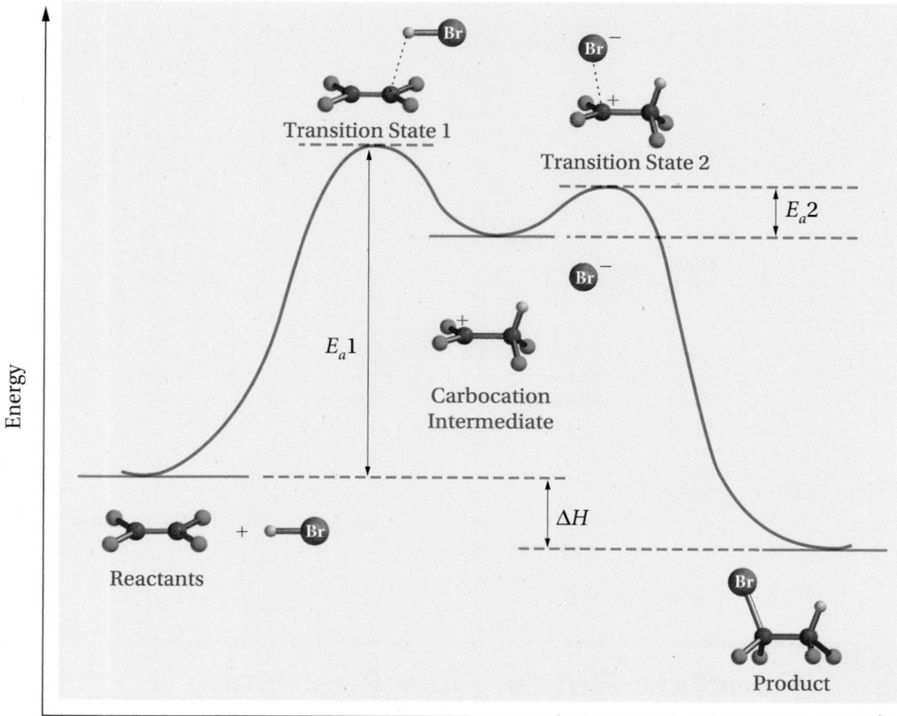

■ **Figure 3.11**
Reaction energy diagram for the addition of HBr to an alkene.

Notice in Figure 3.11 that although the final product of the reaction is lower in energy (ΔH) than the reactants, the reactants must surmount two energy barriers (E_a1 and E_a2), one for each step of the reaction. Between the two transition states, the carbocation intermediate is at an energy minimum that is higher than reactants or products. The first step of the reaction is *endothermic,* because the carbocation intermediate product is higher in energy than the reactants. The second step is exothermic because the product is lower in energy than the carbocation. The overall reaction is *exothermic,* because the product is lower in energy than the reactants. However, the rate of the reaction is determined by the highest energy barrier, E_a1. The second activation energy, E_a2, is very low compared to the activation energy for the first step.

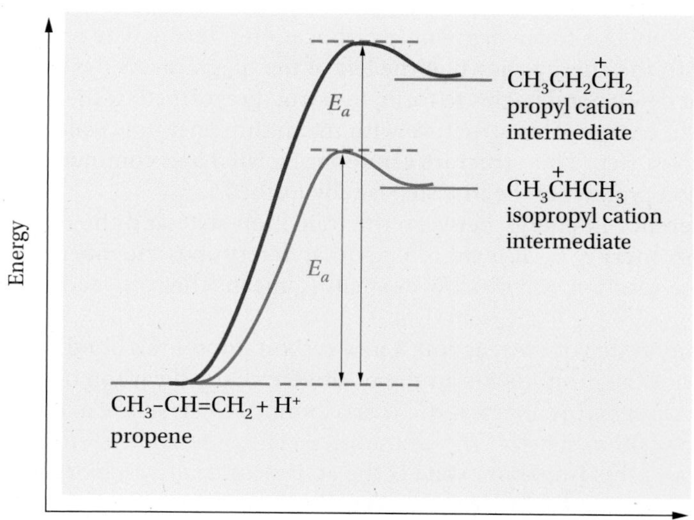

■ **Figure 3.12**
Reaction energy diagram for formation of the isopropyl and propyl cations from propene (eq. 3.21).

Therefore, as described in Section 3.9, the first step is the slower of the two steps, and the rate of the reaction is determined by the rate of this first step.

EXAMPLE 3.5

Sketch a reaction energy diagram for a one-step reaction that is very slow and slightly exothermic.

Solution A very slow reaction has a large E_a, and a slightly exothermic reaction has a small negative ΔH. Therefore, the diagram will look like this:

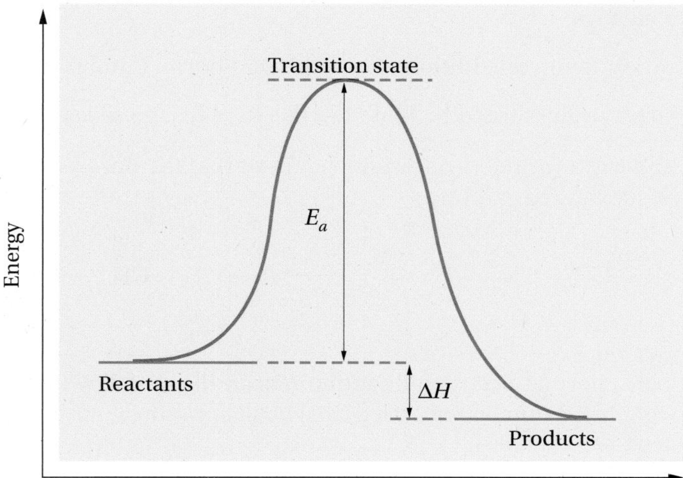

PROBLEM 3.16 Draw a reaction energy diagram for a one-step reaction that is very fast and very exothermic.

PROBLEM 3.17 Draw a reaction energy diagram for a one-step reaction that is very slow and slightly endothermic.

PROBLEM 3.18 Draw a reaction energy diagram for a two-step reaction that has an endothermic first step and an exothermic second step. Label the reactants, transition states, reaction intermediate, activation energies, and enthalpy differences.

Let us see how reaction rates are related to Markovnikov's Rule. In electrophilic addition reactions, more stable carbocations are formed more rapidly than less stable carbocations. This is because more stable carbocations are lower in energy than less stable carbocations, and it follows that the activation energy E_a for the formation of more stable carbocations is also lower. For example, both isopropyl and propyl cations could be formed from propene and H^+ (eq. 3.21), but the isopropyl cation is more stable (i.e., much lower in energy) than the propyl cation (Figure 3.12). Formation of the isopropyl cation, therefore, has a lower activation energy E_a and thus, the isopropyl carbocation is formed more rapidly than the propyl cation. Hence, the regioselectivity of electrophilic additions is the result of competing first steps, in which the more stable carbocation is formed at a faster rate.

Other factors that affect reaction rates are **temperature** and **catalysts**. Heating a reaction generally increases the rate at which the reaction occurs by providing the reactant

Increasing **temperature** or using a **catalyst** increases reaction rates.

molecules with more energy to surmount activation energy barriers. Catalysts speed up a reaction by providing an alternative pathway or mechanism for the reaction, one in which the activation energy is lower. Enzymes play this role in biochemical reactions.

In the next five sections, we will continue our survey of the reactions of alkenes.

3.13 Hydroboration of Alkenes

Hydroboration is the addition of $H{-}B\lessdot$ to an alkene.

Hydroboration was discovered by Professor Herbert C. Brown (1912–2004). This reaction is so useful in synthesis that Brown's work earned him a Nobel Prize in 1979. We will describe here only one practical example of hydroboration, a two-step alcohol synthesis from alkenes.

Hydroboration involves addition of a hydrogen–boron bond to an alkene. From the electronegativity values listed in Table 1.4, the $H{-}B\lessdot$ bond is polarized with the hydrogen $\delta-$ and the boron $\delta+$. Addition occurs so that the boron (the electrophile) adds to the less-substituted carbon.

$$R{-}CH{=}CH_2 + \overset{\delta-}{H}{-}\overset{\delta+}{B}\lessdot \longrightarrow R{-}\underset{\displaystyle H}{\underset{|}{C}H}{-}CH_2{-}B\lessdot \qquad (3.25)$$

Thus, it resembles a normal electrophilic addition to an alkene, following Markovnikov's Rule, even though the addition is concerted (that is, all bond-breaking and bond-making occur in one step).

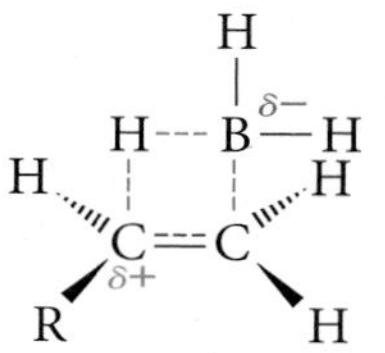

transition state for hydroboration

Because it has three B—H bonds, one molecule of borane, BH_3, can react with three molecules of an alkene. For example, propene gives tri-*n*-propylborane.

$$3\ CH_3CH{=}CH_2 + BH_3 \longrightarrow CH_3CH_2CH_2{-}B(CH_2CH_2CH_3)_2 \qquad (3.26)$$

propene — borane — tri-*n*-propylborane

The trialkylboranes made in this way are usually not isolated but are treated with some other reagent to obtain the desired final product. For example, trialkylboranes can be oxidized by hydrogen peroxide and base to give alcohols.

$$\underset{\text{tri-}n\text{-propylborane}}{(CH_3CH_2CH_2)_3B} + 3\ H_2O_2 + 3\ NaOH \longrightarrow$$

$$\underset{n\text{-propyl alcohol}}{3\ CH_3CH_2CH_2OH} + \underset{\text{sodium borate}}{Na_3BO_3} + 3\ H_2O \qquad (3.27)$$

One great advantage of this hydroboration–oxidation sequence is that it provides a route to alcohols that *cannot* be obtained by the acid-catalyzed hydration of alkenes (review eq. 3.13).

$$\text{R—CH=CH}_2 \begin{cases} \xrightarrow[\text{H}^+]{\text{H—OH}} \underset{\text{Markovnikov product}}{\text{R—}\underset{\text{OH}}{\underset{|}{\text{CH}}}\text{—CH}_3} \\ \xrightarrow[\text{2. H}_2\text{O}_2\text{, OH}^-]{\text{1. BH}_3} \underset{\text{anti-Markovnikov product}}{\text{R—CH}_2\text{—CH}_2\text{OH}} \end{cases} \qquad (3.28)$$

The overall result of the two-step hydroboration sequence *appears* to be the addition of water to the carbon–carbon double bond in the reverse of the usual Markovnikov sense.

EXAMPLE 3.6

What alcohol is obtained from this sequence?

$$\text{CH}_3\text{—}\overset{\text{CH}_3}{\overset{|}{\text{C}}}\text{=CH}_2 \xrightarrow{\text{BH}_3} \xrightarrow[\text{OH}^-]{\text{H}_2\text{O}_2}$$

Solution The boron adds to the less-substituted carbon; oxidation gives the corresponding alcohol. Compare this result with that of eq. 3.15.

$$3\ \text{CH}_3\text{—}\overset{\text{CH}_3}{\overset{|}{\text{C}}}\text{=CH}_2 \xrightarrow{\text{BH}_3} (\text{CH}_3\text{—}\overset{\text{CH}_3}{\overset{|}{\text{CH}}}\text{—CH}_2)_3\text{B} \xrightarrow[\text{OH}^-]{\text{H}_2\text{O}_2} 3\ \text{CH}_3\text{—}\overset{\text{CH}_3}{\overset{|}{\text{CH}}}\text{—CH}_2\text{OH}$$

PROBLEM 3.19 What alcohol is obtained by applying the hydroboration–oxidation sequence to 2-methyl-2-hexene?

PROBLEM 3.20 What alkene is needed to obtain (cyclohexyl)—CH_2CH_2OH via the hydroboration–oxidation sequence? What product would this alkene give with acid-catalyzed hydration?

3.14 Addition of Hydrogen

Hydrogen adds to alkenes in the presence of an appropriate catalyst. The process is called **hydrogenation**.

Hydrogenation is the addition of hydrogen to alkenes in the presence of a catalyst.

$$\text{>C=C<} + \text{H}_2 \xrightarrow{\text{catalyst}} \text{—}\underset{\text{H}}{\underset{|}{\overset{|}{\text{C}}}}\text{—}\underset{\text{H}}{\underset{|}{\overset{|}{\text{C}}}}\text{—} \qquad (3.29)$$

The catalyst is usually a finely divided metal, such as nickel, platinum, or palladium. These metals adsorb hydrogen gas on their surfaces and activate the hydrogen–hydrogen bond. Both hydrogen atoms usually add from the catalyst surface to the same face of the double bond. For example, 1,2-dimethylcyclopentene gives mainly *cis*-1,2-dimethylcyclopentane.

$$\text{1,2-dimethylcyclopentene (}H_3C, CH_3\text{; H, H on catalyst)} \xrightarrow[Pt]{H_2} \text{cis-1,2-dimethylcyclopentane (}H_3C, CH_3\text{; H, H)} \quad \textbf{(3.30)}$$

catalyst catalyst

Catalytic hydrogenation of double bonds is used commercially to convert vegetable oils to margarine and other cooking fats (Sec. 15.2).

PROBLEM 3.21 Write an equation for the catalytic hydrogenation of

a. 2-methyl-2-pentene
b. 4-methylcyclopentene
c. 3-methylcyclohexene
d. vinylcyclobutane

3.15 Additions to Conjugated Systems

3.15.a Electrophilic Additions to Conjugated Dienes

Alternate double and single bonds of conjugated systems have special consequences for their addition reactions. When 1 mole of HBr adds to 1 mole of 1,3-butadiene, a rather surprising result is obtained. Two products are isolated.

$$\overset{1}{C}H_2{=}\overset{2}{C}H{-}\overset{3}{C}H{=}\overset{4}{C}H_2 \xrightarrow{HBr} \begin{cases} CH_2(H){-}CH(Br){-}CH{=}CH_2 \quad \text{(1,2-addition)} \quad \text{3-bromo-1-butene} \\ CH_2(H){-}CH{=}CH{-}CH_2(Br) \quad \text{(1,4-addition)} \quad \text{1-bromo-2-butene} \end{cases} \quad \textbf{(3.31)}$$

1,3-butadiene

In one of these products, HBr has added to one of the two double bonds, and the other double bond is still present in its original position. We call this the product of **1,2-addition**. The other product may at first seem unexpected. The hydrogen and bromine have added to carbon-1 and carbon-4 of the original diene, and a new double bond has appeared between carbon-2 and carbon-3. This process, known as **1,4-addition**, is quite general for electrophilic additions to conjugated systems. How can we explain it?

In **1,2-addition**, a reagent is added to the first and second carbons of a conjugated diene, whereas **1,4-addition** is addition to the first and fourth carbons.

In the first step, the proton adds to the terminal carbon atom, according to Markovnikov's Rule.

$$H^+ + CH_2{=}CH{-}CH{=}CH_2 \longrightarrow CH_3{-}\overset{+}{C}H{-}CH{=}CH_2 \quad \textbf{(3.32)}$$

The resulting carbocation can be stabilized by resonance; in fact, it is a hybrid of two contributing resonance structures (see Sec. 1.12).

$$[CH_3-\overset{+}{C}H-CH=CH_2 \longleftrightarrow CH_3-CH=CH-\overset{+}{C}H_2]$$

The positive charge is delocalized over carbon-2 and carbon-4. When, in the next step, the carbocation reacts with bromide ion (the nucleophile), it can react either at carbon-2 to give the product of 1,2-addition or at carbon-4 to give the product of 1,4-addition.

$$\left.\begin{matrix} \underset{1}{CH_3}-\underset{2}{\overset{+}{C}H}-\underset{3}{CH}=\underset{4}{CH_2} \\ \updownarrow \\ \underset{1}{CH_3}-\underset{2}{CH}=\underset{3}{CH}-\underset{4}{\overset{+}{C}H_2} \end{matrix}\right\} \xrightarrow{Br^-} \begin{matrix} CH_3-\underset{\displaystyle Br}{\underset{|}{C}H}-CH=CH_2 \\ + \\ CH_3-CH=CH-\underset{\displaystyle Br}{\underset{|}{C}H_2} \end{matrix} \qquad (3.33)$$

PROBLEM 3.22 Explain why, in the first step of this reaction, the proton adds to C-1 (eq. 3.32) and not to C-2.

The carbocation intermediate in these reactions is a single species, a resonance hybrid. This type of carbocation, with a carbon–carbon double bond adjacent to the positive carbon, is called an **allylic cation**. The parent allyl cation, shown below as a resonance hybrid, is a primary carbocation, but it is more stable than simple primary ions (such as propyl) because its positive charge is delocalized over the two end carbon atoms as shown in the electrostatic potential map accompanying eq. 3.34.

In an **allylic cation**, a carbon–carbon double bond is adjacent to the positively charged carbon atom.

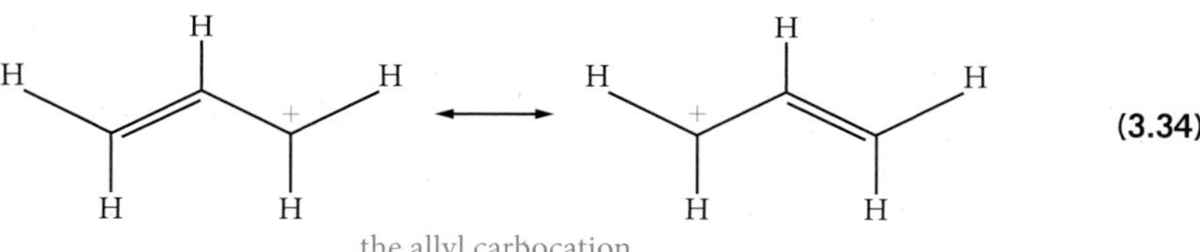

(3.34)

the allyl carbocation

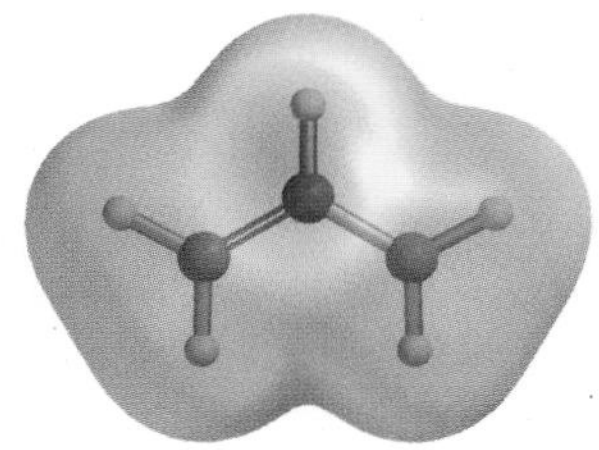

PROBLEM 3.23 Draw the contributors to the resonance hybrid structure of the 3-cyclopentenyl cation [structure].

PROBLEM 3.24 Write an equation for the expected products of 1,2-addition and 1,4-addition of bromine, Br_2, to 1,3-butadiene.

3.15.b Cycloaddition to Conjugated Dienes: The Diels–Alder Reaction

Conjugated dienes undergo another type of 1,4-addition when they react with alkenes (or alkynes). The simplest example is the addition of ethylene to 1,3-butadiene to give cyclohexene.

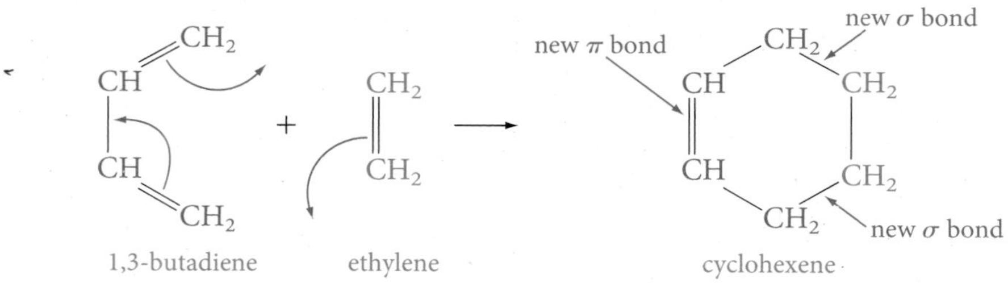

1,3-butadiene + ethylene → cyclohexene (new π bond; new σ bond; new σ bond) **(3.35)**

The **Diels–Alder reaction** is the **cycloaddition reaction** of a conjugated **diene** and a **dienophile** to give a cyclic product in which three π bonds are converted to two σ bonds and a new π bond.

This reaction is an example of a **cycloaddition reaction**, an addition that results in a cyclic product. This cycloaddition, which converts three π bonds to two σ bonds and one new π bond, is called the **Diels–Alder reaction**, after its discoverers, Otto Diels and Kurt Alder. It is so useful for making cyclic compounds that it earned the 1950 Nobel Prize in chemistry for its discoverers. As with hydroboration (Sec. 3.13), this reaction is *concerted.* All bond-breaking and bond-making occur at the same time.

The two reactants are a **diene** and a **dienophile** (diene lover). The simple example in eq. 3.35 is not typical of most Diels–Alder reactions because it proceeds only under pressure and not in good yield. However, this type of reaction gives excellent yields at moderate temperatures if the dienophile has *electron-withdrawing groups** attached, as in the following examples:

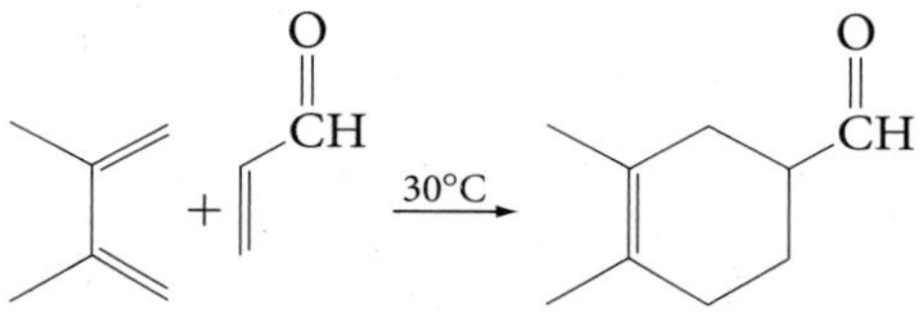

(30°C) **(3.36)**

(0°C) **(3.37)**

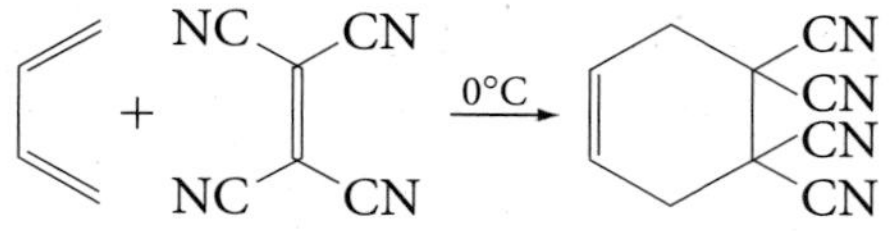

EXAMPLE 3.7

How could a Diels–Alder reaction be used to synthesize the following compound?

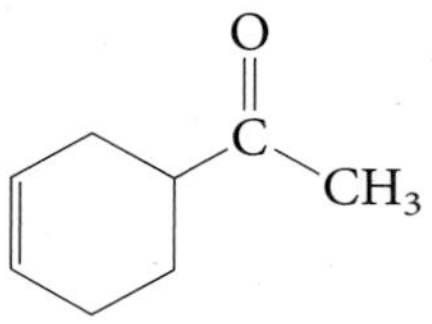

---

*Electron-withdrawing groups are groups of atoms that attract the electrons of the π bond, making the alkene electron poor and therefore more electrophilic toward the diene.

Solution Work backward. The double bond in the product was a single bond in the starting diene. Therefore,

$$\text{1,3-butadiene} + CH_2{=}CH{-}C({=}O){-}CH_3 \longrightarrow \text{cyclohexenyl}{-}C({=}O){-}CH_3$$

PROBLEM 3.25 Show how limonene (Figure 1.13) could be formed by a Diels–Alder reaction of isoprene (2-methyl-1,3-butadiene) with itself.

PROBLEM 3.26 Draw the structure of the product of each of the following cycloaddition reactions.

a. furan (O) + $CH_2{=}CH{-}CN$

b. $CH_2{=}CH{-}CH{=}CH_2 + NC{-}C{\equiv}C{-}CN$

c. $CH_2{=}CH{-}CH{=}CH_2 + H_2C{=}CH{-}NO_2$

3.16 Free-Radical Additions; Polyethylene

Some reagents add to alkenes by a free-radical mechanism instead of by an ionic mechanism. From a commercial standpoint, the most important of these free-radical additions are those that lead to polymers.

A **polymer** is a large molecule, usually with a high molecular weight, built up from small repeating units. The simple molecule from which these repeating units are derived is called a **monomer**, and the process of converting a monomer to a polymer is called **polymerization**.

A **polymer** is a large molecule containing a repeating unit derived from small molecules called **monomers**. The process of polymer formation is called **polymerization**.

The free-radical polymerization of ethylene gives **polyethylene**, a material that is produced on a very large scale (more than 150 billion pounds worldwide annually). The reaction is carried out by heating ethylene under pressure with a catalyst (eq. 3.38). How does this reaction occur?

$$\underset{\text{ethylene}}{CH_2{=}CH_2} \xrightarrow[1000\text{ atm}, >100^\circ C]{ROOR} \underset{\substack{\text{polyethylene} \\ (n\,=\,\text{several thousand})}}{\left(CH_2{-}CH_2\right)_n} \quad (3.38)$$

One common type of catalyst for polymerization is an organic peroxide. The O—O single bond is weak, and on heating, this bond breaks, with one electron going to each of the oxygens.

$$\underset{\text{organic peroxide}}{R{-}O{-}O{-}R} \xrightarrow{\text{heat}} \underset{\text{two radicals}}{2\,R{-}O\cdot} \quad (3.39)$$

A catalyst radical then adds to the carbon–carbon double bond:

$$\underset{\substack{\text{catalyst} \\ \text{radical}}}{RO\cdot} \quad CH_2{=}CH_2 \longrightarrow \underset{\substack{\text{a carbon-centered} \\ \text{free radical}}}{RO{-}CH_2{-}\dot{C}H_2} \quad (3.40)$$

The result of this addition is a carbon-centered free radical, which may add to another ethylene molecule, and another, and another, and so on.

$$\mathrm{RO CH_2\dot{C}H_2} \xrightarrow{\mathrm{CH_2{=}CH_2}} \mathrm{ROCH_2CH_2CH_2\dot{C}H_2} \xrightarrow{\mathrm{CH_2{=}CH_2}} \mathrm{ROCH_2CH_2CH_2CH_2CH_2\dot{C}H_2}\ \text{and so on} \tag{3.41}$$

The carbon chain continues to grow in length until some chain-termination reaction occurs (perhaps a combination of two radicals).

We might think that only a single long chain of carbons will be formed in this way, but this is not always the case. A "growing" polymer chain may abstract a hydrogen atom from its back, so to speak, to cause chain branching.

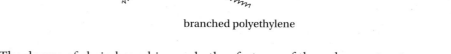

$\cdot CH_2$ H → H — CH_2 $\xrightarrow{CH_2=CH_2}$ CH_3 … CH_2 — $CH_2\cdot$ and so on (3.42)

Fish in polyethylene bag.
Mel Curtis/PhotoDisc, Inc.

A giant molecule with long and short branches is thus formed:

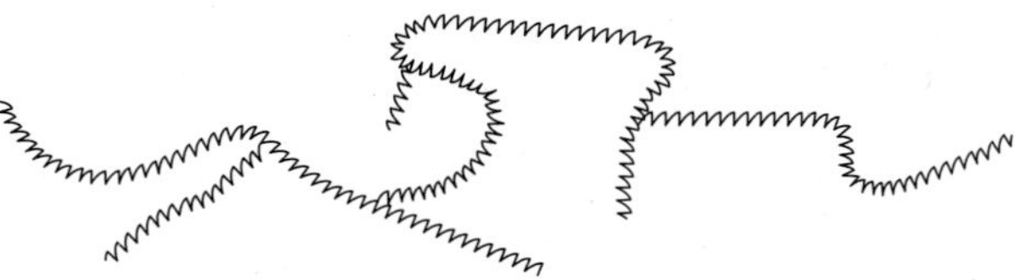
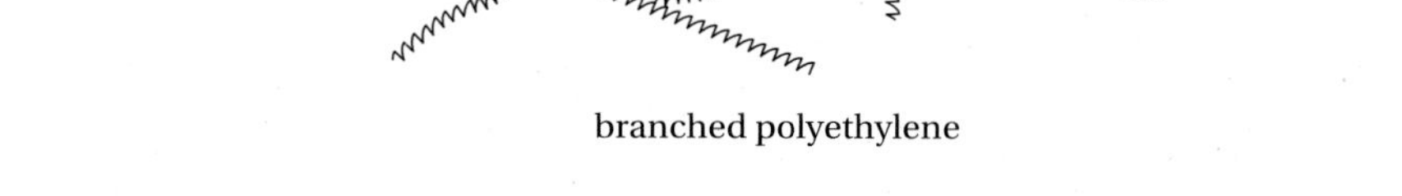
branched polyethylene

The degree of chain branching and other features of the polymer structure can often be controlled by the choice of catalyst and reaction conditions.

A polyethylene molecule is mainly saturated despite its name (polyethyl*ene*) and consists mostly of linked CH_2 groups, but with CH groups at the branch points and CH_3 groups at the ends of the branches. It also contains an OR group from the catalyst at one end, but since the molecular weight is very large, this OR group constitutes a minor and, as far as properties go, relatively insignificant fraction of the molecule.

Polyethylene made in this way is transparent and used in packaging and film (for example, for grocery bags as well as for freezer and sandwich bags).

In Chapter 14, we will describe many other polymers, some made by the process just described for polyethylene and some made by other methods.

3.17 Oxidation of Alkenes

In general, alkenes are more easily oxidized than alkanes by chemical oxidizing agents. These reagents attack the pi electrons of the double bond. The reactions may be useful as chemical tests for the presence of a double bond or for synthesis.

3.17.a Oxidation with Permanganate; a Chemical Test

Alkenes react with alkaline potassium permanganate to form **glycols** (compounds with two adjacent hydroxyl groups).

Glycols are compounds with two hydroxyl groups on adjacent carbons.

$$3\ \rangle C{=}C\langle\ + 2\,K^+MnO_4^- + 4\,H_2O \longrightarrow 3\ {-}\underset{OH}{\underset{|}{\overset{|}{C}}}{-}\underset{OH}{\underset{|}{\overset{|}{C}}}{-}\ + 2\,MnO_2 + 2\,K^+OH^- \quad (3.43)$$

alkene; potassium permanganate (purple); a glycol; manganese dioxide (brown-black)

As the reaction occurs, the purple color of the permanganate ion is replaced by the brown precipitate of manganese dioxide. Because of this color change, the reaction can be used as a chemical test to distinguish alkenes from alkanes, which normally do not react with potassium permanganate.

© Leslie Craine

Hexane does not react with purple $KMnO_4$ (left); cyclohexene (right) reacts, producing a brown-black precipitate of MnO_2.

> **PROBLEM 3.27** Write an equation for the reaction of 2-hexene with potassium permanganate.

3.17.b Ozonolysis of Alkenes

Alkenes react rapidly and quantitatively with ozone, O_3. Ozone is produced naturally in lightning storms, but in the laboratory, ozone is generated by passing oxygen over a high-voltage electric discharge. The resulting gas stream is then bubbled at low temperature into a solution of the alkene in an inert solvent, such as dichloromethane. The first product, a molozonide, is formed by cycloaddition of the oxygen at each end of the ozone molecule to the carbon–carbon double bond. This product then rearranges rapidly to an ozonide. Since these products may be explosive if isolated, they are usually treated directly with a reducing agent, commonly zinc and aqueous acid, to give carbonyl compounds as the isolated products.

$$\rangle C{=}C\langle \xrightarrow{O_3} \left[\text{molozonide}\right] \longrightarrow \text{ozonide} \xrightarrow[H_3O^+]{Zn} \rangle C{=}O + O{=}C\langle \quad (3.44)$$

alkene; molozonide; ozonide; two carbonyl groups

The net result of this reaction is to break the double bond of the alkene and to form two carbon–oxygen double bonds (carbonyl groups), one at each carbon of the original double bond. The overall process is called **ozonolysis**.

Ozonolysis is the oxidation of alkenes with ozone to give carbonyl compounds.

Ozonolysis can be used to locate the position of a double bond. For example, ozonolysis of 1-butene gives two different aldehydes, whereas 2-butene gives a single aldehyde.

$$\underset{\text{1-butene}}{CH_2{=}CHCH_2CH_3} \xrightarrow[2.\ Zn,\ H^+]{1.\ O_3} \underset{\text{formaldehyde}}{CH_2{=}O} \quad \underset{\text{propanal}}{O{=}CHCH_2CH_3} \quad (3.45)$$

$$\underset{\text{2-butene}}{CH_3CH{=}CHCH_3} \xrightarrow[2.\ Zn,\ H^+]{1.\ O_3} \underset{\text{ethanal}}{2\ CH_3CH{=}O} \quad (3.46)$$

Using ozonolysis, one can easily tell which butene isomer is which. By working backward from the structures of ozonolysis products, one can deduce the structure of an unknown alkene.

A WORD ABOUT... Ethylene: Raw Material and Plant Hormone

Cranberry harvest.

Ethylene, the simplest alkene, ranks first among organic chemicals in industrial production. Worldwide, the current annual production of ethylene is over 200 billion pounds. Propene comes in second with about half that amount.

How is all this ethylene produced, and what is it used for? Most hydrocarbons can be "cracked" to give ethylene. (See "A Word About . . . Petroleum, Gasoline, and Octane Number" on pages 102–103.) In the United States, the major raw material for this purpose is ethane.

$$CH_3{-}CH_3 \xrightarrow{700\text{–}900^\circ C} CH_2{=}CH_2 + H_2$$

A substantial fraction of industrial ethylene is, of course, converted to polyethylene, as described in Section 3.16; but ethylene is also a key raw material for the manufacture of other industrial organic chemicals, because of the reactivity of the carbon–carbon double bond. Shown in Figure 3.13 are 9 of the top 50 organic chemicals; each is produced from ethylene.

Ethylene is not only the most important industrial source of organic chemicals, but it also has some biochemical properties that are crucial to agriculture. Ethylene is a **plant hormone** that can cause seeds to sprout, flowers to bloom, fruit to ripen and fall, and leaves and petals to shrivel and turn brown. It is produced naturally by plants from the amino acid *methionine* via an unusual cyclic amino acid, *1-aminocyclopropane-1-carboxylic*

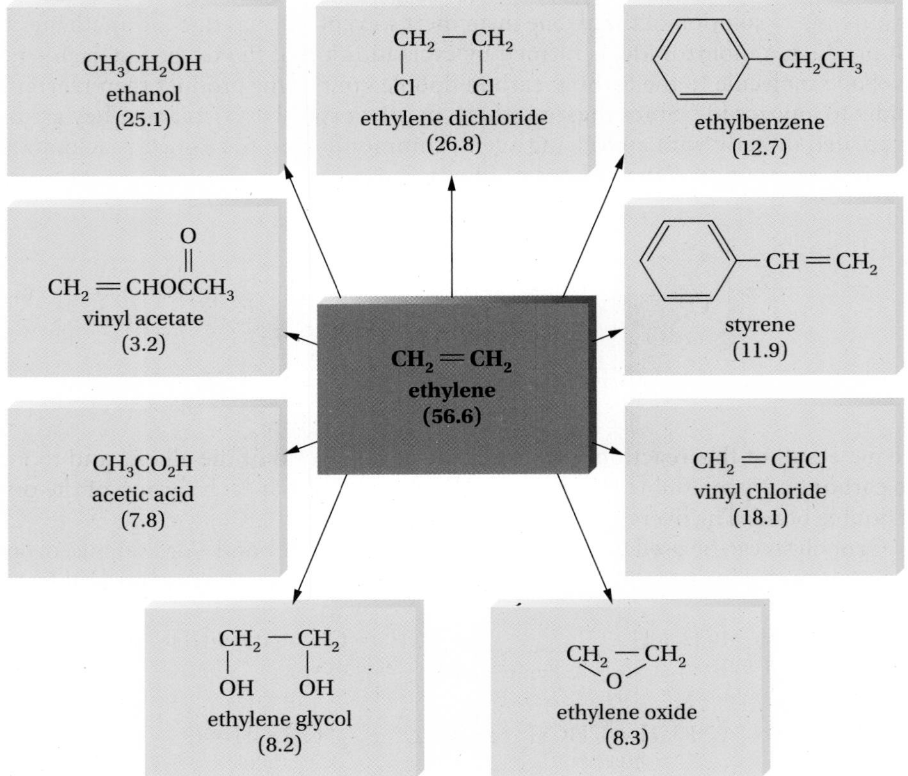

Figure 3.13
Ethylene is central to the manufacture of many industrial organic chemicals. The numbers in parentheses give the recent U.S. production of these chemicals in *billions* of pounds.

acid (ACC), which is then, in several steps, converted to ethylene.

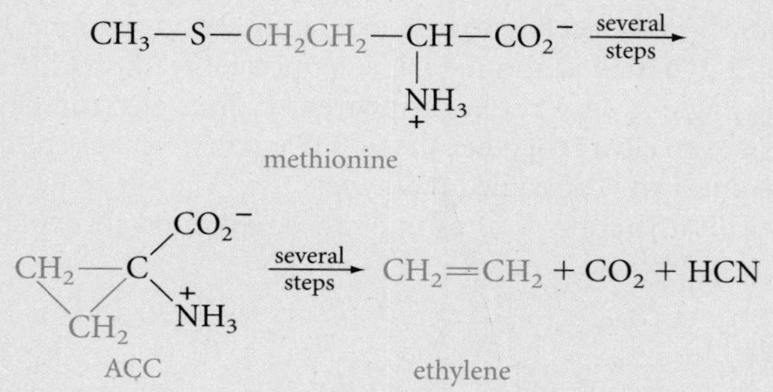

$$CH_3{-}S{-}CH_2CH_2{-}\underset{\overset{|}{\underset{+}{NH_3}}}{CH}{-}CO_2^- \xrightarrow[\text{steps}]{\text{several}}$$

methionine

$$\text{(ACC: cyclopropane ring } CH_2{-}C(CO_2^-)(\overset{+}{N}H_3){-}CH_2\text{)} \xrightarrow[\text{steps}]{\text{several}} CH_2{=}CH_2 + CO_2 + HCN$$

ACC ethylene

The mode by which ethylene functions biologically is still being studied.

Chemists have prepared synthetic compounds that can release ethylene in plants in a controlled manner. One such example is 2-chloroethylphosphonic acid, $ClCH_2CH_2PO(OH)_2$. Sold by Dow Chemical under the trade name *ethrel,* it is water soluble and is taken up by plants, where it breaks down to ethylene, chloride, and phosphate. It has been used commercially to induce fruits, such as pineapples and tomatoes, to ripen uniformly so that an entire field can be harvested efficiently, as shown in the photo on the previous page. Ethylene has also been used to regulate the growth of other crops, such as wheat, apples, cherries, and cotton. Only a small amount need be used, since plants are very sensitive to ethylene and respond to concentrations lower than 0.1 part per million of the gas.

See Problem 3.60.

EXAMPLE 3.8

Ozonolysis of an alkene produces equal amounts of acetone and formaldehyde, $(CH_3)_2C{=}O$ and $CH_2{=}O$, respectively. Deduce the alkene structure.

Solution Connect to each other by a double bond the carbons that are bound to oxygen in the ozonolysis products. The alkene is $(CH_3)_2C{=}CH_2$.

PROBLEM 3.28 Which alkene will give only acetone, $(CH_3)_2C{=}O$, as the ozonolysis product?

3.17.c Other Alkene Oxidations

Various reagents can convert alkenes to epoxides (eq. 3.47).

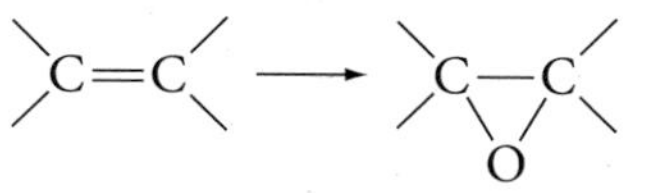

$$\underset{\text{alkene}}{\gt C{=}C\lt} \longrightarrow \underset{\text{epoxide}}{\gt C{-}C\lt \text{ (bridged by O)}} \quad (3.47)$$

This reaction and the chemistry of epoxides are detailed in Chapter 8.

Like alkanes (and all other hydrocarbons), alkenes can be used as fuels. Complete combustion gives carbon dioxide and water.

$$C_nH_{2n} + \tfrac{3n}{2}\,O_2 \longrightarrow nCO_2 + nH_2O \quad (3.48)$$

3.18 Some Facts About Triple Bonds

In the final sections of this chapter, we will describe some of the special features of triple bonds and alkynes.

A carbon atom that is part of a triple bond is directly attached to only *two* other atoms, and the bond angle is 180°. Thus, acetylene is linear, as shown in Figure 3.14. The carbon–carbon triple bond distance is about 1.21 Å, appreciably shorter than that of most double (1.34 Å) or single (1.54 Å) bonds. Apparently, three electron pairs between two carbons draw them even closer together than do two pairs. Because of the linear geometry, no *cis–trans* isomerism is possible for alkynes.

Now let us see how the orbital theory of bonding can be adapted to explain these facts.

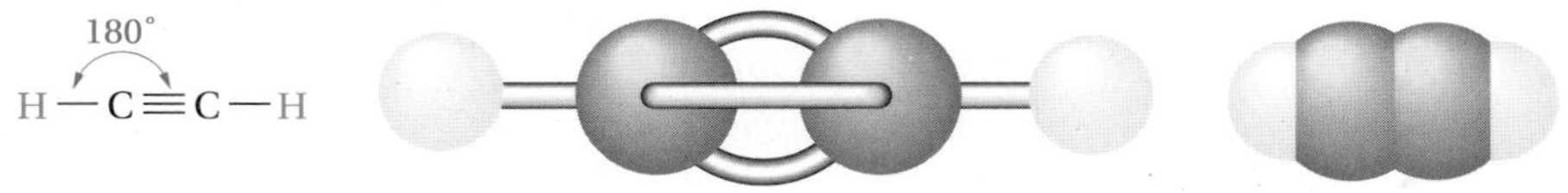

180°

$H-C\equiv C-H$

■ Figure 3.14
Models of acetylene, showing its linearity.

3.19 The Orbital Model of a Triple Bond

***sp*-Hybrid orbitals** are half *s* and half *p* in character. The angle between two *sp* orbitals is 180°.

The carbon atom of an acetylene is connected to only *two* other atoms. Therefore, we combine the 2*s* orbital with only one 2*p* orbital to make two ***sp*-hybrid orbitals** (Figure 3.15). These orbitals extend in opposite directions from the carbon atom. The angle between the two hybrid orbitals is 180° so as to minimize repulsion between any electrons placed in them. One valence electron is placed in each *sp*-hybrid orbital. The remaining two valence electrons occupy two different *p* orbitals that are perpendicular to each other and perpendicular to the hybrid *sp* orbitals.

The formulation of a triple bond from two *sp*-hybridized carbons is shown in Figure 3.16. End-on overlap of two *sp* orbitals forms a sigma bond between the two carbons, and lateral overlap of the properly aligned *p* orbitals forms two pi bonds (designated π_1 and π_2 in the figure). This model nicely explains the linearity of acetylenes.

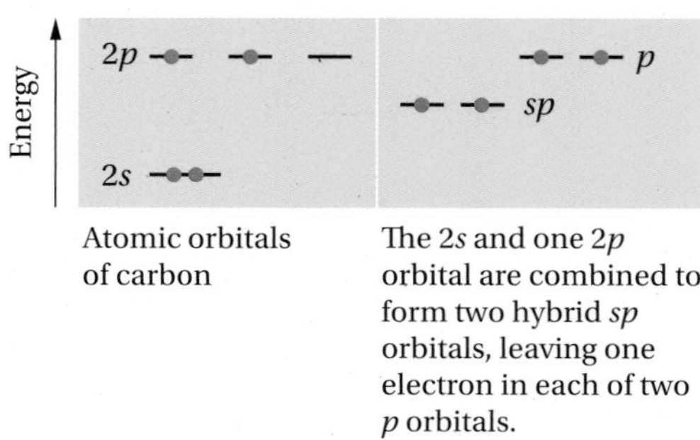
Energy
2p
2s
p
sp
Atomic orbitals of carbon
The 2s and one 2p orbital are combined to form two hybrid sp orbitals, leaving one electron in each of two p orbitals.

■ Figure 3.15
Unhybridized versus *sp*-hybridized orbitals on carbon.

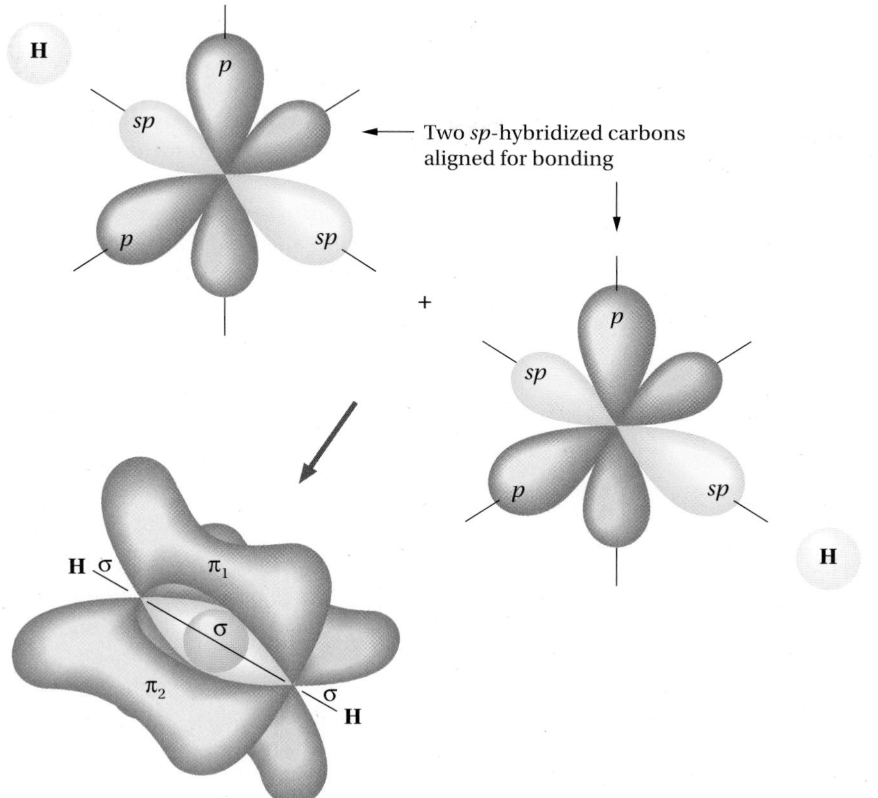

The resulting carbon–carbon triple bond, with a hydrogen atom attached to each remaining *sp* bond. (The orbitals involved in the C—H bonds are omitted for clarity.)

Figure 3.16
A triple bond consists of the end-on overlap of two *sp*-hybrid orbitals to form a σ bond and the lateral overlap of two sets of parallel-oriented *p* orbitals to form two mutually perpendicular π bonds.

3.20 Addition Reactions of Alkynes

As in alkenes, the pi electrons of alkynes are exposed to electrophilic attack (see Figure 3.17). Therefore, many addition reactions described for alkenes also occur, though usually more slowly, with alkynes. For example, bromine adds as follows:

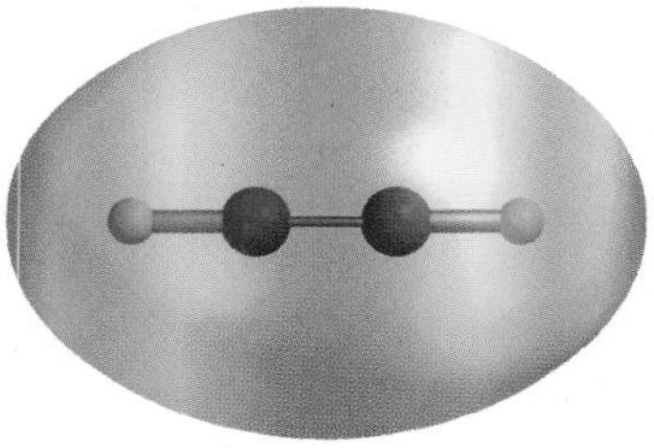

$$\underset{\text{ethyne}}{\mathrm{H{-}C{\equiv}C{-}H}} \xrightarrow{Br_2} \underset{trans\text{-1,2-dibromoethene}}{\mathrm{(H)(Br)C{=}C(Br)(H)}} \xrightarrow{Br_2} \underset{\text{1,1,2,2-tetrabromoethane}}{\mathrm{H{-}CBr_2{-}CBr_2{-}H}} \qquad (3.49)$$

Figure 3.17
Electrostatic potential map of acetylene, showing the exposure of pi electrons (red) to attack by electrophiles.

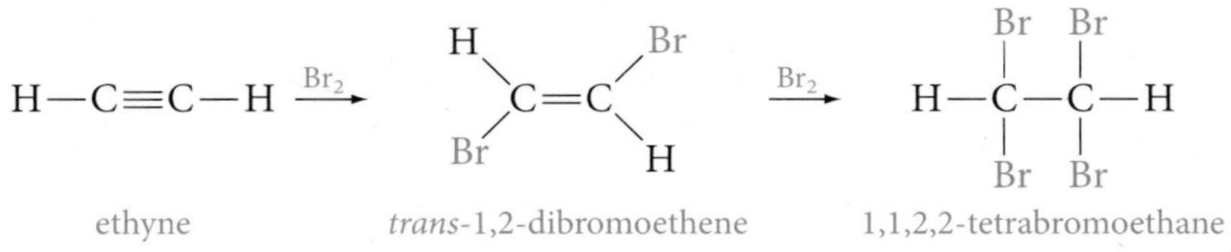

In the first step, the addition occurs mainly *trans*.

With an ordinary nickel or platinum catalyst, alkynes are hydrogenated all the way to alkanes (eq. 3.1). However, a special palladium catalyst (called **Lindlar's catalyst**) can control hydrogen addition so that only 1 mole of hydrogen adds. In this case, the

Lindlar's catalyst limits addition of hydrogen to an alkyne to 1 mole and produces a *cis* alkene.

A WORD ABOUT... Petroleum, Gasoline, and Octane Number

© Peter Menzel/Stock Boston

Petroleum refinery, where crude oil is processed.

Petroleum is at present our most important fossil fuel. The need for petroleum to keep our industrial society going follows close behind our need for food, air, water, and shelter. What is black gold, as petroleum has been called, and how do we use it?

Petroleum is a complex mixture of hydrocarbons formed over eons of time through the gradual decay of buried animal and vegetable matter. **Crude oil** is a viscous black liquid that collects in vast underground pockets in sedimentary rock (the word *petroleum* literally means rock oil, from the Latin *petra*, rock, and *oleum*, oil). It must be brought to the surface via drilling and pumping. To be most useful, the crude oil must be refined.

The first step in petroleum refining is usually **distillation**. The crude oil is heated to about 400°C, and the vapors rise through a tall fractionating column. The lower boiling fractions rise faster and higher in the column before condensing to liquids; higher boiling fractions do not rise as high. By drawing off liquid at various column levels, technicians separate crude oil roughly into the fractions shown in Table 3.3.

The gasoline fraction comprises only about 25% of crude oil. It is the most valuable fraction, however, both as a fuel and as a source material for the petrochemical industry, the industry that furnishes our synthetic fibers, plastics, and many other useful materials. For this reason, many processes have been developed for converting the other fractions into gasoline.

Higher boiling fractions can be "cracked" by heat and catalysts (mainly silica and alumina) to give products with shorter carbon chains and therefore lower boiling points. The carbon chain can break at many points.

$$\underset{\text{alkane}}{C_{10}H_{22}} \longrightarrow \begin{cases} \overset{\text{alkane}}{C_5H_{12}} + \overset{\text{alkene}}{C_5H_{10}} \\ C_8H_{18} + C_2H_4 \\ C_2H_6 + C_8H_{16} \\ C_4H_{10} + (C_4H_8 + C_2H_4) \end{cases}$$

To balance the number of hydrogens, any particular alkane must give at least one alkane and one alkene as products. Thus, catalytic cracking converts larger alkanes into a mixture of smaller alkanes and alkenes and increases the yield of gasoline from petroleum.

During cracking, large amounts of the lower gaseous hydrocarbons—ethene, propene, butanes, and butenes—are formed. Some of these, especially ethene, are used as petrochemical raw materials. To obtain more gasoline, scientists sought methods to convert these low-molecular-weight hydrocarbons to somewhat larger hydrocarbons that boil in the gasoline range. One such process is **alkylation**, the combination of an alkane with an alkene to form a higher-boiling alkane.

$$C_2H_6 + C_4H_8 \xrightarrow{\text{catalyst}} C_6H_{14}$$

$$\underbrace{C_4H_{10} + C_4H_8}_{\text{gases}} \xrightarrow{\text{catalyst}} \underbrace{C_8H_{18}}_{\text{liquids}}$$

Table 3.3 ■ Common Petroleum Fractions

Boiling range, °C	Name	Range of carbon atoms per molecule	Use
<20	gases	C_1 to C_4	heating, cooking, petrochemical raw material
20–200	naphtha; straight-run gasoline	C_5 to C_{12}	fuel; lighter fractions (such as petroleum ether, bp 30°C–60°C) also used as laboratory solvents
200–300	kerosene	C_{12} to C_{15}	fuel
300–400	fuel oil	C_{15} to C_{18}	heating homes, diesel fuel
>400		over C_{18}	lubricating oil, greases, paraffin waxes, asphalt

These processes, which were developed in the 1930s, were important for producing aviation fuel during World War II and are still used to make high-octane gasoline, now for our much-improved automobile engines.

This brings us to **octane number** and why it is important. Some hydrocarbons, especially those with highly branched structures, burn smoothly in an automobile engine and drive the piston forward evenly. Other hydrocarbons, especially those with unbranched carbon chains, tend to explode in the cylinder and drive the piston forward violently. These undesirable explosions produce audible knocks. A scale was set up many years ago to evaluate this important knock property of gasolines. **Isooctane** (2,2,4-trimethylpentane), an excellent fuel with a highly branched structure, was arbitrarily given a rating of 100, and **heptane**, a very poor automotive fuel, was given a rating of 0. A "regular" gasoline with an octane number of 87 has the same "knock" properties as a mixture that is 87% isooctane and 13% heptane.

Small amounts of **tetraethyllead**, $(CH_3CH_2)_4Pb$, in gasoline improve its octane rating but are undesirable for environmental reasons, and most gasoline is now unleaded. However, unleaded gasoline must therefore contain hydrocarbons with a high octane rating. It became important to develop methods that convert straight-chain to branched-chain hydrocarbons, since these have higher octane ratings.

Certain catalysts can produce branched-chain alkanes from straight-chain alkanes. This process, called **isomerization**, is carried out on a large scale commercially.

$$\underset{n\text{-butane}}{CH_3CH_2CH_2CH_3} \xrightarrow[\text{alumina}]{AlCl_3,\ HCl} \underset{\text{isobutane}}{CH_3CH(CH_3)CH_3}$$

Aromatic hydrocarbons, such as benzene and toluene, also have a high octane rating. A platinum catalyst used in a process called **platforming** cyclizes and dehydrogenates alkanes to cycloalkanes and to aromatic hydrocarbons. Of course, large amounts of hydrogen gas are also formed during platforming. Millions of gallons of aromatic hydrocarbons are produced daily by such processes, not only to add to unleaded gasoline to improve its octane rating but also to supply raw materials for many other petrochemical-based products, as we will see in the next chapter.

$$CH_3(CH_2)_5CH_3 \xrightarrow[\text{catalyst}]{Pt} \underset{\text{methylcyclohexane}}{C_6H_{11}CH_3} \xrightarrow[\text{catalyst}]{Pt} \underset{\text{toluene (an aromatic hydrocarbon)}}{C_6H_5CH_3}$$

methylcyclohexane

toluene
(an aromatic hydrocarbon)

See Problem 3.65.

product is a *cis* alkene, because both hydrogens add to the same face of the triple bond from the catalyst surface.

$$\underset{\substack{\text{2-butyne} \\ \text{bp 27°C}}}{CH_3{-}C{\equiv}C{-}CH_3} \xrightarrow[\text{Pd (Lindlar's catalyst)}]{H{-}H} \underset{\substack{cis\text{-2-butene} \\ \text{bp 3.7°C}}}{(CH_3)HC{=}CH(CH_3)} \quad \textbf{(3.50)}$$

With unsymmetric triple bonds and unsymmetric reagents, Markovnikov's Rule is followed in each step, as shown in the following example:

$$CH_3C{\equiv}CH + H{-}Br \longrightarrow CH_3\overset{+}{C}{=}CH_2 + Br^- \longrightarrow CH_3\overset{Br}{C}{=}CH_2 \quad \text{(3.51)}$$
2-bromopropene

$$CH_3\overset{Br}{C}{=}CH_2 + H{-}Br \longrightarrow CH_3\overset{+}{\underset{Br}{C}}{-}CH_3 + Br^- \longrightarrow CH_3{-}\overset{Br}{\underset{Br}{C}}{-}CH_3$$
2,2-dibromopropane

Addition of water to alkynes requires not only an acid catalyst but mercuric ion as well. The mercuric ion forms a complex with the triple bond and activates it for addition. Although the reaction is similar to that of alkenes, the initial product—a **vinyl alcohol** or **enol**—rearranges to a carbonyl compound.

A **vinyl alcohol** or **enol** is an alcohol with a carbon–carbon double bond on the carbon that bears the hydroxyl group.

$$R{-}C{\equiv}CH + H{-}OH \xrightarrow[HgSO_4]{H^+} \left[R{-}\overset{HO}{C}{=}\overset{H}{C}{-}H\right] \longrightarrow R{-}\overset{O}{\overset{\|}{C}}{-}CH_3 \quad \text{(3.52)}$$
a vinyl alcohol, or enol — a methyl ketone

The product is a methyl ketone or, in the case of acetylene itself (R=H), acetaldehyde. We will have more to say about the chemistry of enols and the mechanism of the second step of eq. 3.52 in Chapter 9.

PROBLEM 3.29 Write equations for the following reactions:

a. $CH_3CH_2C{\equiv}CH + Br_2$ (1 mole)
b. $CH_3CH_2C{\equiv}CCH_3 + Cl_2$ (2 moles)
c. 1-hexyne + HBr (1 and 2 moles)
d. 1-butyne + H_2O (Hg^{2+}, H^+)

3.21 Acidity of Alkynes

A hydrogen atom on a triply bonded carbon is weakly acidic and can be removed by a very strong base. Sodium amide, for example, converts acetylenes to acetylides.

$$R{-}C{\equiv}C{-}H + Na^+NH_2^- \xrightarrow{\text{liquid } NH_3} R{-}C{\equiv}C{:}^-Na^+ + NH_3 \quad \text{(3.53)}$$
sodium amide — a sodium acetylide

this hydrogen is weakly acidic

This type of reaction occurs easily with a hydrogen attached to a carbon-carbon triple bond (i.e., RC≡C—H), but less so when the hydrogen is adjacent to a double or single bond. Why? Consider the hybridization of the carbon atom in each type of C—H bond:

$-CH_3$	$=CH_2$	$\equiv C{-}H$
sp^3	sp^2	sp
25% *s*, 75% *p*	$33\frac{1}{3}$% *s*, $66\frac{2}{3}$% *p*	50% *s*, 50% *p*

increasing acidity →

A CLOSER LOOK AT... Petroleum

Conduct research on the Internet to find more information on petroleum and to answer the following questions.

Petroleum Consumption and Extraction

1. How much oil does the world use in a year? What percentage of U.S. energy needs is supplied through oil?
2. What are some of the earliest known examples of oil use by humans? When did the modern oil industry come into being?
3. What are some of the techniques used to extract petroleum (crude oil) from rock? Describe some of the new technologies being developed to extract oil.

Chemical Processing

1. What is cracking? What is the difference between thermal and catalytic cracking?
2. *Unification* and *alteration* are other methods of chemical processing. Define these processes and explain why they are used.

As the hybridization at carbon becomes more s-like and less p-like, the acidity of the attached hydrogen increases. Recall that *s* orbitals are closer to the nucleus than are *p* orbitals. Consequently, the bonding electrons are closest to the carbon nucleus in the ≡C—H bond, making it easiest for a base to remove that type of proton. Sodium amide is a sufficiently strong base for this purpose.*

PROBLEM 3.30 Write an equation for the reaction of 1-hexyne with sodium amide in liquid ammonia.

Though acidic, 1-alkynes are much less so than water. Acetylides can therefore be hydrolyzed to alkynes by water. Internal alkynes have no exceptionally acidic hydrogens.

PROBLEM 3.31 Write an equation for the reaction of a sodium acetylide with water.

PROBLEM 3.32 Will 2-hexyne react with sodium amide? Explain.

*See Table C in the Appendix for the relative acidities of organic functional groups.

REACTION SUMMARY

1. Reactions of Alkenes

a. Addition of Halogens (Sec. 3.7a)

$$\text{>C=C<} + X_2 \longrightarrow -\underset{X}{\overset{|}{\underset{|}{C}}}-\underset{X}{\overset{|}{\underset{|}{C}}}- \quad (X = \text{Cl, Br})$$

b. Addition of Polar Reagents (Sec. 3.7b and Sec. 3.7c)

$$\text{>C=C<} + \text{H—OH} \xrightarrow{H^+} -\underset{H}{\overset{|}{\underset{|}{C}}}-\underset{OH}{\overset{|}{\underset{|}{C}}}-$$

$$\text{>C=C<} + \text{H—X} \longrightarrow -\underset{H}{\overset{|}{\underset{|}{C}}}-\underset{X}{\overset{|}{\underset{|}{C}}}- \quad \left(\begin{matrix} X = \text{F, Cl, Br, I} \\ -OSO_3H \end{matrix}\right)$$

c. Hydroboration–Oxidation (Sec. 3.13)

$$3\ RCH{=}CH_2 \xrightarrow{BH_3} (RCH_2CH_2)_3B \xrightarrow[HO^-]{H_2O_2} 3\ RCH_2CH_2OH$$

d. Addition of Hydrogen (Sec. 3.14)

$$\text{>C=C<} + H_2 \xrightarrow{\text{Pd, Pt, or Ni}} -\underset{H}{\overset{|}{\underset{|}{C}}}-\underset{H}{\overset{|}{\underset{|}{C}}}-$$

e. Addition of X_2 and HX to Conjugated Dienes (Sec. 3.15a)

$$C{=}C{-}C{=}C + X_2 \longrightarrow \underset{\text{1,2-addition}}{\underset{X}{\underset{|}{C}}-\underset{X}{\underset{|}{C}}-C{=}C} + \underset{\text{1,4-addition}}{\underset{X}{\underset{|}{C}}-C{=}C-\underset{X}{\underset{|}{C}}}$$

$$C{=}C{-}C{=}C + \text{H—X} \longrightarrow \underset{\text{1,2-addition}}{\underset{H}{\underset{|}{C}}-\underset{X}{\underset{|}{C}}-C{=}C} + \underset{\text{1,4-addition}}{\underset{H}{\underset{|}{C}}-C{=}C-\underset{X}{\underset{|}{C}}}$$

(X = Cl, Br)

f. Cycloaddition to Conjugated Dienes (Sec. 3.15b)

$$\text{C=C–C=C (diene)} + \text{C=C} \longrightarrow \text{cyclohexene ring (C}_6\text{)}$$

g. Polymerization of Ethylene (Sec. 3.16)

$$nH_2C{=}CH_2 \xrightarrow{\text{catalyst}} \leftarrow\!\!\!\!\!\!\!\;\;CH_2{-}CH_2\!\!\!\rightarrow_n$$

h. Oxidation to Diols or Carbonyl-Containing Compounds (Sec. 3.17)

$$RCH{=}CHR \xrightarrow{KMnO_4} \underset{OH}{RCH}{-}\underset{OH}{CHR} + MnO_2$$

$$\rangle C{=}C\langle \xrightarrow[\text{2. Zn, } H^+]{\text{1. } O_3} \rangle C{=}O + O{=}C\langle$$

2. Reactions of Alkynes

a. Additions to the Triple Bond (Sec. 3.20)

$$R{-}C{\equiv}C{-}R + H_2 \xrightarrow[\text{catalyst}]{\text{Lindlar's}} (R)(H)C{=}C(R)(H)$$ (*cis* addition)

$$R{-}C{\equiv}C{-}H + X_2 \longrightarrow (R)(X)C{=}C(X)(H) \xrightarrow{X_2} RCX_2CHX_2$$

$$R{-}C{\equiv}C{-}H + H{-}X \longrightarrow (R)(X)C{=}C(H)(H) \xrightarrow{H-X} RCX_2CH_3$$

$$R{-}C{\equiv}C{-}H \xrightarrow{H_2O,\ Hg^{2+},\ H^+} R{-}\overset{O}{\overset{\|}{C}}{-}CH_3$$

b. Formation of Acetylide Anions (Sec. 3.21)

$$R{-}C{\equiv}C{-}H + Na^+\,NH_2^- \xrightarrow{NH_3} R{-}C{\equiv}C{:}^-\,Na^+ + NH_3$$

MECHANISM SUMMARY

1. Electrophilic Addition (E^+ = electrophile and $Nu{:}^-$ = nucleophile; Sec. 3.9)

$$\rangle C{=}C\langle \xrightarrow{E^+} {-}\underset{E}{\overset{|}{C}}{-}\overset{|}{\underset{+}{C}}{-} \xrightarrow{Nu{:}^-} {-}\underset{E}{\overset{|}{C}}{-}\underset{Nu}{\overset{|}{C}}{-}$$

carbocation

2. 1,2-Addition and 1,4-Addition (Sec. 3.15a)

$$\mathrm{C{=}C{-}C{=}C} \xrightarrow{E^+} \left[\mathrm{C(E){-}\overset{+}{C}{-}C{=}C} \longleftrightarrow \mathrm{C(E){-}C{=}C{-}\overset{+}{C}}\right] \quad \text{allyl carbocation}$$

$\mathrm{Nu{:}^-}$

$$\underset{\text{1,2-product}}{\mathrm{C(E){-}C(Nu){-}C{=}C}} + \underset{\text{1,4-product}}{\mathrm{E{-}C{-}C{=}C{-}C{-}Nu}}$$

3. Cycloaddition (Sec. 3.15b)

$$\mathrm{C{=}C{-}C{=}C} + \mathrm{C{=}C} \longrightarrow \text{cyclohexene (C}_6\text{ ring)}$$

4. Free-Radical Polymerization of Ethylene (Sec. 3.16)

$$\mathrm{R\cdot} \quad \mathrm{C{=}C} \longrightarrow \mathrm{R{-}C{-}C\cdot}$$

$$\mathrm{R{-}C{-}C\cdot} \quad \mathrm{C{=}C} \longrightarrow \mathrm{R{-}C{-}C{-}C{-}C\cdot} \text{ (and so on)}$$

ADDITIONAL PROBLEMS

OWL Interactive versions of these problems are assignable in OWL.

Alkenes and Alkynes: Nomenclature and Structure

3.33 For the following compounds, write structural formulas and IUPAC names for all possible isomers having the indicated number of multiple bonds:

a. C_4H_6 (one triple bond) b. C_5H_{10} (one double bond) c. C_5H_8 (two double bonds)

3.34 Name the following compounds by the IUPAC system:

a. $CH_3CH{=}C(CH_2CH_2CH_3)_2$

b. $(CH_3)_2CHCH{=}CHCH_3$

c. 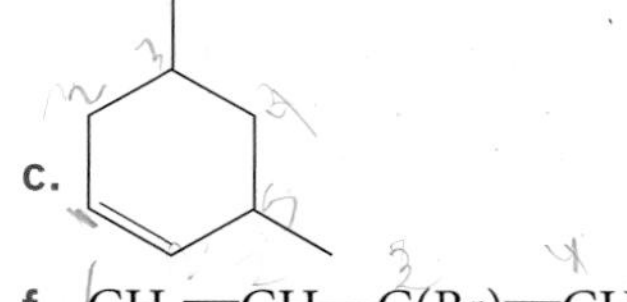

d. $CH_3C{\equiv}CCH(Cl)CH_2CH_3$

e. $CH_3{-}C{\equiv}C{-}CH_2{-}CH{=}CHCH_3$

f. $CH_2{=}CH{-}C(Br){=}CH_2$

g. $CH_3{-}C{\equiv}C{-}CH{=}CH_2$

h.

i.

j. 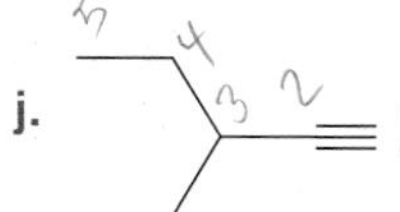

k. 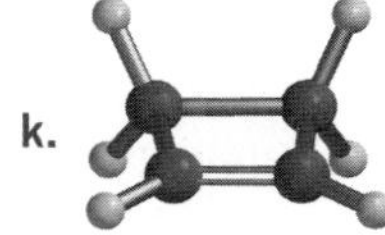

l. 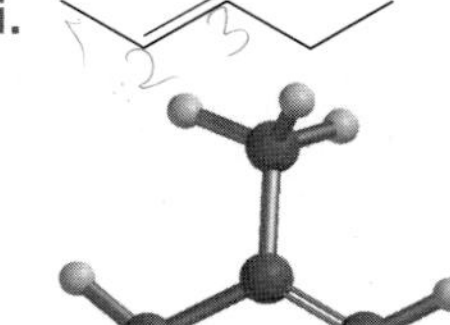

3.35 Write a structural formula for each of the following compounds:

a. 1-bromo-3-heptene
b. 3-methylcyclopentene
c. vinyl bromide
d. vinylcyclohexane
e. 1,3-difluoro-2-butene
f. 4,5-dimethyl-2-heptyne
g. 3-ethyl-1,4-cyclohexadiene
h. 1,3-dimethylcyclopentene
i. vinylcyclopentane
j. allyl chloride
k. 2,3-dichloro-1,3-cyclopentadiene

3.36 Explain why the following names are incorrect and give a correct name in each case:

a. 5-octyne
b. 3-pentene
c. 3-buten-1-yne
d. 1-methyl-2-pentene
e. 2-methylcyclopentene
f. 2-propyl-1-propene
g. 3-pentyne-1-ene
h. 3-ethyl-1,3-butadiene

3.37 a. What are the usual lengths for the single (sp^3–sp^3), double (sp^2–sp^2), and triple (sp–sp) carbon–carbon bonds?
b. The *single* bond in each of the following compounds has the length shown. Suggest a possible explanation for the observed shortening.

$CH_2{=}CH{-}CH{=}CH_2$ (single bond 1.47 Å) $CH_2{=}CH{-}C{\equiv}CH$ (single bond 1.43 Å) $HC{\equiv}C{-}C{\equiv}CH$ (single bond 1.37 Å)

3.38 Which of the following compounds can exist as *cis–trans* isomers? If such isomerism is possible, draw the structures in a way that clearly illustrates the geometry.

a. 3-octene
b. 3-chloropropene
c. 1-hexene
d. 1-bromopropene
e. 4-methylcyclohexene
f. 1,3,5,7-octatetraene
g. 2,3-difluoro-2-butene

3.39 The mold metabolite and antibiotic *mycomycin* has the formula:

$$HC{\equiv}C{-}C{\equiv}C{-}CH{=}C{=}CH{-}CH{=}CH{-}CH{=}CH{-}CH_2{-}\overset{\overset{\large O}{\|}}{C}{-}OH$$

Number the carbon chain, starting with the carbonyl carbon.

a. Which multiple bonds are conjugated?
b. Which multiple bonds are cumulated?
c. Which multiple bonds are isolated?

3.40 Certain tropical plants create extra protection from hungry arthropods, such as caterpillars, by secreting volatile terpenoids, such as nerolidol and 4,8-dimethyl-1,3,7-nonatriene, in order to attract the herbivore's natural enemies:

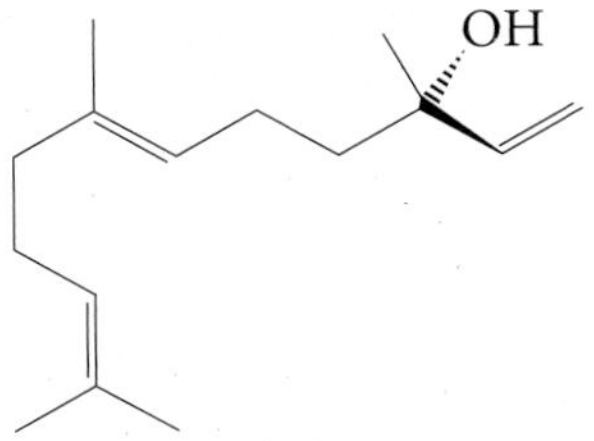

OH

nerolidol

Which double bond(s) in nerolidol and 4,8-dimethyl-1,3,7-nonatriene can exist as *cis–trans* isomers?

Electrophilic Addition to Alkenes

3.41 Write the structural formula and name of the product when each of the following reacts with one mole of bromine.

a. $CH_3CH_2CH{=}CHCH_2CH_3$

b. $CH_2{=}CHCH_2Cl$

c.

d.

e.

f. $CH_3CH{=}C(CH_3){-}CH_2CH_3$

3.42 What reagent will react by addition to what unsaturated hydrocarbon to form each of the following compounds?

a. $CH_3CHBrCHBrCH_3$

b. $(CH_3)_2CHOSO_3H$

c. $(CH_3)_3COH$

d. —Cl

e. $CH_3CH{=}CHCH_2Cl$

f. $CH_3CH_2CCl_2CCl_2CH_3$

g. 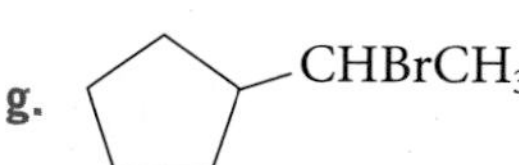

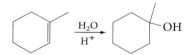

h. 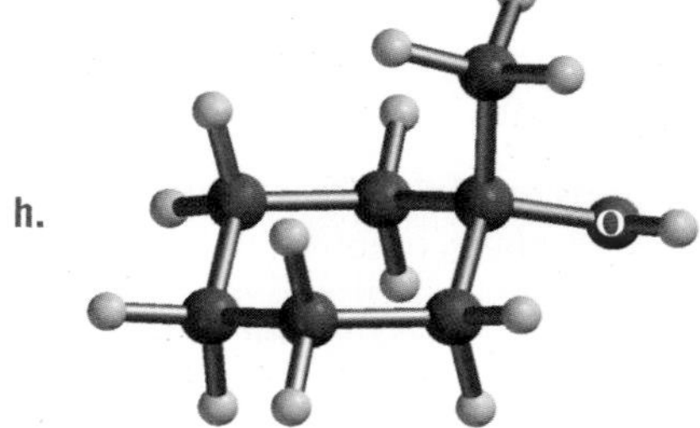

3.43 Which of the following reagents are electrophiles? Which are nucleophiles?

a. H_3O^+
b. HCl
c. Br^-
d. $FeCl_3$
e. H_2SO_4
f. $AlBr_3$
g. HO^-

3.44 The acid-catalyzed hydration of 1-methylcyclohexene gives 1-methylcyclohexanol.

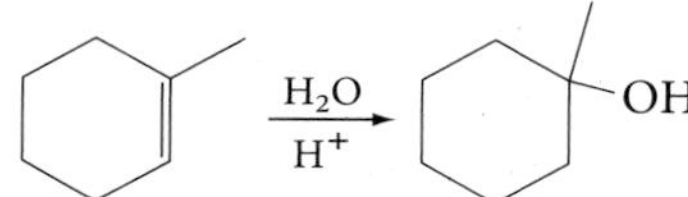

Write every step in the mechanism of this reaction.

3.45 When 2-methylpropene reacts with water and an acid catalyst, only one product alcohol is observed: *tert*-butyl alcohol (2-methyl-2-propanol).

a. Draw the structures of the two intermediate carbocations that could form from the protonation of 2-methylpropene. Which is more stable (has lower energy)? (*Hint:* See eq. 3.21.)

b. Draw a reaction energy diagram for the formation of the two intermediate carbocations in 3.45a from protonation of 2-methylpropene. Use your diagram to explain why only one alcohol is formed (see Figure 3.11).

3.46 *Caryophyllene* is an unsaturated hydrocarbon mainly responsible for the odor of oil of cloves. It has the molecular formula $C_{15}H_{24}$. Hydrogenation of caryophyllene gives a saturated hydrocarbon $C_{15}H_{28}$. Does caryophyllene contain any rings? How many? What else can be learned about the structure of caryophyllene from its hydrogenation?

3.47 Water can act as an electrophile or as a nucleophile. Explain.

3.48 Predict the structures of the two possible monohydration products of limonene (Figure 1.13). These alcohols are called *terpineols*. Predict the structure of the diol (dialcohol) obtained by hydrating both double bonds in limonene. These alcohols are used in the cough medicine "elixir of terpin hydrate."

Reactions of Conjugated Dienes

3.49 Draw the resonance contributors to the carbocation

$$(CH_3)_2CH\overset{+}{C}HCH{=}CHCH(CH_3)_2$$

Does the ion have a symmetric structure?

3.50 Adding 1 mole of hydrogen chloride (HCl) to 1,3-octadiene gives two products. Give their structures, and write all of the steps in a reaction mechanism that explains how each product is formed.

3.51 Predict the product of each of the following Diels–Alder reactions for the diene (left) and dienophile (right) as shown below.

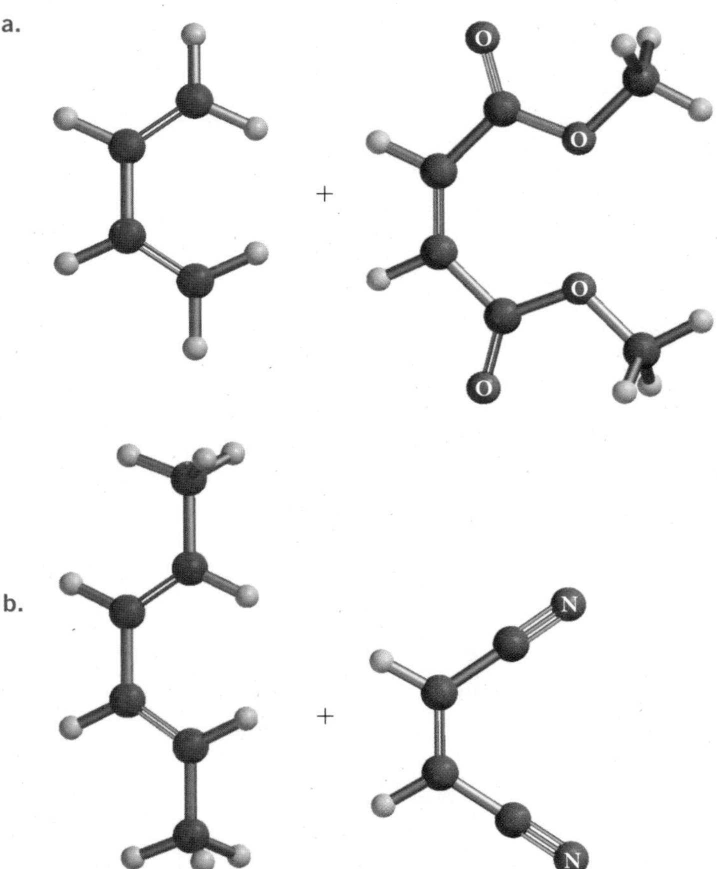

3.52 From what diene and dienophile could each of the following be made?

a. O, OCH_3

b. CN, CN

Other Reactions of Alkenes

3.53 Write an equation that clearly shows the structure of the alcohol obtained from the sequential hydroboration and H_2O_2/OH^- oxidation of

a. 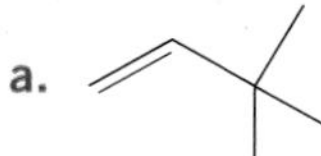

b.

c.

3.54 Write equations to show how $=CH_2$ could be converted to:

a. CH_3 OH

b. $-CH_2OH$

3.55 Describe two simple chemical tests that could be used to distinguish methylcyclohexane from methylcyclohexene. (*Hint:* Both tests produce color changes when alkenes are present.)

3.56 Give the structural formulas of the alkenes that, on ozonolysis, give:

a. $(CH_3)_2C{=}O$ and $CH_2{=}O$
b. only $(CH_3CH_2)_2C{=}O$
c. $CH_3CH{=}O$ and $CH_3CH_2CH{=}O$
d. $O{=}CHCH_2CH_2CH_2CH{=}O$

3.57 Given the information that free-radical stability follows the same order as carbocation stability ($3° > 2° > 1°$), predict the structure of polypropylene produced by the free-radical polymerization of propene. It should help to write out each step in the mechanism, as in eqs. 3.40 and 3.41.

Reactions of Alkynes

3.58 Write equations for the following reactions:

a. 2-octyne + H_2 (1 mole, Lindlar's catalyst)
b. 3-hexyne + Br_2 (2 moles)
c. 1-hexyne + sodium amide in liquid ammonia
d. 1-butyne + H_2O (H^+, Hg^{2+} catalyst)

3.59 Determine what alkyne and what reagent will give

a. 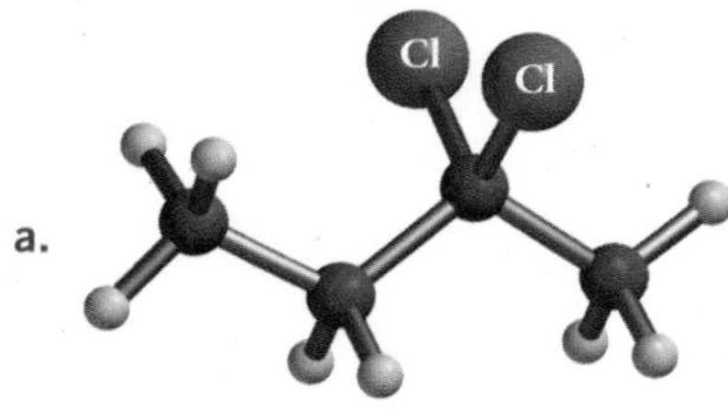

b. 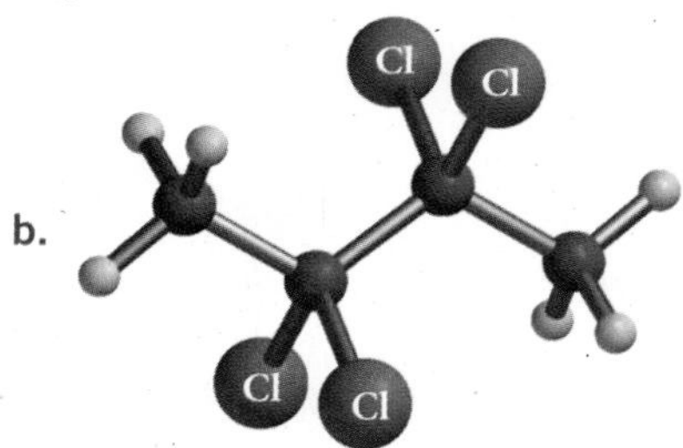

Summary Problems

3.60 As discussed in "A Word About . . . Ethylene" on pp. 98–99, 2-chloroethylphosphonic acid, $ClCH_2CH_2PO(OH)_2$, is a commercially synthesized compound that is used for inducing the ripening of fruit.

a. Write a balanced equation for the conversion of $ClCH_2CH_2PO(OH)_2$ with water to form ethylene, chloride, and phosphate.
b. Which functional group in 2-chloroethylphosphonic acid is responsible for its water solubility?

3.61 Write an equation for the reaction of $CH_2{=}CHCH_2CH_3$ with each of the following reagents:

a. hydrogen chloride
b. bromine
c. hydrogen (Pt catalyst)
d. H_2O, H^+
e. ozone, followed by H^+
f. $KMnO_4$, OH^-
g. BH_3 followed by H_2O_2, OH^-
h. oxygen (combustion)

3.62 Write a complete reaction mechanism for the addition of HCl to 1-butyne (Problem 3.59a).

3.63 For the structures of nerolidol and 4,8-dimethyl-1,3,7-nonatriene in Problem 3.40, provide the products for each compound after:

a. hydrogenation (3 moles H_2, Pt catalyst) b. addition of 1 mole of Br_2 c. ozonolysis

3.64 When 1-butyne is treated with $[(CH_3)_2CHCH(CH_3)]_2BH$ followed by H_2O_2 and OH^-, the product isolated is the aldehyde butanal: $CH_3CH_2CH_2{-}C({=}O){-}H$.

a. What reaction of alkenes is similar to this reaction?
b. Compare the product with the product of Problem 3.58d. Write the structure of an intermediate product (not isolated) that rearranges to form butanal. (*Hint:* See eq. 3.52.)

3.65 As discussed in "A Word About . . . Petroleum, Gasoline, and Octane Number" on p. 103, isomerization is often used to convert one alkane to a more substituted alkane. Suggest the highly branched product obtained from the isomerization of pentane using a mixture of $AlCl_3$ and HCl over an alumina catalyst. Provide the correct IUPAC name for the product.

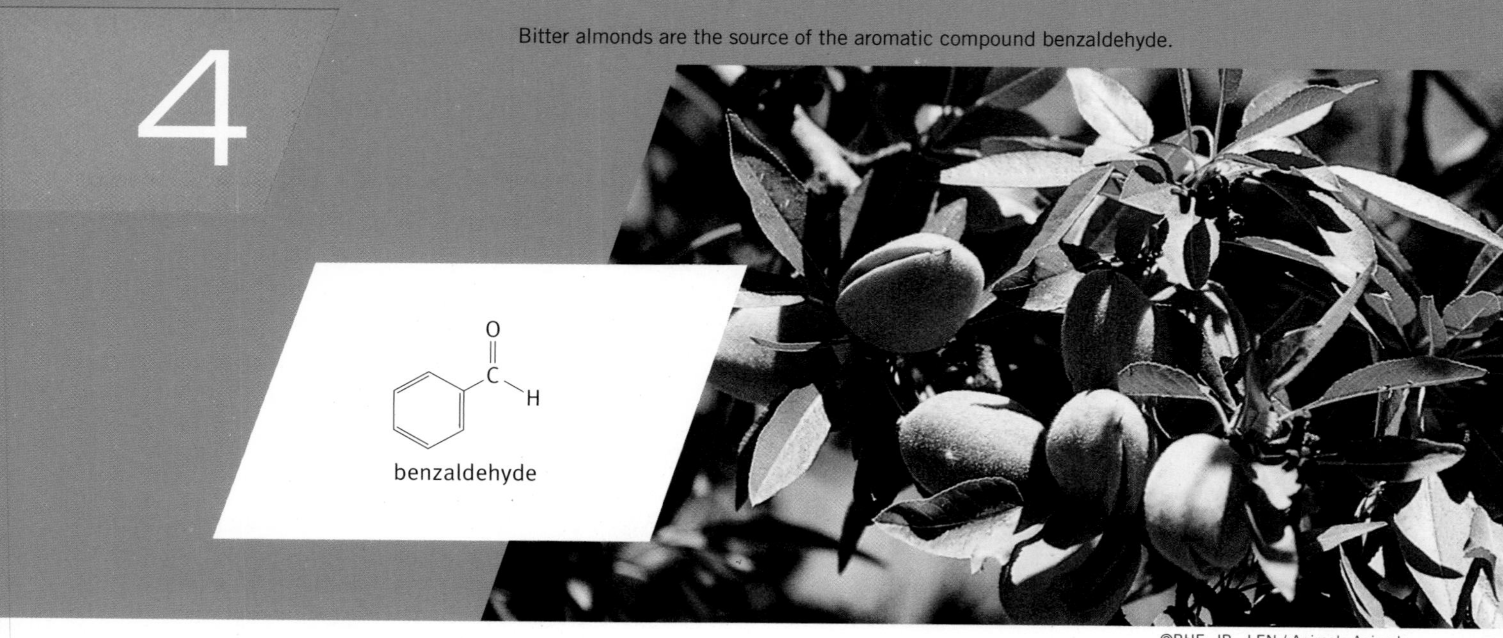

Bitter almonds are the source of the aromatic compound benzaldehyde.

Online homework for this chapter can be assigned in OWL, an online homework assessment tool.

Aromatic Compounds

Spices and herbs have long played a romantic role in the course of history. They bring to mind frankincense and myrrh and the great explorers of past centuries—Vasco da Gama, Christopher Columbus, Ferdinand Magellan, Sir Francis Drake—whose quest for spices helped open up the Western world. Though not without risk, trade in spices was immensely profitable. It was natural, therefore, that spices and herbs were among the first natural products studied by organic chemists. If one could extract from plants the pure compounds with these desirable fragrances and flavors and determine their structures, perhaps one could synthesize them in large quantity, at low cost, and without danger.

It turned out that many of these aromatic substances have rather simple structures. Many contain a six-carbon unit that passes unscathed through various chemical reactions that alter only the rest of the structure. This group, C_6H_5—, is common to many substances, including benzaldehyde (isolated from the oil of bitter almonds), benzyl alcohol (isolated from gum benzoin, a balsam resin obtained from certain Southeast Asian trees), and toluene (a hydrocarbon isolated from tolu balsam). When any of these three compounds is oxidized, the C_6H_5 group remains intact; the product is benzoic acid (another constituent of gum benzoin). The calcium salt of this acid, when heated, yields the parent hydrocarbon C_6H_6 (eq. 4.1).

$$\begin{array}{l} C_6H_5CH{=}O \text{ (benzaldehyde)} \xrightarrow{\text{oxidize}} \\ C_6H_5CH_2OH \text{ (benzyl alcohol)} \xrightarrow{\text{oxidize}} C_6H_5CO_2H \text{ (benzoic acid)} \xrightarrow[\text{2. heat}]{\text{1. CaO}} C_6H_6 \text{ (benzene)} \\ C_6H_5CH_3 \text{ (toluene)} \xrightarrow{\text{oxidize}} \end{array} \quad (4.1)$$

This same hydrocarbon, first isolated from compressed illuminating gas by Michael Faraday in 1825, is now called **benzene.*** It is the parent hydrocarbon of a class of substances that we now call **aromatic compounds**, *not because of their aroma,* but because of their special chemical properties, in particular, their stability. Why is benzene unusually stable, and what chemical reactions will benzene and related aromatic compounds undergo? These are the subjects of this chapter.

Benzene, C_6H_6, is the parent hydrocarbon of the especially stable compounds known as **aromatic compounds**.

4.1 Some Facts About Benzene

The carbon-to-hydrogen ratio in benzene, C_6H_6, suggests a highly unsaturated structure. Compare the number of hydrogens, for example, with that in hexane, C_6H_{14}, or in cyclohexane, C_6H_{12}, both of which also have six carbons but are saturated.

PROBLEM 4.1 Draw at least five isomeric structures that have the molecular formula C_6H_6. Note that all are highly unsaturated or contain small, strained rings.

Despite its molecular formula, benzene for the most part does not behave as if it were unsaturated. For instance, it does not decolorize bromine solutions the way alkenes and alkynes do (Sec. 3.7a), nor is it easily oxidized by potassium permanganate (Sec. 3.17a). It does not undergo the typical addition reactions of alkenes or alkynes. Instead, *benzene reacts mainly by substitution.* For example, when treated with bromine (Br_2) in the presence of ferric bromide as a catalyst, benzene gives bromobenzene and hydrogen bromide as products.

$$\underset{\text{benzene}}{C_6H_6} + Br_2 \xrightarrow[\text{catalyst}]{FeBr_3} \underset{\text{bromobenzene}}{C_6H_5Br} + HBr \quad (4.2)$$

Chlorine, with a ferric chloride catalyst, reacts similarly.

$$\underset{\text{benzene}}{C_6H_6} + Cl_2 \xrightarrow[\text{catalyst}]{FeCl_3} \underset{\text{chlorobenzene}}{C_6H_5Cl} + HCl \quad (4.3)$$

Only *one* monobromobenzene or monochlorobenzene has ever been isolated; that is, no isomers are obtained in either of these reactions. This result requires *all six hydrogens in benzene to be chemically equivalent.* It does not matter which hydrogen is replaced by bromine; we get the same monobromobenzene. This fact has to be accounted for in any structure proposed for benzene.

*Today, benzene is one of the most important commercial organic chemicals. Approximately 17 billion pounds are produced annually in the United States alone. Benzene is obtained mostly from petroleum by catalytic reforming of alkanes and cycloalkanes or by cracking certain gasoline fractions. It is used to make styrene, phenol, acetone, cyclohexane, and other industrial chemicals.

When bromobenzene is treated with a second equivalent of bromine and the same type of catalyst, *three di*bromobenzenes are obtained.

$$C_6H_5Br + Br_2 \xrightarrow[\text{catalyst}]{FeBr_3} \underset{\substack{\text{dibromobenzenes} \\ \text{(three isomers)}}}{C_6H_4Br_2} + HBr \qquad \textbf{(4.4)}$$

The isomers are not formed in equal amounts. Two of them predominate, and only a small amount of the third isomer is formed. The important point is that there are three isomers—no more and no less. Similar results are obtained when chlorobenzene is further chlorinated to give dichlorobenzenes. These facts also have to be explained by any structure proposed for benzene.

The problem of benzene's structure does not sound overwhelming, yet it took decades to solve. Let us examine the main ideas that led to our modern view of its structure.

4.2 The Kekulé Structure of Benzene

Friedrich August Kekulé.

© Bettmann/CORBIS

In 1865, Kekulé proposed a reasonable structure for benzene.* He suggested that six carbon atoms are located at the corners of a regular hexagon, with one hydrogen atom attached to each carbon atom. To give each carbon atom a valence of 4, he suggested that single and double bonds alternate around the ring (what we now call a *conjugated* system of double bonds). But this structure is highly unsaturated. To explain benzene's negative tests for unsaturation (that is, its failure to decolorize bromine or to give a permanganate test), Kekulé suggested that the single and double bonds exchange positions around the ring *so rapidly* that the typical reactions of alkenes cannot take place.

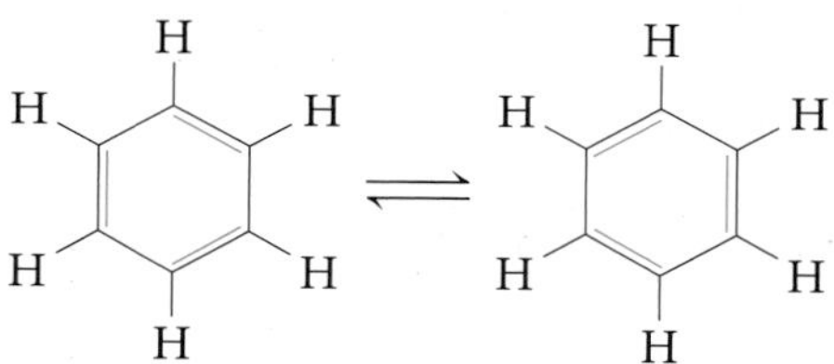

the Kekulé structures for benzene

PROBLEM 4.2 Write out eqs. 4.2 and 4.4 using a Kekulé structure for benzene. Does this model explain the existence of only one monobromobenzene? Only three dibromobenzenes?

PROBLEM 4.3 How might Kekulé explain the fact that there is only one dibromobenzene with the bromines on adjacent carbon atoms, even though we can draw two different structures, with either a double or a single bond between the bromine-bearing carbons?

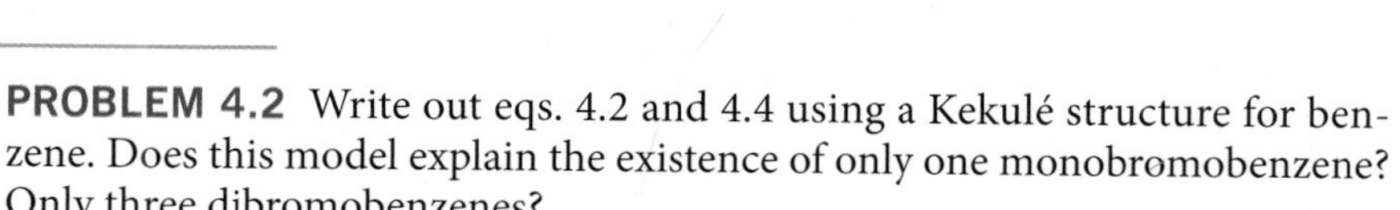

*Friedrich August Kekulé (1829–1896) was a pioneer in the development of structural formulas in organic chemistry. He was among the first to recognize the tetracovalence of carbon and the importance of carbon chains in organic structures. He is best known for his proposal regarding the structure of benzene and other aromatic compounds. It is interesting that Kekulé first studied architecture, and only later switched to chemistry. Judging from his contributions, he apparently viewed chemistry as molecular architecture.

4.3 Resonance Model for Benzene

Kekulé's model for the structure of benzene is nearly, but not entirely, correct. *Kekulé's two structures for benzene differ only in the arrangement of the electrons*; all of the atoms occupy the same positions in both structures. *This is precisely the requirement for resonance* (review Sec. 1.12). Kekulé's formulas represent two identical contributing structures to a *single resonance hybrid* structure of benzene. Instead of writing an equilibrium symbol between them, as Kekulé did, we now write the double-headed arrow (↔) used to indicate a resonance hybrid:

Benzene is a resonance hybrid of these two contributing structures.

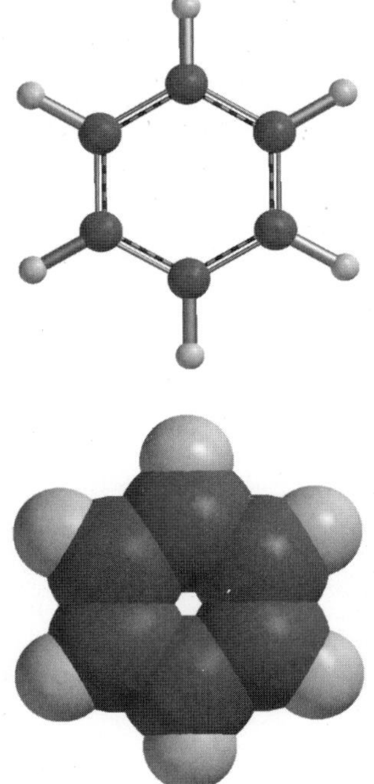

Properties:
colorless liquid
bp 80°C
mp 5.5°C

Figure 4.1
Ball-and-stick and space-filling models of benzene.

To express this model another way, all benzene molecules are identical, and their structure is not adequately represented by either of Kekulé's contributing structures. Being a resonance hybrid, benzene is more stable than either of its contributing Kekulé structures. There are no single or double bonds in benzene—only one type of carbon–carbon bond, which is of some intermediate type. Consequently, it is not surprising that benzene does not react chemically exactly like alkenes.

Modern physical measurements support this model for the benzene structure. *Benzene is planar, and each carbon atom is at the corner of a regular hexagon. All of the carbon–carbon bond lengths are identical*: 1.39 Å, intermediate between typical single (1.54 Å) and double (1.34 Å) carbon–carbon bond lengths. Figure 4.1 shows a space-filling model of the benzene molecule.*

PROBLEM 4.4 How does the resonance model for benzene explain the fact that there are only three isomers of dibromobenzene?

4.4 Orbital Model for Benzene

Orbital theory, which is so useful in rationalizing the geometries of alkanes, alkenes, and alkynes, is also useful in explaining the structure of benzene. Each carbon atom in benzene is connected to only *three* other atoms (two carbons and a hydrogen). Each carbon is therefore sp^2-hybridized, as in ethylene. Two sp^2 orbitals of each carbon atom overlap with similar orbitals of adjacent carbon atoms to form the sigma bonds of the hexagonal ring. The third sp^2 orbital of each carbon overlaps with a hydrogen 1*s* orbital to form the C—H sigma bonds. Perpendicular to the plane of the three sp^2 orbitals at each carbon is a *p* orbital containing one electron, the fourth valence electron. The *p* orbitals on all six carbon atoms can overlap laterally to form pi orbitals that create a ring or cloud of electrons above and below the plane of the ring. The construction of a benzene ring from six sp^2-hybridized carbons is shown schematically in Figure 4.2. This model explains nicely the planarity of benzene. It also explains its hexagonal shape, with H—C—C and C—C—C angles of 120°.

*Notice the difference in the shapes of benzene and cyclohexane (Figure 2.6).

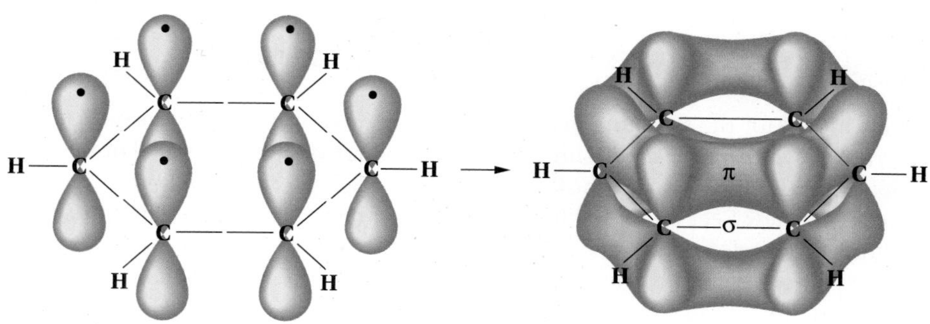

Figure 4.2
An orbital representation of the bonding in benzene. Sigma (σ) bonds are formed by the end-on overlap of sp^2 orbitals. In addition, each carbon contributes one electron to the pi (π) system by lateral overlap of its p orbital with the p orbitals of its two neighbors.

4.5 Symbols for Benzene

Two symbols are used to represent benzene. One is the Kekulé structure, and the other is a hexagon with an inscribed circle, to represent the idea of a delocalized pi electron cloud.

Kekulé delocalized pi cloud

Regardless of which symbol is used, the hydrogens are usually not written explicitly, but we must remember that one hydrogen atom is attached to the carbon at each corner of the hexagon.

The symbol with the inscribed circle emphasizes the fact that the electrons are distributed evenly around the ring, and in this sense, it is perhaps the more accurate of the two. The Kekulé symbol, however, reminds us very clearly that there are six pi electrons in benzene. For this reason, it is particularly useful in allowing us to keep track of the valence electrons during chemical reactions of benzene. In this book, we will use the Kekulé symbol. However, we must keep in mind that the "double bonds" are not fixed in the positions shown, nor are they really double bonds at all.

4.6 Nomenclature of Aromatic Compounds

Because aromatic chemistry developed in a haphazard fashion many years before systematic methods of nomenclature were developed, common names have acquired historic respectability and are accepted by IUPAC. Examples include:

CH_3 CH_3CHCH_3 $CH{=}CH_2$ OH

benzene toluene cumene styrene phenol

OCH_3 $CH{=}O$ $CH_3C{=}O$ CO_2H NH_2

anisole benzaldehyde acetophenone benzoic acid aniline

EXAMPLE 4.1

Write the structural formula for benzaldehyde (eq. 4.1).

Solution One hydrogen in the formula for benzene is replaced by the aldehyde group.

PROBLEM 4.5 Write the formulas for benzyl alcohol, toluene, and benzoic acid (eq. 4.1).

Monosubstituted benzenes that do not have common names accepted by IUPAC are named as derivatives of benzene.

bromobenzene chlorobenzene nitrobenzene ethylbenzene propylbenzene

When two substituents are present, three isomeric structures are possible. They are designated by the prefixes ***ortho*-**, ***meta*-**, and ***para*-**, which are usually abbreviated as ***o*-**, ***m*-**, and ***p*-**, respectively. If substituent X is attached (by convention) to carbon 1, then *o*-groups are on carbons 2 and 6, *m*-groups are on carbons 3 and 5, and *p*-groups are on carbon 4.*

Specific examples are

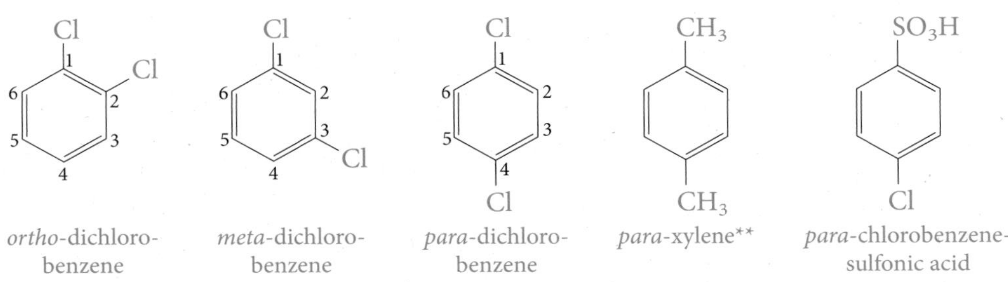

ortho-dichlorobenzene *meta*-dichlorobenzene *para*-dichlorobenzene *para*-xylene** *para*-chlorobenzenesulfonic acid

*Note that X can be on any carbon of the ring. It is the *position of the second substituent relative to that of X* that is important.

**The common and IUPAC name is xylene, *not p*-methyltoluene.

PROBLEM 4.6 Draw the structures for *ortho*-xylene and *meta*-xylene.

The prefixes *ortho*-, *meta*-, and *para*- are used even when the two substituents are not identical.

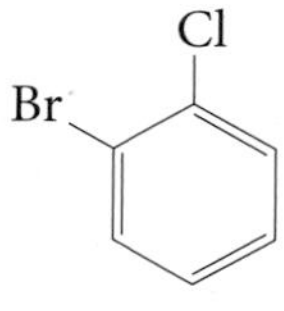

o-bromochlorobenzene (note alphabetical order) 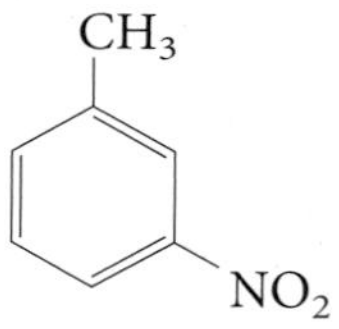 *m*-nitrotoluene 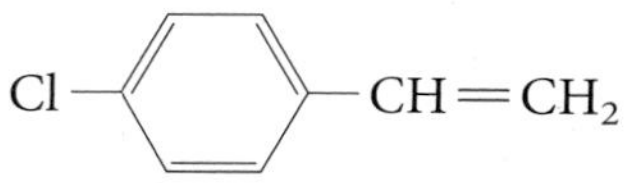 *p*-chlorostyrene 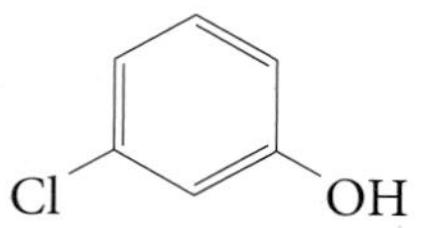 *m*-chlorophenol 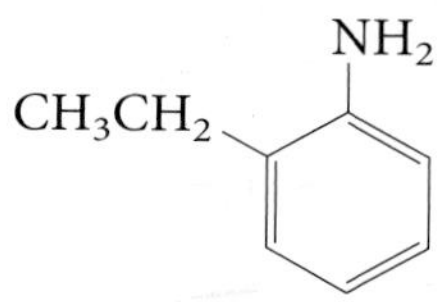 *o*-ethylaniline

Although *o*-, *m*-, and *p*- designations are commonly used in naming disubstituted benzenes, position numbers of substituents can also be used. For example, *o*-dichlorobenzene can also be named 1,2-dichlorobenzene, and *m*-chlorophenol can also be named 3-chlorophenol.

When more than two substituents are present, their positions are designated by numbering the ring.

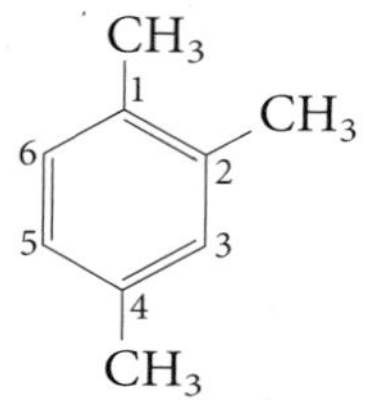

1,2,4-trimethylbenzene 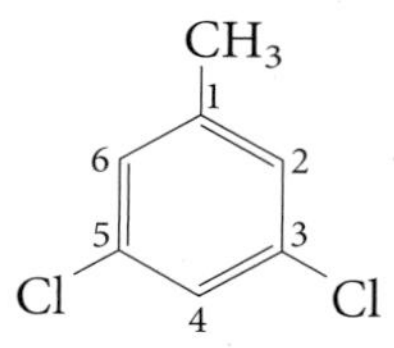 3,5-dichlorotoluene 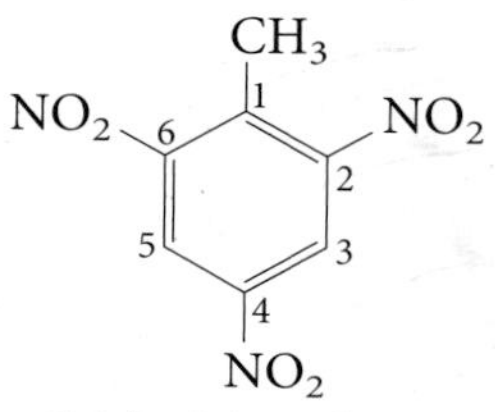 2,4,6-trinitrotoluene (TNT)

PROBLEM 4.7 Draw the structure of

a. *o*-dinitrobenzene
b. *m*-nitrotoluene
c. *p*-bromophenol
d. 4-chloroaniline
e. *p*-vinylbenzaldehyde
f. 1,4-dichlorobenzene

PROBLEM 4.8 Draw the structure of

a. 1,3,5-trimethylphenol
b. 4-ethyl-2,6-difluorotoluene

Aromatic hydrocarbons are called **arenes**. An aromatic substituent is called an **aryl group, Ar**.

Aromatic hydrocarbons, as a class, are called **arenes**. The symbol **Ar** is used for an **aryl group**, just as the symbol R is used for an alkyl group. Therefore, the formula Ar—R would represent any arylalkane.

Two groups with special names occur frequently in aromatic compounds. They are the **phenyl group** and the **benzyl group**.

C_6H_5— or

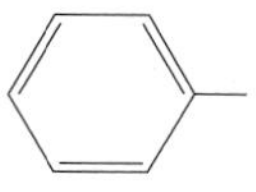

phenyl group

$C_6H_5CH_2$— or —CH_2—

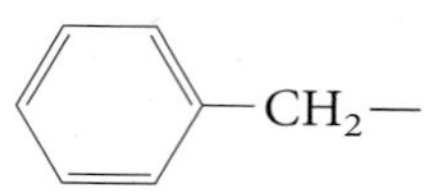

benzyl group

The symbol Ph is sometimes used as an abbreviation for the phenyl group. The use of these group names is illustrated in the following examples:

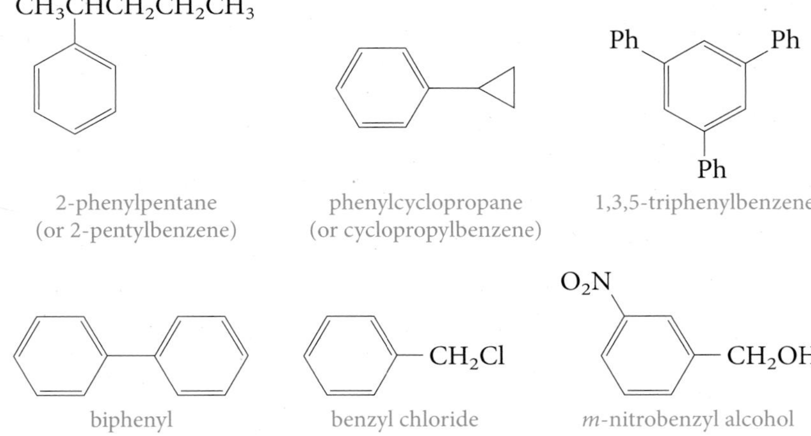

PROBLEM 4.9 Draw the structure of

a. dibenzyl b. cyclobutylbenzene
c. benzyl iodide d. *p*-phenylstyrene

PROBLEM 4.10 Name the following structures:

a.

b. OH, CH_2

4.7 The Resonance Energy of Benzene

We have asserted that a resonance hybrid is always more stable than any of its contributing structures. Fortunately, in the case of benzene, this assertion can be proved experimentally, and we can even measure how much more stable benzene is than the hypothetical molecule 1,3,5-cyclohexatriene (the IUPAC name for one Kekulé structure).

Hydrogenation of a carbon–carbon double bond is an exothermic reaction. The amount of energy (heat) released is about 26 to 30 kcal/mol for each double bond (eq. 4.5). (The exact value depends on the substituents attached to the double bond.) When two double bonds in a molecule are hydrogenated, twice as much heat is evolved, and so on.

$$\text{>C=C<} + \text{H—H} \longrightarrow \text{—}\underset{\text{H}}{\overset{|}{\text{C}}}\text{—}\underset{\text{H}}{\overset{|}{\text{C}}}\text{—} + \text{heat (26–30 kcal/mol)} \tag{4.5}$$

Hydrogenation of cyclohexene releases 28.6 kcal/mol (Figure 4.3). We expect that the complete hydrogenation of 1,3-cyclohexadiene should release twice that amount of heat, or $2 \times 28.6 = 57.2$ kcal/mol; experimentally, the value is close to what we expect (Figure 4.3). It seems reasonable, therefore, to expect that the heat of hydrogenation of

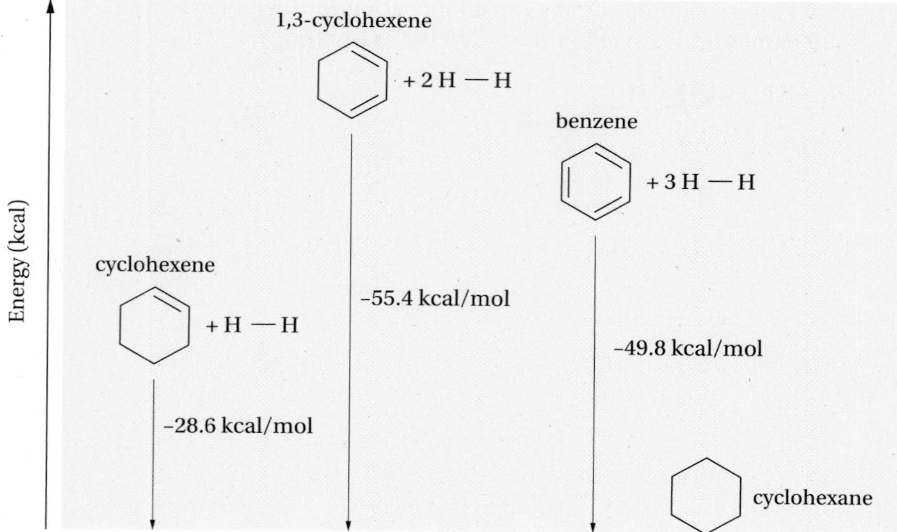

Figure 4.3
Comparison of heat released by the hydrogenation of cyclohexene, 1,3-cyclohexadiene, and benzene to produce cyclohexane.

a Kekulé structure (the *hypothetical* triene 1,3,5-cyclohexatriene) should correspond to that for *three* double bonds, or about 84 to 86 kcal/mol. However, we find experimentally that benzene is more difficult to hydrogenate than simple alkenes, and the heat evolved when benzene is hydrogenated to cyclohexane is *much lower* than expected: only 49.8 kcal/mol (Figure 4.3).

We conclude that *real benzene molecules are more stable than the contributing resonance structures* (the hypothetical molecule 1,3,5-cyclohexatriene) *by about 36 kcal/mol* ($86 - 50 = 36$).

The **stabilization energy**, or **resonance energy**, of a substance is the difference between the energy of the real molecule and the calculated energy of the most stable contributing structure.

We define the **stabilization energy**, or **resonance energy**, of a substance as the difference between the actual energy of the real molecule (the resonance hybrid) and the calculated energy of the most stable contributing structure. For benzene, this value is about 36 kcal/mol. This is a substantial amount of energy. Consequently, as we will see, *benzene and other aromatic compounds usually react in such a way as to preserve their aromatic structure and therefore retain their resonance energy.*

4.8 Electrophilic Aromatic Substitution

The most common reactions of aromatic compounds involve substitution of other atoms or groups for a ring hydrogen on the aromatic unit. Here are some typical substitution reactions of benzene.

$$C_6H_6 + Cl_2 \xrightarrow{FeCl_3} C_6H_5Cl + HCl$$ chlorination **(4.6)**

$$C_6H_6 + Br_2 \xrightarrow{FeBr_3} C_6H_5Br + HBr$$ bromination **(4.7)**

$$C_6H_6 + HNO_3\ (HONO_2) \xrightarrow{H_2SO_4} C_6H_5NO_2 + H_2O$$ nitration **(4.8)**

$$C_6H_6 + H_2SO_4 \ (HOSO_3H) \xrightarrow{SO_3} C_6H_5SO_3H + H_2O \quad \text{sulfonation} \tag{4.9}$$

$$C_6H_6 + RCl \xrightarrow{AlCl_3} C_6H_5R + HCl \quad \text{alkylation} \tag{4.10}$$

(R = an alkyl group such as CH_3—, CH_3CH_2—)

$$C_6H_6 + CH_2{=}CH_2 \xrightarrow{H_2SO_4} C_6H_5CH_2CH_3 \quad \text{alkylation} \tag{4.11}$$

$$C_6H_6 + R{-}\overset{O}{\overset{\|}{C}}Cl \xrightarrow{AlCl_3} C_6H_5\overset{O}{\overset{\|}{C}}{-}R + HCl \quad \text{acylation} \tag{4.12}$$

Most of these reactions are carried out at temperatures between about 0°C and 50°C, but these conditions may have to be milder or more severe if other substituents are already present on the benzene ring. Also, the conditions can usually be adjusted to introduce more than one substituent, if desired.

How do these reactions take place? And why do we observe substitution instead of addition? And what role does the catalyst play? In the next sections, we will try to answer these questions.

4.9 The Mechanism of Electrophilic Aromatic Substitution

Much evidence indicates that all of the substitution reactions listed in the previous section involve initial attack on the benzene ring by an electrophile. Consider chlorination (eq. 4.6) as a specific example. The reaction of benzene with chlorine is exceedingly slow without a catalyst, but it occurs quite briskly with one. What does the catalyst do? It acts as a Lewis acid and converts chlorine to a strong electrophile by forming a complex and polarizing the Cl—Cl bond.

$$:\ddot{\underset{..}{Cl}}-\ddot{\underset{..}{Cl}}: + FeCl_3 \rightleftharpoons \overset{\delta+}{Cl}\cdots\cdots\overset{\delta-}{Cl}\cdots\cdots FeCl_3 \tag{4.13}$$

weak electrophile — strong electrophile

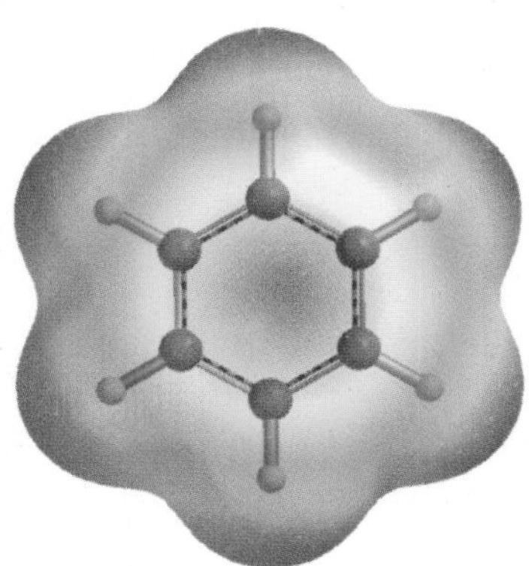

■ **Figure 4.4**
Electrostatic potential map of benzene illustrates the pi electron density (red-yellow) of the aromatic ring.

The reason why a *strong* electrophile is required will become apparent shortly.

The electrophile bonds to one carbon atom of the benzene ring, using two of the pi electrons from the pi cloud to form a sigma bond with a ring carbon atom. This carbon atom becomes sp^3-hybridized. The benzene ring acts as a pi-electron donor, or nucleophile, toward the electrophilic reagent (Figure 4.4).

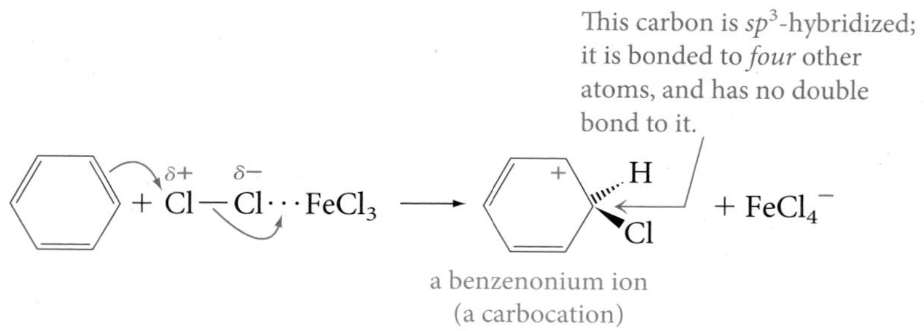

This carbon is sp^3-hybridized; it is bonded to *four* other atoms, and has no double bond to it.

$$\text{C}_6\text{H}_6 + \overset{\delta+}{\text{Cl}}-\overset{\delta-}{\text{Cl}}\cdots\text{FeCl}_3 \longrightarrow \text{C}_6\text{H}_6\text{Cl}^+ + \text{FeCl}_4^- \quad (4.14)$$

a benzenonium ion (a carbocation)

The resonance-stabilized **benzenonium ion**, a carbocation, is the intermediate in electrophilic aromatic substitution reactions.

The resulting carbocation is a **benzenonium ion**, in which the positive charge is delocalized by resonance to the carbon atoms *ortho* and *para* to the carbon to which the chlorine atom became attached; that is, *ortho* and *para* to the sp^3 carbon atom.

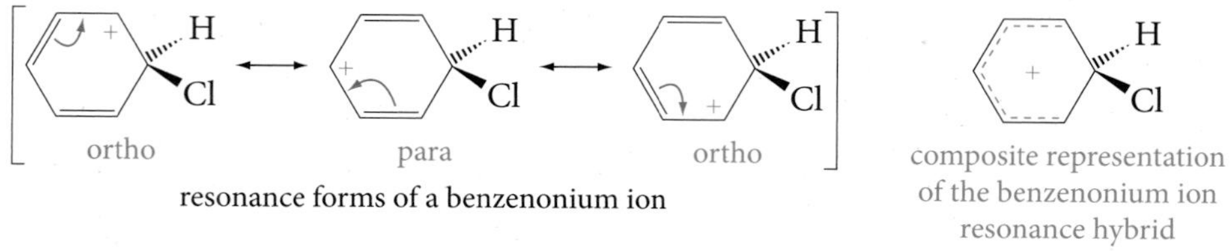

ortho ⟷ para ⟷ ortho

resonance forms of a benzenonium ion

composite representation of the benzenonium ion resonance hybrid

A benzenonium ion is similar to an allylic carbocation (Sec. 3.15.a), but the positive charge is delocalized over three carbon atoms instead of only two. Although stabilized by resonance compared with other carbocations, its resonance energy is much less than that of the starting benzene ring.

Substitution is completed by loss of a proton from the sp^3 carbon atom, the same atom to which the electrophile became attached.*

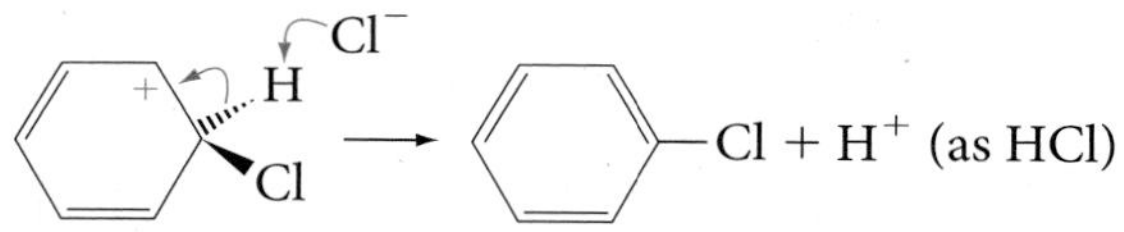

$$\text{C}_6\text{H}_6\text{Cl}^+ + \text{Cl}^- \longrightarrow \text{C}_6\text{H}_5-\text{Cl} + \text{H}^+ \text{ (as HCl)} \quad (4.15)$$

We can generalize this two-step mechanism for all the electrophilic aromatic substitutions in Section 4.8 with the following equation:

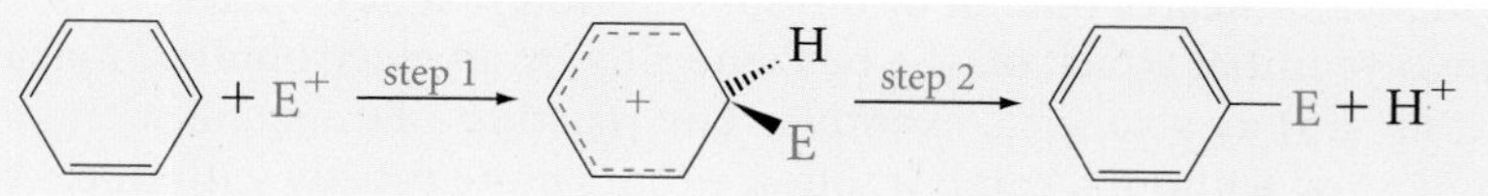

$$\text{C}_6\text{H}_6 + \text{E}^+ \xrightarrow{\text{step 1}} \text{C}_6\text{H}_6\text{E}^+ \xrightarrow{\text{step 2}} \text{C}_6\text{H}_5-\text{E} + \text{H}^+ \quad (4.16)$$

The reason why a strong electrophile is important, and why we observe substitution instead of addition, now becomes clear. In step 1, the stabilization energy (resonance energy) of the aromatic ring is lost, due to disruption of the aromatic pi system. This disruption, caused by addition of the electrophile to one of the ring carbons, requires energy and a strong electrophile. In step 2, the aromatic resonance energy is regained by loss of a proton. This would not be the case if the intermediate carbocation added a nucleophile (as in electrophilic *additions* to double bonds, Sec. 3.9).

*Generally, some base present in solution assists the removal of the proton. The base does not have to be strong, since the regeneration of the aromatic ring is highly favored due to its stability.

Nu, H, H, E — addition of Nu : (aromaticity lost) ← [cyclohexadienyl cation with H and E] — substitution (aromaticity restored) → [benzene]—E + H^+ **(4.17)**

The first step in eq. 4.16 is usually slow or rate determining because it requires a substantial activation energy to disrupt the aromatic system. The second step has a low activation energy and is usually fast because it regenerates the aromatic system. This is illustrated in Figure 4.5.

PROBLEM 4.11 Electrophilic aromatic substitutions to benzene and electrophilic additions to alkenes both involve a slow first step and a fast second step. Using Figure 4.5 as a guide, draw a reaction energy diagram for the reaction shown in eqs. 4.14 and 4.15.

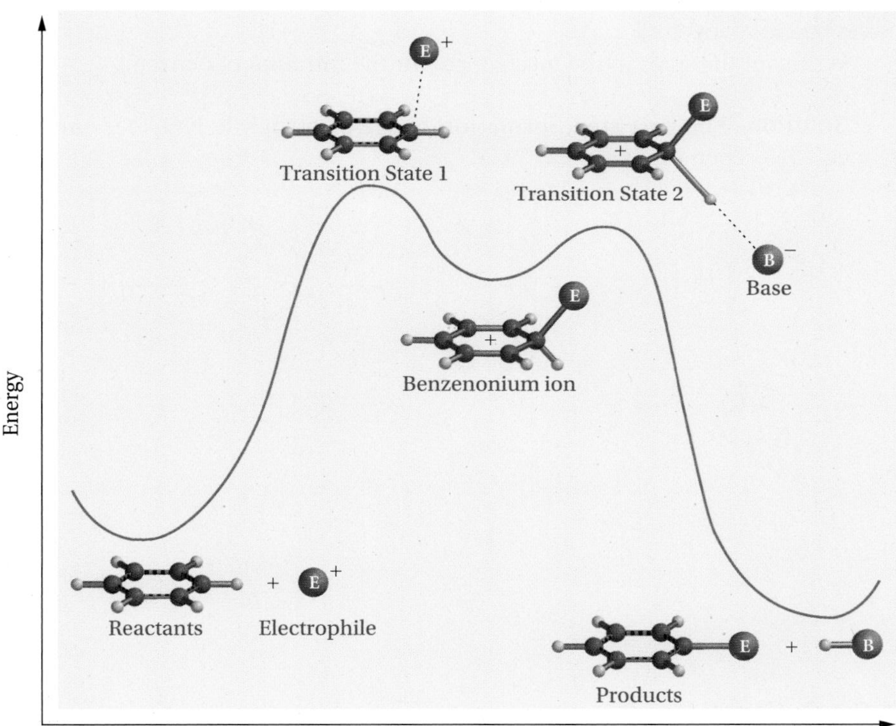

Figure 4.5 A reaction energy diagram for electrophilic aromatic substitution. The electrophile adds to benzene in an endothermic first step, with loss of aromaticity. A proton at the carbon attacked by the electrophile is lost in an exothermic second step, with restoration of aromaticity.

Now let us briefly consider separately each of the various types of electrophilic aromatic substitutions listed in Section 4.8.

4.9.a Halogenation

Chlorine or bromine is introduced into an aromatic ring by using the halogen together with the corresponding iron halide as a catalyst (that is, $Cl_2 + FeCl_3$ or $Br_2 + FeBr_3$). Usually the reaction is carried out by adding the halogen slowly to a mixture of the aromatic compound and iron filings. The iron reacts with the halogen to form the iron halide, which then catalyzes the halogenation.

Direct fluorination or iodination of aromatic rings is also possible but requires special methods.

4.9.b Nitration

The positively charged **nitronium ion** is the electrophile in the nitration of benzene.

In aromatic nitration reactions (eq. 4.18), the sulfuric acid catalyst* protonates the nitric acid, which then loses water to generate the **nitronium ion** (NO_2^+), which contains a positively charged nitrogen atom.

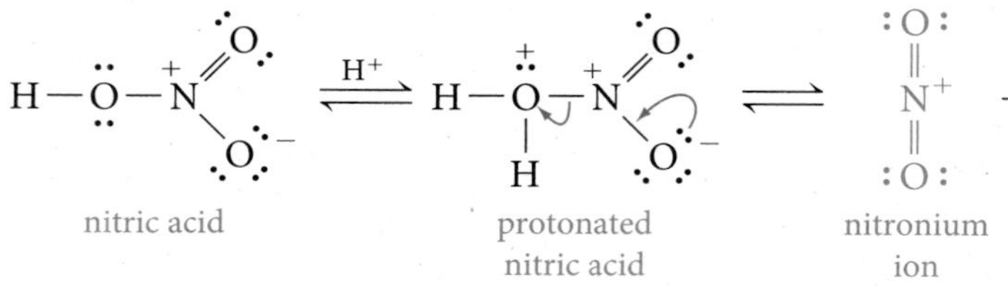

$$H{-}\ddot{O}{-}\overset{+}{N}(=O)O^- \underset{}{\overset{H^+}{\rightleftharpoons}} H{-}\overset{+}{O}(H){-}\overset{+}{N}(=O)O^- \rightleftharpoons O{=}N^+{=}O + H_2O \quad \textbf{(4.18)}$$

nitric acid — protonated nitric acid — nitronium ion

The nitronium ion, a strong electrophile, is then attacked by the aromatic ring.

EXAMPLE 4.2

Write out the steps in the mechanism for the nitration of benzene.

Solution The first step, formation of the electrophile NO_2^+, is shown in eq. 4.18. Then

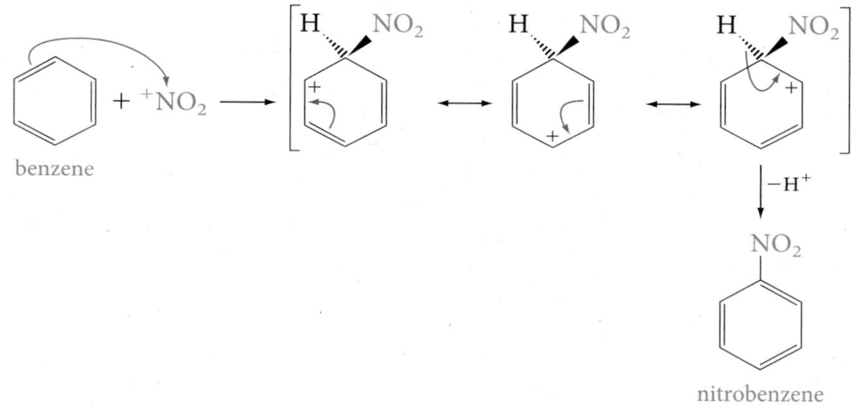

benzene + $^+NO_2$ ⟶ [H, NO_2 ⟷ H, NO_2 ⟷ H, NO_2] $\xrightarrow{-H^+}$ NO_2 nitrobenzene

4.9.c Sulfonation

In aromatic sulfonation reactions (eq. 4.19), we use either concentrated or fuming sulfuric acid, and the electrophile may be sulfur trioxide, SO_3, or protonated sulfur trioxide, $^+SO_3H$. The following resonance structures demonstrate that SO_3 is a strong electrophile at sulfur.

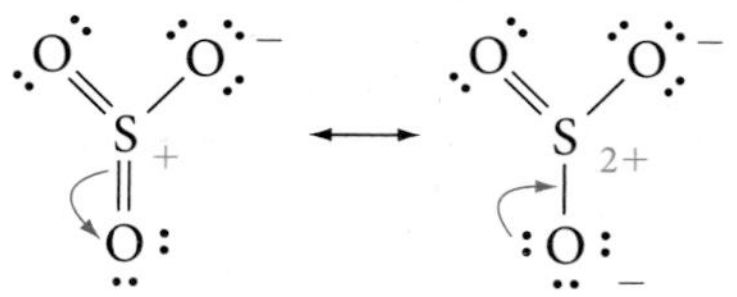

$$O{=}S^{+}(O^-){=}O \longleftrightarrow O{=}S^{2+}(O^-){-}O^-$$

*Sulfuric acid provides a proton catalyst as shown in the following equilibrium: $H{-}OSO_3H \rightleftharpoons H^+ + {}^-OSO_3H$.

The products, sulfonic acids, are strong organic acids. Also, they can be converted to phenols by reaction with base at high temperatures.

$$\text{benzene} \xrightarrow[H_2SO_4]{SO_3} \underset{\text{benzenesulfonic acid}}{C_6H_5SO_3H} \xrightarrow[200°C]{NaOH} \underset{\text{phenol}}{C_6H_5OH} \quad (4.19)$$

PROBLEM 4.12 Write out the steps in the mechanism for the sulfonation of benzene.

4.9.d Alkylation and Acylation

Alkylation of aromatic compounds (eqs. 4.10 and 4.11) is referred to as the **Friedel–Crafts reaction**, after Charles Friedel (French) and James Mason Crafts (American), who first discovered the reaction in 1877. The electrophile is a carbocation, which can be formed either by removing a halide ion from an alkyl halide with a Lewis acid catalyst (for example, $AlCl_3$) or by adding a proton to an alkene. For example, the synthesis of ethylbenzene can be carried out as follows:

The alkylation or acylation of an aromatic compound is referred to as a **Friedel–Crafts reaction.**

$$AlCl_3 + ClCH_2CH_3 \rightleftharpoons AlCl_4^- + \underset{\text{ethyl cation}}{\overset{+}{C}H_2CH_3} \xleftarrow{H^+} CH_2{=}CH_2 \quad (4.20)$$

$$\underset{\text{benzene}}{C_6H_6} + {}^+CH_2CH_3 \longrightarrow \left[\text{arenium ion resonance structures (H, } CH_2CH_3 \text{)} \right] \xrightarrow{-H^+} \underset{\text{ethylbenzene}}{C_6H_5CH_2CH_3} \quad (4.21)$$

PROBLEM 4.13 Which product would you expect if propene were used in place of ethene in eq. 4.11 (or eqs. 4.20 and 4.21): propylbenzene or isopropylbenzene? Explain.

The Friedel–Crafts alkylation reaction has some limitations. It cannot be applied to an aromatic ring that already has on it a nitro or sulfonic acid group, because these groups form complexes with and deactivate the aluminum chloride catalyst.

Friedel–Crafts acylations (eq. 4.12) occur similarly. The electrophile is an acyl cation generated from an acid derivative, usually an acyl halide. The reaction provides a useful general route to aromatic ketones.

$$\underset{\text{acetyl choride}}{CH_3\overset{O}{\overset{\|}{C}}Cl} + AlCl_3 \rightleftharpoons \underset{\text{acetyl cation}}{CH_3\overset{+}{C}{=}O} + AlCl_4^- \quad (4.22)$$

$$C_6H_6 + CH_3\overset{+}{C}{=}O \rightleftharpoons \left[C_6H_6^+(H)(\overset{O}{\overset{\|}{C}}CH_3) \right] \xrightarrow{-H^+} \underset{\text{acetophenone}}{C_6H_5\overset{O}{\overset{\|}{C}}CH_3} \quad (4.23)$$

A WORD ABOUT... Vitamin E: Tocopherols and Tocotrienols

The vitamin E family includes eight structurally similar molecules, four tocopherol (TCP) isomers (α, β, γ, and δ) and four tocotrienol (TCT) isomers (α, β, γ, and δ). Vitamin E that is purchased on a store shelf is usually TCP. The eight compounds of the TCP and TCT family contain an aromatic chromanol ring with methyl substitutions at different positions. While TCTs and TCPs have an isoprenoid chain, TCT has an unsaturated side chain with three double bonds, but TCP has a saturated alkyl chain. TCPs are predominantly found in corn, soybean, and olive oil, and TCTs are particularly prominent in palm, rice bran, and barley oil.

Within the vitamin E family, there has been significant focus on TCPs relative to that of TCTs. Current developments in vitamin E research show that each member of the vitamin E family is unique with respect to its biological functions. Researchers have discovered that α-TCT, in particular, has different biological functions than α-TCP. For example, α-TCT has been shown to have more potent anti-oxidative properties than α-TCP. Moreover, it has been established that TCTs possess better neuroprotective, anticancer, and cholesterol-lowering properties than TCPs. TCTs are also connected with anti-carcinogenic properties and anti-angiogenic functions. However, the reason(s) for the enhanced biological activity of TCTs is not known, and researchers are actively exploring this question.

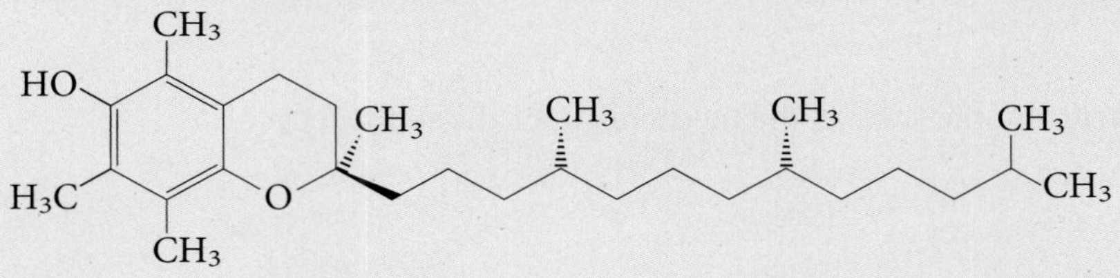

α-tocopherol (α-TCP)

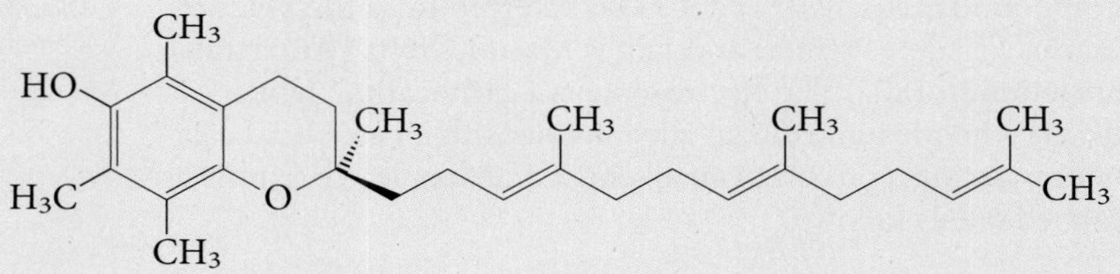

α-tocotrienol (α-TCT)

See Problem 4.42.

For more information, see Sen, C., Khanna, S., and Roy, S., *Vitamin E and Health*, **2004**, *1031*, 127.

4.10 Ring-Activating and Ring-Deactivating Substituents

In this section and the next, we will present experimental evidence that supports the electrophilic aromatic substitution mechanism just described. We will do this by examining how substituents already present on an aromatic ring affect further substitution reactions.

For example, consider the relative nitration rates of the following compounds, all under the same reaction conditions:

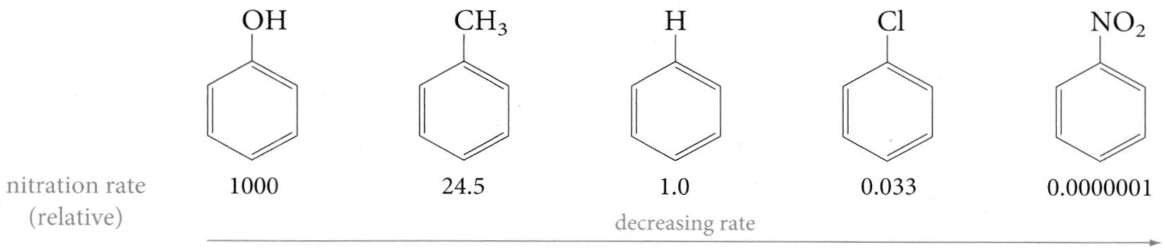

Taking benzene as the standard, we see that some substituents (for example, OH and CH_3) speed up the reaction, and other substituents (Cl and NO_2) retard the reaction. We know from other evidence that hydroxyl and methyl groups are more electron donating than hydrogen, whereas chloro and nitro groups are more electron withdrawing than hydrogen. The electrostatic potential maps in Figure 4.6 illustrate the effects of electron-donating and electron-withdrawing groups on the electron density in the benzene ring.

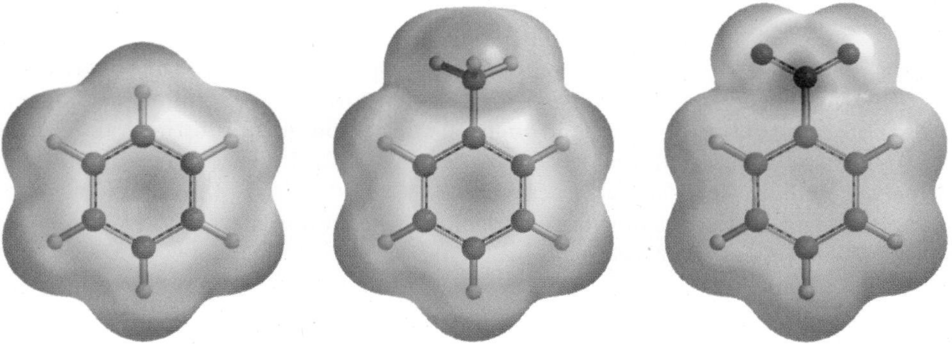

Figure 4.6
Electrostatic potential maps of benzene (left), toluene (middle), and nitrobenzene (right). The pi electron clouds in benzene and toluene are electron-rich (red), while the pi electron cloud in nitrobenzene is electron-deficient (blue).

These observations support the electrophilic mechanism for substitution. If the reaction rate depends on electrophilic (that is, electron-seeking) attack on the aromatic ring, then substituents that donate electrons to the ring will increase its electron density and, hence, speed up the reaction; substituents that withdraw electrons from the ring will decrease electron density in the ring and therefore slow down the reaction. This reactivity pattern is exactly what is observed, not only with nitration, but also with all electrophilic aromatic substitution reactions.

4.11 *Ortho,Para*-Directing and *Meta*-Directing Groups

Substituents already present on an aromatic ring determine the position taken by a new substituent. For example, nitration of toluene gives mainly a mixture of *o*- and *p*-nitrotoluene.

CH_3 (toluene) $\xrightarrow{HONO_2}$ CH_3, NO_2 (*ortho* isomer, bp 222°C, 59%) + CH_3, NO_2 (*para* isomer, bp 238°C, mp 51°C, 37%) (+ 4% *meta* isomer) **(4.24)**

On the other hand, nitration of nitrobenzene under similar conditions gives mainly the *meta* isomer.

NO_2 nitrobenzene $\xrightarrow{HONO_2}$ NO_2, NO_2 *meta* isomer mp 89°C 93% (+ 7% *ortho* isomer) **(4.25)**

This pattern is also followed for other electrophilic aromatic substitutions—chlorination, bromination, sulfonation, and so on. Toluene undergoes mainly *ortho,para* substitution, whereas nitrobenzene undergoes *meta* substitution.

An ***ortho,para*-directing** substituent on a benzene ring directs a second electrophile to positions *ortho* and *para* to it on the ring. A ***meta*-directing** substituent directs a second electrophile to a position *meta* to it.

In general, groups fall into one of two categories. Certain groups are ***ortho,para* directing**, and others are ***meta* directing**. Table 4.1 lists some of the common groups in each category. Let us see how the electrophilic substitution mechanism accounts for the behavior of these two classes of substituents.

4.11.a *Ortho,Para*-Directing Groups

Consider the nitration of toluene. In the first step, the nitronium ion may attack a ring carbon that is *ortho, meta,* or *para* to the methyl group.

Ortho,para attack

CH_3 toluene; *ortho* → [CH_3 H NO_2 ⟷ CH_3 H NO_2 ⟷ CH_3 H NO_2]; *para* → [CH_3 H NO_2 ⟷ CH_3 H NO_2 ⟷ CH_3 H NO_2] **(4.26)**

Meta attack

CH_3 toluene $\xrightarrow{meta}$ [CH_3 H NO_2 ⟷ CH_3 H NO_2 ⟷ CH_3 H NO_2] **(4.27)**

In one of the three resonance contributors to the benzenonium ion intermediate for *ortho* or *para* substitution (shown in blue dashed boxes), the positive charge is on the methyl-bearing carbon. That contributor is a *tertiary* carbocation and more stable than the other contributors, which are secondary carbocations. However, with *meta* attack, *all* of the resonance contributors are secondary carbocations; the positive charge in the intermediate benzenonium ion is never adjacent to the methyl substituent. Therefore, the methyl group is *ortho,para* directing, so that the reaction can proceed via the most stable carbocation intermediate. Similarly, all other alkyl groups are *ortho,para* directing.

Consider now the other *ortho,para*-directing groups listed in Table 4.1. *In each of them, the atom attached to the aromatic ring has an unshared electron pair.*

$$-\ddot{\underset{\cdot\cdot}{F}}: \qquad -\ddot{\underset{\cdot\cdot}{O}}H \qquad -\ddot{N}H_2$$

This unshared electron pair can stabilize an adjacent positive charge. Let us consider, as an example, the bromination of phenol.

Table 4.1 ■ Directing and Activating Effects of Common Functional Groups (Groups are Listed in Decreasing Order of Activation)

	Substituent group	Name of group	
Ortho,Para-Directing	$-\ddot{N}H_2$, $-\ddot{N}HR$, $-\ddot{N}R_2$	amino	Activating
	$-\ddot{\underset{\cdot\cdot}{O}}H$, $-\ddot{\underset{\cdot\cdot}{O}}CH_3$, $-\ddot{\underset{\cdot\cdot}{O}}R$	hydroxy, alkoxy	
	$-\ddot{N}H\overset{O}{\overset{\Vert}{C}}-R$	acylamino	
	$-CH_3$, $-CH_2CH_3$, $-R$	alkyl	
	$-\ddot{\underset{\cdot\cdot}{F}}:$, $-\ddot{\underset{\cdot\cdot}{Cl}}:$, $-\ddot{\underset{\cdot\cdot}{Br}}:$, $-\ddot{\underset{\cdot\cdot}{I}}:$	halo	Deactivating
Meta-Directing	$-\overset{:O:}{\overset{\Vert}{C}}-R$ $-\overset{:O:}{\overset{\Vert}{C}}-\ddot{\underset{\cdot\cdot}{O}}H$	acyl, carboxy	
	$-\overset{:O:}{\overset{\Vert}{C}}-\ddot{N}H_2$ $-\overset{:O:}{\overset{\Vert}{C}}-\ddot{\underset{\cdot\cdot}{O}}R$	carboxamido, carboalkoxy	
	$-\overset{:O:}{\overset{\Vert}{\underset{\underset{:O:}{\Vert}}{S}}}-\ddot{\underset{\cdot\cdot}{O}}H$	sulfonic acid	
	$-C\equiv N:$	cyano	
	$-\overset{+}{N}(=\ddot{O}:)-\ddot{\underset{\cdot\cdot}{O}}:^-$	nitro	

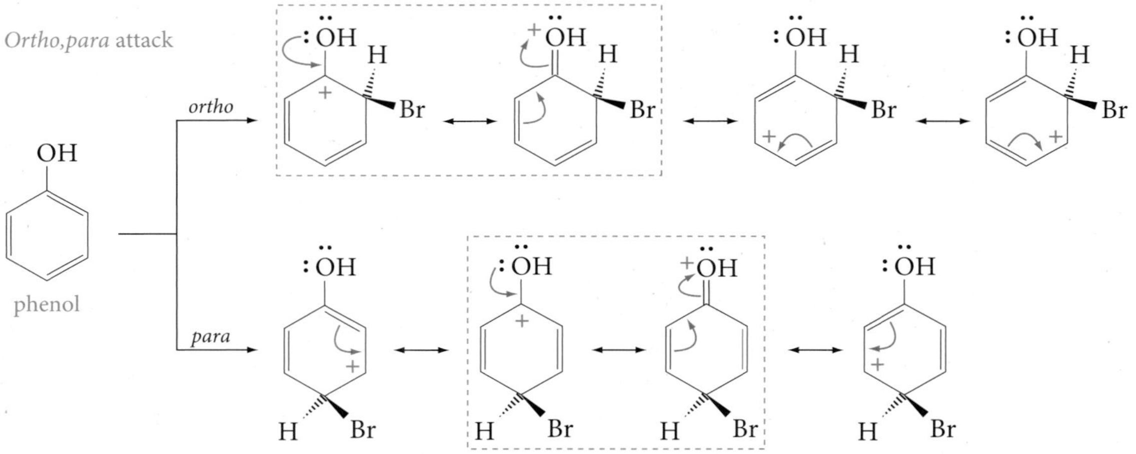

Ortho,para attack

ortho

para

phenol

(4.28)

Meta attack

meta

phenol

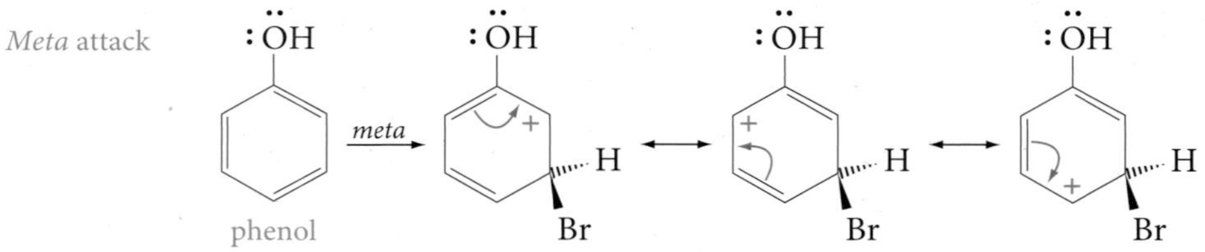

(4.29)

In the case of *ortho* or *para* attack, one of the contributors to the intermediate benzenonium ion places the positive charge on the hydroxyl-bearing carbon. *Shift of an unshared electron pair from the oxygen to the positive carbon allows the positive charge to be delocalized even further, onto the oxygen* (see the structures in the blue dashed boxes). No such structures are possible for *meta* attack. Therefore, the hydroxyl group is *ortho,para* directing.

We can generalize this observation. *All groups with unshared electrons on the atom attached to the ring are* ortho,para *directing.*

PROBLEM 4.14 Draw the important resonance contributors for the benzenonium intermediate in the bromination of aniline, and explain why *ortho,para* substitution predominates.

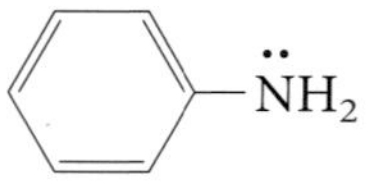

aniline

4.11.b *Meta*-Directing Groups

Now let us examine the nitration of nitrobenzene in the same way, to see if we can explain the *meta*-directing effect of the nitro group. In nitrobenzene, the nitrogen has a formal charge of +1, as shown on the structures. The equations for forming the intermediate benzenonium ion are

Ortho,para attack

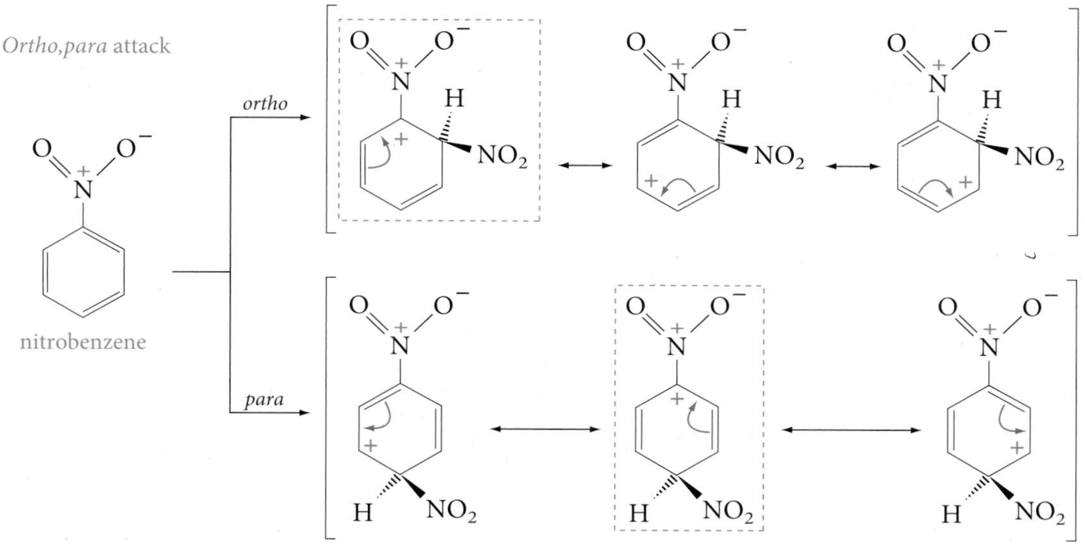

ortho

para

nitrobenzene

(4.30)

Meta attack

meta

nitrobenzene

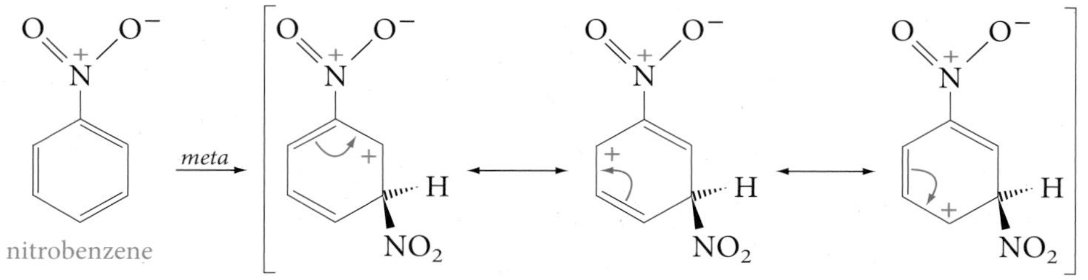

(4.31)

In eq. 4.30, one of the contributors to the resonance hybrid intermediate for *ortho* or *para* substitution (shown in the blue dashed boxes) has *two adjacent positive charges,* a highly *undesirable* arrangement, because like charges repel each other. No such intermediate is present for *meta* substitution (eq. 4.31). For this reason, *meta* substitution is preferred.

Can we generalize this explanation to the other *meta*-directing groups in Table 4.1? Notice that each *meta*-directing group is connected to the aromatic ring by an atom that is part of a double or triple bond, at the other end of which is an atom more electronegative than carbon (for example, an oxygen or nitrogen atom). In such cases, the *atom directly attached to the benzene ring will carry a partial positive charge* (like the nitrogen in the nitro group). This is because of resonance contributors, such as

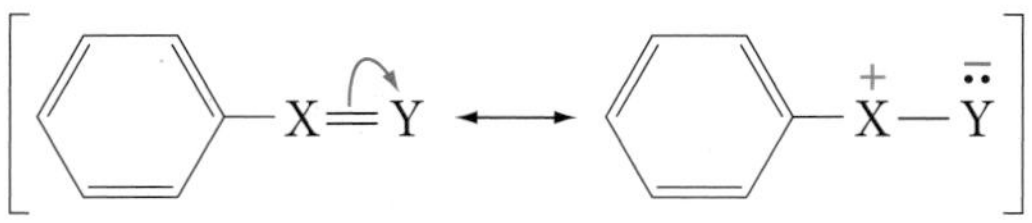

Y is an electron-withdrawing atom such as oxygen or nitrogen; atom X carries a positive charge in one of the resonance contributors.

All such groups will be *meta* directing for the same reason that the nitro group is *meta* directing: to avoid having two adjacent positive charges in the intermediate benzenonium ion. We can generalize as follows: *All groups in which the atom directly attached to the aromatic ring is positively charged or is part of a multiple bond to a more electronegative element will be* meta *directing.*

PROBLEM 4.15 Compare the intermediate benzenonium ions for *ortho, meta* and *para* bromination of benzoic acid, and explain why the main product is *m*-bromobenzoic acid.

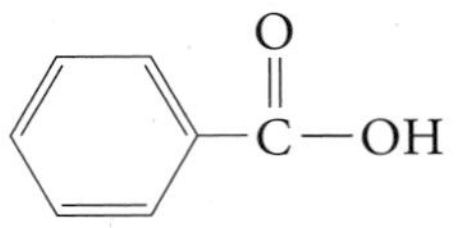

benzoic acid

4.11.c Substituent Effects on Reactivity

A substituent is **activating** if the rate of electrophilic aromatic substitution is faster for the substituted benzene than for unsubstituted benzene. Likewise, a substituent is **deactivating** if the rate of reaction is slower than for benzene.

Substituents not only affect the position of substitution, they also affect the *rate* of substitution, whether it will occur slower or faster than for benzene. A substituent is considered to be **activating** if the rate is faster and **deactivating** if the rate is slower (see Table 4.1) than for benzene. Is this rate effect related to the orientation effect?

In all *meta*-directing groups, the atom connected to the ring carries a full or partial positive charge and will therefore withdraw electrons from the ring. *All* meta-*directing groups are therefore ring-deactivating groups.* On the other hand, ortho,para-*directing groups in general supply electrons to the ring and are therefore ring activating.* With the halogens (F, Cl, Br, and I), two opposing effects bring about the only important exception to these rules. *Because they are strongly electron withdrawing, the halogens are ring deactivating; but because they have unshared electron pairs, they are* ortho,para *directing.*

4.12 The Importance of Directing Effects in Synthesis

When designing a multistep synthesis involving electrophilic aromatic substitution, we must keep in mind the directing and activating effects of the groups involved. Consider, for example, the bromination and nitration of benzene to make bromonitrobenzene. If we brominate first and then nitrate, we will get a mixture of the *ortho* and *para* isomers.

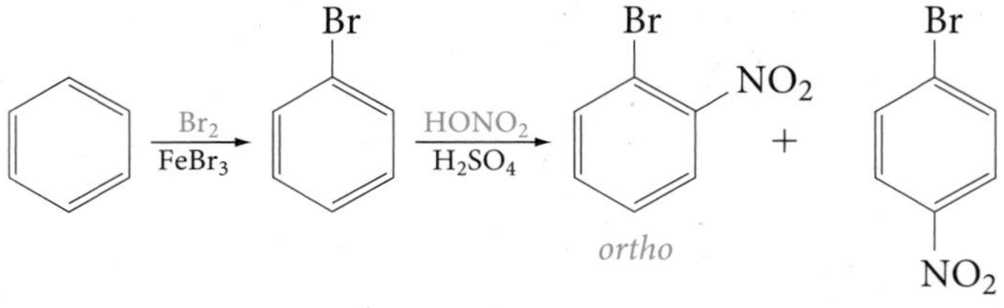

Br_2 / $FeBr_3$; $HONO_2$ / H_2SO_4; Br; Br, NO_2 + Br, NO_2; *ortho*; *para* **(4.32)**

This is because the bromine atom in bromobenzene is *ortho,para* directing. On the other hand, if we nitrate first and then brominate, we will get mainly the *meta isomer* because the nitro group is *meta* directing.

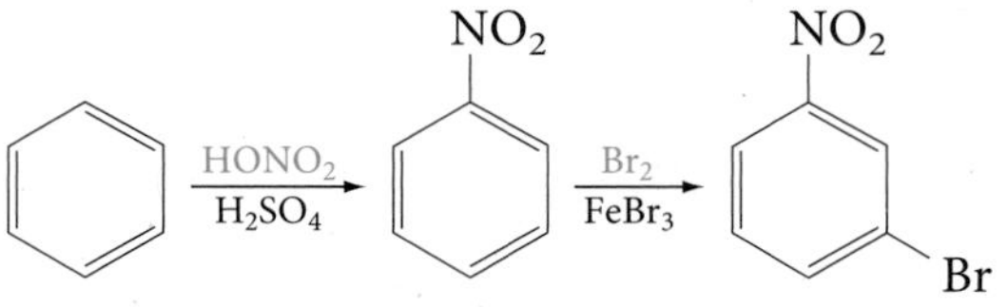

$HONO_2$ / H_2SO_4; NO_2; Br_2 / $FeBr_3$; NO_2, Br; *meta* **(4.33)**

The sequence in which we carry out the reactions of bromination and nitration is therefore very important. It determines which type of product is formed.

PROBLEM 4.16 Devise a synthesis for each of the following, starting with benzene:

a. *m*-bromobenzenesulfonic acid
b. *p*-nitrotoluene

PROBLEM 4.17 Explain why it is *not* possible to prepare *m*-bromochlorobenzene or *p*-nitrobenzenesulfonic acid by carrying out two successive electrophilic aromatic substitutions.

4.13 Polycyclic Aromatic Hydrocarbons

The concept of **aromaticity** is the unusual stability of certain fully conjugated **cyclic** systems.

The concept of **aromaticity**—the unusual stability of certain fully conjugated **cyclic** systems—can be extended well beyond benzene itself or simple substituted benzenes.*

Coke, required in huge quantities for the manufacture of steel, is obtained by heating coal in the absence of air. A by-product of this conversion of coal to coke is a distillate called coal tar, a complex mixture containing many aromatic hydrocarbons (including benzene, toluene, and xylenes). Naphthalene, $C_{10}H_8$, was the *first* pure compound to be obtained from the higher boiling fractions of coal tar. Naphthalene was easily isolated because it sublimes from the tar as a beautiful colorless crystalline solid, mp 80°C. Naphthalene is a planar molecule with two fused benzene rings. The two rings share two carbon atoms.

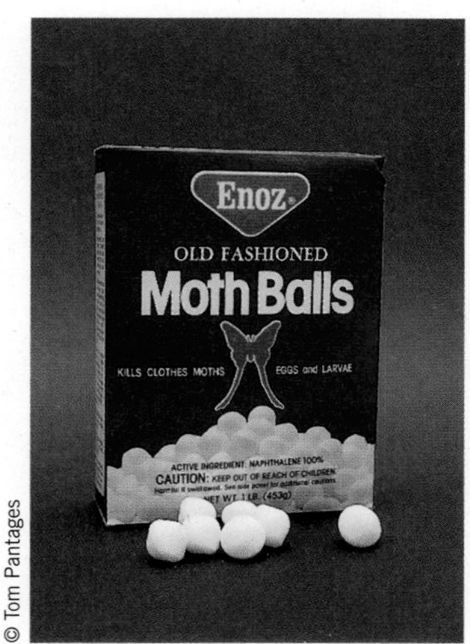

© Tom Pantages

These mothballs are made of naphthalene.

8 1 8a 7 2 6 3 4a 5 4

naphthalene
mp 80°C

1.42 Å 1.36 Å 1.39 Å 1.40 Å

bond lengths in naphthalene

The bond lengths in naphthalene are not all identical, but they all approximate the bond length in benzene (1.39 Å). Although it has two six-membered rings, naphthalene has a resonance energy somewhat less than twice that of benzene, about 60 kcal/mol. Because of its symmetry, naphthalene has three sets of equivalent carbon atoms: C-4a and C-8a; C-1, C-4, C-5, and C-8; and C-2, C-3, C-6, and C-7. Like benzene, naphthalene undergoes electrophilic substitution reactions (halogenation, nitration, and so on), usually under somewhat milder conditions than benzene. Although two monosubstitution products are possible, substitution at C-1 usually predominates.

$$\text{naphthalene} \xrightarrow[50^\circ C]{HNO_3} \text{1-nitronaphthalene} + \text{2-nitronaphthalene} \quad \text{(ratio 10:1)} \tag{4.34}$$

NO_2 NO_2

1-nitronaphthalene 2-nitronaphthalene
(ratio 10:1)

*In general, ring systems containing 4n + 2 pi electrons (n = 0, 1, 2, . . .) in adjacent p orbitals are aromatic. This is known as Hückel's Rule. Although the theoretical basis for this rule is too advanced to be treated in this book, your instructor can provide you with additional information.

A WORD ABOUT... Polycyclic Aromatic Hydrocarbons and Cancer

Certain polycyclic aromatic hydrocarbons (PAHs) are carcinogenic (that is, they produce cancers). They can produce a tumor on mice in a short time when only trace amounts are painted on the skin. These carcinogenic hydrocarbons are present not only in coal tar but also in soot and tobacco smoke and can be formed in barbecuing meat. Their biological effect was noted as long ago as 1775 when soot was identified as the cause of the high incidence of scrotal cancer in chimney sweeps. A similar occurrence of lung and lip cancer is common in habitual smokers.

The way these carcinogens produce cancer is now fairly well understood. Unsubstituted PAHs are not very soluble, so to eliminate hydrocarbons, the body usually oxidizes them to render them more water soluble and then excretes them in urine. The metabolic oxidation products seem to be the real culprits in causing cancer. For example, one of the most potent carcinogens of this type is benzo[a]pyrene. Enzymatic oxidation converts it to the diol-epoxide shown below. The diol-epoxide reacts with cellular DNA, causing mutations that eventually prevent the cells from reproducing normally.

Benzene itself is quite toxic to humans and can cause severe liver damage; toluene, although not harmless, is much less toxic. How can such different behavior of two very similar compounds be possible? To eliminate benzene from the body, the aromatic ring must be oxidized, but intermediates in this oxidation are damaging. However, the *methyl side chain* of toluene can be oxidized to give benzoic acid, which can be excreted. None of the intermediates in this process causes health problems. Conversely, nitro-substituted PAHs, especially of benzo[a]pyrene, are often about 10 times more carcinogenic than their unsubstituted analogues, and studies are currently underway to understand this effect.

Although some chemicals may cause cancer, others can help prevent or cure it. Many substances inhibit cancer growth, and the study of cancer chemotherapy has contributed substantially to human health.

See Problems 4.27, 4.28, and 4.50.

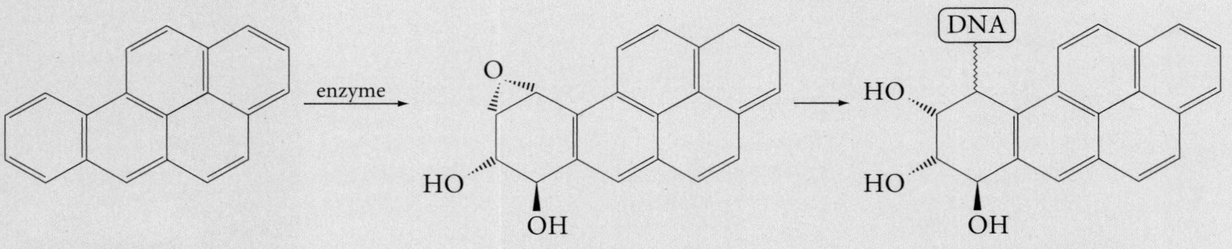

enzyme

O

HO

OH

DNA

HO

HO

OH

benzo[a]pyrene diol-epoxide DNA-adduct

EXAMPLE 4.3

Draw the resonance contributors for the carbocation intermediate in nitration of naphthalene at C-1; include only structures that retain benzenoid aromaticity in the unsubstituted ring.

Solution Four such contributors are possible.

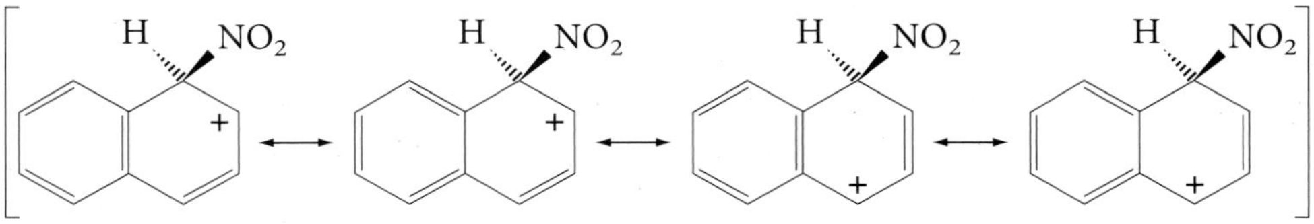

H NO_2 + ⟷ H NO_2 + ⟷ H NO_2 + ⟷ H NO_2 +

PROBLEM 4.18 Repeat Example 4.3 for nitration at C-2. Can you suggest why substitution at C-1 is preferred?

A CLOSER LOOK AT... Polycyclic Aromatic Hydrocarbons

Conduct research on the Internet to find more information on PAHs and to answer the following questions.

Sources of PAHs

1. What are some natural sources of PAHs?
2. What human activities lead to the production of PAHs?
3. What are some ways in which PAHs in the environment can be broken down?

PAHs and Health Concerns

1. Which PAHs are intentionally synthesized, and what are their uses?
2. What are some consumer products that may contain PAHs?
3. What immediate health effects can occur from exposure to PAHs?

Naphthalene is the parent compound of a series of **fused polycyclic hydrocarbons**, a few other examples of which are

Fused polycyclic hydrocarbons contain at least two benzene rings; each ring shares two carbon atoms with at least one other ring.

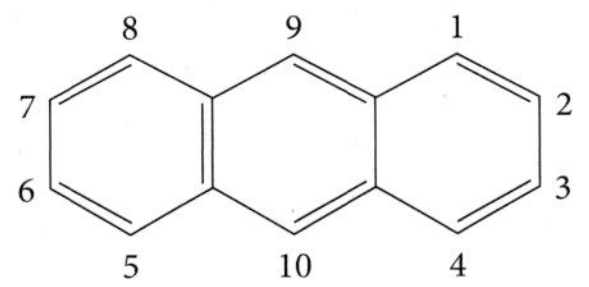

anthracene
mp 217°C

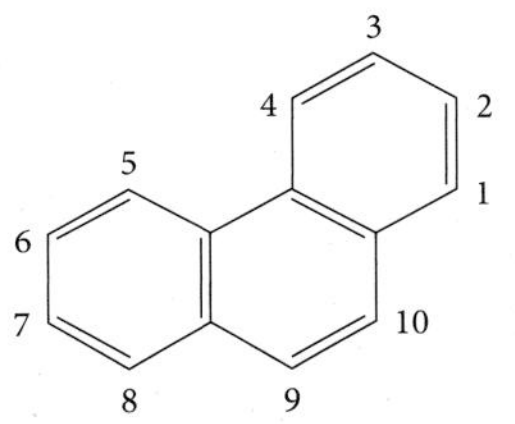

phenanthrene
mp 98°C

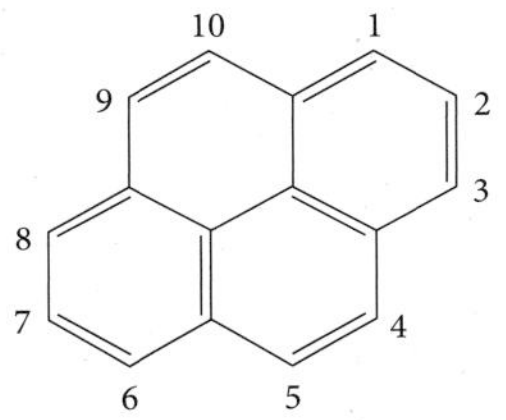

pyrene
mp 156°C

Infinite extension of such rings leads to sheets of hexagonally arranged carbons, the structure of graphite (a form of elemental carbon).

PROBLEM 4.19 Calculate the ratio of carbons to hydrogens in benzene, naphthalene, anthracene, and pyrene. Notice that as the number of fused rings increases, the proportion of carbon also increases so that extrapolation to a very large number of such rings leads to graphite (see "A Word about C_{60}, an Aromatic Sphere: The Fullerenes," pages 138–139).

PAHs comprise a large percentage of the carbon found in interstellar space. They have even been observed in interstellar ice (Halley's comet, for example). It has been shown that ultraviolet irradiation of PAHs in ice produces aromatic ketones (Chapter 9), alcohols (Chapter 7), and other compounds, suggesting a role of PAHs in prebiotic chemistry (see "A Word about Methane, Marsh Gas, and Miller's Experiment," page 60).

A WORD ABOUT... C_{60}, an Aromatic Sphere: The Fullerenes

No recent chemical discovery has been as dramatic nor stimulated such an explosive burst of research activity as that of C_{60}. Until the mid-1980s, only two major allotropic forms of elemental carbon were common: *diamond* (page 51) and *graphite.* But in September 1985, scientists interested in small carbon fragments blown out of giant red stars in interstellar space were trying to produce these same particles here on earth for closer study. To do so, they subjected graphite to a high-energy pulsed laser beam, and passed the vaporized fragments into a mass spectrometer for analysis. Although they did find some of the small fragments they sought, they also saw, to their great surprise, an exceptionally intense mass spectral peak at the very high mass of 720, corresponding to C_{60} ($C = 12 \times 60 = 720$). Certain other less-intense high-mass peaks were also observed (for example, C_{70}, at mass 840). They proposed a unique structure for C_{60} to account for its remarkable stability; it turned out to be correct, and five years later, methods were developed for producing this new form of carbon in quantities sufficient to study its chemistry. What is C_{60} and why is it (and not C_{59} or C_{61}) so stable?

First, let us consider how the carbon atoms are arranged in graphite, C_{60}'s precursor. Graphite consists of layers of planar hexagonal carbon rings. Each carbon atom (except for the few at the outer edges) is connected to three other carbon atoms in the same layer, by bonds that are approximately the same length (1.42 Å) as the carbon–carbon bonds in benzene (1.39 Å).

In a very real sense, then, the sheets in graphite are like an infinite number of fused aromatic rings. There are no covalent bonds between carbon atoms in different layers, which are 3.4 Å apart. Only weak forces hold the layers together. The lubricant properties of graphite are thought to result from sliding of the layers with respect to each other.

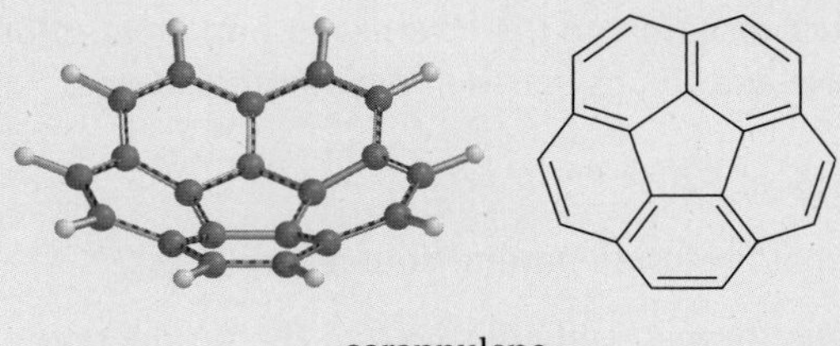

corannulene

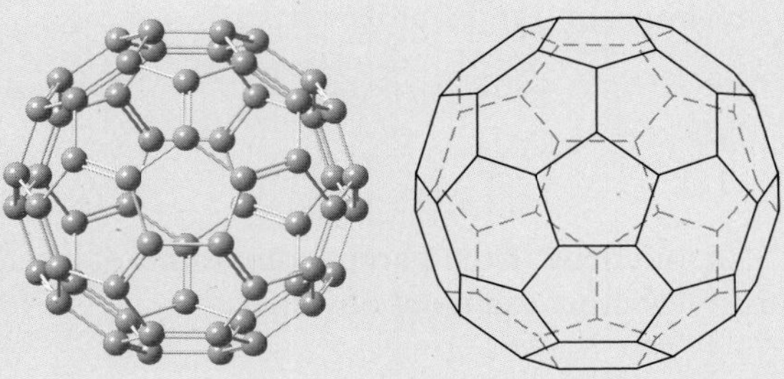

C_{60} (the pi bonds are not shown)

Consider what happens if a carbon atom is expelled from one of the rings, reducing one ring from a hexagon to a pentagon. *The structure will no longer be flat, but will curve!* The known aromatic hydrocarbon **corannulene**, for example, with five benzene rings around a pentagon, is saucer shaped, not flat (see ball-and-stick model). Ejection of additional carbon atoms from the graphite layers leads to increasing curvature and eventual closure to spherelike structures. Although the relationship between graphite and C_{60} is clear, the details of how C_{60} is formed from vaporized graphite are still being investigated.*

The structure originally guessed at and later verified for C_{60} is the same as for a soccer ball. It is a polygon

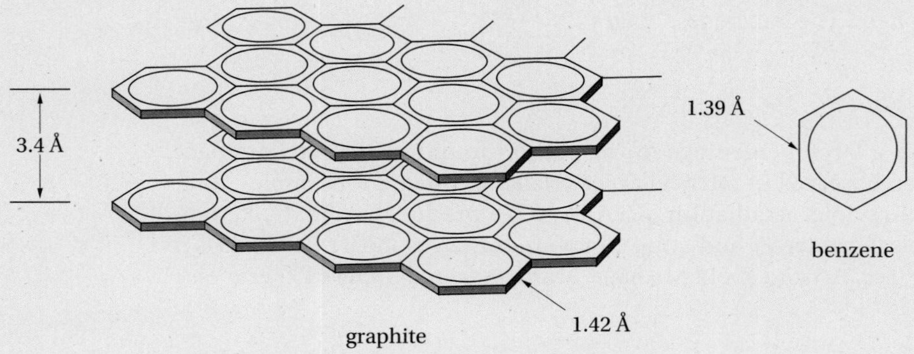

graphite

benzene

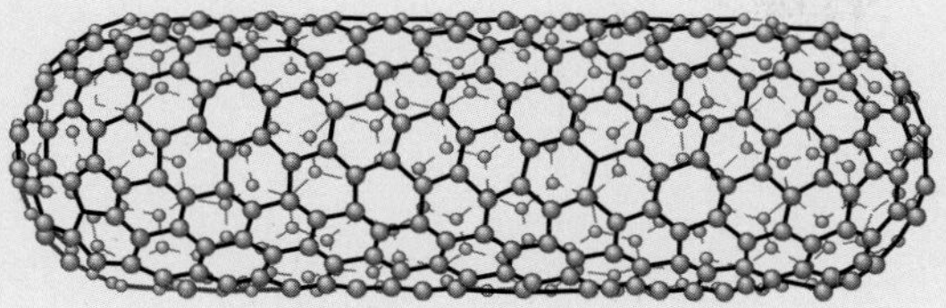

Carbon nanotube[1]

Fullerene derivative used in a photovoltaic device[2]

(technically, a truncated icosahedron) with 60 vertices, one carbon at each. There are 32 faces, 12 of which are *pentagons* (12 pentagons × 5 carbons each = 60 carbons) and 20 of which are hexagons (20 hexagons × 6 carbons each = 120 carbons ÷ 2 = 60 carbons; we must divide by 2 because each carbon atom is shared by *two* hexagons). Each pentagon is surrounded by five hexagons—no two pentagons are adjacent. However, the six bonds of each hexagon are fused alternately to three pentagons and three hexagons.

C_{60} was trivially named "buckminsterfullerene" after R. Buckminster Fuller, the engineer, architect, and philosopher who used similar shapes to construct geodesic domes. Due to their spherical shape, C_{60} molecules are sometimes colloquially called "buckyballs." Compounds derived from C_{60} and related carbon clusters that enclose space (such as C_{70}, which is egg shaped, C_{76}, C_{84}, and others) are called "fullerenes." Even carbon "nanotubes" are now known; some carbon nanotubes are even being used as "storage containers" to deliver anticancer drugs into cells for chemotherapy applications.

Why is C_{60} so stable? As with benzene and graphite, each carbon atom in C_{60} is attached to three other atoms and therefore is sp^2-hybridized. The fourth valence electron of each carbon lies in a *p* orbital that is perpendicular to the spherical surface. These orbitals overlap to form a pi cloud *outside* and *inside* the sphere, like the pi cloud above and below the plane of a benzene ring. Hence the structure is in a sense aromatic and exceptionally stable. Indeed, C_{60} was recently detected in certain black, lustrous rocks where it may have been trapped more than half a billion years ago!

C_{60} can now be made in gram quantities. It is extracted by organic solvents (in which it is sparingly soluble) from a specially prepared soot, and is separated chromatographically from higher fullerenes that are simultaneously formed. Its solutions are beautifully colored, for example, magenta in hexane.

What about its reactions? Since substitution is impossible (there are no hydrogens in C_{60}), it reacts mainly by addition. Although all 60 carbons are equivalent (just like the six carbons in benzene), the bonds are not. C_{60} has two types of bonds, the 6–6 bonds shared by adjacent hexagons, and the 5–6 bonds shared by a pentagon and a hexagon. The 6–6 bonds are a bit shorter (1.39 Å) than the 5–6 bonds (1.43 Å) and are more like double bonds. Therefore, most additions to C_{60} occur across a 6–6 bond. Multiple additions are possible.

We have lessons to learn from the serendipitous and entirely unanticipated discovery of C_{60}. This advance, which opened up a whole new field of chemistry, was the unexpected result of studies in fundamental science. Yet new types of polymers, materials that conduct electricity or store energy from sunlight, fascinating structures with metals or other atoms (even helium) trapped inside carbon clusters, new catalysts, new probes for electron microscopy, and even pharmaceuticals—all of these and other commercial possibilities not yet imagined—are emerging from this exciting discovery. The C_{60} story illustrates once more why it is so important, in a technological world, to support research in the fundamental sciences. Where the research will lead cannot be predicted with certainty, but experience shows that the eventual practical benefits that follow, even from only a small fraction of fundamental discoveries, compensate many times over for the initial investment. (Richard E. Smalley, Robert F. Curl, and Harold Kroto shared the 1996 Nobel Prize in chemistry for their discovery of fullerenes.)

See Problem 4.48.

---

*The first rational synthesis of C_{60} has been described by Lawrence T. Scott and his coworkers: *Science*, **2002**, *295*, 1500. This development may provide C_{60} derivatives that could not be prepared from C_{60} itself.

[1]The Lawrence Berkeley National Laboratory.

[2]Reprinted with permission from *Chemical and Engineering News*, April 12, 1999.

REACTION SUMMARY

1. Catalytic Hydrogenation (Sec. 4.7)

benzene + $3\ H_2$ —(metal catalyst)→ cyclohexane

cyclohexane

2. Electrophilic Aromatic Substitution (Sec. 4.8)

a. Halogenation

benzene + X_2 —(FeX_3)→ C_6H_5X + HX

X = Cl, Br

b. Nitration

benzene + $HONO_2$* —(H_2SO_4)→ $C_6H_5NO_2$ + H_2O

c. Sulfonation

benzene + $HOSO_3H$ —(SO_3, heat)→ $C_6H_5SO_3H$ + H_2O

d. Alkylation (Friedel–Crafts)

benzene + RCl —($AlCl_3$)→ C_6H_5R + HCl

R = alkyl group

e. Alkylation

benzene + $CH_2{=}CH_2$ —(H^+)→ $C_6H_5CH_2CH_3$

*Nitric acid, HNO_3

f. **Acylation (Friedel–Crafts)**

$$\text{C}_6\text{H}_6 + \text{R}-\overset{\text{O}}{\overset{\|}{\text{C}}}-\text{Cl} \xrightarrow{\text{AlCl}_3} \text{C}_6\text{H}_5-\overset{\text{O}}{\overset{\|}{\text{C}}}-\text{R} + \text{HCl}$$

MECHANISM SUMMARY

Electrophilic Aromatic Substitution (Sec. 4.9)

$$\text{C}_6\text{H}_6 + \text{E}^+ \rightleftharpoons \text{benzenonium ion} \xrightarrow{-\text{H}^+} \text{C}_6\text{H}_5\text{E}$$

benzenonium ion

ADDITIONAL PROBLEMS

OWL Interactive versions of these problems are assignable in OWL.

Aromatic Compounds: Nomenclature and Structural Formulas

4.20 Write structural formulas for the following compounds:

a. *p*-chlorostyrene
b. 2,3,5-trifluoroaniline
c. *o*-chlorophenol
d. allylbenzene
e. *p*-isopropylphenol
f. *p*-dimethylbenzene
g. 1,3-diphenylhexane
h. 2-iodo-4-methyl-3-nitroanisole
i. *p*-chlorobenzenesulfonic acid
j. *m*-nitrophenol
k. benzyl bromide
l. 2,4,6-triethyltoluene
m. *m*-bromocumene
n. *m*-chlorobenzaldehyde
o. 1,4-dicyclohexylbenzene
p. *o*-methylacetophenone

4.21 Name the following compounds:

a.

b. CH=O, Br

c. Cl Cl

d. Cl, CH_3, Cl

e.

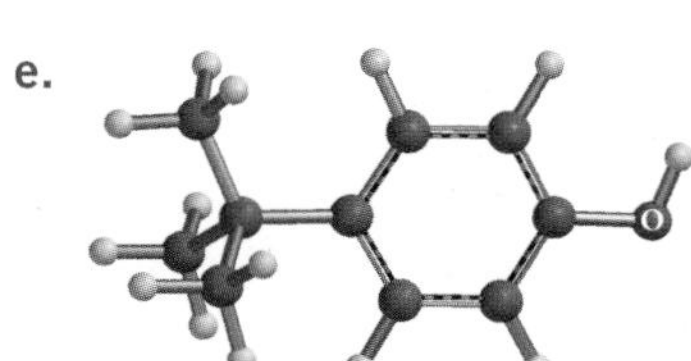

f. CH_3, NO_2

g. F, F, F, F, F, F

h. $CH{=}CH_2$, Br, Br

i.

4.22 Give the structures and names for all possible

a. trimethylbenzenes
b. dibromobenzoic acids
c. dinitroanisoles
d. dichloronitrobenzenes

4.23 There are three dibromobenzenes (*o*-, *m*-, and *p*-). Suppose we have samples of each in separate bottles, but we don't know which is which. Let us call them A, B, and C. On nitration, compound A (mp 87°C) gives only *one* nitrodibromobenzene. What is the structure of A? B and C are both liquids. On nitration, B gives *two* nitrodibromobenzenes, and C gives *three* nitrodibromobenzenes (of course, not in equal amounts). What are the structures of B and C? Of their mononitration products? (This method, known as Körner's method, was used years ago to assign structures to isomeric benzene derivatives.) (*Hint:* Start by drawing the structures of the three dibromobenzenes. Then figure out where they can be nitrated.)

4.24 Give the structure and name of each of the following aromatic hydrocarbons:

a. C_8H_{10}; has two possible ring-substituted monobromo derivatives
b. C_9H_{12}; can give only one mononitro product after nitration
c. C_9H_{12}; can give four mononitro derivatives after nitration

Aromaticity and Resonance

4.25 The observed amount of heat evolved when 1,3,5,7-cyclooctatetraene is hydrogenated is 110 kcal/mol. What does this tell you about the possible resonance energy of this compound?

4.26 The structure of the nitro group ($—NO_2$) is usually shown as $—\overset{+}{N}(=O)—O^-$, yet experiments show that the two nitrogen–oxygen bonds have the same length of 1.21 Å. This length is intermediate between 1.36 Å for the N—O single bond and 1.18 Å for the N═O double bond. Draw structural formulas that explain this observation.

4.27 Draw all reasonable resonance structures for naphthalene ($C_{10}H_8$, Sec. 4.13) and rationalize the different bond lengths for the C—C bonds.

4.28 When naphthalene is hydrogenated, the heat released is about 80 kcal/mol. Using an isolated cyclohexene unit (Figure 4.3) for comparison, estimate the resonance energy of naphthalene. Why is the resonance energy of naphthalene less than twice that of benzene? (*Hint:* Look at your answers to Problem 4.27.)

4.29 Draw all reasonable electron-dot formulas for the nitronium ion (NO_2^+), the electrophile in aromatic nitrations. Show any formal charges. Which structure is favored and why? (See Secs. 1.11 and 1.12.)

Mechanism of Electrophilic Aromatic Substitution

4.30 Write out all steps in the mechanism for the reaction of

a. $H_3C—C_6H_4—CH_3$ + nitric acid (H_2SO_4 catalyst)
b. $C_6H_5—CH_3$ + *t*-butyl chloride + $AlCl_3$

4.31 Draw all possible contributing structures to the carbocation intermediate in the chlorination of chlorobenzene. Explain why the major products are *o*- and *p*-dichlorobenzene. (*Note: p*-Dichlorobenzene is produced commercially this way, for use against clothes moths.)

4.32 Repeat Problem 4.31 for the chlorination of acetophenone, and explain why the product is *m*-chloroacetophenone.

4.33 When benzene is treated with propene and sulfuric acid (see structures of reactants below), two different monoalkylation products are possible. Draw their structures. Which one do you expect to be the major product? Why? (*Hint:* See eq. 3.21, p. 84.)

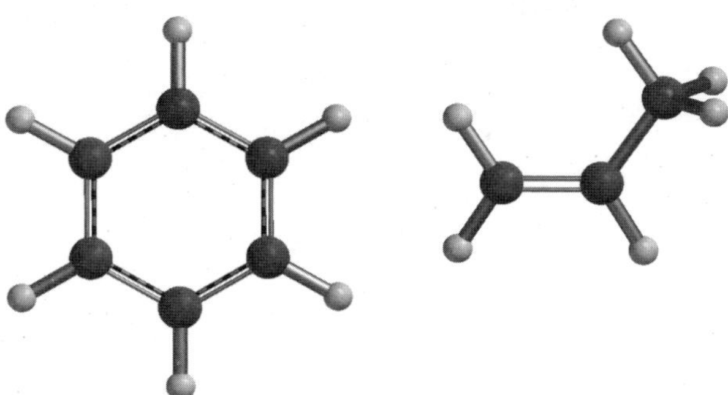

4.34 In aromatic chlorinations, we use $FeCl_3$ with Cl_2 as the reagents, and for aromatic brominations, we use $FeBr_3$ with Br_2. Suggest a reason as to why the iron halide (FeX_3) always has the same halide (X) as the halogenating agent (X_2).

4.35 When benzene is treated with excess D_2SO_4 at room temperature, the hydrogens on the benzene ring are gradually replaced by deuterium. Write a mechanism that explains this observation. (*Hint:* D_2SO_4 is a form of the acid H_2SO_4 in which deuterium has been substituted for hydrogen.)

4.36 Draw a molecular orbital picture for the resonance hybrid benzenonium ion shown in eq. 4.16, and describe the hybridization of each ring carbon atom. (*Hint:* Examine the molecular orbital structure of benzene in Figure 4.2.)

Reactions of Substituted Benzenes: Activating and Directing Effects

4.37 Predict whether the following substituents on the benzene ring are likely to be *ortho,para* directing or *meta* directing and whether they are likely to be ring activating or ring deactivating:

a. $-\overset{+}{N}H(CH_3)_2$
b. $-\overset{\overset{\displaystyle O}{\|}}{C}-OCH_3$
c. $-SCH_3$
d. $-N{=}O$
e. $-C{\equiv}N$
f. $-Br$
g. $-OCH(CH_3)_2$

4.38 For each of the monosubstituted benzenes shown below,

(1) indicate whether the substituent is *ortho,para* directing or *meta* directing.
(2) draw the structure of the main *mono*substitution product for each of the reactions indicated.

a. OCH_3 (benzene ring) + chlorine (Fe catalyst)

b. NO_2 (benzene ring) + bromine (Fe catalyst)

c. CH_3 (benzene ring) + bromine (Fe catalyst)

d. Br (benzene ring) + concentrated sulfuric acid (heat) + SO_3

e. 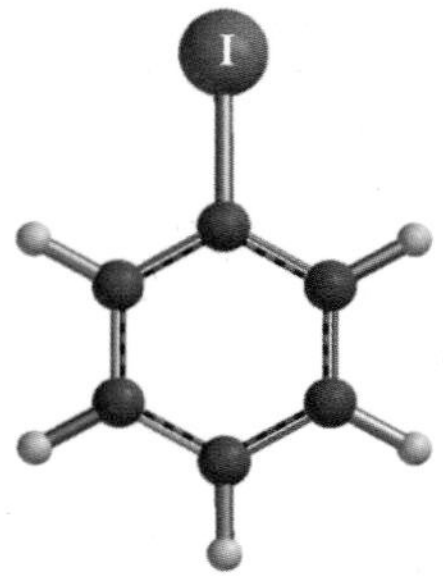 + chlorine (Fe catalyst)

f. 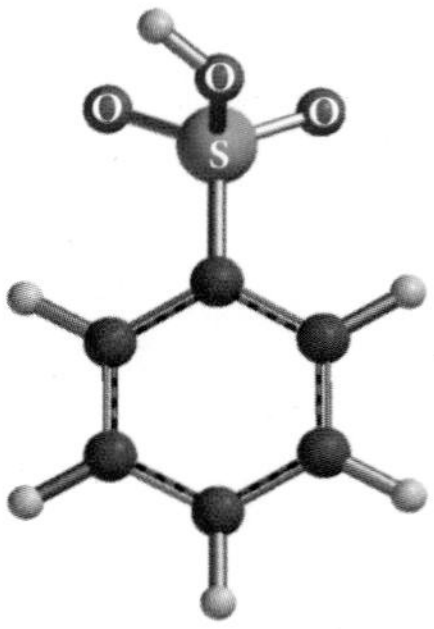 + concentrated nitric acid (heat) (H_2SO_4 catalyst)

g. 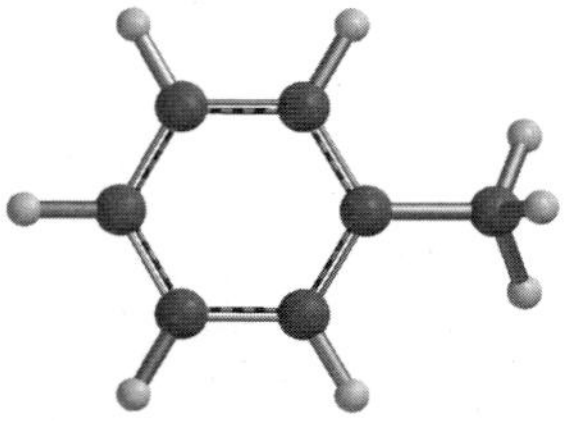 + acetyl chloride ($AlCl_3$ catalyst)

h. 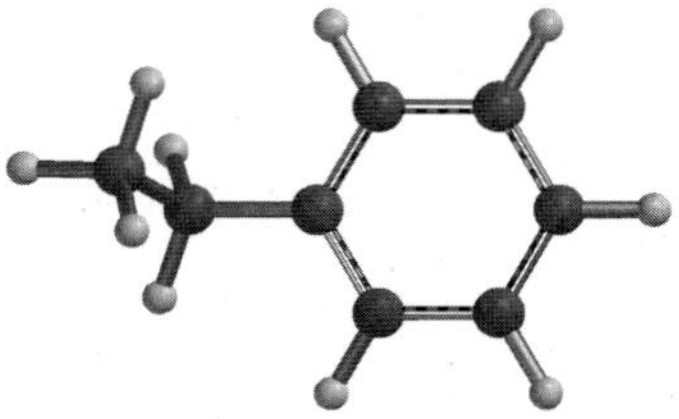 + concentrated nitric acid (H_2SO_4 catalyst)

4.39 When toluene ($C_6H_5CH_3$) is treated with 2-methyl-2-butene and sulfuric acid, two products are observed. What are the structures of the two products? Provide a mechanism for the formation of each product.

4.40 Which compound is more reactive toward electrophilic substitution (for example, nitration)?

a. 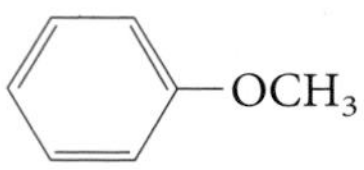 C_6H_5–OCH_3 or C_6H_5–C(=O)–OH   b. 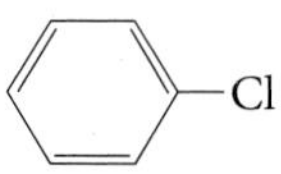 C_6H_5–Cl or C_6H_5–CH_2CH_3

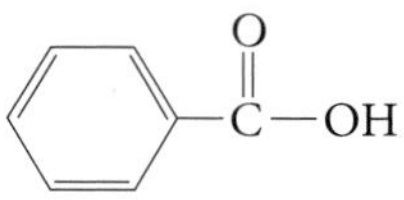

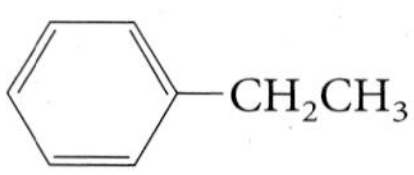

4.41 The explosive TNT (2,4,6-trinitrotoluene) can be made by nitrating toluene with a mixture of nitric and sulfuric acids, but the reaction conditions must gradually be made more severe as the nitration proceeds. Explain why.

4.42 As discussed in "A Word About . . . Vitamin E—Tocopherols and Tocotrienols" on page 128, these compounds can be obtained from different sources. How would you convert α-tocotrienol to α-tocopherol?

Electrophilic Aromatic Substitution Reactions in Synthesis

4.43 For the compounds named below,

(1) draw the structure of each compound.
(2) using benzene or toluene as the only aromatic starting material, devise a synthesis of each compound.

a. *p*-bromotoluene
b. *p*-nitroethylbenzene
c. *p*-bromonitrobenzene
d. *p*-isopropylbenzenesulfonic acid
e. *tert*-butylcyclohexane

4.44 Using benzene or toluene as the only aromatic organic starting material, devise a synthesis for each of the following compounds. Name the product.

a. 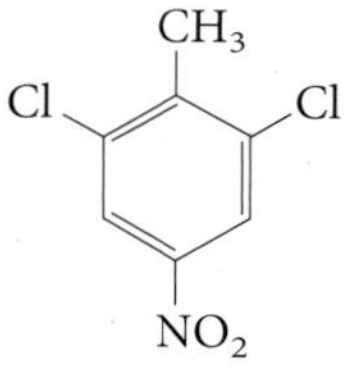 ring with CH_3, Cl, Cl, NO_2
b. 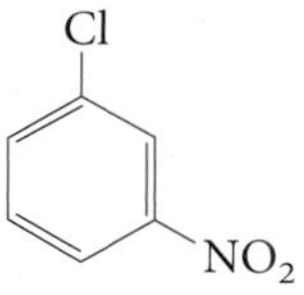 ring with Cl, NO_2
c. 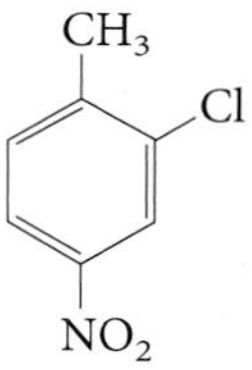 ring with CH_3, Cl, NO_2
d. 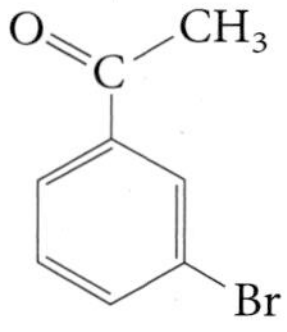 ring with C(=O)CH_3, Br

4.45 For a one-step synthesis of 3-bromo-5-nitrobenzoic acid, which is the better starting material: 3-bromobenzoic acid or 3-nitrobenzoic acid? Why?

3-bromo-5-nitrobenzoic acid

4.46 Vinclozolin (shown below) is a fungicide and has been reported to cause developmental defects in male fertility.

vinclozolin

A reasonable intermediate in the synthesis of vinclozolin is 1-chloro-3-nitrobenzene. Starting from benzene, suggest a reasonable synthesis of 1-chloro-3-nitrobenzene.

4.47 Show how pure 3,5-dinitrochlorobenzene can be prepared, starting from a disubstituted benzene.

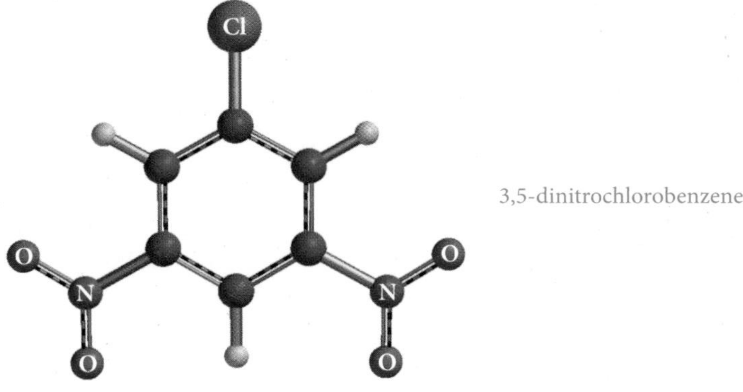

3,5-dinitrochlorobenzene

Polycyclic Aromatic Hydrocarbons

4.48 In the "A Word About . . . C_{60}, An Aromatic Sphere: The Fullerenes" on pages 138–139, corannulene was described as a nonplanar, curved, aromatic hydrocarbon. If corannulene would undergo dibromination, how many unique isomers could result? Show their structures.

4.49 How many possible monosubstitution products are there for each of the following? (See page 137 for structures.)

a. anthracene   b. phenanthrene

4.50 In the "A Word About . . . Polycyclic Aromatic Hydrocarbons and Cancer" on page 136, it was noted that some polycyclic aromatic hydrocarbons are used to treat cancer. Indeed, daunorubicin (below) is a member of the anthracycline family, and these compounds are some of the most effective chemotherapy tools used in anticancer treatment. A compound like *m*-hydroxybenzaldehyde could be considered as an intermediate in the synthesis of daunorubicin. Using benzaldehyde as a starting reagent, suggest a synthesis of *m*-hydroxybenzaldehyde.

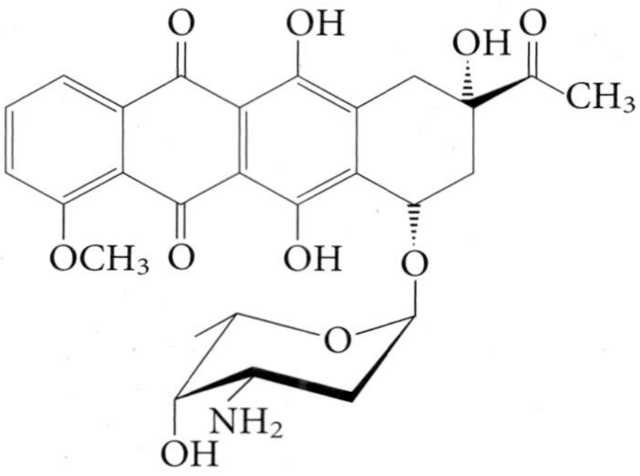

O OH OH O CH3 OCH3 O OH O O NH2 OH

daunorubicin

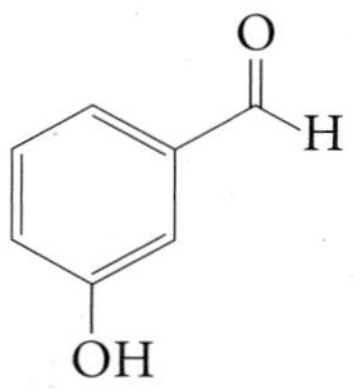

O H OH

m-hydroxybenzaldehyde

4.51 Bromination of anthracene gives mainly 9-bromoanthracene. Write out the steps in the mechanism of this reaction.

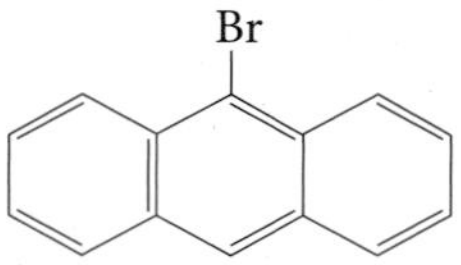

Br

9-bromoanthracene

The odors of caraway seeds (left) and mint leaves (right) arise from the enantiomers of carvone, which differ in arrangement of the atoms attached to the indicated (*) carbon.

© PhotoDisc, Inc.   Okapia/Photo Researchers

5

Stereoisomerism

Stereoisomers have the same order of attachment of the atoms, but different arrangements of the atoms in space. The differences between stereoisomers are more subtle than those between structural isomers. Yet stereoisomerism is responsible for significant differences in chemical properties of molecules. The effectiveness of a drug often depends on which stereoisomer is used, as does the presence or absence of side effects (see "A Word About . . . Enantiomers and Biological Activity," pp. 172–173). The chemistry of life itself is affected by the natural predominance of particular stereoisomers in biological molecules such as carbohydrates (Chapter 16), amino acids (Chapter 17), and nucleic acids (Chapter 18).

We have already seen that stereoisomers can be characterized according to the ease with which they can be interconverted (see Sec. 2.11 and Figure 2.8). That is, they may be **conformers**, which can be interconverted by rotation about a single bond, or they may be **configurational isomers**, which can be interconverted only by breaking and remaking covalent bonds. Here we will consider other useful ways to categorize stereoisomers, ways that are particularly helpful in describing their properties.

Online homework for this chapter can be assigned in OWL, an online homework assessment tool.

Stereoisomers have the same order of attachment of atoms but different spatial arrangements of atoms. **Conformers** and **configurational isomers** are two classes of stereoisomers (Sec. 2.11).

5.1 Chirality and Enantiomers

Consider the difference between a pair of gloves and a pair of socks. A sock, like its partner, can be worn on either the left or the right foot. But a left-hand glove, unlike its partner, cannot be worn on the right hand. Like a pair of gloves, certain molecules possess this property of "handedness," which affects their chemical behavior. Let us examine the idea of molecular handedness.

Chiral molecules possess the property of handedness, whereas **achiral** molecules do not.

A molecule (or object) is either **chiral** or **achiral**. The word *chiral*, pronounced "kairal" to rhyme with spiral, comes from the Greek *χειρ* (*cheir*, hand). A chiral molecule (or object) is one that exhibits the property of handedness. An achiral molecule does not have this property.

What test can we apply to tell whether a molecule (or object) is chiral or achiral? We examine the molecule (or object) *and its mirror image. The mirror image of a chiral molecule cannot be superimposed on the molecule itself. The mirror image of an achiral molecule, however, is identical to or superimposable on the molecule itself.*

Let us apply this test to some specific examples. Figure 5.1 shows one of the more obvious examples. The mirror image of a left hand is not another left hand, but a right hand. A hand and its mirror image are *not* superimposable. A hand is chiral. But the mirror image of a ball (sphere) is also a ball (sphere), so a ball (sphere) is achiral.

PROBLEM 5.1 Which of the following objects are chiral and which are achiral?

a. golf club b. teacup c. football d. corkscrew
e. tennis racket f. shoe g. portrait h. pencil

▰ **Figure 5.1**
The mirror-image relationships of chiral and achiral objects.

The mirror image of a left hand is not a left hand, but a right hand.

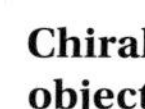
Chiral object

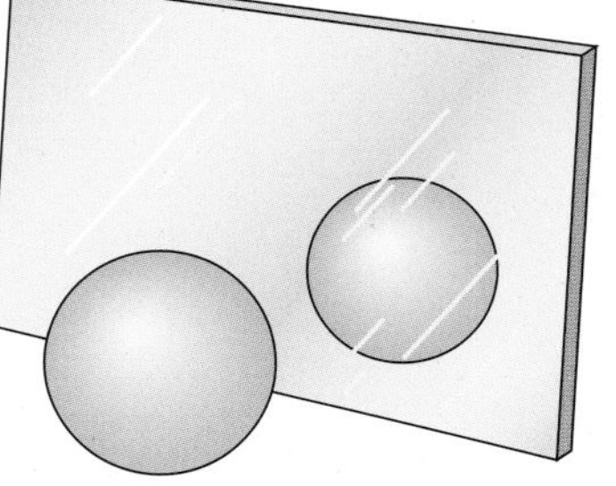
The mirror image of a ball is identical with the object itself.

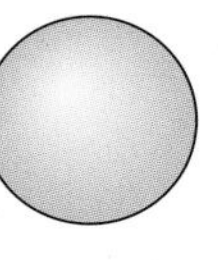
Achiral object

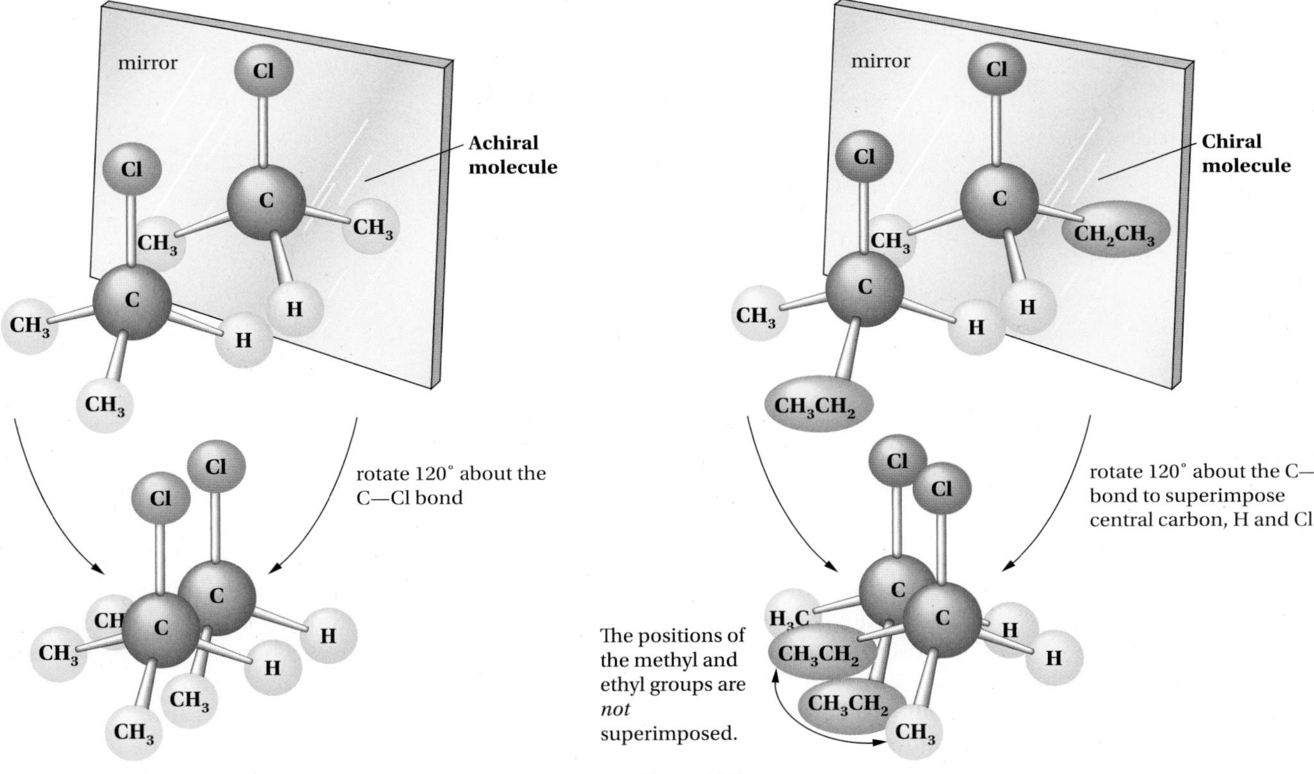

■ **Figure 5.2**
Model of 2-chloropropane and its mirror image. The mirror image is superimposable on the original molecule.

■ **Figure 5.3**
Model of 2-chlorobutane and its mirror image. The mirror image is *not* superimposable on the original molecule. The two forms of 2-chlorobutane are enantiomers.

Now let us look at two molecules, 2-chloropropane and 2-chlorobutane, and their mirror images.* Figure 5.2 shows that 2-chloropropane is achiral. Its mirror image is superimposable on the molecule itself. Therefore 2-chloropropane has only one possible structure.

On the other hand, as Figure 5.3 shows, 2-chlorobutane has two possible structures, related to one another as nonsuperimposable mirror images. We call a pair of molecules that are related as nonsuperimposable mirror images **enantiomers**. Every molecule, of course, has a mirror image. Only those that are *nonsuperimposable* are called enantiomers.

Enantiomers are a pair of molecules related as nonsuperimposable mirror images.

5.2 Stereogenic Centers; the Stereogenic Carbon Atom

What is it about their structures that leads to chirality in 2-chlorobutane but not in 2-chloropropane? Notice that, in 2-chlorobutane, carbon atom 2, the one marked with an asterisk, has four different groups attached to it (Cl, H, CH_3, and CH_2CH_3). A carbon atom with four different groups attached to it is called a **stereogenic carbon atom**. This type of carbon is also called a **stereogenic center** because it gives rise to stereoisomers.

A **stereogenic carbon atom** or **stereogenic center** is a carbon atom with four different groups attached to it.

$$CH_3-\overset{\overset{\large Cl}{|}}{\underset{\underset{\large H}{|}}{\overset{*}{C}}}-CH_2CH_3$$

*Build 3D models of these molecules to visualize them better. In general, using 3D models while reading this chapter will help you understand the concepts.

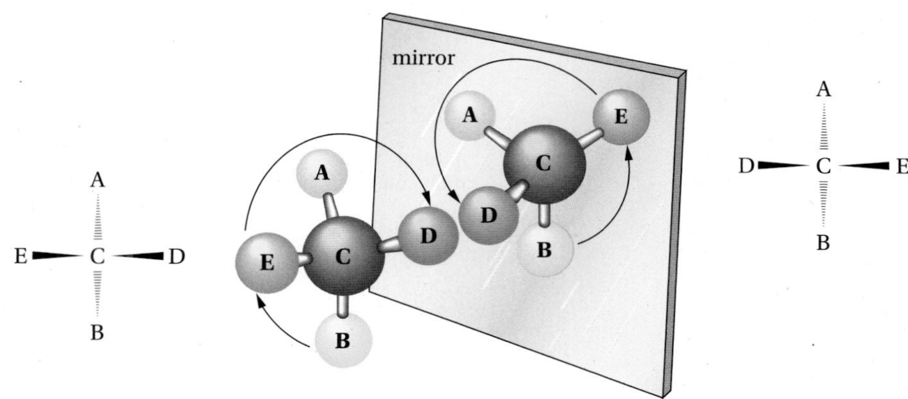

Figure 5.4
The chirality of enantiomers. Looking down the C—A bond, we have to read clockwise to spell BED for the model on the left, but we must read counterclockwise for its mirror image.

Let us examine the more general case of a carbon atom with any four different groups attached; let us call the groups A, B, D, and E. Figure 5.4 shows such a molecule and its mirror image. That the molecules on each side of the mirror in Figure 5.4 are nonsuperimposable mirror images (enantiomers) becomes clear by examining Figure 5.5. (We strongly urge you to use molecular models when studying this chapter. It is sometimes difficult to visualize three-dimensional structures when they are drawn on a two-dimensional surface [this page or a blackboard], though with experience, your ability to do so will improve.) The handedness of these molecules is also illustrated in Figure 5.4, where the clockwise or counterclockwise arrangement of the groups (we might call them right- or left-handed arrangements) is apparent.

What happens when all four of the groups attached to the central carbon atom are *not* different from one another? Suppose two of the groups are identical—say, A, A, B, and D. Figure 5.6 describes this situation. The molecule and its mirror image are now *identical,* and the molecule is achiral. This is exactly the situation with 2-chloropropane, where two of the four groups attached to carbon-2 are identical (CH_3, CH_3, H, and Cl).

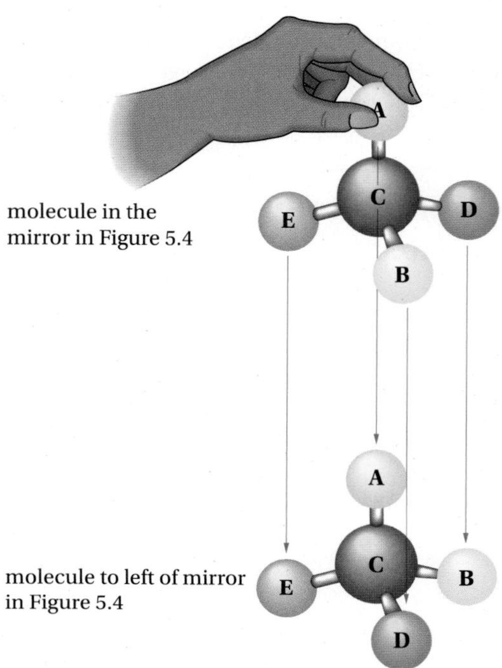

Figure 5.5
When the four different groups attached to a stereogenic carbon atom are arranged to form mirror images, the molecules are not superimposable. The models may be twisted or turned in any direction, but as long as no bonds are broken, only two of the four attached groups can be made to coincide.

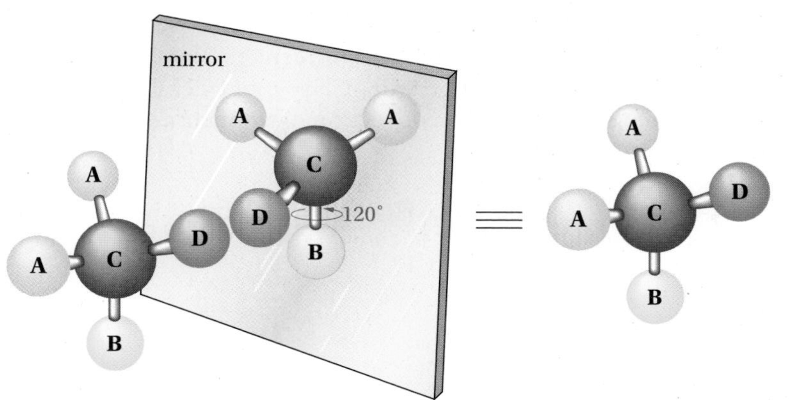

Figure 5.6
The tetrahedral model at the left has two corners occupied by identical groups (A). It has a plane of symmetry that passes through atoms B, C, and D and bisects angle ACA. Its mirror image is identical to itself, seen by a 120° rotation of the mirror image about the C—B bond. Hence the model is achiral.

Notice that the molecule in Figure 5.6 has a plane of symmetry. This plane passes through atoms B, C, and D and bisects the ACA angle. On the other hand, the molecule in Figure 5.4 does *not* have a symmetry plane.

A **plane of symmetry** (sometimes called a mirror plane) is a plane that passes through a molecule (or object) in such a way that what is on one side of the plane is the exact reflection of what is on the other side. *Any molecule with a plane of symmetry is achiral. Chiral molecules do not have a plane of symmetry.* Seeking a plane of symmetry is usually one quick way to tell whether a molecule is chiral or achiral.

What is on one side of a **plane of symmetry**, or mirror plane, is the exact reflection of what is on the other side.

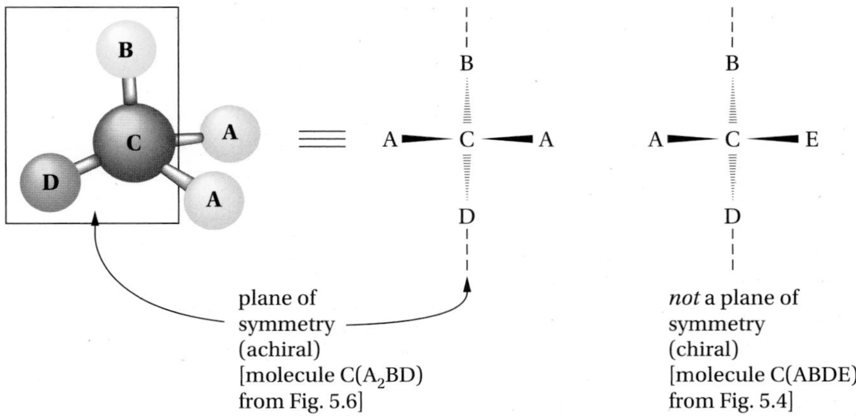

To summarize, a molecule with a stereogenic center (in our examples, the stereogenic center is a carbon atom with four different groups attached to it) can exist in two stereoisomeric forms, that is, as a pair of enantiomers. Such a molecule does not have a symmetry plane. Compounds with a symmetry plane are achiral.

EXAMPLE 5.1

Locate the stereogenic center in 3-methylhexane.

Solution Draw the structure, and look for a carbon atom with four different groups attached.

$$\overset{1}{C}H_3\overset{2}{C}H_2\overset{3}{C}H\overset{4}{C}H_2\overset{5}{C}H_2\overset{6}{C}H_3$$
$$\text{(with } CH_3 \text{ attached to C-3)}$$

All of the carbons except carbon-3 have at least two hydrogens (two identical groups) and therefore cannot be stereogenic centers. But carbon-3 has four different groups attached (H, CH_3—, CH_3CH_2—, and $CH_3CH_2CH_2$—) and is therefore a stereogenic center. By convention, we sometimes mark such centers with an asterisk.

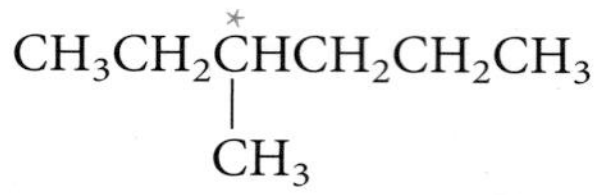

$$CH_3CH_2\overset{*}{C}HCH_2CH_2CH_3 \text{ (with } CH_3 \text{ on C-3)}$$

EXAMPLE 5.2

Draw the two enantiomers of 3-methylhexane.

Solution There are many ways to do this. Here are two of them. First draw carbon-3 with four tetrahedral bonds.

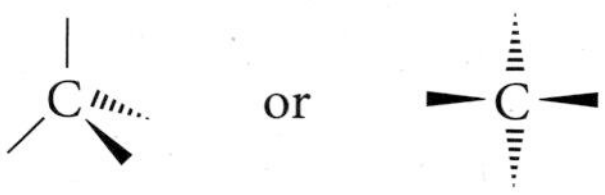

C (tetrahedral, wedge/dash) or —C— (wedge/dash cross)

Then attach the four different groups, in any order.

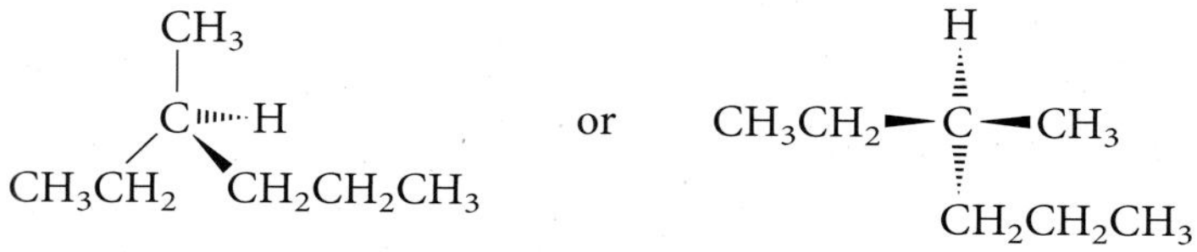

CH_3 above C, H on dashed bond, CH_3CH_2 and $CH_2CH_2CH_3$ below; or CH_3CH_2—C—CH_3 with H above and $CH_2CH_2CH_3$ below

Now draw the mirror image, or interchange the positions of any two groups.

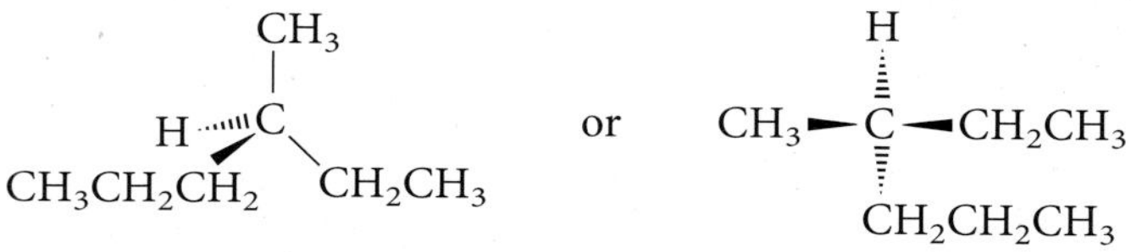

CH_3 above C, H on dashed bond, $CH_3CH_2CH_2$ and CH_2CH_3 below; or CH_3—C—CH_2CH_3 with H above and $CH_2CH_2CH_3$ below

To convince yourself that the *interchange of any two groups at a stereogenic center produces the enantiomer,* work with molecular models.

PROBLEM 5.2 Find the stereogenic centers in

a. $CH_3CH_2CHBrCH_2CH_2CH_2CH_3$
b. 3-methylcyclohexene
c. $ClFCHCH_3$
d. 2,3-dibromobutane

PROBLEM 5.3 Which of the following compounds is chiral?

a. 1-bromo-1-phenylethane
b. 1-bromo-2-phenylethane

PROBLEM 5.4 Draw three-dimensional structures for the two enantiomers of the chiral compound in Problem 5.3.

PROBLEM 5.5 Locate the planes of symmetry in the eclipsed conformation of ethane. In this conformation, is ethane chiral or achiral?

PROBLEM 5.6 Does the staggered conformation of ethane have planes of symmetry? In this conformation, is ethane chiral or achiral? *(Careful!)*

PROBLEM 5.7 Locate the planes of symmetry in *cis*- and *trans*-1,2-dichloroethene. Are these molecules chiral or achiral? *(Careful!)*

5.3 Configuration and the *R-S* Convention

Enantiomers differ in the arrangement of the groups attached to the stereogenic center. This arrangement of groups is called the **configuration** of the stereogenic center. *Enantiomers are another type of configurational isomer; they are said to have opposite configurations.*

The arrangement of four groups attached to a stereogenic center is called the **configuration** of that center.

When referring to a particular enantiomer, we would like to be able to specify which configuration we mean without having to draw the structure. A convention for doing this is known as the *R-S* or Cahn–Ingold–Prelog* system. Here is how it works.

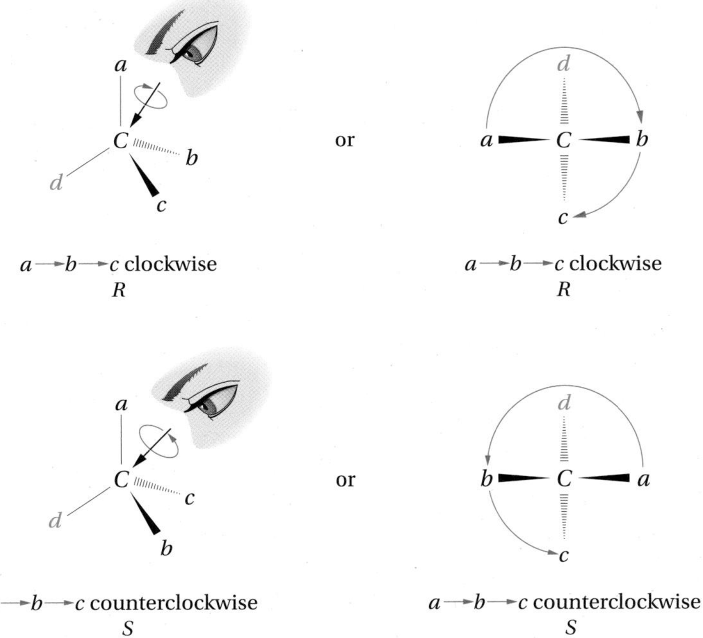

The four groups attached to the stereogenic center are placed in a priority order (by a system we will describe next), $a \longrightarrow b \longrightarrow c \longrightarrow d$. The stereogenic center is then observed *from the side opposite the lowest priority group, d.* If the remaining three groups ($a \longrightarrow b \longrightarrow c$) form a clockwise array, the configuration is designated *R* (from the Latin *rectus*, right).** If they form a *counterclockwise* array, the configuration is designated as *S* (from the Latin *sinister*, left).

The priority order of the four groups is set in the following way:

Rule 1

The atoms directly attached to the stereogenic center are ranked according to *atomic number:* the higher the atomic number, the higher the priority.

$$\text{Cl} > \text{O} > \text{C} > \text{H}$$

high priority — low priority

*Named after R. S. Cahn and C. K. Ingold, both British organic chemists, and V. Prelog, a Swiss chemist and Nobel Prize winner.

**More precisely, *rectus* means "right" in the sense of "correct, or proper," and not in the sense of direction (which is *dexter* = right, opposite to left). It may not be entirely coincidental that the initials of one of the inventors of this system are *R. S.*

If one of the four groups is H, it always has the lowest priority, and one views the stereogenic center looking down the C—H bond from C to H.

Rule 2

If a decision cannot be reached with rule 1 (that is, if two or more of the directly attached atoms are the same), work outward from the stereogenic center until a decision is reached. For example, the ethyl group has a higher priority than the methyl group, because at the first point of difference, working outward from the stereogenic center, we come to a *carbon* (higher priority) in the ethyl group and a *hydrogen* (lower priority) in the methyl group.

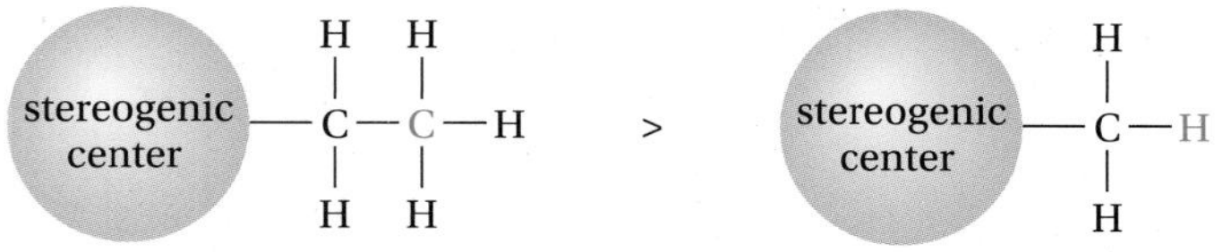

stereogenic center —C(H)(H)—C(H)(H)—H > stereogenic center —C(H)(H)—H

ethyl methyl

EXAMPLE 5.3

Assign a priority order to the following groups: —H, —Br, $—CH_2CH_3$, and $—CH_2OCH_3$.

Solution $—Br > —CH_2OCH_3 > —CH_2CH_3 > —H$

The atomic numbers of the directly attached atoms are ordered Br > C > H. To prioritize the two carbon groups, we must continue along the chain until a point of difference is reached.

$$—CH_2OCH_3 > —CH_2CH_3 \qquad (O > C)$$

PROBLEM 5.8 Assign a priority order to each of the following sets of groups:

a. $—CH(CH_3)_2$, $—CH_3$, —H, $—NH_2$
b. —OH, —Br, $—CH_3$, $—CH_2OH$
c. $—OCH_3$, $—NH(CH_3)_2$, $—CH_2NH_2$, —OH
d. $—CH_2CH_2CH_3$, $—CH_2CH_3$, $—C(CH_3)_3$, $—CH(CH_3)_2$

For stereogenic centers in cyclic compounds, the same rule for assigning priorities is followed. For example, in 1,1,3-trimethylcyclohexane, the four groups attached to carbon-3 in order of priority are $—CH_2C(CH_3)_2CH_2 > —CH_2CH_2 > —CH_3 > —H$.

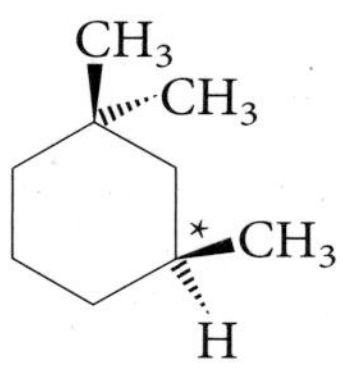

CH_3 CH_3 * CH_3 H

1,1,3-trimethylcyclohexane

A third, somewhat more complicated, rule is required to handle double or triple bonds and aromatic rings (which are written in the Kekulé fashion).

Rule 3

Multiple bonds are treated as if they were an equal number of single bonds. For example, the vinyl group $—CH{=}CH_2$ is counted as

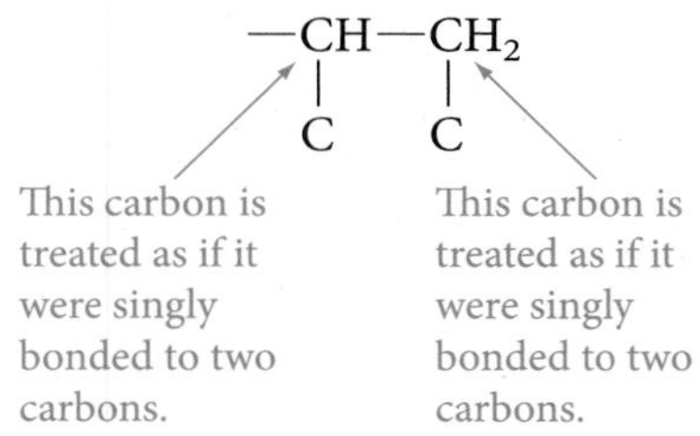

$$\begin{array}{cc} —CH— & CH_2 \\ | & | \\ C & C \end{array}$$

This carbon is treated as if it were singly bonded to two carbons.

This carbon is treated as if it were singly bonded to two carbons.

Similarly,

$$—C{\equiv}CH \quad \text{is treated as} \quad \begin{array}{ccc} C & C & \\ | & | & \\ —C— & C & —H \\ | & | & \\ C & C & \end{array}$$

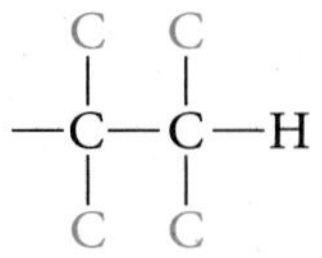

and

$$—CH{=}O \quad \text{is treated as} \quad \begin{array}{cc} H & \\ | & \\ —C— & O \\ | & | \\ O & C \end{array}$$

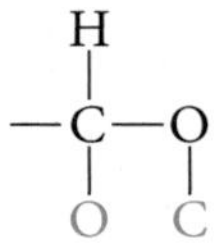

EXAMPLE 5.4

Which group has the higher priority, isopropyl or vinyl?

Solution The vinyl group has the higher priority. We continue along the chain until we reach a difference, shown in color.

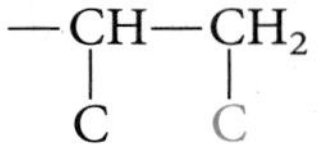

$$\underset{\text{vinyl}}{—CH{=}CH_2} \equiv \begin{array}{cc} —CH— & CH_2 \\ | & | \\ C & C \end{array}$$

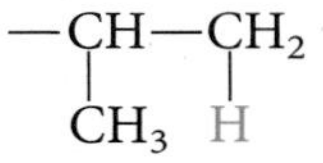

$$\underset{\text{isopropyl}}{—CH(CH_3)_2} \equiv \begin{array}{cc} —CH— & CH_2 \\ | & | \\ CH_3 & H \end{array}$$

PROBLEM 5.9 Assign a priority order to

a. $—C{\equiv}CH$ and $—CH{=}CH_2$ b. $—CH{=}CH_2$ and —(phenyl)

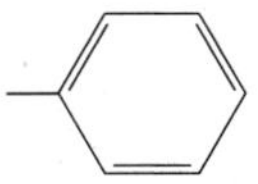

c. $—CH{=}O, —CH{=}CH_2, —CH_2CH_3,$ and $—CH_2OH$

Now let us see how these rules are applied.

EXAMPLE 5.5

Assign the configuration (*R* or *S*) to the following enantiomer of 3-methyl-hexane (see Example 5.2).

CH_3, C, H, CH_3CH_2, $CH_2CH_2CH_3$

Solution First assign the priority order to the four different groups attached to the stereogenic center.

$$—CH_2CH_2CH_3 > —CH_2CH_3 > —CH_3 > —H$$

Now view the molecule *from the side opposite the lowest-priority group* (—H) and determine whether the remaining three groups, from high to low priority, form a clockwise (*R*) or counterclockwise (*S*) array.

CH_3, C, H, CH_3CH_2, $CH_2CH_2CH_3$

R (clockwise)

We write the name (*R*)-3-methylhexane.

If we view the other representation of this molecule shown in Example 5.2, we come to the same conclusion.

H, CH_3CH_2, C, CH_3, $CH_2CH_2CH_3$

view down the C····H bond; the configuration is *R*

PROBLEM 5.10 Determine the configuration (*R* or *S*) at the stereogenic center in

a. CH=O, H, HO, CH_3

b. H, CH_3—C—(phenyl), NH_2

EXAMPLE 5.6

Draw the structure of (*R*)-2-bromobutane.

Solution First, write out the structure and prioritize the groups attached to the stereogenic center.

$$CH_3\overset{*}{C}HCH_2CH_3 \text{ (Br on C*)}$$

$$Br— > CH_3CH_2— > CH_3— > H—$$

Now make the drawing with the H (lowest priority group) “away” from you, and place the three remaining groups (Br → CH_3CH_2 → CH_3) in a clockwise (*R*) array.

Br — C(CH$_3$)(CH$_2$CH$_3$)H ... or ... CH$_3$—C(H)(Br)—CH$_2$CH$_3$

Of course, we could have started with the top-priority group at either of the other two bonds to give the following structures, which are equivalent to the previous structures:

CH$_3$, CH$_3$CH$_2$, Br, H or CH$_2$CH$_3$, Br, CH$_3$, H and CH$_3$CH$_2$—C(H)(Br)—CH$_3$ or Br—C(H)(CH$_3$)—CH$_2$CH$_3$

PROBLEM 5.11 Draw the structure of

a. (*S*)-3-phenylheptane
b. (*S*)-3-methylcyclopentene
c. (*R*)-4-methyl-2-hexene

5.4 The *E-Z* Convention for *Cis–Trans* Isomers

Before we continue with other aspects of chirality, let us digress briefly to describe a useful extension of the Cahn–Ingold–Prelog system of nomenclature to *cis–trans* isomers. Although we can easily use *cis–trans* nomenclature for 1,2-dichloroethene or 2-butene (see Sec. 3.5), that system is sometimes ambiguous, as in the following examples:

F, Cl C=C Br, I — *cis* or *trans?*

CH$_3$CH$_2$, CH$_3$ C=C Cl, Br — *cis* or *trans?*

The system we have just discussed for stereogenic centers has been extended to double-bond isomers. We use exactly the same priority rules. *The two groups attached to each carbon of the double bond are assigned priorities.* If the two higher-priority groups are on *opposite* sides of the double bond, the prefix *E* (from the German *entgegen,* opposite) is used. If the two higher-priority groups are on the *same* side of the double bond, the prefix is *Z* (from the German *zusammen,* together). The higher-priority groups for the previous examples are shown here in color, and the correct names are given below the structures.

F, Cl C=C Br, I

(*Z*)-1-bromo-2-chloro-2-fluoro-1-iodoethene

CH$_3$CH$_2$, CH$_3$ C=C Cl, Br

(*E*)-1-bromo-1-chloro-2-methyl-1-butene

PROBLEM 5.12 Name each compound by the *E-Z* system.

a. CH_3(H)C=C(H)(CH_2CH_3) b. F(Br)C=C(Cl)(H)

PROBLEM 5.13 Write the structure for

a. (*E*)-1,3-hexadiene b. (*Z*)-2-butene

5.5 Polarized Light and Optical Activity

The concept of molecular chirality follows logically from the tetrahedral geometry of carbon, as developed in Sections 5.1 and 5.2. Historically, however, these concepts were developed in the reverse order; how this happened is one of the most elegant and logically beautiful stories in the history of science. The story began in the early eighteenth century with the discovery of polarized light and with studies on how molecules placed in the path of such a light beam affect it.

Plane-polarized light is a light beam consisting of waves that vibrate in parallel planes.

An ordinary light beam consists of waves that vibrate in all possible planes perpendicular to its path. However, if this light beam is passed through certain types of substances, the waves of the transmitted beam will all vibrate in parallel planes. Such a light beam, said to be **plane polarized**, is illustrated in Figure 5.7. One convenient way to polarize light is to pass it through a device composed of Iceland spar (crystalline calcium carbonate) called a Nicol prism (invented in 1828 by the British physicist William Nicol). A more recently developed polarizing material is Polaroid, invented by the American E. H. Land. It contains a crystalline organic compound properly oriented and embedded in a transparent plastic. Sunglasses, for example, are often made from Polaroid.

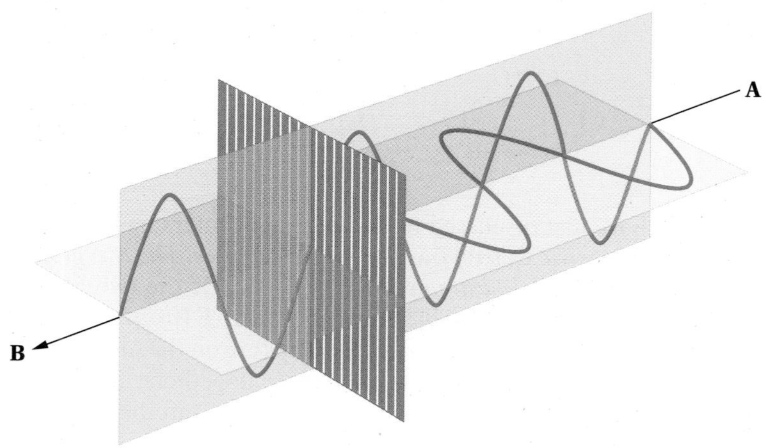

Figure 5.7
A beam of light, AB, initially vibrating in all directions, passes through a polarizing substance that "strains" the light so that only the vertical component emerges.

A light beam will pass through *two* samples of polarizing material only if their polarizing axes are aligned. If the axes are perpendicular, no light will pass through. This result, illustrated in Figure 5.8, is the basis of an instrument used to study the effect of various substances on plane-polarized light.

▰ **Figure 5.8**
The two sheets of polarizing material shown have their axes aligned perpendicularly. Although each disk alone is almost transparent, the area where they overlap is opaque. You can duplicate this effect using two pairs of Polaroid sunglasses. Try it! (Courtesy of the Polaroid Corporation.)

A **polarimeter** is shown schematically in Figure 5.9. Here is how it works. With the light on and the sample tube empty, the analyzer prism is rotated so that the light beam that has been polarized by the polarizing prism is completely blocked and the field of view is dark. At this point, the prism axes of the polarizer prism and the analyzer prism are perpendicular to one another. Now the sample is placed in the sample tube. If the substance is **optically inactive**, nothing changes. The field of view remains dark. But if an **optically active** substance is placed in the tube, it rotates the plane of polarization, and some light passes through the analyzer to the observer. By turning the analyzer prism clockwise or counterclockwise, the observer can again block the light beam and restore the dark field.

A **polarimeter**, or **spectropolarimeter**, is an instrument used to detect optical activity. An **optically active** substance rotates plane-polarized light, whereas an **optically inactive** substance does not.

The angle through which the analyzer prism must be rotated in this experiment is called α, the **observed rotation**. It is equal to the number of degrees that the optically active substance rotated the beam of plane-polarized light. If the analyzer must be rotated to the *right* (clockwise), the optically active substance is said to be **dextrorotatory** (+); if rotated to the *left* (counterclockwise), the substance is **levorotatory** (−).*

An optically active substance that is **dextrorotatory** would rotate plane-polarized light to the right (clockwise), while a **levorotatory** compound would rotate plane-polarized light to the left (counterclockwise).

The observed rotation, α, of a sample of an optically active substance depends on its molecular structure and also on the number of molecules in the sample tube, the

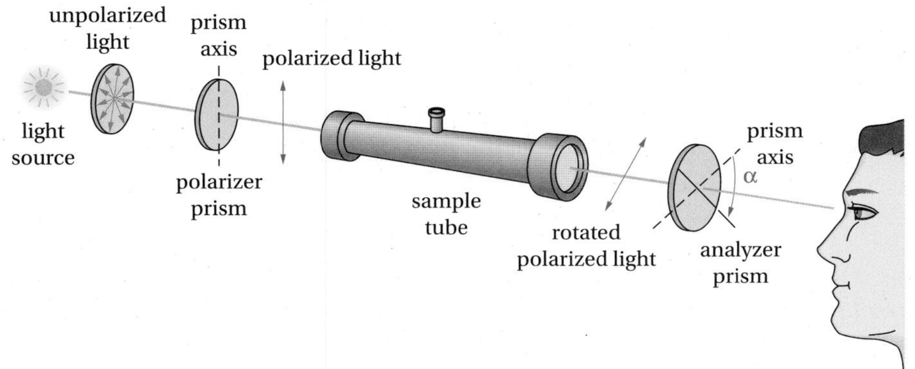

▰ **Figure 5.9**
Diagram of a polarimeter.

*It is not possible to tell from a single measurement whether a rotation is + or −. For example, is a reading +10° or −350°? We can distinguish between these alternatives by, for example, increasing the sample concentration by 10%. Then a +10° reading would change to +11°, and a −350° reading would change to −385° (that is, −25°).

A WORD ABOUT... Pasteur's Experiments and the van't Hoff–LeBel Explanation

Louis Pasteur in his laboratory.

Louis Pasteur (1822–95) in his Laboratory, 1885 (oil on canvas), Edelfelt, Albert Gustaf Aristides (1854–1905)/Musee d'Orsay, Paris, France/Giraudon/ The Bridgeman Art Library

The great French scientist Louis Pasteur (1822–1895) was the first to recognize that optical activity is related to what we now call chirality. He realized that similar molecules that rotate plane-polarized light through equal angles but in opposite directions must be related to one another as an object and its nonsuperimposable mirror image (that is, as a pair of enantiomers). Here is how he came to that conclusion.

Working in the mid-nineteenth century in a country famous for its wine industry, Pasteur was aware of two *isomeric* acids that deposit in wine casks during fermentation. One of these, called **tartaric acid**, was optically active and dextrorotatory. The other, at the time called **racemic acid**, was optically inactive.

Pasteur prepared various salts of these acids. He noticed that *crystals* of the sodium ammonium salt of *tartaric acid* were *not* symmetric (that is, they were not identical to their mirror images). In other words, they exhibited the property of handedness (chirality). Let us say that the crystals were all left-handed.

When Pasteur next examined crystals of the same salt of *racemic acid,* he found that they, too, were chiral but that some of the crystals were left-handed and others were right-handed. The crystals were related to one another as an object and its nonsuperimposable mirror image, and they were present in equal amounts. With a magnifying lens and a pair of tweezers, Pasteur carefully separated these crystals into two piles: the left-handed ones and the right-handed ones.

Then Pasteur made a crucial observation. When he *dissolved* the two types of crystals *separately* in water and placed the solutions in a polarimeter, he found that each solution was optically active (remember, he obtained these crystals from racemic acid, which was optically inactive). One solution had a specific rotation identical to that of the sodium ammonium salt of tartaric acid! The other had an equal but *opposite* specific rotation. This meant that it must be the mirror image, or levorotatory tartaric acid. Pasteur correctly concluded that racemic acid was not really a single substance, but a 50:50 mixture of (+) and (−)-tartaric acids. Racemic acid was optically inactive because it contained equal amounts of two enantiomers. We now define a **racemic mixture** as a 50:50 mixture of enantiomers, and of course, such a mixture is optically inactive because the rotations of the two enantiomers cancel out.

Pasteur recognized that optical activity must be due to some property of tartaric acid molecules themselves,

length of the tube, the wavelength of the polarized light, and the temperature. All of these have to be standardized if we want to compare the optical activity of different substances. This is done using the **specific rotation** $[\alpha]$, defined as follows:

The **specific rotation** of an optically active substance (a standardized version of its **observed rotation**) is a characteristic physical property of the substance.

$$\text{Specific rotation} = [\alpha]_\lambda^t = \frac{\alpha}{l \times c}\text{(solvent)}$$

where l is the length of the sample tube in *decimeters, c* is the concentration in *grams per milliliter, t* is the temperature of the solution, and λ is the wavelength of light. The solvent used is indicated in parentheses. Measurements are usually made at room temperature, and the most common light source is the D-line of a sodium vapor lamp (λ = 589.3 nm), although modern instruments called **spectropolarimeters** allow the light wavelength to be varied at will. The specific rotation of an optically active substance at a particular wavelength is as definite a property of the substance as its melting point, boiling point, or density.

not to some property of the crystals, because the crystalline shape was lost when the crystals were dissolved in water in order to measure the specific rotation. However, the precise explanation in terms of molecular structure eluded Pasteur and was not to come for another 25 years.

Pasteur's experiments were performed at about the same time that Kekulé in Germany was developing his theories about organic structures. Kekulé recognized that carbon is tetravalent, and there is even a hint in some of his writings (about 1867) and also in the writings of the Russian chemist A. M. Butlerov (1862) and the Italian E. Paterno (1869) that carbon might be tetrahedral. But it was not until 1874 that the Dutch physical chemist J. H. van't Hoff (1852–1911) and the Frenchman J. A. LeBel (1847–1930) simultaneously, but quite independently, made a bold hypothesis about carbon that would explain the optical activity of some organic molecules and the optical inactivity of others.

These scientists knew their solid geometry. They knew that four different objects can be arranged in two different ways at the corners of a tetrahedron, and that these two ways are related to one another as an object and its nonsuperimposable mirror image. They also knew that this arrangement resulted in right- and left-handedness, as shown in the drawing at the top of the next column.

They made the bold hypothesis that the four valences of carbon were directed toward the corners of a tetrahedron, and that optically active molecules would contain at least one carbon atom with four different groups attached to it. This idea would explain

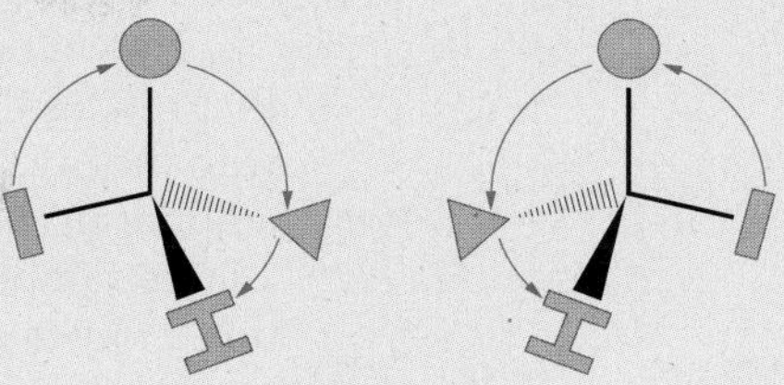

why Pasteur's (+) and (−)-tartaric acids rotated plane-polarized light to equalextents but in opposite (right- and left-handed) directions. Optically inactive organic substances would either contain no asymmetric carbon atom (stereogenic carbon) or be a 50:50 mixture of enantiomers.

This, then, is how the tetrahedral geometry of carbon was first recognized.* The boldness and brilliance of this proposal is remarkable when one realizes that neither the electron nor the nucleus of an atom had yet been discovered and that almost nothing was then known about the physical nature of chemical bonds. At the time they made their proposal, van't Hoff and LeBel were relatively unknown chemists. Their hypothesis was ridiculed by at least one establishment chemist at the time, but it was soon generally accepted, has survived many tests, and can now be regarded as fact. In 1901, van't Hoff received the first Nobel Prize in chemistry.

See Problem 5.60.

*Actually, LeBel developed his ideas based on symmetry considerations, whereas van't Hoff based his (at least at first) on the idea of an asymmetric carbon atom. For a stimulating article on these differences, see R. B. Grossman, *J. Chem. Educ.* **1989**, *66*, 30–33.

PROBLEM 5.14 Camphor is optically active. A camphor sample (1.5 g) dissolved in ethanol (optically inactive) to a total volume of 50 mL, placed in a 5-cm polarimeter sample tube, gives an observed rotation of +0.66° at 20°C (using the sodium D-line). Calculate and express the specific rotation of camphor.

In the early nineteenth century, the French physicist Jean Baptiste Biot (1774–1862) studied the behavior of a great many substances in a polarimeter. Some, such as turpentine, lemon oil, solutions of camphor in alcohol, and solutions of cane sugar in water, were optically active. Others, such as water, alcohol, and solutions of salt in water, were optically inactive. Later, many natural products (carbohydrates, proteins, and steroids, to name just a few) were added to the list of optically active compounds. What is it about the structure of molecules that causes some to be optically active and others inactive?

When plane-polarized light passes through a single molecule, the light and the electrons in the molecule interact. This interaction causes the plane of polarization to rotate slightly.* But when we place a substance in a polarimeter, *we do not place a single molecule there; we place a large collection of molecules there* (recall that even as little as one-thousandth of a mole contains 6×10^{20} molecules).

Now, if the substance is achiral, then for every single molecule in one conformation that rotates the plane of polarization in one direction, there will be another molecule with the mirror-image conformation that will rotate the plane of polarization an equal amount in the opposite direction. Every conceivable conformation will be present in a large sample of achiral molecules, and the effects of these conformers will cancel one another out. The result is that the light beam passes through a sample of achiral molecules without any net change in the plane of polarization. *Achiral molecules are optically inactive.*

But for chiral molecules, the situation is different. A 50:50 mixture of a pair of enantiomers will not rotate the plane of polarization for the same reason that achiral molecules do not rotate it. Consider, however, a sample of *one enantiomer* (say, *R*) of a chiral molecule. For any molecule with a given configuration in the sample, *there can be no mirror-image configuration* (because the mirror image gives a different molecule, the *S* enantiomer). Therefore, the rotation in the polarization plane caused by one molecule is *not* canceled by any other molecule, and the light beam passes through the sample with a net change in the plane of polarization. *Single enantiomers of chiral molecules are optically active.***

5.6 Properties of Enantiomers

In what properties do enantiomers differ from one another? Enantiomers differ only with respect to chirality. In all other respects, they are identical. For this reason, they differ from one another *only in properties that are also chiral.* Let us illustrate this idea first with familiar objects.

A left-handed baseball player (chiral) can use the same ball or bat (achiral) as can a right-handed player. But, of course, a left-handed player (chiral) can use only a left-handed baseball glove (chiral). A bolt with a right-handed thread (chiral) can use the same washer (achiral) as a bolt with a left-handed thread, but it can only fit into a nut (chiral) with a right-handed thread. To generalize, *the chirality of an object is most significant when the object interacts with another chiral object.*

Enantiomers have identical achiral properties, such as melting point, boiling point, density, and various types of spectra. Their solubilities in an ordinary, achiral solvent are also identical. However, *enantiomers have different chiral properties,* one of which is the *direction* in which they rotate plane-polarized light (clockwise or counterclockwise). Although enantiomers rotate plane-polarized light in opposite directions, they have specific rotations of the same magnitude (but with opposite signs), because the *number of degrees* is not a chiral property. Only the *direction* of rotation is a chiral property. Here is a specific example.

Lactic acid is an optically active hydroxyacid that is important in several biological processes. It has one stereogenic center. Its structure and some of its properties are shown in Figure 5.10. Note that both enantiomers have identical melting points and, *except for sign,* identical specific rotations.

---

*This is because the electric and magnetic fields that result from electronic motions in the molecule affect the electric and magnetic fields of the light. Even achiral molecules have this effect, because they can have chiral conformations (for example, the enantiomeric gauche conformations of 1,2-dichloroethane).

**Mixtures of enantiomers that contain more of one enantiomer than the other are also optically active.

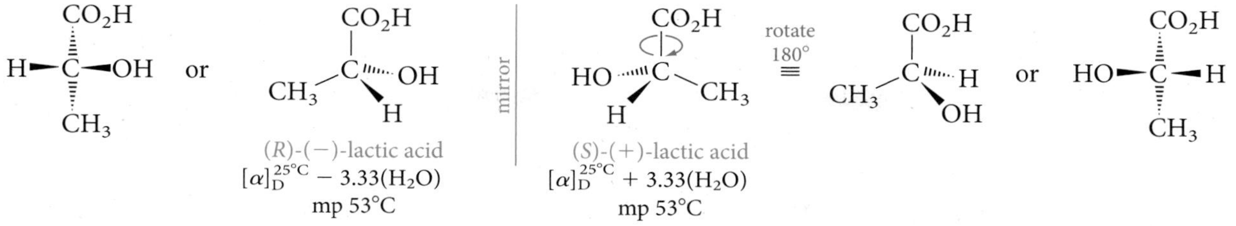

(R)-(−)-lactic acid
$[\alpha]_D^{25°C}$ − 3.33(H_2O)
mp 53°C

mirror

(S)-(+)-lactic acid
$[\alpha]_D^{25°C}$ + 3.33(H_2O)
mp 53°C

rotate 180°

■ **Figure 5.10**
The structures and properties of the lactic acid enantiomers.

There is no obvious relationship between configuration (*R* or *S*) and sign of rotation (+ or −). For example, (*R*)-lactic acid is levorotatory. When (*R*)-lactic acid is converted to its methyl ester (eq. 5.1), the configuration is unchanged because none of the bonds to the stereogenic carbon is involved in the reaction. Yet the sign of rotation of the product, a physical property, changes from − to +.

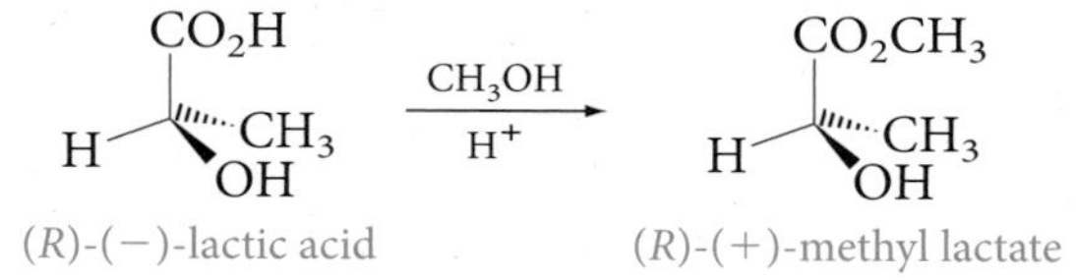

(R)-(−)-lactic acid $\xrightarrow[H^+]{CH_3OH}$ (R)-(+)-methyl lactate **(5.1)**

Enantiomers often behave differently in a biological setting because these properties usually involve a reaction with another chiral molecule. For example, the enzyme *lactic acid dehydrogenase* will oxidize (+)-lactic acid to pyruvic acid, but it will *not* oxidize (−)-lactic acid (eq. 5.2).

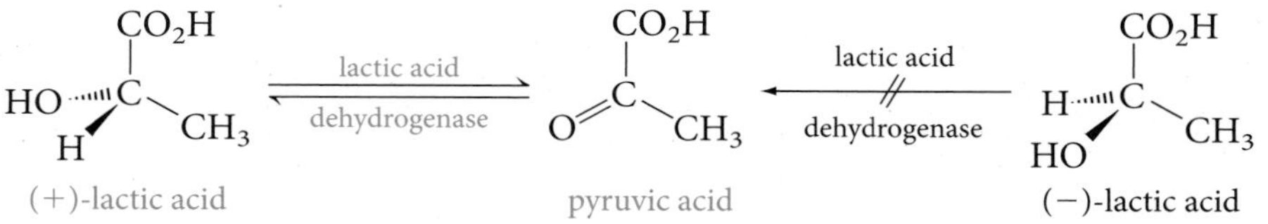

(+)-lactic acid $\underset{\text{dehydrogenase}}{\overset{\text{lactic acid}}{\rightleftharpoons}}$ pyruvic acid $\underset{\text{dehydrogenase}}{\overset{\text{lactic acid}}{\not\leftarrow}}$ (−)-lactic acid **(5.2)**

Why? The enzyme itself is chiral and can distinguish between right- and left-handed lactic acid molecules, just as a right hand distinguishes between left-handed and right-handed gloves.

Enantiomers differ in many types of biological activity. One enantiomer may be a drug, whereas its enantiomer may be ineffective. For example, only (−)-adrenaline is a cardiac stimulant; (+)-adrenaline is ineffective. One enantiomer may be toxic, another harmless. One may be an antibiotic, the other useless. One may be an insect sex attractant, the other without effect or perhaps a repellant. Chirality is of paramount importance in the biological world.

5.7 Fischer Projection Formulas

Instead of using dashed and solid wedges to show the three-dimensional arrangements of groups in a chiral molecule, it is sometimes convenient to have a two-dimensional way of doing so. A useful way to do this was devised many years ago by Emil Fischer*; the formulas are called **Fischer projections**.

A **Fischer projection** is a type of two-dimensional formula of a molecule used to represent the three-dimensional configurations of stereogenic centers.

*Emil Fischer (1852–1919), who devised these formulas, was one of the early giants in the field of organic chemistry. He did much to elucidate the structures of carbohydrates, proteins, and other natural products and received the 1902 Nobel Prize in chemistry.

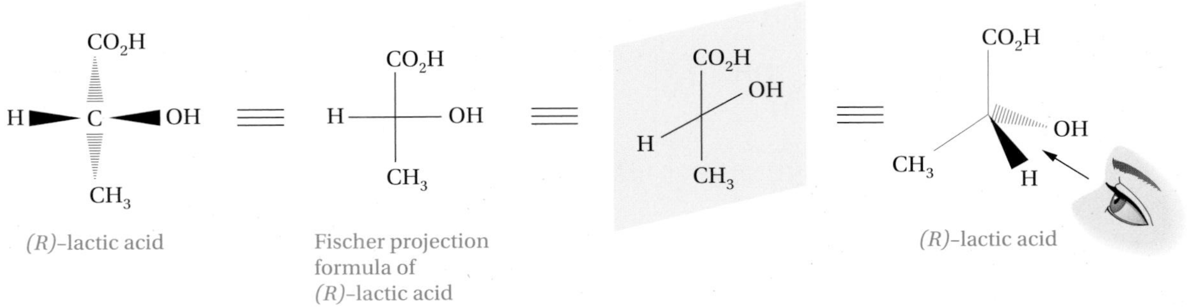

(R)-lactic acid

Fischer projection formula of (R)-lactic acid

(R)-lactic acid

▰ Figure 5.11
Projecting the model at the right onto a plane gives the Fischer projection formula.

Consider the formula for (*R*)-lactic acid, to the left of the mirror in Figure 5.10. If we project that three-dimensional formula onto a plane, as illustrated in Figure 5.11, we obtain the flattened Fischer projection formula.

There are two important things to notice about Fischer projection formulas. First, the C for the stereogenic carbon atom is omitted and is represented simply as the crossing point of the horizontal and vertical lines. Second, horizontal lines connect the stereogenic center to groups that project *above* the plane of the page, *toward* the viewer; vertical lines lead to groups that project *below* the plane of the page, *away* from the viewer. As with other stereorepresentations, interchange of any two groups always gives the enantiomer.

PROBLEM 5.15 Draw a Fischer projection formula for (*S*)-lactic acid.

The following example demonstrates how the absolute configuration of a stereocenter is determined from its Fischer projection.

EXAMPLE 5.7

Determine the absolute (*R* or *S*) configuration of the stereoisomer of 2-chlorobutane shown in the following Fischer projection.

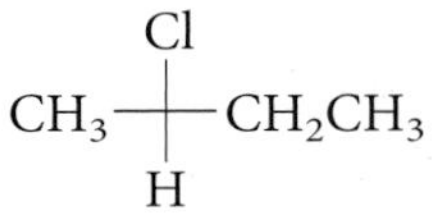

Solution First prioritize the four groups attached to the central carbon according to the Cahn–Ingold–Prelog rules:

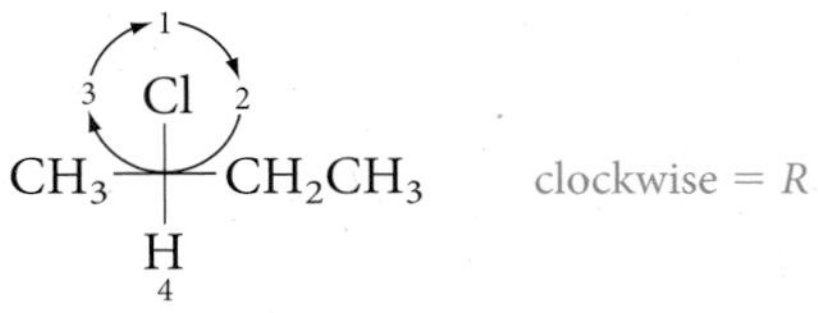

clockwise = *R*

Because atoms attached to vertical arms project away from the viewer, you can now determine whether the sequence 1→2→3 is clockwise or counterclockwise. In this case, it is clockwise, so the absolute configuration of this enantiomer of 2-chlorobutane is *R*.

Note that if the lowest-priority group is on a horizontal arm, you can still determine the absolute configuration of the stereocenter. This can be done by *rotating three* of the four groups so that the lowest-priority group is located on a vertical arm and then proceeding as above:

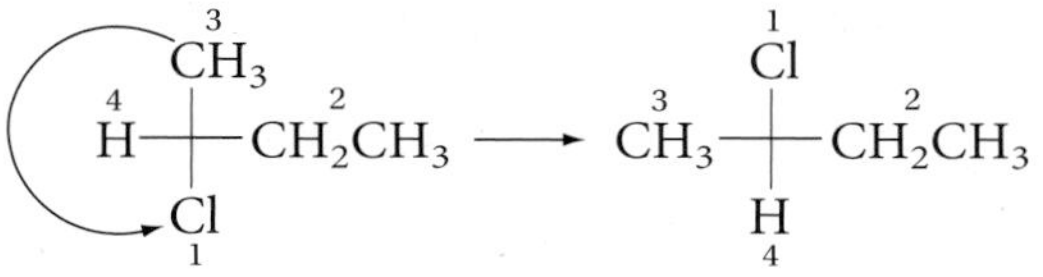

$$\begin{array}{c} \overset{3}{CH_3} \\ \overset{4}{H}-\!\!\!+\!\!\!-\overset{2}{CH_2CH_3} \\ \underset{1}{Cl} \end{array} \longrightarrow \begin{array}{c} \overset{1}{Cl} \\ \overset{3}{CH_3}-\!\!\!+\!\!\!-\overset{2}{CH_2CH_3} \\ \underset{4}{H} \end{array}$$

It can also be done by remembering that a substituent on a horizontal arm is pointing out at the viewer. If the sequence of the three priority groups 1→2→3 is counterclockwise, as in this case, the absolute configuration is *R*. (If the direction were clockwise, the absolute configuration would be *S*.)

PROBLEM 5.16 Determine the absolute configuration of the following enantiomer of 2-butanol from its Fischer projection:

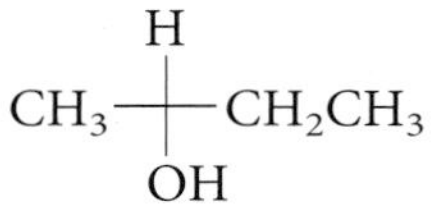

$$\begin{array}{c} H \\ CH_3-\!\!\!+\!\!\!-CH_2CH_3 \\ OH \end{array}$$

Fischer projections are used extensively in biochemistry and in carbohydrate chemistry, where compounds frequently contain more than one stereocenter. In the next section, you will see how useful Fischer projections are in dealing with compounds containing more than one stereogenic center.

5.8 Compounds with More Than One Stereogenic Center; Diastereomers

Compounds may have more than one stereogenic center, so it is important to be able to determine how many isomers exist in such cases and how they are related to one another. Consider the molecule 2-bromo-3-chlorobutane.

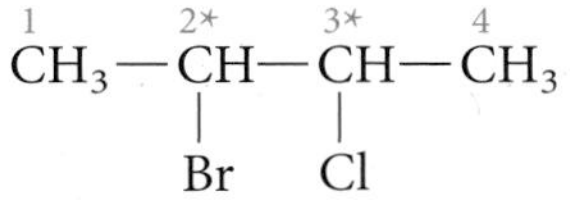

$$\overset{1}{CH_3}-\underset{\displaystyle Br}{\overset{2*}{CH}}-\underset{\displaystyle Cl}{\overset{3*}{CH}}-\overset{4}{CH_3}$$

2-bromo-3-chlorobutane

As indicated by the asterisks, the molecule has two stereogenic centers. Each of these could have the configuration *R* or *S*. Thus, four isomers in all are possible: (2*R*,3*R*), (2*S*,3*S*), (2*R*,3*S*), and (2*S*,3*R*). We can draw these four isomers as shown in Figure 5.12. Note that there are two pairs of enantiomers. The (2*R*,3*R*) and (2*S*,3*S*) forms are nonsuperimposable mirror images, and the (2*R*,3*S*) and (2*S*,3*R*) forms are another such pair.

Let us see how to use Fischer projection formulas for these molecules. Consider the (2*R*,3*R*) isomer, the one at the left in Figure 5.12. The solid-dashed wedge drawing has horizontal groups projecting out of the plane of the paper toward us and vertical

▪ **Figure 5.12**
The four stereoisomers of 2-bromo-3-chlorobutane, a compound with two stereogenic centers.

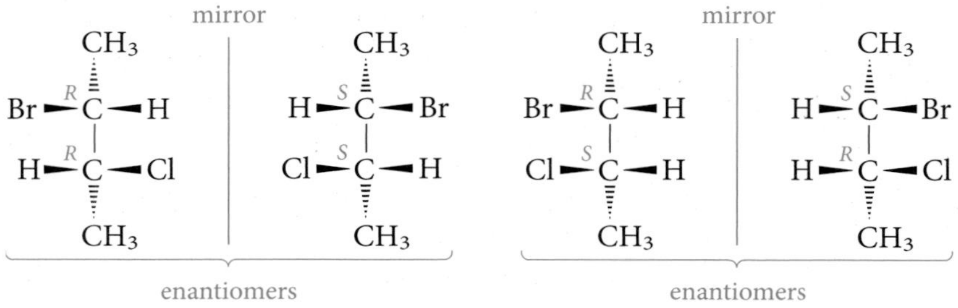

groups going away from us, behind the paper. These facts are expressed in the equivalent Fischer projection formula as shown.*

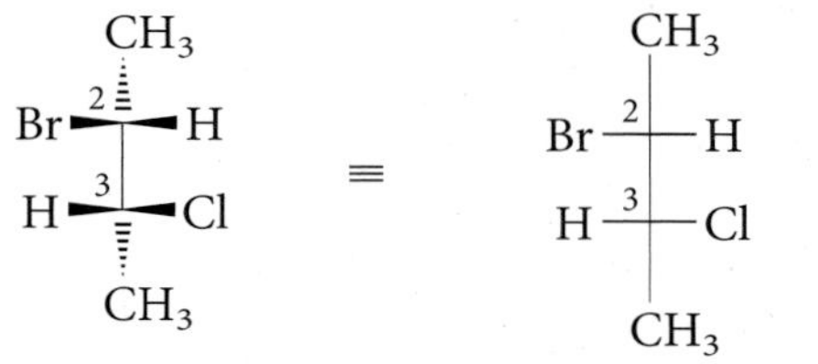

solid-dashed wedge formula ≡ Fischer projection formula

PROBLEM 5.17 Draw the Fischer projection formulas for the remaining stereoisomers of 2-bromo-3-chlorobutane shown in Figure 5.12.

Diastereomers are stereoisomers that are not mirror images of each other.

Now we come to an extremely important new idea. Consider the relationship between, for example, the (2*R*,3*R*) and (2*R*,3*S*) forms of the isomers in Figure 5.12. These forms are *not* mirror images because they have the *same* configuration at carbon-2, though they have opposite configurations at carbon-3. They are certainly stereoisomers, but they are not enantiomers. For such pairs of stereoisomers, we use the term **diastereomers**. Diastereomers are stereoisomers that are not mirror images of one another.

*Make 3D models of the molecules in Figure 5.12 to help you visualize them. Notice that these structures are derived from an eclipsed conformation of the molecule, viewed from above so that horizontal groups project toward the viewer. The actual molecule is an equilibrium mixture of several staggered conformations, one of which is shown below. Fischer formulas are used to represent the correct *configurations,* but not necessarily the lowest energy *conformations* of a molecule.

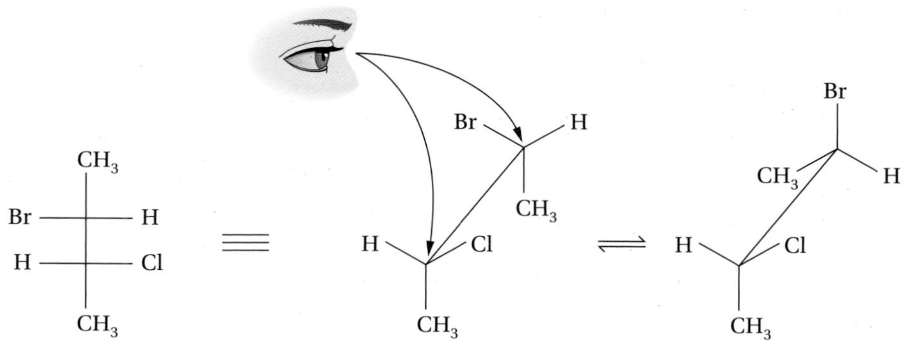

eclipsed

staggered

There is an important, fundamental difference between enantiomers and diastereomers. Because they are mirror images, enantiomers differ *only* in mirror-image (chiral) properties. They have the same achiral properties, such as melting point, boiling point, and solubility in ordinary solvents. Enantiomers cannot be separated from one another by methods that depend on achiral properties, such as recrystallization or distillation. On the other hand, diastereomers are *not* mirror images. They may differ in *all* properties, whether chiral or achiral. As a consequence, diastereomers may differ in melting point, boiling point, and solubility, not only in direction but also in the number of degrees that they rotate plane-polarized light—in short, they behave as two different chemical substances.

PROBLEM 5.18 How do you expect the specific rotations of the (2*R*,3*R*) and (2*S*,3*S*) forms of 2-bromo-3-chlorobutane to be related? Answer the same question for the (2*R*,3*R*) and (2*S*,3*R*) forms.

Can we generalize about the number of stereoisomers possible when a larger number of stereogenic centers is present? Suppose, for example, that we add a third stereogenic center to the compounds shown in Figure 5.12 (say, 2-bromo-3-chloro-4-iodopentane). The new stereogenic center added to each of the four structures can once again have either an *R* or an *S* configuration, so that with three different stereogenic centers, eight stereoisomers are possible. The situation is summed up in a single rule: If a molecule has *n* different stereogenic centers, it may exist in a maximum of 2^n stereoisomeric forms. There will be a maximum of $2^n/2$ pairs of enantiomers.

Polarized light micrograph of glucose crystals.

PROBLEM 5.19 The Fischer projection formula for glucose (blood sugar, Sec. 16.4, p. 466) is

```
        CH=O
     H ─┼─ OH
    HO ─┼─ H
     H ─┼─ OH
     H ─┼─ OH
        CH2OH
       glucose
```

Altogether, how many stereoisomers of this sugar are possible?

Actually, the number of isomers predicted by this rule is the *maximum* number possible. Sometimes, certain structural features reduce the actual number of isomers. In the next section, we examine a case of this type.

5.9 *Meso* Compounds; the Stereoisomers of Tartaric Acid

Consider the stereoisomers of 2,3-dichlorobutane. There are two stereogenic centers.

$$\overset{1}{CH_3}-\underset{\underset{Cl}{|}}{\overset{2*}{CH}}-\underset{\underset{Cl}{|}}{\overset{3*}{CH}}-\overset{4}{CH_3}$$

2,3-dichlorobutane

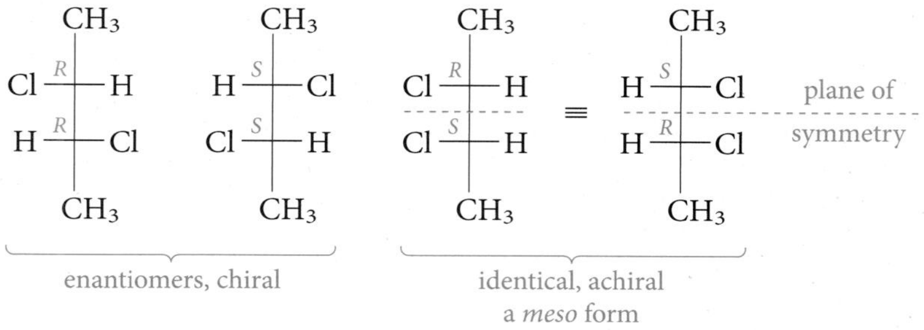

Figure 5.13
Fischer projections of the stereoisomers of 2,3-dichlorobutane.

We can write out the stereoisomers just as we did in Figure 5.12; they are shown in Figure 5.13. Once again, the (*R,R*) and (*S,S*) isomers constitute a pair of nonsuperimposable mirror images, or enantiomers. *However, the other "two" structures, (R,S) and (S,R), in fact, now represent a single compound.*

Look more closely at the structures to the right in Figure 5.13. Notice that they have a plane of symmetry that is perpendicular to the plane of the paper and bisects the central C—C bond. The reason is that each stereogenic center has the *same* four groups attached. The structures are identical, superimposable mirror images and therefore *achiral.* We call such a structure a ***meso* compound**. A *meso* compound is an achiral diastereomer of a compound with stereogenic centers. Its stereogenic centers have opposite configurations. Being achiral, *meso* compounds are optically inactive.*

A ***meso* compound** is an achiral diastereomer of a compound with stereogenic centers.

Now let us take a look at tartaric acid, the compound whose optical activity was so carefully studied by Louis Pasteur (see "A Word About . . . Pasteur's Experiments" on pages 160–161). It has two identical stereogenic centers.

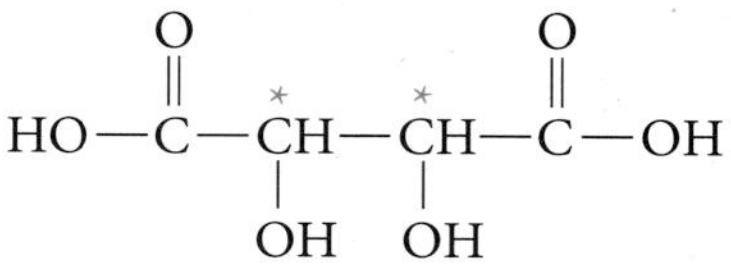

$$HO-\overset{O}{\overset{\|}{C}}-\underset{OH}{\underset{|}{\overset{*}{C}H}}-\underset{OH}{\underset{|}{\overset{*}{C}H}}-\overset{O}{\overset{\|}{C}}-OH$$

tartaric acid

The structures of these three stereoisomers and two of their properties are shown in Figure 5.14. Note that the enantiomers have identical properties except for the *sign* of the specific rotation, whereas the *meso* form, being a diastereomer of each enantiomer, differs from them in both properties.

For about 100 years after Pasteur's research, it was still not possible to determine the configuration associated with a particular enantiomer of tartaric acid. For example, it was not known whether (+)-tartaric acid had the (*R,R*) or the (*S,S*)

Figure 5.14
The stereoisomers of tartaric acid.

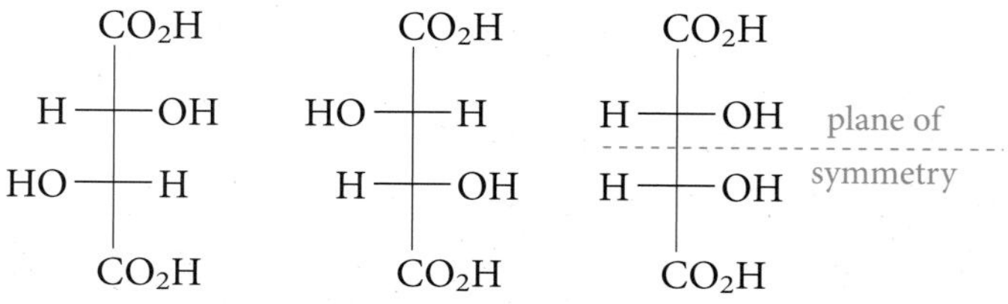

Configuration	(*R,R*)	(*S,S*)	*meso* (*R,S*)
$[\alpha]_D^{20°}$ (H_2O)	+12	−12	0
Melting point, °C	170	170	140

*Fischer projections may be turned 180° in the plane of the paper without changing configuration. You can see that such an operation on the enantiomeric pair in Figure 5.13 does *not* interconvert them, but when performed on the *meso* form, it does.

configuration. It was known that (+)-tartaric acid had to have one of these two configurations and that (−)-tartaric acid had to have the opposite configuration, but which isomer had which?

In 1951, the Dutch scientist J. M. Bijvoet developed a special x-ray technique that solved the problem. Using this technique on crystals of the sodium rubidium salt of (+)-tartaric acid, Bijvoet showed that it had the (*R*,*R*) configuration. So this was the tartaric acid studied by Pasteur, and racemic acid was a 50:50 mixture of the (*R*,*R*) and (*S*,*S*) isomers. The *meso* form was not studied until later.

Since tartaric acid had been converted chemically into other chiral compounds and these in turn into still others, it became possible as a result of Bijvoet's work to assign **absolute configurations** (that is, the correct *R* or *S* configuration for each stereocenter) to many pairs of enantiomers.

The correctly assigned (*R* or *S*) configuration of a stereocenter in a molecule is called the **absolute configuration** of the stereocenter.

PROBLEM 5.20 Show that *trans*-1,2-dimethylcyclopentane can exist in chiral, enantiomeric forms.

PROBLEM 5.21 Is *cis*-1,2-dimethylcyclopentane chiral or achiral? What stereochemical term can we give to it?

5.10 Stereochemistry: A Recap of Definitions

We have seen here and in Section 2.11 that *stereoisomers* can be classified in three different ways. They may be either *conformers* or *configurational isomers;* they may be *chiral* or *achiral;* and they may be *enantiomers* or *diastereomers.*

A	*Conformers:*	interconvertible by rotation about single bonds
	Configurational Isomers:	not interconvertible by rotation, only by breaking and making bonds
B	*Chiral:*	mirror image not superimposable on itself
	Achiral:	molecule and mirror image are identical
C	*Enantiomers:*	mirror images; have opposite configurations at all stereogenic centers
	Diastereomers:	stereoisomers but not mirror images; have same configuration at one or more centers, but differ at the remaining stereogenic centers

Various combinations of these three sets of terms can be applied to any pair of stereoisomers. Here are a few examples:

1. *Cis-* and *trans*-2-butene (*Z-* and *E*-2-butene).

$$\begin{array}{c}CH_3 \quad\quad CH_3\\ C{=}C\\ H \quad\quad\quad H\end{array} \quad \text{and} \quad \begin{array}{c}CH_3 \quad\quad H\\ C{=}C\\ H \quad\quad\quad CH_3\end{array}$$

These isomers are *configurational* (not interconverted by rotation about single bonds), *achiral* (the mirror image of each is superimposable on the original), and *diastereomers* (although they are stereoisomers, they are *not* mirror images of one another; hence they must be diastereomers).

2. Staggered and eclipsed ethane.

These are *achiral conformers*. They are *diastereomeric conformers* (but without stereogenic centers) because they are not mirror images.

3. (*R*)- and (*S*)-lactic acid.

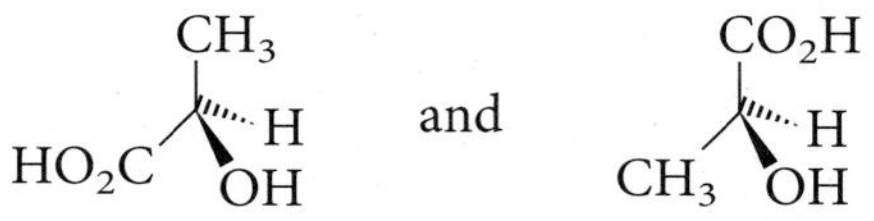

These isomers are *configurational*, each is *chiral*, and they constitute a pair of *enantiomers*.

4. *Meso*- and (*R*,*R*)-tartaric acids.

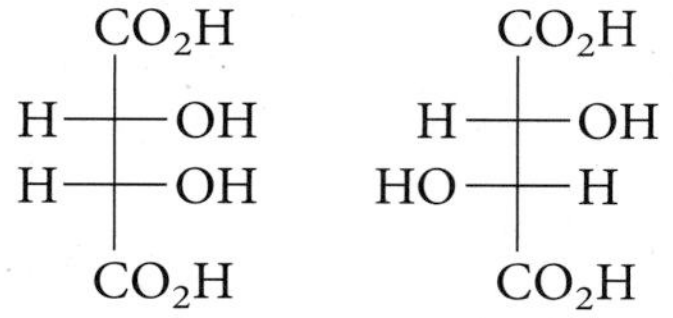

meso (*R*,*R*)

These isomers are *configurational* and *diastereomers*. One is *achiral*, and the other is *chiral*.

Enantiomers, such as (*R*)- and (*S*)-lactic acid, differ only in chiral properties and therefore cannot be separated by ordinary achiral methods such as distillation or recrystallization. Diastereomers differ in all properties, chiral or achiral. *If* they are also configurational isomers [such as *cis*- and *trans*-2-butene, or *meso*- and (*R*,*R*)-tartaric acid], they can be separated by ordinary achiral methods, such as distillation or recrystallization. *If*, on the other hand, they are conformers (such as staggered and eclipsed ethane], they may interconvert so readily by bond rotation as to not be separable.

PROBLEM 5.22 Draw the two stereoisomers of 1,3-dimethylcyclobutane, and classify the pair according to the categories listed in A, B, and C above.

5.11 Stereochemistry and Chemical Reactions

How important is stereochemistry in chemical reactions? The answer depends on the nature of the reactants. First, consider the formation of a chiral product from achiral reactants; for example, the addition of hydrogen bromide to 1-butene to give 2-bromobutane in accord with Markovnikov's Rule.

$$CH_3CH_2CH{=}CH_2 + HBr \longrightarrow CH_3CH_2\overset{*}{C}HCH_3 \quad (\text{Br on C-2}) \tag{5.3}$$

1-butene → 2-bromobutane

The product has one stereogenic center, marked with an asterisk, but both enantiomers are formed in exactly equal amounts. The product is a **racemic mixture**. Why? Although this result will be obtained *regardless* of the reaction mechanism, let us consider the generally accepted mechanism.

A **racemic mixture** is a 50:50 mixture of a pair of enantiomers.

$$CH_3CH_2CH{=}CH_2 + H^+ \longrightarrow CH_3CH_2\overset{+}{C}HCH_3 \xrightarrow{Br^-} CH_3CH_2CHCH_3 \quad (\text{Br on C-2}) \tag{5.4}$$

2-butyl cation

The intermediate 2-butyl cation obtained by adding a proton to the end carbon is planar, and bromide ion can combine with it from the "top" or "bottom" side with exactly equal probability.

(5.5)

2-butyl cation + :Br:⁻ → top: (*S*)-2-bromobutane; bottom: (*R*)-2-bromobutane

Therefore, the product is a racemic mixture, an optically inactive 50:50 mixture of the two enantiomers.

We can generalize this result. *When chiral products are obtained from achiral reactants, both enantiomers are formed at the same rates, in equal amounts.*

PROBLEM 5.23 Show that, if the mechanism of addition of HBr to 1-butene involved *no* intermediates, but *simultaneous one-step* addition (in the Markovnikov sense), the product would still be racemic 2-bromobutane.

PROBLEM 5.24 Show that the chlorination of butane at carbon-2 will give a 50:50 mixture of enantiomers.

Now consider a different situation, the reaction of a *chiral* molecule with an achiral reagent to create a second stereogenic center. Consider, for example, the addition of HBr to 3-chloro-1-butene.

$$CH_3\overset{*}{C}HCH{=}CH_2 + HBr \longrightarrow CH_3\overset{*}{C}H{-}\overset{*}{C}HCH_3 \tag{5.6}$$

3-chloro-1-butene (Cl on C-3) → 2-bromo-3-chlorobutane (Cl and Br on the starred carbons)

A WORD ABOUT... Enantiomers and Biological Activity

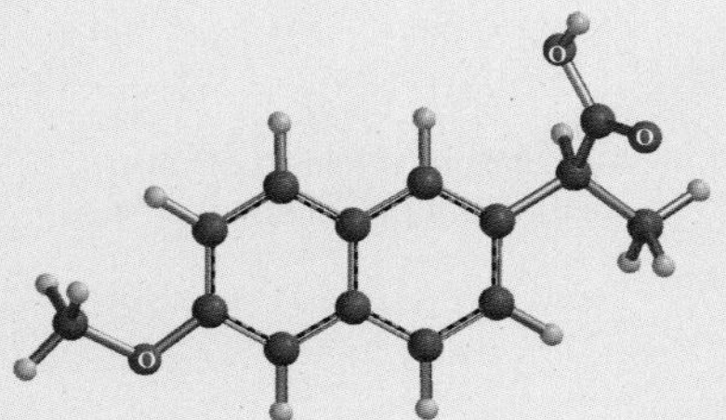
Ball-and-stick structure of (*S*)-naproxen.

Enantiomers of chiral molecules can elicit vastly different biological responses when ingested by living organisms. The tastes, odors, medicinal properties, toxicity, bactericidal, fungicidal, insecticidal, and other properties of enantiomers often differ widely. Here are some examples: The amino acid (*R*)-asparagine tastes sweet, while (*S*)-asparagine tastes bitter; (*R*)-carvone has the odor of spearmint, whereas (*S*)-carvone is responsible for the smell of caraway; (*S*)-naproxen is an important anti-inflammatory drug, while its enantiomer is a liver toxin; (*R,R*)-chloramphenicol is a useful antibiotic but its enantiomer is harmless to bacteria; (*R,R*)-paclobutrazol is a fungicide, while its enantiomer is a plant growth regulator; and (*R*)-thalidomide is a sedative and hypnotic, while its enantiomer is a potent teratogen.

How can it be that two molecules as similar in structure as enantiomers have such different biological activities? The reason is that biological activity is initiated by binding of the small molecule to a receptor molecule in the living organism to form a small molecule–receptor complex. The receptors are usually chiral, nonracemic molecules such as proteins, complex carbohydrates, or nucleic acids that bind well to only one of a pair of enantiomers. Because of their different three-dimensional shapes, (*R*)-asparagine binds to a receptor in humans that triggers a sensation of sweetness, whereas (*S*)-asparagine does not bind to that receptor. Instead, it binds to a differently shaped receptor that results in a bitter sensation.

A useful analogy for this selective binding is the manner in which your left shoe (receptor) interacts with your left and right feet (a pair of enantiomers). The left foot fits (binds) into the shoe comfortably (one response) whereas the right foot does not, or only uncomfortably (a different response).

There are now a number of examples where one enantiomer of a chiral molecule elicits a desirable biological response, while the other enantiomer results in a detrimental response (naproxen, for example). Therefore, it has become increasingly important for chiral pharmaceutical and agrochemical compounds to be marketed as single enantiomers rather than as racemic mixtures. This has recently stimulated the development of new synthetic

Suppose we start with one pure enantiomer of 3-chloro-1-butene, say, the *R* isomer. What can we say about the stereochemistry of the products? One way to see the answer quickly is to draw Fischer projections.

$$\begin{array}{c} CH_3 \\ Cl \overset{R}{-\!\!\!+\!\!\!-} H \\ CH \\ \| \\ CH_2 \end{array} \xrightarrow{HBr} \begin{array}{c} CH_3 \\ Cl \overset{R}{-\!\!\!+\!\!\!-} H \\ H \overset{R}{-\!\!\!+\!\!\!-} Br \\ CH_3 \end{array} + \begin{array}{c} CH_3 \\ Cl \overset{R}{-\!\!\!+\!\!\!-} H \\ Br \overset{S}{-\!\!\!+\!\!\!-} H \\ CH_3 \end{array} \quad (5.7)$$

(*R*)-3-chloro-1-butene (2*R*,3*R*)-2-bromo-3-chlorobutane (2*S*,3*R*)-2-bromo-3-chlorobutane

The configuration where the chloro substituent is located remains *R* and unchanged, but the new stereogenic center can be either *R* or *S*. Therefore, the products are *diastereomers.* Are they formed in equal amounts? No. Looking at the starting material in eq. 5.7, we can see that it has no plane of symmetry. Approach of the bromine to the double bond from the H side or from the Cl side of the stereogenic center should not occur with equal ease.

We can generalize this result. *Reaction of a chiral reagent with an achiral reagent, when it creates a new stereogenic center, leads to diastereomeric products at different rates and in unequal amounts.*

methods that provide only a single enantiomer of a chiral molecule, a process called *asymmetric synthesis.* In fact, the 2001 Nobel Prize in chemistry was awarded to William S. Knowles (Monsanto), K. Barry Sharpless (Scripps Institute), and Ryoji Noyori (Nagoya University) for their seminal and practical contributions to this field of research.

The need for enantiomerically pure compounds has also led to the development of new methods for separating racemic mixtures into their enantiomeric components, a process called resolution.

See Problem 5.57.

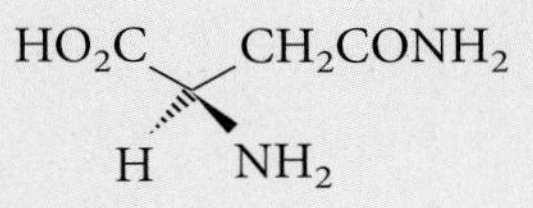

(*R*)-asparagine

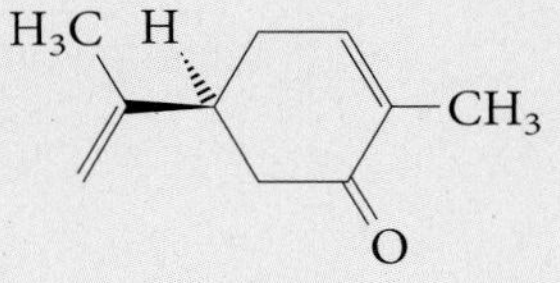

(*S*)-carvone

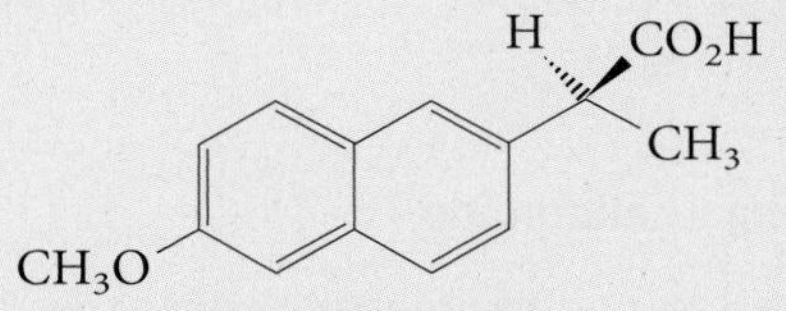

(*S*)-naproxen

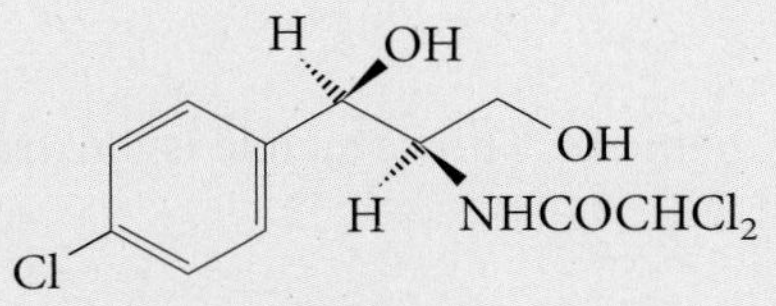

(*R,R*)-chloramphenicol

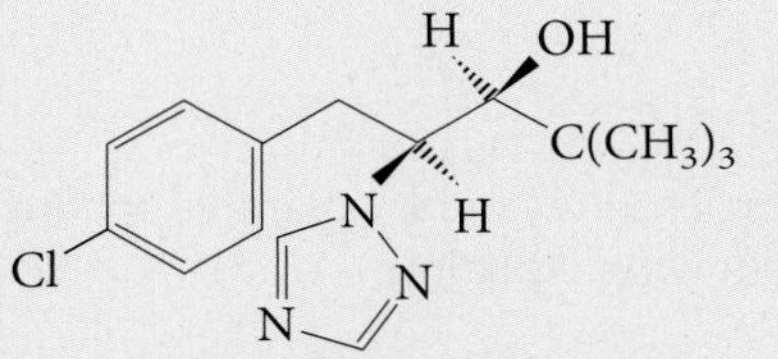

(*S,S*)-paclobutrazol

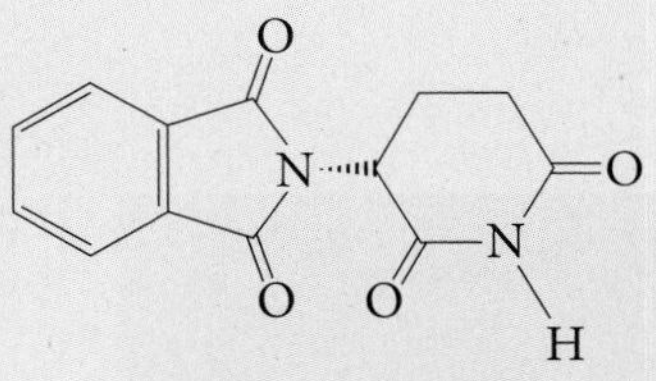

(*R*)-thalidomide

Enantiomers of Some Biologically Active Compounds

PROBLEM 5.25 Let us say that the (2*R*,3*R*) and (2*S*,3*R*) products in eq. 5.7 are formed in a 60:40 ratio. What products would be formed and in what ratio by adding HBr to pure (*S*)-3-chloro-1-butene? By adding HBr to a racemic mixture of (*R*)- and (*S*)-3-chloro-1-butene?

5.12 Resolution of a Racemic Mixture

We have just seen (eq. 5.5) that, when reaction between two achiral reagents leads to a chiral product, it always gives a racemic (50:50) mixture of enantiomers. Suppose we want to obtain each enantiomer pure and free of the other. The process of separating a racemic mixture into its enantiomers is called **resolution**. Since enantiomers have identical achiral properties, how can we resolve a racemic mixture into its components? The answer is to convert them to diastereomers, separate the *diastereomers,* and then reconvert the now-separated diastereomers back to enantiomers.

The process of separating the two enantiomers of a racemic (50:50) mixture is called **resolution**.

To separate two enantiomers, we first let them react with a chiral reagent. The product will be a pair of *diastereomers.* These, as we have seen earlier, differ in all types of

A CLOSER LOOK AT... Thalidomide

Conduct research on the Internet to find more information on thalidomide and to answer the following questions.

The History of Thalidomide

1. When and where was the drug thalidomide first used?
2. For what purpose was it used?
3. What are the unintended consequences of its use by pregnant women?

Patients Taking Thalidomide

1. During what period of pregnancy does this drug cause damage?
2. In addition to its teratogenic effects, thalidomide also can give rise to other side effects. What are some of these side effects?

Why We Still Use Thalidomide

1. What are some diseases for which thalidomide is proving to be an effective medication?
2. What are other potential uses of this drug?

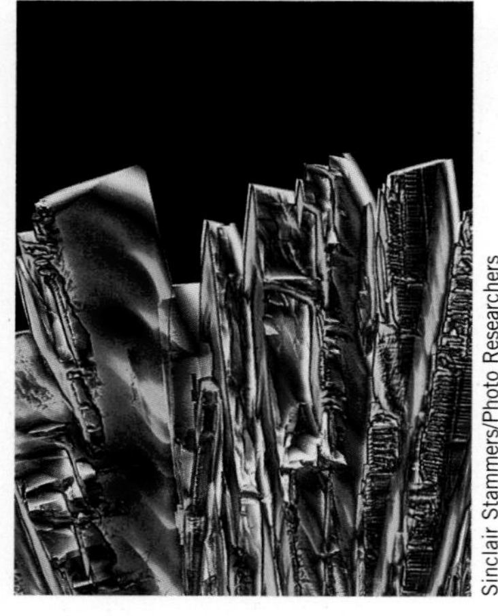
Sinclair Stammers/Photo Researchers

Tartaric acid crystals under polarized light.

properties and can be separated by ordinary methods. This principle is illustrated in the following equation:

$$\left\{\begin{matrix} R \\ S \end{matrix}\right\} + R \longrightarrow \left\{\begin{matrix} R—R \\ S—R \end{matrix}\right\} \quad (5.8)$$

pair of enantiomers (not separable) — chiral reagent — diastereomeric products (separable)

After the diastereomers are separated, we then carry out reactions that regenerate the chiral reagent and the separated enantiomers.

$$R—R \longrightarrow R + R \quad \text{and} \quad S—R \longrightarrow S + R \quad (5.9)$$

Louis Pasteur was the first to resolve a racemic mixture when he separated the sodium ammonium salts of (+)- and (−)-tartaric acid. In a sense, he was the chiral reagent, since he could distinguish between the right- and left-handed crystals. In Chapter 11, we will see a specific example of how this is done chemically.

The principle behind the resolution of racemic mixtures is the same as the principle involved in the specificity of many biological reactions. That is, a chiral reagent (in a cell, usually an enzyme) can discriminate between enantiomers.

A WORD ABOUT... Green Chemistry: L-DOPA

Enantiomers often have very different properties when interacting with biological enzymes that are chiral. As we just discussed, sometimes we use a resolution to convert an enantiomeric mixture to a set of diastereomers in order to separate them. This technique of resolving one enantiomer from another is a practical approach, but it is not necessarily good for the environment. Additional steps are needed to prepare and to separate the diastereomers, and then the enantiomers need to be regenerated as discussed in Section 5.12. Each of these steps would need solvents and, of course, would generate some waste. A greener approach is to prepare the desired enantiomer in its pure form and without formation of the undesired enantiomer.

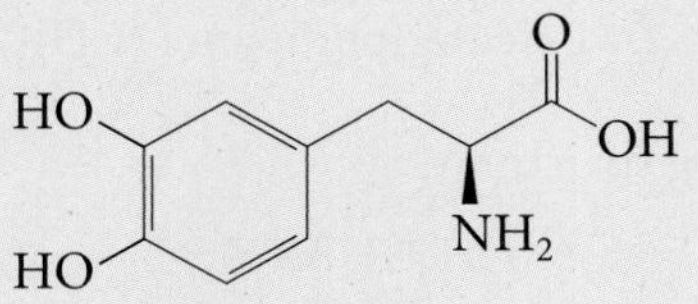

HO, HO, O, OH, NH_2

L-DOPA

L-DOPA is an amino acid that is found in some foods and herbs. L-DOPA is the precursor to a number of neurotransmitters, including dopamine and adrenaline (page 163). In 2000, Arvid Carlsson won a Nobel Prize in physiology or medicine for his work with L-DOPA that showed its ability to increase dopamine concentrations in the brains of Parkinson's patients.

L-DOPA has a stereogenic center and is chiral. Biochemically, it can be synthesized from other essential amino acids, but only the L-enantiomer is biologically useful. Many researchers have investigated the chemistry and biochemistry of L-DOPA. In 2001, the Nobel Prize in chemistry was awarded to work related to L-DOPA; one-fourth of the prize went to William Knowles for his work at the Monsanto company on the synthesis of this compound by an asymmetric hydrogenation reaction.

Why was a Nobel Prize awarded for the synthesis of L-DOPA? In an attempt to avoid many steps in resolution, Knowles and his coworkers prepared L-DOPA in an enantiomerically pure manner by using a chiral hydrogenation reaction to synthesize L-DOPA in 95% enantiomerically pure form. In this rhodium-catalyzed hydrogenation, they used a chiral and enantiomerically pure (*R*,*R*)-DiPAMP phosphine ligand. Since the phosphine ligand that surrounded the Rh metal was chiral, the formation of each of the two products occurred through a different transition state (an analogous situation to Figure 3.12).

Therefore, there was a lower activation barrier that created a kinetic preference for one enantiomer to be made over the other. The preferred enantiomeric product from that reaction could then be converted to the desired L-DOPA with a simple aqueous hydrolysis step with acid.

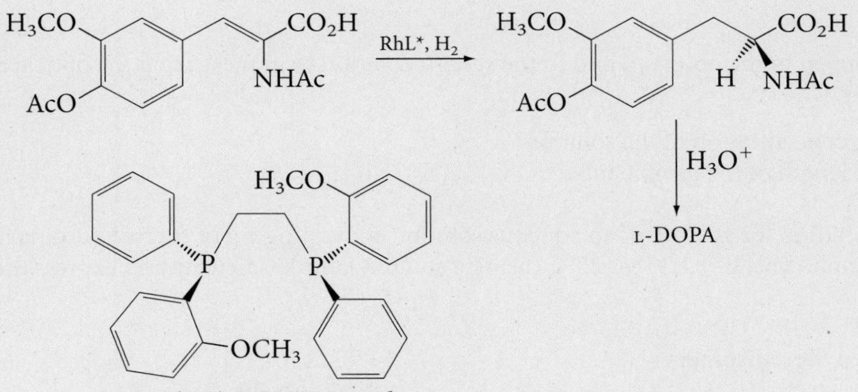

H_3CO, AcO, CO_2H, NHAc —(RhL^*, H_2)→ H_3CO, AcO, CO_2H, H, NHAc —(H_3O^+)→ L-DOPA

H_3CO, P, P, OCH_3

L = (*R*,*R*)-DiPAMP Ligand

This synthetic process avoided a number of purification steps and is an excellent example of environmentally efficient green chemistry.

See Problem 5.59.

ADDITIONAL PROBLEMS

OWL Interactive versions of these problems are assignable in OWL.

Stereochemistry: Definitions and Stereogenic Centers

5.26 Define or describe the following terms.

a. diastereomers
b. plane of symmetry
c. stereogenic center
d. *meso* form
e. racemic mixture
f. plane-polarized light
g. specific rotation
h. chiral molecule
i. enantiomers
j. resolution

5.27 Which of the following substances contain stereogenic centers? (*Hint:* Drawing the structures will help you answer this question.)

a. 2,2-dibromobutane
b. 3-methylcyclopentene
c. 1,2-difluoropropane
d. 2,3-dimethylheptane
e. methylcyclobutane
f. 1-deuteriopropanol (CH_3CH_2CHDOH)

5.28 Locate with an asterisk the stereogenic centers (if any) in the following structures.

a. $HOCH_2CH(OH)CH(OH)CHO$
b. $C_6H_5CH_2CH(OH)CO_2H$
c. $CH_3CHBrCF_3$

d. 

e. 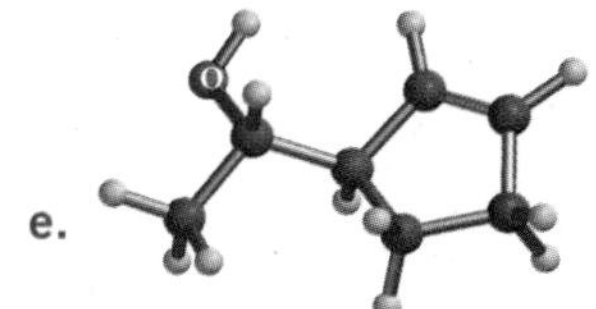

f. 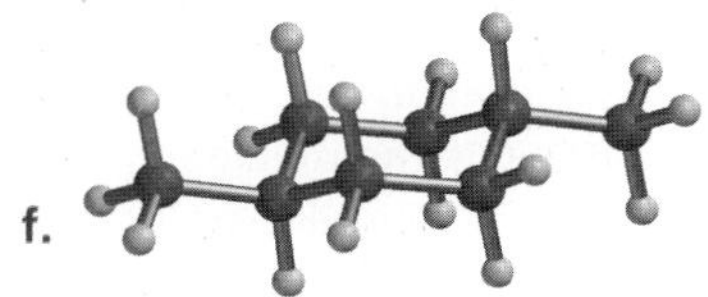

Optical Activity

5.29 What would happen to the observed and to the *specific* rotation if, in measuring the optical activity of a solution of sugar in water, we

a. doubled the concentration of the solution?
b. doubled the length of the sample tube?

5.30 The observed rotation for 100 mL of an aqueous solution containing 1 g of sucrose (ordinary sugar), placed in a 2-decimeter sample tube, is +1.33° at 25°C (using a sodium lamp). Calculate and express the specific rotation of sucrose.

Relationships Between Stereoisomers

5.31 Tell whether the following structures are identical or enantiomers:

a. 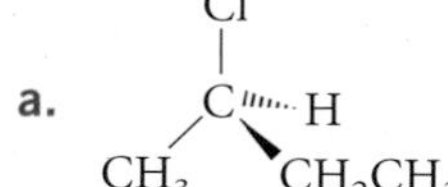

and 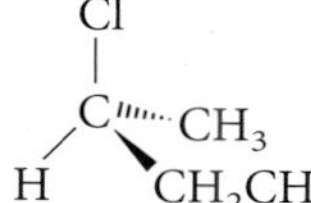

b. and 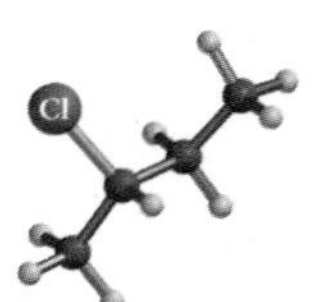

5.32 Draw a structural formula for an optically active compound with the molecular formula

a. C_6H_{14}
b. $C_5H_{11}Br$
c. $C_4H_{10}O$
d. $C_4H_8Cl_2$

5.33 Draw the formula of an unsaturated bromide, C_5H_9Br, that can show

a. neither *cis–trans* isomerism nor optical activity
b. *cis–trans* isomerism but no optical activity
c. no *cis–trans* isomerism but optical activity
d. *cis–trans* isomerism and optical activity

The *R-S* and *E-Z* Conventions

5.34 Place the members of the following groups in order of decreasing priority according to the *R-S* convention:

a. $CH_3—$, $H—$, $C_6H_5—$, $CH_3CH_2—$ (see page 120 for structure of $C_6H_5—$ group)
b. $CH_3CH_2—$, $CH_3CH_2CH_2—$, $CH_2{=}CH—$, $O{=}CH—$
c. $CH_3CH_2—$, $HS—$, $H—$, $Br—$
d. $CH_3—$, $HS—$, $BrCH_2—$, $HOCH_2—$

5.35 Assume that the four groups in each part of Problem 5.34 are attached to one carbon atom.

a. Draw a three-dimensional formula for the *R* configuration of the molecule in 5.34a and 5.34b.
b. Draw a three-dimensional formula for the *S* configuration of the molecule in 5.34c and 5.34d.

5.36 Tell whether the stereogenic centers marked with an asterisk in the following structures have the *R* or the *S* configuration:

a. CH_3, H, O, H, $CH(CH_3)_2$

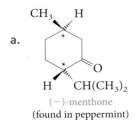

(−)-menthone
(found in peppermint)

b. CO_2H, $H_2N—C^*—H$, CH_2OH

(−)-serine
(an amino acid found in proteins)

c.

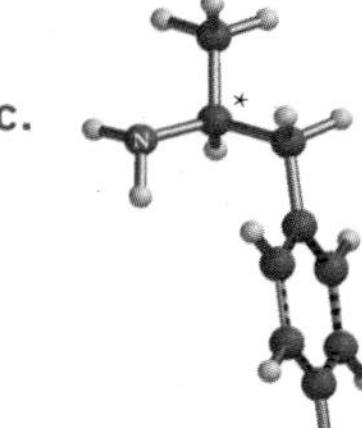

(+)-amphetamine
(central nervous system stimulant)

5.37 Determine the configuration, *R* or *S*, of (+)-carvone (page 147), the compound responsible for the odor of caraway seeds.

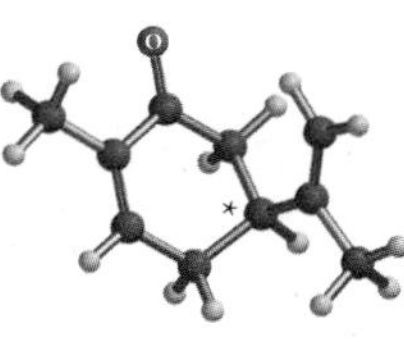

(+)-carvone

5.38 In a recent collaboration, French and American chemists found that (−)-bromochlorofluoromethane (CHBrClF), one of the simplest chiral molecules, has the *R* configuration. Draw a three-dimensional structural formula for (*R*)-(−)-bromochlorofluoromethane.

5.39 Name the following compounds, using *E-Z* notation:

a. 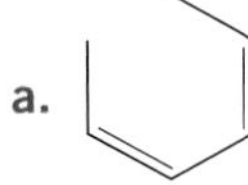

b.

c.

d.

5.40 Two possible isomers of 1,2-bromoethene are:

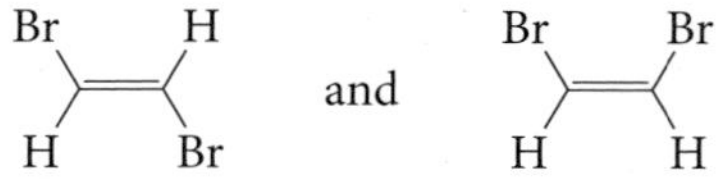

Br, H / H, Br (alkene) and Br, Br / H, H (alkene)

Classify them fully, according to the discussion in Section 5.10.

5.41 4-Chloro-2-pentene has a double bond that can have either the *E* or the *Z* configuration and a stereogenic center that can have either the *R* or the *S* configuration. How many stereoisomers are possible altogether? Draw the structure of each, and group the pairs of enantiomers.

5.42 How many stereoisomers are possible for each of the following structures? Draw them, and name each by the *R-S* and *E-Z* conventions. (See Problem 5.41.)

a. 2,5-dichloro-3-hexene
b. 2-chloro-5-fluoro-3-hexene
c. 3-methyl-1,4-pentadiene
d. 3-methyl-1,4-heptadiene

Fischer and Newman Projections

5.43 Which of the following Fischer projection formulas have the same configuration as **A**, and which are the enantiomer of **A?**

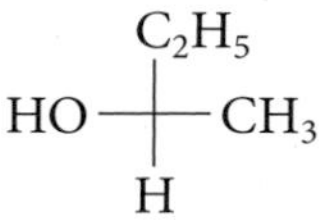

A: top C_2H_5; left HO; right CH_3; bottom H

a. top CH_3; left H; right OH; bottom C_2H_5

b. 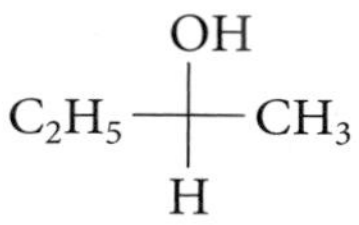 top OH; left C_2H_5; right CH_3; bottom H

c. top C_2H_5; left H; right CH_3; bottom OH

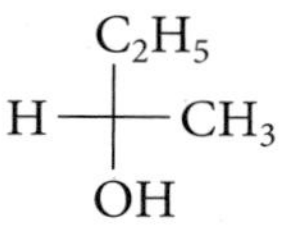

5.44 Following are Newman projections for the three tartaric acids (*R,R*), (*S,S*), and *meso*. Which is which?

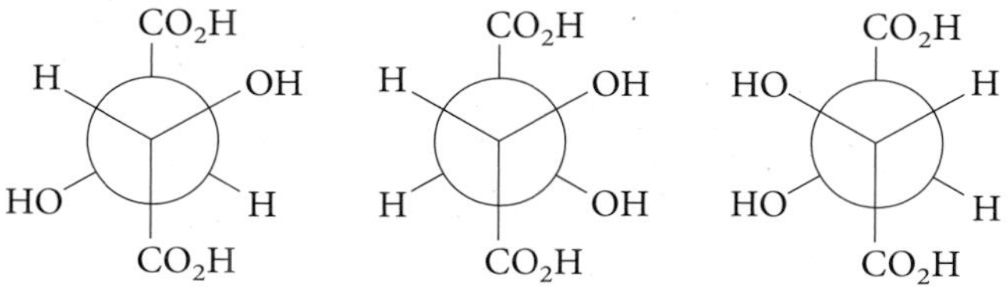

(1) front: CO_2H (top), H (upper left), OH (upper right); back: HO (lower left), H (lower right), CO_2H (bottom)

(2) front: CO_2H (top), H (upper left), OH (upper right); back: H (lower left), OH (lower right), CO_2H (bottom)

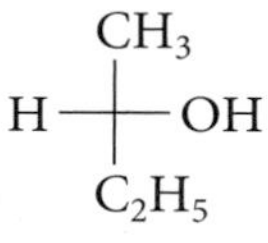

(3) front: CO_2H (top), HO (upper left), H (upper right); back: HO (lower left), H (lower right), CO_2H (bottom)

5.45 Convert the following sawhorse formula for one isomer of tartaric acid to a Fischer projection formula. Which isomer of tartaric acid is it?

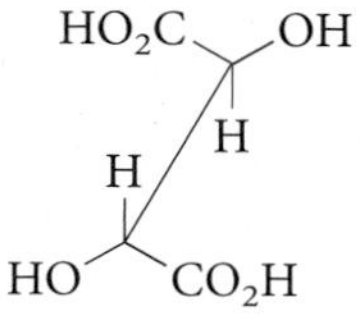

Upper carbon: HO_2C, OH, H; lower carbon: H, HO, CO_2H

5.46 Two possible isomeric structures of 1,2-dichloroethane are

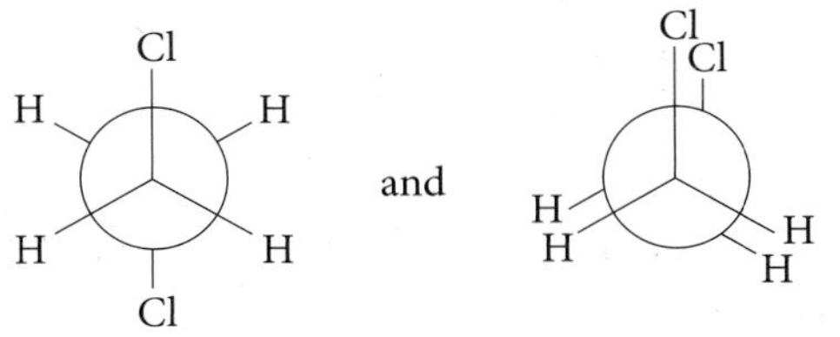

Classify them fully, according to the discussion in Section 5.10.

5.47 Two possible configurations for a molecule with three different stereogenic centers are (*R,R,R*) and its mirror image (*S,S,S*). What are all of the remaining possibilities? Repeat for a compound with four different stereogenic centers.

Stereochemistry: Natural and Synthetic Applications

5.48 The formula for muscarine, a toxic constituent of poisonous mushrooms, is

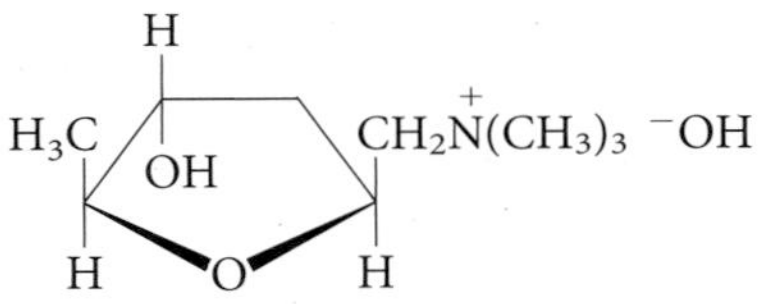

Is it chiral? How many stereoisomers of this structure are possible? An interesting murder mystery, which you might enjoy reading and which depends for its solution on the distinction between optically active and racemic forms of this poison, is Dorothy L. Sayers's *The Documents in the Case,* published in paperback by Avon Books. (See an article by H. Hart, "Accident, Suicide, or Murder? A Question of Stereochemistry," *J. Chem. Educ.,* **1975**, *52,* 444.)

5.49 Chloramphenicol is an antibiotic that is particularly effective against typhoid fever. Its structure is

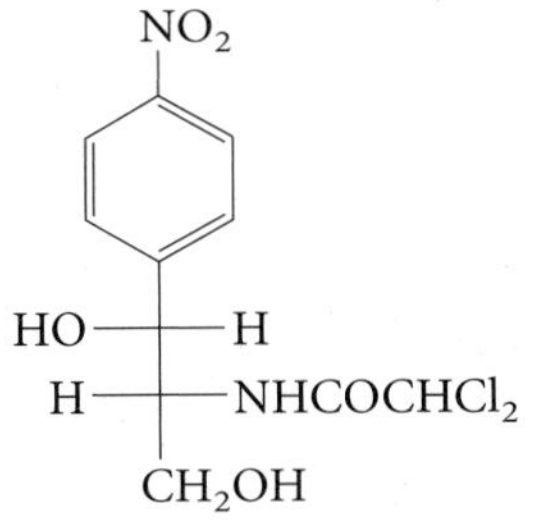

What is the configuration (*R,S*) at each stereogenic center?

5.50 Methoprene (marketed as Precor), an insect juvenile hormone mimic used in flea control products for pets, works by preventing the development of flea eggs and larvae. The effective form of methoprene, shown here, is optically active. Locate the stereogenic center and determine its configuration (*R,S*).

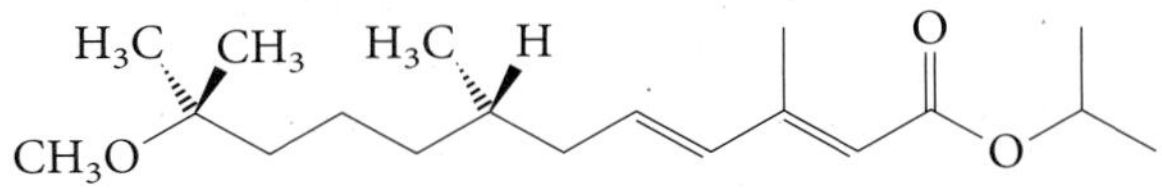

methoprene

© Stephen Dalton/Animals Animals

Adult cat flea jumping.

5.51 A South Korean research group has isolated and synthesized "daumone," the pheromone that induces hibernation in *Caenorhabditis elegans* worms when food becomes scarce, thus extending their life span (*Nature,* **2005**, *433,* 541). Following is a ball-and-stick representation of daumone.

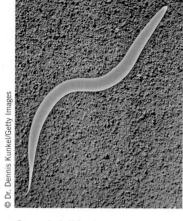

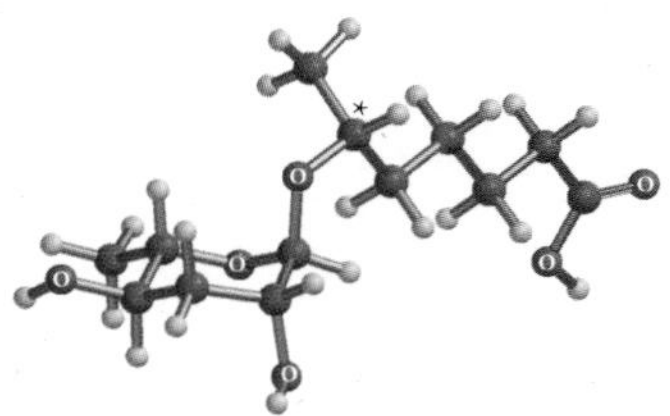

a. What is the absolute configuration, *R* or *S*, of the stereogenic carbon marked with an asterisk?

b. There are four additional stereogenic carbons in daumone. Locate these carbons and determine their absolute configurations.

c. Draw a representation of the enantiomer of daumone.

d. Do you think the enantiomer of daumone would induce hibernation in *Caenorhabditis elegans* worms? Why or why not?

Caenorhabditis elegans **worms.**

5.52 Mature crocodiles secrete from their skin glands the compound with the following structure. This compound is thought to be a communication pheromone for nesting or mating.

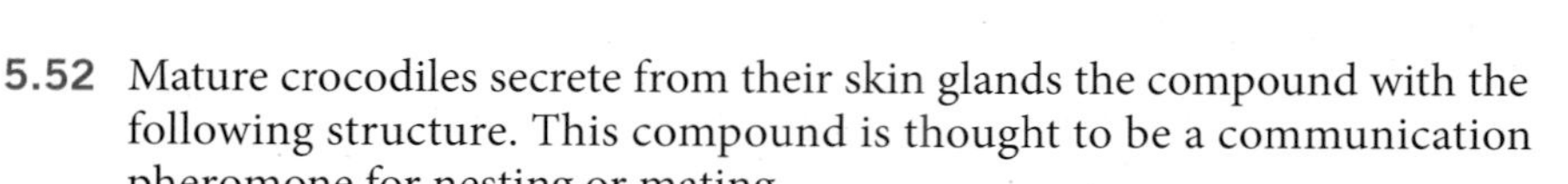

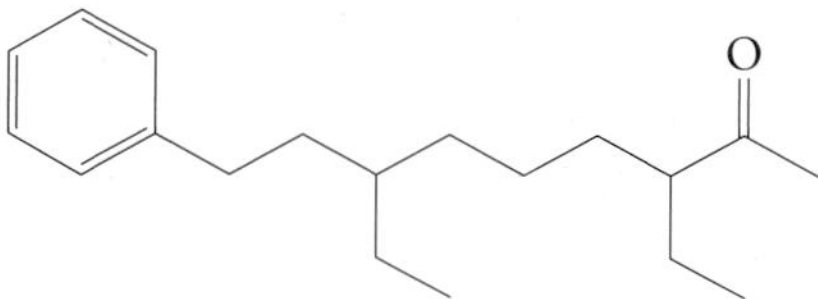

Adult crocodiles.

a. How many stereogenic centers are there in this compound? Mark them with an asterisk.

b. Two stereoisomers of this compound have been isolated from crocodile skin gland secretions. How many possible stereoisomers of this compound are there?

5.53 The structure of nerolidol is shown in Problem 3.40. What is the absolute configuration, *R* or *S*, of the stereogenic carbon in nerolidol?

5.54 Extract of *Ephedra sinica,* a Chinese herbal treatment for asthma, contains the compound ephedrine, which dilates the air passages of the lungs. The naturally occurring stereoisomer is levorotatory and has the structure shown here. (a) What is the configuration (*R,S*) at each stereogenic center? (b) How many stereoisomers of ephedrine are possible altogether? (c) Compare the structure of (−)-ephedrine to that of (−)-epinephrine. How are they similar and how do they differ?

H, $NHCH_3$, HO, H, CH_3

(−)-ephedrine

H, OH, HO, C, *, CH_2NHCH_3, HO

(−)-epinephrine
(adrenalin)

5.55 The structure of kavain, a natural relaxant popular in the South Pacific, is shown here. Draw a three-dimensional structure for the *R* configuration of the molecule.

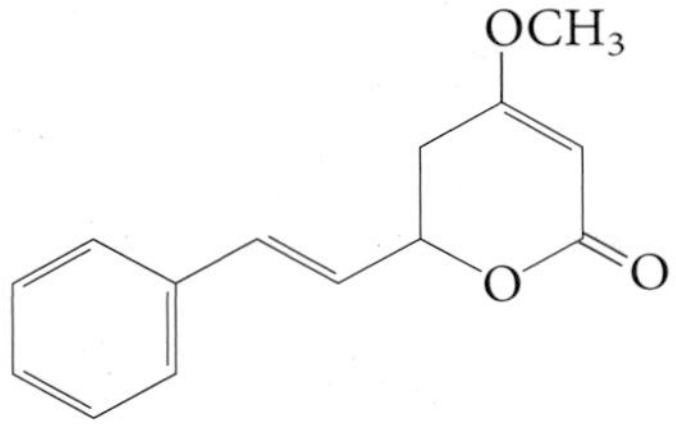

OCH_3

kavain

Stereochemistry and Chemical Reactions

5.56 When (*R*)-2-chlorobutane is chlorinated, we obtain some 2,3-dichlorobutane. It consists of 71% *meso* isomer and 29% racemic isomers. Explain why the mixture need not be 50:50 *meso* and (2*R*,3*R*)-2,3-dichlorobutane.

5.57 (+)- and (−)-Carvone [see Problem 5.37 for the structure of (+)-carvone] are enantiomers that have very different odors and are responsible for the odors of caraway seeds and spearmint, respectively. Suggest a possible explanation.

5.58 What can you say about the stereochemistry of the products in the following reactions? (See Sec. 5.11 and eq. 3.28.)

a. 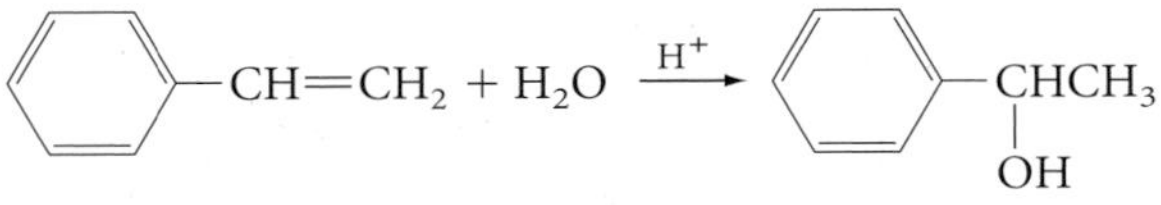 $C_6H_5-CH{=}CH_2 + H_2O \xrightarrow{H^+} C_6H_5-CH(OH)CH_3$

b. $CH_3CH(OH)CH{=}CH_2 + H_2O \xrightarrow{H^+} CH_3CH(OH)-CH(OH)CH_3$

(*R*-enantiomer)

5.59 As discussed in the "A Word About . . . Green Chemistry: L-DOPA" on page 175, the (*R*,*R*)-DiPAMP phosphine ligand was used to prepare the precursor to L-DOPA. What is the absolute configuration, *R* or *S*, of the stereogenic carbon in L-DOPA? What do you expect to be the product of the analogous hydrogenation reaction if (*S*,*S*)-DiPAMP was used as the phosphine ligand?

5.60 As discussed in the "A Word About . . . Pasteur's Experiments and the van't Hoff-LeBel Explanation" on pages 160–161, a critical experiment and observation made by Pasteur was that when he dissolved pure forms of the left- and right-handed crystals, he found that each solution was optically active, even though each set of crystals was obtained from crystallization of the racemic acid. The optical rotations of (*R*,*R*)- and (*S*,*S*)-tartaric acid are provided in Figure 5.14. Predict the specific rotation of a solution that contains 75% (*R*,*R*) and 25% (*S*,*S*) crystals of tartaric acid.

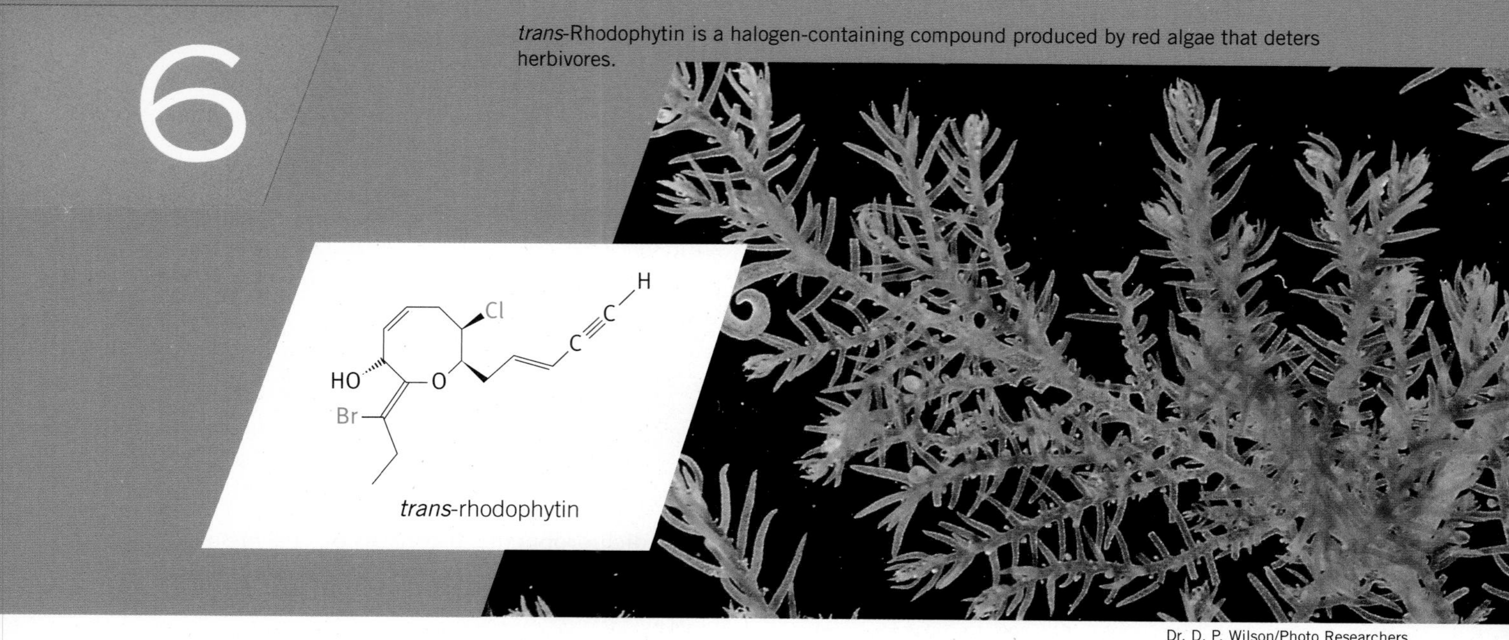

trans-Rhodophytin is a halogen-containing compound produced by red algae that deters herbivores.

Dr. D. P. Wilson/Photo Researchers

Organic Halogen Compounds; Substitution and Elimination Reactions

Chlorine- and bromine-containing natural products have been isolated from various species that live in the sea—sponges, mollusks, and other ocean creatures that adapted to their environment by metabolizing inorganic chlorides and bromides that are prevalent there. With these exceptions, most organic halogen compounds are creatures of the laboratory.

Halogen compounds are important for several reasons. Simple alkyl and aryl halides, especially chlorides and bromides, are versatile reagents in syntheses. Through *substitution reactions*, which we will discuss in this chapter, halogens can be replaced by many other functional groups. Organic halides can be converted to unsaturated compounds through dehydrohalogenation. Also, some halogen compounds, especially those that contain two or more halogen atoms per molecule, have practical uses; for example, as solvents, insecticides, herbicides, fire retardants, cleaning fluids, and refrigerants, and in polymers such as Teflon. In this chapter, we will discuss all of the aspects of halogen compounds described.

Online homework for this chapter can be assigned in OWL, an online homework assessment tool.

6.1 Nucleophilic Substitution

Let us look at a typical **nucleophilic substitution reaction**. Ethyl bromide (bromoethane) reacts with hydroxide ion to give ethyl alcohol and bromide ion.*

Alkyl halides undergo **nucleophilic substitution reactions**, in which a nucleophile displaces the halide **leaving group** from the alkyl halide **substrate**.

$$\underset{}{HO^-} + \underset{\text{ethyl bromide}}{CH_3CH_2-Br} \xrightarrow{H_2O} \underset{\text{ethanol}}{CH_3CH_2-OH} + Br^- \tag{6.1}$$

Hydroxide ion is the *nucleophile* (Sec. 3.9). It reacts with the **substrate** (ethyl bromide) and displaces bromide ion. The bromide ion is called the **leaving group**.

In reactions of this type, one covalent bond is broken, and a new covalent bond is formed. In this example, the carbon–bromine bond is broken and the carbon–oxygen bond is formed. The leaving group (bromide) takes with it *both* of the electrons from the carbon–bromine bond, and the nucleophile (hydroxide ion) supplies *both* electrons for the new carbon–oxygen bond.

These ideas are generalized in the following equations for a nucleophilic substitution reaction:

$$\underset{\substack{\text{nucleophile}\\\text{(neutral)}}}{Nu{:}} + \underset{\text{substrate}}{R{:}L} \longrightarrow \underset{\text{product}}{R{:}\overset{+}{Nu}} + \underset{\substack{\text{leaving}\\\text{group}}}{{:}L^-} \tag{6.2}$$

$$\underset{\substack{\text{nucleophile}\\\text{(anion)}}}{Nu{:}^-} + \underset{\text{substrate}}{R{:}L} \longrightarrow \underset{\text{product}}{R{:}Nu} + \underset{\substack{\text{leaving}\\\text{group}}}{{:}L^-} \tag{6.3}$$

If the nucleophile and substrate are neutral (eq. 6.2), the product will be positively charged. If the nucleophile is a negative ion and the substrate is neutral (eq. 6.3), the product will be neutral. In either case, an unshared electron pair on the nucleophile supplies the electrons for the new covalent bond.

In principle, of course, these reactions may be reversible because the leaving group also has an unshared electron pair that can be used to form a covalent bond. However, we can use various methods to force the reactions to go in the forward direction. For example, we can choose the nucleophile so that it is a *stronger* nucleophile than the leaving group. Or we can shift the equilibrium by using a large excess of one reagent or by removing one of the products as it is formed. Nucleophilic substitution is a versatile reaction, widely used in organic synthesis.

6.2 Examples of Nucleophilic Substitutions

Nucleophiles can be classified according to the kind of atom that forms a new covalent bond. For example, the hydroxide ion in eq. 6.1 is an *oxygen* nucleophile. In the product, a new carbon–*oxygen* bond is formed. *The most common nucleophiles are oxygen, nitrogen, sulfur, halogen, and carbon nucleophiles.* Table 6.1 shows some examples of nucleophiles and the products that they form when they react with an alkyl halide.

Let us consider a few specific examples of these reactions, to see how they may be used in synthesis.

*The nomenclature of alkyl halides was discussed in Section 2.4.

Table 6.1 ■ Reactions of Common Nucleophiles with Alkyl Halides (Eqs. 6.2 and 6.3)

Nu		R—Nu		
Formula	**Name**	**Formula**	**Name**	**Comments**
Oxygen nucleophiles				
1. $HO\ddot{}:^-$	hydroxide	R—OH	alcohol	
2. $RO:^-$	alkoxide	R—OR	ether	
3. HOH	water	$R-\overset{+}{O}H_2$	alkyloxonium ion	These ions lose a proton and the products are alcohols and ethers. $\xrightarrow{-H^+}$ ROH (alcohol)
4. ROH	alcohol	$R-\overset{+}{O}(R)H$	dialkyloxonium ion	$\xrightarrow{-H^+}$ ROR (ether)
5. $R-C(=O)O:^-$	carboxylate	$R-OC(=O)-R$	ester	
Nitrogen nucleophiles				
6. NH_3	ammonia	$R-\overset{+}{N}H_3$	alkylammonium ion	With a base, these ions readily lose a proton to give amines. $\xrightarrow{-H^+}$ RNH_2
7. RNH_2	primary amine	$R-\overset{+}{N}H_2R$	dialkylammonium ion	$\xrightarrow{-H^+}$ R_2NH
8. R_2NH	secondary amine	$R-\overset{+}{N}HR_2$	trialkylammonium ion	$\xrightarrow{-H^+}$ $R_3N:$
9. R_3N	tertiary amine	$R-\overset{+}{N}R_3$	tetraalkylammonium ion	
Sulfur nucleophiles				
10. $HS:^-$	hydrosulfide	R—SH	thiol	
11. $RS:^-$	mercaptide	R—SR	thioether (sulfide)	
12. $R_2S:$	thioether	$R-\overset{+}{S}R_2$	trialkylsulfonium ion	
Halogen nucleophiles				
13. $:I:^-$	iodide	R—I:	alkyl iodide	The usual solvent is acetone. Sodium iodide is soluble in acetone, but sodium bromide and sodium chloride are not.
Carbon nucleophiles				
14. $^-:C\equiv N:$	cyanide	$R-C\equiv N:$	alkyl cyanide (nitrile)	Sometimes the isonitrile, $R-\overset{+}{N}\equiv\overset{-}{C}:$, is formed.
15. $^-:C\equiv CR$	acetylide	$R-C\equiv CR$	alkyne	

EXAMPLE 6.1

Use Table 6.1 to write an equation for the reaction of sodium ethoxide with bromoethane.

Solution

$$\underset{\text{sodium ethoxide}}{CH_3CH_2\ddot{\underset{..}{O}}{:}^-\ Na^+} + \underset{\text{bromoethane}}{CH_3CH_2Br} \longrightarrow \underset{\text{diethyl ether}}{CH_3CH_2\ddot{\underset{..}{O}}CH_2CH_3} + Na^+Br^-$$

Ethoxide is the nucleophile (entry 2 in Table 6.1), bromoethane is the substrate, and bromide ion is the leaving group. The product is diethyl ether, which is used as an anesthetic. Notice that the counterion of the nucleophile, Na^+, is merely a spectator during the reaction. It is present at the beginning and end of the reaction.

EXAMPLE 6.2

Devise a synthesis for propyl cyanide in which a nucleophilic substitution reaction is used.

Solution First, write the structure of the desired product.

$$\underset{\text{propyl cyanide}}{CH_3CH_2CH_2{-}C{\equiv}N{:}}$$

If we use cyanide ion as the nucleophile (entry 14 in Table 6.1), the alkyl halide must have the halogen (Cl, Br, or I) attached to a propyl group. The equation is

$$^-{:}C{\equiv}N{:} + CH_3CH_2CH_2Br \longrightarrow CH_3CH_2CH_2C{\equiv}N{:} + Br^-$$

Sodium cyanide or potassium cyanide can be used to supply the nucleophile.

EXAMPLE 6.3

Show how 1-butyne could be converted to 3-hexyne using a nucleophilic substitution reaction.

Solution Compare the starting material with the product.

$$\underset{\text{1-butyne}}{CH_3CH_2C{\equiv}CH} \qquad \underset{\text{3-hexyne}}{CH_3CH_2C{\equiv}CCH_2CH_3}$$

From Table 6.1, entry 15, we see that acetylides react with alkyl halides to give acetylenes. We therefore need to convert 1-butyne to an acetylide (review eq. 3.53), then treat it with a 2-carbon alkyl halide.

$$CH_3CH_2C{\equiv}CH + NaNH_2 \xrightarrow{NH_3} CH_3CH_2C{\equiv}C{:}^-\ Na^+$$

$$CH_3CH_2C{\equiv}C{:}^-\ Na^+ + CH_3CH_2Br \longrightarrow CH_3CH_2C{\equiv}CCH_2CH_3 + Na^+Br^-$$

EXAMPLE 6.4

Complete the following equation:

$$:NH_3 + CH_3CH_2CH_2Br \longrightarrow$$

Solution Ammonia is a nitrogen nucleophile (Table 6.1, entry 6). Since both reactants are neutral, the product has a positive charge (the formal +1 charge is on the nitrogen—check it out!).

$$:NH_3 + CH_3CH_2CH_2Br \longrightarrow CH_3CH_2CH_2{-}\overset{+}{N}H_3 + Br^-$$

PROBLEM 6.1 Using Table 6.1, write complete equations for the following nucleophilic substitution reactions:

a. $Na^{+}\,{}^{-}OH + CH_3CH_2CH_2Br$

b. $(CH_3CH_2)_3N: + CH_3CH_2Br$

c. $Na^{+}\,{}^{-}SH$ + (benzene ring)$-CH_2Br$

PROBLEM 6.2 Write an equation for the preparation of each of the following compounds, using a nucleophilic substitution reaction. In each case, label the nucleophile, the substrate, and the leaving group.

a. $(CH_3CH_2)_3N$

b. $CH_3CH_2CH_2OH$

c. $(CH_3)_2CHCH_2C{\equiv}N$

d. $CH_3CH_2CH_2CH_2OCH_3$

e. $(CH_3CH_2)_3S^{+}Br^{-}$

f. $CH_2{=}CHCH_2I$

The substitution reactions in Table 6.1 have some limitations with respect to the structure of the *R* group in the alkyl halide. For example, these are reactions of *alkyl* halides (halogen bonded to sp^3-hybridized carbon). *Aryl* halides and *vinyl* halides, in which the halogen is bonded to sp^2-hybridized carbon, do not undergo this type of nucleophilic substitution reaction. Another important limitation often occurs when the nucleophile is an anion or a base or both. For example,

$$\underset{\text{anion}}{{}^{-}CN} + \underset{\text{primary alkyl halide*}}{CH_3CH_2CH_2CH_2Br} \longrightarrow CH_3CH_2CH_2CH_2CN + Br^- \qquad \textbf{(6.4)}$$

but

$$\underset{\text{anion}}{{}^{-}CN} + \underset{\text{tertiary alkyl halide}}{CH_3{-}\underset{Br}{\overset{CH_3}{C}}{-}CH_3} \longrightarrow \underset{\text{methylpropene}}{CH_3{-}\overset{CH_2}{\overset{\|}{C}}{-}CH_3} + HCN + Br^- \qquad \textbf{(6.5)}$$

Another example is

$$\underset{\text{neutral, not very basic}}{H_2O} + \underset{\text{tertiary alkyl halide}}{CH_3{-}\underset{Br}{\overset{CH_3}{C}}{-}CH_3} \longrightarrow \underset{\text{(about 80\%; some methylpropene is also formed)}}{CH_3{-}\underset{OH}{\overset{CH_3}{C}}{-}CH_3} + H^+ + Br^- \qquad \textbf{(6.6)}$$

*For the definition of primary, secondary, and tertiary alkyl groups, review Section 3.10.

but

$$\underset{\substack{\text{strong}\\\text{base}}}{^-OH} + \underset{\text{tertiary alkyl halide}}{CH_3-\overset{\overset{\displaystyle CH_3}{|}}{\underset{\underset{\displaystyle Br}{|}}{C}}-CH_3} \longrightarrow \underset{\text{methylpropene}}{CH_3-\overset{\overset{\displaystyle CH_2}{\|}}{C}-CH_3} + \underset{(H-OH)}{H_2O} + Br^- \quad (6.7)$$

To understand these differences, we must consider the mechanisms by which the substitutions in Table 6.1 take place.

6.3 Nucleophilic Substitution Mechanisms

As a result of experiments that began more than 70 years ago, we now understand the mechanisms of nucleophilic substitution reactions rather well. We use the plural because such *nucleophilic substitutions occur by more than one mechanism.* The mechanism observed in a particular case depends on the structures of the nucleophile and the alkyl halide, the solvent, the reaction temperature, and other factors.

There are two main nucleophilic substitution mechanisms. These are described by the symbols $\mathbf{S_N2}$ and $\mathbf{S_N1}$, respectively. The S_N part of each symbol stands for "substitution, nucleophilic." The meaning of the numbers 2 and 1 will become clear as we discuss each mechanism.

6.4 The S_N2 Mechanism

The **S_N2 mechanism** is a one-step process, represented by the following equation:

The **S_N2 mechanism** is a one-step process in which the bond to the leaving group begins to break as the bond to the nucleophile begins to form.

$$\underset{\text{nucleophile}}{Nu:^-} + \underset{\text{substrate}}{>C-L} \longrightarrow \underset{\text{transition state}}{\left[\overset{\delta-}{Nu}\cdots C\cdots\overset{\delta-}{L}\right]^{TS}} \longrightarrow \underset{\text{product}}{Nu-C<} + \underset{\substack{\text{leaving}\\\text{group}}}{:L^-} \quad (6.8)$$

The nucleophile attacks from the *back*side of the C—L bond (remember, there is a small "back" lobe to an sp^3 hybrid bond orbital; see Figure 1.7). At some stage (the transition state), the nucleophile *and* the leaving group are *both* partly bonded to the carbon at which substitution occurs. As the leaving group departs *with its electron pair,* the nucleophile supplies another electron pair to the carbon atom.

The number 2 is used in describing this mechanism because the reaction is *bi*molecular. That is, two molecules—the nucleophile and the substrate—are involved in the key step (the *only* step) in the reaction mechanism. The reaction shown in eq. 6.1 occurs by an S_N2 mechanism. A reaction energy diagram is shown in Figure 6.1.

PROBLEM 6.3 Draw a reaction energy diagram for the reaction between $CH_3CH_2CH_2Br$ and sodium cyanide (NaCN). Label the energy of activation (E_a) and ΔH for the reaction. (Refer to Sec. 3.12 if you need help.)

How can we recognize when a particular nucleophile and substrate react by the S_N2 mechanism? There are several telltale signs.

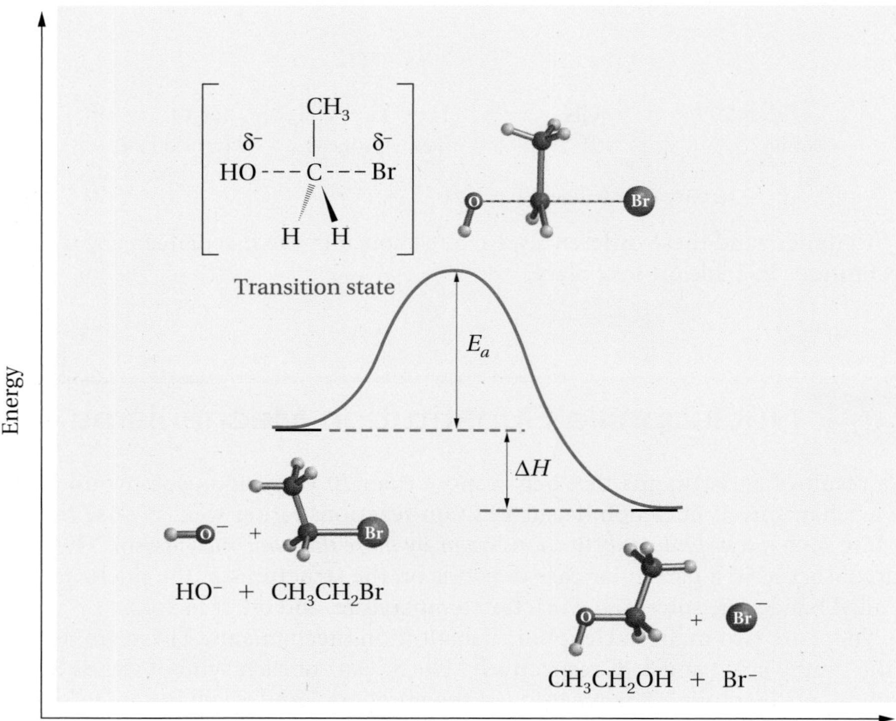

Figure 6.1
Reaction energy diagram for an S_N2 reaction.

1. *The rate of the reaction depends on both the nucleophile and the substrate concentrations.* The reaction of hydroxide ion with ethyl bromide (eq. 6.1) is an example of an S_N2 reaction. If we double the nucleophile concentration (HO^-), the reaction goes twice as fast. The same thing happens if we double the ethyl bromide concentration. We will see shortly that this rate behavior is *not* observed in the S_N1 mechanism.
2. *Every S_N2 displacement occurs with inversion of configuration.* For example, if we treat (*R*)-2-bromobutane with sodium hydroxide, we obtain (*S*)-2-butanol.

$$HO^- + \underset{(R)\text{-2-bromobutane}}{H(CH_3)(CH_2CH_3)C{-}Br} \longrightarrow \underset{(S)\text{-2-butanol}}{HO{-}C(CH_3)(H)(CH_2CH_3)} + Br^- \quad \textbf{(6.9)}$$

This experimental result, which at first came as a surprise to chemists, meant that the OH group did *not* take the exact position occupied by the Br. If it had, the configuration would have been retained; (*R*)-bromide would have given (*R*)-alcohol. What is the only reasonable explanation? The hydroxide ion must attack the C—Br bond from the rear. As substitution occurs, the three groups attached to the sp^3 carbon *invert,* somewhat like an umbrella caught in a strong wind.*
3. The reaction is fastest when the alkyl group of the substrate is methyl or primary and slowest when it is tertiary. Secondary alkyl halides react at an intermediate rate. The reason for this reactivity order is fairly obvious if we think about the S_N2 mechanism. The rear side of the carbon, where displacement occurs, is more crowded if more alkyl groups are attached to it, thus slowing down the reaction rate.**

*Because HO^- is both an anion and a base, a competing reaction is the formation of alkenes (see eqs. 6.5 and 6.7). We will see this again in Section 6.7.

**In eqs. 6.10 and 6.11, the leaving group L (see eq. 6.8) is a halide, which is given the general symbol X. We will see later (Chapter 7) that groups other than halides can behave as leaving groups.

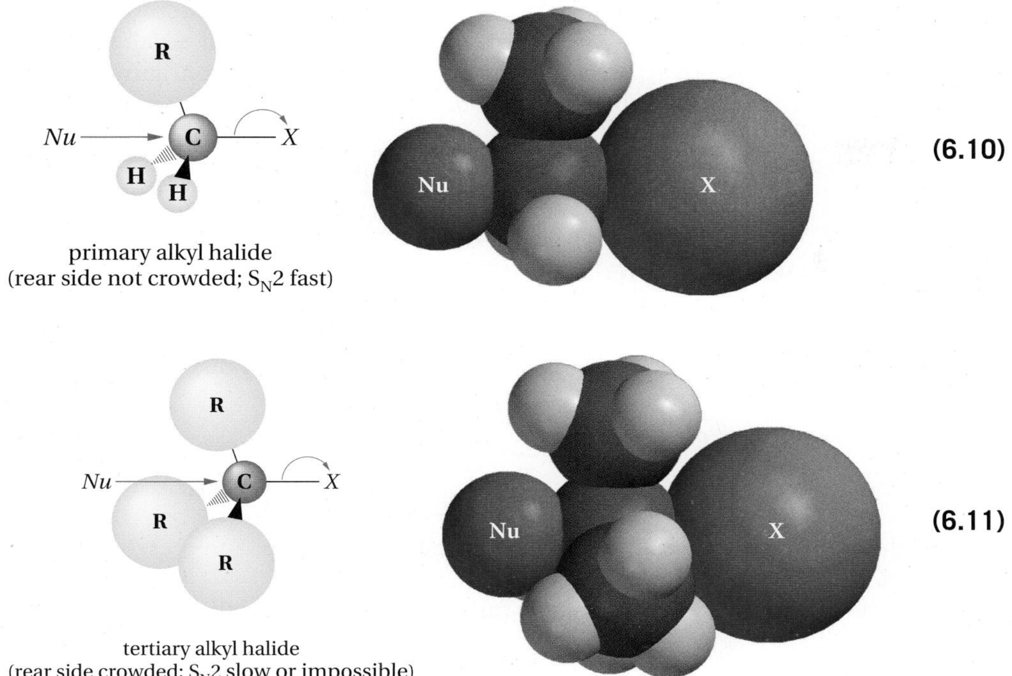

(6.10)

primary alkyl halide (rear side not crowded; S_N2 fast)

(6.11)

tertiary alkyl halide (rear side crowded; S_N2 slow or impossible)

EXAMPLE 6.5

Predict the product from the S_N2 reaction of *cis*-4-methylcyclohexyl bromide with cyanide ion.

Solution

cis (CH_3, Br, H) $\xrightarrow{^-C\equiv N}$ [*transition state*: CH_3, $\delta-$ Br, H, $C\equiv N$ $\delta-$] $\longrightarrow$ *trans* (CH_3, H, $C\equiv N$) + Br^-

Cyanide ion attacks the C—Br bond from the rear and therefore the cyano group ends up *trans* to the methyl group.

PROBLEM 6.4 Predict the product from the S_N2 reaction of

a. *trans*-4-methylcyclohexyl bromide with cyanide ion.
b. (*S*)-2-bromopentane with cyanide ion.
c. (*R*)-2-chlorobutane with NaSH.

PROBLEM 6.5 Arrange the following compounds in order of *decreasing* S_N2 reactivity toward sodium ethoxide:

CH_3CH_2CHBr (with CH_3 on the CHBr carbon) CH_3CHCH_2Br (with CH_3 on the CH carbon) $CH_3CH_2CH_2CH_2Br$

To summarize, the S_N2 mechanism is a one-step process favored for methyl and primary halides. It occurs more slowly with secondary halides and usually not at all with tertiary halides. An S_N2 reaction occurs with inversion of configuration, and its rate depends on the concentration of *both* the nucleophile and the substrate (the alkyl halide).

Now let us see how these features differ for the S_N1 mechanism.

6.5 The S_N1 Mechanism

The S_N1 **mechanism** is a two-step process: the bond between the carbon and the leaving group breaks first and then the resulting carbocation combines with the nucleophile.

The S_N**1 mechanism** is a two-step process. In the first step, which is slow, the bond between the carbon and the leaving group breaks as the substrate dissociates (ionizes).

$$\text{substrate } C{-}L \underset{}{\overset{\text{slow}}{\rightleftharpoons}} \text{carbocation } C^{+} + {}^{-}{:}L \text{ leaving group} \quad (6.12)$$

The electrons of the C—L bond go with the leaving group, and a carbocation is formed.

In the second step, which is fast, the carbocation combines with the nucleophile to give the product.

$$\text{carbocation } C^{+} + {:}Nu \text{ (nucleophile)} \xrightarrow{\text{fast}} C{-}\overset{+}{Nu} \text{ and } \overset{+}{Nu}{-}C \quad (6.13)$$

When the nucleophile is a neutral molecule, such as water or an alcohol, loss of a proton from the nucleophilic oxygen, in a third step, gives the final product.

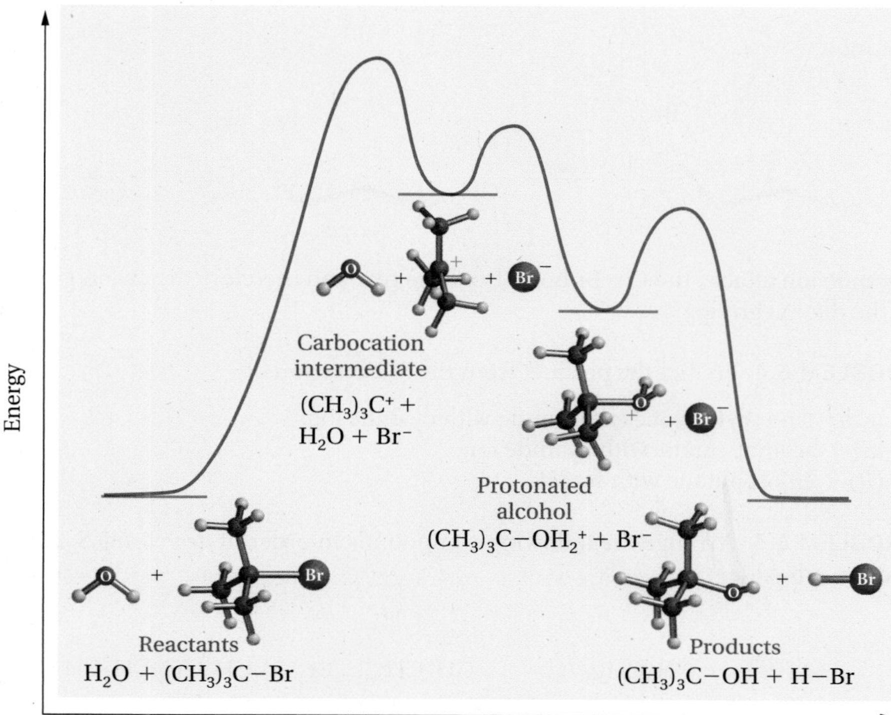

■ **Figure 6.2**
Reaction energy diagram for an S_N1 reaction.

The number 1 is used to designate this mechanism because the slow, or rate-determining, step involves *only one* of the two reactants: the substrate (eq. 6.12). It does *not* involve the nucleophile at all. That is, the first step is *uni*molecular. The reaction shown in eq. 6.6 occurs by an S_N1 mechanism, and a reaction energy diagram for that reaction is shown in Figure 6.2. Notice that the energy diagram for this reaction, and all S_N1 reactions, resembles that of an electrophilic addition to an alkene (Figure 3.11), another reaction that has a carbocation intermediate. Also notice that the energy of activation for the first step (the rate-determining step) is much greater than for subsequent steps. This first step forms the highest energy (most unstable) species in the reaction energy diagram.

PROBLEM 6.6 What are the products expected from the reaction of $(CH_3)_3C—Cl$ with $CH_3—OH$? Draw a reaction energy diagram for the reaction.

How can we recognize when a particular nucleophile and substrate react by the S_N1 mechanism? Here are the signs:

1. *The rate of the reaction does not depend on the concentration of the nucleophile.* The first step is rate-determining, and the nucleophile is not involved in this step. Therefore, the bottleneck in the reaction rate is the rate of formation of the carbocation, not the rate of its reaction with the nucleophile, which is nearly instantaneous.
2. If the carbon bearing the leaving group is stereogenic, the reaction occurs mainly with loss of optical activity (that is, with racemization). In carbocations, only three groups are attached to the positively charged carbon. Therefore, the positively charged carbon is sp^2-hybridized and planar. As shown in eq. 6.13, the nucleophile can react at either "face" of the carbocation to give a 50:50 mixture of two enantiomers, a racemic mixture. For example, the reaction of (*R*)-3-bromo-3-methylhexane with water gives the racemic alcohol.

Br; CH_3CH_2, $CH_3CH_2CH_2$, CH_3 on C
(*R*)-3-bromo-3-methylhexane

aqueous acetone →

CH_3CH_2, $CH_3CH_2CH_2$ on C(+)—CH_3 + Br^-
carbocation intermediate

(see item 3 in Table 6.1)

↓ H_2O
↓ $-H^+$ **(6.14)**

CH_3CH_2, $CH_3CH_2CH_2$, CH_3 on C—OH
50% *S*
(product from attack on the bottom face of the carbocation)

+

OH on C with CH_3CH_2, $CH_3CH_2CH_2$, CH_3
50% *R*
(product from attack on the top face of the carbocation)

The intermediate carbocation is planar and achiral. Combination with H_2O from the "top" or "bottom" of the carbocation intermediate is equally probable, giving the *R* and *S* alcohols, respectively, in equal amounts.

3. The reaction is fastest when the alkyl group of the substrate is tertiary and slowest when it is primary. The reason is that S_N1 reactions proceed via carbocations, so the reactivity order corresponds to that of carbocation stability (3° > 2° > 1°).

That is, the easier it is to form the carbocation, the faster the reaction will proceed. For this reason, S_N1 reactivity is also favored for resonance-stabilized carbocations, such as allylic carbocations (see Sec. 3.15). Likewise, S_N1 reactivity is disfavored for aryl and vinyl halides because aryl and vinyl carbocations are unstable and not easily formed.

PROBLEM 6.7 Which of the following bromides will react faster with methanol (via an S_N1 reaction)? What are the reaction products in each case?

a. 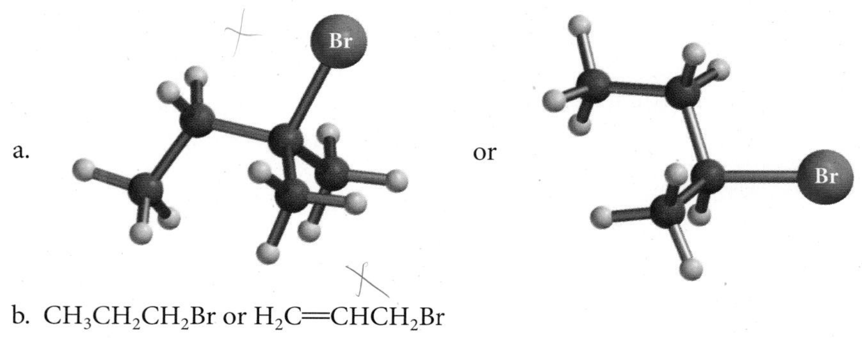

b. $CH_3CH_2CH_2Br$ or $H_2C{=}CHCH_2Br$

To summarize, the S_N1 mechanism is a two-step process and is favored when the alkyl halide is tertiary. Primary halides normally do not react by this mechanism. The S_N1 process occurs with racemization, and its rate is independent of the nucleophile's concentration.

6.6 The S_N1 and S_N2 Mechanisms Compared

How can we tell whether a particular nucleophilic substitution reaction will proceed by an S_N2 or an S_N1 mechanism? And why do we care? Well, we care for several reasons. When we perform a reaction in the laboratory, we want to be sure that the reaction will proceed at a rate fast enough to obtain the product in a reasonable time. If the reaction has stereochemical consequences, we want to know in advance what that outcome will be: inversion or racemization.

Table 6.2 should be helpful. It summarizes what we have said so far about the two substitution mechanisms, and it compares them with respect to two other variables, solvent and nucleophile structure, which we will discuss here.

Primary halides almost always react by the S_N2 mechanism, whereas tertiary halides react by the S_N1 mechanism. Only with secondary halides are we likely to encounter both possibilities.

Polar protic solvents are solvents such as water or alcohols that can donate protons.

One experimental variable that we can use to help control the mechanism is the solvent polarity. Water and alcohols are **polar protic solvents** (protic because of the proton-donating ability of the hydroxyl groups). How will such solvents affect S_N1 and S_N2 reactions?

The first step of the S_N1 mechanism involves the formation of ions. Since polar solvents can solvate ions, the rate of S_N1 processes is enhanced by polar solvents. On the other hand, solvation of nucleophiles ties up their unshared electron pairs. Therefore, S_N2 reactions, whose rates depend on nucleophile effectiveness, are usually retarded by polar protic solvents. Polar but *aprotic* solvents [examples are acetone, dimethyl sulfoxide, $(CH_3)_2S{=}O$, or dimethylformamide, $(CH_3)_2NCHO$] solvate cations preferentially. These solvents *accelerate* S_N2 reactions because, by solvating the cation (say, K^+ in $K^{+-}CN$), they leave the anion more "naked" or unsolvated, thus improving its nucleophilicity.

Table 6.2 ▪ Comparison of S_N2 and S_N1 Substitutions

Variables	S_N2	S_N1
Halide structure		
Primary or CH_3	Common	Rarely*
Secondary	Sometimes	Sometimes
Tertiary	Rarely	Common
Stereochemistry	Inversion	Racemization
Solvent	Rate is retarded by polar protic solvents and increased by polar aprotic solvents	Because the intermediates are ions, the rate is increased by polar solvents
Nucleophile	Rate depends on nucleophile concentration; mechanism is favored when the nucleophile is an anion	Rate is independent of nucleophile concentration; mechanism is more likely with neutral nucleophiles

*Allyl and benzyl substrates are the common exceptions (see Problem 6.7b).

A WORD ABOUT... S_N2 Reactions in Nature: Biological Methylations

Substitution and elimination reactions are so useful that it is not surprising that they occur in nature. Alkyl halides, however, are not compatible with cytoplasm because they are hydrocarbon-like and therefore insoluble in water. In the cell, alkyl phosphates play the role that alkyl halides do in the laboratory. **Adenosine triphosphate** (ATP) is an example of a biologically important alkyl phosphate. We will abbreviate its structure here as follows (the full structure is given in Sec. 18.12).

$$\text{Ad}-\text{O}-\underset{\text{OH}}{\overset{\text{O}}{\overset{\|}{\underset{|}{\text{P}}}}}-\text{O}-\underset{\text{OH}}{\overset{\text{O}}{\overset{\|}{\underset{|}{\text{P}}}}}-\text{O}-\underset{\text{OH}}{\overset{\text{O}}{\overset{\|}{\underset{|}{\text{P}}}}}-\text{OH} \quad \text{or} \quad \text{Ad}-\text{OPPP}$$

Ad = adenosyl (p. 546)

Many compounds in nature have a methyl group attached to an oxygen ($RO—CH_3$) or nitrogen atom ($R_2N—CH_3$).

Examples include mescaline, an hallucinogen derived from the peyote cactus, and morphine (p. 405), the pain-relieving drug derived from opium. How do the methyl groups get there? Two steps are involved, both of them nucleophilic substitutions.

The methyl carrier in most biochemical methylations is a sulfur-containing amino acid called **methionine**. In the first step, methionine is alkylated by ATP to form ***S*-adenosylmethionine**. This reaction is a biological example of reaction 12 in Table 6.1. The methionine acts as a sulfur nucleophile in an S_N2 reaction and displaces the triphosphate ion. In the second step, the oxygen or nitrogen atom to be methylated acts as a nucleophile. The *S*-adenosylmethionine acts just like a methyl halide. Indeed, it has been shown that these methylation reactions take place with inversion of configuration.

$$\underset{\text{methionine}}{CH_3-\ddot{\underset{\cdot\cdot}{S}}-CH_2CH_2CH(NH_2)CO_2H} \xrightarrow[\text{Ad—OPPP(ATP)}]{S_N2} \underset{S\text{-adenosylmethionine}}{CH_3-\overset{\text{Ad}}{\overset{|}{\underset{+}{S}}}-CH_2CH_2CH(NH_2)CO_2H} \quad {}^{-}\text{OPPP}$$

$$\downarrow \; R\ddot{O}H \text{ or } R_2\ddot{N}H, \; S_N2$$

$$RO-CH_3 \text{ or } R_2N-CH_3 + :\overset{\text{Ad}}{\overset{|}{\underset{\cdot\cdot}{S}}}-CH_2CH_2CH(NH_2)CO_2H + H-OPPP$$

See Problem 6.19.

Now let us consider the other variable in Table 6.2—the nucleophile. As we have seen, the rate of an S_N2 reaction (but *not* an S_N1 reaction) depends on the nucleophile. If the nucleophile is *strong*, the S_N2 mechanism will be favored. How can we tell whether a nucleophile is strong or weak, or whether one nucleophile is stronger than another? Here are a few useful generalizations.

1. *Negative ions are more nucleophilic, or better electron suppliers, than the corresponding neutral molecules.* Thus,

$$HO^- > HOH \qquad RS^- > RSH \qquad RO^- > ROH$$

2. *Elements low in the periodic table tend to be more nucleophilic than elements above them in the same column.* Thus,

$$HS^- > HO^- \qquad I^- > Br^- > Cl^- > F^- \qquad \text{(in protic solvents)}$$

3. *Across a row in the periodic table, more electronegative elements (that is, the more tightly an element holds electrons to itself) tend to be less nucleophilic.* Thus,

$$R_3C^- > R_2N^- > R{-}O^- > F^- \qquad \text{and} \qquad H_3N\!: > H_2\ddot{O}\!: > H\ddot{F}\!:$$

Can we juggle all of these factors to make some predictions about particular substitution reactions? Here are some examples.

EXAMPLE 6.6

Which mechanism, S_N1 or S_N2, would you predict for this reaction?

$$(CH_3)_3CBr + CH_3OH \longrightarrow (CH_3)_3COCH_3 + HBr$$

Solution S_N1, because the substrate is a tertiary alkyl halide. Also, methanol is a weak, neutral nucleophile and, if used as the reaction solvent, rather polar. Thus, it favors ionization.

EXAMPLE 6.7

Which mechanism, S_N1 or S_N2, would you predict for this reaction?

$$CH_3CH_2{-}I + NaOCH_3 \longrightarrow CH_3CH_2{-}OCH_3 + NaI$$

Solution S_N2, because the substrate is a primary halide, and methoxide (CH_3O^-), an anion, is a rather strong nucleophile.

PROBLEM 6.8 Which mechanism, S_N1 or S_N2, would you predict for each of the following reactions?

a. $CH_3\underset{\displaystyle Br}{\underset{|}{C}}HCH_2CH_2CH_3 + Na^{+\,-}SH \longrightarrow CH_3\underset{\displaystyle SH}{\underset{|}{C}}HCH_2CH_2CH_3 + NaBr$

b. $CH_3\underset{\displaystyle Br}{\underset{|}{C}}HCH_2CH_2CH_3 + CH_3OH \longrightarrow CH_3\underset{\displaystyle OCH_3}{\underset{|}{C}}HCH_2CH_2CH_3 + HBr$

6.7 Dehydrohalogenation, an Elimination Reaction; the E2 and E1 Mechanisms

We have seen several examples of reactions in which two reactants give not a single product but mixtures. Examples include halogenation of alkanes (eq. 2.13), addition to double bonds (eq. 3.31), and electrophilic aromatic substitutions (Sec. 4.11), where more than one isomer may be formed from the same two reactants. Even in nucleophilic substitution, more than one substitution product may form. For example, hydrolysis of a single alkyl bromide gives a mixture of two alcohols in eq. 6.14. But sometimes we find two entirely different reaction types occurring at the same time between the same two reactants, to give two (or more) entirely different types of products. Let us consider one example.

In **elimination** (or **dehydrohalogenation**) reactions of alkyl halides, a hydrogen atom and a halogen atom from adjacent carbons are eliminated and a carbon–carbon double bond is formed.

When an alkyl halide with a hydrogen attached to the carbon *adjacent* to the halogen-bearing carbon reacts with a nucleophile, two competing reaction paths are possible: substitution or **elimination**.

$$\text{H}-\overset{|}{\underset{|}{\text{C}}}_2-\overset{|}{\underset{|}{\text{C}}}_1-\text{X} + \text{Nu:}^- \xrightarrow{\text{substitution (S)}} \text{H}-\overset{|}{\underset{|}{\text{C}}}-\overset{|}{\underset{|}{\text{C}}}-\text{Nu} + \text{X}^- \quad (6.15)$$

$$\xrightarrow{\text{elimination (E)}} \text{C}=\text{C} + \text{Nu}-\text{H} + \text{X}^- \quad (6.16)$$

In the substitution reaction, the nucleophile replaces the halogen X. In the elimination reaction, the nucleophile acts as a base and removes a proton from carbon-2, the carbon next to the one that bears the halogen X. The halogen X and the hydrogen from the *adjacent* carbon atom are *eliminated,* and a new bond (a pi bond) is formed between carbons-1 and -2.* The symbol E is used to designate an elimination process. Since in this case a hydrogen halide is eliminated, the reaction is called **dehydrohalogenation**. Elimination reactions provide a useful way to prepare compounds with double or triple bonds.

Often substitution and elimination reactions occur simultaneously with the same set of reactants—a nucleophile and a substrate. One reaction type or the other may predominate, depending on the structure of the nucleophile, the structure of the substrate, and other reaction conditions. As with substitution reactions, *there are two main mechanisms for elimination reactions, designated E2 and E1.* To learn how to control these reactions, we must first understand each mechanism.

The **E2 mechanism** is a process in which HX is eliminated and a C═C bond is formed in the same step.

Like the S_N2 mechanism, the **E2 mechanism** is a one-step process. The nucleophile, acting as a base, removes the proton (hydrogen) on a carbon atom adjacent to the one that bears the leaving group. At the same time, the leaving group departs and a double bond is formed. The bond breaking and bond making that occurs during an E2 reaction is shown by the curved arrows:

(6.17)

*For a discussion of pi bonds and bonding in alkenes, review Section 3.4.

The preferred conformation for the substrate in an E2 reaction is also shown in eq. 6.17. The H—C—C—L atoms lie in a single plane, with H and L in an *anti*-arrangement. The reason for this preference is that the C—H and C—L bonds are parallel in this conformation. This alignment is needed to form the new pi bond as the C—H and C—L bonds break.

The **E1 mechanism** is a two-step process with the same first step as an S_N1 reaction.

The **E1 mechanism** is a two-step process and has the same first step as the S_N1 mechanism, the slow and rate-determining ionization of the substrate to give a carbocation (compare with eq. 6.12).

$$\underset{\text{substrate}}{-\overset{H}{\overset{|}{\underset{|}{C}}}-\overset{|}{\underset{|}{C}}-L} \xrightleftharpoons{\text{slow}} \underset{\text{carbocation}}{-\overset{H}{\overset{|}{\underset{|}{C}}}-\overset{|}{\underset{|}{C^+}}} + :L^- \quad (6.18)$$

Two reactions are then possible for the carbocation. It may combine with a nucleophile (the S_N1 process), or it may lose a proton from a carbon atom adjacent to the positive carbon, as shown by the curved arrow, to give an alkene (the E1 process).

$$\underset{\text{carbocation}}{-\overset{H}{\overset{|}{\underset{|}{C}}}-\overset{|}{\underset{|}{C^+}}} \begin{cases} \xrightarrow{\text{Nu}:} -\overset{H}{\overset{|}{\underset{|}{C}}}-\overset{|}{\underset{|}{C}}-\text{Nu} & S_N1 \\ \xrightarrow{-H^+} \;\rangle C=C\langle \; + H^+ & \text{E1} \end{cases} \quad (6.19)$$

6.8 Substitution and Elimination in Competition

Now we can consider how substitution and elimination reactions compete with one another. Let us consider the options for each class of alkyl halide.

6.8.a Tertiary Halides

Substitution can only occur by the S_N1 mechanism, but elimination can occur by either the E1 or the E2 mechanism. With weak nucleophiles and polar solvents, the S_N1 and E1 mechanisms compete with each other. For example,

$$\underset{t\text{-butyl bromide}}{(CH_3)_3CBr} \xrightleftharpoons{H_2O} (CH_3)_3C^+ + Br^- \begin{cases} \xrightarrow{H_2O,\ S_N1} \underset{\text{(about 80\%)}}{(CH_3)_3COH} \\ \xrightarrow{\text{E1}} \underset{\text{(about 20\%)}}{(CH_3)_2C=CH_2 + H^+} \end{cases} \quad (6.20)$$

If we use a strong nucleophile (which can act as a base) instead of a weak one, and if we use a less polar solvent, we favor elimination by the E2 mechanism. Thus, with OH^- or CN^- as nucleophiles, only elimination occurs (eqs. 6.5 and 6.7), and the exclusive product is the alkene.

$$HO^- + \underset{t\text{-butyl bromide}}{H_2C(H)-C(CH_3)_2Br} \xrightarrow{\text{E2}} \underset{\text{methylpropene (100\%)}}{H_2C=C(CH_3)_2} + \underset{(H-OH)}{H_2O} + Br^- \quad (6.21)$$

Because the tertiary carbon is too hindered sterically for S_N2 attack (eq. 6.11), substitution does not compete with elimination.

6.8.b Primary Halides

Only the S_N2 and E2 mechanisms are possible, because ionization to a primary carbocation, the first step required for the S_N1 or E1 mechanisms, does not occur.

With most nucleophiles, primary halides give mainly substitution products (S_N2). Only with very bulky, strongly basic nucleophiles do we see that the E2 process is favored. For example,

$$\underset{\text{1-bromobutane}}{CH_3CH_2CH_2CH_2Br} \xrightarrow[\text{in ethanol}]{CH_3CH_2O^-\ Na^+} \underset{\substack{\text{butyl ethyl ether}\\ (S_N2;\ 90\%)}}{CH_3CH_2CH_2CH_2OCH_2CH_3} + \underset{\substack{\text{1-butene}\\ (\text{E2};\ 10\%)}}{CH_3CH_2CH{=}CH_2} \tag{6.22}$$

$$\underset{\text{1-bromobutane}}{CH_3CH_2CH_2CH_2Br} \xrightarrow[\text{in } t\text{-butyl alcohol}]{(CH_3)_3C{-}O^-\ K^+} \underset{\substack{\text{butyl } t\text{-butyl ether}\\ (S_N2;\ 15\%)}}{CH_3CH_2CH_2CH_2OC(CH_3)_3} + \underset{\substack{\text{1-butene}\\ (\text{E2};\ 85\%)}}{CH_3CH_2CH{=}CH_2}$$

Potassium *t*-butoxide is a bulky base (Figure 6.3). Hence substitution is retarded, and the main reaction is elimination.

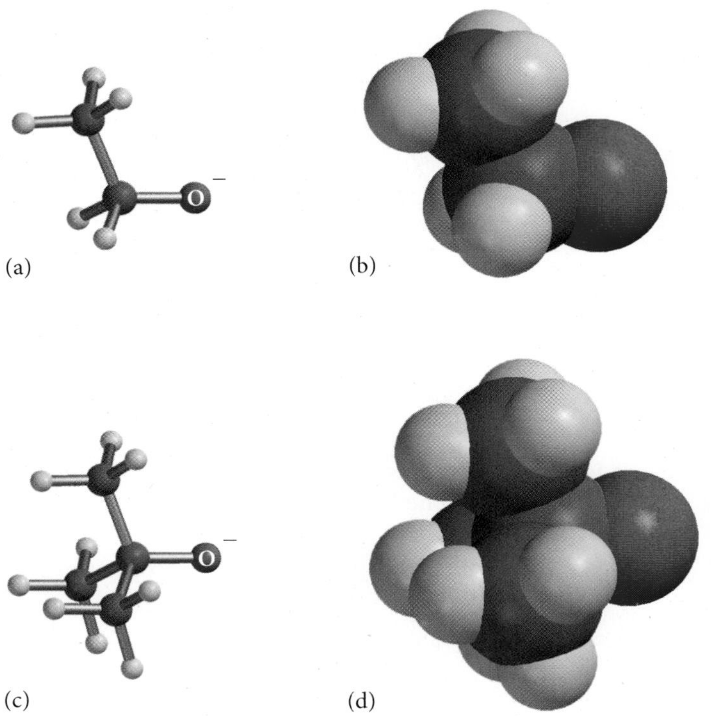

Figure 6.3
Ball-and-stick (a) and space-filling (b) structure for ethoxide ($CH_3CH_2O^-$) and for *t*-butoxide [$(CH_3)_3CO^-$] as ball-and-stick (c) and space-filling (d) models.

6.8.c Secondary Halides

All four mechanisms, S_N2 and E2 as well as S_N1 and E1, are possible. The product composition is sensitive to the nucleophile (its strength as a nucleophile and as a base) and to the reaction conditions (solvent, temperature). In general, substitution is favored with stronger nucleophiles that are not strong bases (S_N2) or by weaker nucleophiles in polar solvents (S_N1), but elimination is favored by strong bases (E2).

$$CH_3CHCH_3 \text{ (Br on C2; 2-bromopropane)} \xrightarrow[\text{strong nucleophile}]{CH_3CH_2S^- \ Na^+} CH_3CH(SCH_2CH_3)CH_3 \quad (S_N2)$$

$$\xrightarrow[\text{weak nucleophile}]{CH_3CH_2OH} \underset{(S_N1;\ \text{major})}{CH_3CH(OCH_2CH_3)CH_3} + \underset{(E1;\ \text{minor})}{CH_3CH{=}CH_2} \qquad (6.23)$$

$$\xrightarrow[\text{strong base}]{CH_3CH_2O^- \ Na^+} \underset{(S_N2;\ \text{minor})}{CH_3CH(OCH_2CH_3)CH_3} + \underset{(E2;\ \text{major})}{CH_3CH{=}CH_2}$$

EXAMPLE 6.8

Predict the product of the reaction of 1-bromo-1-methylcyclohexane with

a. sodium ethoxide in ethanol.
b. refluxing ethanol.

Solution The alkyl bromide is tertiary

CH_3 Br

a. The first set of conditions favors the E2 process, because sodium ethoxide is a strong base. Two elimination products are possible, depending on whether the base attacks a hydrogen on an adjacent CH_2 or CH_3 group.

[1-methylcyclohexene, $-CH_3$] and [methylenecyclohexane, $=CH_2$]

b. This set of conditions favors ionization, because the ethanol is neutral (hence a weak nucleophile) and, as a solvent, fairly polar. The S_N1 process predominates, and the main product is the ether.

CH_3 OCH_2CH_3

Some of the above alkenes will also be formed by the E1 mechanism.

PROBLEM 6.9 Draw structures for *all* possible elimination products obtainable from 2-chloro-2-methylpentane.

PROBLEM 6.10 Treatment of the alkyl halide in Problem 6.9 with KOH in methanol gives mainly a mixture of the alkenes whose structures you drew. But treatment with only methanol gives a different product. What is it, and by what mechanism is it formed?

6.9 Polyhalogenated Aliphatic Compounds

Because of their useful properties, many polyhalogenated compounds are produced commercially. As industrial chemicals, they are usually given common names.

Chlorinated methanes* are made by the chlorination of methane (eqs. 2.10 and 2.12). Carbon tetrachloride (CCl_4, bp 77°C), chloroform ($CHCl_3$, bp 62°C), and methylene chloride (CH_2Cl_2, bp 40°C) are all *insoluble* in water, but they are effective solvents for organic compounds. Also important for this purpose are trichloroethylenes and tetrachloroethylenes, used in dry cleaning and as degreasing agents in metal and textile processing.

$Cl_2C{=}CHCl$ — trichloroethylene, bp 87°C

$Cl_2C{=}CCl_2$ — tetrachloroethylene, bp 121°C

Because some of these chlorinated compounds are suspected carcinogens, adequate ventilation is essential when they are used as solvents.

Tetrafluoroethylene is the raw material for **Teflon**, a polymer related to polyethylene (Sec. 3.16) but with all of the hydrogens replaced by fluorine atoms.

$$n\,CF_2{=}CF_2 \xrightarrow[\text{catalyst}]{\text{peroxide}} \underset{\text{Teflon}}{+\!CF_2CF_2\!+_n} \quad (6.24)$$

Teflon has exceptional properties. It is resistant to almost all chemicals and is widely used as a nonstick coating for pots, pans, and other cooking utensils. Another use of Teflon is in Gore-Tex-like fabrics, materials with as many as nine billion pores per square inch. These pores are the right size to transmit water vapor, but not liquid water. Thus, perspiration vapor can pass through the fabric, but wind, rain, and snow cannot. Gore-Tex has revolutionized cold- and wet-weather gear for both military and civilian uses. It is used in skiwear, boots, sleeping bags, tents, and other rugged outdoor gear.

Nonpolymeric perfluorochemicals (hydrocarbons, ethers, or amines in which all of the hydrogens are replaced by fluorine atoms) also have fascinating and useful properties. For example, perfluorochemicals such as perfluorotributylamine, $(CF_3CF_2CF_2CF_2)_3N$, can dissolve as much as 60% oxygen by volume. By contrast, whole blood dissolves only about 20%, and blood plasma about 0.3%. Because of this property, these perfluorochemicals are important components of artificial blood.

Other polyhalogenated compounds that contain two or three different halogens per molecule are commercially important. The best known are the **chlorofluorocarbons** (**CFCs**, formerly known as **Freons**). The two that have been produced on the largest scale are CFC-11 and CFC-12, made by fluorination of carbon tetrachloride.

Chlorofluorocarbons (**CFCs**, also known as **Freons**) are polyhalogenated compounds containing chlorine and fluorine. Bromine-containing compounds of this type are called **Halons**.

$$\underset{\text{(bp 77°C)}}{CCl_4} \xrightarrow[SbF_5]{HF} \underset{\substack{\text{trichlorofluoromethane}\\ \text{(CFC-11)}\\ \text{bp 24°C}}}{CCl_3F} \xrightarrow[SbF_5]{HF} \underset{\substack{\text{dichlorodifluoromethane}\\ \text{(CFC-12)}\\ \text{bp −30°C}}}{CCl_2F_2} \quad (6.25)$$

They are used as refrigerants, as blowing agents in the manufacture of foams, as cleaning fluids, and as aerosol propellants. They are exceptionally stable. Because of this stability, they accumulate in the upper stratosphere, where they damage the earth's ozone layer. Consequently, their use for nonessential propellant purposes is now banned in most countries, and the search is on for replacements (see "A Word About... CFCs, the Ozone Layer, and Trade-Offs.")

*The analogous F, Br, and I compounds are also known, but are more expensive and not commercially important.

A WORD ABOUT... CFCs, the Ozone Layer, and Trade-Offs

The story of chlorofluorocarbons (CFCs) has many lessons for us. Fluorine's reputation as one of the most reactive of the elements was widespread. It was quite remarkable and unexpected, therefore, that fluorocarbons and CFCs should be so extremely *un*reactive. (Indeed, Thomas Midgley, the American who discovered CFCs, used to demonstrate their nontoxicity, nonflammability, and noncorrosiveness by inhaling them and then puffing out a candle.)

Because of these and other properties, at least *four* major commercial uses of CFCs were developed. Their low boiling points (bp's) and other heat properties make them excellent *refrigerants,* far superior to ammonia, sulfur dioxide, and other rather difficult-to-handle refrigerants. Thus CFCs became widely used in freezers, refrigerators, and air conditioners. CFCs make excellent *blowing agents* for rigid foams (such as those used for ice chests, fast-food take-out boxes, and other packaging materials) and for flexible foams (like those used to make pillows and furniture cushions). Their low surface tension and low viscosity give them excellent wetting properties, which led to their use as *cleaning fluids* for printed computer circuits, artificial limbs, and other products. Finally, they were used as *propellants* in aerosol sprays. CFCs were manufactured on such a scale that they constituted a multibillion-dollar industry.

But their extreme stability led to a major world problem. CFCs are so stable that, when they are released into the atmosphere, they do not decompose in the lower atmosphere, as do most other industrial chemicals. Instead, they eventually rise to the stratosphere where, through ultraviolet radiation, the C—Cl bonds are broken and chlorine atoms are released. These chlorine atoms initiate a chain of reactions that damages the ozone layer, which is needed to protect life on earth from harmful ultraviolet rays.*

How do we solve the problem? Ban all use of CFCs? To do so could bring about a major crisis for civilization. How, without refrigeration, could we ensure safe and adequate food supplies to urban areas, or delivery of heat-sensitive medical supplies? A return to old-fashioned refrigerants would also be fraught with environmental hazards. So the solution is not simple.

What has been done is to cut back or ban some less essential uses gradually. For example, in the United States (but *not* worldwide), all nonessential aerosol use of CFCs is banned. In 1987, 24 countries signed the Montreal Protocol, an agreement that called for cutting CFC use to half of the 1986 level by 1998. This was later expanded to a complete ban on CFC production and use by 1996.

Chemists and other scientists have been seeking replacements to fill the gap left by elimination of CFC production and use. The properties of polyhalogenated compounds are so uniquely useful that for the moment the best chance to find CFC substitutes is among this type of compound. For example, it was found that introducing one or more hydrogen atoms into the molecule substantially increases its decomposition rate in the lower atmosphere, where no damage to the ozone layer is possible. Hydrochlorofluorocarbon (HCFC) and hydrofluorocarbon (HFC) compounds such as CF_3CHCl_2 (HCFC-123, bp 28°C) and CF_3CH_2F (HFC-134a, bp –26°C) are now being used as replacements for CFC-11 and CFC-12, respectively.

The Montreal Protocol has brought results. Worldwide production of CFCs has declined from more than 1 million metric tons per year in 1986 to about 150,000 metric tons in 1997. But significant problems remain. CFC production is still legal in some countries and a black market trade in CFCs has developed because of the cost associated with replacing CFC equipment, such as chillers and air conditioners, with HCFC and HFC equipment. Nonetheless, the market for CFC replacements is rapidly growing, and this conversion should have positive environmental consequences.

CFCs are just one of many examples of the trade-offs between beneficial and possible harmful effects of a new research product. During World War II, the use of CFCs as propellants for the insecticide DDT saved the lives of many troops in the Pacific zone who were suffering more casualties from malaria than from enemy action. But later, indiscriminate use of CFCs as propellants for all sorts of trivial purposes led to problems with their environmental accumulation. Chemicals (in this case, CFCs) are neither good nor evil, but we must exercise good judgment in how we use them.

---

*For an account of the discovery and development of the CFC–ozone problem by F. Sherwood Rowland and Mario J. Molina, chemists who initiated ozone depletion research, see *Chemical and Engineering News,* **1994**, *77* (33), 8–13. Rowland, Molina, and Paul Crutzen shared the 1995 Nobel Prize in chemistry for their contributions to atmospheric chemistry. For one overview of the situation, see *Chemical and Engineering News,* **1997**, *75* (37), 24.

See Problem 6.30.

Bromine-containing aliphatic compounds are now widely used to extinguish fires. Called **Halons**, the best known are

$CBrClF_2$	$CBrF_3$
bromochloro-difluoromethane	bromotrifluoro-methane
(Halon-1211)	(Halon-1301)

Halons are much more effective than carbon tetrachloride. They are very important in air safety because of their ability to douse fires within seconds, a particularly important feature on airplanes. However, compared to chlorine- and fluorine-containing CFCs, bromine-containing compounds are even more effective at ozone depletion in the stratosphere. While chemists search for replacements of bromine-containing halons, many of these compounds are still in use for fire suppression applications.

Halon fire extinguisher.

A WORD ABOUT... Halogenated Organic Compounds from the Sea

Terrestrial plants and animals have long served as a source of substances for treatment of various maladies. For example, extracts of willow bark, now known to contain salicylic acid—the active component in aspirin—were used more than three millennia ago to treat pain. Marine organisms have also been used as a source of folk medicines, but only recently have chemists begun to determine the structures of biologically active compounds from the sea. It turns out that many of these substances are halogenated organic compounds, perhaps not surprising given the abundance of bromide and chloride ions in seawater. A few examples are shown in Figure 6.4.

Halomon, a polyhalogenated terpene produced by red algae, kills a variety of tumor cells and is being investigated as a lead compound for the development of new anticancer agents.* Moloka'iamine, a halogenated aromatic compound that inhibits the growth of barnacles, has been isolated from a marine sponge. Such compounds might be of interest as antifouling agents to the boating industry. It is possible that the two bromines in moloka'iamine are introduced by an electrophilic aromatic substitution reaction in which bromide ion is oxidized to provide an electrophilic source of bromine. 3,4,5-Tribromopyrrole-2-carboxylic acid has been isolated from sponges and structurally related hexahalogenated bipyrroles (HHBPs) have been isolated from seabird eggs, leading to speculation that halogenated organic compounds produced by marine organisms might accumulate in terrestrial animals that live by the sea. These are but a few examples of the many fascinating halogenated compounds that come from the sea.

*See Section 15.9 for the definition of *terpene*.

halomon

$H_2NCH_2CH_2CH_2O$—(dibromophenylene)—$CH_2CH_2NH_2$

moloka'iamine

3,4,5-tribromopyrrole-2-carboxylic acid

HHBP

Figure 6.4

Halogenated organic compounds from the sea. See Problems 6.15 and 6.17.

REACTION SUMMARY

1. Nucleophilic Substitutions (S_N1 and S_N2)

Alkyl halides react with a variety of nucleophiles to give alcohols, ethers, alkyl halides, alkynes, and other families of compounds. Examples are shown in Table 6.1 and Section 6.2.

$$\text{Nu:} + \text{R—X} \longrightarrow \text{R—Nu}^+ + \text{X}^- \qquad \text{Nu:}^- + \text{R—X} \longrightarrow \text{R—Nu} + \text{X}^-$$

2. Elimination Reactions (E1 and E2)

Alkyl halides react with bases to give alkenes (Sec. 6.7).

$$\text{H—C—C—X} \xrightarrow{\text{B:}^-} \text{C=C} + \text{BH} + \text{X}^-$$

MECHANISM SUMMARY

1. S_N2: Bimolecular Nucleophilic Substitution (Sec. 6.4)

$$\text{Nu:}^- + \text{C—L} \longrightarrow \left[\overset{\delta-}{\text{Nu}}\cdots\text{C}\cdots\overset{\delta-}{\text{L}}\right]^{\text{TS}} \longrightarrow \text{Nu—C} + \text{:L}^-$$

(nucleophile) (substrate) (leaving group)

2. S_N1: Unimolecular Nucleophilic Substitution (Sec. 6.5)

$$\text{C—L} \xrightleftharpoons{\text{slow}} \text{C}^+ + \text{:L}^- \xrightarrow[\text{Nu:}^-]{\text{fast}} \text{C—Nu} \text{ and } \text{Nu—C}$$

(substrate) (carbocation)

3. E2: Bimolecular Elimination (Sec. 6.7)

$$\text{B:} + \text{H—C—C—L} \longrightarrow \text{BH}^+ + \text{C=C} + \text{:L}^-$$

4. E1: Unimolecular Elimination (Sec. 6.7)

$$\text{H—C—C—L} \rightleftharpoons \text{H—C—C}^+ + \text{:L}^- \longrightarrow \text{C=C} + \text{H}^+$$

ADDITIONAL PROBLEMS

OWL Interactive versions of these problems are assignable in OWL.

Alkyl Halide Structure

6.11 Draw the structure of

a. a primary alkyl chloride, C_4H_9Cl.
b. a tertiary alkyl bromide, $C_5H_{11}Br$.
c. a secondary alkyl iodide, $C_6H_{11}I$.

Nucleophilic Substitution Reactions of Alkyl Halides

6.12 Using Table 6.1, write an equation for each of the following substitution reactions:

a. *p*-methylbenzyl chloride + sodium acetylide
b. *n*-propyl bromide + sodium cyanide
c. 2-iodopropane + sodium hydrosulfide
d. 1-bromopentane + sodium iodide
e. 2-chlorobutane + sodium ethoxide
f. *t*-butyl bromide + ethanol
g. 1,6-dibromohexane + sodium cyanide (excess)
h. allyl chloride + ammonia (2 equivalents)
i. 1-methyl-1-chlorocyclohexane + water

6.13 Select an alkyl halide and a nucleophile that will give each of the following products:

a. $CH_3OCH_2CH_2CH_2CH_3$
b. $HC{\equiv}CCH_2CH_2CH_3$
c. $CH_3CH_2CH_2NH_2$
d. $CH_3CH_2CH_2SCH_2CH_2CH_3$
e.

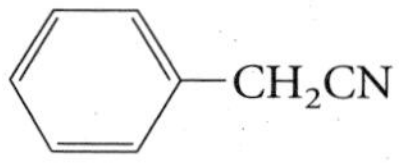

f.

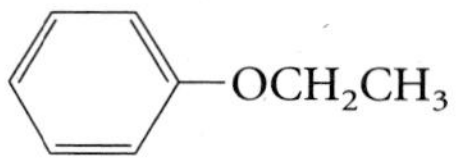

Stereochemistry of Nucleophilic Substitution Reactions

6.14 Draw each of the following equations in a way that shows clearly the stereochemistry of the reactants and products.

a. (*R*)-2-bromobutane + sodium methoxide (in methanol) $\xrightarrow{S_N2}$ 2-methoxybutane
b. (*S*)-3-bromo-3-methylhexane + methanol $\xrightarrow{S_N1}$ 3-methoxy-3-methylhexane
c. *cis*-1-bromo-4-methylcyclohexane + NaSH $\longrightarrow$ 4-methylcyclohexanethiol

6.15 The (+) enantiomer of the inhalation anesthetic desflurane ($CF_3CHFOCHF_2$) has the *S* configuration. Draw a three-dimensional representation of (*S*)-(+)-desflurane.

6.16 When treated with sodium iodide, a solution of (*R*)-2-iodooctane in acetone gradually loses all of its optical activity. Explain.

6.17 In Figure 6.4 and in the "A Word About…Halogenated Organic Compounds from the Sea," halomon, a polyhalogenated terpene produced by red algae, is shown.

a. Assign *R* or *S* configuration to the different stereogenic centers in halomon.
b. Of the two carbon centers bonded to Br, which one would be more reactive with a nucleophile like cyanide ion? Explain.

6.18 Draw a Fischer projection formula for the product of this S_N2 reaction:

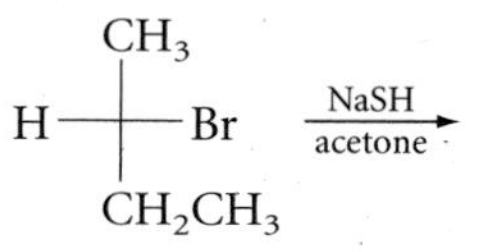

6.19 Predict the product of the reaction if CH_3–S–$CH_2CH_2CH_2CH_2$–Br is heated in a polar organic solvent, such as methanol. Similarly, what would be the product for hexyl bromide in methanol? (In hexyl bromide, note that the S has been replaced by a CH_2 group.) Explain why the reaction of CH_3–S–$CH_2CH_2CH_2CH_2$–Br is much faster. (*Hint*: see the "A Word About…S_N2 Reactions in Nature: Biological Methylations" on page 193.)

Nucleophilic Substitution and Elimination Reaction Mechanisms

6.20 Equation 6.20 shows that hydrolysis of *t*-butyl bromide gives about 80% $(CH_3)_3COH$ and 20% $(CH_3)_2C{=}CH_2$. The same ratio of alcohol to alkene is obtained whether the starting halide is *t*-butyl chloride or *t*-butyl iodide. Explain.

6.21 Determine the order of reactivity for $(CH_3)_2CHCH_2Br$, $(CH_3)_3CBr$, and $CH_3\underset{\underset{Br}{|}}{C}HCH_2CH_3$ in substitution reactions with

a. sodium cyanide.
b. 50% aqueous acetone.

6.22 Tell what product you expect, and by what mechanism it is formed, for each of the following reactions:

a. 1-chloro-1-methylcyclohexane + ethanol
b. 1-chloro-1-methylcyclohexane + sodium ethoxide (in ethanol)

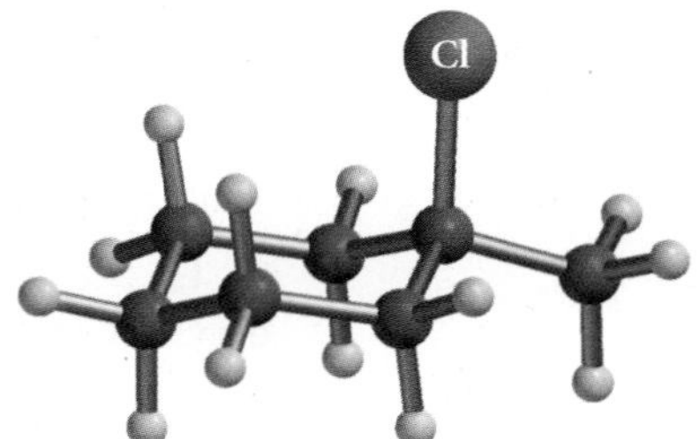

1-chloro-1-methylcyclohexane

6.23 Give the structures of all possible products when 2-chloro-2-methylhexane reacts by the E1 mechanism.

6.24 Explain the different products of the following two reactions by considering the mechanism by which each reaction proceeds. As part of your explanation, use the curved arrow formalism to draw a mechanism for each reaction.

$$CH_2{=}CH{-}\underset{\underset{Br}{|}}{CH}{-}CH_3 + Na^{+}\,{}^{-}OCH_3 \xrightarrow{CH_3OH} CH_2{=}CH{-}\underset{\underset{OCH_3}{|}}{CH}{-}CH_3$$

$$CH_2{=}CH{-}\underset{\underset{Br}{|}}{CH}{-}CH_3 + CH_3OH \longrightarrow CH_2{=}CH{-}\underset{\underset{OCH_3}{|}}{CH}{-}CH_3 + \underset{\underset{OCH_3}{|}}{CH_2}CH{=}CHCH_3$$

Nucleophilic Substitution and Elimination Reactions in Organic Synthesis

6.25 Provide equations for the synthesis of the following compounds from 1-bromo-1-phenylethane.

a. OCH_2CH_3 (1-phenylethyl structure)
b. $N(CH_3)_2$ (1-phenylethyl structure)
c. $C{\equiv}N$ (1-phenylethyl structure)
d.
e.
f. I (1-phenylethyl structure)

6.26 Devise a synthesis of

a. $CH_3OCH_2CH_2CH_3$ from an alkoxide and an alkyl halide.
b. $CH_3OC(CH_3)_3$ from an alcohol and an alkyl halide.

6.27 Provide an equation for the preparation of the following alkene from an alkyl halide. Do you anticipate problems with formation of other elimination or substitution reactions? Explain.

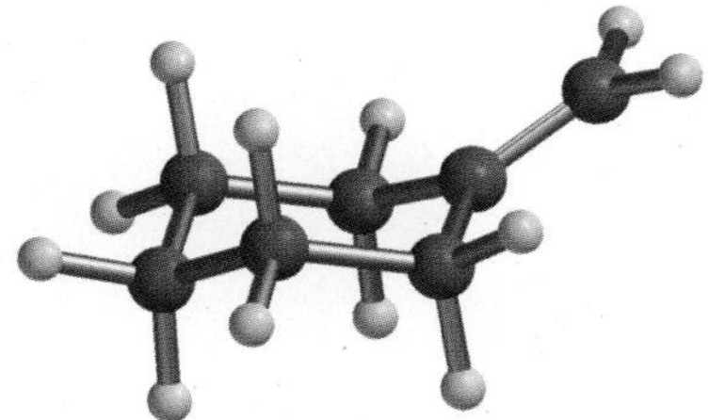

6.28 Combine the reaction in eq. 3.53 with a nucleophilic substitution to devise

a. a two-step synthesis of $CH_3C{\equiv}C{-}CH_2{-}C_6H_5$ from $C_6H_5{-}CH_2Br$.
b. a four-step synthesis of $CH_3C{\equiv}CCH_2CH_3$ from acetylene and appropriate alkyl halides.

6.29 Combine a nucleophilic substitution with a catalytic hydrogenation to synthesize

a. *cis*-3-heptene from butyne and bromopropane.
b. $CH_3CH_2CH_2OH$ from $CH_2{=}CHCH_2Br$.

6.30 HCFC-123 (CF_3CHCl_2) and HFC-134a (CF_3CH_2F) are replacements for chlorofluorocarbons (CFCs), which are known to deplete the stratospheric ozone layer, as described in the "A Word About…CFCs, the Ozone Layer, and Trade-Offs" on page 200. Predict the products from the base-induced elimination reactions of HCFC-123 and HFC-134a.

The alcohol ethanol is obtained from the fermentation of carbohydrates (Chapter 16) contained in fruits and grains.

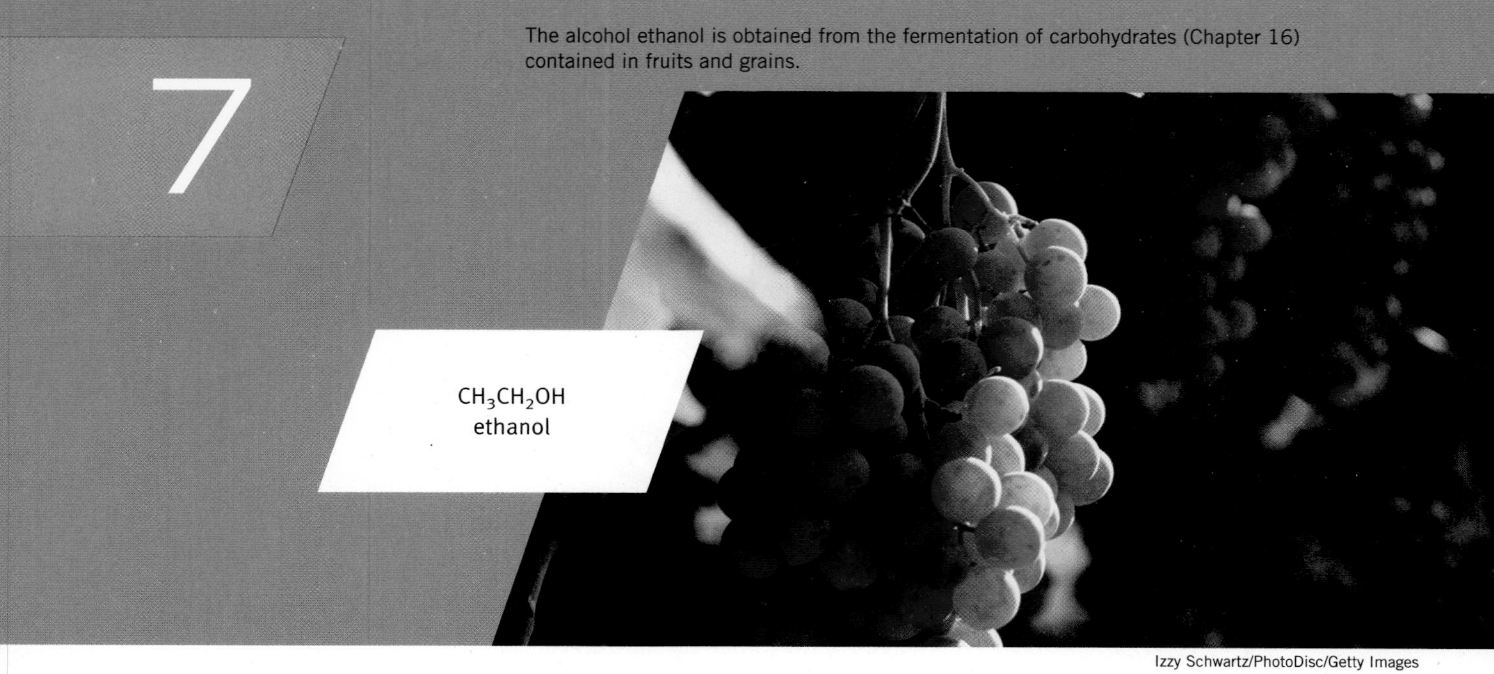

Izzy Schwartz/PhotoDisc/Getty Images

Alcohols, Phenols, and Thiols

The word *alcohol* immediately brings to mind ethanol, the intoxicating compound in wine and beer. But ethanol is just one member of a family of organic compounds called alcohols that abound in nature. Naturally occurring alcohols include 2-phenylethanol, the compound responsible for the intoxicating smell of a rose; cholesterol, a tasty alcohol with which many of us have developed a love–hate relationship; sucrose, a sugar we use to satisfy our sweet tooth; and many others. In this chapter, we will discuss the structural and physical properties as well as the main chemical reactions of alcohols and their structural relatives, phenols and thiols.

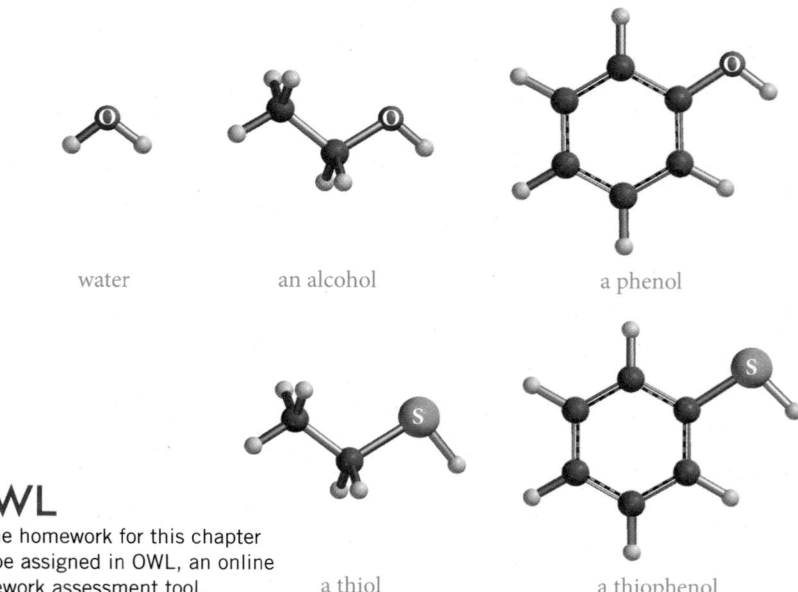

OWL

Online homework for this chapter can be assigned in OWL, an online homework assessment tool.

Alcohols have the general formula R—OH and are characterized by the presence of a **hydroxyl group, —OH.** They are structurally similar to water, but with one of the hydrogens replaced by an alkyl group. **Phenols** have a hydroxyl group attached directly to an aromatic ring. **Thiols** and thiophenols are similar to alcohols and phenols, except the oxygen is replaced by sulfur.

Alcohols contain the **hydroxyl (—OH) group.** In **phenols,** the hydroxyl group is attached to an aromatic ring, and in **thiols,** oxygen is replaced by sulfur.

7.1 Nomenclature of Alcohols

In the IUPAC system, the hydroxyl group in alcohols is indicated by the ending **-ol.** In common names, the separate word *alcohol* is placed after the name of the alkyl group. The following examples illustrate the use of IUPAC rules, with common names given in parentheses.

CH_3OH	CH_3CH_2OH	$\overset{3}{C}H_3\overset{2}{C}H_2\overset{1}{C}H_2OH$	$\overset{1}{C}H_3\overset{2}{C}H(OH)\overset{3}{C}H_3$
methanol (methyl alcohol)	ethanol (ethyl alcohol)	1-propanol (*n*-propyl alcohol)	2-propanol (isopropyl alcohol)

$CH_3CH_2CH_2CH_2OH$	$CH_3CH(OH)CH_2CH_3$	$CH_3CH(CH_3)CH_2OH$	$(CH_3)_3C—OH$
1-butanol (*n*-butyl alcohol)	2-butanol (*sec*-butyl alcohol)	2-methyl-1-propanol (isobutyl alcohol)	2-methyl-2-propanol (*tert*-butyl alcohol)

$CH_2{=}CHCH_2OH$	cyclohexane ring bearing H and OH	benzene ring—CH_2OH
2-propen-1-ol (allyl alcohol)	cyclohexanol (cyclohexyl alcohol)	phenylmethanol (benzyl alcohol)

With unsaturated alcohols, two endings are needed: one for the double or triple bond and one for the hydroxyl group (see the IUPAC name for allyl alcohol). In these cases, the *-ol* suffix comes last and takes precedence in numbering.

EXAMPLE 7.1

Name the following alcohols by the IUPAC system:

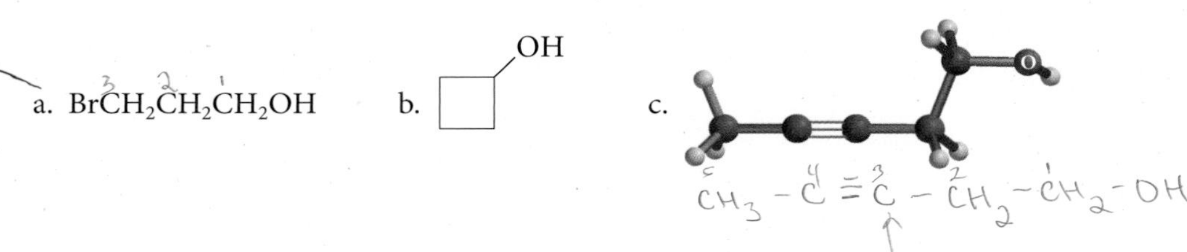

Solution

a. 3-bromopropanol (number from the hydroxyl-bearing carbon)
b. cyclobutanol
c. 3-pentyne-1-ol (*not* 2-pentyne-5-ol)

PROBLEM 7.1 Name these alcohols by the IUPAC system:

a. $ClCH_2CH_2OH$ b. H OH (cyclopentane ring) c.

PROBLEM 7.2 Write a structural formula for

a. 3-methylcyclohexanol b. 2-phenylpropanol c. 3-pentyn-2-ol

A WORD ABOUT... Industrial Alcohols

The lower alcohols (those with up to four carbon atoms) are manufactured on a large scale. They are used as raw materials for other valuable chemicals and also have important uses in their own right.

Methanol was at one time produced from wood by distillation and is still sometimes called wood alcohol. At present, however, methanol is manufactured from carbon monoxide and hydrogen.

$$CO + 2H_2 \xrightarrow[400°C,\ 150\ atm]{ZnO—Cr_2O_3} CH_3OH$$

The U.S. production of methanol is 1.4 billion gallons per year. Most of it is used to produce formaldehyde and other chemicals, but some is used as a solvent and as an antifreeze. Methanol is highly toxic and can cause permanent blindness because when taken internally, it is oxidized to formaldehyde ($CH_2{=}O$), which binds to opsin, preventing formation of rhodopsin, the light-sensitive pigment needed for vision (see "A Word About . . . The Chemistry of Vision," pages 76–77).

Ethanol is prepared from multiple sources, and the total U.S. production of ethanol is 9.0 billion gallons per year. Ethanol is prepared by the fermentation of blackstrap molasses, the residue that results from the purification of cane sugar.

$$\underset{\text{cane sugar}}{C_{12}H_{22}O_{11}} + H_2O \xrightarrow{\text{yeast}} \underset{\text{ethanol}}{4\ CH_3CH_2OH} + 4\ CO_2$$

The starch in corn, sugar beets, grain, potatoes, and rice can be fermented similarly to produce ethanol, sometimes called grain alcohol.

Besides fermentation, ethanol is also manufactured by the acid-catalyzed hydration of ethylene (eq. 3.7). This method, using sulfuric acid or other acid catalysts, results in an annual U.S. production of 170 million gallons.

Commercial alcohol is a constant-boiling mixture containing 95% ethanol and 5% water and cannot be further purified by distillation. To remove the remaining water to obtain **absolute alcohol**, one adds quicklime (CaO), which reacts with water to form calcium hydroxide but does not react with ethanol.

Since earlier times, ethanol has been known as an ingredient in fermented beverages (beer, wine, and whiskey). The term *proof*, as used in the United States in reference to alcoholic beverages, is approximately twice the volume percentage of alcohol present. For example 100-proof whiskey contains 50% ethanol.

Ethanol is used as a solvent, as a topical antiseptic (for example, when drawing blood), and as a starting material for the manufacture of ethers (Chapter 8) and ethyl esters (Chapter 10). It also can be used as a fuel, often as a blend with gasoline (e.g., E85: 85% ethanol, 15% gasoline). In many states, ethanol is added to gasoline to provide an EPA-mandated oxygen content, in order to improve air quality in the troposphere.

2-Propanol (isopropyl alcohol) is manufactured commercially by the acid-catalyzed hydration of propene (eq. 3.13). It is the main component of rubbing alcohol and is used in many household and personal care products.

See Problem 7.45.

7.2 Classification of Alcohols

Alcohols are classified as primary (1°), secondary (2°), or tertiary (3°), depending on whether one, two, or three organic groups are connected to the hydroxyl-bearing carbon atom.

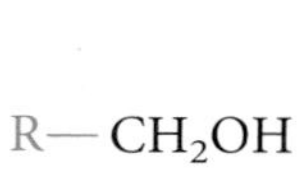

R—CH_2OH — primary (1°)

R_2CHOH (R—CHOH with a second R on carbon) — secondary (2°)

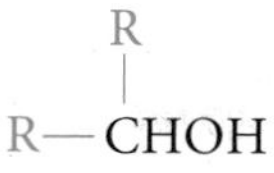

R_3C—OH (R—C—OH with R above and below) — tertiary (3°)

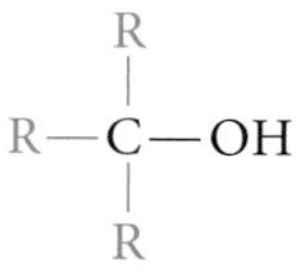

Methyl alcohol, which is not strictly covered by this classification, is usually grouped with the primary alcohols. This classification is similar to that for carbocations (Sec. 3.10). We will see that the chemistry of an alcohol sometimes depends on its class.

PROBLEM 7.3 Classify as 1°, 2°, or 3° the eleven alcohols listed in Section 7.1.

7.3 Nomenclature of Phenols

Phenols are usually named as derivatives of the parent compounds.

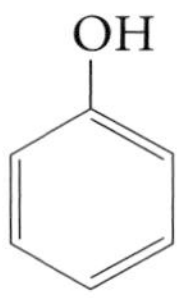

phenol    *p*-chlorophenol 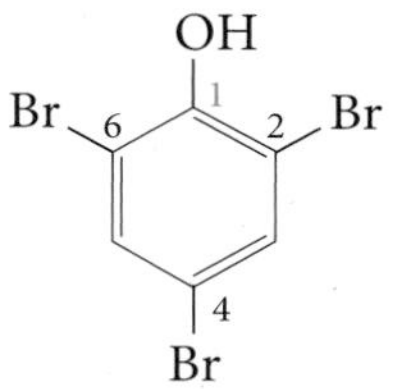   2,4,6-tribromophenol

The hydroxyl group is named as a substituent when it occurs in the same molecule with carboxylic acid, aldehyde, or ketone functionalities, which have priority in naming. Examples are

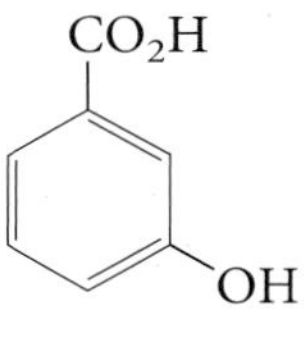

m-hydroxy benzoic acid 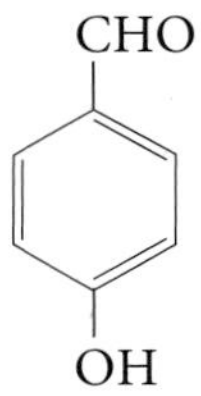   *p*-hydroxybenzaldehyde   but   *p*-nitrophenol (*not p*-hydroxynitrobenzene)

PROBLEM 7.4 Write the structure for

a. pentachlorophenol (an insecticide for termite control, and a fungicide)
b. *m*-isopropylphenol
c. *o*-hydroxyacetophenone (for the structure of acetophenone, see Sec. 4.6)
d. 3-hydroxy-5-nitrobenzaldehyde

7.4 Hydrogen Bonding in Alcohols and Phenols

The boiling points (bp's) of alcohols are much higher than those of ethers or hydrocarbons with similar molecular weights.

	CH_3CH_2OH	CH_3OCH_3	$CH_3CH_2CH_3$
mol wt	46	46	44
bp	+78.5°C	−24°C	−42°C

Why? Because alcohols form *hydrogen bonds* with one another (see Sec. 2.7). The O—H bond is polarized by the high electronegativity of the oxygen atom. This polarization places a partial positive charge on the hydrogen atom and a partial negative charge on the oxygen atom (Figure 7.1). Because of its small size and partial positive charge, the hydrogen atom can link two electronegative atoms such as oxygen.

$$\underbrace{\overset{R}{\overset{\delta-}{O}}-\overset{\delta+}{H} + \overset{R}{\overset{\delta-}{O}}-\overset{\delta+}{H}}_{\text{two separate alcohol molecules}} \rightleftharpoons \overset{R}{\overset{\delta-}{O}}-\underbrace{\overset{\delta+}{H}\cdots\overset{R}{\overset{\delta-}{O}}}_{\text{a hydrogen bond}}-\overset{\delta+}{H} \tag{7.1}$$

Two or more alcohol molecules thus become loosely bonded to one another through hydrogen bonds.

Hydrogen bonds are weaker than ordinary covalent bonds.* Nevertheless, their strength is significant, about 5 to 10 kcal/mol (20 to 40 kJ/mol). Consequently, alcohols and phenols have relatively high boiling points because we must not only supply enough heat (energy) to vaporize each molecule but must also supply enough heat to break the hydrogen bonds before each molecule can be vaporized.

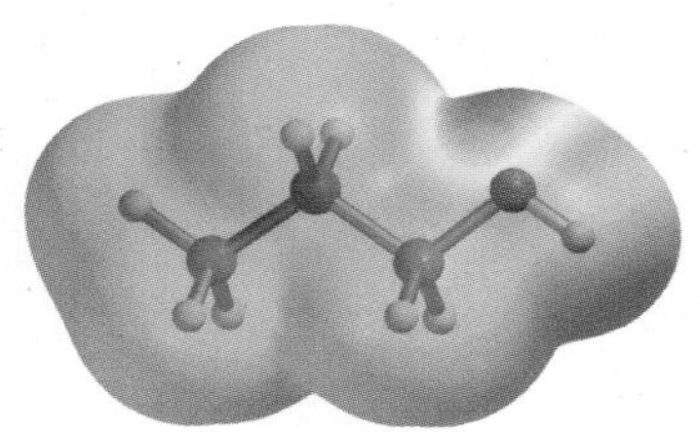

Figure 7.1
Electrostatic potential map for 1-propanol, revealing the electron-rich (red) nature of the oxygen and the electron-poor (blue) nature of the attached proton.

Water, of course, is also a hydrogen-bonded liquid (see Figure 2.2). The lower-molecular-weight alcohols can readily replace water molecules in the hydrogen-bonded network.

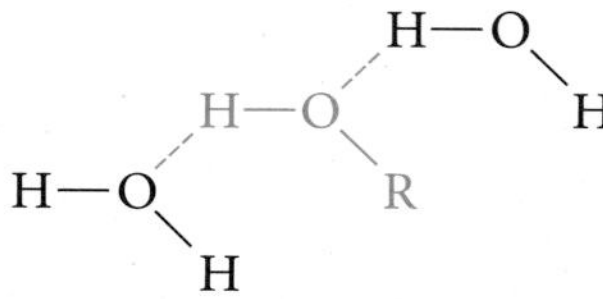

This accounts for the complete miscibility of the lower alcohols with water. However, as the organic chain lengthens and the alcohol becomes relatively more hydrocarbon-like, its water solubility decreases. Table 7.1 illustrates these properties.

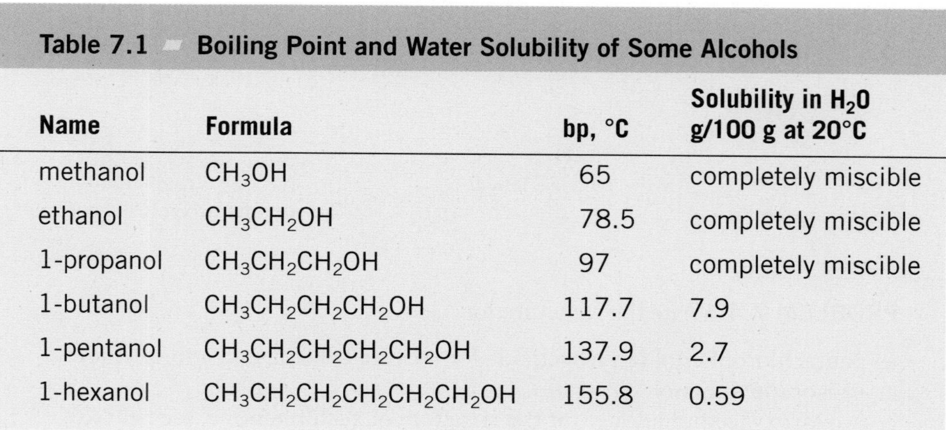

Table 7.1 ▪ Boiling Point and Water Solubility of Some Alcohols

Name	Formula	bp, °C	Solubility in H_2O g/100 g at 20°C
methanol	CH_3OH	65	completely miscible
ethanol	CH_3CH_2OH	78.5	completely miscible
1-propanol	$CH_3CH_2CH_2OH$	97	completely miscible
1-butanol	$CH_3CH_2CH_2CH_2OH$	117.7	7.9
1-pentanol	$CH_3CH_2CH_2CH_2CH_2OH$	137.9	2.7
1-hexanol	$CH_3CH_2CH_2CH_2CH_2CH_2OH$	155.8	0.59

*Covalent O—H bond strengths are about 120 kcal/mol (480 kJ/mol).

7.5 Acidity and Basicity Reviewed

The acid–base behavior of organic compounds often helps to explain their chemistry; this is certainly true of alcohols. It is a good idea, therefore, to review the fundamental concepts of acidity and basicity.

Acids and **bases** are defined in two ways. According to the **Brønsted–Lowry definition**, an acid is a proton donor, and a base is a proton acceptor. For example, in eq. 7.2, which represents what occurs when hydrogen chloride dissolves in water, the water accepts a proton from the hydrogen chloride.

A **Brønsted–Lowry acid** is a proton donor, whereas a **Brønsted–Lowry base** is a proton acceptor.

$$\underset{\text{base}}{H_2\ddot{O}:} + \underset{\text{acid}}{H-\ddot{Cl}:} \rightleftharpoons \underset{\text{conjugate acid of water}}{H_3\ddot{O}^+} + \underset{\text{conjugate base of hydrogen chloride}}{:\ddot{Cl}:^-} \quad (7.2)$$

Here water acts as a base or proton acceptor, and hydrogen chloride acts as an acid or proton donor. The products of this proton exchange are called the *conjugate acid* and the *conjugate base*.

The strength of an acid (in water) is measured quantitatively by its **acidity constant,** or **ionization constant, K_a**. For example, any acid dissolved in water is in equilibrium with hydronium ions and its conjugate base A^-:

The **acidity** (or **ionization**) **constant, K_a**, of an acid is a quantitative measure of its strength in water.

$$HA + H_2O \rightleftharpoons H_3O^+ + A^- \quad (7.3)$$

K_a is related to the equilibrium constant (see page 86) for this reaction and is defined as follows*:

$$K_a = \frac{[H_3O^+][A^-]}{[HA]} \quad (7.4)$$

The stronger the acid, the more this equilibrium is shifted to the right, thus increasing the concentration of H_3O^+ and the value of K_a.

For water, these expressions are

$$H_2O + H_2O \rightleftharpoons H_3O^+ + HO^- \quad (7.5)$$

and

$$K_a = \frac{[H_3O^+][HO^-]}{[H_2O]} = 1.8 \times 10^{-16} \quad (7.6)$$

PROBLEM 7.5 Verify from eq. 7.6 and from the molarity of water (55.5 *M*) that the concentrations of both H_3O^+ and HO^- in water are 10^{-7} moles per liter.

*The square brackets used in the expression for K_a indicate concentration, at equilibrium, of the enclosed species in moles per liter. The acidity constant K_a is related to the equilibrium constant for the reaction shown in eq. 7.3; only the concentration of water $[H_2O]$ is omitted from the denominator of the expression since it remains nearly constant at 55.5 *M*, very large compared to the concentrations of the other three species. For a discussion of reaction equilibria and equilibrium constants, see Section 3.11.

The **pK_a** of an acid is the negative logarithm of the acidity constant.

To avoid using numbers with negative exponents, such as those we have just seen for the acidity constant K_a for water, we often express acidity as **pK_a**, the negative logarithm of the acidity constant.

$$\mathrm{p}K_a = -\log K_a \tag{7.7}$$

The pK_a of water is

$$-\log(1.8 \times 10^{-16}) = -\log 1.8 - \log 10^{-16} = -0.26 + 16 = +15.74$$

The mathematical relationship between the values for K_a and pK_a means that *the smaller K_a or the larger pK_a, the weaker the acid.*

It is useful to keep in mind that there is an inverse relationship between the strength of an acid and the strength of its conjugate base. In eq. 7.2, for example, hydrogen chloride is a *strong* acid since the equilibrium is shifted largely to the right. It follows that the chloride ion must be a *weak* base, since it has relatively little affinity for a proton. Similarly, since water is a *weak* acid, its conjugate base, hydroxide ion, must be a *strong* base.

A **Lewis acid** is an electron pair acceptor; a **Lewis base** is an electron pair donor.

Another way to define acids and bases was first proposed by G. N. Lewis. A **Lewis acid** is a substance that can accept an electron pair, and a **Lewis base** is a substance that can donate an electron pair. According to this definition, a proton is considered to be a Lewis acid because it can accept an electron pair from a donor (a Lewis base) to fill its 1*s* shell.

$$\underset{\text{Lewis acid}}{H^+} + \underset{\text{Lewis base}}{:\ddot{O}H_2} \rightleftharpoons H-\overset{+}{\ddot{O}}(H)-H \tag{7.8}$$

Any atom with an unshared electron pair can act as a Lewis base.

Compounds with an element whose valence shell is incomplete also act as Lewis acids. For example,

$$\underset{\text{Lewis acid}}{BF_3} + \underset{\text{Lewis base}}{:\ddot{F}:^-} \rightleftharpoons F-\overset{-}{B}(F)(F)-\ddot{F}: \tag{7.9}$$

Similarly, when $FeCl_3$ or $AlCl_3$ acts as a catalyst for electrophilic aromatic chlorination (eqs. 4.13 and 4.14) or the Friedel–Crafts reaction (eqs. 4.20 and 4.22), they are acting as Lewis acids; the metal atom accepts an electron pair from chlorine or from an alkyl or acyl chloride to complete its valence shell of electrons.

Finally, some substances can act as either an acid or a base, depending on the other reactant. For example, in eq. 7.2, water acts as a base (a proton acceptor). However, in its reaction with ammonia, water acts as an acid (a proton donor).

$$\underset{\substack{\text{water}\\\text{(acid)}}}{H-\ddot{O}-H} + \underset{\substack{\text{ammonia}\\\text{(base)}}}{:NH_3} \rightleftharpoons \underset{\substack{\text{hydroxide ion}\\\text{(conjugate base)}}}{H-\ddot{O}:^-} + \underset{\substack{\text{ammonium ion}\\\text{(conjugate acid)}}}{H-\overset{+}{N}H_3} \tag{7.10}$$

An **amphoteric** substance can act as an acid or as a base.

Water acts as a base toward acids that are stronger than itself (HCl) and as an acid toward bases that are stronger than itself (NH_3). Substances that can act as either an acid or a base are said to be **amphoteric**.

PROBLEM 7.6 The K_a for ethanol is 1.0×10^{-16}. What is its pK_a?

PROBLEM 7.7 The pK_a's of hydrogen cyanide and acetic acid are 9.2 and 4.7, respectively. Which is the stronger acid?

PROBLEM 7.8 Which of the following are Lewis acids and which are Lewis bases?

a. Mg^{2+}	b. $(CH_3)_3C{:}^-$	c. CH_3NH_2
d. Zn^{2+}	e. $CH_3CH_2OCH_2CH_3$	f. $(CH_3)_3C^+$
g. $(CH_3)_3B$	h. $(CH_3)_3N$	i. $H{:}^-$

PROBLEM 7.9 In eq. 3.53, is the amide ion, NH_2^-, functioning as an acid or as a base?

7.6 The Acidity of Alcohols and Phenols

Like water, alcohols and phenols are weak acids. The hydroxyl group can act as a proton donor, and dissociation occurs in a manner similar to that for water:

$$\underset{\text{alcohol}}{R\ddot{\underset{\cdot\cdot}{O}}-H} \rightleftharpoons \underset{\text{alkoxide ion}}{R\ddot{\underset{\cdot\cdot}{O}}{:}^-} + H^+ \tag{7.11}$$

The conjugate base of an alcohol is an **alkoxide ion** (for example, *meth*oxide ion from *meth*anol, *eth*oxide ion from *eth*anol, and so on).

The conjugate base of an alcohol is an **alkoxide ion**.

Table 7.2 lists pK_a values for selected alcohols and phenols.* Methanol and ethanol have nearly the same acid strength as water; bulky alcohols such as *t*-butyl alcohol are somewhat weaker because their bulk makes it difficult to solvate the corresponding alkoxide ion (see Figure 6.3).

Phenol is a much stronger acid than ethanol. How can we explain this acidity difference between alcohols and phenols, since in both types of compounds, the proton donor is a hydroxyl group?

Phenols are stronger acids than alcohols mainly because the corresponding phenoxide ions are stabilized by resonance. The negative charge of an alkoxide ion is concentrated on the oxygen atom, but the negative charge on a phenoxide ion can be delocalized to the *ortho* and *para* ring positions through resonance.

$R-\ddot{\underset{\cdot\cdot}{O}}{:}^-$

charge localized on the oxygen atom in alkoxide ions

charge delocalized in phenoxide ion

Because phenoxide ions are stabilized in this way, the equilibrium for their formation is more favorable than that for alkoxide ions. Thus, phenols are stronger acids than alcohols.

*To compare the acidity of alcohols and phenols with that of other organic compounds, see Table C in the Appendix.

Table 7.2 ■ pK_a's of Selected Alcohols and Phenols in Aqueous Aolution

Name	Formula	pK_a
water	HO—H	15.7
methanol	CH_3O—H	15.5
ethanol	CH_3CH_2O—H	15.9
t-butyl alcohol	$(CH_3)_3CO$—H	18
2,2,2-trifluoroethanol	CF_3CH_2O—H	12.4
phenol	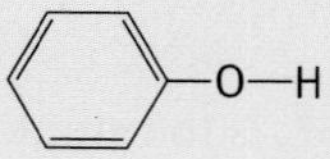 C_6H_5—O—H	10.0
p-nitrophenol	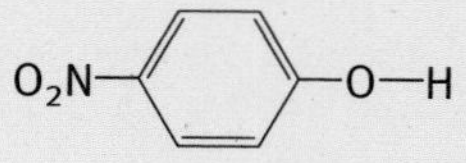 O_2N—C_6H_4—O—H	7.2
picric acid	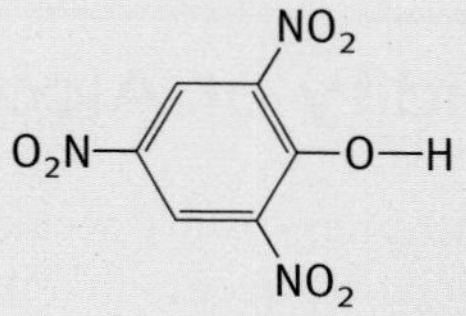 2,4,6-$(NO_2)_3C_6H_2$—O—H	0.25

We see in Table 7.2 that 2,2,2-trifluoroethanol is a much stronger acid than ethanol. How can we explain this effect of fluorine substitution? Again, think about the stabilities of the respective anions. Fluorine is a strongly electronegative element, so each C—F bond is polarized, with the fluorine partially negative and the carbon partially positive.

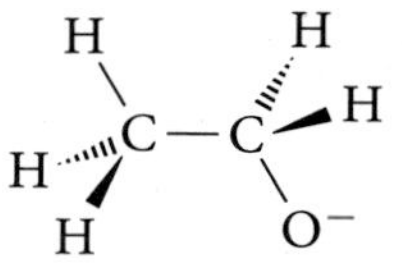

ethoxide ion 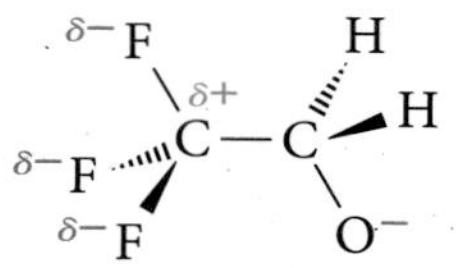 2,2,2-trifluoroethoxide ion

Polar bonds that place a partial positive charge near the negative charge on an alkoxide ion stabilize the ion by an **inductive effect**.

The positive charge on the carbon is located near the negative charge on the nearby oxygen atom, where it can partially neutralize and hence stabilize it. This **inductive effect**, as it is called, is absent in ethoxide ion.

The acidity-increasing effect of fluorine seen here is not a special case, but a general phenomenon. *All electron-withdrawing groups increase acidity* by stabilizing the conjugate base. *Electron-donating groups decrease acidity* because they destabilize the conjugate base.

Here is another example. *p*-Nitrophenol (Table 7.2) is a much stronger acid than phenol. In this case, the nitro group acts in two ways to stabilize the *p*-nitrophenoxide ion.

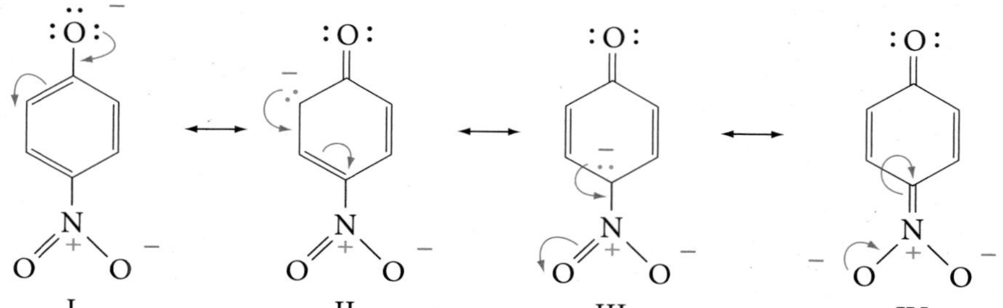

I II III IV

p-nitrophenoxide ion resonance contributors

First, the nitrogen atom has a formal positive charge and is therefore strongly electron withdrawing. Thus, it increases the acidity of *p*-nitrophenol through an inductive effect. Second, the negative charge on the oxygen of the phenoxide can be delocalized through resonance, not only to the *ortho* and *para* ring carbons, as in phenoxide itself, but to the oxygen atoms of the nitro group as well (structure IV). Both the inductive and the resonance effects of the nitro group are acid strengthening.

Additional nitro groups on the benzene ring further increase phenolic acidity. Picric acid (2,4,6-trinitrophenol) is an even stronger acid than *p*-nitrophenol.

PROBLEM 7.10 Draw the resonance contributors for the 2,4,6-trinitrophenoxide (picrate) ion, and show that the negative charge can be delocalized to every oxygen atom.

PROBLEM 7.11 Rank the following five compounds in order of increasing acid strength: 2-chloroethanol, *p*-chlorophenol, *p*-methylphenol, ethanol, and phenol.

Alkoxides, the conjugate bases of alcohols, are strong bases just like hydroxide ion. They are ionic compounds and are frequently used as strong bases in organic chemistry. They can be prepared by the reaction of an alcohol with sodium or potassium metal (eq. 7.12) or with a metal hydride (eq. 7.13). These reactions proceed irreversibly to give the metal alkoxides that can frequently be isolated as white solids.

$$\underset{\text{alcohol}}{2\,\text{R}\ddot{\underset{..}{\text{O}}}\text{—H}} + 2\,\text{K} \longrightarrow \underset{\text{potassium alkoxide}}{2\,\text{R}\ddot{\underset{..}{\text{O}}}\text{:}^-\ \text{K}^+} + \text{H}_2 \tag{7.12}$$

$$\text{R}\ddot{\underset{..}{\text{O}}}\text{—H} + \underset{\text{sodium hydride}}{\text{NaH}} \longrightarrow \underset{\text{sodium alkoxide}}{\text{R}\ddot{\underset{..}{\text{O}}}\text{:}^-\ \text{Na}^+} + \text{H—H} \tag{7.13}$$

PROBLEM 7.12 Write the equation for the reaction of *t*-butyl alcohol with potassium metal. Name the product.

Ordinarily, treatment of alcohols with sodium hydroxide does not convert them to their alkoxides. This is because alkoxides are stronger bases than hydroxide ion, so the reaction goes in the reverse direction. Phenols, however, can be converted to phenoxide ions in this way.

$$\text{ROH} + \text{Na}^+\text{HO}^- \not\rightleftharpoons \text{RO}^-\text{Na}^+ + \text{H}_2\text{O} \tag{7.14}$$

$$\underset{\text{phenol}}{\text{C}_6\text{H}_5\text{—OH}} + \text{Na}^+\text{HO}^- \longrightarrow \underset{\text{sodium phenoxide}}{\text{C}_6\text{H}_5\text{—O}^-\text{Na}^+} + \text{HOH} \tag{7.15}$$

PROBLEM 7.13 Write an equation for the reaction, if any, between

a. *p*-nitrophenol and aqueous potassium hydroxide
b. cyclohexanol and aqueous potassium hydroxide

7.7 The Basicity of Alcohols and Phenols

Alcohols (and phenols) function not only as weak acids but also as weak bases. They have unshared electron pairs on the oxygen and are therefore Lewis bases. They can be protonated by strong acids. The product, analogous to the oxonium ion, H_3O^+, is an alkyloxonium ion.

$$\underset{\text{alcohol acting as a base}}{R-\ddot{O}-H} + H^+ \rightleftharpoons \underset{\text{alkyloxonium ion}}{R-\overset{+}{\ddot{O}}(H)-H} \qquad (7.16)$$

This protonation is the first step in two important reactions of alcohols that are discussed in the following two sections: their dehydration to alkenes and their conversion to alkyl halides.

7.8 Dehydration of Alcohols to Alkenes

Alcohols can be dehydrated by heating them with a strong acid. For example, when ethanol is heated at 180°C with a small amount of concentrated sulfuric acid, a good yield of ethylene is obtained.

$$\underset{\text{ethanol}}{H-CH_2CH_2-OH} \xrightarrow{H^+,\ 180°C} \underset{\text{ethylene}}{CH_2=CH_2} + H-OH \qquad (7.17)$$

This type of reaction, which can be used to prepare alkenes, is the reverse of hydration (Sec. 3.7.b). It is an *elimination reaction* and can occur by either an E1 or an E2 mechanism, depending on the class of the alcohol.

Tertiary alcohols dehydrate by the E1 mechanism. *t*-Butyl alcohol is a typical example. The first step involves rapid and reversible protonation of the hydroxyl group.

$$(CH_3)_3C-\ddot{O}H + H^+ \rightleftharpoons (CH_3)_3C-\overset{+}{\ddot{O}}(H)-H \qquad (7.18)$$

Ionization (the rate-determining step), with water as the leaving group, occurs readily because the resulting carbocation is tertiary.

$$(CH_3)_3C-\overset{+}{\ddot{O}}(H)-H \rightleftharpoons \underset{t\text{-butyl cation}}{(CH_3)_3C^+} + H_2O \qquad (7.19)$$

Proton loss from a carbon atom adjacent to the positive carbon completes the reaction.

$$H-CH_2-\overset{CH_3}{\underset{CH_3}{C^+}} \longrightarrow CH_2=C(CH_3)_2 + H^+ \qquad (7.20)$$

The overall dehydration reaction is the sum of all three steps.

$$\underset{t\text{-butyl alcohol}}{\overset{\text{H}\quad \text{CH}_3}{\text{CH}_2-\underset{\text{CH}_3}{\text{C}}-\text{OH}}} \xrightarrow[\text{heat}]{\text{H}^+} \underset{\substack{\text{2-methylpropene}\\ \text{(isobutylene)}}}{\text{CH}_2{=}\text{C}(\text{CH}_3)_2} + \text{H}-\text{OH} \tag{7.21}$$

With a primary alcohol, a primary carbocation intermediate is avoided by combining the last two steps of the mechanism. The loss of water and an adjacent proton occurs simultaneously in an E2 mechanism.

$$\text{CH}_3\text{CH}_2\ddot{\text{O}}\text{H} + \text{H}^+ \rightleftharpoons \text{CH}_3\text{CH}_2-\overset{+}{\ddot{\text{O}}}-\text{H} \quad (\text{with H on O}) \tag{7.22}$$

$$\overset{\text{H}}{\text{CH}_2}-\text{CH}_2-\overset{+}{\ddot{\text{O}}}(\text{H})-\text{H} \longrightarrow \text{CH}_2{=}\text{CH}_2 + \text{H}^+ + \text{H}_2\text{O} \tag{7.23}$$

The important things to remember about alcohol dehydrations are that (1) they all begin by protonation of the hydroxyl group (that is, the alcohol acts as a base) and (2) the ease of alcohol dehydration is 3° > 2° > 1° (the same as the order of carbocation stability).

Sometimes a single alcohol gives two or more alkenes because the proton lost during dehydration can come from any carbon atom that is directly attached to the hydroxyl-bearing carbon. For example, 2-methyl-2-butanol can give two alkenes.

$$\underset{\text{2-methyl-2-butanol}}{\text{CH}_2(\text{H})-\text{C}(\text{OH})(\text{CH}_3)-\text{CH}(\text{H})-\text{CH}_3} \xrightarrow[\text{heat}\ -\text{H}_2\text{O}]{\text{H}^+} \underset{\text{2-methyl-1-butene}}{\text{CH}_2{=}\text{C}(\text{CH}_3)-\text{CH}_2\text{CH}_3} \text{ and/or } \underset{\text{2-methyl-2-butene}}{\text{CH}_3-\text{C}(\text{CH}_3){=}\text{CHCH}_3} \tag{7.24}$$

In these cases, *the alkene with the most substituted double bond usually predominates.* By "most substituted," we mean the alkene with the greatest number of alkyl groups on the doubly bonded carbons. Thus, in the example shown, the major product is 2-methyl-2-butene.

PROBLEM 7.14 Write the structure for all possible dehydration products of

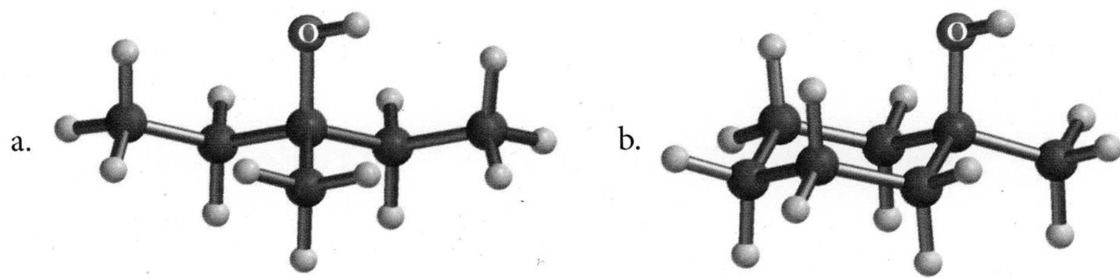

In each case, which product do you expect to predominate?

7.9 The Reaction of Alcohols with Hydrogen Halides

Alcohols react with hydrogen halides (HCl, HBr, and HI) to give alkyl halides (chlorides, bromides, and iodides).

$$\underset{\text{alcohol}}{R{-}OH} + H{-}X \longrightarrow \underset{\text{alkyl halide}}{R{-}X} + H{-}OH \quad (7.25)$$

This substitution reaction provides a useful general route to alkyl halides. Because halide ions are good nucleophiles, we obtain mainly substitution products instead of dehydration. Once again, the reaction rate and mechanism depend on the class of alcohol (tertiary, secondary, or primary).

Tertiary alcohols react the fastest. For example, we can convert *t*-butyl alcohol to *t*-butyl chloride simply by shaking it for a few minutes at room temperature (rt) with concentrated hydrochloric acid.

$$\underset{t\text{-butyl alcohol}}{(CH_3)_3COH} + H{-}Cl \xrightarrow[15\text{ min}]{\text{rt}} \underset{t\text{-butyl chloride}}{(CH_3)_3C{-}Cl} + H{-}OH \quad (7.26)$$

The reaction occurs by an S_N1 mechanism and involves a carbocation intermediate. The first two steps in the mechanism are identical to those shown in eqs. 7.18 and 7.19. The final step involves capture of the *t*-butyl carbocation by chloride ion.

$$(CH_3)_3C^+ + Cl^- \xrightarrow{\text{fast}} (CH_3)_3CCl \quad (7.27)$$

On the other hand, 1-butanol, a primary alcohol, reacts slowly and must be heated for several hours with a mixture of concentrated hydrochloric acid and a Lewis acid catalyst such as zinc chloride to accomplish the same type of reaction.

$$\underset{\text{1-butanol}}{CH_3CH_2CH_2CH_2OH} + H{-}Cl \xrightarrow[\text{several hours}]{\text{heat, }ZnCl_2} \underset{\text{1-chlorobutane}}{CH_3CH_2CH_2CH_2{-}Cl} + H{-}OH \quad (7.28)$$

The reaction occurs by an S_N2 mechanism. In the first step, the alcohol is protonated by the acid.

$$CH_3CH_2CH_2CH_2{-}\ddot{\underset{\cdot\cdot}{O}}H + H^+ \rightleftharpoons CH_3CH_2CH_2CH_2{-}\overset{+}{\ddot{O}}{-}H \ (\text{with } {-}H \text{ on O}) \quad (7.29)$$

In the second step, chloride ion displaces water in a typical S_N2 process. The zinc chloride is a good Lewis acid and can serve the same role as a proton in sharing an electron pair of the hydroxyl oxygen. It also increases the chloride ion concentration, thus speeding up the S_N2 displacement.

$$Cl^- \curvearrowright (CH_3CH_2CH_2)(H)(H)C{-}\overset{+}{\ddot{O}}(H){-}H \longrightarrow CH_3CH_2CH_2CH_2Cl + H_2O \quad (7.30)$$

Secondary alcohols react at intermediate rates by both S_N1 and S_N2 mechanisms.

EXAMPLE 7.2

Explain why *t*-butyl alcohol reacts at equal rates with HCl, HBr, and HI (to form, in each case, the corresponding *t*-butyl halide).

Solution *t*-Butyl alcohol is a tertiary alcohol; thus it reacts by an S_N1 mechanism. As in all S_N1 reactions, the rate-determining step involves formation of a carbocation; in this case, the *t*-butyl carbocation. The rate of this step does not depend on which acid is used, so all of the reactions proceed at equal rates.

$$(CH_3)_3COH + H^+ \rightleftharpoons (CH_3)_3C-\ddot{O}^{+}(H)-H \xrightarrow[\text{slow step}]{S_N1} \underset{t\text{-butyl cation}}{(CH_3)_3C^+} + H_2O$$

The reaction of the carbocation with Cl^-, Br^-, or I^- is then fast.

PROBLEM 7.15 Explain why 1-butanol reacts with hydrogen halides in the rate order HI > HBr > HCl (to form, in each case, the corresponding butyl halide).

PROBLEM 7.16 Write equations for reactions of the following alcohols with concentrated HBr.

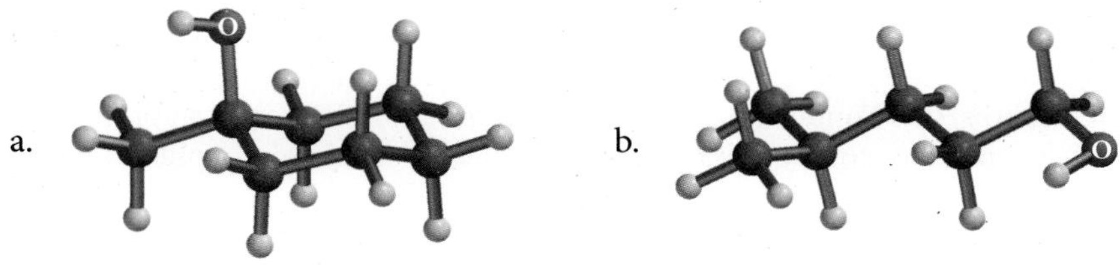

7.10 Other Ways to Prepare Alkyl Halides from Alcohols

Since alkyl halides are extremely useful in synthesis, it is not surprising that chemists have devised several ways to prepare them from alcohols. For example, thionyl chloride (eq. 7.31) reacts with alcohols to give alkyl chlorides. The alcohol is first converted to a chlorosulfite ester intermediate, a step that converts the hydroxyl group into a good leaving group. This is followed by a nucleophilic substitution whose mechanism (S_N1 or S_N2) depends on whether the alcohol is primary, secondary, or tertiary.

$$R-OH + \underset{\text{thionyl chloride}}{Cl-\overset{O}{\overset{\|}{S}}-Cl} \xrightarrow{\text{heat}} \underset{\substack{\text{chlorosulfite ester}\\ \text{intermediate}}}{\left[R-O-\overset{O}{\overset{\|}{S}}-Cl\right]} \; (Cl^-\ H^+) \longrightarrow R-Cl + SO_2\uparrow + HCl\uparrow \qquad (7.31)$$

One advantage of this method is that two of the reaction products, hydrogen chloride and sulfur dioxide, are gases and evolve from the reaction mixture (indicated by the upward pointing arrows), leaving behind only the desired alkyl chloride. The method is not effective, however, for preparing low-boiling alkyl chlorides (in which R has only a few carbon atoms), because they easily boil out of the reaction mixture with the gaseous products.

Phosphorus halides (eq. 7.32) also convert alcohols to alkyl halides.

$$3\,ROH + \underset{\text{phosphorus halide}}{PX_3} \longrightarrow 3\,RX + H_3PO_3\ (X = Cl \text{ or } Br) \tag{7.32}$$

In this case, the other reaction product, phosphorous acid, has a rather high boiling point. Thus, the alkyl halide is usually the lowest boiling component of the reaction mixture and can be isolated by distillation.

Both of these methods are used mainly with primary and secondary alcohols, whose reaction with hydrogen halides is slow.

PROBLEM 7.17 Write balanced equations for the preparation of the following alkyl halides from the corresponding alcohol and $SOCl_2$, PCl_3, or PBr_3.

a. cyclohexyl–CH_2Br b. cyclohexyl–Cl

7.11 A Comparison of Alcohols and Phenols

Because they have the same functional group, alcohols and phenols have many similar properties. But whereas it is relatively easy, with acid catalysis, to break the C—OH bond of alcohols, this bond is difficult to break in phenols. Protonation of the phenolic hydroxyl group can occur, but loss of a water molecule would give a phenyl cation.

$$C_6H_5\text{–}\overset{+}{O}H_2 \not\longrightarrow \underset{\text{a phenyl cation}}{C_6H_5^+} + H_2O \tag{7.33}$$

With only two attached groups, the positive carbon in a phenyl cation should be *sp*-hybridized and linear. But this geometry is prevented by the structure of the benzene ring, so *phenyl cations are energetically unstable and are exceedingly difficult to form.* Consequently, phenols cannot undergo replacement of the hydroxyl group by an S_N1 mechanism. Neither can phenols undergo displacement by the S_N2 mechanism. (The geometry of the ring makes the usual inversion mechanism impossible.) Therefore, hydrogen halides, phosphorus halides, or thionyl halides cannot readily cause replacement of the hydroxyl group by halogens in phenols.

PROBLEM 7.18 Compare the reactions of cyclohexanol and phenol with

a. HBr b. H_2SO_4, heat

7.12 Oxidation of Alcohols to Aldehydes, Ketones, and Carboxylic Acids

Alcohols with at least one hydrogen attached to the hydroxyl-bearing carbon can be oxidized to carbonyl compounds. Primary alcohols give aldehydes, which may be further oxidized to carboxylic acids. Secondary alcohols give ketones. Notice that as an alcohol is oxidized to an aldehyde or ketone and then to a carboxylic acid, the number of bonds between the reactive carbon atom and oxygen atoms increases from one to two to three. In other words, we say that the oxidation state of that carbon increases as we go from an alcohol to an aldehyde or from a ketone to a carboxylic acid.

$$\underset{\text{primary alcohol}}{R-\overset{\overset{\large OH}{|}}{\underset{\underset{\large H}{|}}{C}}-H} \xrightarrow[\text{agent}]{\text{oxidizing}} \underset{\text{aldehyde}}{R-\overset{\overset{\large O}{||}}{C}-H} \xrightarrow[\text{agent}]{\text{oxidizing}} \underset{\text{carboxylic acid}}{R-\overset{\overset{\large O}{||}}{C}-OH} \quad (7.34)$$

$$\underset{\text{secondary alcohol}}{R-\overset{\overset{\large OH}{|}}{\underset{\underset{\large H}{|}}{C}}-R'} \xrightarrow[\text{agent}]{\text{oxidizing}} \underset{\text{ketone}}{R-\overset{\overset{\large O}{||}}{C}-R'} \quad (7.35)$$

Tertiary alcohols, having no hydrogen atom on the hydroxyl-bearing carbon, do not undergo this type of oxidation.

A common laboratory oxidizing agent for alcohols is chromic anhydride, CrO_3, dissolved in aqueous sulfuric acid (**Jones' reagent**). Acetone is used as a solvent in such oxidations. Typical examples are

Jones' reagent is an oxidizing agent composed of CrO_3 dissolved in aqueous H_2SO_4.

$$\underset{\text{cyclohexanol}}{C_6H_{10}(H)(OH)} \xrightarrow[\substack{H^+,\ \text{acetone} \\ \text{(Jones' reagent)}}]{CrO_3} \underset{\text{cyclohexanone}}{C_6H_{10}{=}O} \quad (7.36)$$

$$\underset{\text{1-octanol}}{CH_3(CH_2)_6CH_2OH} \xrightarrow[\text{reagent}]{\text{Jones'}} \underset{\text{octanoic acid}}{CH_3(CH_2)_6CO_2H} \quad (7.37)$$

With primary alcohols, oxidation can be stopped at the aldehyde stage by special reagents, such as pyridinium chlorochromate (PCC), shown in eq. 7.38.*

$$\underset{\text{1-octanol}}{CH_3(CH_2)_6CH_2OH} \xrightarrow[CH_2Cl_2,\ 25°C]{PCC} \underset{\text{octanal}}{CH_3(CH_2)_6\overset{\overset{\large O}{||}}{C}-H} \quad (7.38)$$

PCC is prepared by dissolving CrO_3 in hydrochloric acid and then adding pyridine:

$$CrO_3 + HCl + \underset{\text{pyridine}}{C_5H_5N:} \longrightarrow \underset{\text{pyridinium chlorochromate (PCC)}}{C_5H_5N^+{-}H\ \ CrO_3Cl^-} \quad (7.39)$$

PROBLEM 7.19 Write an equation for the oxidation of

a. 4-methyl-1-octanol with Jones' reagent
b. 4-methyl-1-octanol with PCC
c. 4-phenyl-2-butanol with Jones' reagent
d. 4-phenyl-2-butanol with PCC

*In the oxidation reactions shown in eqs. 7.36–7.38, the chromium is reduced from Cr^{6+} to Cr^{3+}. Aqueous solutions of Cr^{6+} are orange, whereas aqueous solutions of Cr^{3+} are green. This color change has been used as the basis for detecting ethanol in Breathalyzer tests.

A WORD ABOUT... Biologically Important Alcohols and Phenols

The hydroxyl group appears in many biologically important molecules.

Four metabolically important unsaturated primary alcohols are 3-methyl-2-buten-1-ol, 3-methyl-3-buten-1-ol, geraniol, and farnesol.

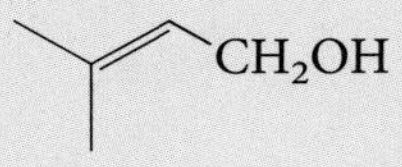

CH_2OH CH_2OH

3-methyl-2-buten-1-ol 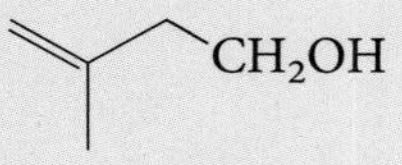 3-methyl-3-buten-1-ol

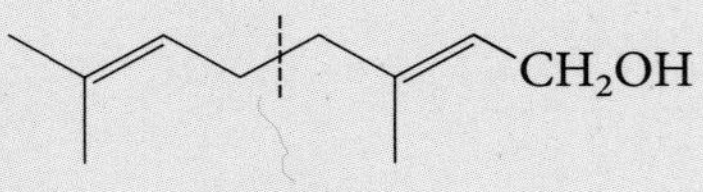

CH_2OH

geraniol

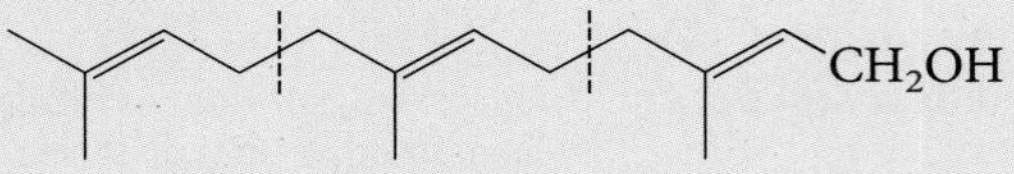

CH_2OH

farnesol

The two smaller alcohols contain a five-carbon unit, called an **isoprene unit**, that is present in many natural products (see page 452). This unit consists of a four-carbon chain with a one-carbon branch at carbon-2. These five-carbon alcohols can combine to give geraniol (10 carbons), which then can add yet another five-carbon unit to give farnesol (15 carbons). Note the isoprene units, marked off by dotted lines, in the structures of geraniol and farnesol.

Compounds of this type are called **terpenes**. Terpenes occur in many plants and flowers. They have 10, 15, 20, or more carbon atoms and are formed by linking isoprene units in various ways.

Geraniol, as its name implies, occurs in oil of geranium but also constitutes about 50% of rose oil, the extract of rose petals. **Farnesol**, which occurs in the essential oils of rose and cyclamen, has a pleasing lily-of-the-valley odor. Both geraniol and farnesol are used in making perfumes.

Combination of two farnesol units leads to **squalene**, a 30-carbon hydrocarbon present in small amounts in the livers of most higher animals. Squalene is the biological precursor of steroids.

Cholesterol, a typical steroidal alcohol, has the following structure:

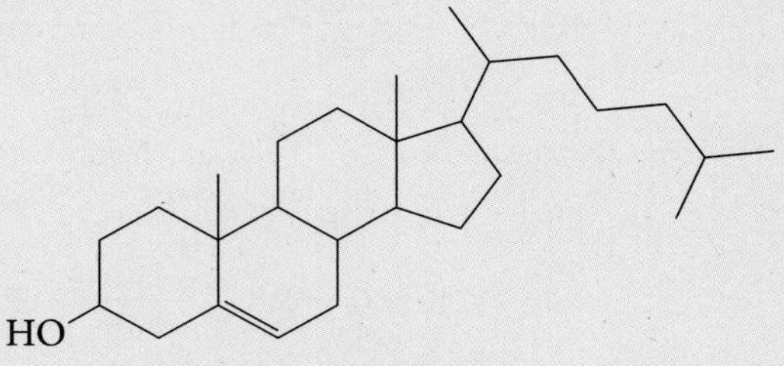

HO

cholesterol

Although it has 27 carbon atoms (instead of 30) and is therefore not strictly a terpene, cholesterol is synthesized in the body from the terpene squalene through a complex process that, in its final stages, involves the loss of three carbon atoms.

Phenols are less involved than alcohols in fundamental metabolic processes. Three phenolic alcohols do, however, form the basic building blocks of **lignins**, complex polymeric substances that, together with cellulose, form the woody parts of trees and shrubs. They have very similar structures.

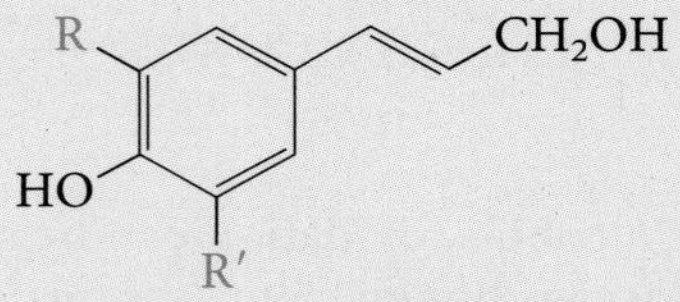

R CH_2OH HO R′

coniferyl alcohol ($R = OCH_3$, $R' = H$)
sinapyl alcohol ($R = R' = OCH_3$)
p-coumaryl alcohol ($R = R' = H$)

Some phenolic natural products to be avoided are **urushiols**, the active allergenic ingredients in poison ivy and poison oak.

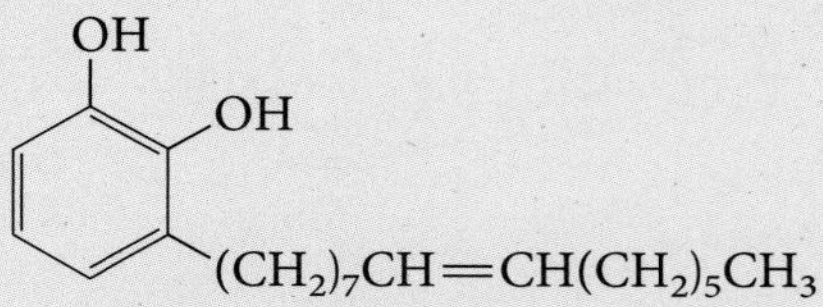

OH OH $(CH_2)_7CH{=}CH(CH_2)_5CH_3$

an urushiol

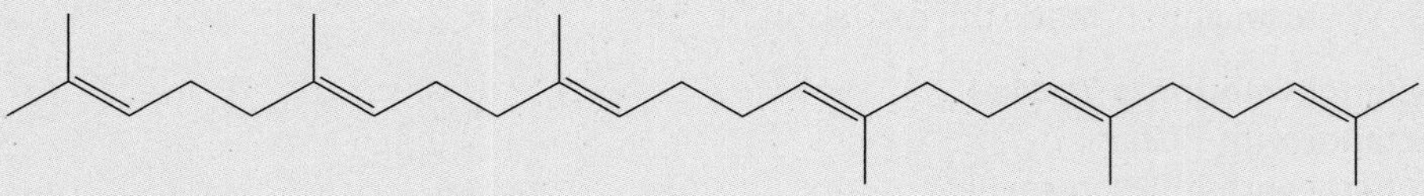

squalene

See Problems 7.30, 7.44, and 7.52.

In the body, similar oxidations are accomplished by enzymes, together with a rather complex coenzyme called nicotinamide adenine dinucleotide, NAD^+ (for its structure, see page 547). Oxidation occurs in the liver and is a key step in the body's attempt to rid itself of imbibed alcohol.

$$\underset{\text{ethanol}}{CH_3CH_2OH} + NAD^+ \underset{\text{dehydrogenase}}{\overset{\text{alcohol}}{\rightleftharpoons}} \underset{\text{acetaldehyde}}{CH_3\overset{\overset{O}{\|}}{C}-H} + NADH \quad (7.40)$$

The resulting acetaldehyde—also toxic—is further oxidized in the body to acetic acid and eventually to carbon dioxide and water.

7.13 Alcohols with More Than One Hydroxyl Group

Compounds with two adjacent alcohol groups are called *glycols* (page 97). The most important example is **ethylene glycol**.* Compounds with more than two hydroxyl groups are also known, and several, such as **glycerol** and **sorbitol**, are important commercial chemicals.

Some important polyols are **ethylene glycol, glycerol,** and **sorbitol.**

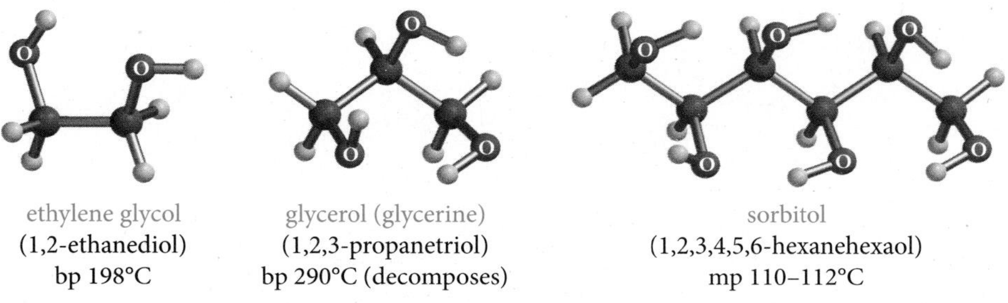

ethylene glycol (1,2-ethanediol) bp 198°C

glycerol (glycerine) (1,2,3-propanetriol) bp 290°C (decomposes)

sorbitol (1,2,3,4,5,6-hexanehexaol) mp 110–112°C

Ethylene glycol is used as the "permanent" antifreeze in automobile radiators and as a raw material in the manufacture of Dacron. Ethylene glycol is completely miscible with water. Because of its increased capacity for hydrogen bonding, ethylene glycol has an exceptionally high boiling point for its molecular weight—much higher than that of ethanol.

Glycerol is a syrupy, colorless, water-soluble, high-boiling liquid with a distinctly sweet taste. Its soothing qualities make it useful in shaving and toilet soaps and in cough drops and syrups. Triesters of glycerol are fats and oils, whose chemistry is discussed in Chapter 15.

Nitration of glycerol gives glyceryl trinitrate (nitroglycerine), a powerful and shock-sensitive explosive. Alfred Nobel, who invented dynamite in 1866, found that glyceryl trinitrate could be controlled by absorbing it on an inert porous material. Dynamite contains about 15% glyceryl (and glycol) nitrate along with other explosive materials. Dynamite is used mainly in mining and construction.

$$\underset{\text{glycerol}}{\begin{array}{l} CH_2OH \\ | \\ CHOH \\ | \\ CH_2OH \end{array}} + \underset{\text{nitric acid}}{3\ HONO_2} \xrightarrow{H_2SO_4} \underset{\text{glyceryl trinitrate (nitroglycerine)}}{\begin{array}{l} CH_2ONO_2 \\ | \\ CHONO_2 \\ | \\ CH_2ONO_2 \end{array}} + 3\ H_2O \quad (7.41)$$

Nitroglycerine is also used in medicine as a vasodilator, to prevent heart attacks in patients who suffer with angina.

*Notice that despite the *-ene* ending in the common name of ethylene glycol, there is *no double bond* between the carbons.

Sorbitol, with its many hydroxyl groups, is water soluble. It is almost as sweet as cane sugar and is used in candy making and as a sugar substitute for diabetics. In Chapter 16, we will see that carbohydrates, for example, sucrose (table sugar), starch, and cellulose, have many hydroxyl groups.

7.14 Aromatic Substitution in Phenols

Now we will examine some reactions that occur with phenols, but not with alcohols. Phenols undergo electrophilic aromatic substitution under very mild conditions because the hydroxyl group is strongly ring activating. For example, phenol can be nitrated with *dilute aqueous* nitric acid.

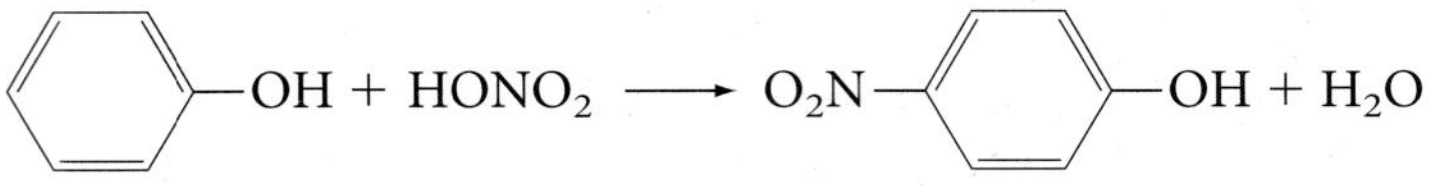

$$C_6H_5\text{—OH} + HONO_2 \longrightarrow O_2N\text{—}C_6H_4\text{—OH} + H_2O \qquad (7.42)$$

phenol — dilute nitric acid — *p*-nitrophenol

Phenol is also brominated rapidly with *bromine* in water, to produce 2,4,6-tribromophenol.

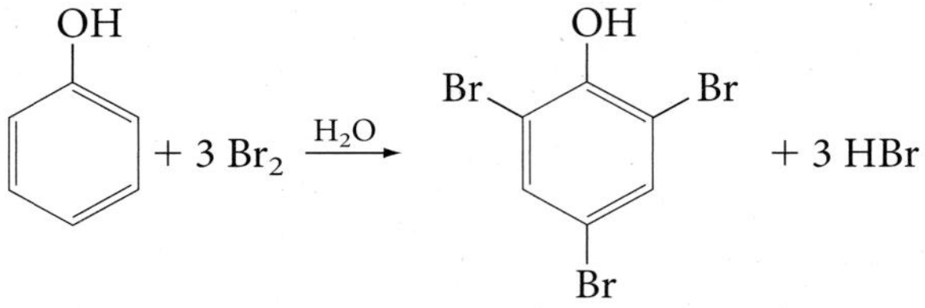

$$C_6H_5OH + 3\ Br_2 \xrightarrow{H_2O} 2,4,6\text{-}Br_3C_6H_2OH + 3\ HBr \qquad (7.43)$$

phenol — 2,4,6-tribromophenol

EXAMPLE 7.3

Draw the intermediate in electrophilic aromatic substitution *para* to a hydroxyl group, and show how the intermediate benzenonium ion is stabilized by the hydroxyl group.

Solution

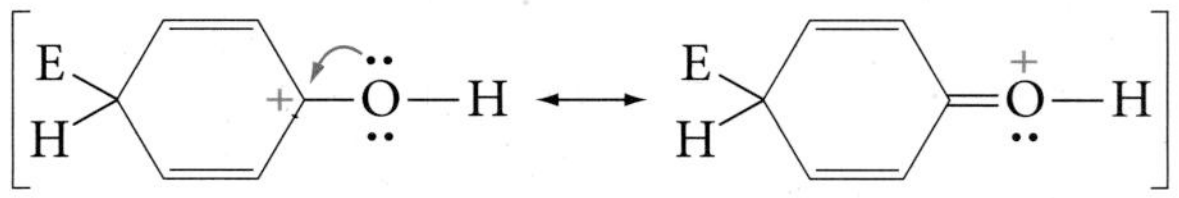

[Resonance structures: benzenonium ion (E, H on sp³ carbon) with + para to $\ddot{O}$—H ⟷ structure with C=$\overset{+}{O}$—H]

An unshared electron pair on the oxygen atom helps to delocalize the positive charge.

PROBLEM 7.20 Explain why phenoxide ion undergoes electrophilic aromatic substitution even more easily than does phenol.

PROBLEM 7.21 Write an equation for the reaction of

a. *p*-methylphenol + $HONO_2$ (1 mole)
b. *o*-chlorophenol + Br_2 (1 mole)

A WORD ABOUT... Quinones and the Bombardier Beetle

Tom Eisner and Dan Aneshansley, Cornell University

Bombardier beetle spraying.

The bombardier beetle has a curious way of defending itself from attack by a predator. It sprays the attacker with a hot mixture of noxious chemicals. These are discharged with remarkable accuracy and an audible pop from a pair of glands at the tip of its abdomen. The components of the spray are 1,4-benzoquinone, 2-methyl-1,4-benzoquinone, and hot water. The propellant is oxygen. This surprising and unpleasant event deters predators and provides the beetle with a chance to retreat to cover. How does the bombardier beetle accomplish this feat?

Each abdominal gland of the bombardier beetle consists of two connected chambers. The inner chamber contains a solution of 1,4-hydroquinone, 2-methyl-1,4-hydroquinone, and hydrogen peroxide dissolved in water. This is connected by a tube to an outer chamber. A one-way valve in the tube prevents the contents of the two chambers from mixing when closed, and allows flow only from the inner chamber to the outer chamber when opened. The outer chamber contains a water solution of enzymes called **peroxidases** and **catalases**. The peroxidase promotes a reaction between the hydrogen peroxide and hydroquinones to produce **quinones**, while the catalase converts hydrogen peroxide to water and oxygen. These reactions are exothermic enough to rapidly bring water to its boiling point. When under attack, the bombardier beetle opens the valve, and the contents of the inner compartment flow into the outer compartment. The aforementioned chemical reactions occur and the quinones and hot water are, by pressure produced by the oxygen, ejected as a pulsating spray through a vent in the outer chamber, accompanied by the audible pop.

$$\text{R-hydroquinone} + H_2O_2 \xrightarrow{\text{peroxidase}} \text{R-benzoquinone} + 2\,H_2O$$

R = H 1,4-hydroquinone
R = CH_3 2-methyl-1,4-hydroquinone

R = H 1,4-benzoquinone
R = CH_3 2-methyl-1,4-benzoquinone

$$2\,H_2O_2 \xrightarrow{\text{catalase}} 2\,H_2O + O_2$$

Although the delivery system used by the bombardier beetle is particularly impressive, quinones are also used by other beetles, and certain earwigs, cockroaches, and spiders, as components of defensive secretions. Furthermore, two-chambered glands related to that of the bombardier beetle are used by other organisms to deliver defensive substances (Sec. 9.10). For some fascinating reading, see W. Agosta, *Bombardier Beetles and Fever Trees: A Close-Up Look at Chemical Warfare and Signals in Animals and Plants,* Addison-Wesley (1996). Also see T. Eisner and J. Meinwald, *Science*, **1966**, *153*, 1341.

See Problem 7.56.

7.15 Oxidation of Phenols

As indicated in the "A Word About . . . Quinones and the Bombardier Beetle" box on this page 225, phenols are easily oxidized. Samples that stand exposed to air for some time often become highly colored due to the formation of oxidation products. With hydroquinone (1,4-dihydroxybenzene), the reaction is easily controlled to give 1,4-benzoquinone (commonly called *quinone*).

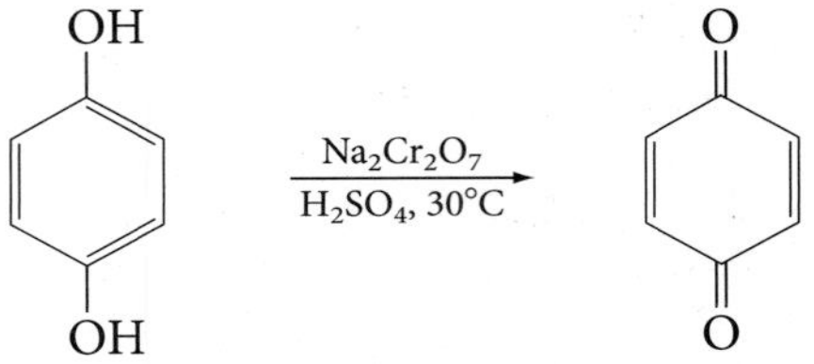

$$\text{hydroquinone (HO–C}_6\text{H}_4\text{–OH)} \xrightarrow[H_2SO_4,\ 30°C]{Na_2Cr_2O_7} \text{1,4-benzoquinone (O=C}_6\text{H}_4\text{=O)} \quad (7.44)$$

hydroquinone
colorless, mp 171°C

1,4-benzoquinone
yellow, mp 116°C

Hydroquinone and related compounds are used in photographic developers. They reduce silver ion that has not been exposed to light to metallic silver (and, in turn, they are oxidized to quinones). The oxidation of hydroquinones to quinones is reversible; this interconversion plays an important role in several biological oxidation–reduction reactions.

7.16 Phenols as Antioxidants

Phenols are **antioxidants,** preventing oxidation of substances sensitive to air oxidation.

Substances that are sensitive to air oxidation, such as foods and lubricating oils, can be protected by phenolic additives. Phenols function as **antioxidants**. They react with and destroy peroxy (ROO·)and hydroxy (HO·) radicals, which otherwise react with the alkenes present in foods and oils to cause their degradation. The peroxy and hydroxy radicals abstract the phenolic hydrogen atom to produce more stable phenoxy radicals that cause less damage to the alkenes (eq. 7.45).

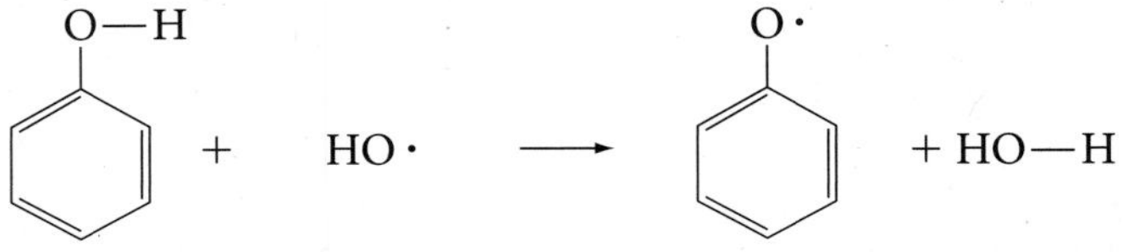

$$C_6H_5\text{–O–H} + HO\cdot \longrightarrow C_6H_5\text{–O}\cdot + HO\text{–}H \quad (7.45)$$

hydroxy radical phenoxy radical

PROBLEM 7.22 Write resonance structures that indicate how the unpaired electron in the phenoxy radical can be delocalized to the *ortho* and *para* positions (reason by analogy with the resonance structures for the phenoxide anion on page 215).

Two commercial phenolic antioxidants are BHA (butylated hydroxyanisole) and BHT (butylated hydroxytoluene). BHA is used as an antioxidant in foods, especially meat products. BHT is used not only in foods, animal feeds, and vegetable oils, but also in lubricating oils, synthetic rubber, and various plastics.

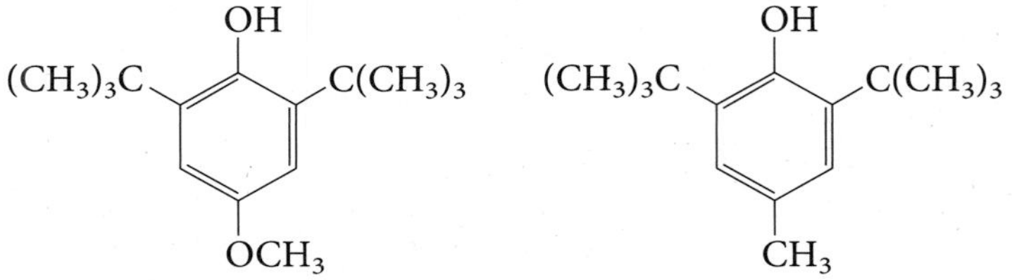

OH, $(CH_3)_3C$, $C(CH_3)_3$, OCH_3
BHA

OH, $(CH_3)_3C$, $C(CH_3)_3$, CH_3
BHT

PROBLEM 7.23 Write an equation for the reaction of BHT with hydroxy radical.

Human beings and other animals also use antioxidants for protection against biological sources of peroxy and hydroxy radicals. **Vitamin E** (α-tocopherol) is a natural phenolic antioxidant that protects the body against free radicals; see also "A Word About . . . Vitamin E: Tocopherols and Tocotrienols," page 128. Vitamin E is obtained largely through dietary sources such as leafy vegetables, egg yolks, wheat germ, vegetable oil, and legumes. Deficiencies of vitamin E can cause eye problems in premature infants and nerve damage in older children. Resveratrol is another phenolic natural product that is a common constituent of the human diet. It is found in a number of foods, including peanuts and grapes. Resveratrol is also an antioxidant and has been studied as a possible cancer chemopreventive agent.

Vitamin E is a natural phenolic antioxidant that protects the body against free radicals.

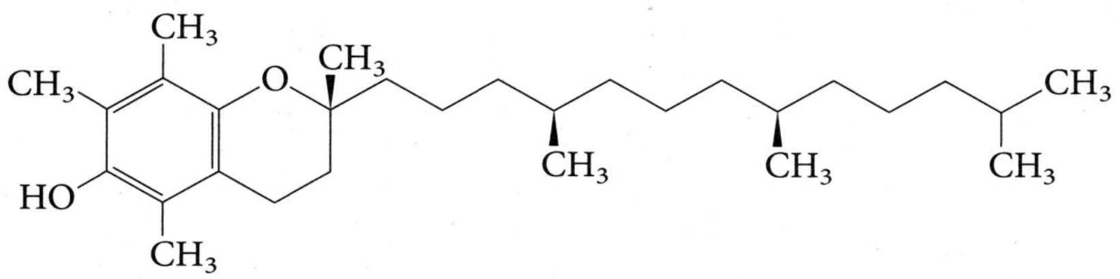

vitamin E (α-tocopherol)

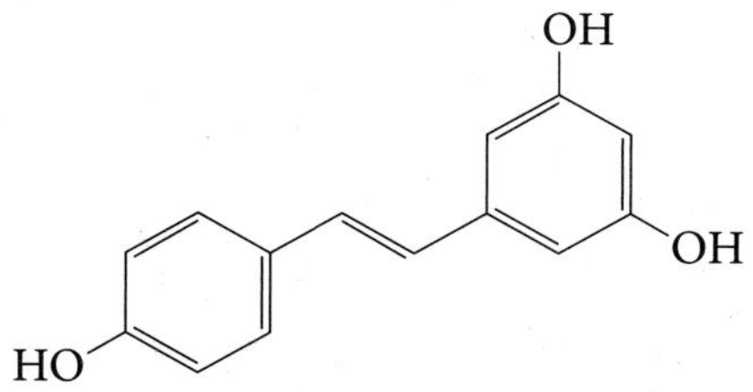

resveratrol

7.17 Thiols, the Sulfur Analogs of Alcohols and Phenols

Sulfur is immediately beneath oxygen in the periodic table and can often take its place in organic structures. The —**SH** group, called the **sulfhydryl group**, is the functional group of thiols (page 207). Thiols are named as follows:

The **sulfhydryl group, —SH,** is the functional group of thiols.

CH_3SH	$CH_3CH_2CH_2CH_2SH$	C_6H_5—SH
methanethiol (methyl mercaptan)	1-butanethiol (*n*-butyl mercaptan)	thiophenol (phenyl mercaptan)

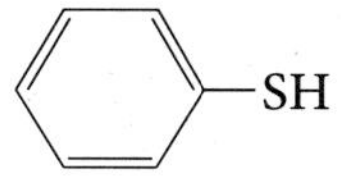

Thiols are sometimes called **mercaptans**, a name that refers to their reaction with mercuric ion to form mercury salts, called **mercaptides**.

Thiols are also called **mercaptans**; their mercury salts are called **mercaptides**.

$$2\ RSH + HgCl_2 \longrightarrow \underset{\text{a mercaptide}}{(RS)_2Hg} + 2\ HCl \quad \textbf{(7.46)}$$

PROBLEM 7.24 Draw the structure for

a. cyclohexyl mercaptan b. 2-hexanethiol c. isobutyl mercaptan

Alkyl thiols can be made from alkyl halides by nucleophilic displacement with sulfhydryl ion (Table 6.1, entry 10).

$$R{-}X + {}^-SH \longrightarrow R{-}SH + X^- \quad \textbf{(7.47)}$$

Perhaps the most distinctive feature of thiols is their intense and disagreeable odor. The thiols $CH_3CH{=}CHCH_2SH$ and $(CH_3)_2CHCH_2CH_2SH$, for example, are responsible for the odor of a skunk. The structurally related thiol $(CH_3)_2C{=}CHCH_2SH$ has recently been shown to be responsible for the skunky odor and taste of beer that has been exposed to light.

A WORD ABOUT... Hair—Curly or Straight

Hair consists of a fibrous protein called **keratin**, which, as proteins go, contains an unusually large percentage of the sulfur-containing amino acid **cystine**. Horse hair, for example, contains about 8% cystine:

$$\underset{\displaystyle NH_2}{HO_2CCHCH_2S}-\underset{\displaystyle NH_2}{SCH_2CHCO_2H}$$

cystine (CyS—SCy)

The disulfide link in cystine serves to cross-link the chains of amino acids that make up the protein (Figure 7.2).

The chemistry used in waving or straightening hair involves the oxidation–reduction chemistry of the disulfide bond (eq. 7.49). First, the hair is treated with a reducing agent, which breaks the S—S bonds, converting each sulfur to an —SH group. This breaks the cross-links between the long protein chains. The reduced hair can now be shaped as desired, either waved or straightened. Finally, the reduced and rearranged hair is treated with an oxidizing agent to reform the disulfide cross-links. The new disulfide bonds, no longer in their original positions, hold the hair in its new shape.

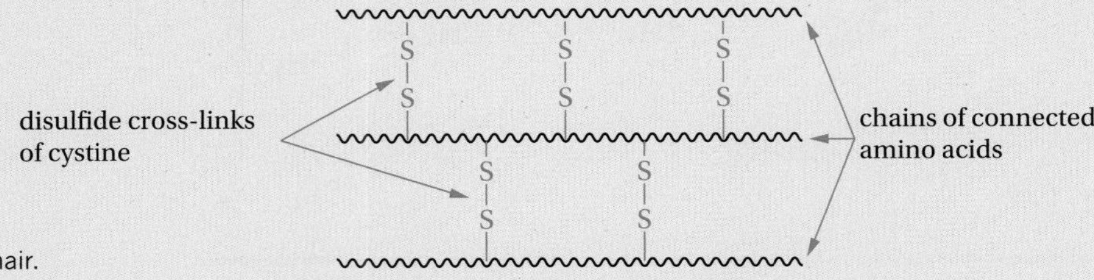

■ **Figure 7.2**
Schematic structure of hair.

The striped skunk (*Mephitis mephitis*) sprays a foul mixture of thiols at its enemies.

Thiols are more acidic than alcohols. The pK_a of ethanethiol, for example, is 10.6 whereas that of ethanol is 15.9. Hence, thiols react with aqueous base to give thiolates.

$$RSH + Na^+OH^- \longrightarrow \underset{\text{a sodium thiolate}}{RS^-Na^+} + HOH \qquad (7.48)$$

PROBLEM 7.25 Write an equation for the reaction of propanethiol ($CH_3CH_2CH_2SH$) with

a. NaH
b. $CH_3CH_2O^-Na^+$
c. KOH
d. $HgCl_2$

Disulfides are compounds containing an S—S bond.

Thiols are easily oxidized to **disulfides,** compounds containing an S—S bond, by mild oxidizing agents such as hydrogen peroxide or iodine. A naturally occurring disulfide whose smell you are probably familiar with is diallyl disulfide ($CH_2{=}CHCH_2S{-}SCH_2CH{=}CH_2$), which is responsible for the odor of fresh garlic.*

$$\underset{\text{thiol}}{2\ RS-H} \underset{\text{reduction}}{\overset{\text{oxidation}}{\rightleftharpoons}} \underset{\text{disulfide}}{RS-SR} \qquad (7.49)$$

The reaction shown in eq. 7.49 can be reversed with a variety of reducing agents. Since proteins contain disulfide links, these reversible oxidation–reduction reactions can be used to manipulate the three-dimensional shapes of proteins.

*Garlic belongs to the plant family *Allium*, from which the *allyl* group gets its name.

REACTION SUMMARY

1. Alcohols

a. Conversion to Alkoxides (Sec. 7.6)

$$2\,RO{-}H + 2\,Na \longrightarrow 2\,RO^- Na^+ + H_2$$

$$RO{-}H + NaH \longrightarrow RO^- Na^+ + H_2$$

b. Dehydration to Alkenes (Sec. 7.8)

$$H{-}C{-}C{-}OH \xrightarrow{H^+ \text{ (cat)}} C{=}C + H{-}OH$$

c. Conversion to Alkyl Halides (Secs. 7.9–7.10)

$$R{-}OH + HX \longrightarrow R{-}X + H_2O \; (X{=}Cl, Br, I)$$

$$R{-}OH + SOCl_2 \longrightarrow R{-}Cl + HCl + SO_2$$

$$3\,R{-}OH + PX_3 \longrightarrow 3\,R{-}X + H_3PO_3 \; (X{=}Cl, Br)$$

d. Oxidation (Sec. 7.12)

$$\underset{\text{primary}}{RCH_2OH} \xrightarrow{PCC} R{-}\overset{O}{\overset{\|}{C}}{-}H \quad \text{(aldehyde)}$$

$$\underset{\text{primary}}{RCH_2OH} \xrightarrow[H^+]{CrO_3} R{-}\overset{O}{\overset{\|}{C}}{-}OH \quad \text{(carboxylic acid)}$$

$$\underset{\text{secondary}}{R_2CHOH} \xrightarrow{PCC \text{ or } CrO_3,\, H^+} R{-}\overset{O}{\overset{\|}{C}}{-}R \quad \text{(ketone)}$$

2. Phenols

a. Preparation of Phenoxides (Sec. 7.6)

$$ArO{-}H + NaOH \longrightarrow ArO^- Na^+ + H_2O$$

b. Electrophilic Aromatic Substitution (Sec. 7.14)

$$C_6H_5{-}OH \xrightarrow{HNO_3} O_2N{-}C_6H_4{-}OH$$

$$C_6H_5{-}OH \xrightarrow{Br_2} \text{2,4,6-}Br_3C_6H_2{-}OH$$

c. Oxidation to Quinones (Sec. 7.15)

$$HO{-}C_6H_4{-}OH \xrightarrow[H_2SO_4,\, H_2O]{Na_2Cr_2O_7} O{=}C_6H_4{=}O$$

quinone

3. Thiols

a. Conversion to Thiolates (Sec. 7.17)

$$\underset{\text{thiol}}{RS{-}H} + NaOH \longrightarrow \underset{\text{thiolate}}{RS^- Na^+} + H_2O$$

b. Oxidation to Disulfides (Sec. 7.17)

$$\underset{\text{thiol}}{2\,RSH} \xrightarrow{\text{oxidation}} \underset{\text{disulfide}}{RS{-}SR}$$

ADDITIONAL PROBLEMS

OWL Interactive versions of these problems are assignable in OWL.

Nomenclature and Structure of Alcohols

7.26 Write a structural formula for each of the following compounds:

a. 1-methylcyclohexanol
b. *p*-bromophenol
c. 2,3-butanediol
d. 2,2-dimethyl-1-butanol
e. sodium ethoxide
f. *cis*-2-methylcyclopentanol
g. (*S*)-2-butanethiol
h. 2-phenylpropanol
i. 2-methyl-2-propen-1-ol
j. 2-cyclohexenol

7.27 Name each of the following alcohols:

a. $CH_3CH(Cl)CH(OH)CH_2CH_3$
b. $CH_3CH(OH)CH_2CH_3$
c. $(CH_3)_2C(Cl)CH(OH)CH_2CH_3$
d. $CH_3CH(Cl)CH_2CH(OH)CH_3$

7.28 Name each of the following compounds:

a. $HOCH_2CH(OH)CH(OH)CH_2OH$
b. $(CH_3)_3CO^-K^+$
c. 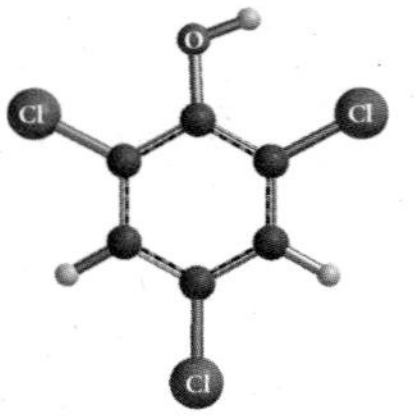 [ball-and-stick model: Cl, Cl, Cl, O]
d. 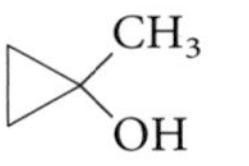 [cyclopropane with CH_3 and OH]
e. 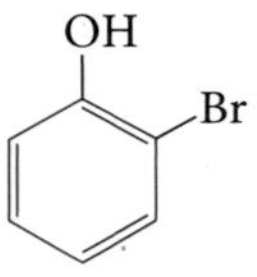 [benzene ring with OH and Br]
f. 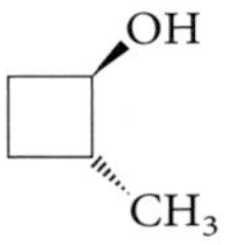 [cyclobutane with OH and CH_3]
g. $CH_3CH{=}CHCH_2OH$
h. 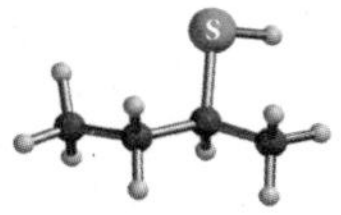 [ball-and-stick model: S]
i. $CH_3CH_2CH(OH)CH_3$
j. $CH_3CHBrC(CH_3)_2OH$

7.29 Explain why each of the following names is unsatisfactory, and give a correct name:

a. 2-ethyl-1-propanol
b. 2,2-dimethyl-3-butanol
c. 1-propene-3-ol
d. 2-chloro-4-pentanol
e. 3,6-dibromophenol

7.30 Members of the fungus-like genus *Phytophthora* are responsible for many crop diseases that threaten plants around the world. Sexual reproduction for this genus is modulated by a mating hormone, shown below:

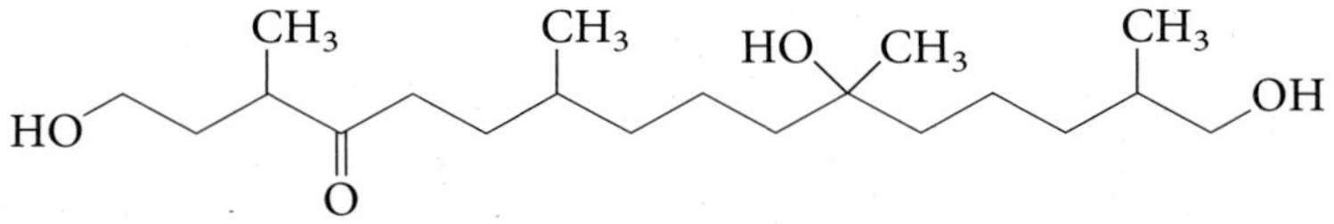

a. How many stereogenic centers are present in this compound?
b. Draw structures for the unique stereoisomers of this hormone and provide the name of each structure.

7.31 The odor of a human's armpit is due to a number of molecules, including some sulfur-containing compounds. Indeed, the major component in the odor of sweat is (*S*)-3-methyl-3-mercapto-1-hexanol. Draw the structure of this thiol.

7.32 Thymol is an antibacterial oil obtained from thyme (*Thymus vulgaris*). The IUPAC name of this compound is 2-isopropyl-5-methylphenol. Draw the structure of thymol.

© Squared Studios/Photodisc/Getty Images

Thyme (*Thymus vulgaris*), source of the antibacterial oil thymol.

Properties of Alcohols

7.33 Classify the alcohols in parts a, d, f, h, i, and j of Problem 7.26 as primary, secondary, or tertiary.

7.34 Arrange the compounds in each of the following groups in order of increasing solubility in water, and briefly explain your answers:

a. 1-octanol; ethanol; ethyl chloride
b. $HOCH_2(CHOH)_3CH_2OH$; 1,5-pentanediol; 1-pentanol

Acid–Base Reactions of Alcohols and Thiols

7.35 The following classes of organic compounds are Lewis bases. Write an equation that shows how each class might react with H^+.

a. amine, R_3N: b. ether, $R\ddot{\underset{\cdot\cdot}{O}}R$ c. ketone, $R_2C{=}\ddot{\underset{\cdot\cdot}{O}}$

7.36 Arrange the following compounds in order of increasing acidity, and explain the reasons for your choice of order: cyclopentanol, phenol, *p*-nitrophenol, and 2-chlorocyclopentanol.

7.37 Which is the stronger base, potassium *t*-butoxide or potassium ethoxide? (*Hint:* Use the data in Table 7.2.)

7.38 Complete each of the following equations:

a. $CH_3CH(OH)CH_2CH_3 + K \longrightarrow$
b. $(CH_3)_2CHOH + NaH \longrightarrow$
c. Cl–⟨benzene ring⟩–OH + NaOH ⟶
d. cyclopentane ring bearing H and OH on the same carbon + NaOH ⟶
e. $CH_3CH{=}CHCH_2SH + NaOH \longrightarrow$

7.39 Explain why your answers to parts c, d, and e of Problem 7.38 are consistent with the pK_a's of the starting acids and product acids (see eqs. 7.14, 7.15, and 7.48).

Acid-Catalyzed Dehydration of Alcohols

7.40 Show the structures of all possible acid-catalyzed dehydration products of the following. If more than one alkene is possible, predict which one will be formed in the largest amount.

a. 3-methylcyclopentanol
b. 1-methylcyclopentanol
c. 2-phenylethanol
d. 2-hexanol

7.41 Explain why the reaction shown in eq. 7.19 occurs much more easily than the reaction $(CH_3)_3C{-}OH \rightleftharpoons (CH_3)_3C^+ + HO^-$. (That is, why is it necessary to protonate the alcohol before ionization can occur?)

7.42 Draw a reaction energy diagram for the dehydration of *tert*-butyl alcohol (eq. 7.21). Include the steps shown in eqs. 7.18–7.20 in your diagram.

7.43 Write out all of the steps in the mechanism for eq. 7.24, showing how each product is formed.

7.44 The sesquiterpene, (3*S*)-(*E*)-nerolidol (structure shown in Problem 3.40), is derived from multiple isoprene units as are some of the biologically important alcohols discussed on page 222. What products would result from the acid-catalyzed dehydration of (3*S*)-(*E*)-nerolidol?

7.45 2-Propanol is prepared commercially from the acid-catalyzed hydration of propene (see "A Word About . . . Industrial Alcohols" on page 208). Write out all of the steps in this acid-catalyzed mechanism.

Alkyl Halides from Alcohols

7.46 Although the reaction shown in eq. 7.26 occurs faster than that shown in eq. 7.28, the yield of product is lower. The yield of *t*-butyl chloride is only 80%, whereas the yield of *n*-butyl chloride is nearly 100%. What by-product is formed in eq. 7.26, and by what mechanism is it formed? Why is a similar by-product *not* formed in eq. 7.28?

7.47 Treatment of 3-buten-2-ol with concentrated hydrochloric acid gives a mixture of two products, 3-chloro-1-butene and 1-chloro-2-butene. Write a reaction mechanism that explains how both products are formed.

Synthesis and Reactions of Alcohols

7.48 Write an equation for each of the following reactions:

a. 2-methyl-2-butanol + HCl
b. 3-pentanol + Na
c. cyclohexanol + PBr_3
d. 2-phenylethanol + $SOCl_2$
e. 1-methylcyclopentanol + H_2SO_4, heat
f. ethylene glycol + $HONO_2$
g. 1-octanol + HBr + $ZnBr_2$
h. 1-pentanol + aqueous NaOH
i. 1-pentanol + CrO_3, H^+
j. 2-cyclohexylethanol + PCC

7.49 Write an equation for each of the following two-step syntheses:

a. cyclohexene to cyclohexanone
b. 1-chlorobutane to butanal
c. 1-butanol to 1-butanethiol

Oxidation Reactions of Alcohols, Phenols, and Thiols

7.50 The alcohol citronellol is a terpene found in rose oil. The product formed when citronellol is oxidized with pyridinium chlorochromate (PCC) is a constituent of lemon oil. Draw the structure of the product when citronellol is oxidized with PCC.

© Squared Studios/Photodisc/Getty Images

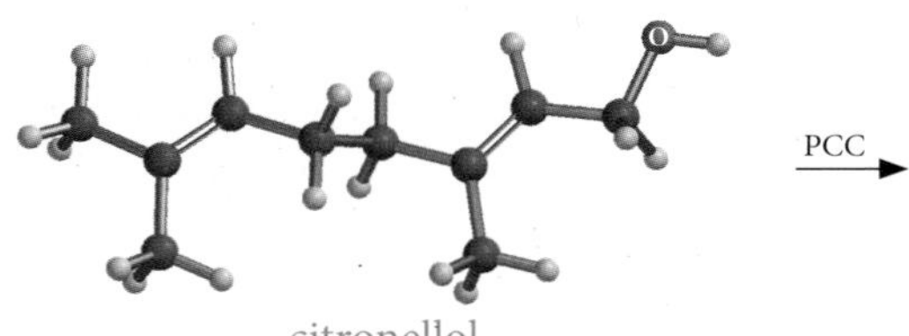

7.51 What product do you expect from the oxidation of cholesterol with CrO_3 and H^+? (See page 222 for the formula of cholesterol.)

7.52 Linoleic acid (Table 15.1) is an unsaturated vegetable oil found in soybean, sunflower, and corn oils. Often these unsaturated vegetable oils are considered to be good for one's heart since unsaturated vegetable oils do not contribute to levels of bad cholesterol (see "A Word About . . . Biologically Important Alcohols and Phenols," page 222). However, it has been reported that heating linoleic acid at frying temperatures, such as 365°F for 30 minutes or longer, leads to the formation of a toxic aldehyde, (2*E*)-4-hydroxy-2-nonenal. What is the structure of (2*E*)-4-hydroxy-2-nonenal? If (2*E*)-4-hydroxy-2-nonenal was oxidized by PCC, what would be the expected product? If (2*E*)-4-hydroxy-2-nonenal was oxidized by Jones' reagent, what would be the expected product?

7.53 Draw the structure of the quinone expected from the oxidation of

a. (structure: OH, OH)

b. (structure: OH, OH)

c. (structure: OH, CH_3, OH)

7.54 Dimethyl disulfide, $CH_3S—SCH_3$, found in the vaginal secretions of female hamsters, acts as a sexual attractant for the male hamster. Write an equation for its synthesis from methanethiol.

7.55 The disulfide shown below is a component of the odorous secretion of mink. Describe a synthesis of this disulfide, starting with 3-methyl-1-butanol.

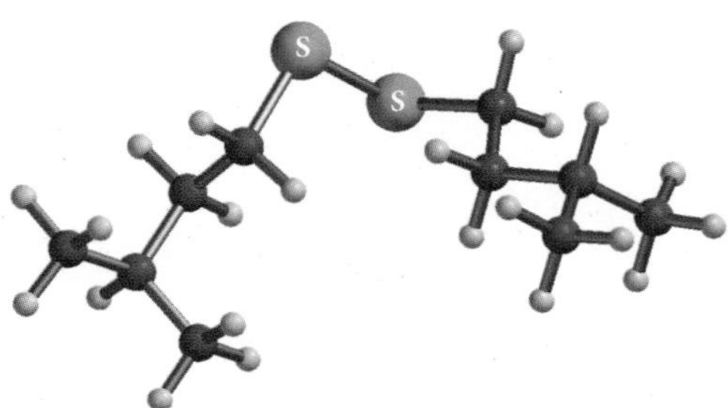

7.56 2,3,6-Trimethyl-1,4-naphthoquinone (TMNQ) is a quinone that was recently isolated from tobacco leaves and was shown to slow the metabolism of dopamine, a neurotransmitter whose depletion can lead to Parkinson's disease. What is the structure of the hydroquinone whose oxidation gives TMNQ?

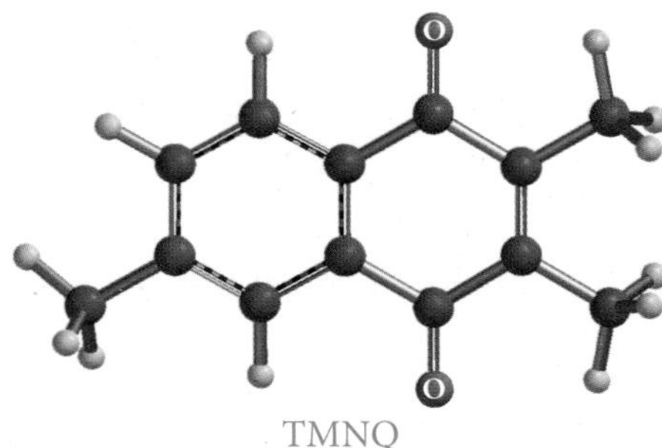

TMNQ

8

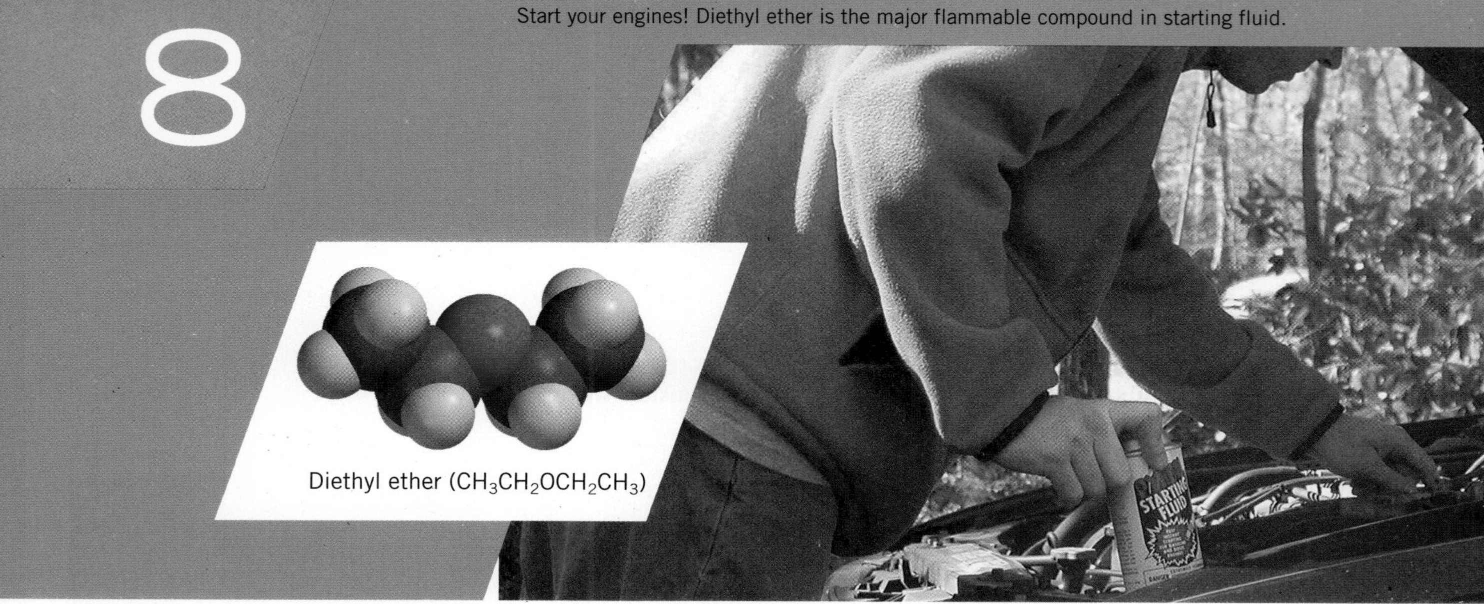

Start your engines! Diethyl ether is the major flammable compound in starting fluid.

Ethers and Epoxides

For many people the word *ether* is associated with the well-known general anesthetic—(see "A Word About... Ether and Anesthesia" later in this chapter). *Ether* is also a common ingredient in starter fluid for car engines. However, this is only one member of a class of compounds known as **ethers**. The natural antibiotic monensin and the pheromone disparlure (see "A Word About... The Gypsy Moth's Epoxide" later in this chapter) both contain the ether functional group. The drug tetrahydrocannabinol (THC), the active ingredient in marijuana, contains a cyclic ether. Synthetic ethers include the gasoline additive MTBE (page 241) and ethylene oxide, the industrial precursor to ethylene glycol, which is used as antifreeze and as a starting material in the manufacture of polyesters (Chapter 14).

tetrahydrocannabinol (THC)

monensin

OWL

Online homework for this chapter can be assigned in OWL, an online homework assessment tool.

All ethers are compounds in which two organic groups are connected to a single oxygen atom. The general formula for an ether is R—O—R′, where R and R′ may be identical or different, and they may be alkyl or aryl groups. In the common anesthetic, both R and R′ are ethyl groups, $CH_3CH_2—O—CH_2CH_3$. In this chapter, we will describe the physical and chemical properties of ethers. Their excellent solvent properties are applied in the preparation of Grignard reagents, organometallic compounds with a carbon–magnesium bond. We will give special attention to **epoxides**, cyclic three-membered ring ethers that have important industrial utility.

Ethers are compounds that have two organic groups connected to a single oxygen atom. **Epoxides** are cyclic, three-membered ring ethers.

8.1 Nomenclature of Ethers

Ethers are usually named by giving the name of each alkyl or aryl group, in alphabetical order, followed by the word *ether.*

$CH_3CH_2—O—CH_3$
ethyl methyl ether

$CH_3CH_2—O—CH_2CH_3$
diethyl ether (the prefix *di-* is sometimes omitted)

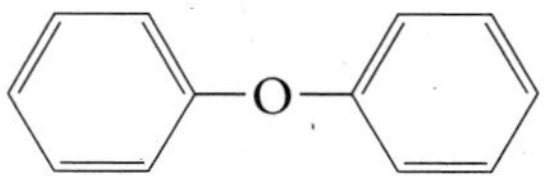

C₆H₅—O—C₆H₅
diphenyl ether

For ethers with more complex structures, it may be necessary to name the —OR group as an alkoxy group. In the IUPAC system, the smaller alkoxy group is named as a substituent.

$CH_3CHCH_2CH_2CH_3$ with OCH_3 on C-2
2-methoxypentane

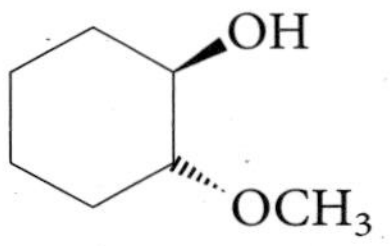

OH, OCH_3 on cyclohexane
trans-2-methoxycyclohexanol

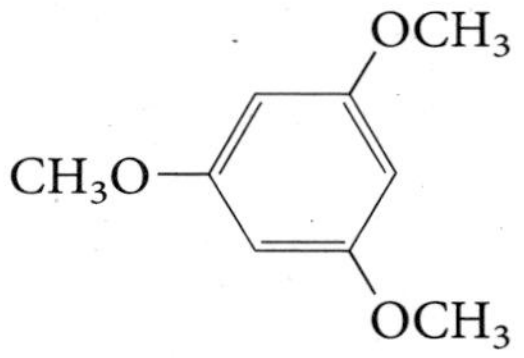

CH_3O, OCH_3, OCH_3 on benzene
1,3,5-trimethoxybenzene

EXAMPLE 8.1

Give a correct name for $CH_3CHCH(CH_3)_2$ with OCH_2CH_3 on C-2.

Solution

$\overset{1}{C}H_3\overset{2}{C}H\overset{3}{C}H\overset{4}{C}H_3$ with CH_3 on C-3 and OCH_2CH_3 on C-2

2-ethoxy-3-methylbutane

PROBLEM 8.1 Give a correct name for

a. $(CH_3)_2CHCH_2OCH_2CH_3$ b. (cyclohexyl)$-O-CH_2CH_2CH_3$

c.

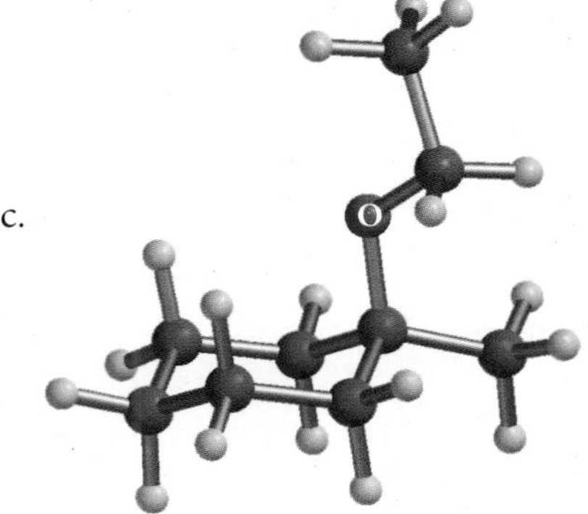

PROBLEM 8.2 Write the structural formula for

a. dicyclopropyl ether b. dibenzyl ether c. 2-methoxyhexane

8.2 Physical Properties of Ethers

Ethers are colorless compounds with characteristic, relatively pleasant odors. They have lower boiling points (bp's) than alcohols with an equal number of carbon atoms. In fact, an ether has nearly the same bp as the corresponding hydrocarbon in which a $-CH_2-$ group replaces the ether's oxygen. The data in Table 8.1 illustrate these facts:

Table 8.1 ■ **Properties of Alcohols, Ethers, and Hydrocarbons of Similar Molecular Weight**

Compound	Formula	bp	mol wt	Water solubility (g/100 mL, 20°C)
1-butanol	$CH_3CH_2CH_2CH_2OH$	118°C	74	7.9
diethyl ether	$CH_3CH_2-O-CH_2CH_3$	35°C	74	7.5
pentane	$CH_3CH_2-CH_2-CH_2CH_3$	36°C	72	0.03

Because of their structures (no O—H bonds), ether molecules cannot form hydrogen bonds with one another. This is why they boil at much lower temperatures than their isomeric alcohols.*

PROBLEM 8.3 Write structures for each of the following *isomers* and arrange them in order of decreasing boiling point: 3-methoxy-1-propanol, 1,2-dimethoxyethane, and 1,4-butanediol.

*Although ethers are slightly more polar than alkanes, the major attractive forces between ether molecules are van der Waals attractions (see Sec. 2.7).

Although ethers cannot form hydrogen bonds with one another, they do form hydrogen bonds with alcohols:

$$\underset{\displaystyle R}{\underset{|}{R-\ddot{O}:}}\cdots\cdots \underset{\displaystyle R}{\underset{|}{H-\ddot{O}:}}$$

For this reason, alcohols and ethers are usually mutually soluble. Low-molecular-weight ethers, such as dimethyl ether, are quite soluble in water. Likewise, the modest solubility of diethyl ether in water is similar to that of its isomer 1-butanol (see Table 8.1) because each can form a hydrogen bond to water. Ethers are less dense than water.

8.3 Ethers as Solvents

Ethers are relatively inert compounds. They do not usually react with dilute acids, with dilute bases, or with common oxidizing and reducing agents. They do not react with metallic sodium—a property that distinguishes them from alcohols. This general inertness, coupled with the fact that most organic compounds are ether-soluble, makes ethers excellent solvents in which to carry out organic reactions.

Ethers are also used frequently to extract organic compounds from their natural sources. Diethyl ether is particularly good for this purpose. Its low boiling point makes it easy to remove from an extract and easy to recover by distillation. It is highly flammable, however, and must not be used if there are any flames in the laboratory.

Ethers that have been in a laboratory for a long time, exposed to air, may contain organic peroxides as a result of oxidation.

$$CH_3CH_2OCH_2CH_3 + O_2 \longrightarrow \underset{\text{an ether hydroperoxide}}{CH_3CH_2O\underset{\displaystyle OOH}{\underset{|}{C}}HCH_3} \tag{8.1}$$

These peroxides are explosive and must be removed before the ether can be used safely. Test papers are used to detect peroxides, and shaking with aqueous ferrous sulfate ($FeSO_4$) destroys these peroxides by reduction.

8.4 The Grignard Reagent; an Organometallic Compound

One of the most striking examples of the solvating power of ethers is in the preparation of **Grignard reagents**. These reagents, which are exceedingly useful in organic synthesis, were discovered by the French organic chemist Victor Grignard [pronounced "greenyar(d)"]. In 1912, he received a Nobel Prize for his contribution to organic synthesis.*

Grignard reagents are alkyl- or arylmagnesium halides.

Grignard found that when magnesium turnings are stirred with an ether solution of an alkyl or aryl halide, an exothermic reaction occurs. The magnesium, which is insoluble in ether, disappears as it reacts with the halide to give solutions of ether-soluble Grignard reagents.

$$R-X + Mg \xrightarrow{\text{dry ether}} \underset{\text{a Grignard reagent}}{R-MgX} \tag{8.2}$$

The carbon–halogen bond is broken, and both the alkyl group and halogen become bonded to the magnesium.

---

*For a brief account of how Grignard discovered these reagents, see D. Hodson, *Chemistry in Britain*, **1987**, 141–142.

Although the ether used as a solvent for this reaction is normally not shown as part of the Grignard reagent structure, it does play an important role. The unshared electron pairs on the ether oxygen help to stabilize the magnesium through coordination.

$$\begin{array}{c} R_2\ddot{O}: \\ R'-Mg-X \\ R_2\ddot{O}: \end{array}$$

Acting as a Lewis base, ether stabilizes a Grignard reagent.

The two ethers most commonly used in Grignard preparations are diethyl ether and the cyclic ether tetrahydrofuran, abbreviated THF (page 247). The Grignard reagent will not form unless the ether is scrupulously dry, free of traces of water or alcohols.

Grignard reagents are named as shown in the following equations:

$$\underset{\text{methyl iodide}}{CH_3-I} + Mg \xrightarrow{\text{ether}} \underset{\text{methylmagnesium iodide}}{CH_3MgI} \qquad \textbf{(8.3)}$$

$$\underset{\text{bromobenzene}}{C_6H_5-Br} + Mg \xrightarrow{\text{ether}} \underset{\text{phenylmagnesium bromide}}{C_6H_5-MgBr} \qquad \textbf{(8.4)}$$

Notice that there is no space between the name of the organic group and magnesium, but that there is a space before the halide name.

Grignard reagents usually react as if the alkyl or aryl group is negatively charged (a carbanion) and the magnesium atom is positively charged.

$$\overset{\delta-}{R}-\overset{\delta+}{MgX}$$

A **carbanion** is an alkyl or aryl group with a negatively charged carbon atom. Carbanions are strong bases.

***Carbanions** are strong bases.* They are the conjugate bases of hydrocarbons, which are very weak acids. It is not surprising, then, that Grignard reagents react vigorously with even such a weak acid as water, or with any other compound with an OH, SH, or NH bond.

$$\underset{\text{stronger base}}{\overset{\delta-}{R}-MgX} + \underset{\text{stronger acid}}{\overset{\delta+}{H}-OH} \longrightarrow \underset{\text{weaker acid}}{R-H} + \underset{\text{weaker base}}{Mg^{2+}(OH)^-X^-} \qquad \textbf{(8.5)}$$

This is why the ether used as a solvent for the Grignard reagent *must be scrupulously free of water or alcohol.*

PROBLEM 8.4 Write an equation for the reaction between

a. ethylmagnesium iodide and water
b. phenylmagnesium bromide and ethanol

EXAMPLE 8.2

Is it possible to prepare a Grignard reagent from $HOCH_2CH_2CH_2Br$ and magnesium?

Solution No! Any Grignard reagent that forms is immediately destroyed by protons from the OH group. Grignard and hydroxyl groups in the same molecule are incompatible.

PROBLEM 8.5 Is it possible or impossible to prepare a Grignard reagent from $CH_3OCH_2CH_2CH_2Br$? Explain.

The reaction of a Grignard reagent with water can be put to useful purpose. For example, if heavy water (D_2O) is used, deuterium (an isotope of hydrogen) can be substituted for a halogen.

$$CH_3-C_6H_4-Br \xrightarrow[\text{ether}]{\text{Mg}} CH_3-C_6H_4-MgBr \xrightarrow{D_2O} CH_3-C_6H_4-D \quad \textbf{(8.6)}$$

p-bromotoluene — *p*-tolylmagnesium bromide — *p*-deuteriotoluene

This is a useful way to introduce an isotopic label into an organic compound.*

EXAMPLE 8.3

Show how to prepare CH_3CHDCH_3 from $CH_2{=}CHCH_3$.

Solution

$$CH_2{=}CHCH_3 \xrightarrow{HBr} CH_3\underset{\underset{Br}{|}}{C}HCH_3 \xrightarrow[\text{ether}]{\text{Mg}} CH_3\underset{\underset{MgBr}{|}}{C}HCH_3 \xrightarrow{D_2O} CH_3CHDCH_3$$

PROBLEM 8.6 Show how to prepare CH_3CHDCH_3 from $(CH_3)_2CHOH$.

Grignard reagents are **organometallic compounds**; they contain a carbon–metal bond. Many other types of organometallic compounds are known. Acetylides (eq. 3.53), for example, are similar to Grignard reagents in their reactions. **Organolithium compounds**, which can be prepared in a manner similar to that for Grignard reagents, are also useful in synthesis.

Organometallic compounds are organic compounds that contain a carbon–metal bond. **Organolithium compounds** are organic compounds that contain a carbon–lithium bond.

$$\overset{\delta+}{R}-\overset{\delta-}{X} + 2\,Li \xrightarrow{\text{ether}} \overset{\delta-}{R}-\overset{\delta+}{Li} + Li^+X^- \quad \textbf{(8.7)}$$

an alkyllithium

*An isotope label serves as a "tag" that allows chemists to obtain information about such things as reaction mechanisms and rates. For a detailed example of how deuterium labels have been used to determine reaction mechanisms, see Section 9.16.

PROBLEM 8.7 Write an equation for the preparation of

a. propyllithium b. the acetylide of 1-hexyne

and for the reaction of each with D_2O.

Later in this chapter and elsewhere in this book, we will see examples of the synthetic utility of organometallic reagents.

8.5 Preparation of Ethers

The most important commercial ether is diethyl ether. It is prepared from ethanol and sulfuric acid.

$$\underset{\text{ethanol}}{CH_3CH_2OH + HOCH_2CH_3} \xrightarrow[140°C]{H_2SO_4} \underset{\text{diethyl ether}}{CH_3CH_2OCH_2CH_3} + H_2O \quad (8.8)$$

Note that ethanol can be dehydrated by sulfuric acid to give either ethylene (eq. 7.17) or diethyl ether (eq. 8.8). Of course, the reaction conditions are different in each case. These reactions provide a good example of how important it is to control reaction conditions and to specify them in equations.

PROBLEM 8.8 The reaction in eq. 7.17 occurs by an E2 mechanism (review eqs. 7.22 and 7.23). By what mechanism does the reaction in eq. 8.8 occur?

Although it can be adapted to other ethers, the alcohol–sulfuric acid method is most commonly used to make symmetric ethers from primary alcohols.

PROBLEM 8.9 Write an equation for the synthesis of propyl ether from 1-propanol.

The commercial production of *t*-butyl methyl ether has become important in recent years. In 2002, worldwide consumption of MTBE* was about 7 billion gallons. With an octane value of 110, it is used as an octane number enhancer in unleaded gasolines.** It is prepared by the acid-catalyzed addition of methanol to 2-methylpropene. The reaction is related to the hydration of alkenes (Sec. 3.7.b). The only difference is that an alcohol, methanol, is used as the nucleophile instead of water.

$$\underset{\text{methanol}}{CH_3OH} + \underset{\text{2-methylpropene}}{CH_2{=}C(CH_3)_2} \xrightarrow{H^+} \underset{t\text{-butyl methyl ether}}{CH_3O{-}\overset{\displaystyle CH_3}{\underset{\displaystyle CH_3}{\overset{|}{\underset{|}{C}}}}{-}CH_3} \quad (8.9)$$

PROBLEM 8.10 Write out the steps in the mechanism for eq. 8.9 (see eqs. 3.18 and 3.20).

***t*-Butyl methyl ether is incorrectly referred to as methyl *t*-butyl ether (MTBE) by the general public.

**See "A Word About... Petroleum, Gasoline, and Octane Number," pages 102–103.

A CLOSER LOOK AT... MTBE

Conduct research on the Internet to find more information on MTBE and to answer the following questions. Investigate different points of view concerning the use of MTBE.

FAQs about MTBE from the EPA

1. What is MTBE and how is it made?
2. Why is MTBE used in gasoline? What is reformulated gasoline (RFG)? To what extent is MTBE used in RFG?
3. Does the use of MTBE improve air quality?
4. Is MTBE hazardous to humans when inhaled? When ingested in drinking water? What evidence is there with regard to the health dangers presented by MTBE?
5. How does MTBE get into the environment? What properties of MTBE make it likely to be found far from the original site of a spill or leak?
6. How can MTBE be removed from the environment?
7. MTBE is now being phased out for use in gasoline. What is its replacement?

Most important for the laboratory synthesis of unsymmetric ethers is the Williamson synthesis, named after the British chemist who devised it. This method has two steps, both of which we have already discussed. In the first step, an alcohol is converted to its alkoxide by treatment with a reactive metal (sodium or potassium) or metal hydride (review eqs. 7.12 and 7.13). In the second step, an S_N2 displacement is carried out between the alkoxide and an alkyl halide (see Table 6.1, entry 2). The Williamson synthesis is summarized by the general equations

$$2\,ROH + 2\,Na \longrightarrow 2\,RO^-Na^+ + H_2 \quad (8.10)$$

$$RO^-Na^+ + R'{-}X \longrightarrow ROR' + Na^+X^- \quad (8.11)$$

Since the second step is an S_N2 reaction, it works best if R′ in the alkyl halide is primary and not well at all if R′ is tertiary.

EXAMPLE 8.4

Write an equation for the synthesis of $CH_3OCH_2CH_2CH_3$ using the Williamson method.

Solution There are two possibilities, depending on which alcohol and which alkyl halide are used:

$$\underset{\text{from } CH_3O^-Na^+ + XCH_2CH_2CH_3}{CH_3OCH_2CH_2CH_3} \quad \text{or} \quad \underset{\text{from } CH_3X + Na^{+-}OCH_2CH_2CH_3}{CH_3OCH_2CH_2CH_3}$$

The equations are

$$2\,CH_3OH + 2\,Na \longrightarrow 2\,CH_3O^-Na^+ + H_2$$

$$CH_3O^-Na^+ + CH_3CH_2CH_2X \longrightarrow CH_3OCH_2CH_2CH_3 + Na^+X^-$$

or

$$2\,CH_3CH_2CH_2OH + 2\,Na \longrightarrow 2\,CH_3CH_2CH_2O^-Na^+ + H_2$$

$$CH_3CH_2CH_2O^-Na^+ + CH_3X \longrightarrow CH_3CH_2CH_2OCH_3 + Na^+X^-$$

X is usually Cl, Br, or I.

PROBLEM 8.11 Write equations for the synthesis of the following ethers by the Williamson method:

a. C_6H_5—OCH_3 (benzene ring)—OCH_3 b. $(CH_3)_3COCH_3$ c. $C_6H_5CH_2OC(CH_3)_3$

(*Reminder:* The second step proceeds by an S_N2 mechanism.)

8.6 Cleavage of Ethers

Ethers have unshared electron pairs on the oxygen atom and are therefore Lewis bases (Figure 8.1). They react with strong proton acids and with Lewis acids such as the boron halides.

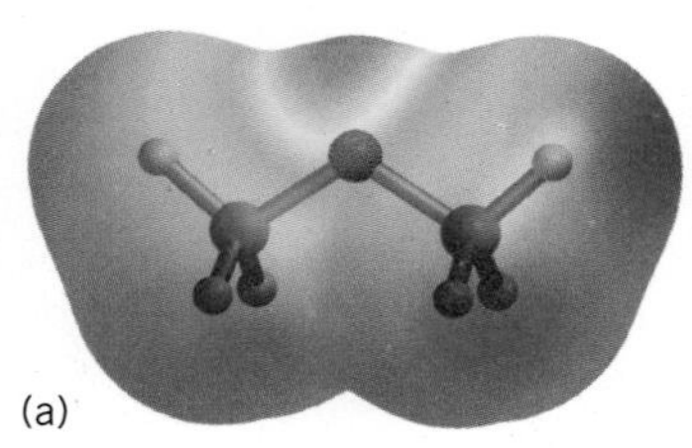
(a)

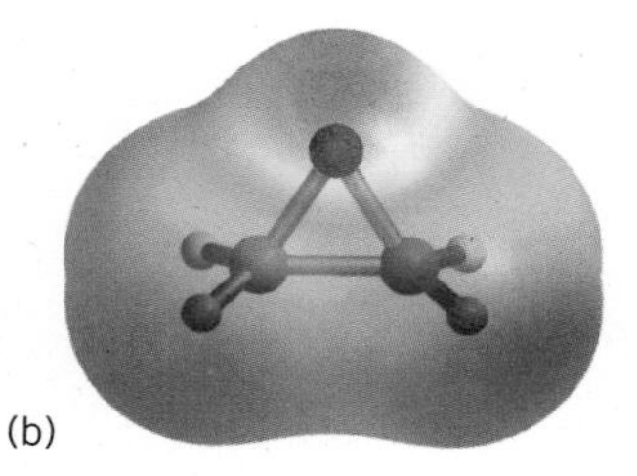
(b)

▪ **Figure 8.1**
Electrostatic potential map for (a) dimethyl ether and (b) ethylene oxide, showing the electron-rich (red) character of the oxygen atom.

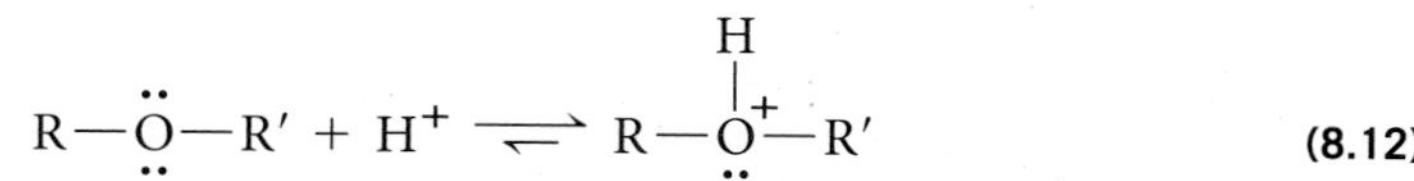

$$R-\ddot{O}-R' + H^+ \rightleftharpoons R-\overset{H}{\overset{|}{\dot{O}^+}}-R' \quad (8.12)$$

$$R-\ddot{O}-R' + Br-\underset{Br}{\underset{|}{B}}-Br \rightleftharpoons R-\overset{+}{\dot{O}}(-\overset{-}{B}Br_3)-R' \quad (8.13)$$

These reactions are similar to the reaction of alcohols with strong acids (eq. 7.18). If the alkyl groups R and/or R′ are primary or secondary, the bond to oxygen can be broken by reaction with a strong nucleophile such as I^- or Br^- (by an S_N2 process). For example,

$$\underset{\text{ethyl isopropyl ether}}{CH_3CH_2OCH(CH_3)_2} + HI \xrightarrow{\text{heat}} \underset{\text{ethyl iodide}}{CH_3CH_2I} + \underset{\text{isopropyl alcohol}}{HOCH(CH_3)_2} \quad (8.14)$$

$$\underset{\text{anisole}}{C_6H_5-OCH_3} + BBr_3 \xrightarrow[\text{2. } H_2O]{\text{1. heat}} \underset{\text{phenol}}{C_6H_5-OH} + \underset{\text{methyl bromide}}{CH_3Br} \quad (8.15)$$

If R or R′ is tertiary, a strong nucleophile is not required since reaction will occur by an S_N1 (or E1) mechanism.

$$\underset{t\text{-butyl phenyl ether}}{C_6H_5-OC(CH_3)_3} \xrightarrow[H_2O]{H^+} \underset{\text{phenol}}{C_6H_5-OH} + \underset{t\text{-butyl alcohol (and } (CH_3)_2C=CH_2)}{(CH_3)_3COH} \quad (8.16)$$

The net result of these reactions is cleavage of the ether at one of the C—O bonds. Ether cleavage is a useful reaction for determining the structure of a complex, naturally occurring ether because it allows one to break the large molecule into more easily handled, smaller fragments.

A WORD ABOUT... Ether and Anesthesia

Prior to the 1840s, pain during surgery was relieved by various methods (asphyxiation, pressure on nerves, and administration of narcotics or alcohol), but on the whole, it was almost worse torture to undergo an operation than to endure the disease. Modern use of anesthesia during surgery has changed all that. Anesthesia stems from the work of several physicians in the mid-nineteenth century. The earliest experiments used nitrous oxide, ether, or chloroform. Perhaps the best known of these experiments was the removal of a tumor from the jaw of a patient anesthetized by ether, performed by Boston dentist William T. G. Morton in 1846.

Anesthetics fall into two major categories, general and local. **General anesthetics** are usually administered to accomplish three ends: insensitivity to pain (analgesia), loss of consciousness, and muscle relaxation. Gases, such as nitrous oxide and cyclopropane, as well as volatile liquids, such as ether, are administered by inhalation, but other general anesthetics, such as barbiturates, are injected intravenously.

The exact mechanism by which anesthetics affect the central nervous system is not completely known. Unconsciousness may result from several factors: changes in the properties of nerve cell membranes, suppression of certain enzymatic reactions, and solubility of the anesthetic in lipid membranes (Chapter 15).

A good inhalation anesthetic should vaporize readily and have appropriate solubility in the blood and tissues. It should also be stable, inert, nonflammable, potent, and minimally toxic. It should have an acceptable odor and cause minimal side effects such as nausea or vomiting. No anesthetic that meets *all* of these specifications has yet been developed. Although **diethyl ether** is perhaps the best-known general anesthetic to the layperson, it fails on several counts (flammability, side effects of nausea or vomiting, and relatively slow action). It is quite potent, however, and produces good analgesia and muscle relaxation. The use of ether at present is rather limited, mainly because of its undesirable side effects. **Halothane**, $CF_3CHBrCl$, comes closest to an ideal inhalation anesthetic at present, but halogenated ethers such as **enflurane**, $CF_2H{-}O{-}CF_2CHClF$, are also used.

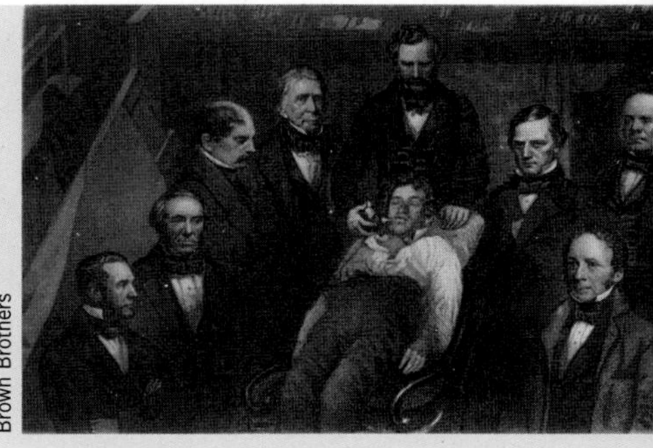

Brown Brothers

Dr. William T. G. Morton making the first public demonstration of anesthetization with ether.

Local anesthetics are either applied to body surfaces or injected near nerves to desensitize a particular region of the body to pain. The best known of these anesthetics is procaine (Novocain), an aromatic amino-ester (see "A Word About... Morphine and Other Nitrogen-Containing Drugs," pages 405–406).

The discovery of anesthetics enabled physicians to perform surgery with deliberation and care, leading to many of the advances of modern medicine.

See Problem 6.15.

EXAMPLE 8.5

Write out the steps in the mechanism for eq. 8.14.

Solution The ether is first protonated by the acid.

$$CH_3CH_2\ddot{\underset{\cdot\cdot}{O}}CH(CH_3)_2 \overset{H^+}{\rightleftharpoons} CH_3CH_2\underset{\cdot\cdot}{\overset{H}{\overset{|+}{O}}}CH(CH_3)_2$$

oxonium ion

The resulting oxonium ion is then cleaved by S_N2 attack of iodide ion at the primary carbon (recall that $1° > 2°$ in S_N2 reactions).

$$I^- + CH_3CH_2{-}\overset{H}{\overset{|+}{O}}CH(CH_3)_2 \longrightarrow CH_3CH_2I + HOCH(CH_3)_2$$

PROBLEM 8.12 Write out the steps in the mechanism for formation of *t*-butyl alcohol in eq. 8.16. Which C—O bond cleaves, the one to the phenyl or the one to the *t*-butyl group? What kind of mechanism is operative in the C—O bond cleavage step?

8.7 Epoxides (Oxiranes)

Epoxides (or oxiranes) are cyclic ethers with a three-membered ring containing one oxygen atom.

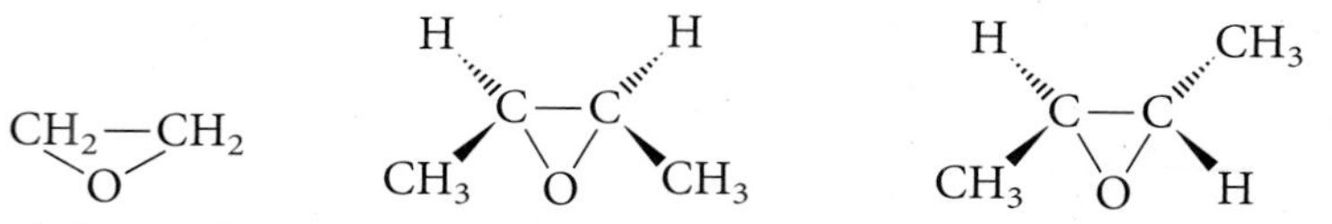

ethylene oxide (oxirane) bp 13.5°C

cis-2-butene oxide (*cis*-2,3-dimethyloxirane) bp 60°C

trans-2-butene oxide (*trans*-2,3-dimethyloxirane) bp 54°C

The most important commercial epoxide is ethylene oxide, produced by the silver-catalyzed air oxidation of ethylene.

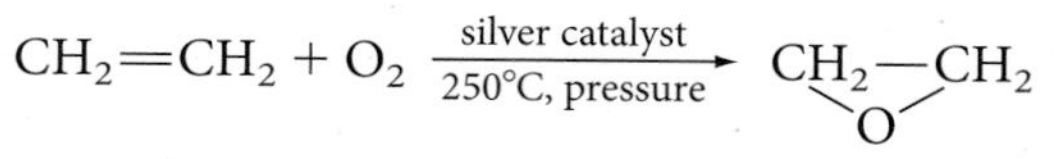

$$CH_2{=}CH_2 + O_2 \xrightarrow[250°C,\ pressure]{silver\ catalyst} \underset{ethylene\ oxide}{CH_2{-}CH_2\ (O)} \qquad (8.17)$$

Worldwide production of ethylene oxide exceeds 30 billion pounds per year. Only rather small amounts are used directly (for example, as a fumigant in grain storage). Most of the ethylene oxide constitutes a versatile raw material for the manufacture of other products, the main one being ethylene glycol.

The reaction in eq. 8.17 is suitable only for ethylene oxide. Other epoxides are usually prepared by the reaction of an alkene with an organic peroxyacid (often called simply a peracid).

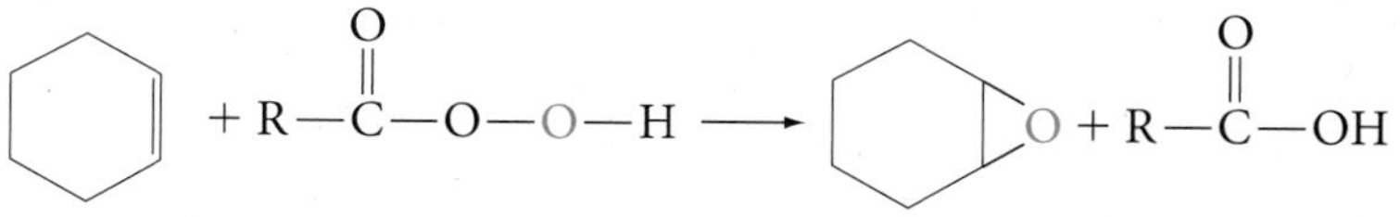

$$\text{cyclohexene} + R{-}\overset{O}{\overset{\|}{C}}{-}O{-}O{-}H \longrightarrow \text{cyclohexene oxide} + R{-}\overset{O}{\overset{\|}{C}}{-}OH \qquad (8.18)$$

cyclohexene; organic peroxy acid; cyclohexene oxide; organic acid

Peroxy acids are structurally related to hydrogen peroxide H—O—O—H, and peroxy acids are good oxidizing agents. On a large scale, peracetic acid (eq. 8.18, R = CH_3) is used, whereas in the laboratory, organic peroxy acids such as *m*-chloroperbenzoic acid (R = *m*-chlorophenyl) are frequently used.

PROBLEM 8.13 Write an equation for the reaction of cyclohexene with *m*-chloroperbenzoic acid,

$$m\text{-}ClC_6H_4{-}\overset{O}{\overset{\|}{C}}{-}O{-}O{-}H$$

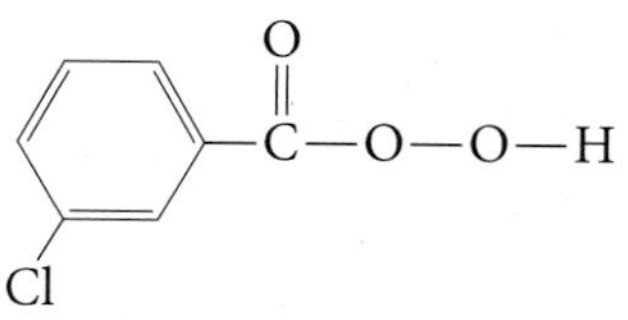

A WORD ABOUT... The Gypsy Moth's Epoxide

The main mode of communication among insects is via the emission and detection of specific chemical substances. These substances are called **pheromones**. The word is from the Greek *pherein,* to carry, and *horman,* to excite. Even though they are emitted and detected in exceedingly small amounts, pheromones have profound biological effects. One of their main effects is sexual attraction and stimulation, but they are also used as alarm substances to alert members of the same species to danger, as aggregation substances to call together both sexes of a species, and as trail substances to lead members of a species to food.

Often pheromones are chemically simple compounds—alcohols, esters, aldehydes, ketones, ethers, epoxides, or even hydrocarbons. Two examples are **muscalure** and **bombykol**, the sex attractants of the common housefly and the silkworm moth, respectively.

$$\underset{\text{muscalure}}{\overset{CH_3(CH_2)_7}{\underset{H}{}}C{=}C\overset{(CH_2)_{12}CH_3}{\underset{H}{}}}$$

$$\underset{\text{bombykol}}{CH_3(CH_2)_2CH{=}CH{-}CH{=}CH(CH_2)_8CH_2OH}$$

Their molecular weights are low enough that the substances are volatile, yet not so low that they disperse too rapidly. Also, their molecular structures must be distinctive to make them species-specific; survival of the species would not be served by attracting another species. Often this specificity is attained through stereoisomerism (at double bonds and/or at chiral centers), but it can also be achieved by using specific ratios of two or more pheromones for a particular communication purpose.

© Color Pic/Animals Animals
Male (top) and female (bottom) gypsy moths.

Let us consider a specific pheromone, **disparlure**, the sex attractant of the gypsy moth (*Lymantria dispar*). The gypsy moth is a serious despoiler of forest and shade trees as well as fruit orchards. Gypsy moth larvae, which hatch each spring, are voracious eaters and can strip a tree bare of leaves in just a few weeks.

The abdominal tips (last two segments) of the virgin female moth contain the sex attractant. Extraction of 78,000 tips led to isolation of the main sex attractant, which was the following *cis*-epoxide.

$$(CH_3)_2CH(CH_2)_4\overset{7}{C}H\!-\!\overset{8}{C}H(CH_2)_9CH_3 \quad (\text{epoxide ring, O bridging C7–C8; both H on the same side})$$

(7*R*,8*S*)-(+)-7,8-epoxy-2-methyloctadecane
(disparlure)

The active isomer has the *R* configuration at carbon-7 and the *S* configuration at carbon-8. This isomer can be detected by the male gypsy moth at a concentration as low as 10^{-10} g/mL; its enantiomer is inactive in solutions that are a million times more concentrated.

Disparlure has been synthesized in the laboratory. The synthetic material can be used to lure the male to traps and in that way to control the insect population. This form of insect control sometimes has advantages over spraying with insecticides.

For more on pheromones, chemical ecology, and organic synthesis, see K. Mori, *Acc. Chem. Res.*, **2000**, *33*, 102–110.

See Problem 8.38.

8.8 Reactions of Epoxides

Because of the strain in the three-membered ring, epoxides are much more reactive than ordinary ethers and give products in which the ring has opened. For example, with water they undergo acid-catalyzed ring opening to give glycols. [Remember that treating alkenes with alkaline $KMnO_4$ (Sec. 3.17.a) also produces glycols.]

$$\underset{\text{ethylene oxide}}{\underset{\diagdown O \diagup}{CH_2{-}CH_2}} + H{-}OH \xrightarrow{H^+} \underset{\text{ethylene glycol}}{\underset{OH\quad OH}{CH_2{-}CH_2}} \qquad (8.19)$$

In this way, over 8 billion pounds of ethylene glycol are produced annually in the United States alone. Approximately half of it is used in automobile cooling systems as antifreeze. Most of the rest is used to prepare polyester fibers.

EXAMPLE 8.6

Write equations that show the mechanism for eq. 8.19.

Solution The first step is reversible protonation of the epoxide oxygen, as in eq. 8.12.

$$\underset{\diagdown\ \ddot{\text{O}}\ \diagup}{CH_2{-}CH_2} + H^+ \rightleftharpoons \underset{\underset{\text{H}}{|}}{\underset{\diagdown\ \ddot{\text{O}}^+\ \diagup}{CH_2{-}CH_2}}$$

The second step is a nucleophilic S_N2 displacement on the primary carbon, with water as the nucleophile. Then proton loss yields the glycol (see Table 6.1, entry 3).

$$H_2\ddot{O}: + \underset{\underset{\text{H}}{|}}{\underset{\diagdown\ \ddot{\text{O}}^+\ \diagup}{CH_2{-}CH_2}} \longrightarrow H{-}\overset{+}{\ddot{\underset{\underset{\text{H}}{|}}{O}}}{-}CH_2{-}CH_2{-}OH \rightleftharpoons HO{-}CH_2CH_2{-}OH + H^+$$

PROBLEM 8.14 Write an equation for the acid-catalyzed reaction of cyclohexene oxide with water. Predict the stereochemistry of the product.

Other nucleophiles add to epoxides in a similar way.

$$\underset{\diagdown\ \text{O}\ \diagup}{CH_2{-}CH_2} \xrightarrow{H^+} \begin{cases} \xrightarrow{CH_3OH} HOCH_2CH_2OCH_3 \quad \text{2-methoxyethanol} \\ \xrightarrow{HOCH_2CH_2OH} HOCH_2CH_2OCH_2CH_2OH \quad \text{diethylene glycol} \end{cases} \qquad (8.20)$$

2-Methoxyethanol is an additive for jet fuels, used to prevent water from freezing in fuel lines. Being both an alcohol and an ether, it is soluble in both water and organic solvents. *Diethylene glycol* is useful as a plasticizer (softener) in cork gaskets and tiles.

Grignard reagents and organolithium compounds are strong nucleophiles capable of opening the ethylene oxide ring. The initial product is a magnesium alkoxide or lithium alkoxide, but after hydrolysis, we obtain a primary alcohol with two more carbon atoms than the organometallic reagent.

$$\overset{\delta-}{R}{-}\overset{\delta+}{Mg}X + \underset{\diagdown\ \underset{\delta-}{O}\ \diagup}{H_2\overset{\delta+}{C}{-}CH_2} \longrightarrow \underset{\text{a magnesium alkoxide}}{RCH_2CH_2OMgX} \xrightarrow{H-OH} RCH_2CH_2OH + Mg(OH)X \qquad (8.21)$$

$$\overset{\delta-}{R}{-}\overset{\delta+}{Li} + \underset{\diagdown\ \underset{\delta-}{O}\ \diagup}{H_2\overset{\delta+}{C}{-}CH_2} \longrightarrow \underset{\text{a lithium alkoxide}}{RCH_2CH_2OLi} \xrightarrow{H-OH} RCH_2CH_2OH + LiOH \qquad (8.22)$$

PROBLEM 8.15 Write an equation for the reaction between ethylene oxide and

a. $CH_3CH_2CH_2CH_2MgCl$ followed by hydrolysis
b. $C_6H_5CH_2MgBr$ followed by hydrolysis
c. $H_2C{=}CHLi$ followed by hydrolysis
d. $CH_3C{\equiv}C^-Na^+$ followed by hydrolysis

8.9 Cyclic Ethers

Cyclic ethers whose rings are larger than the three-membered epoxides are known. The most common are five- or six-membered rings. Some examples are

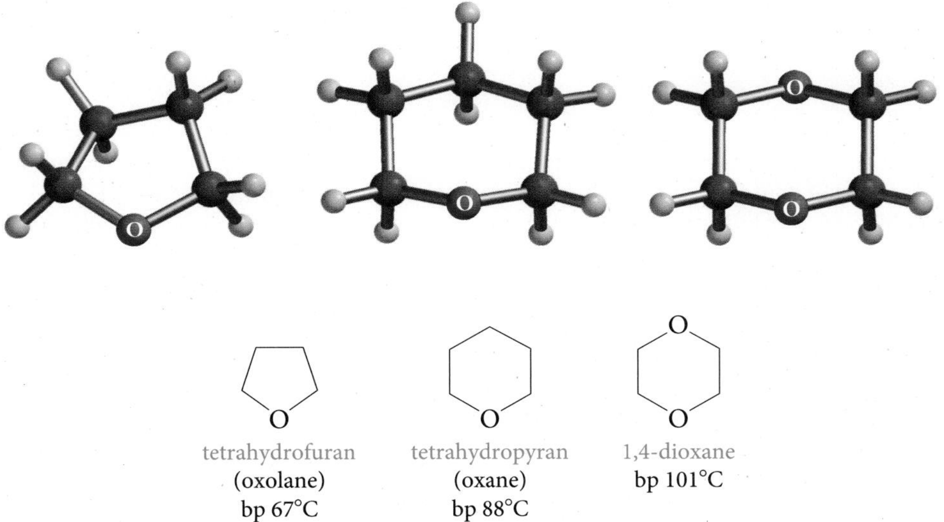

Tetrahydrofuran (THF) is a particularly useful solvent that not only dissolves many organic compounds but is miscible with water. THF is an excellent solvent—often superior to diethyl ether—in which to prepare Grignard reagents. Although it has the same number of carbon atoms as diethyl ether, they are "pinned back" in a ring. The oxygen in THF is therefore less hindered and better at coordinating with the magnesium in a Grignard reagent. **Tetrahydropyran** and **1,4-dioxane** are also soluble in both water and organic solvents.

Tetrahydrofuran (THF), **tetrahydropyran**, and **1,4-dioxane** are important cyclic ethers.

In recent years, there has been much interest in macrocyclic (large-ring) polyethers. Some examples are

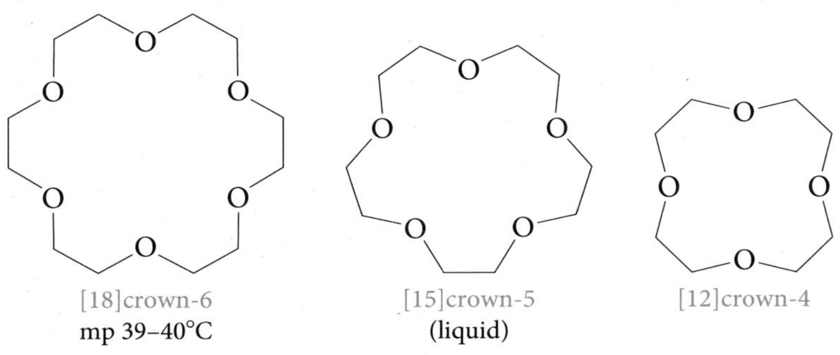

Crown ethers are crown-shaped macrocyclic polyethers that form complexes with positive ions.

These compounds are called **crown ethers** because their molecules have a crownlike shape. The bracketed number in their common names gives the ring size, and the terminal number gives the number of oxygens. The oxygens are usually separated from one another by two carbon atoms.

Crown ethers have the unique property of forming complexes with positive ions (Na^+, K^+, and so on). The positive ions fit within the macrocyclic rings selectively, depending on the sizes of the particular ring and ion. For example, [18]crown-6 binds K^+ more tightly than it does the smaller Na^+ (too loose a fit) or the larger Cs^+ (too large to fit in the hole). Similarly, [15]crown-5 binds Na^+, and [12]crown-4 binds Li^+. The crown ethers act as hosts for their ionic guests.

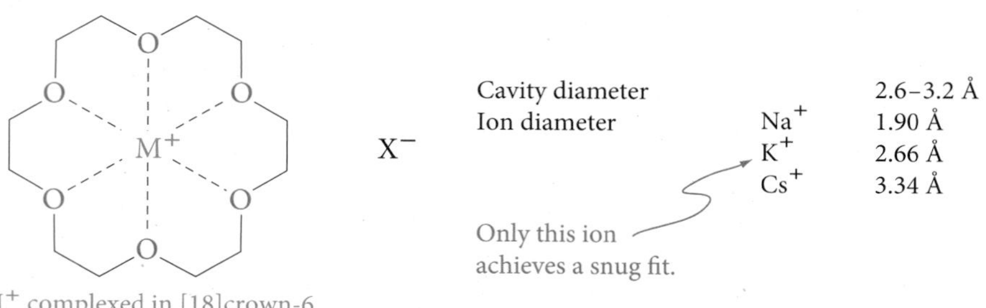

M$^+$ complexed in [18]crown-6

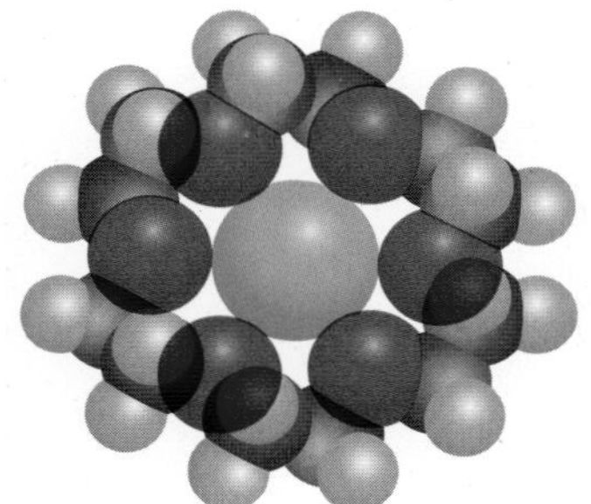

Model of [18]crown-6 complex with K$^+$.

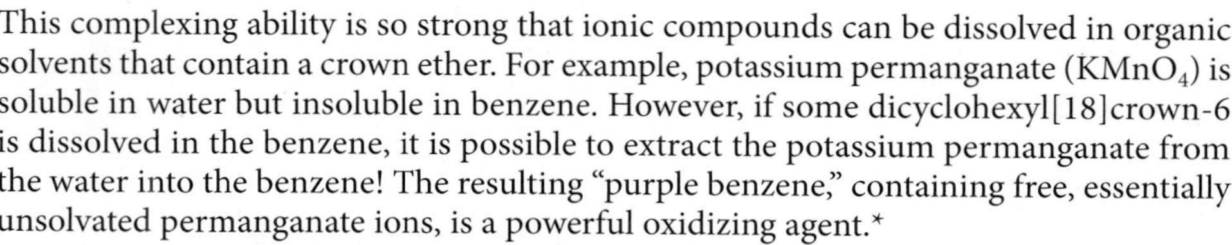

This complexing ability is so strong that ionic compounds can be dissolved in organic solvents that contain a crown ether. For example, potassium permanganate ($KMnO_4$) is soluble in water but insoluble in benzene. However, if some dicyclohexyl[18]crown-6 is dissolved in the benzene, it is possible to extract the potassium permanganate from the water into the benzene! The resulting "purple benzene," containing free, essentially unsolvated permanganate ions, is a powerful oxidizing agent.*

The selective binding of metallic ions by macrocyclic compounds is important in nature. Several antibiotics, such as **nonactin**, have large rings that contain regularly spaced oxygen atoms. Nonactin (which contains four tetrahydrofuran rings joined by four ester links) selectively binds K^+ (in the presence of Na^+) in aqueous media, thus allowing the selective transport of K^+ (but not Na^+) through cell membranes.

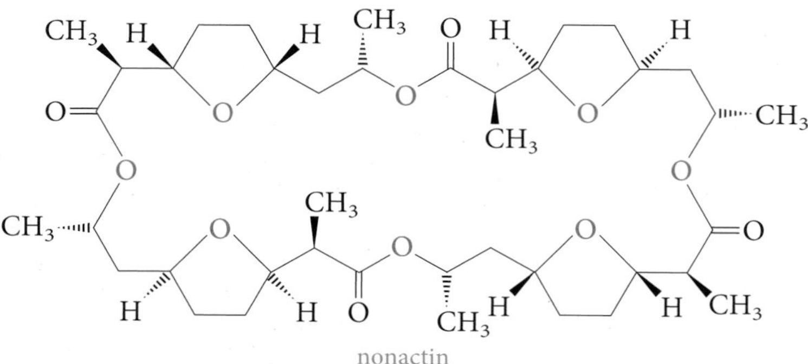

nonactin

*Crown ethers were discovered by Charles J. Pedersen, working at DuPont Company. This discovery had broad implications for a field now known as molecular recognition, or host–guest chemistry. Pedersen, Donald J. Cram (U.S.), and Jean-Marie Lehn (France) shared the 1987 Nobel Prize in chemistry for their imaginative development of this field. You might enjoy Pedersen's personal account of this discovery (*Journal of Inclusion Phenomena,* **1988,** *6,* 337–350); the same journal contains the Nobel lectures by Cram and Lehn on their work.

REACTION SUMMARY

1. Organometallic Compounds

a. Preparation of Grignard Reagents (Sec. 8.4)

$$R{-}X + Mg \xrightarrow{\text{ether or THF}} R{-}MgX$$
$X = Cl, Br, I$

b. Preparation of Organolithium Reagents (Sec. 8.4)

$$R{-}X + 2\,Li \longrightarrow R{-}Li + LiX$$
$X = Cl, Br, I$

c. Hydrolysis of Organometallics to Alkanes (Sec. 8.4)

$$R{-}MgX + H_2O \longrightarrow R{-}H + Mg(OH)X$$
$$R{-}Li + H_2O \longrightarrow R{-}H + LiOH$$

2. Ethers

a. Preparation by Dehydration of Alcohols (Sec. 8.5)

$$2\,R{-}OH \xrightarrow[\Delta]{H_2SO_4} R{-}O{-}R + H_2O$$

b. Preparation from Alkenes and Alcohols (Sec. 8.5)

$$>C{=}C< \xrightarrow[H^+ \text{ (catalyst)}]{ROH} H{-}\overset{|}{\underset{|}{C}}{-}\overset{|}{\underset{|}{C}}{-}OR$$

c. Preparation from Alcohols and Alkyl Halides (Sec. 8.5)

$$2\,ROH + 2\,Na \longrightarrow 2\,RO^-Na^+ + H_2$$
$$\text{or } ROH + NaH \longrightarrow RO^-Na^+ + H_2$$
$$RO^-Na^+ + R'{-}X \longrightarrow RO{-}R' + Na^+X^-$$
(best for R' = primary)

d. Cleavage by Hydrogen Halides (Sec. 8.6)

$$R{-}O{-}R + HX \longrightarrow R{-}X + R{-}OH$$

e. Cleavage by Boron Tribromide (Sec. 8.6)

$$R{-}O{-}R \xrightarrow[2.\ H_2O]{1.\ BBr_3} RBr + ROH$$

3. Epoxides

a. Preparation from Alkenes (Sec. 8.7)

$$>C{=}C< + RCO_3H \longrightarrow >\overset{O}{C{-}C}< + RCO_2H$$

b. Reaction with Water and Alcohols (Sec. 8.8)

$$>\overset{O}{C{-}C}< \xrightarrow[H^+]{H{-}OH} {-}\overset{HO}{\underset{|}{C}}{-}\overset{OH}{\underset{|}{C}}{-}$$

$$>\overset{O}{C{-}C}< \xrightarrow[H{-}OR]{H^+} {-}\overset{RO}{\underset{|}{C}}{-}\overset{OH}{\underset{|}{C}}{-}$$

c. Reaction with Organometallic Reagents (Sec. 8.8)

$$>\overset{O}{C{-}C}< \xrightarrow{\text{RMgX or RLi}} {-}\overset{R}{\underset{|}{C}}{-}\overset{O^-M^+}{\underset{|}{C}}{-} \xrightarrow{H_2O} {-}\overset{R}{\underset{|}{C}}{-}\overset{OH}{\underset{|}{C}}{-} + M{-}OH$$
M = MgX or Li

ADDITIONAL PROBLEMS

OWL Interactive versions of these problems are assignable in OWL.

Ethers and Epoxides: Structure, Nomenclature, and Properties

8.16 Write a structural formula for each of the following compounds:

a. diethyl ether
b. propylene oxide
c. 2-methoxypropane
d. benzyl propyl ether
e. 1-methoxyhexene
f. *m*-chlorophenyl ethyl ether
g. *trans*-2-ethoxycyclohexanol
h. ethyl isopropyl ether
i. propylene glycol
j. *p*-isopropoxyanisole

8.17 Name each of the following compounds:

a. $CH_3CH_2CH_2OCH_2CH(CH_3)_2$
b. $CH_3OCH_2CH(CH_3)_2$
c. $CH_3CH{-}CH_2$ (epoxide; CH and CH_2 bridged by O)

d. 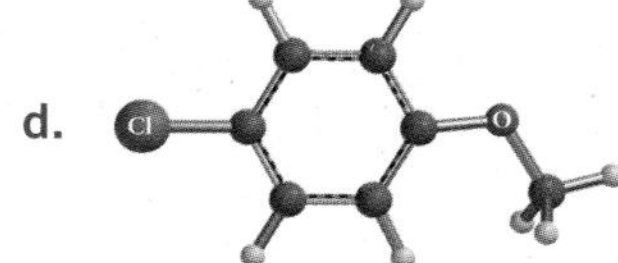

e. $CH_3CH_2CH(OCH_3)CH_2CH_2CH_2OH$

f. 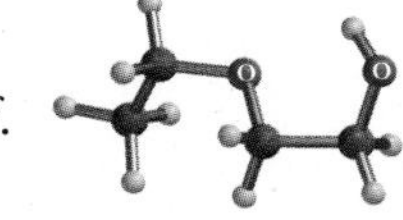

g. (five-membered ring with O, substituted with CH_3 on the carbon adjacent to O)
O CH$_3$

h. $CH_3OCH_2CH_2C{\equiv}CH$

8.18 Ethers and alcohols can be isomeric (see Sec. 1.8). Write the structures and give the names for all possible isomers with the molecular formula $C_4H_{10}O$.

8.19 Consider four compounds that have nearly the same molecular weights: 1,2-dimethoxyethane, ethyl propyl ether, hexane, and 1-pentanol.

a. Draw the structural formulas of these four molecules.
b. Which would you expect to have the highest boiling point? Which would be most soluble in water? Explain the reasons for your choices.

8.20 Capsaicin is the alkaloid that provides the hot (perhaps, more like blistering) taste of a habañero chili pepper. Commercially, capsaicin is also used in repellent sprays to ward off mammalian pests from gardens.

H_3CO O N H HO

capsaicin

H_3CO CH_3 HO

A reasonable intermediate for the synthesis of capsaicin would be the phenol shown to the right above. What is the IUPAC name for the phenol?

Preparation and Reaction of Grignard Reagents

8.21 Write equations for the reaction of each of the following with (1) Mg in ether followed by (2) addition of D_2O to the resulting solution:

a. $CH_3CH_2CH_2CH_2I$
b. $CH_3CH_2OCH_2CH_2Br$
c. $C_6H_5CH_2Br$
d. allyl bromide

8.22 The following steps can be used to convert anisole to *o*-*t*-butylanisole. Give the reagent for each step. Explain why the overall result cannot be achieved in one step by a Friedel–Crafts alkylation. (*Hint:* Sections 4.10 and 4.11 may be helpful for the second part of this question.)

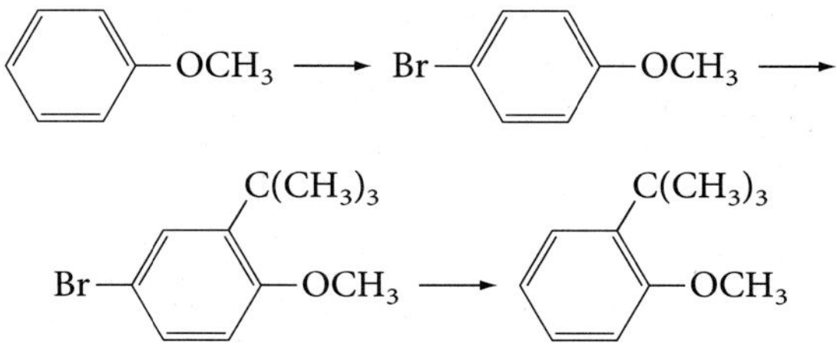

$$C_6H_5{-}OCH_3 \longrightarrow Br{-}C_6H_4{-}OCH_3 \longrightarrow$$

$$Br{-}C_6H_3(C(CH_3)_3){-}OCH_3 \longrightarrow C_6H_4(C(CH_3)_3){-}OCH_3$$

Preparation of Ethers

8.23 Write equations for the best method to prepare each of the following ethers:

a.

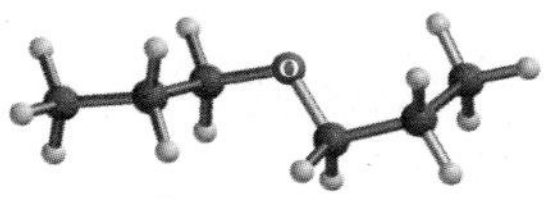

b.

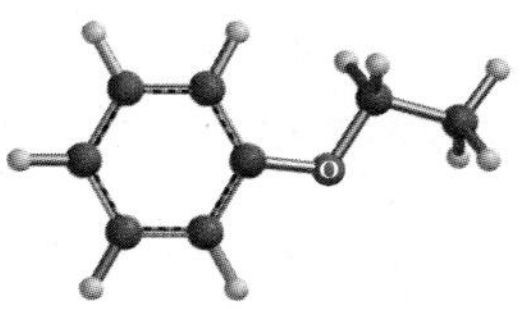

c.

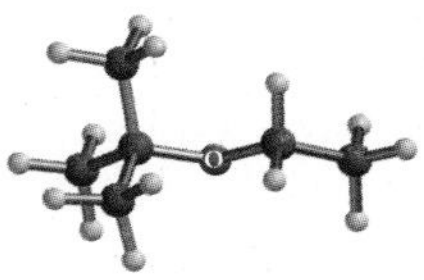

8.24 Explain why the Williamson synthesis cannot be used to prepare diphenyl ether.

8.25 Provide *two* synthetic routes to prepare $C_6H_5CH_2OC(CH_3)_3$, starting from $C_6H_5CH_2OH$ in one case and $HOC(CH_3)_3$ for the other. What general mechanistic pathway (S_N1, S_N2, etc.) is being used for each of your two proposed synthetic routes?

8.26 Several companies are developing the manufacture and use of dimethyl ether, CH_3OCH_3, as an efficient and clean alternative to diesel fuel [*Chemical and Engineering News,* **1995** (May 29), 37–39]. Much of this interest in dimethyl ether and other related compounds is focused on minimizing C—C bonds in fuels in order to limit formation of polycyclic aromatic hydrocarbons; see "A Word About... Polycyclic Aromatic Hydrocarbons and Cancer" on page 136. Show how dimethyl ether could be synthesized from methanol.

Behavior of Ethers in Acids and Bases

8.27 Write an equation for each of the following reactions. If no reaction occurs, say so.

a. methyl propyl ether + excess HBr (hot) $\longrightarrow$

b. dibutyl ether + boiling aqueous NaOH $\longrightarrow$

c. ethyl ether + cold concentrated H_2SO_4 $\longrightarrow$

d. dipropyl ether + Na $\longrightarrow$

e. ethyl phenyl ether $\xrightarrow[2.\ H_2O]{1.\ BBr_3}$

8.28 Ethers are soluble in cold, concentrated sulfuric acid, but alkanes are not. This difference can be used as a simple chemical test to distinguish between these two classes of compounds. What chemistry (show an equation) is the basis for this difference? (*Hint:* See Sec. 8.6.)

8.29 When heated with excess HBr, a cyclic ether gave 1,4-dibromobutane as the only organic product. Write a structure for the ether and an equation for the reaction. (*Hint:* See Secs. 8.6 and 7.9.)

Preparation and Reactions of Epoxides

8.30 Using the peroxyacid epoxidation of an alkene and the ring opening of an epoxide, devise a two-step synthesis of 1,2-butanediol from 1-butene.

8.31 Write an equation for the reaction of ethylene oxide with

a. 1 mole of HCl
b. excess HCl
c. phenol + H^+
d. phenylmagnesium bromide

8.32 $CH_3CH_2OCH_2CH_2OH$ (ethyl cellosolve) and $CH_3CH_2OCH_2CH_2OCH_2CH_2OH$ (ethyl carbitol) are solvents used in the formulation of lacquers. They are produced commercially from ethylene oxide and certain other reagents. Show with equations how this might be done.

8.33 2-Phenylethanol, which has the aroma of oil of roses, is used in perfumes. Write equations to show how 2-phenylethanol can be synthesized from bromobenzene and ethylene oxide, using a Grignard reagent.

8.34 2,2-Dimethyloxirane dissolved in excess methanol and treated with a little acid yields the product 2-methoxy-2-methyl-1-propanol (and no 1-methoxy-2-methyl-2-propanol). What reaction mechanism explains this result? (*Hint:* See Secs. 6.5 and 3.10.)

8.35 Write an equation for the reaction of ammonia with ethylene oxide. The product is a water-soluble organic base used to absorb and concentrate CO_2 in the manufacture of dry ice.

8.36 Design a synthesis of 3-pentyn-1-ol using propyne and ethylene oxide as the only sources of carbon atoms.

8.37 Write out the steps in the reaction mechanisms for the reactions given in eq. 8.20.

8.38 In "A Word About... The Gypsy Moth's Epoxide," the structure of this moth's sex attractant, disparlure (page 245), was shown. Using a multistep synthetic sequence including a catalytic hydrogenation with Lindlar's catalyst and a subsequent epoxidation with a peroxyacid, suggest a synthesis of racemic disparlure, starting from the alkyne, $(CH_3)_2CH(CH_2)_4C{\equiv}CH$, and the alkyl bromide, $CH_3(CH_2)_8CH_2Br$.

Cyclic Ethers

8.39 The cyclic polyether *nonactin* selectively binds and transports K^+ ions, while the acyclic polyether *monensin* (see page 234 for the structure) selectively binds and transports Na^+ ions. Use information in Section 8.9 to help you answer the following questions.

a. Which atoms in these two compounds are likely to be involved in binding the metal cations? Explain.
b. Suggest a reason why *nonactin* is selective for K^+ while *monensin* is selective for Na^+.

8.40 What halogen-containing products would be prepared when 1,4-dioxane, a cyclic ether (Sec. 8.9, structure shown on page 247), is heated with excess HBr?

Puzzles

8.41 An organic compound with the molecular formula $C_4H_{10}O_3$ shows properties of both an alcohol and an ether. When treated with an excess of hydrogen bromide, it yields only one organic compound, 1,2-dibromoethane. Draw a structural formula for the original compound.

8.42 What chemical test will distinguish between the compounds in each of the following pairs? Indicate what is visually observed with each test. (*Hint:* Make a list of tests you have learned from this chapter and previous chapters for identifying the functional groups in these compounds.)

a. phenol and anisole
b. 1-butanol and methyl propyl ether
c. dipropyl ether and hexane
d. ethyl phenyl ether and allyl phenyl ether

The civet cat (*Viverra civetta*) is the original source of civetone, a sweet and pungent ketone, now produced synthetically and used as a fixative in perfumery.

© Nigel J. Dennis / Photo Researchers

Aldehydes and Ketones

Aldehydes and ketones are a large family of organic compounds that permeate our everyday lives. They are responsible for the fragrant odors of many fruits and fine perfumes. For example, cinnamaldehyde (an aldehyde) provides the smell we associate with cinnamon, and civetone (a ketone) is used to provide the musky odor of many perfumes. Formaldehyde is a component of many building materials we use to construct our houses. The ketones testosterone and estrone are known to many as hormones responsible for our sexual characteristics. And the chemistry of aldehydes and ketones plays a role in how we digest food and even in how we can see the words on this page (see "A Word About... The Chemistry of Vision" on pages 76–77). So what are aldehydes and ketones?

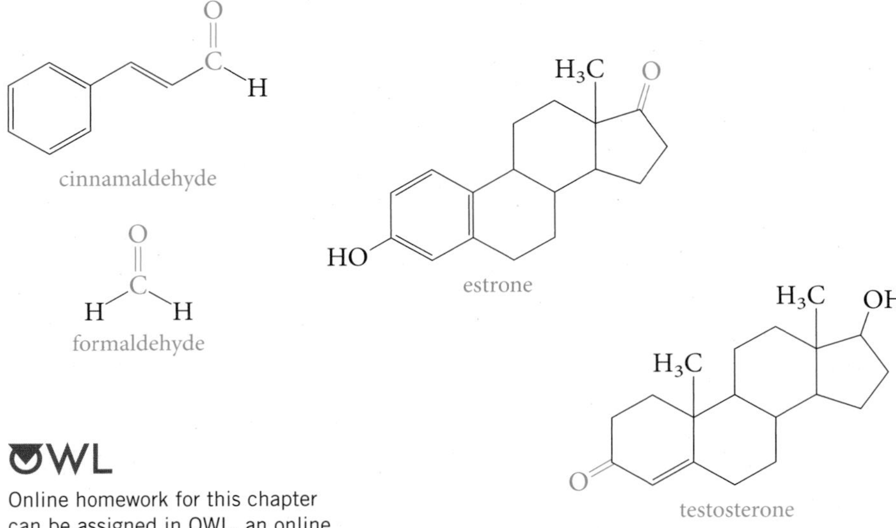

OWL
Online homework for this chapter can be assigned in OWL, an online homework assessment tool.

The **carbonyl group** is a C=O unit. **Aldehydes** have at least one hydrogen atom attached to the carbonyl carbon atom.
In **ketones**, the carbonyl carbon atom is connected to two other carbon atoms.

Aldehydes and ketones are characterized by the presence of the **carbonyl group**, perhaps the most important functional group in organic chemistry. Aldehydes have at least one hydrogen atom attached to the carbonyl carbon atom. The remaining group may be another hydrogen atom or any aliphatic or aromatic organic group. The —CH=O group characteristic of aldehydes is often called a **formyl group**. In ketones, the carbonyl carbon atom is connected to two other carbon atoms.

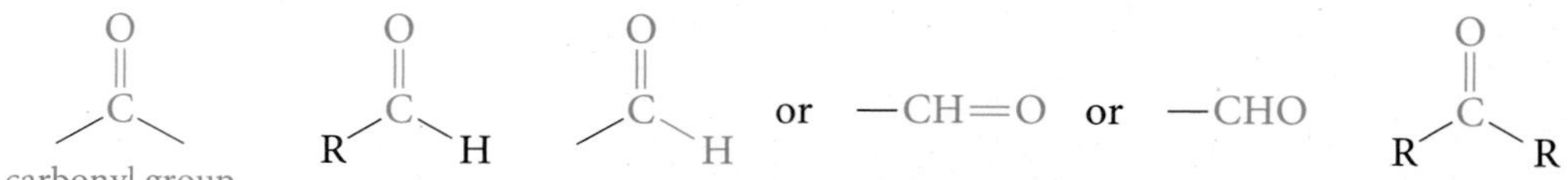

$$\begin{array}{ccccc} \overset{O}{\overset{\|}{\diagup C \diagdown}} & \overset{O}{\overset{\|}{R\diagup C \diagdown H}} & \overset{O}{\overset{\|}{\diagup C \diagdown H}} \text{ or } -CH{=}O \text{ or } -CHO & \overset{O}{\overset{\|}{R\diagup C \diagdown R}} \\ \text{carbonyl group} & \text{aldehyde} & \text{aldehyde group or formyl group} & \text{ketone} \end{array}$$

We will see that the carbonyl group appears in many other organic compounds, including carboxylic acids and their derivatives (Chapter 10). This chapter, however, will focus only on aldehydes and ketones.

9.1 Nomenclature of Aldehydes and Ketones

In the IUPAC system, the characteristic ending for aldehydes is *-al* (from the first syllable of *aldehyde*). The following examples illustrate the system:

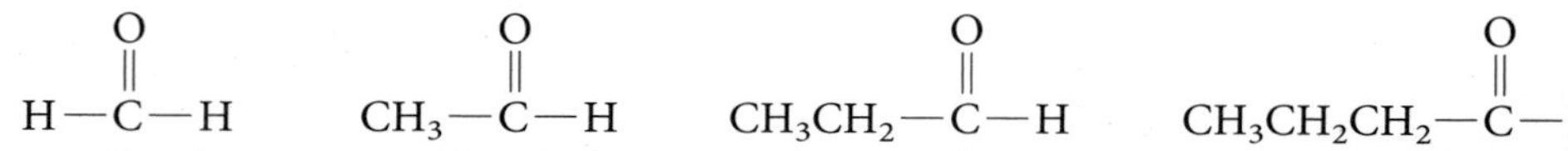

$$\begin{array}{cccc} H-\overset{O}{\overset{\|}{C}}-H & CH_3-\overset{O}{\overset{\|}{C}}-H & CH_3CH_2-\overset{O}{\overset{\|}{C}}-H & CH_3CH_2CH_2-\overset{O}{\overset{\|}{C}}-H \\ \text{methanal} & \text{ethanal} & \text{propanal} & \text{butanal} \\ \text{(formaldehyde)} & \text{(acetaldehyde)} & \text{(propionaldehyde)} & (n\text{-butyraldehyde}) \end{array}$$

The common names shown below the IUPAC names are frequently used, so you should learn them.

For substituted aldehydes, we number the chain starting with the aldehyde carbon, as the following examples illustrate:

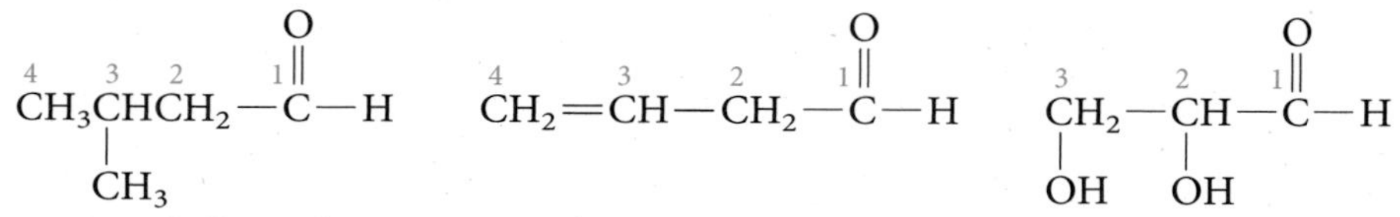

$$\begin{array}{ccc} \overset{4}{C}H_3\overset{3}{C}H(CH_3)\overset{2}{C}H_2-\overset{O}{\overset{\|}{\overset{1}{C}}}-H & \overset{4}{C}H_2{=}\overset{3}{C}H-\overset{2}{C}H_2-\overset{O}{\overset{\|}{\overset{1}{C}}}-H & \underset{OH}{\overset{3}{C}H_2}-\underset{OH}{\overset{2}{C}H}-\overset{O}{\overset{\|}{\overset{1}{C}}}-H \\ \text{3-methylbutanal} & \text{3-butenal} & \text{2,3-dihydroxypropanal} \\ & & \text{(glyceraldehyde)} \end{array}$$

Notice from the last two examples that an aldehyde group has priority over a double bond or a hydroxyl group, not only in numbering but also as the suffix. For cyclic aldehydes, the suffix *-carbaldehyde* is used. Aromatic aldehydes often have common names:

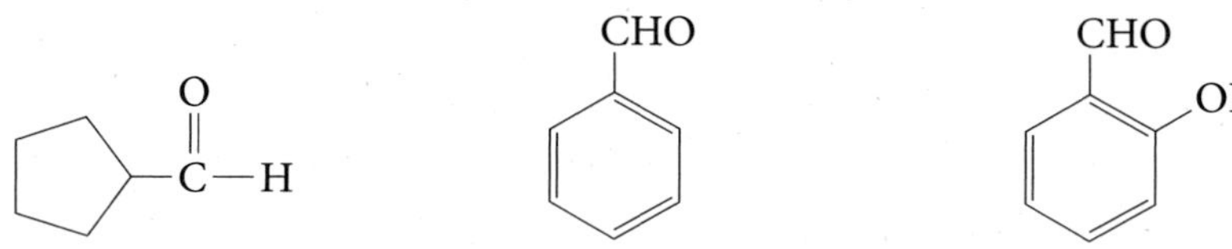

cyclopentanecarbaldehyde (formylcyclopentane) | benzaldehyde (benzenecarbaldehyde) | salicylaldehyde (2-hydroxybenzenecarbaldehyde)

In the IUPAC system, the ending for ketones is *-one* (from the last syllable of *ketone*). The chain is numbered so that the carbonyl carbon has the lowest possible number. Common names of ketones are formed by adding the word *ketone* to the names of the alkyl or aryl groups attached to the carbonyl carbon. In still other cases, traditional names are used. The following examples illustrate these methods:

$CH_3-\overset{\overset{O}{\|}}{C}-CH_3$
propanone
(acetone)

$\overset{1}{CH_3}-\overset{\overset{O}{\|}}{\underset{2}{C}}-\overset{3}{CH_2}\overset{4}{CH_3}$
2-butanone
(ethyl methyl ketone)

$\overset{1}{CH_3}\overset{2}{CH_2}-\overset{\overset{O}{\|}}{\underset{3}{C}}-\overset{4}{CH_2}\overset{5}{CH_3}$
3-pentanone
(diethyl ketone)

cyclohexanone

2-methylcyclopentanone

$\overset{4}{CH_2}=\overset{3}{CH}-\overset{\overset{O}{\|}}{\underset{2}{C}}-\overset{1}{CH_3}$
3-buten-2-one
(methyl vinyl ketone)

acetophenone
(methyl phenyl ketone)

benzophenone
(diphenyl ketone)

dicyclopropyl ketone

PROBLEM 9.1 Using the examples as a guide, write a structure for

a. hexanal
b. *o*-chlorobenzaldehyde
c. 4-heptanone
d. isopropyl methyl ketone
e. cyclobutanecarbaldehyde
f. 3-octyn-2-one

PROBLEM 9.2 Using the examples as a guide, write a correct name for

a. $(CH_3)_2C(Br)CH_2CH{=}O$
b. $CH_3CH{=}CHCH{=}O$

c. 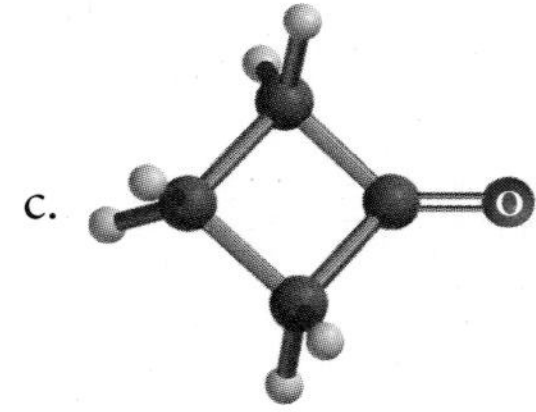

d. 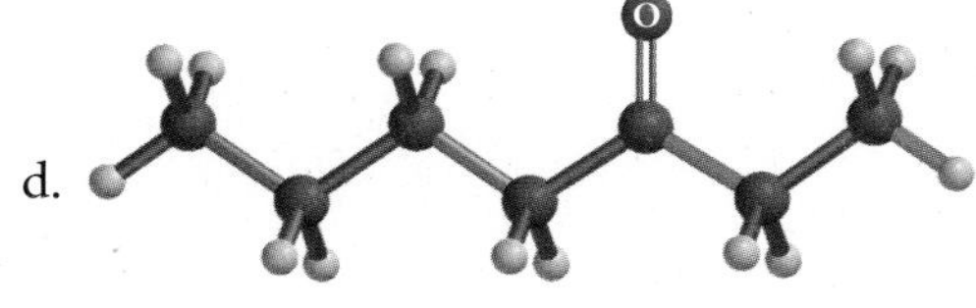

9.2 Some Common Aldehydes and Ketones

Formaldehyde ($HCH{=}O$), the simplest aldehyde, is manufactured on a very large scale by the oxidation of methanol.

$$CH_3OH \xrightarrow[600\text{–}700°C]{\text{Ag catalyst}} \underset{\text{formaldehyde}}{CH_2{=}O} + H_2 \quad (9.1)$$

Annual world production is more than 46 billion pounds. Formaldehyde is a gas (bp −21°C), but it cannot be stored in a free state because it polymerizes readily.*

*The polymer derived from formaldehyde is a long chain of alternating CH_2 and oxygen units, which can be described by the structure $(CH_2O)_n$. See Chapter 14 for a discussion of polymers.

Normally it is supplied as a 37% aqueous solution called formalin. In this form, it is used as a disinfectant and preservative, but formaldehyde is mostly used in the manufacture of plastics, building insulation, particleboard, and plywood.

Acetaldehyde ($CH_3CH{=}O$) boils close to room temperature (bp 20°C). It is manufactured mainly by the oxidation of ethylene over a palladium–copper catalyst, and about 1 billion pounds are produced worldwide each year.

$$2\,CH_2{=}CH_2 + O_2 \xrightarrow[100\text{–}300^\circ C]{Pd\text{–}Cu} 2\,CH_3CH{=}O \quad (9.2)$$

About half of the acetaldehyde produced annually is oxidized to acetic acid. The rest is used for the production of 1-butanol and other commercial chemicals.

Acetone ($(CH_3)_2C{=}O$), the simplest ketone, is also produced on a large scale—about 4 billion pounds annually. The most common methods for its commercial synthesis are the oxidation of propene (analogous to eq. 9.2), the oxidation of isopropyl alcohol (eq. 7.35, $R = R' = CH_3$), and the oxidation of isopropylbenzene (eq. 9.3).

$$C_6H_5C(H)(CH_3)_2 \xrightarrow{O_2} C_6H_5C(OOH)(CH_3)_2 \xrightarrow[H_2O]{dil\ H_2SO_4} C_6H_5OH + CH_3{-}\overset{O}{\overset{\|}{C}}{-}CH_3 \quad (9.3)$$

phenol acetone

About 30% of the acetone is used directly, because it is not only completely miscible with water but is also an excellent solvent for many organic substances (resins, paints, dyes, and nail polish). The rest is used to manufacture other commercial chemicals, including bisphenol-A for epoxy resins (Sec. 14.9).

$$2\,HO{-}C_6H_5 + CH_3\overset{O}{\overset{\|}{C}}CH_3 \xrightarrow[-H_2O]{H^+} HO{-}C_6H_4{-}C(CH_3)_2{-}C_6H_4{-}OH \quad (9.4)$$

phenol acetone bisphenol-A

Quinones constitute a unique class of carbonyl compounds. They are cyclic conjugated diketones. The simplest example is 1,4-benzoquinone (eq. 7.44). All quinones are colored, and many are naturally occurring pigments that are used as dyes. Alizarin is an orange–red quinone that was used to dye the red coats of the British army during the American Revolution. Vitamin K is a quinone that is required for the normal clotting of blood.

alizarin
mp 290°C

vitamin K
mp −20°C

9.3 Synthesis of Aldehydes and Ketones

We have already seen, in previous chapters, several ways to prepare aldehydes and ketones. One of the most useful is the oxidation of alcohols.

$$\underset{\text{OH}}{\overset{\text{H}}{>\text{C}<}} \xrightarrow[\text{agent}]{\text{oxidizing}} >\text{C}{=}\text{O} \tag{9.5}$$

Oxidation of a primary alcohol gives an aldehyde, and oxidation of a secondary alcohol gives a ketone. Chromium reagents, such as pyridinium chlorochromate (PCC), are commonly used in the laboratory for this purpose (review Sec. 7.12).

PROBLEM 9.3 Give the product expected from treatment of

a. cyclohexanol with Jones' reagent (see Sec. 7.12)
b. 3-methylcyclopentanol with pyridinium chlorochromate
c. 5-methyl-1-heptanol with Jones' reagent
d. 5-methyl-1-heptanol with pyridinium chlorochromate

EXAMPLE 9.1

Write an equation for the oxidation of an appropriate alcohol to $(CH_3)_2CHCH_2CHO$ (3-methylbutanal).

Solution Aldehydes are prepared by oxidation of 1° alcohols (RCH_2OH) with PCC. First, find the carbonyl carbon in 3-methylbutanal (marked with an asterisk). Convert this carbon to a primary alcohol. A proper equation is:

$$(CH_3)_2CHCH_2{-}CH_2{-}OH \xrightarrow{\text{PCC}} (CH_3)_2CHCH_2{-}\underset{*}{\overset{\text{O}}{\overset{\|}{\text{C}}}}{-}H$$

PROBLEM 9.4 Write an equation for the oxidation of an appropriate alcohol to

a. cyclopentyl–CHO b. cyclopentanone (cyclopentane=O)

Aromatic ketones can be prepared by Friedel–Crafts acylation of an aromatic ring (review eq. 4.12 and Sec. 4.9.d). For example,

$$C_6H_6 + C_6H_5{-}\overset{\text{O}}{\overset{\|}{\text{C}}}\text{Cl} \xrightarrow{\text{AlCl}_3} C_6H_5{-}\overset{\text{O}}{\overset{\|}{\text{C}}}{-}C_6H_5 + \text{HCl} \tag{9.6}$$

benzene, benzoyl chloride, benzophenone

PROBLEM 9.5 Complete the following equation and name the product:

$$\text{C}_6\text{H}_6 + \text{CH}_3\overset{\displaystyle\text{O}}{\overset{\|}{\text{C}}}\text{Cl} \xrightarrow{\text{AlCl}_3}$$

Cloves, source of 2-heptanone.

Methyl ketones can be prepared by hydration of terminal alkynes, catalyzed by acid and mercuric ion (review eq. 3.52). For example,

$$\underset{\text{1-octyne}}{\text{CH}_3(\text{CH}_2)_5\text{C}{\equiv}\text{CH}} \xrightarrow[\text{Hg}^{2+}]{\text{H}^+,\ \text{H}_2\text{O}} \underset{\text{2-octanone}}{\text{CH}_3(\text{CH}_2)_5\overset{\displaystyle\text{O}}{\overset{\|}{\text{C}}}\text{CH}_3} \qquad \textbf{(9.7)}$$

PROBLEM 9.6 What alkyne would be useful for the synthesis of 2-heptanone (oil of cloves)?

Vanilla beans, source of vanillin.

9.4 Aldehydes and Ketones in Nature

Aldehydes and ketones occur very widely in nature. Figures 1.12 and 1.13 show three examples, and Figure 9.1 gives several more. Many aldehydes and ketones have pleasant odors and flavors and are used for these properties in perfumes and other consumer products (soaps, bleaches, and air fresheners, for example). However, the gathering and extraction of these fragrant substances from flowers, plants, and animal glands are extremely expensive. Chanel No. 5, introduced to the perfume market in 1921, was the first fine fragrance to use *synthetic* organic chemicals. Today most fragrances do.

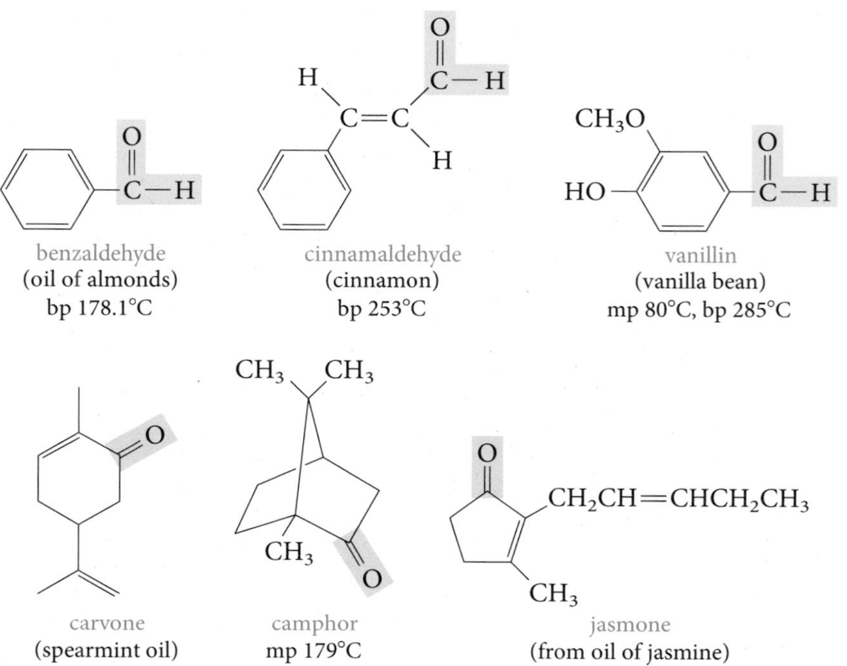

■ **Figure 9.1**
Some naturally occurring aldehydes and ketones.

9.5 The Carbonyl Group

To best understand the reactions of aldehydes, ketones, and other carbonyl compounds, we must first appreciate the structure and properties of the carbonyl group.

The carbon–oxygen double bond consists of a sigma bond and a pi bond (Figure 9.2). The carbon atom is sp^2-hybridized. *The three atoms attached to the carbonyl carbon lie in a plane with bond angles of 120°.* The pi bond is formed by overlap of a *p* orbital on carbon with an oxygen *p* orbital. There are also two unshared electron pairs on the oxygen atom. The C=O bond distance is 1.24 Å, shorter than the C—O distance in alcohols and ethers (1.43 Å).

Oxygen is much more electronegative than carbon. Therefore, the electrons in the C=O bond are attracted to the oxygen, producing a highly polarized bond. This effect is especially pronounced for the pi electrons and can be expressed in the following ways:

$$\left[\; \mathrm{C{=}\ddot{O}\!:} \longleftrightarrow \overset{+}{\mathrm{C}}{-}\mathrm{\ddot{\underset{..}{O}}\!:^{-}} \right] \qquad \overset{\delta+}{\mathrm{C}}{=}\overset{\delta-}{\mathrm{O}} \qquad \mathrm{C{=}O}$$

resonance contributors to the carbonyl group

polarization of the carbonyl group

As a consequence of this polarization, *most carbonyl reactions involve* nucleophilic attack *at the carbonyl carbon,* often accompanied by addition of a proton to the oxygen, consistent with the electrostatic potential map (Figure 9.2c) showing that the oxygen is electron rich (red).

attack here by a nucleophile ⟶ $\overset{\delta+}{\mathrm{C}}{=}\overset{\delta-}{\mathrm{O}}$ ⟵ may react with a proton

C=O bonds are quite different, then, from C=C bonds, which are not polarized and where attack at carbon is usually by an electrophile (Sec. 3.9).

In addition to its effect on reactivity, polarization of the C=O bond influences the physical properties of carbonyl compounds. For example, carbonyl compounds boil at higher temperatures than hydrocarbons, but at lower temperatures than alcohols of comparable molecular weight.

$CH_3(CH_2)_4CH_3$	$CH_3(CH_2)_3CH{=}O$	$CH_3(CH_2)_3CH_2OH$
hexane (bp 69°C)	pentanal (bp 102°C)	pentanol (bp 118°C)

Why is this so? Unlike hydrocarbon molecules, which can be temporarily polarized, molecules of carbonyl compounds have *permanently polar C=O bonds* and thus have a stronger tendency to associate. *The positive part of one molecule is attracted to the negative part of another molecule.* These intermolecular forces of attraction, called **dipole–dipole interactions**, are generally stronger than van der Waals attractions (Sec. 2.7) but not as strong as hydrogen bonds (Sec. 7.4). Carbonyl compounds such

Dipole–dipole interactions are opposite pole attractions between polar molecules.

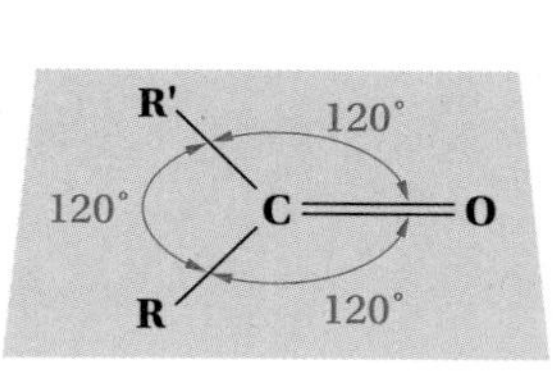

(a)

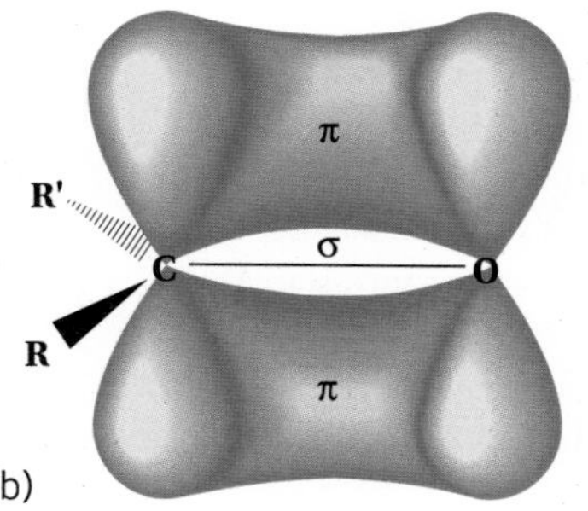

(b)

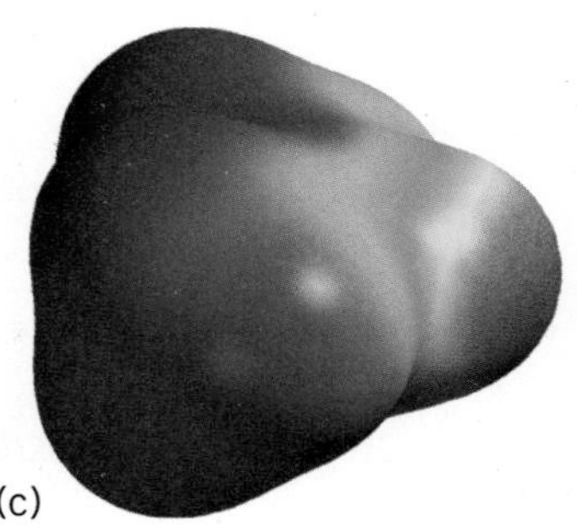

(c)

▰ **Figure 9.2**
Bonding in the carbonyl group: (a) carbonyl carbon is sp^2-hybridized, (b) C=O group consists of sigma and pi bonds, (c) electrostatic potential map illustrates the electron-poor nature (blue) of the carbonyl carbon and the electron-rich nature (red) of the oxygen.

as aldehydes and ketones that have a C═O bond, but no O—H bond, cannot form hydrogen bonds with one another, as can alcohols. Consequently, carbonyl compounds require more energy (heat) than hydrocarbons of comparable molecular weight to overcome intermolecular attractive forces when converted from liquid to vapor, but less than alcohols.

The polarity of the carbonyl group also affects the solubility properties of aldehydes and ketones. For example, carbonyl compounds with low molecular weights are soluble in water. Although they cannot form hydrogen bonds with themselves, they can form hydrogen bonds with O—H or N—H compounds.

$$\overset{\delta+}{C}=\overset{\delta-}{\ddot{O}}:\cdots\overset{\delta+}{H}-\overset{\delta-}{O}-H$$

PROBLEM 9.7 Arrange benzaldehyde (mol. wt. 106), benzyl alcohol (mol. wt. 108), and *p*-xylene (mol. wt. 106) in order of

a. increasing boiling point b. increasing water solubility

9.6 Nucleophilic Addition to Carbonyl Groups: An Overview

Nucleophiles attack the carbon atom of a carbon–oxygen double bond because that carbon has a partial positive charge. The pi electrons of the C═O bond move to the oxygen atom, which, because of its electronegativity, can easily accommodate the negative charge that it acquires. When these reactions are carried out in a hydroxylic solvent such as alcohol or water, the reaction is usually completed by addition of a proton to the negative oxygen. The overall reaction involves addition of a nucleophile and a proton across the pi bond of the carbonyl group.

$$\mathrm{Nu}:^- + \underset{\text{trigonal reactant}}{C=\ddot{O}:} \rightleftharpoons \underset{\text{tetrahedral intermediate}}{\mathrm{Nu}-C-\ddot{O}:^-} \xrightleftharpoons{H_2O} \underset{\text{tetrahedral product}}{\mathrm{Nu}-C-\ddot{O}H} \quad (9.8)$$

The carbonyl carbon, which is trigonal and sp^2-hybridized in the starting aldehyde or ketone, becomes tetrahedral and sp^3-hybridized in the reaction product.

Because of the unshared electron pairs on the oxygen atom (Figure 9.2c), carbonyl compounds are weak Lewis bases and can be protonated. *Acids can catalyze the addition of weak nucleophiles to carbonyl compounds* by protonating the carbonyl oxygen atom.

$$C=\ddot{O}: + H^+ \rightleftharpoons \underset{\text{a resonance-stabilized carbocation}}{\left[C=\overset{+}{\ddot{O}}H \longleftrightarrow \overset{+}{C}-\ddot{O}H\right]} \xrightarrow{\mathrm{Nu}:^-} \mathrm{Nu}-C-\ddot{O}H \quad (9.9)$$

This converts the carbonyl carbon to a carbocation and enhances its susceptibility to attack by nucleophiles.

Nucleophiles can be classified as those that add reversibly to the carbonyl carbon and those that add irreversibly. Nucleophiles that add reversibly are also good leaving groups. In other words, they are the conjugate bases of relatively strong acids. Nucleophiles that add irreversibly are poor leaving groups, the conjugate bases of weak acids. This classification will be useful when we consider the mechanism of carbonyl additions.

In general, *ketones are somewhat less reactive than aldehydes toward nucleophiles.* There are two main reasons for this reactivity difference. The first reason is *steric.* The carbonyl carbon atom is more crowded in ketones (two organic groups) than in aldehydes (one organic group and one hydrogen atom). In nucleophilic addition, we bring these attached groups closer together because the hybridization changes from sp^2 to sp^3 and the bond angles decrease from 120° to 109.5° (eq. 9.8). Less strain is involved in additions to aldehydes than in additions to ketones because one of the groups (H) is small. The second reason is *electronic.* As we have already seen in connection with carbocation stability, alkyl groups are usually electron-donating compared to hydrogen (Sec. 3.10). Therefore, they tend to neutralize the partial positive charge on the carbonyl carbon, decreasing its reactivity toward nucleophiles. Ketones have two such alkyl groups; aldehydes have only one. If, however, the attached groups are strongly electron-withdrawing (contain halogens, for example), they can have the opposite effect and increase carbonyl reactivity toward nucleophiles.

In the following discussion, we will classify nucleophilic additions to aldehydes and ketones according to the type of new bond formed to the carbonyl carbon. We will consider oxygen, carbon, and nitrogen nucleophiles, in that sequence.

9.7 Addition of Alcohols: Formation of Hemiacetals and Acetals

The reactions discussed in this section are extremely important because they are crucial to understanding the chemistry of carbohydrates, which we will discuss in Chapter 16.

Alcohols are oxygen nucleophiles. They add to the C=O bond, the OR group becoming attached to the carbon and the proton becoming attached to the oxygen:

$$\underset{\text{alcohol}}{\text{ROH}} + \underset{\text{aldehyde}}{\text{R}'(\text{H})\text{C=O}} \underset{}{\overset{\text{H}^+}{\rightleftharpoons}} \underset{\text{hemiacetal}}{\text{RO–C(R')(H)–OH}} \qquad \textbf{(9.10)}$$

Because alcohols are *weak* nucleophiles, an acid catalyst is required.* The product is a **hemiacetal**; it contains both alcohol and ether functional groups on the same carbon atom. The addition is reversible.

A **hemiacetal** contains both alcohol and ether functional groups on the same carbon atom.

The mechanism of hemiacetal formation involves three steps. First, the carbonyl oxygen is protonated by the acid catalyst. The alcohol oxygen then attacks the carbonyl carbon, and a proton is lost from the resulting positive oxygen. *Each step is reversible.* In terms of acid–base reactions, the starting acid in each step is converted to a product acid of similar strength (Sec. 7.5).

$$\underset{\text{aldehyde}}{\text{R}'(\text{H})\text{C=}\ddot{\text{O}}\text{:}} \underset{-\text{H}^+}{\overset{\text{H}^+}{\rightleftharpoons}} \underset{\text{protonated aldehyde}}{\text{R}'(\text{H})\text{C=}\overset{+}{\ddot{\text{O}}}\text{H}} \underset{-\text{ROH}}{\overset{\text{RÖH}}{\rightleftharpoons}} \underset{\text{protonated hemiacetal}}{\text{RO}^+\text{(H)–C(R')(H)–}\ddot{\text{O}}\text{H}} \underset{\text{H}^+}{\overset{-\text{H}^+}{\rightleftharpoons}} \underset{\text{hemiacetal}}{\text{RO–C(R')(H)–}\ddot{\text{O}}\text{H}} \qquad \textbf{(9.11)}$$

*Many acid catalysts can be used. Sulfuric acid and *p*-toluenesulfonic acid are commonly used in the laboratory.

PROBLEM 9.8 Write an equation for the formation of a hemiacetal from propanal (CH_3CH_2CHO), methanol (CH_3OH), and H^+. Show each step in the reaction mechanism.

An **acetal** has two ether functional groups on the same carbon atom.

In the presence of *excess alcohol,* hemiacetals react further to form **acetals**.

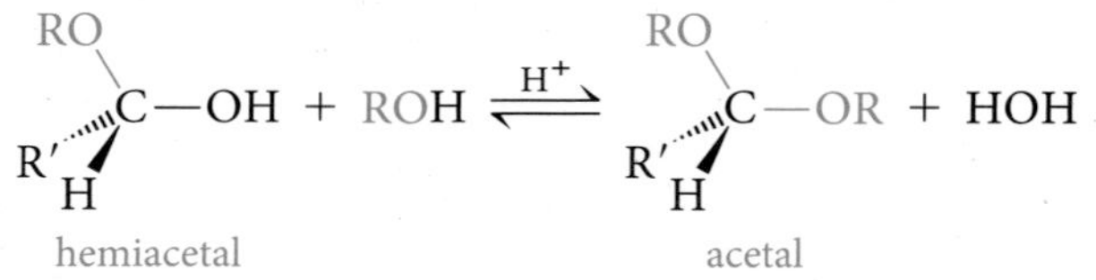

$$\text{hemiacetal } (RO)(R')(H)C\text{—OH} + ROH \underset{}{\overset{H^+}{\rightleftharpoons}} \text{acetal } (RO)(R')(H)C\text{—OR} + HOH \quad (9.12)$$

The hydroxyl group of the hemiacetal is replaced by an alkoxyl group. Acetals have two ether functions at the same carbon atom.

The mechanism of acetal formation involves the following steps.

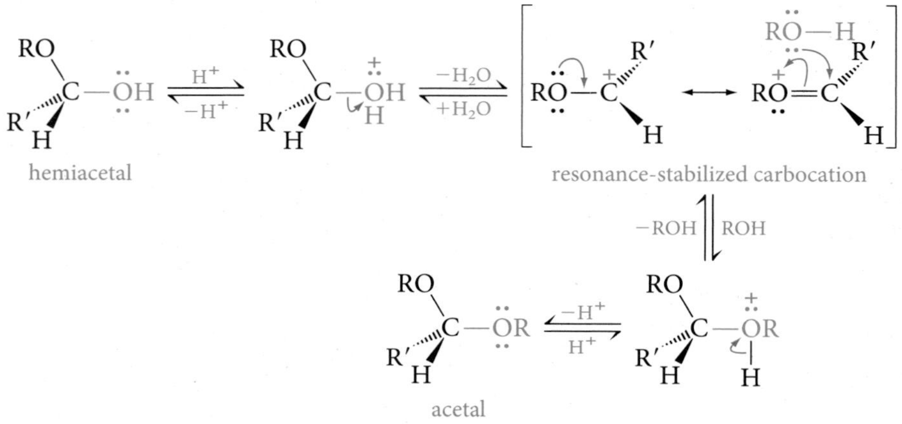

$$\text{hemiacetal } (RO)(R')(H)C\text{—}\ddot{O}H \underset{-H^+}{\overset{H^+}{\rightleftharpoons}} (RO)(R')(H)C\text{—}\overset{+}{O}H_2 \underset{+H_2O}{\overset{-H_2O}{\rightleftharpoons}} \left[R\ddot{O}\text{—}\overset{+}{C}(R')(H) \longleftrightarrow R\overset{+}{O}\text{=}C(R')(H) \right] \text{ resonance-stabilized carbocation}$$

$$\underset{-ROH}{\overset{ROH}{\rightleftharpoons}} (RO)(R')(H)C\text{—}\overset{+}{O}(R)H \underset{H^+}{\overset{-H^+}{\rightleftharpoons}} (RO)(R')(H)C\text{—}\ddot{O}R \text{ acetal} \quad (9.13)$$

Either oxygen of the hemiacetal can be protonated. When the hydroxyl oxygen is protonated, loss of water leads to a resonance-stabilized carbocation. Reaction of this carbocation with the alcohol, which is usually the solvent and is present in large excess, gives (after proton loss) the acetal. The mechanism is like an S_N1 reaction. *Each step is reversible.*

PROBLEM 9.9 Write an equation for the reaction of the hemiacetal

$$CH_3\underset{}{\overset{OH}{\overset{|}{C}}}HOCH_2CH_3$$

with excess ethanol and H^+. Show each step in the mechanism.

In a **cyclic hemiacetal**, the ether functional group is cyclic.

Aldehydes that have an appropriately located hydroxyl group *in the same molecule* may exist in equilibrium with a **cyclic hemiacetal**, formed by *intramolecular* nucleophilic addition. For example, 5-hydroxypentanal exists mainly in the cyclic hemiacetal form:

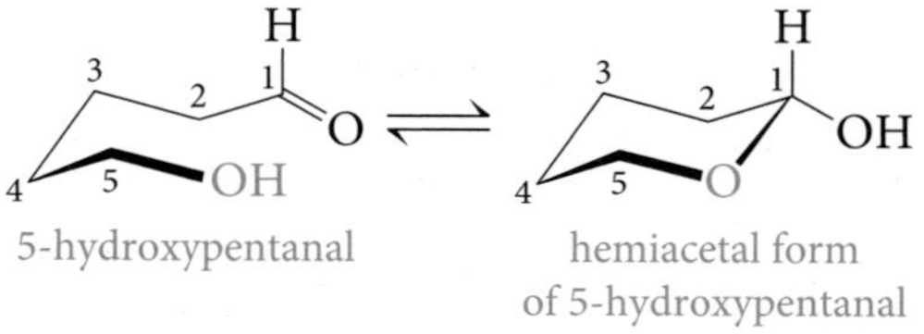

$$\text{5-hydroxypentanal} \rightleftharpoons \text{hemiacetal form of 5-hydroxypentanal} \quad (9.14)$$

The hydroxyl group is favorably located to act as a nucleophile toward the carbonyl carbon, and cyclization occurs by the following mechanism:

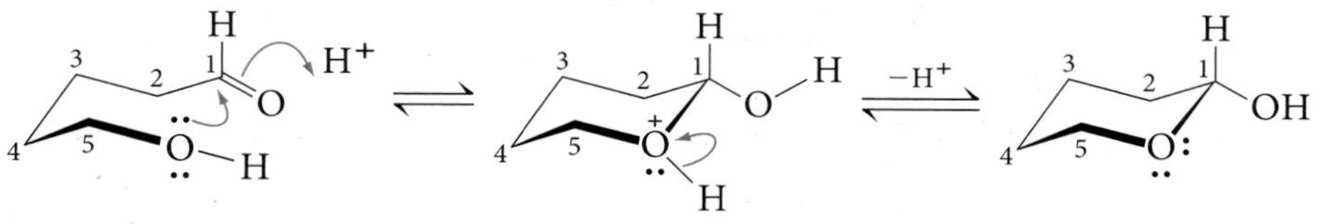

(9.15)

Compounds with a hydroxyl group that is four or five carbons away from the aldehyde group tend to form cyclic hemiacetals and acetals because the ring size (five- or six-membered) is relatively strain free. As we will see in Chapter 16, these structures are crucial to the chemistry of carbohydrates. For example, glucose is an important carbohydrate that exists as a cyclic hemiacetal.

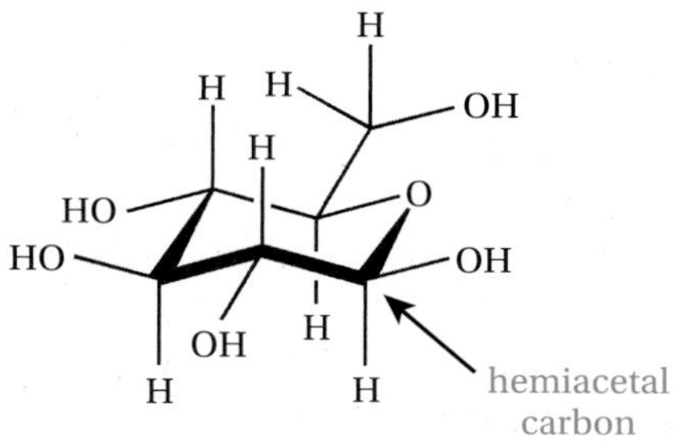

hemiacetal carbon

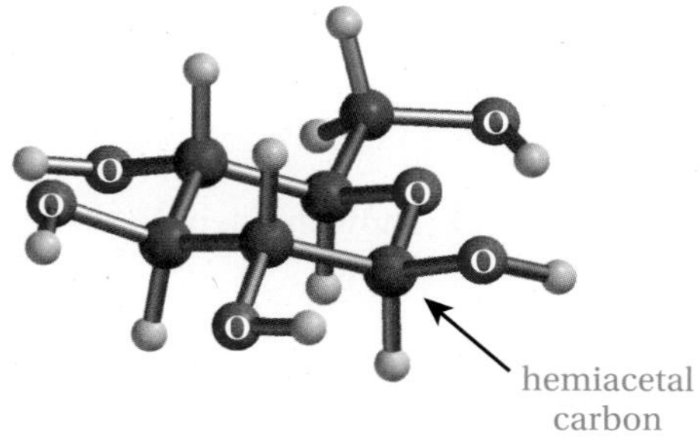

hemiacetal carbon

Ketones also form acetals. If, as in the following example, a glycol is used as the alcohol, the product will be cyclic.

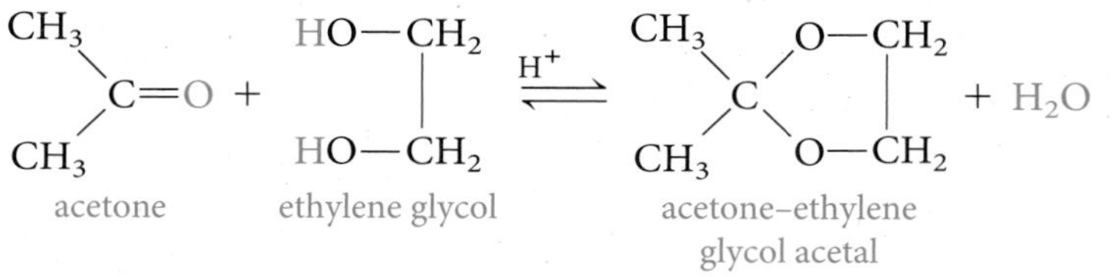

$$(CH_3)_2C{=}O + HO{-}CH_2{-}CH_2{-}OH \underset{}{\overset{H^+}{\rightleftharpoons}} (CH_3)_2C(O{-}CH_2)_2 + H_2O \quad (9.16)$$

acetone · ethylene glycol · acetone–ethylene glycol acetal

To summarize, aldehydes and ketones react with alcohols to form, first, hemiacetals and then, if excess alcohol is present, acetals.

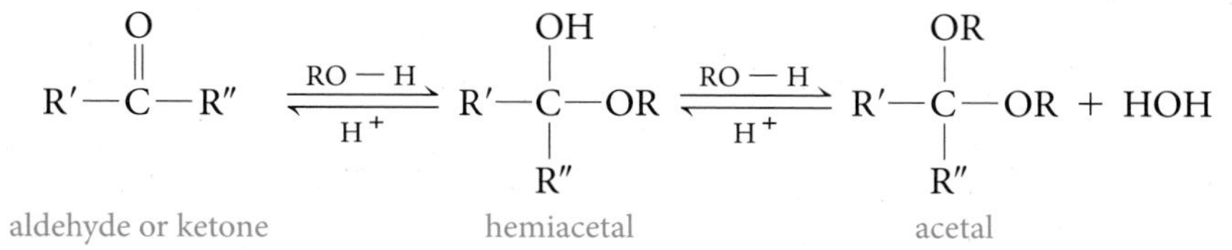

$$R'{-}\overset{O}{\overset{\|}{C}}{-}R'' \underset{H^+}{\overset{RO-H}{\rightleftharpoons}} R'{-}\underset{R''}{\overset{OH}{C}}{-}OR \underset{H^+}{\overset{RO-H}{\rightleftharpoons}} R'{-}\underset{R''}{\overset{OR}{C}}{-}OR + HOH \quad (9.17)$$

aldehyde or ketone · hemiacetal · acetal

EXAMPLE 9.2

Write an equation for the reaction of benzaldehyde with excess methanol and an acid catalyst.

Solution

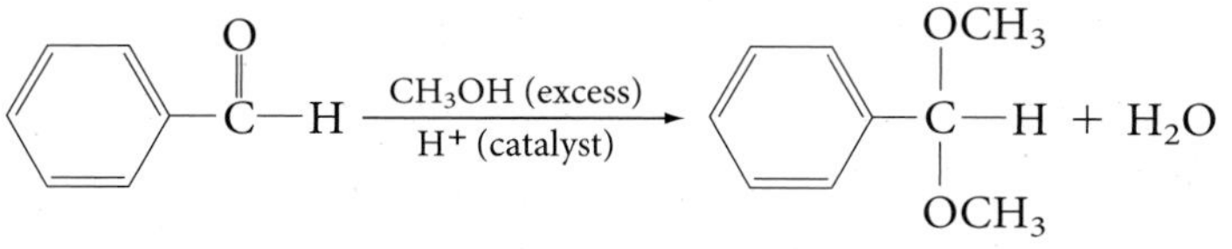

$$C_6H_5{-}\overset{O}{\overset{\|}{C}}{-}H \xrightarrow[H^+ \text{ (catalyst)}]{CH_3OH \text{ (excess)}} C_6H_5{-}\underset{OCH_3}{\overset{OCH_3}{C}}{-}H + H_2O \quad (9.18)$$

PROBLEM 9.10 Show the steps in the mechanism for eq. 9.18.

PROBLEM 9.11 Write an equation for the acid catalyzed reactions between cyclohexanone and

a. excess ethanol

b. excess ethylene glycol ($HOCH_2CH_2OH$)

Notice that acetal formation is a reversible process that involves a series of equilibria (eq. 9.17). How can these reactions be driven in the forward direction? One way is to use a large excess of alcohol. Another way is to remove water, a product of the forward reaction, as it is formed.* The reverse of acetal formation, called **acetal hydrolysis**, cannot proceed without water. On the other hand, an acetal can be hydrolyzed to its aldehyde or ketone and alcohol components by treatment with *excess water* in the presence of an acid catalyst. The hemiacetal intermediate in both the forward and reverse processes usually cannot be isolated when R′ and R″ are simple alkyl or aryl groups.

Acetal hydrolysis is the reverse of acetal formation.

EXAMPLE 9.3

Write an equation for the reaction of benzaldehyde dimethylacetal with aqueous acid.

Solution

$$C_6H_5CH(OCH_3)_2 \xrightarrow[H^+]{H_2O} C_6H_5CH{=}O + 2\,CH_3OH \quad (9.19)$$

PROBLEM 9.12 Show the steps in the mechanism for eq. 9.19.

The acid-catalyzed cleavage of acetals occurs much more readily than the acid-catalyzed cleavage of simple ethers (Sec. 8.6) because the intermediate carbocation is resonance stabilized. However, acetals, like ordinary ethers, are stable toward bases.

9.8 Addition of Water; Hydration of Aldehydes and Ketones

Water, like alcohols, is an oxygen nucleophile and can add reversibly to aldehydes and ketones. For example, formaldehyde in aqueous solution exists mainly as its hydrate.

$$H_2C{=}O + H{-}OH \rightleftharpoons H_2C(OH)_2 \quad (9.20)$$

formaldehyde — formaldehyde hydrate

With most other aldehydes or ketones, however, the hydrates cannot be isolated because they readily lose water to reform the carbonyl compound; that is, the equilibrium constant (Sec. 3.11) is less than 1. An exception is trichloroacetaldehyde (chloral), which forms a stable crystalline hydrate, $CCl_3CH(OH)_2$. Chloral hydrate is used in medicine as a sedative and in veterinary medicine as a narcotic and anesthetic** for horses, cattle, swine, and poultry. The potent drink known as a Mickey Finn is a combination of alcohol and chloral hydrate.

PROBLEM 9.13 Hydrolysis of $CH_3CBr_2CH_3$ with sodium hydroxide does *not* give $CH_3C(OH)_2CH_3$. Instead, it gives acetone. Explain.

---

*In the laboratory, this can be accomplished in several ways. One method involves distilling the water from the reaction mixture. Another method involves trapping the water with molecular sieves, inorganic materials with cavities of the size and shape required to absorb water molecules.

**See also "A Word about Ether and Anesthesia" in Chapter 8, page 243.

9.9 Addition of Grignard Reagents and Acetylides

Grignard reagents act as carbon nucleophiles toward carbonyl compounds. The R group of the Grignard reagent adds irreversibly to the carbonyl carbon, forming a new carbon–carbon bond. In terms of acid–base reactions, the addition is favorable because the product (an alkoxide) is a much weaker base than the starting carbanion (Grignard reagent). The alkoxide can be protonated to give an alcohol.

$$\text{C=O} + \text{RMgX} \xrightarrow{\text{ether}} \text{R–C–}\overset{-}{\text{O}}\overset{+}{\text{MgX}} \xrightarrow[\text{HCl}]{H_2O} \text{R–C–OH} + Mg^{2+}X^-Cl^- \quad (9.21)$$

intermediate addition product (a magnesium alkoxide) — an alcohol

The reaction is normally carried out by slowly adding an ether solution of the aldehyde or ketone to an ether solution of the Grignard reagent. After all of the carbonyl compound is added and the reaction is complete, the resulting magnesium alkoxide is hydrolyzed with aqueous acid.

The reaction of a Grignard reagent with a carbonyl compound provides a useful route to alcohols. The type of carbonyl compound chosen determines the class of alcohol produced. *Formaldehyde gives primary alcohols.*

$$\text{R–MgX} + \text{H–}\overset{O}{\overset{\|}{C}}\text{–H} \longrightarrow \text{R–}\underset{H}{\overset{H}{C}}\text{–OMgX} \xrightarrow[H^+]{H_2O} \text{R–}\underset{H}{\overset{H}{C}}\text{–OH} \quad (9.22)$$

formaldehyde — a primary alcohol

Other aldehydes give secondary alcohols.

$$\text{R–MgX} + \text{R}'\text{–}\overset{O}{\overset{\|}{C}}\text{–H} \longrightarrow \text{R–}\underset{H}{\overset{R'}{C}}\text{–OMgX} \xrightarrow[H^+]{H_2O} \text{R–}\underset{H}{\overset{R'}{C}}\text{–OH} \quad (9.23)$$

aldehyde — a secondary alcohol

Ketones give tertiary alcohols.

$$\text{R–MgX} + \text{R}'\text{–}\overset{O}{\overset{\|}{C}}\text{–R}'' \longrightarrow \text{R–}\underset{R''}{\overset{R'}{C}}\text{–OMgX} \xrightarrow[H^+]{H_2O} \text{R–}\underset{R''}{\overset{R'}{C}}\text{–OH} \quad (9.24)$$

ketone — a tertiary alcohol

Note that only *one* of the R groups (shown in black) attached to the hydroxyl-bearing carbon of the alcohol comes from the Grignard reagent. The rest of the alcohol's carbon skeleton comes from the carbonyl compound.

EXAMPLE 9.4

What is the product expected from the reaction between ethylmagnesium bromide and 3-pentanone followed by hydrolysis?

Solution 3-Pentanone is a ketone. Following eq. 9.24 as an example, the product is 3-ethyl-3-pentanol.

$$CH_3CH_2-\overset{\overset{\large O}{\|}}{C}-CH_2CH_3 + CH_3CH_2MgBr \longrightarrow CH_3CH_2-\underset{\underset{\large CH_2CH_3}{|}}{\overset{\overset{\large OMgBr}{|}}{C}}-CH_2CH_3 \xrightarrow[H^+]{H_2O} CH_3CH_2-\underset{\underset{\large CH_2CH_3}{|}}{\overset{\overset{\large OH}{|}}{C}}-CH_2CH_3$$

PROBLEM 9.14 Provide the products expected from the reaction of

a. formaldehyde with benzylmagnesium bromide followed by hydrolysis
b. pentanal with phenylmagnesium bromide followed by hydrolysis
c. benzaldehyde with ethylmagnesium bromide followed by hydrolysis
d. cyclohexanone with methylmagnesium bromide followed by hydrolysis

EXAMPLE 9.5

Show how the following alcohol can be synthesized from a Grignard reagent and a carbonyl compound:

$$C_6H_5-\overset{\overset{\large OH}{|}}{C}HCH_3$$

Solution The alcohol is secondary, so the carbonyl compound must be an aldehyde. We can use either a methyl or a phenyl Grignard reagent.

$$C_6H_5-\overset{\overset{\large OH}{|}}{C}H(-CH_3) \leftarrow CH_3MgBr \qquad (C_6H_5)-\overset{\overset{\large OH}{|}}{C}H-CH_3 \leftarrow C_6H_5-MgBr$$

The equations are

$$\underset{\text{methylmagnesium bromide}}{CH_3-MgBr} + \underset{\text{benzaldehyde}}{C_6H_5-\overset{\overset{\large O}{\|}}{C}H}$$

$$\downarrow$$

$$\underset{\text{alkoxide}}{CH_3-\overset{\overset{\large OMgBr}{|}}{C}H-C_6H_5} \xrightarrow[H^+]{H_2O} CH_3-\overset{\overset{\large OH}{|}}{C}H-C_6H_5 \qquad (9.25)$$

$$\uparrow$$

$$\underset{\text{phenylmagnesium bromide}}{C_6H_5-MgBr} + \underset{\text{acetaldehyde}}{CH_3-\overset{\overset{\large O}{\|}}{C}-H}$$

The choice between the possible sets of reactants may be made by availability or cost of reactants, or for chemical reasons (for example, the more reactive aldehyde or ketone might be selected).

PROBLEM 9.15 Show how each of the following alcohols can be made from a Grignard reagent and a carbonyl compound:

a.

b.

Other organometallic reagents, such as organolithium compounds and acetylides, react with carbonyl compounds in a similar fashion to Grignard reagents. For example,

$$\text{cyclopentanone (a ketone)} + Na^{+}\,{}^{-}C{\equiv}CH \longrightarrow \text{1-ethynylcyclopentyl } Na^{+}O^{-} \xrightarrow[H_2O]{H^+} \text{1-ethynylcyclopentanol (a tertiary acetylenic alcohol)} \quad (9.26)$$

a ketone; sodium acetylide; a tertiary acetylenic alcohol

PROBLEM 9.16 Write the structural formula of the product expected from the reaction of $CH_3C{\equiv}C^{-}Na^{+}$ with 3-pentanone followed by H_3O^{+}.

9.10 Addition of Hydrogen Cyanide; Cyanohydrins

Cyanohydrins are compounds with a hydroxyl and a cyano group attached to the same carbon.

Hydrogen cyanide adds reversibly to the carbonyl group of aldehydes and ketones to form **cyanohydrins**, compounds with a hydroxyl and a cyano group attached to the same carbon. A basic catalyst is required.

$$\text{>C=O} + \text{HCN} \xrightarrow{\text{KOH}} \text{(NC)C–OH} \quad \text{(9.27)}$$

a cyanohydrin

Acetone, for example, reacts as follows:

$$CH_3-\overset{O}{\overset{\|}{C}}-CH_3 + HCN \xrightarrow{KOH} CH_3-\underset{CN}{\overset{OH}{C}}-CH_3 \quad \text{(9.28)}$$

acetone — acetone cyanohydrin

Hydrogen cyanide has no unshared electron pair on its carbon, so it cannot function as a carbon nucleophile. The base converts some of the hydrogen cyanide to cyanide ion (NC^-), however, which then acts as a carbon nucleophile.

$$\text{>C=\ddot{O}:} + {}^-\text{:C≡N:} \rightleftharpoons \text{(NC)C–\ddot{O}:}^- \underset{}{\overset{\text{HCN}}{\rightleftharpoons}} \text{(NC)C–\ddot{O}H} + {}^-\text{CN} \quad \text{(9.29)}$$

cyanohydrin

PROBLEM 9.17 Write an equation for the addition of HCN to

a. propanal b. cyclopentanecarbaldehyde c. benzophenone (Sec. 9.1)

Cyanohydrin chemistry plays a central role in the defense system of *Apheloria corrugata.* This millipede uses a two-chamber gland much like that used by the bombardier beetle (see "A Word About . . . Quinones and the Bombardier Beetle" in Chapter 7) to deliver a secretion that contains hydrogen cyanide. *Apheloria* stores benzaldehyde cyanohydrin and, when threatened, converts it to a mixture of benzaldehyde and hydrogen cyanide, which is then secreted. The hydrogen cyanide gas that emanates from the secretion is an effective deterrent of predators.

$$C_6H_5CH(OH)CN \xrightarrow{\text{enzyme catalyst}} C_6H_5CHO + HCN \quad \text{(9.30)}$$

benzaldehyde cyanohydrin — benzaldehyde

9.11 Addition of Nitrogen Nucleophiles

Ammonia, amines, and certain related compounds have an unshared electron pair on the nitrogen atom and act as nitrogen nucleophiles toward the carbonyl carbon atom. For example, primary amines react as follows:

$$>C=O + \ddot{N}H_2-R \rightleftharpoons \left[\begin{matrix} HO \\ >C-NHR \end{matrix} \right] \xrightarrow{-HOH} >C=NR \quad (9.31)$$

primary amine; tetrahedral addition product; imine

The tetrahedral addition product that is formed first is similar to a hemiacetal, but with an NH group in place of one of the oxygens. These addition products are normally not stable. They eliminate water to form a product with a carbon–nitrogen double bond. With primary amines, the products are called **imines**. Imines are like carbonyl compounds, except that the O is replaced by NR. They are important intermediates in some biochemical reactions, particularly in binding carbonyl compounds to the free amino groups that are present in most enzymes.

Imines are compounds containing a carbon–nitrogen double bond.

$$\text{enzyme}-NH_2 + O=C< \text{(substrate)} \longrightarrow \text{enzyme}-N=C< + H_2O \quad (9.32)$$

enzyme-substrate compound

For example, retinal (see "A Word About . . . The Chemistry of Vision" in Chapter 3) binds to the protein opsin in this way to form rhodopsin.

PROBLEM 9.18 Write an equation for the reaction of benzaldehyde with aniline (the formula of which is $C_6H_5NH_2$).

Other ammonia derivatives containing an $-NH_2$ group react with carbonyl compounds similarly to primary amines. Table 9.1 lists some specific examples. Notice that in each of these reactions, the two hydrogens attached to nitrogen and the oxygen of the carbonyl group are eliminated as water.

Table 9.1 ■ Nitrogen Derivatives of Carbonyl Compounds

Formula of ammonia derivative	Name	Formula of carbonyl derivative	Name
RNH_2 or $ArNH_2$	primary amine	$>C=NR$ or $>C=NAr$	imine
NH_2OH	hydroxylamine	$>C=NOH$	oxime
NH_2NH_2	hydrazine	$>C=NNH_2$	hydrazone
$NH_2NHC_6H_5$	phenylhydrazine	$>C=NNHC_6H_5$	phenylhydrazone

EXAMPLE 9.6

Using Table 9.1 as a guide, write an equation for the reaction of hydrazine with cyclohexanone.

Solution

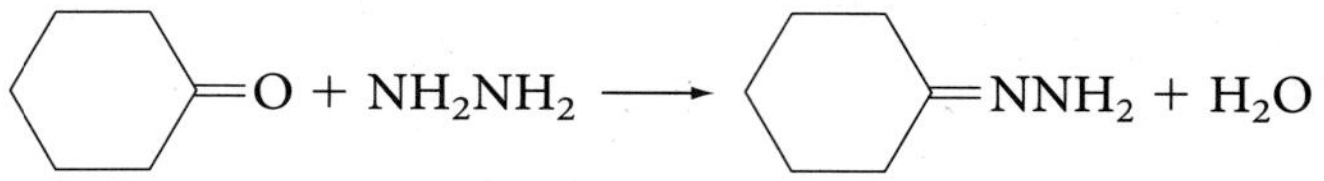

$$\text{cyclohexanone}{=}O + NH_2NH_2 \longrightarrow \text{cyclohexane}{=}NNH_2 + H_2O$$

The product is a hydrazone.

PROBLEM 9.19 Using Table 9.1 as a guide, write an equation for the reaction of pentanal ($CH_3CH_2CH_2CH_2CH{=}O$) with

a. hydroxylamine (NH_2OH)
b. ethylamine ($CH_3CH_2NH_2$)
c. hydrazine (NH_2NH_2)
d. phenylhydrazine ($NH_2NHC_6H_5$)

9.12 Reduction of Carbonyl Compounds

Aldehydes and ketones are easily reduced to primary and secondary alcohols, respectively. Reduction can be accomplished in many ways, most commonly by metal hydrides.

The most common metal hydrides used to reduce carbonyl compounds are lithium aluminum hydride ($LiAlH_4$) and sodium borohydride ($NaBH_4$). The metal–hydride bond is polarized, with the metal positive and the hydrogen negative. Therefore, the reaction involves irreversible nucleophilic attack of the hydride (H^-) at the carbonyl carbon:

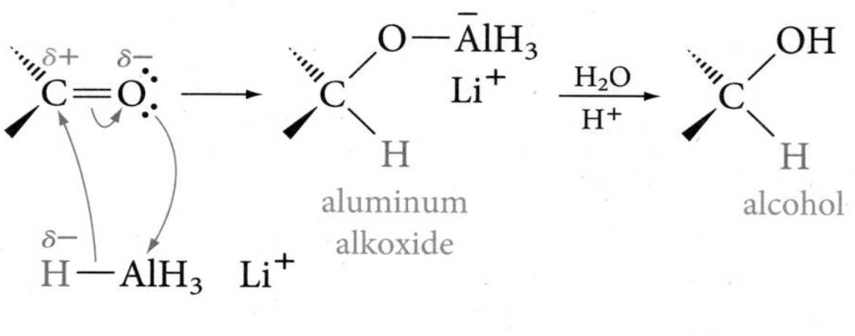

$$\overset{\delta+}{C}{=}\overset{\delta-}{O}: \; + \; \overset{\delta-}{H}{-}AlH_3 \; Li^+ \longrightarrow \underset{\text{aluminum alkoxide}}{H{-}C{-}O{-}\bar{A}lH_3 \; Li^+} \xrightarrow[H^+]{H_2O} \underset{\text{alcohol}}{H{-}C{-}OH} \tag{9.33}$$

The initial product is an aluminum alkoxide, which is subsequently hydrolyzed by water and acid to give the alcohol. The net result is addition of hydrogen across the carbon–oxygen double bond. A specific example is

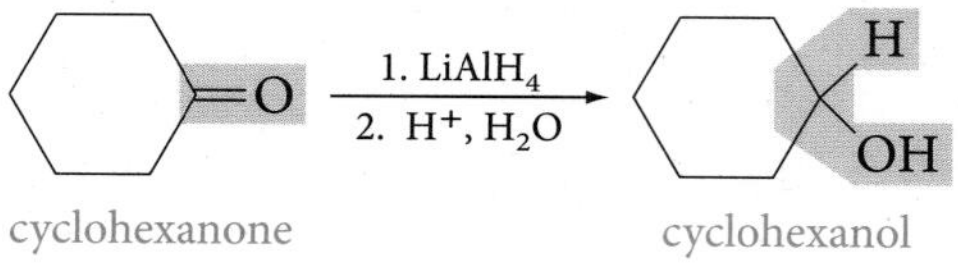

$$\underset{\text{cyclohexanone}}{C_6H_{10}{=}O} \xrightarrow[2.\ H^+,\ H_2O]{1.\ LiAlH_4} \underset{\text{cyclohexanol}}{C_6H_{10}(H)OH} \tag{9.34}$$

PROBLEM 9.20 Show how the following alcohols can be made from lithium aluminum hydride and a carbonyl compound:

a. 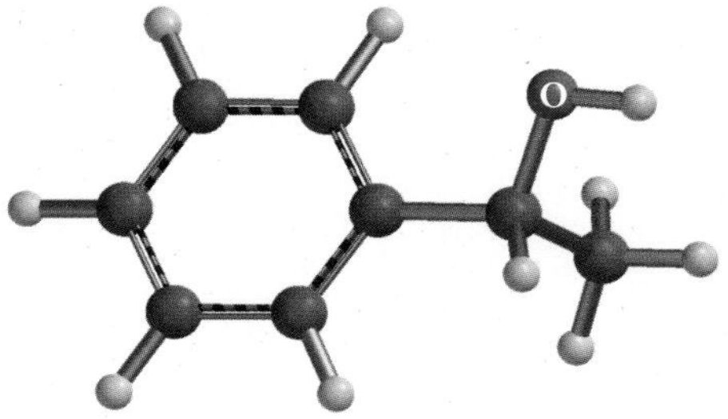 b.

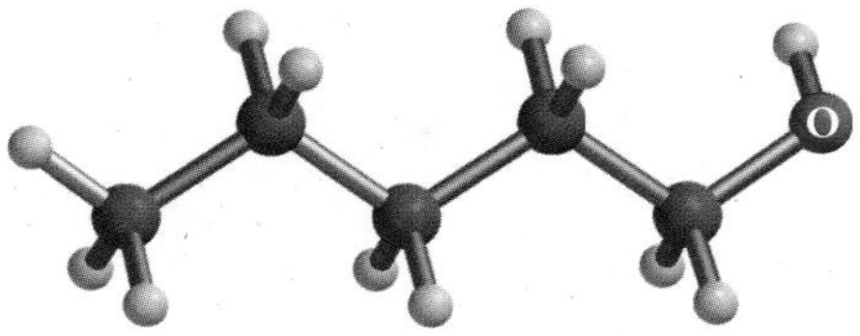

Because a carbon–carbon double bond is not readily attacked by nucleophiles, metal hydrides can be used to reduce a carbon–oxygen double bond to the corresponding alcohol without reducing a carbon–carbon double bond present in the same compound.

$$CH_3{-}CH{=}CH{-}\overset{O}{\overset{\|}{C}}H \xrightarrow{NaBH_4} CH_3{-}CH{=}CH{-}CH_2OH \quad (9.35)$$

2-butenal → 2-buten-1-ol

PROBLEM 9.21 Show how

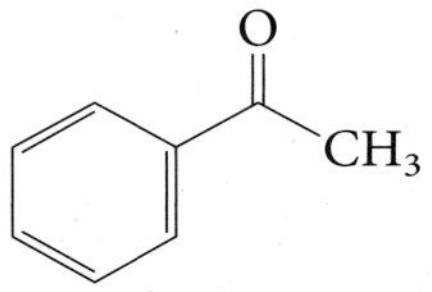

$C_6H_5C(=O)CH_3$ can be reduced to $C_6H_5CH(OH)CH_3$

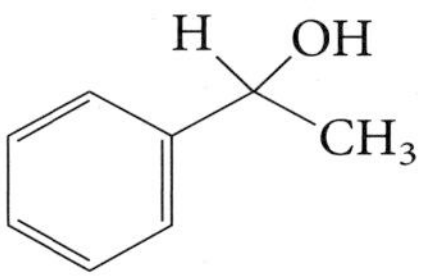

9.13 Oxidation of Carbonyl Compounds

Aldehydes are more easily oxidized than ketones. Oxidation of an aldehyde gives a carboxylic acid with the same number of carbon atoms.

$$R{-}\overset{O}{\overset{\|}{C}}{-}H \xrightarrow{\text{oxidizing agent}} R{-}\overset{O}{\overset{\|}{C}}{-}OH \quad (9.36)$$

aldehyde → acid

Because the reaction occurs easily, many oxidizing agents, such as $KMnO_4$, CrO_3, Ag_2O, and peracids (see eq. 8.18), will work. Specific examples are

$$CH_3(CH_2)_5CH{=}O \xrightarrow[\text{(Jones' reagent)}]{CrO_3,\ H^+} CH_3(CH_2)_5CO_2H \quad (9.37)$$

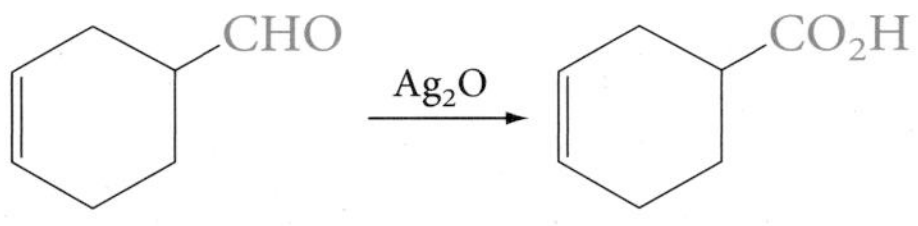

$$\text{3-cyclohexenecarbaldehyde (CHO)} \xrightarrow{Ag_2O} \text{3-cyclohexenecarboxylic acid } (CO_2H) \quad (9.38)$$

Silver ion as an oxidant is expensive but has the virtue that it selectively oxidizes aldehydes to carboxylic acids in the presence of alkenes (eq. 9.38).

The symbol ↓ indicates formation of a precipitate; the symbol ↑ indicates the formation of a gas.

A laboratory test that distinguishes aldehydes from ketones takes advantage of their different ease of oxidation. In the Tollens' silver mirror test, the silver–ammonia complex ion is reduced by aldehydes (but not by ketones) to metallic silver.* The equation for the reaction may be written as follows:

$$\underset{\text{aldehyde}}{\mathrm{R}\overset{\displaystyle \mathrm{O} \atop \|}{\mathrm{C}}\mathrm{H}} + \underset{\substack{\text{silver–ammonia}\\ \text{complex ion}\\ \text{(colorless)}}}{2\,\mathrm{Ag(NH_3)_2}^+} + 3\,\mathrm{HO^-} \longrightarrow \underset{\substack{\text{acid}\\ \text{anion}}}{\mathrm{R}\overset{\displaystyle \mathrm{O} \atop \|}{\mathrm{C}}\mathrm{—O^-}} + \underset{\substack{\text{silver}\\ \text{mirror}}}{2\,\mathrm{Ag}\downarrow} + 4\,\mathrm{NH_3}\uparrow + 2\,\mathrm{H_2O} \qquad \textbf{(9.39)}$$

If the glass vessel in which the test is performed is thoroughly clean, the silver deposits as a mirror on the glass surface. This reaction is also employed to silver glass, using the inexpensive aldehyde formaldehyde.

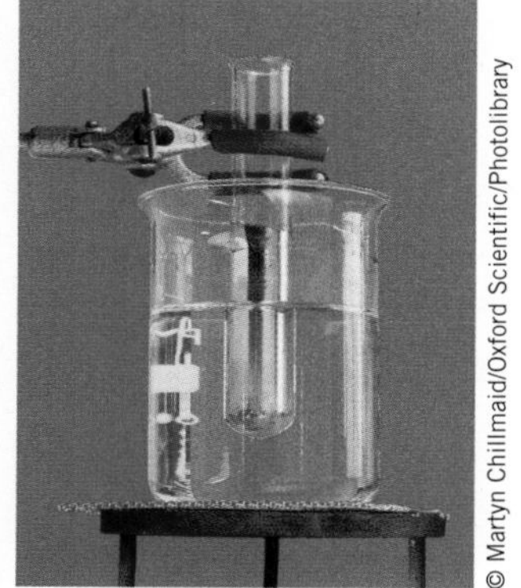

© Martyn Chillmaid/Oxford Scientific/Photolibrary

Tollens' silver mirror test (eq. 9.39).

PROBLEM 9.22 Write an equation for the formation of a silver mirror from formaldehyde and Tollens' reagent.

Aldehydes are so easily oxidized that stored samples usually contain some of the corresponding acid. This contamination is caused by air oxidation.

$$2\,\mathrm{RCHO} + \mathrm{O_2} \longrightarrow 2\,\mathrm{RCO_2H} \qquad \textbf{(9.40)}$$

Ketones also can be oxidized, but require special oxidizing conditions. For example, cyclohexanone is oxidized commercially to adipic acid, an important industrial chemical used to manufacture nylon. Over 5 billion pounds of adipic acid are produced each year.

$$\underset{\text{cyclohexanone}}{\text{cyclohexanone}} + \mathrm{HNO_3} \xrightarrow{\mathrm{V_2O_5}} \underset{\text{adipic acid}}{\mathrm{HO—}\overset{\displaystyle \mathrm{O} \atop \|}{\mathrm{C}}\mathrm{—CH_2CH_2CH_2CH_2—}\overset{\displaystyle \mathrm{O} \atop \|}{\mathrm{C}}\mathrm{—OH}} \qquad \textbf{(9.41)}$$

one of these C—C bonds is cleaved in the oxidation

9.14 Keto–Enol Tautomerism

Aldehydes and ketones may exist as an equilibrium mixture of two forms, called the **keto form** and the **enol form**. The two forms differ in the location of a proton and a double bond.

$$\underset{\text{keto form}}{\mathrm{—\overset{\displaystyle H \atop |}{\underset{|}{C}}—\overset{\displaystyle O \atop \|}{C}—}} \rightleftharpoons \underset{\text{enol form}}{\mathrm{\gt C{=}C\lt^{OH}}} \qquad \textbf{(9.42)}$$

*Silver hydroxide is insoluble in water, so the silver ion must be complexed with ammonia to keep it in solution in a basic medium.

This type of structural isomerism is called tautomerism (from the Greek *tauto*, the same, and *meros*, part). The two forms of the aldehyde or ketone are called **tautomers**.

Tautomers are structural isomers that differ in the location of a proton and a double bond. The **keto** and **enol** forms of an aldehyde or ketone are tautomers.

EXAMPLE 9.7

Write formulas for the keto and enol forms of acetone.

Solution

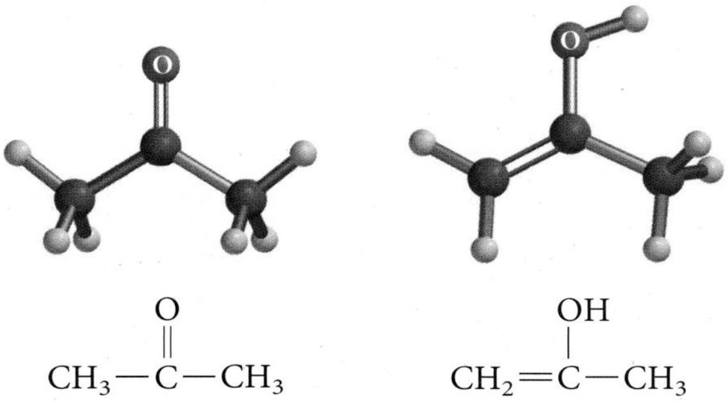

$CH_3-C(=O)-CH_3$
keto form

$CH_2=C(OH)-CH_3$
enol form

PROBLEM 9.23 Draw the structural formula for the enol form of

a. cyclohexanone
b. propanal (CH_3CH_2CHO)
c. 4-methoxycyclohexanone

Tautomers are structural isomers, *not* contributors to a resonance hybrid. They readily equilibrate, and we indicate that fact by using the equilibrium symbol $\rightleftharpoons$ between their structures.

To be capable of existing in an enol form, a carbonyl compound must have a hydrogen atom attached to the carbon atom adjacent to the carbonyl group. This hydrogen is called an **α-hydrogen** and is attached to the **α-carbon atom** (from the first letter of the Greek alphabet, α, or alpha).

An **α-hydrogen** is attached to the **α-carbon atom**, the carbon atom adjacent to a carbonyl group.

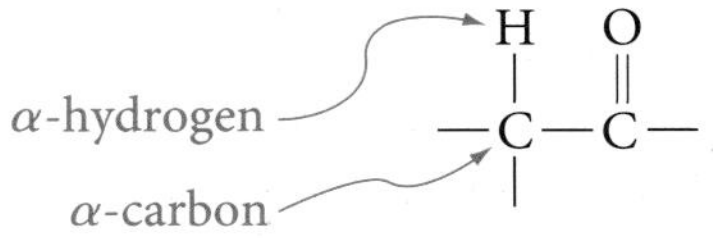

α-hydrogen → H on $-C(H)-C(=O)-$
α-carbon → C

Most simple aldehydes and ketones exist mainly in the keto form. Acetone, for example, is 99.9997% in the keto form, with only 0.0003% of the enol present. The main reason for the greater stability of the keto form is that the C=O plus C—H bond energy present in the keto form is greater than the C=C plus O—H bond energy of the enol form. We have already encountered some molecules, however, that have mainly the enol structure—the *phenols*. In this case, the resonance stabilization of the aromatic ring is greater than the usual energy difference that favors the keto over the enol form. Aromaticity would be destroyed if the molecule existed in the keto form; therefore, the enol form is preferred.

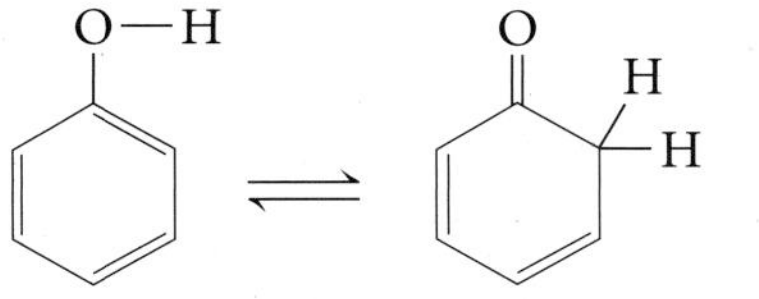

O—H (phenol) $\rightleftharpoons$ O, H, H (cyclohexadienone) **(9.43)**

enol form of phenol — keto form of phenol

A WORD ABOUT... Tautomerism and Photochromism

Some carbonyl compounds exist predominantly in the enol form due to favorable interactions between the tautomeric enol's OH group and an adjacent lone pair via a intermolecular hydrogen bond. For example, the equilibrium constant for 2,4-pentanedione in water lies in favor of the enol form and is about 4. Often, the equilibrium constant for the keto–enol tautomerism is very solvent dependent, and the keto form is more favored in less polar solvents, such as cyclohexane.

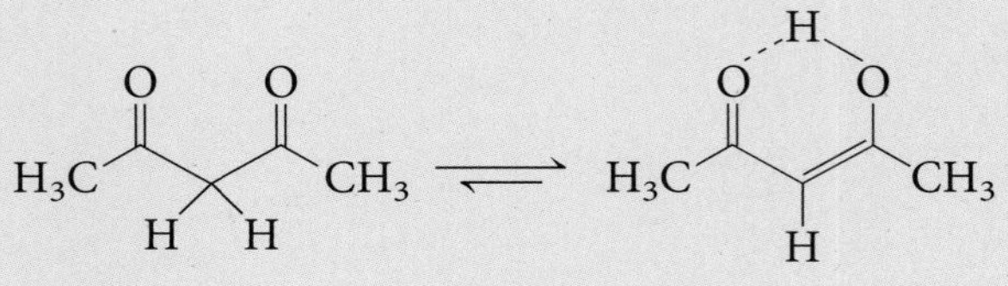

keto form ⇌ enol form

The concept of tautomerism can be expanded beyond keto and enol forms to include any pair or group of isomers that can be easily interconverted by the relocation of an atom and/or bonds. For example, imines and enamines (unsaturated amines) are tautomers whose relationship is similar to that of keto and enol forms.

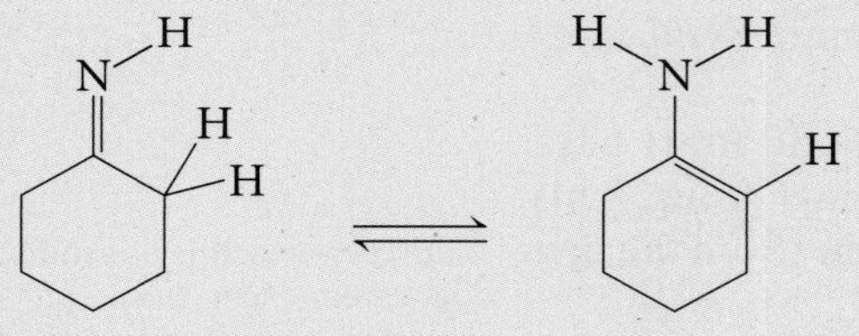

an imine ⇌ an enamine

For some pairs of tautomers, one can be converted to the other photochemically—that is, by the absorption of light. Irradiation of the following pale yellow phenol-imine causes the hydrogen atom to shift from the oxygen to the nitrogen, with appropriate rebonding.

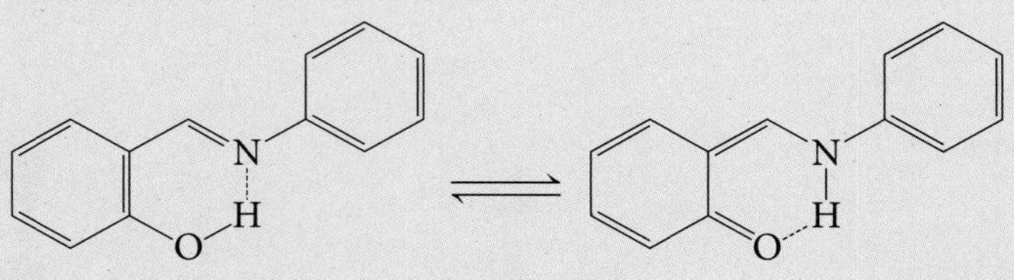

a phenol-imine
pale yellow
(both rings are aromatic)

a keto-enamine
red
(only one ring is aromatic)

If the keto-enamine product of this photochemical reaction is allowed to remain in the dark, it gradually changes back to the more stable phenol-imine.

Of what use can such a cycle of reactions, with no net change, possibly be? In this example, one tautomer is pale yellow, and the other is red. This phenomenon, in which two compounds undergo a reversible photochemical color change, is called **photochromism**. Photochromic substances have many practical uses. Glasses coated with these substances become darker when exposed to sunlight, because their lenses are impregnated with a photochromic material. When the sunlight dims or when taken indoors, the colored photochromic substance gradually changes back to its colorless form. Photochromic substances can also be used for data storage and display (as in digital watches), as second-order nonlinear optical materials in laser and telecommunication applications, as chemical switches in computers, for micro images (microfilm), for protection against sudden light flashes, and in many other creative ways. (See the special issue in "Photochromism: Memories and Switches," *Chemical Reviews*, **2000**, *100*, no. 5.)

See Problems 9.55 and 9.58.

Photochromism is the phenomenon in which two compounds undergo a reversible photochemical color change.

Carbonyl compounds that do not have an α-hydrogen cannot form enols and exist only in the keto form. Examples are

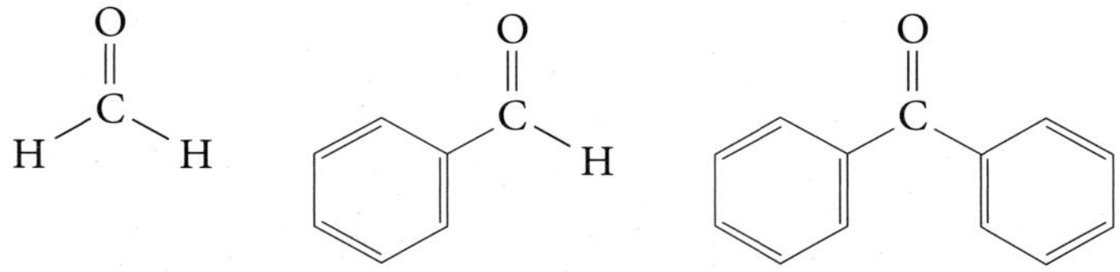

formaldehyde benzaldehyde benzophenone

9.15 Acidity of α-Hydrogens; the Enolate Anion

The α-hydrogen in a carbonyl compound is more acidic than a normal hydrogen bonded to a carbon atom. Table 9.2 shows the pK_a values for a typical aldehyde and ketone as well as for reference compounds. The result of placing a carbonyl group adjacent to methyl protons is truly striking, an increase in their acidity of over 10^{30}! (Compare acetaldehyde or acetone with propane.) Indeed, these compounds are almost as acidic as the O—H protons in alcohols. Why is this?

There are two reasons. First, the carbonyl carbon carries a partial positive charge. Bonding electrons are displaced toward the carbonyl carbon and away from the α-hydrogen (shown by the red arrows below), making it easy for a base to remove the α-hydrogen as a proton (that is, without its bonding electrons).

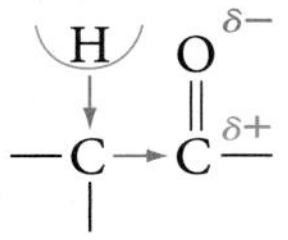

H O δ− δ+ —C→C—

Second, the resulting anion is stabilized by resonance.

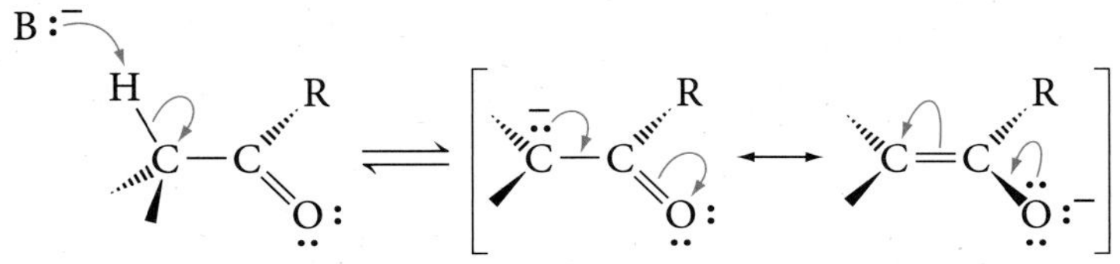

B:⁻ H R C—C O: ⇌ [C—C O: ⟷ C=C O:⁻] (9.44)

resonance structures of an enolate anion

The anion is called an **enolate anion**. Its negative charge is distributed between the α-carbon and the carbonyl oxygen atom.

An **enolate anion** is formed by removal of the α-hydrogen of a ketone or aldehyde.

Table 9.2 ■ Acidity of α-Hydrogens

Compound	Name	pK_a
$CH_3CH_2CH_3$	propane	~50
$CH_3C(=O)CH_3$	acetone	19
$CH_3C(=O)H$	acetaldehyde	17
CH_3CH_2OH	ethanol	16

EXAMPLE 9.8

Draw the formula for the enolate anion of acetone.

Solution

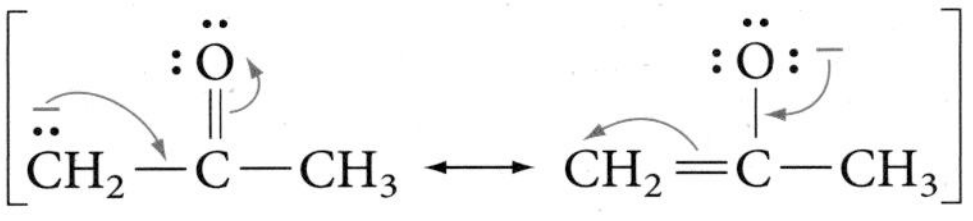

$$\left[\overset{-}{C}H_2—C(=O)—CH_3 \longleftrightarrow CH_2=C(—O^-)—CH_3\right]$$

An enolate anion is a resonance hybrid of two contributing structures that differ *only* in the arrangement of the electrons.

PROBLEM 9.24 Draw the resonance contributors to the enolate anion of

a. propanal b. cyclopentanone

9.16 Deuterium Exchange in Carbonyl Compounds

Even though its concentration is very low, the presence of the enol form of ordinary aldehydes and ketones can be demonstrated experimentally. For example, the α-hydrogens can be exchanged for deuterium by placing the carbonyl compound in a solvent such as D_2O or CH_3OD. The exchange is catalyzed by acid or base. *Only the α-hydrogens exchange,* as illustrated by the following examples:

cyclohexanone $\xrightarrow[\text{CH}_3\text{OD (excess)}]{\text{Na}^{+\,-}\text{OCH}_3}$ 2,2,6,6-tetradeuteriocyclohexanone (9.45)

$$CH_3CH_2CH_2CH{=}O \xrightarrow[D^+]{D_2O} CH_3CH_2CD_2CH{=}O \quad (9.46)$$

butanal → 2,2-dideuteriobutanal

The mechanism of the base-catalyzed exchange of the α-hydrogens (eq. 9.45) involves two steps.

cyclohexanone + $^-OCH_3$ ⇌ [enolate anion resonance structures] + CH_3OH

enolate anion

$\downarrow\uparrow$ CH_3OD

2-deuteriocyclohexanone + CH_3O^- (9.47)

The base (methoxide ion) removes an α-proton to form the enolate anion. Reprotonation, but with CH_3OD, replaces the α-hydrogen with deuterium. With excess CH_3OD, all four α-hydrogens are eventually exchanged.

The mechanism of the acid-catalyzed exchange of the α-hydrogens (eq. 9.46) also involves several steps. The keto form is first protonated and, by loss of an α-hydrogen, converted to its enol.

$$\underset{\text{keto form}}{CH_3CH_2CH_2\overset{\overset{\ddot{O}:}{\|}}{C}H} \underset{}{\overset{D^+}{\rightleftharpoons}} CH_3CH_2\overset{\overset{H}{|}}{\underset{\underset{H}{|}}{C}}-\overset{\overset{+\ddot{O}-D}{\|}}{C}H \overset{-H^+}{\rightleftharpoons} \underset{\text{enol form}}{CH_3CH_2CH=\overset{\overset{:\ddot{O}-D}{|}}{C}H} \quad (9.48)$$

In the reversal of these equilibria, the enol then adds D^+ at the α-carbon.

$$D^+ \quad CH_3CH_2CH=\overset{\overset{:\ddot{O}-D}{|}}{C}H \rightleftharpoons CH_3CH_2\overset{\overset{D}{|}}{C}H-\overset{\overset{:\ddot{O}}{\|}}{C}H + D^+ \quad (9.49)$$

Repetition of this sequence results in exchange of the other α-hydrogen.

PROBLEM 9.25 Identify the hydrogens that are readily exchanged for deuterium in

a. 2-ethylcyclopentanone (cyclopentanone ring with C=O and CH_2CH_3 substituent on adjacent carbon)

b. $(CH_3)_3C\overset{\overset{O}{\|}}{C}CH_3$

9.17 The Aldol Condensation

Enolate anions may act as carbon nucleophiles. They add reversibly to the carbonyl group of another aldehyde or ketone molecule in a reaction called the **aldol condensation**, an extremely useful carbon–carbon bond-forming reaction.

An enolate anion adds to the carbonyl group of an aldehyde or ketone in an **aldol condensation**. An **aldol** is a 3-hydroxyaldehyde or 3-hydroxyketone.

The simplest example of an aldol condensation is the combination of two acetaldehyde molecules, which occurs when a solution of acetaldehyde is treated with catalytic amounts of aqueous base.

$$\underset{\text{acetaldehyde}}{CH_3\overset{\overset{O}{\|}}{C}H + CH_3\overset{\overset{O}{\|}}{C}H} \overset{HO^-}{\rightleftharpoons} \underset{\substack{\text{3-hydroxybutanal} \\ \text{(an aldol)}}}{CH_3\overset{\overset{OH}{|}}{C}H-CH_2\overset{\overset{O}{\|}}{C}H} \quad (9.50)$$

The product is called an **aldol** (so named because the product is both an *ald*ehyde and an alcoh*ol*).

The aldol condensation of acetaldehyde occurs according to the following three-step mechanism:

Step 1.

$$\overset{\alpha}{CH_3}-\overset{\overset{\ddot{O}:}{\|}}{C}-H + HO^- \rightleftharpoons \underset{\text{enolate anion}}{\overset{..}{\bar{C}}H_2-\overset{\overset{\ddot{O}:}{\|}}{C}-H} + HOH \quad (9.51)$$

Step 2.

$$CH_3-\overset{\overset{\ddot{O}:}{\|}}{C}H + \underset{\text{nucleophile}}{\overset{..}{\bar{C}}H_2-\overset{\overset{\ddot{O}:}{\|}}{C}H} \rightleftharpoons \underset{\text{an alkoxide ion}}{CH_3\overset{\overset{:\ddot{O}:^-}{|}}{C}H-CH_2\overset{\overset{\ddot{O}:}{\|}}{C}H} \quad (9.52)$$

Step 3. $$CH_3\overset{\overset{:\ddot{O}:^-}{|}}{C}H-CH_2\overset{\overset{\ddot{O}:}{\|}}{C}H + HOH \rightleftharpoons CH_3\overset{\overset{:\ddot{O}H}{|}}{C}H-\overset{\alpha}{C}H_2\overset{\overset{\ddot{O}:}{\|}}{C}H + HO^- \quad \text{(aldol)} \tag{9.53}$$

In step 1, the base removes an α-hydrogen to form the enolate anion. In step 2, this anion adds to the carbonyl carbon of *another* acetaldehyde molecule, forming a new carbon–carbon bond. Ordinary bases convert only a small fraction of the carbonyl compound to the enolate anion so that a substantial fraction of the aldehyde is still present in the un-ionized carbonyl form needed for this step. In step 3, the alkoxide ion formed in step 2 accepts a proton from the solvent, thus regenerating the hydroxide ion needed for the first step.

In the aldol condensation, the α-carbon of one aldehyde molecule becomes connected to the carbonyl carbon of another aldehyde molecule.

$$RCH_2\overset{\overset{O}{\|}}{C}H + R\overset{\alpha}{C}H_2\overset{\overset{O}{\|}}{C}H \xrightarrow[H_2O]{HO^-} RCH_2\overset{\overset{OH}{|}}{\underset{3}{C}}H-\underset{\underset{R}{|}}{\overset{\alpha}{\underset{2}{C}}}H\overset{\overset{O}{\|}}{\underset{1}{C}}H \quad \text{(an aldol)} \tag{9.54}$$

Aldols are therefore 3-hydroxyaldehydes. *Since it is always the α-carbon that acts as a nucleophile, the product always has just one carbon atom between the aldehyde and alcohol carbons,* regardless of how long the carbon chain is in the starting aldehyde.

EXAMPLE 9.9

Give the structure of the aldol that is obtained by treating propanal ($CH_3CH_2CH{=}O$) with base.

Solution Rewriting eq. 9.54 with R = CH_3, the product is

$$CH_3CH_2\overset{\overset{OH}{|}}{C}H-\underset{\underset{CH_3}{|}}{C}H\overset{\overset{O}{\|}}{C}H$$

PROBLEM 9.26 Write out the steps in the mechanism for formation of the product in Example 9.9.

9.18 The Mixed Aldol Condensation

The aldol condensation is versatile in that the enolate anion of *one* carbonyl compound can be made to add to the carbonyl carbon of *another,* provided that the reaction partners are carefully selected. Consider, for example, the reaction between acetaldehyde and benzaldehyde, when treated with base. Only acetaldehyde can form an enolate anion (benzaldehyde has no α-hydrogen). If the enolate ion of acetaldehyde adds to the benzaldehyde carbonyl group, a mixed aldol condensation occurs.

$$C_6H_5\overset{\overset{O}{\|}}{C}H + \overset{\alpha}{C}H_3\overset{\overset{O}{\|}}{C}H \overset{HO^-}{\rightleftharpoons} C_6H_5\overset{\overset{OH}{|}}{C}H-CH_2\overset{\overset{O}{\|}}{C}H \xrightarrow[-H_2O]{heat} C_6H_5CH{=}CH\overset{\overset{O}{\|}}{C}H \tag{9.55}$$

a mixed aldol → cinnamaldehyde

A WORD ABOUT... Water Treatment and the Chemistry of Enols/Enolates

In a city with a population of about 35,000, around 1.2 billion gallons of drinking water is delivered annually to homes. This water is usually purified before use to remove contaminants such as viruses and bacteria; organic chemicals, including pesticides and herbicides; and inorganic chemicals, including radioactive substances. These contaminants come from a variety of industrial and natural sources. The cleanup process involves several steps, including removal of water-insoluble material by sedimentation, coagulation, and filtration; softening of the water using lime (calcium oxide, CaO) and soda ash (sodium carbonate, Na_2CO_3); adjustment of the pH to 7.5–8.0 with carbon dioxide; and disinfection with chlorine (Cl_2). A typical water quality report will show that while the water delivered to homes is quite clean, α-halogenated ketones, halogenated carboxylic acids, and trihalomethanes (such as chloroform) are sometimes present at very low levels due to chemistry that occurs during the treatment process. What are these compounds, and how do they get into the water during the treatment process?

$$\underset{\text{acetone (2-propanone)}}{CH_3COCH_3} \xrightarrow{\text{enolization}} \underset{\text{enol of acetone}}{CH_3C(OH){=}CH_2} \xrightarrow{Cl-Cl} \underset{\alpha\text{-halogenated ketone}}{CH_3COCH_2Cl} + HCl \rightleftharpoons \underset{\text{1,1,1-trichloroacetone}}{CH_3COCCl_3}$$

When ketones are treated with halogens (Cl_2, Br_2, I_2), they undergo α-halogenation. This is illustrated above for acetone (2-propanone) and chlorine. Enolization of the acetone is followed by the reaction of enol with chlorine to provide α-chloroacetone. The enol (or enolate) of acetone behaves as a nucleophile and the chlorine behaves as an electrophile in this reaction. Repetition of this process several times provides polychlorinated compounds such as 1,1,1-trichloroacetone. Polychlorinated ketones can then undergo further reactions that provide chlorinated carboxylic acids and chloroform. Because acetone is a common water-soluble organic solvent and is also an end product of fat metabolism in humans and other animals, it is not surprising that it shows up in water treatment plants. The normal chemistry of ketones (via their enols/enolates) that takes place during the water treatment process, therefore, is responsible in part for the presence of low levels of chlorinated organic compounds in our drinking water. Whereas one could argue that water treatment introduces potentially toxic compounds into the drinking water supply, this is balanced by the removal of far more dangerous materials by the treatment process.

Some recent alternatives for water treatment include using the dissolved organic carbon, nitrates, and other components in the water as a food source to grow algae. Then such algal biomass can be harvested in order to capture the cell wall constituents, which include lipids and triglycerides (Chapter 15). The lipids can be converted to biodiesel and used as an alternative fuel source to petroleum-derived diesel (petrodiesel) in diesel-combustion engines (see Chapter 3 and "A Word About . . . Petroleum, Gasoline, and Octane Number," pages 102–103). For additional information, see http://www.nrel.gov/biomass/.

See Problem 9.52.

In this particular example, the resulting mixed aldol eliminates water on heating to give cinnamaldehyde (the flavor constituent of cinnamon).

EXAMPLE 9.10

Write the structure of the mixed aldol obtained from acetone and formaldehyde.

Solution Of the two reactants, only acetone has α-hydrogens.

$$H{-}\overset{O}{\overset{\|}{C}}{-}H + CH_3\overset{O}{\overset{\|}{C}}CH_3 \xrightarrow{\text{base}} H{-}\underset{H}{\underset{|}{\overset{OH}{\overset{|}{C}}}}{-}CH_2\overset{O}{\overset{\|}{C}}CH_3$$

PROBLEM 9.27 Using eqs. 9.51 through 9.53 as a guide, write out the steps in the mechanism for eq. 9.55.

PROBLEM 9.28 Write the structure of the mixed aldol obtained from propanal and benzaldehyde. What structure is obtained from dehydration of this mixed aldol?

9.19 Commercial Syntheses via the Aldol Condensation

Aldols are useful in synthesis. For example, acetaldehyde is converted commercially to crotonaldehyde, 1-butanol, and butanal using the aldol condensation.

$$\underset{\text{acetaldehyde}}{2\,CH_3\overset{\overset{O}{\|}}{C}H} \xrightarrow{HO^-} \underset{\text{aldol}}{CH_3\overset{\overset{OH}{|}}{C}HCH_2\overset{\overset{O}{\|}}{C}H} \xrightarrow[-H_2O]{H^+} \underset{\text{crotonaldehyde}}{CH_3CH{=}CH\overset{\overset{O}{\|}}{C}H} \xrightarrow[\text{catalyst}]{H_2}$$

$$\underset{\text{butanal}}{CH_3CH_2CH_2\overset{\overset{O}{\|}}{C}H} \quad \text{or} \quad \underset{\text{1-butanol}}{CH_3CH_2CH_2CH_2OH} \qquad (9.56)$$

The particular product obtained in the hydrogenation step depends on the catalyst and reaction conditions.

Butanal is the starting material for the synthesis of the mosquito repellent "6-12" (2-ethylhexane-1,3-diol). The first step is an aldol condensation, and the second step is reduction of the aldehyde group to a primary alcohol.

$$\underset{\text{butanal}}{2\,CH_3CH_2CH_2\overset{\overset{O}{\|}}{C}H} \xrightarrow{HO^-} \underset{\text{butanal aldol}}{CH_3CH_2CH_2\overset{\overset{OH}{|}}{C}H\underset{\underset{CH_3CH_2}{|}}{C}H\overset{\overset{O}{\|}}{C}H} \xrightarrow[Ni]{H_2} \underset{\text{2-ethylhexane-1,3-diol ("6-12")}}{CH_3CH_2CH_2\overset{\overset{OH}{|}}{C}H\underset{\underset{CH_3CH_2}{|}}{C}HCH_2OH} \qquad (9.57)$$

The aldol condensation is also used in nature to build up (and, in the case of *reverse* aldol condensations, to break down) carbon chains.

PROBLEM 9.29 2-Ethylhexanol, used commercially in the manufacture of plasticizers and synthetic lubricants, is synthesized from butanal via its aldol product. Devise a route to it.

REACTION SUMMARY

1. Preparation of Aldehydes and Ketones

a. Oxidation of Alcohols (Sec. 9.3)

$$^{*}R{-}CH_2{-}OH \xrightarrow{PCC} R{-}\overset{O}{\overset{\|}{C}}{-}H$$

1 alcohol → aldehyde

$$R{-}\underset{}{\overset{OH}{\overset{|}{CH}}}{-}R \xrightarrow{PCC} R{-}\overset{O}{\overset{\|}{C}}{-}R$$

2 alcohol → ketone

b. Friedel–Crafts Acylation (Sec. 9.3)

$$C_6H_6 + R{-}\overset{O}{\overset{\|}{C}}{-}Cl \xrightarrow{AlCl_3} C_6H_5{-}\overset{O}{\overset{\|}{C}}{-}R$$

c. Hydration of Alkynes (Sec. 9.3)

$$R{-}C{\equiv}C{-}H \xrightarrow[Hg^{2+}]{H_3O^+} R{-}\overset{O}{\overset{\|}{C}}{-}CH_3$$

2. Reactions of Aldehydes and Ketones

a. Formation and Hydrolysis of Acetals (Sec. 9.7)

$$ROH + \underset{R'\quad R''}{\overset{O}{\overset{\|}{C}}} \underset{}{\overset{H^+}{\rightleftharpoons}} R'{-}\underset{OR}{\underset{|}{\overset{OH}{\overset{|}{C}}}}{-}R'' \underset{H^+}{\overset{ROH}{\rightleftharpoons}} R'{-}\underset{OR}{\underset{|}{\overset{OR}{\overset{|}{C}}}}{-}R'' + H_2O$$

alcohol, carbonyl → hemiacetal → acetal

b. Addition of Grignard Reagents (Sec. 9.9)

$$R'{-}MgX + R{-}\overset{O}{\overset{\|}{C}}{-}R \longrightarrow R{-}\underset{R'}{\underset{|}{\overset{OMgX}{\overset{|}{C}}}}{-}R \xrightarrow{H_3O^+} R{-}\underset{R'}{\underset{|}{\overset{OH}{\overset{|}{C}}}}{-}R$$

Formaldehyde gives 1° alcohols; other aldehydes give 2° alcohols; and ketones give 3° alcohols.

c. Formation of Cyanohydrins (Sec. 9.10)

$$HC{\equiv}N + R{-}\overset{O}{\overset{\|}{C}}{-}R \underset{}{\overset{\text{NaOH catalyst}}{\rightleftharpoons}} R{-}\underset{CN}{\underset{|}{\overset{OH}{\overset{|}{C}}}}{-}R$$

R=H, alkyl → cyanohydrin

*For all syntheses and reactions of carbonyl compounds in this summary, R can be alkyl or aryl.

d. Addition of Nitrogen Nucleophiles (Sec. 9.11 and Table 9.1)

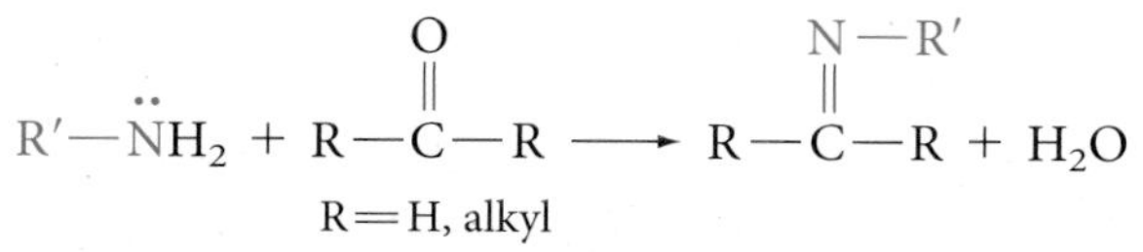

The product is an imine when R′ is an alkyl group.

e. Reduction to Alcohols (Sec. 9.12)

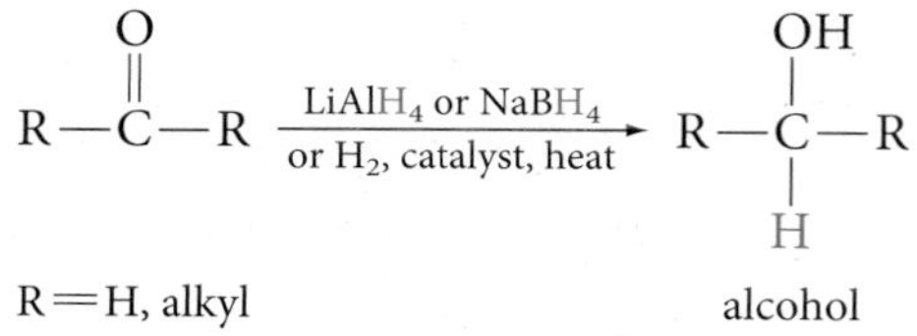

f. Oxidation to Carboxylic Acids (Sec. 9.13)

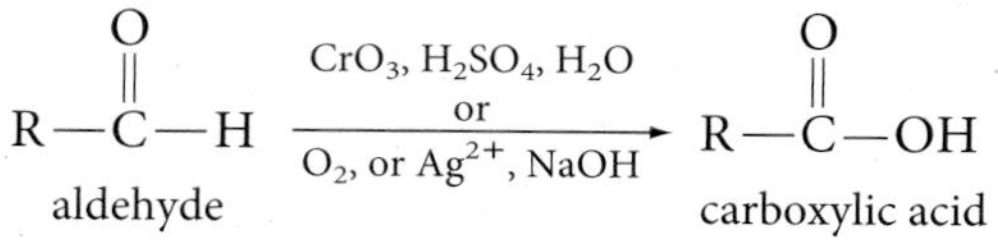

g. Aldol condensation (Sec. 9.17)

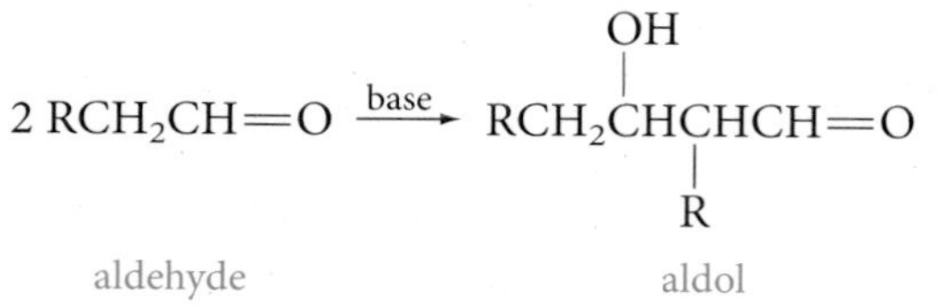

MECHANISM SUMMARY

Nucleophilic Addition (Sec. 9.6)

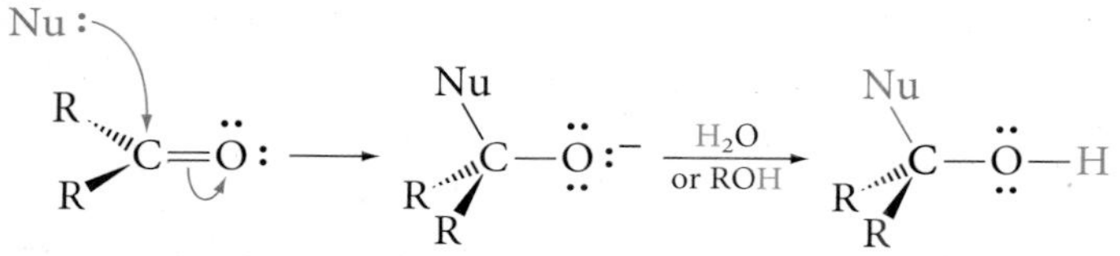

ADDITIONAL PROBLEMS

OWL Interactive versions of these problems are assignable in OWL.

Nomenclature, Structure, and Properties of Aldehydes and Ketones

9.30 Name each of the following compounds.

a. 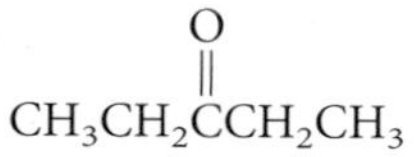

b. $CH_3(CH_2)_5CH{=}O$

c. $(C_6H_5)_2C{=}O$

d. 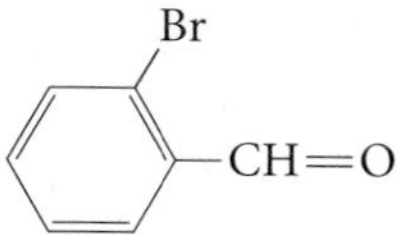

e. 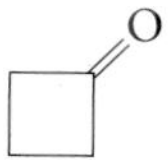

f. 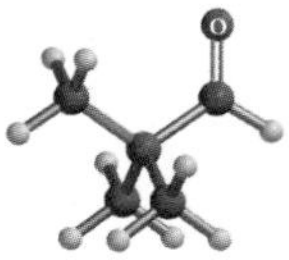

g. 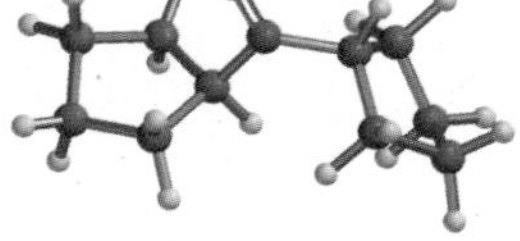

h. 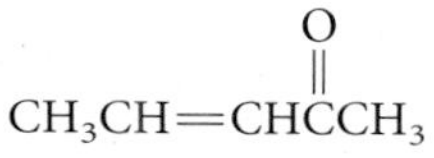

i. 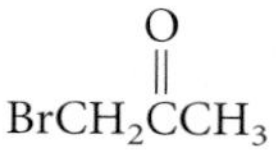

9.31 Write a structural formula for each of the following:

a. 3-heptanone
b. 3-isopropylheptanal
c. *p*-bromobenzaldehyde
d. (*S*)-2-methylcyclohexanone
e. 3-hexenal
f. benzyl *p*-methylphenyl ketone
g. 2,2-dichlorohexanal
h. 4-phenyl-2-butanone
i. cyclobutanone
j. *p*-tolualdehyde

9.32 Give an example of each of the following:

a. acetal
b. hemiacetal
c. cyanohydrin
d. imine
e. oxime
f. phenylhydrazone
g. enol
h. aldehyde with no α-hydrogen
i. enolate
j. hydrazone

9.33 The boiling points of the isomeric carbonyl compounds heptanal, 4-heptanone, and 2,4-dimethyl-3-pentanone are 155°C, 144°C, and 124°C, respectively. Suggest a possible explanation for the observed order. (*Hint:* Recall the effect of chain branching on the boiling points of isomeric alkanes, and how steric effects can influence the association of the molecules.)

Synthesis of Ketones and Aldehydes

9.34 Write an equation for the synthesis of 2-octanone by

a. oxidation of an alcohol
b. hydration of an alkyne

9.35 Write an equation for the synthesis of hexanal from an alcohol.

9.36 Write an equation, using the Friedel–Crafts reaction, for the preparation of

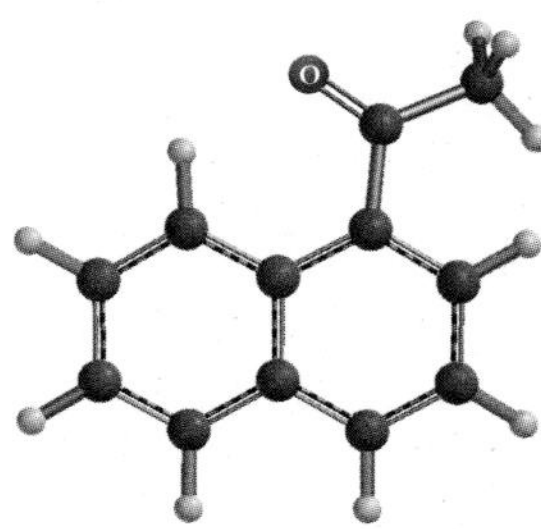

9.37 Using 1-hexyne as a starting reagent, suggest a synthesis of

a. 2-hexanone
b. 2-hexyl alcohol
c. a mixture of 1-hexene and 2-hexene
d. 1-hexene (and without 2-hexene present)

Reactions of Aldehydes and Ketones

9.38 Write an equation for the reaction, if any, of *p*-bromobenzaldehyde with each of the following reagents, and name the organic product.

Br–C_6H_4–CHO

p-bromobenzaldehyde

a. methylmagnesium bromide, then H_3O^+
b. methylamine (CH_3NH_2)
c. ethylene glycol, H^+
d. phenylhydrazine
e. HCN (catalytic NaOH)
f. Tollens' reagent

g. CrO_3, H^+
h. phenylmagnesium bromide, then H_3O^+
i. excess methanol, dry HCl
j. hydroxylamine
k. sodium borohydride

9.39 What simple chemical test can distinguish between the members of the following pairs of compounds? (*Hint:* Think of a reaction that one compound will undergo and the other will not.)

a. pentanal and 2-pentanone
b. benzyl alcohol and benzaldehyde
c. cyclohexanone and 2-cyclohexenone

9.40 Use the structures shown in Figure 9.1 to write equations for the following reactions of natural products:

a. benzaldehyde + Jones' reagent
b. cinnamaldehyde + Tollens' reagent
c. vanillin + hydrazine
d. jasmone + sodium borohydride
e. carvone + lithium aluminum hydride
f. camphor + (1) methylmagnesium bromide and (2) H_3O^+
g. vanillin + excess methanol, dry HCl

9.41 Complete each of the following equations:

a. butanal + excess methanol, $H^+ \longrightarrow$

b. $CH_3CH(OCH_3)_2 + H_2O, H^+ \longrightarrow$

c. 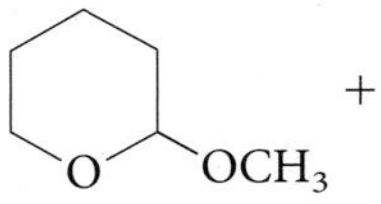 [tetrahydropyran ring with O and OCH_3] $+ H_2O, H^+ \longrightarrow$

d. 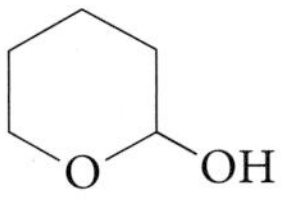 [tetrahydropyran ring with O and OH] + excess $CH_3CH_2OH, H^+ \longrightarrow$

9.42 Estrone (page 253) can be readily converted to ethynylestradiol:

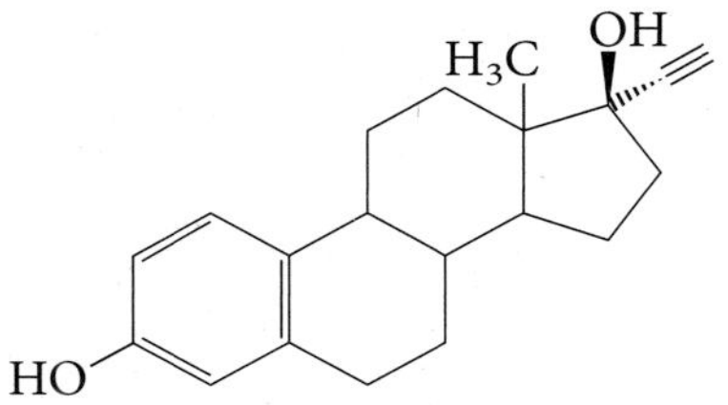

[structure: ethynylestradiol, labels H_3C, OH, HO]

Ethynylestradiol was the first synthetic estrogen to be used in formulations of oral contraceptive pills. Suggest a synthesis of ethynylestradiol from estrone.

9.43. Two equivalents of phenol and one equivalent of acetone are reacted with an acid catalyst to yield bisphenol-A (eq. 9.4), which is used in a number of commercial applications. Bisphenol-A has been also shown to mimic estrogen biologically. Provide a mechanism for the formation of bisphenol-A from phenol and acetone. (*Hint*: See Sec. 4.9.d.)

Reactions with Grignard Reagents and Other Nucleophiles

9.44 Write an equation for the reaction of each of the following with ethylmagnesium bromide, followed by hydrolysis with aqueous acid:

a. formaldehyde
b. hexanal
c. acetophenone
d. ethylene oxide
e. cyclohexanone

9.45 Using a Grignard reagent and the appropriate aldehyde or ketone, show how each of the following can be prepared:

a. methylcyclohexanol
b. 3-octanol
c. 2-methyl-2-pentanol
d. 1-cyclopentylcyclopentanol
e. 1-phenyl-1-propanol
f. 3-butene-2-ol

9.46 Complete the equation for the reaction of

a. cyclohexanone + $Na^{+-}C{\equiv}CH \longrightarrow \xrightarrow[H^+]{H_2O}$

b. cyclopentanone + HCN $\xrightarrow{KOH}$

c. 2-butanone + $NH_2OH \xrightarrow{H^+}$

d. benzaldehyde + benzylamine $\longrightarrow$

e. propanal + phenylhydrazine $\longrightarrow$

9.47 The equilibrium constant for hydrate formation from acetaldehyde (CH_3CHO) is 1 (Sec. 9.8), and at equilibrium, there is an equal amount of acetaldehyde and its hydrate ($CH_3CH(OH)_2$) in solution. The equilibrium constant for forming the hydrate of chloral (Cl_3CCHO), however, is much larger, and at equilibrium, only the chloral hydrate ($Cl_3CCH(OH)_2$) is observed. As noted above, chloral hydrate has been used as an anesthetic and as a sedative (also, see "A Word About . . . Ether and Anesthesia," page 243). Provide an explanation for the difference in hydrate formation for acetaldehyde and chloral. (*Hint:* See Sec. 9.6.)

Oxidations and Reductions

9.48 Give the structure of each product.

a. $CH_3C(=O)-C_6H_5 \xrightarrow[2.\ H_2O,\ H^+]{1.\ LiAlH_4}$

b. $C_6H_5-CH{=}CH-CH{=}O \xrightarrow[Ni,\ heat]{excess\ H_2}$

c. 2-methylcyclohex-2-en-1-one (ring with CH_3 and C=O) $\xrightarrow[2.\ H_2O,\ H^+]{1.\ NaBH_4}$

d. $C_6H_5-CH_2CH_2CH{=}O \xrightarrow[reagent]{Jones'}$

e. $CH_3CH{=}CHCHO \xrightarrow{Ag_2O}$

Enols, Enolates, and the Aldol Reaction

9.49 Write the structural formulas for all possible enols of

a. $CH_3CH_2-C(=O)-CH_3$

b. $C_6H_5-CH_2CHO$

c. $CH_3C(=O)CH_2C(=O)CH_3$

9.50 Complete the reaction shown below by drawing the structure of the product.

$$CH_3CH_2-C(=O)-C_6H_5 \xrightarrow[CH_3OD\ (excess)]{CH_3O^-Na^+}$$

9.51 How many hydrogens are replaced by deuterium when each of the following compounds is treated with NaOD in D_2O?

a. 3-methylcyclopentanone

b. 3-methylhexanal

9.52 As described in "A Word About . . . Water Treatment and the Chemistry of Enols/Enolates," carbonyl compounds can become halogenated by treatment with a base and a halogen source, such as Cl_2. As shown on page 279, base-catalyzed chlorination of acetone leads to 1,1,1-trichloroacetone. For acetone, provide a mechanism that rationalizes the formation of 1,1,1-trichloroacetone vs. chlorination on both sides of the carbonyl group.

9.53 Write out the steps in the mechanism for the aldol condensation of butanal (the first step in the synthesis of the mosquito repellent "6-12," eq. 9.57).

9.54 Explain why the enol form of phenol is more stable than the keto form of phenol (eq. 9.43).

9.55 Draw the keto and enol tautomers of 1,3-diphenyl-1,3-propanedione.

9.56 Lily aldehyde, used in perfumes, can be made starting with a mixed aldol condensation between two different aldehydes. Provide their structures.

$(CH_3)_3C{-}C_6H_4{-}CH_2CH(CH_3)CH{=}O$

lily aldehyde

Puzzles

9.57 Excess benzaldehyde reacts with acetone and base to give a yellow crystalline product, $C_{17}H_{14}O$. Deduce its structure and explain how it is formed.

9.58 For the phenol-imine (pale yellow) and keto-enamine (red) compounds discussed in the "A Word About . . . Tautomerism and Photochromism," page 274, suggest what is unique about the two structures that would give each compound a unique color. (*Hint:* See Chapter 4.)

9.59 Vitamin B_6 (an aldehyde) reacts with an enzyme (partial structure shown below) to form a *coenzyme* that catalyzes the conversion of α-amino acids (Chapter 17) to α-keto acids.

$^{-}O{-}P(=O)(=O){-}OCH_2$–(pyridinium ring bearing $HC{=}O$, OH, CH_3, and $\overset{+}{N}{-}H$) $+ \; H_2N{-}\text{enzyme} \longrightarrow \text{coenzyme}$

vitamin B_6

a. Draw the structure of the coenzyme.
b. α-Amino acids have the general structure shown below. Draw the structure of an α-keto acid.

$R{-}CH(NH_2){-}C(=O){-}OH$ α-amino acid

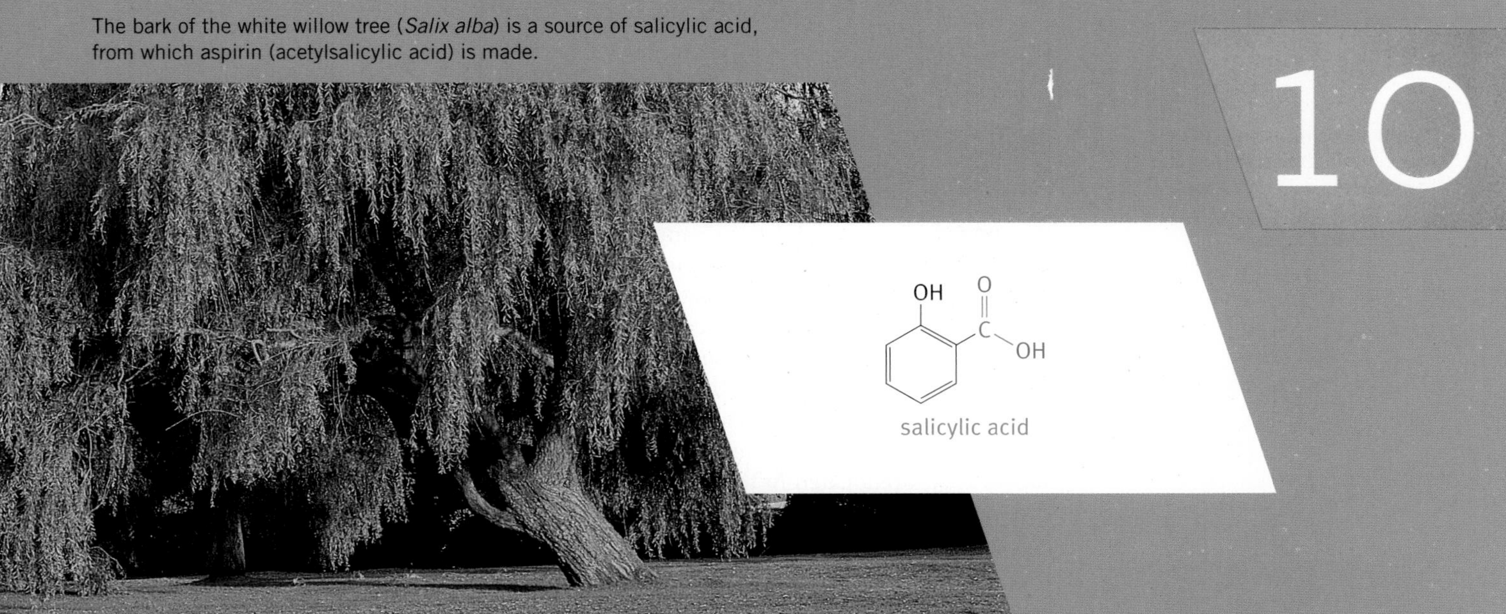

The bark of the white willow tree (*Salix alba*) is a source of salicylic acid, from which aspirin (acetylsalicylic acid) is made.

Carboxylic Acids and Their Derivatives

The taste of vinegar, the sting of an ant, the rancid smell of butter, and the relief derived from aspirin or ibuprofen—all of these are due to compounds that belong to the most important family of organic acids, the **carboxylic acids**. The resilience of polyester and nylon fabrics, the remarkable properties of Velcro, the softness of silk, the no-calorie sugar substitutes, the strength of bacterial cell walls, and the strength of our own cell membranes—all of these are due to properties of derivatives of carboxylic acids. The functional group common to all carboxylic acids is the **carboxyl group**. The name is a contraction of the parts: the *carb*onyl and hydr*oxyl* groups. The general formula for a carboxylic acid can be written in expanded or abbreviated forms.

$$-\overset{\overset{\large O}{\|}}{C}-OH \qquad R-\overset{\overset{\large O}{\|}}{C}-OH \quad \text{or} \quad RCOOH \quad \text{or} \quad RCO_2H \qquad R-\overset{\overset{\large O}{\|}}{C}-X$$

carboxyl group — three ways to represent a carboxylic acid — acid derivatives

In this chapter, we will describe the structures, properties, preparation, and reactions of carboxylic acids and will also discuss some common **carboxylic acid derivatives**, in which the hydroxyl group of an acid is replaced by other functional groups.

Online homework for this chapter can be assigned in OWL, an online homework assessment tool.

Carboxylic acids are organic acids that contain the **carboxyl group**. In **carboxylic acid derivatives**, the —OH group is replaced by other groups.

10.1 Nomenclature of Acids

Because of their abundance in nature, carboxylic acids were among the earliest classes of compounds studied by organic chemists. It is not surprising, then, that many of them have common names. These names usually come from some Latin or Greek word that indicates the original source of the acid. Table 10.1 lists the first ten unbranched carboxylic acids, with their common and IUPAC names. To obtain the IUPAC name of a carboxylic acid, we replace the final *e* in the name of the corresponding alkane with the suffix *-oic* and add the word *acid.* Substituted acids are named in two ways. In the IUPAC system, the chain is numbered beginning with the carboxyl carbon atom, and substituents are located in the usual way. If the common name of the acid is used, substituents are located with Greek letters, beginning with the α-carbon atom. IUPAC and common naming systems should not be mixed.

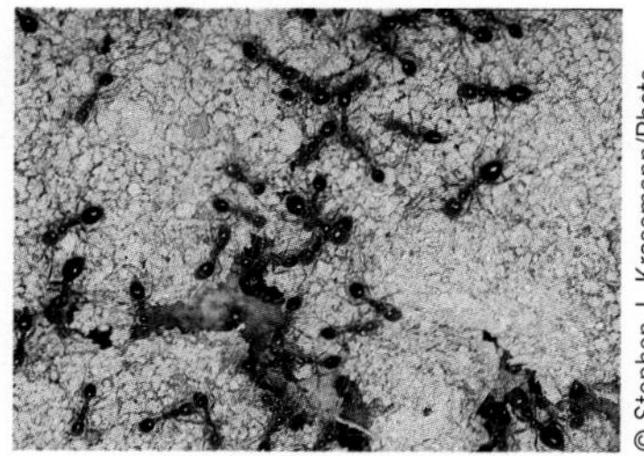

Stinging ants, source of formic acid, HCOOH.

$$\overset{\beta}{\overset{3}{CH_3}}-\overset{\alpha}{\overset{2}{\underset{Br}{\underset{|}{CH}}}}-\overset{1}{CO_2H}$$
2-bromopropanoic acid
(α-bromopropionic acid)

$$\overset{3}{CH_2}=\overset{2}{C}H\overset{1}{C}O_2H$$
propenoic acid
(acrylic acid)

$$\overset{\gamma}{\overset{4}{CH_3}}\overset{\beta}{\overset{3}{\underset{OH}{\underset{|}{CH}}}}\overset{\alpha}{\overset{2}{CH_2}}\overset{1}{CO_2H}$$
3-hydroxybutanoic acid
(β-hydroxybutyric acid)

The carboxyl group has priority over alcohol, aldehyde, or ketone functionality in naming. In the latter cases, the prefix *oxo-* is used to locate the carbonyl group of the aldehyde or ketone, as in these examples:

$$\underset{3}{H\overset{O}{\overset{\|}{C}}}-\overset{2}{CH_2}\overset{1}{CO_2H}$$
3-oxopropanoic acid

$$\overset{5}{CH_3}\underset{4}{\overset{O}{\overset{\|}{C}}}\overset{3}{CH_2}\overset{2}{\underset{Br}{\underset{|}{CH}}}\overset{1}{CO_2H}$$
2-bromo-4-oxopentanoic acid

Table 10.1 ■ Aliphatic Carboxylic Acids

Carbon atoms	Formula	Source	Common name	IUPAC name
1	HCOOH	ants (Latin, *formica*)	formic acid	methanoic acid
2	CH_3COOH	vinegar (Latin, *acetum*)	acetic acid	ethanoic acid
3	CH_3CH_2COOH	milk (Greek, *protos pion*, first fat)	propionic acid	propanoic acid
4	$CH_3(CH_2)_2COOH$	butter (Latin, *butyrum*)	butyric acid	butanoic acid
5	$CH_3(CH_2)_3COOH$	valerian root (Latin, *valere*, to be strong)	valeric acid	pentanoic acid
6	$CH_3(CH_2)_4COOH$	goats (Latin, *caper*)	caproic acid	hexanoic acid
7	$CH_3(CH_2)_5COOH$	vine blossom (Greek, *oenanthe*)	enanthic acid	heptanoic acid
8	$CH_3(CH_2)_6COOH$	goats (Latin, *caper*)	caprylic acid	octanoic acid
9	$CH_3(CH_2)_7COOH$	pelargonium (an herb with stork-shaped seed capsules; Greek, *pelargos*, stork)	pelargonic acid	nonanoic acid
10	$CH_3(CH_2)_8COOH$	goats (Latin, *caper*)	capric acid	decanoic acid

PROBLEM 10.1 Write the structure for

a. 3-hydroxyhexanoic acid
b. 2-iodo-2-methyloctanoic acid
c. 2-butynoic acid
d. 5-ethyl-6-oxoheptanoic acid

PROBLEM 10.2 Give an IUPAC name for

a. C_6H_5—CH_2CO_2H (benzene ring)
b. $Br_2CHCH_2CO_2H$
c. $CH_3CH{=}CHCO_2H$
d. $(CH_3)_3CCH_2CH_2CO_2H$

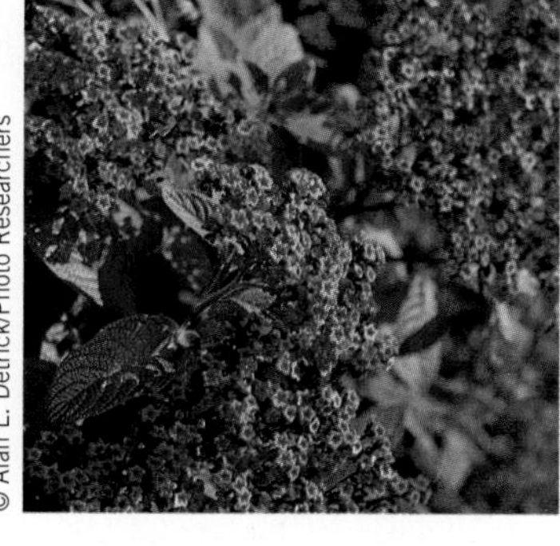

© Alan L. Detrick/Photo Researchers

The root of Garden Heliotrope is a source of valeric acid, $CH_3(CH_2)_3COOH$.

When the carboxyl group is attached to a ring, the ending *-carboxylic acid* is added to the name of the parent cycloalkane.

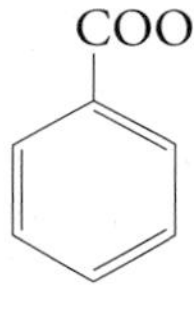

cyclopentanecarboxylic acid

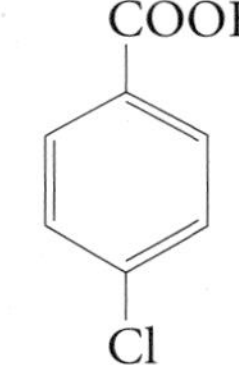

trans-3-chlorocyclobutanecarboxylic acid

Aromatic acids are named by attaching the suffix *-oic acid* or *-ic acid* to an appropriate prefix derived from the aromatic hydrocarbon.

COOH

benzoic acid
(benzenecarboxylic acid)

COOH / Cl

p-chlorobenzoic acid
(4-chlorobenzenecarboxylic acid)

COOH / CH_3

o-toluic acid
(2-methylbenzenecarboxylic acid)

COOH

2-naphthoic acid
(2-naphthalenecarboxylic acid)

Mircea Bezergheanu/Shutterstock

Goats, source of caproic, caprylic, and capric acids: $CH_3(CH_2)_nCOOH$, n = 4, 6, 8.

PROBLEM 10.3 Write the structure for

a. *cis*-3-isopropylcyclohexanecarboxylic acid
b. *o*-nitrobenzoic acid

PROBLEM 10.4 Give the correct name for

a.

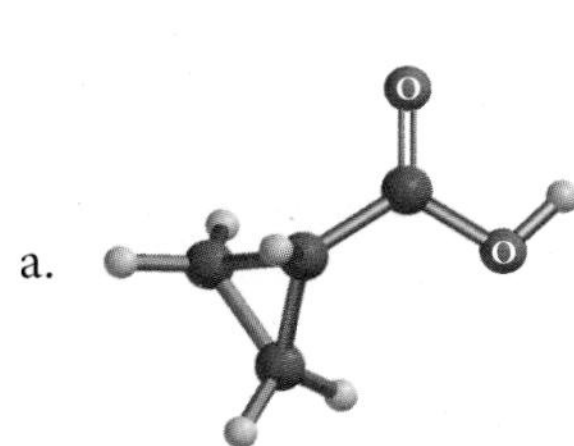

b.

Table 10.2 ■ Aliphatic Dicarboxylic Acids

Formula	Common name	Source	IUPAC name
HOOC—COOH	oxalic acid	plants of the *oxalic* family (for example, sorrel)	ethanedioic acid
HOOC—CH_2—COOH	malonic acid	apple (Gk. *malon*)	propanedioic acid
HOOC—$(CH_2)_2$—COOH	succinic acid	amber (L. *succinum*)	butanedioic acid
HOOC—$(CH_2)_3$—COOH	glutaric acid	gluten	pentanedioic acid
HOOC—$(CH_2)_4$—COOH	adipic acid	fat (L. *adeps*)	hexanedioic acid
HOOC—$(CH_2)_5$—COOH	pimelic acid	fat (Gk. *pimele*)	heptanedioic acid

Juliette Wade/Garden Picture Library/Photolibrary

Rhubarb, a source of oxalic acid, HOOCCOOH.

Aliphatic dicarboxylic acids are given the suffix *-dioic acid* in the IUPAC system. For example,

$$\overset{1}{HO_2C}-\overset{2}{CH_2}\overset{3}{CH_2}-\overset{4}{CO_2H} \qquad HO_2C-C\equiv C-CO_2H$$

butanedioic acid   butynedioic acid

Many dicarboxylic acids occur in nature and go by their common names, which are based on their source. Table 10.2 lists some common aliphatic diacids.* The most important commercial compound in this group is adipic acid, used to manufacture nylon.

The two butenedioic acids played a historic role in the discovery of *cis–trans* isomerism and are usually known by their common names maleic** and fumaric*** acid.

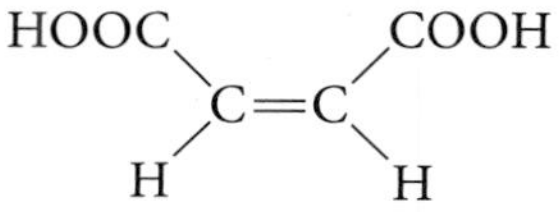

maleic acid
(*cis*-2-butenedioic acid)

and

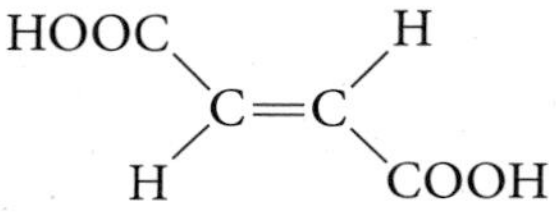

fumaric acid
(*trans*-2-butenedioic acid)

The three benzenedicarboxylic acids are generally known by their common names.

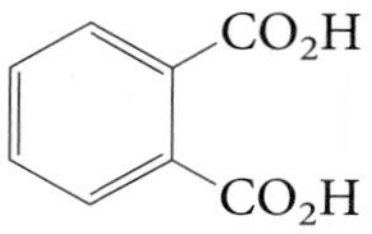

phthalic acid

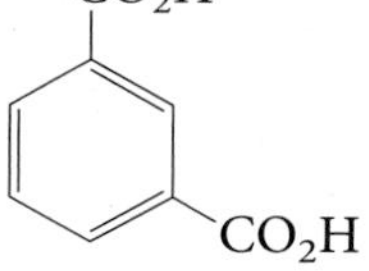

isophthalic acid

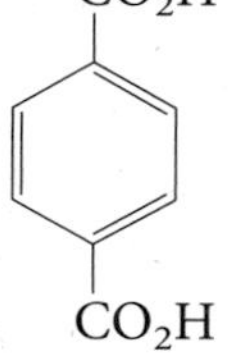

terephthalic acid

*The first letter of each word in the sentence "Oh my, such good apple pie" gives, in order, the first letters of the common names of these acids and can help you to remember them.

**From the Latin *malum* (apple). Malic acid (2-hydroxybutanedioic acid), found in apples, can be dehydrated on heating to give maleic acid.

***Found in fumitory, an herb of the genus *Fumaria*.

All three are important commercial chemicals, used to make polymers and other useful materials.

Finally, it is useful to have a name for an **acyl group**. Particular acyl groups are named from the corresponding acid by changing the *-ic* ending to *-yl.*

$R-\overset{O}{\overset{\|}{C}}-$	$H-\overset{O}{\overset{\|}{C}}-$	$CH_3-\overset{O}{\overset{\|}{C}}-$	$CH_3CH_2\overset{O}{\overset{\|}{C}}-$	$C_6H_5-\overset{O}{\overset{\|}{C}}-$
an acyl group	formyl (methanoyl)	acetyl (ethanoyl)	propanoyl	benzoyl

PROBLEM 10.5 Write the formula for

a. 4-formylbenzoic acid
b. benzoyl bromide
c. octanoyl bromide
d. acetylcyclopentane

10.2 Physical Properties of Acids

The first members of the carboxylic acid series are colorless liquids with sharp or unpleasant odors. Acetic acid, which constitutes about 4% to 5% of vinegar, provides the characteristic odor and flavor. Butyric acid gives rancid butter its disagreeable odor, and the goat acids (caproic, caprylic, and capric in Table 10.1) smell like goats. 3-Methyl-2-hexenoic acid, produced by bacteria, is responsible for the offensive odor of human armpits. Table 10.3 lists some physical properties of selected carboxylic acids.

Carboxylic acids are polar. Like alcohols, they form hydrogen bonds with themselves or with other molecules (Sec. 7.4). Therefore, they have high boiling points for their molecular weights—higher even than those of comparable alcohols. For example, acetic acid and propyl alcohol, which have the same formula weights (60 g/mol), boil at 118°C and 97°C, respectively. Carboxylic acids form dimers, with

Table 10.3 ■ Physical Properties of Some Carboxylic Acids

Name	bp, °C	mp, °C	Solubility, g/100 g H_2O at 25°C
formic acid	101	8	miscible (∞)
acetic acid	118	17	miscible (∞)
propanoic acid	141	−22	miscible (∞)
butanoic acid	164	−8	miscible (∞)
hexanoic acid	205	−1.5	1.0
octanoic acid	240	17	0.06
decanoic acid	270	31	0.01
benzoic acid	249	122	0.4 (but 6.8 at 95°C)

the individual units neatly held together by *two* hydrogen bonds between the electron-rich oxygens and the electron-poor hydrogens (see Sec. 7.4).

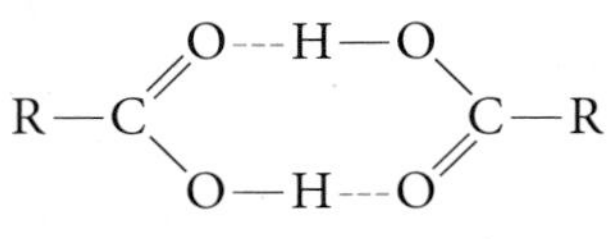

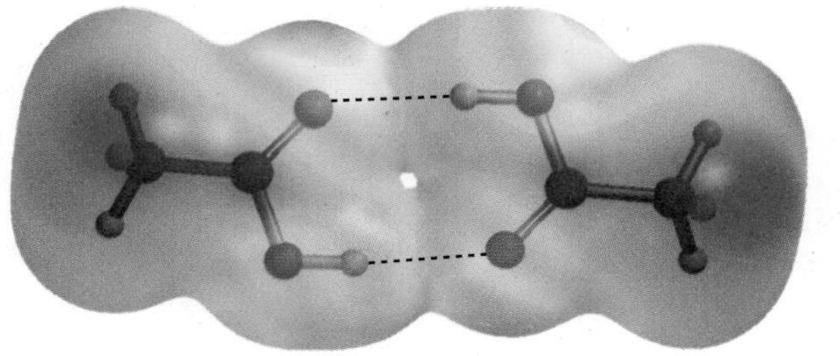

Hydrogen bonding also explains the water solubility of the lower molecular weight carboxylic acids.

10.3 Acidity and Acidity Constants

Carboxylic acids (RCO_2H) dissociate in water, yielding a carboxylate anion (RCO_2^-) and a hydronium ion.

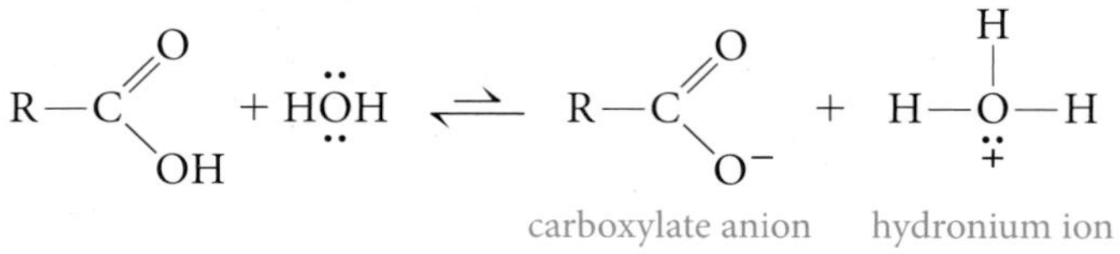

$$R{-}C(=O)OH + H\ddot{O}H \rightleftharpoons R{-}C(=O)O^- + H_3O^+ \quad (10.1)$$

carboxylate anion hydronium ion

Their acidity constants K_a in water are given by the expression

$$K_a = \frac{[RCO_2^-][H_3O^+]}{[RCO_2H]} \quad (10.2)$$

(Before proceeding further, it would be a good idea for you to review Secs. 7.5 and 7.6.)

Table 10.4 lists the acidity constants for some carboxylic and other acids. In comparing data in this table, remember that the larger the value of K_a or the smaller the value of pK_a, the stronger the acid.

EXAMPLE 10.1

Which is the stronger acid, formic or acetic, and by how much?

Solution Formic acid is stronger; it has the larger K_a. The ratio of acidities is

$$\frac{2.1 \times 10^{-4}}{1.8 \times 10^{-5}} = 1.17 \times 10^1 = 11.7$$

This means that formic acid is 11.7 times stronger than acetic acid.

PROBLEM 10.6 Using the data given in Table 10.4, determine which is the stronger acid, acetic or dichloroacetic, and by how much.

Before we can explain the acidity differences in Table 10.4, we must examine the structural features that make carboxylic acids acidic.

Table 10.4 The Ionization Constants of Some Acids

Name	Formula	K_a	pK_a
formic acid	HCOOH	2.1×10^{-4}	3.68
acetic acid	CH_3COOH	1.8×10^{-5}	4.74
propanoic acid	CH_3CH_2COOH	1.4×10^{-5}	4.85
butanoic acid	$CH_3CH_2CH_2COOH$	1.6×10^{-5}	4.80
chloroacetic acid	$ClCH_2COOH$	1.5×10^{-3}	2.82
dichloroacetic acid	$Cl_2CHCOOH$	5.0×10^{-2}	1.30
trichloroacetic acid	Cl_3CCOOH	2.0×10^{-1}	0.70
2-chlorobutanoic acid	$CH_3CH_2CHClCOOH$	1.4×10^{-3}	2.85
3-chlorobutanoic acid	$CH_3CHClCH_2COOH$	8.9×10^{-5}	4.05
benzoic acid	C_6H_5COOH	6.6×10^{-5}	4.18
o-chlorobenzoic acid	*o*-Cl—C_6H_4COOH	12.5×10^{-4}	2.90
m-chlorobenzoic acid	*m*-Cl—C_6H_4COOH	1.6×10^{-4}	3.80
p-chlorobenzoic acid	*p*-Cl—C_6H_4COOH	1.0×10^{-4}	4.00
p-nitrobenzoic acid	*p*-NO_2—C_6H_4COOH	4.0×10^{-4}	3.40
phenol	C_6H_5OH	1.0×10^{-10}	10.00
ethanol	CH_3CH_2OH	1.0×10^{-16}	16.00
water	HOH	1.8×10^{-16}	15.74

10.4 What Makes Carboxylic Acids Acidic?

You might wonder why carboxylic acids are so much more acidic than alcohols, since each class ionizes by losing H^+ from a hydroxyl group. There are two reasons, which can best be illustrated with a specific example.

From Table 10.4, we see that acetic acid is approximately 10^{11}, or 100,000 million, times stronger an acid than ethanol.

$$CH_3CH_2\ddot{\underset{..}{O}}H \rightleftharpoons CH_3CH_2\ddot{\underset{..}{O}}\!:^- + H^+ \qquad K_a = 10^{-16} \tag{10.3}$$

ethoxide ion

$$CH_3\overset{\overset{\delta-}{:O:}\,\|}{\underset{\delta+}{C}}—\ddot{\underset{..}{O}}H \rightleftharpoons CH_3\overset{:O:\,\|}{C}—\ddot{\underset{..}{O}}\!:^- + H^+ \qquad K_a = 10^{-5} \tag{10.4}$$

acetate ion

The only difference between the structures of acetic acid and ethanol is the replacement of a CH_2 group (in ethanol) by a carbonyl group (in acetic acid). But we saw (Sec. 9.5) that a carbonyl carbon atom carries a substantial *positive* charge ($\delta+$). This charge makes it much easier to place a *negative* charge on the adjacent oxygen atom, which is exactly what happens when we ionize a proton from the hydroxyl group.

In ethoxide ion, *the negative charge is localized on a single oxygen atom.* In acetate ion, on the other hand, *the negative charge can be delocalized through resonance.*

$$\left[CH_3{-}C({=}\ddot{O}{:}){-}\ddot{O}{:}^- \longleftrightarrow CH_3{-}C^+({-}\ddot{O}{:}^-){-}\ddot{O}{:}^- \longleftrightarrow CH_3{-}C({-}\ddot{O}{:}^-){=}\ddot{O}{:} \right] \quad \text{or} \quad CH_3{-}C\left\{ \begin{matrix} O \\ O \end{matrix} \right\}^-$$

resonance in a carboxylate ion (acetate ion)

The negative charge is spread *equally* over the two oxygens so that each oxygen in the carboxylate ion carries only half the negative charge (Figure 10.1). The acetate ion is stabilized by resonance compared to the ethoxide ion, and this stabilization helps to drive the equilibrium more to the right in eq. 10.4 than in eq. 10.3. Consequently, more H^+ is formed from acetic acid than from ethanol.

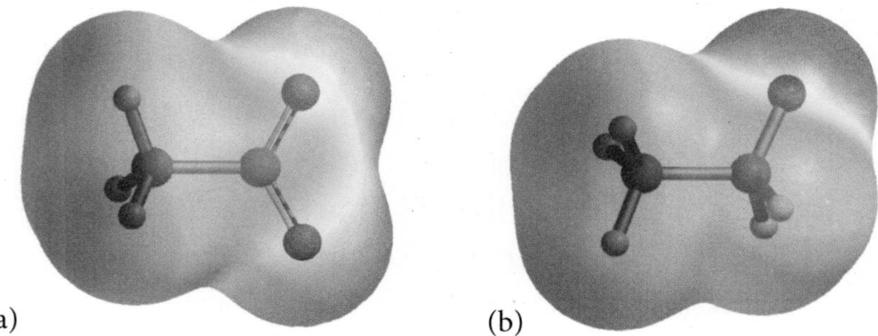

Figure 10.1 Electrostatic potential map of (a) acetate ion and (b) ethoxide ion. Note the negative (red) character for both oxygens in (a).

For both these reasons, the positive charge on the carbonyl carbon and delocalization of the carboxylate ion, carboxylic acids are much more acidic than alcohols.

EXAMPLE 10.2

Phenoxide ions are also stabilized by resonance (Sec. 7.6). Why are phenols weaker acids than carboxylic acids?

Solution First, the carbon atom to which the hydroxyl group is attached in a phenol is not as positive as a carbonyl carbon. Second, charge delocalization is not as great in phenoxide ions as in carboxylate ions because the contributors to the resonance hybrid are not equivalent. Some of them put the negative charge on carbon instead of on oxygen and disrupt aromaticity.

PROBLEM 10.7 Write two resonance structures for the benzoate ion ($C_6H_5CO_2^-$) that show how the negative charge is delocalized over the two oxygens. Can the negative charge in the benzoate ion be delocalized into the aromatic ring?

Physical data support the importance of resonance in carboxylate ions. In formic acid molecules, the two carbon–oxygen bonds have different lengths. But in sodium formate, both carbon–oxygen bonds of the formate ion are identical, and their length is between those of normal double and single carbon–oxygen bonds.

$$H{-}C({=}O)({-}O{-}H) \qquad H{-}C\left\{ \begin{matrix} O \\ O \end{matrix} \right\}^- Na^+$$

formic acid: C=O 1.23 Å; C–O 1.36 Å

sodium formate: C–O 1.27 Å; C–O 1.27 Å

10.5 Effect of Structure on Acidity; the Inductive Effect Revisited

The data in Table 10.4 show that even among carboxylic acids (where the ionizing functional group is kept constant), acidities can vary depending on what other groups are attached to the molecule. Compare, for example, the K_a of acetic acid with those of mono-, di-, and trichloroacetic acids, and note that the acidity varies by a factor of 10,000.

The most important factor operating here is the inductive effect of the groups close to the carboxyl group. This effect relays charge through bonds, by displacing bonding electrons toward electronegative atoms, or away from electropositive atoms. Recall that *electron-withdrawing groups enhance acidity, and electron-releasing groups reduce acidity* (see Sec. 7.6).

Let us examine the carboxylate ions formed when acetic acid and its chloro derivatives ionize:

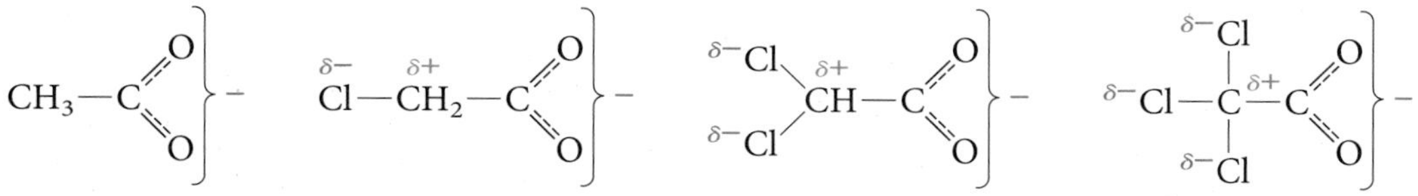

$$CH_3{-}CO_2^-$$ acetate

$$\overset{\delta-}{Cl}{-}\overset{\delta+}{CH_2}{-}CO_2^-$$ chloroacetate

$$(\overset{\delta-}{Cl})_2\overset{\delta+}{CH}{-}CO_2^-$$ dichloroacetate

$$(\overset{\delta-}{Cl})_3\overset{\delta+}{C}{-}CO_2^-$$ trichloroacetate

Because chlorine is more electronegative than carbon, the C—Cl bond is polarized with the chlorine partially negative and the carbon partially positive. Thus, electrons are pulled away from the carboxylate end of the ion toward the chlorine. The effect tends to spread the negative charge over more atoms than in acetate ion itself and thus stabilizes the ion. The more chlorines, the greater the effect and the greater the strength of the acid.

EXAMPLE 10.3

Explain the acidity order in Table 10.4 for butanoic acid and its 2- and 3-chloro derivatives.

Solution The 2-chloro substituent increases the acidity of butanoic acid substantially, due to its inductive effect. In fact, the effect is about the same as for chloroacetic and acetic acids. The 3-chloro substituent exerts a similar *but much smaller* effect, because the C—Cl bond is now farther away from the carboxylate group. *Inductive effects fall off rapidly with distance.*

PROBLEM 10.8 Account for the relative acidities of benzoic acid and its *ortho, meta,* and *para* chloro derivatives (Table 10.4).

We saw in Example 10.1 that formic acid is a substantially stronger acid than acetic acid. This suggests that the methyl group is more electron-releasing (hence anion-destabilizing and acidity-reducing) than hydrogen. This observation is consistent with what we have already learned about carbocation stabilities—that alkyl groups are more effective than hydrogen atoms at releasing electrons to, and therefore stabilizing, a positive carbon atom (see Sec. 3.10). A similar effect was seen for the relative acidity of ethanol and *t*-butanol in water (see Sec. 7.6).

10.6 Conversion of Acids to Salts

Carboxylic acids, when treated with a strong base, form carboxylate salts. For example,

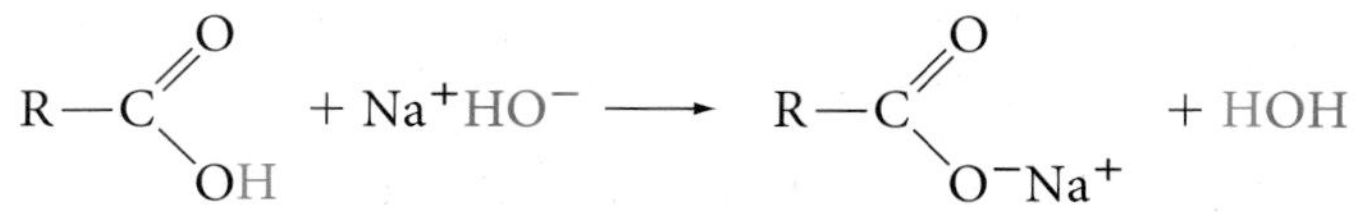

$$\underset{\substack{\text{carboxylic acid}\\ pK_a\ 3\text{–}5}}{R{-}C({=}O){-}OH} + \underset{\substack{\text{strong}\\ \text{base}}}{Na^+HO^-} \longrightarrow \underset{\substack{\text{a sodium carboxylate}\\ \text{(weak base)}}}{R{-}C({=}O){-}O^-Na^+} + \underset{\substack{\text{water}\\ pK_a\ 16}}{HOH} \qquad (10.5)$$

The salt can be isolated by evaporating the water. As we will see in Chapter 15, carboxylate salts of certain acids are useful as soaps and detergents.

Carboxylate salts are named as shown in the following examples:

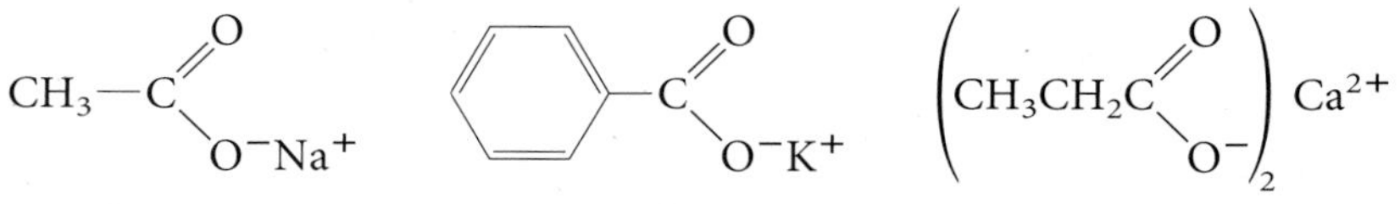

$$\underset{\substack{\text{sodium acetate}\\ \text{(sodium ethanoate)}}}{CH_3{-}C({=}O){-}O^-Na^+} \qquad \underset{\text{potassium benzoate}}{C_6H_5{-}C({=}O){-}O^-K^+} \qquad \underset{\text{calcium propanoate}}{\left(CH_3CH_2C({=}O){-}O^-\right)_2\ Ca^{2+}}$$

The cation is named first, followed by the name of the carboxylate ion, which is obtained by changing the *-ic* ending of the acid to *-ate*.

EXAMPLE 10.4

Name the following carboxylate salt:

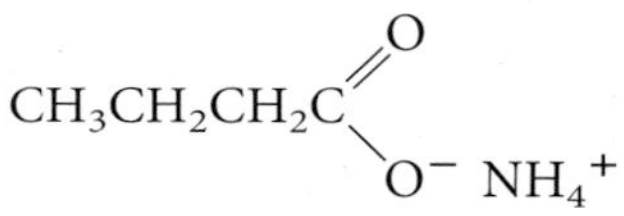

$$CH_3CH_2CH_2C({=}O){-}O^-\ NH_4^+$$

Solution The salt is ammonium butanoate (IUPAC) or ammonium butyrate (common).

PROBLEM 10.9 Write an equation, analogous to eq. 10.5, for the preparation of potassium 3-bromooctanoate from the corresponding acid.

10.7 Preparation of Acids

Organic acids can be prepared in many ways, four of which are described here: (1) oxidation of primary alcohols or aldehydes, (2) oxidation of alkyl side chains on aromatic rings, (3) reaction of Grignard reagents with carbon dioxide, and (4) hydrolysis of alkyl cyanides (nitriles).

10.7.a Oxidation of Primary Alcohols and Aldehydes

The oxidation of primary alcohols (Sec. 7.12) and aldehydes (Sec. 9.13) to carboxylic acids has already been mentioned. It is easy to see that these are oxidation reactions because going from an alcohol to an aldehyde to an acid requires replacement of C—H bonds by C—O bonds.

$$\text{R—CH}_2\text{—OH} \longrightarrow \text{RCH=O} \longrightarrow \text{RCOOH} \quad (10.6)$$

alcohol (one C—O bond); aldehyde (two C—O bonds); acid (three C—O bonds)

The most commonly used oxidizing agents for these purposes are potassium permanganate ($KMnO_4$), chromic acid anhydride (CrO_3), nitric acid (HNO_3), and, with aldehydes only, silver oxide (Ag_2O). For specific examples, see eqs. 7.37, 9.37, 9.38, and 9.41.

10.7.b Oxidation of Aromatic Side Chains

Aromatic acids can be prepared by oxidizing an alkyl side chain on an aromatic ring.

$$\text{C}_6\text{H}_5\text{—CH}_3 \xrightarrow[\text{heat}]{KMnO_4} \text{C}_6\text{H}_5\text{—COOH} \quad (10.7)$$

toluene → benzoic acid

This reaction illustrates the striking stability of aromatic rings; it is the alkane-like methyl group, not the aromatic ring, that is oxidized. The reaction involves attack of the oxidant at a C—H bond adjacent to the benzene ring. Longer side chains are also oxidized to a carboxyl group.

$$\text{C}_6\text{H}_5\text{—CH}_2\text{CH}_2\text{CH}_3 \xrightarrow[\text{heat}]{KMnO_4} \text{C}_6\text{H}_5\text{—CO}_2\text{H} \quad (10.8)$$

If no C—H bond is in the benzylic position, however, the aromatic ring is oxidized, although only under severe reaction conditions.

$$(CH_3)_3C\text{—C}_6\text{H}_5 \xrightarrow[\text{heat}]{KMnO_4} (CH_3)_3CCO_2H \quad (10.9)$$

With oxidants other than potassium permanganate, this reaction is commercially important. For example, terephthalic acid (Sec. 10.1), one of the two raw materials needed to manufacture Dacron, is produced in this way, using a cobalt catalyst and air for the oxidation.

$$CH_3\text{—C}_6\text{H}_4\text{—}CH_3 \xrightarrow[CH_3CO_2H]{O_2,\ Co(III)} \text{HOOC—C}_6\text{H}_4\text{—COOH} \quad (10.10)$$

p-xylene → terephthalic acid

Phthalic acid, used for making plasticizers, resins, and dyestuffs, is manufactured by similar oxidations, starting with *o*-xylene.

$$o\text{-C}_6\text{H}_4(CH_3)_2 \xrightarrow[CH_3CO_2H]{O_2,\ Co(III)} o\text{-C}_6\text{H}_4(COOH)_2 \quad (10.11)$$

o-xylene → phthalic acid

10.7.c Reaction of Grignard Reagents with Carbon Dioxide

As we saw previously, Grignard reagents add to the carbonyl groups of aldehydes or ketones to give alcohols. In a similar way, they add irreversibly to the carbonyl group of carbon dioxide to give acids, after protonation of the intermediate carboxylate salt with a mineral acid like aqueous HCl.

$$\overset{\delta-}{O}{=}\overset{\delta+}{C}{=}\overset{\delta-}{O} + \overset{\delta-}{R}{-}\overset{\delta+}{MgX} \longrightarrow R{-}\overset{O}{\overset{\|}{C}}{-}O^-\overset{+}{M}gX \xrightarrow{H_3O^+} R{-}\overset{O}{\overset{\|}{C}}{-}OH + H_2O \quad (10.12)$$

This reaction gives good yields and is an excellent laboratory method for preparing both aliphatic and aromatic acids. Note that the acid obtained has one more carbon atom than the alkyl or aryl halide from which the Grignard reagent is prepared, so the reaction provides a way to increase the length of a carbon chain.

EXAMPLE 10.5

Show how $(CH_3)_3CBr$ can be converted to $(CH_3)_3CCO_2H$.

Solution $(CH_3)_3CBr \xrightarrow[\text{ether}]{\text{Mg}} (CH_3)_3CMgBr \xrightarrow[2.\ H_3O^+]{1.\ CO_2} (CH_3)_3CCO_2H$

PROBLEM 10.10 Show how 4-methylcyclohexyl chloride can be converted to 4-methylcyclohexanecarboxylic acid.

PROBLEM 10.11 Devise a synthesis of butanoic acid ($CH_3CH_2CH_2CO_2H$) from 1-propanol ($CH_3CH_2CH_2OH$).

10.7.d Hydrolysis of Cyanides (Nitriles)

The carbon–nitrogen triple bond of organic cyanides can be hydrolyzed to a carboxyl group. The reaction requires either acid or base. In acid, the nitrogen atom of the cyanide is converted to an ammonium ion.

$$\underset{\text{a cyanide, or nitrile}}{R{-}C{\equiv}N} + 2\,H_2O \xrightarrow{HCl} \underset{\text{an acid}}{R{-}\overset{O}{\overset{\|}{C}}{-}OH} + \underset{\text{ammonium ion}}{\overset{+}{N}H_4} + Cl^- \quad (10.13)$$

In base, the nitrogen is converted to ammonia and the organic product is the carboxylate salt, which must be neutralized in a separate step to give the acid.

$$R{-}C{\equiv}N + 2\,H_2O \xrightarrow{NaOH} \underset{\text{a carboxylate salt}}{R{-}\overset{O}{\overset{\|}{C}}{-}O^-Na^+} + \underset{\text{ammonia}}{NH_3} \quad (10.14)$$

$$\xrightarrow{H^+} R{-}\overset{O}{\overset{\|}{C}}{-}OH$$

A WORD ABOUT... Green Chemistry and Ibuprofen: A Case Study

From our increasing need to respect nature and protect our environment has come a new field of chemistry—"green chemistry"—the design and development of chemistry that is environmentally friendly, chemistry that avoids pollution. This presents many challenges to organic synthesis. One notion of an ideal synthesis is one that provides a useful compound in one step with formation of no disposable by-products by a process that consumes little energy. Such a synthesis would certainly be environmentally friendly! This goal is seldom met, but general principles can be applied to try to approach this ideal.

First let us consider some reactions we have learned. Addition reactions (catalytic hydrogenation and Diels–Alder reactions, for example) do not create any by-products. The same can be said for isomerization reactions. Such reactions are said to be "atom economical"—all of the atoms in the reactants appear in the product.* On the other hand, elimination reactions and substitution reactions necessarily produce by-products. This does not mean that they are bad, but if a synthesis can be devised that focuses on addition and isomerization reactions, less attention will have to be devoted to disposing of, or developing uses for, by-products.

Some other general strategies for the development of green chemistry are to use catalysts to accomplish reactions (rather than stoichiometric reagents), to minimize the use of heavy metals as stoichiometric oxidants (for example, chromium), to focus on the use of molecular oxygen and hydrogen peroxide as oxidants, and to minimize the use of solvents in reactions.

Let us examine a synthesis that does a good job of meeting the goals of green chemistry. Ibuprofen is a very important anti-inflammatory drug. It is the active ingredient of many over-the-counter drugs used to relieve pain from headaches and arthritis. Approximately 25 million pounds of this simple carboxylic acid were produced by synthesis in 2000! A large number of ibuprofen syntheses have been developed, and several of these have been commercialized. The following synthesis begins with the reaction of isobutylbenzene with acetic anhydride using HF as the solvent. This is a variation of the Friedel–Crafts acylation in which the anhydride serves as the source of an acylium ion (see Sec. 4.9.d). Through clever engineering processes, the reaction solvent (HF) serves as both the acid catalyst and solvent (recyclable) for the reaction, and water is the only major reaction by-product. The second step is an addition reaction, catalytic hydrogenation of a ketone to an alcohol (see eq. 9.57). The final step is a reaction we have not discussed that involves palladium-catalyzed "insertion" of carbon monoxide into a benzylic C—O bond to give the carboxylic acid (ibuprofen). This reaction is clearly atom economical. Finally, the chemical yields of all of these reactions are very high, and very little chemical waste is produced.** Although this synthesis is an excellent example of green chemistry in action, there is room for improvement. For example, this produces a racemic mixture of ibuprofen, whereas only the (*S*)-enantiomer is biologically active (see "A Word About . . . Enantiomers and Biological Activity" in Chapter 5).

It is clear that green chemistry will play an important role in the twenty-first century. This has been recognized by the Presidential Green Chemistry Challenge, initiated by President Clinton in 1995 to reward the development of environmentally benign chemistry. Although biological processes seldom meet the notion of an ideal synthesis, to strive for this ideal can only lead to new and better chemistry.

isobutylbenzene $\xrightarrow[\text{HF}]{(CH_3CO)_2O}$ (ketone) $\xrightarrow{H_2,\ \text{Pd (cat)}}$ (alcohol, OH) $\xrightarrow[Ph_3P]{PdCl_2\ \text{(cat), CO}}$ ibuprofen (CO_2H)

See Problem 10.63.

*For more on "atom economy," see B. M. Trost, *Angew. Chem. Int. Ed. Engl.* **1995**, *34*, 259. For more on green chemistry, see W. Leiner, *Science* **1999**, *284*, 1780, and visit the following Web site: http://www.epa.gov/greenchemistry.

**For an overview of other syntheses of ibuprofen, consult B. G. Reuben and H. A. Wittcoff, *Pharmaceutical Chemicals in Perspective,* John Wiley and Sons, New York, 1989.

The mechanism of nitrile hydrolysis involves acid or base promoted addition of water across the triple bond. This gives an intermediate imidate that tautomerizes to an amide. The amide is then hydrolyzed to the carboxylic acid. The addition of water to the nitrile resembles the hydration of an alkyne (eq. 3.52). The oxygen of water behaves as a nucleophile and bonds to the electrophilic carbon of the nitrile. Amide hydrolysis will be discussed in Section 10.20.

$$\underset{\text{nitrile}}{\overset{\delta+\ \ \delta-}{R{-}C{\equiv}N}} \xrightarrow[H^+ \text{ or } HO^-]{H_2O} \underset{\text{imidate}}{R{-}\overset{OH}{\overset{|}{C}}{=}NH} \xrightarrow{\text{tautomerization}} \underset{\text{amide}}{R{-}\overset{O}{\overset{\|}{C}}{-}NH_2} \xrightarrow[H^+ \text{ or } HO^-]{\text{hydrolysis}} \underset{\text{acid}}{R{-}\overset{O}{\overset{\|}{C}}{-}OH} \quad (10.15)$$

Alkyl cyanides are generally made from the corresponding alkyl halide (usually primary) and sodium cyanide by an S_N2 displacement, as shown in this synthesis of an acid:

$$\underset{\substack{\text{propyl bromide}\\ \text{(1-bromopropane)}}}{CH_3CH_2CH_2Br} \xrightarrow{NaCN} \underset{\substack{\text{butyronitrile}\\ \text{(butanenitrile)}}}{CH_3CH_2CH_2CN} \xrightarrow[H^+]{H_2O} \underset{\substack{\text{butyric acid}\\ \text{(butanoic acid)}}}{CH_3CH_2CH_2CO_2H} + NH_4^+ \quad (10.16)$$

PROBLEM 10.12 Why is it *not* possible to convert bromobenzene to benzoic acid by the nitrile method? Instead, how could this conversion be accomplished?

Organic cyanides are commonly named after the corresponding acid, by changing the *-ic* or *-oic* suffix to *-onitrile* (hence, butyronitrile in eq. 10.16). In the IUPAC system, the suffix *-nitrile* is added to the name of the hydrocarbon with the same number of carbon atoms (hence butanenitrile in eq. 10.16).

Note that with the hydrolysis of nitriles, as with the Grignard method, the acid obtained has one more carbon atom than the alkyl halide from which the cyanide is prepared. Consequently, both methods provide ways of increasing the length of a carbon chain.

PROBLEM 10.13 Write equations for synthesizing phenylacetic acid ($C_6H_5CH_2CO_2H$) from benzyl bromide ($C_6H_5CH_2Br$) by two routes.

10.8 Carboxylic Acid Derivatives

Carboxylic acid derivatives are compounds in which the hydroxyl part of the carboxyl group is replaced by various other groups. All acid derivatives can be hydrolyzed to the corresponding carboxylic acid. In the remainder of this chapter, we will consider the preparation and reactions of the more important of these acid derivatives. Their general formulas are as follows:

$$\underset{\text{ester}}{R{-}\overset{O}{\overset{\|}{C}}{-}OR'} \qquad \underset{\text{acyl halide}}{R{-}\overset{O}{\overset{\|}{C}}{-}X} \left(\substack{\text{X is usually}\\ \text{Cl or Br}}\right) \qquad \underset{\text{acid anhydride}}{R{-}\overset{O}{\overset{\|}{C}}{-}O{-}\overset{O}{\overset{\|}{C}}{-}R} \qquad \underset{\text{primary amide}}{R{-}\overset{O}{\overset{\|}{C}}{-}NH_2}$$

Esters and amides occur widely in nature. Anhydrides, however, are uncommon in nature, and acyl halides are strictly creatures of the laboratory.

10.9 Esters

Esters are derived from acids by replacing the —OH group by an —OR group. They are named in a manner analogous to carboxylic acid salts. The R part of the —OR group is named first, followed by the name of the acid, with the *-ic* ending changed to *-ate.*

An **ester** is a carboxylic acid derivative in which the O—H group is replaced by an —OR group.

$$\underset{\text{methyl acetate (methyl ethanoate) bp 57°C}}{CH_3\overset{\overset{O}{\|}}{C}-OCH_3} \qquad \underset{\text{ethyl acetate (ethyl ethanoate) bp 77°C}}{CH_3\overset{\overset{O}{\|}}{C}-OCH_2CH_3} \qquad \underset{\text{methyl butanoate bp 102.3°C}}{CH_3CH_2CH_2\overset{\overset{O}{\|}}{C}-OCH_3}$$

Notice the different names of the following pair of isomeric esters, where the R and R′ groups are interchanged.

$$\underset{\text{phenyl acetate bp 195.7°C}}{CH_3\overset{\overset{O}{\|}}{C}-O-C_6H_5} \qquad \underset{\text{methyl benzoate bp 196.6°C}}{C_6H_5-\overset{\overset{O}{\|}}{C}-OCH_3}$$

Esters are named as two words that are *not* run together.

EXAMPLE 10.6

Name $CH_3CH_2CO_2CH(CH_3)_2$.

Solution The related acid is $CH_3CH_2CO_2H$, so the last part of the name is *propanoate* (change the *-ic* of propanoic to *-ate*). The alkyl group that replaces the hydrogen is *isopropyl,* or *2-propyl,* so the correct name is *isopropyl propanoate,* or *2-propyl propanoate.*

PROBLEM 10.14 Write the IUPAC name for

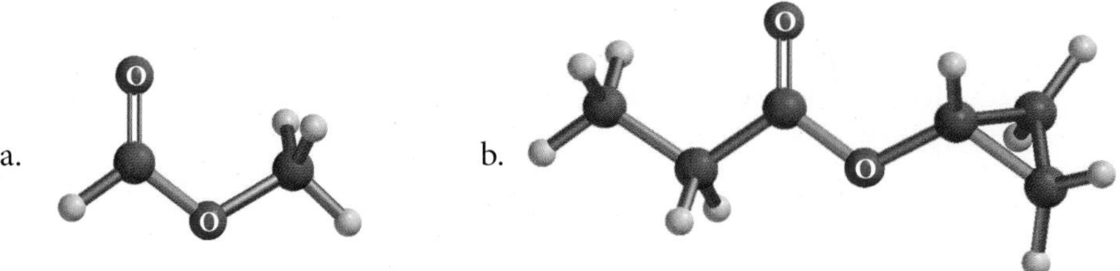

PROBLEM 10.15 Write the structure of

a. 3-pentyl butanoate

b. methyl 2-methylhexanoate

© PhotoDisc/Getty Images

Female elephants release the ester (*Z*)-7-dodecen-1-yl acetate to attract mates.

Many esters are rather pleasant-smelling substances and are responsible for the flavor and fragrance of many fruits and flowers. Among the more common are pentyl acetate (bananas), octyl acetate (oranges), ethyl butanoate (pineapples), and pentyl butanoate (apricots). Natural flavors can be exceedingly complex. For example, no fewer than 53 esters have been identified among the volatile constituents of Bartlett pears! Mixtures of esters are used as perfumes and artificial flavors. Low-molecular-weight esters are also used by insects and animals to transmit signals. Female elephants release (*Z*)-7-dodecen-1-yl acetate to signal their readiness to mate. Many moths release the same ester to attract mates.

10.10 Preparation of Esters; Fischer Esterification

When a carboxylic acid and an alcohol are heated in the presence of an acid catalyst (usually HCl or H_2SO_4), an equilibrium is established with the ester and water.

$$\underset{\text{acid}}{R-\overset{O}{\overset{\|}{C}}-OH} + \underset{\text{alcohol}}{HO-R'} \overset{H^+}{\rightleftharpoons} \underset{\text{ester}}{R-\overset{O}{\overset{\|}{C}}-OR'} + H_2O \quad (10.17)$$

Fischer esterification is the acid-catalyzed condensation of a carboxylic acid and an alcohol.

The process is called **Fischer esterification** after Emil Fischer (page 163), who developed the method. Although the reaction is an equilibrium, it can be shifted to the right in several ways. If either the alcohol or the acid is inexpensive, a large excess can be used. Alternatively, the ester and/or water may be removed as formed (by distillation, for example), thus driving the reaction forward.

PROBLEM 10.16 Following eq. 10.17, write an equation for the preparation of ethyl pentanoate from the correct acid and alcohol.

10.11 The Mechanism of Acid-Catalyzed Esterification; Nucleophilic Acyl Substitution

We can ask the following simple mechanistic question about Fischer esterification: Is the water molecule formed from the hydroxyl group of the acid and the hydrogen of the alcohol (as shown in color in eq. 10.17) or from the hydrogen of the acid and the hydroxyl group of the alcohol? This question may seem rather trivial, but the answer provides a key to understanding much of the chemistry of acids, esters, and their derivatives.

This question was resolved using isotopic labeling. For example, Fischer esterification of benzoic acid with methanol that had been enriched with the ^{18}O isotope of oxygen gave labeled methyl benzoate.*

$$C_6H_5-\overset{O}{\overset{\|}{C}}-OH + H^{18}OCH_3 \overset{H^+}{\rightleftharpoons} \underset{\text{methyl benzoate}}{C_6H_5-\overset{O}{\overset{\|}{C}}-{}^{18}OCH_3} + HOH \quad (10.18)$$

*^{18}O is oxygen with two additional neutrons in its nucleus. It is two mass units heavier than ^{16}O. ^{18}O can be distinguished from ^{16}O by mass spectrometry (see Chapter 12).

None of the ^{18}O appeared in the water. Thus it is clear that *the water was formed using the hydroxyl group of the acid and the hydrogen of the alcohol.* In other words, in Fischer esterification, the —OR group of the alcohol replaces the —OH group of the acid.

How can we explain this experimental fact? A mechanism consistent with this result is as follows (the oxygen atom of the alcohol is shown in color so that its path can be traced):

$$R{-}C({=}O){-}OH \underset{①}{\overset{H^+}{\rightleftharpoons}} R{-}C({=}\overset{+}{O}H){-}OH \overset{②}{\rightleftharpoons} R{-}C(OH)(OH){-}\overset{+}{O}(H)R' \underset{③}{\overset{-H^+}{\rightleftharpoons}} \boxed{R{-}C(OH)(OH){-}OR'}$$

$$\boxed{R{-}C(OH)(OH){-}OR'} \overset{H^+}{\underset{④}{\rightleftharpoons}} R{-}C(OH)(OR'){-}\overset{+}{O}H_2 \underset{⑤}{\overset{-H_2O}{\rightleftharpoons}} R{-}C({=}\overset{+}{O}H){-}OR' \underset{⑥}{\overset{-H^+}{\rightleftharpoons}} R{-}C({=}O){-}OR' \quad (10.19)$$

nucleophilic addition (step 2); elimination step (step 5)

Let us go through this mechanism, which looks more complicated than it really is, one step at a time.

Step 1. The carbonyl group of the acid is reversibly protonated. This step explains how the acid catalyst works. Protonation increases the positive charge on the carboxyl carbon and enhances its reactivity toward nucleophiles (recall the similar effect of acid catalysts with aldehydes and ketones, eq. 9.9). Note that the carbonyl oxygen gets protonated because it is the more basic oxygen.

Step 2. *This is the crucial step.* The alcohol, as a nucleophile, attacks the carbonyl carbon of the protonated acid. This is the step in which the new C—O bond (the ester bond) is formed.

Steps 3 and 4. These steps are equilibria in which oxygens lose or gain a proton. Such acid–base equilibria are reversible and rapid and go on constantly in any acidic solution of an oxygen-containing compound. In step 4, it does not matter which —OH group is protonated since these groups are equivalent.

Step 5. This is the step in which water, one product of the overall reaction, is formed. For this step to occur, an —OH group must be protonated to improve its leaving-group capacity. (This step is similar to the reverse of step 2.)

Step 6. This deprotonation step gives the ester and regenerates the acid catalyst. (This step is similar to the reverse of step 1.)

Some other features of the mechanism in eq. 10.19 are worth examining. The reaction begins with a carboxylic acid, in which the carboxyl carbon is trigonal and sp^2-hybridized. The end product is an ester; the ester carbon is also trigonal and sp^2-hybridized. However, the reaction proceeds through a neutral **tetrahedral intermediate** (shown in a box in eq. 10.19 and in color in eq. 10.20), in which the carbon atom has

A **tetrahedral intermediate** has an sp^3-hybridized carbon atom.

four groups attached to it and is thus sp^3-hybridized. If we omit all of the proton-transfer steps in eq. 10.19, we can focus on this feature of the reaction:

$$\underset{sp^2}{\text{R(HO)C=O}} + \text{R'OH} \rightleftharpoons \underset{\substack{sp^3 \\ \text{tetrahedral intermediate}}}{\text{R(HO)(R'O)C-OH}} \rightleftharpoons \underset{sp^2}{\text{R(R'O)C=O}} + H_2O \qquad (10.20)$$

Nucleophilic acyl substitution is substitution of another group for the —OH group of a carboxylic acid.

The net result of this process is substitution of the —OR′ group of the alcohol for the —OH group of the acid. Hence the reaction is referred to as **nucleophilic acyl substitution**. But the reaction is not a direct substitution. Instead, it occurs in two steps: (1) nucleophilic addition, followed by (2) elimination. We will see in the next and subsequent sections of this chapter that this is a general mechanism for nucleophilic substitutions at the carbonyl carbon atoms of carboxylic acid derivatives.

PROBLEM 10.17 Following eq. 10.19, write out the steps in the mechanism for the acid-catalyzed preparation of ethyl acetate from ethanol and acetic acid. In the United States, this method is used commercially to produce more than 100 million pounds of ethyl acetate annually, mainly for use as a solvent in the paint industry, but also as a solvent for nail polish and various glues.

10.12 Lactones

Hydroxy acids contain a hydroxyl group and a carboxyl group.

Lactones are **cyclic esters**.

Hydroxy acids contain both functional groups required for ester formation. If these groups can come in contact through bending of the chain, they may react with one another to form **cyclic esters** called **lactones**. For example,

$$\overset{\gamma\ \ \beta\ \ \alpha}{\underset{\text{OH}}{\overset{4\ \ 3\ \ 2\ \ 1}{CH_2CH_2CH_2CO_2H}}} \xrightarrow[\text{or heat}]{H^+} \underset{\gamma\text{-butyrolactone}}{\text{(five-membered lactone ring, C1=O, C2, C3, C4, O)}} + H_2O \qquad (10.21)$$

Most common lactones have five- or six-membered rings, although lactones with smaller or larger rings are known. Two examples of six-membered lactones from nature are coumarin, which is responsible for the pleasant odor of newly mown hay, and nepetalactone, the compound in catnip that excites cats. Erythromycin, widely used as an antibiotic, is an example of a macrocyclic lactone.*

*The R and R′ groups in erythromycin are carbohydrate units (see Chapter 16).

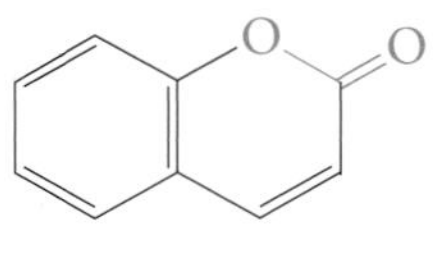

coumarin 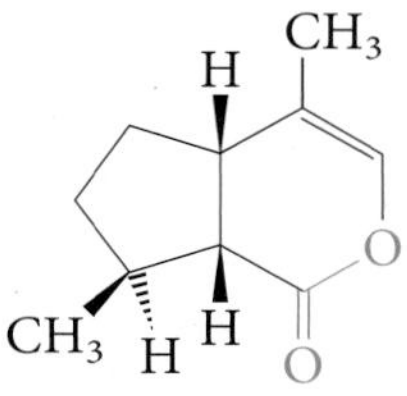 nepetalactone 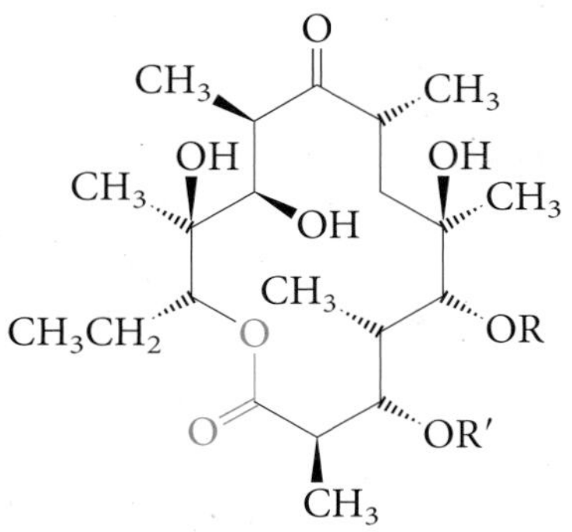 erythromycin

PROBLEM 10.18 Write the steps in the mechanism for the acid-catalyzed reaction in eq. 10.21.

10.13 Saponification of Esters

Esters are commonly hydrolyzed with base. The reaction is called **saponification** (from the Latin *sapon,* soap) because this type of reaction is used to make soaps from fats (Chapter 15). The general reaction is as follows:

Saponification is the hydrolysis of an ester with a base.

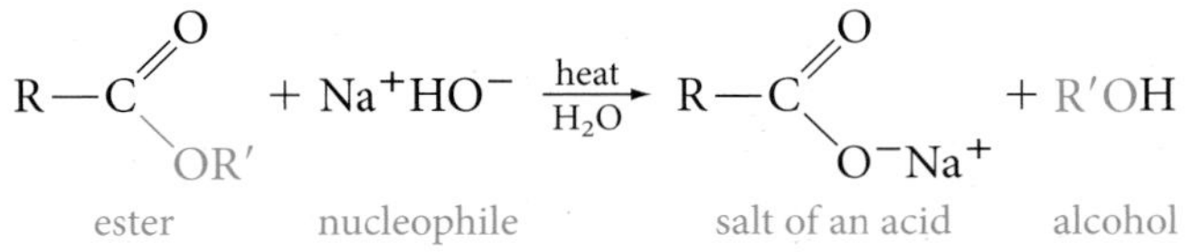

$$\underset{\text{ester}}{R{-}C(=O){-}OR'} + \underset{\text{nucleophile}}{Na^+HO^-} \xrightarrow[H_2O]{\text{heat}} \underset{\text{salt of an acid}}{R{-}C(=O){-}O^-Na^+} + \underset{\text{alcohol}}{R'OH} \quad \textbf{(10.22)}$$

The mechanism is another example of a nucleophilic acyl substitution. It involves nucleophilic attack by hydroxide ion, a strong nucleophile, on the carbonyl carbon of the ester.

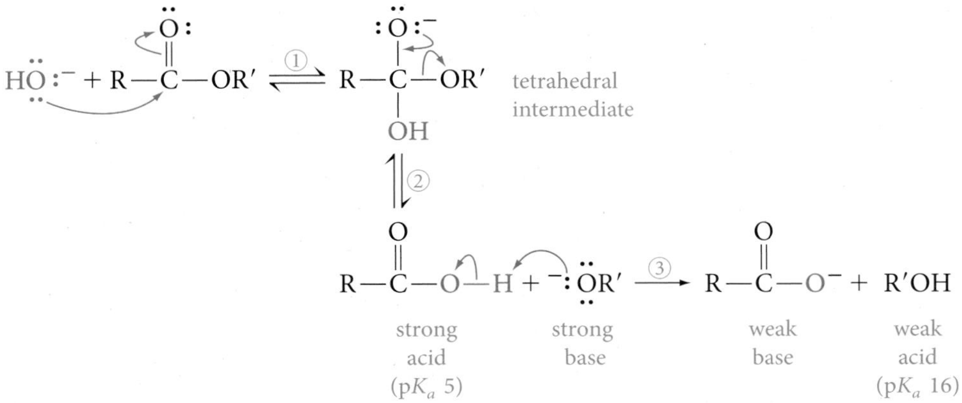

$$HO^- + R{-}C(=O){-}OR' \overset{①}{\rightleftharpoons} \underset{\text{tetrahedral intermediate}}{R{-}C(O^-)(OH){-}OR'} \overset{②}{\rightleftharpoons} \underset{\substack{\text{strong acid}\\(pK_a\ 5)}}{R{-}C(=O){-}O{-}H} + \underset{\text{strong base}}{^-OR'} \xrightarrow{③} \underset{\text{weak base}}{R{-}C(=O){-}O^-} + \underset{\substack{\text{weak acid}\\(pK_a\ 16)}}{R'OH} \quad \textbf{(10.23)}$$

The key step is nucleophilic addition to the carbonyl group (step 1). The reaction proceeds via a tetrahedral intermediate, but the reactant and the product are trigonal. *Saponification is not reversible;* in the final step (3), the strongly basic alkoxide ion removes a proton from the acid to form a carboxylate ion and an alcohol molecule—a step that proceeds completely in the forward direction.

Saponification is especially useful for breaking down an unknown ester, perhaps isolated from a natural source, into its component acid and alcohol for structural determination.

PROBLEM 10.19 Following eq. 10.22, write an equation for the saponification of methyl benzoate.

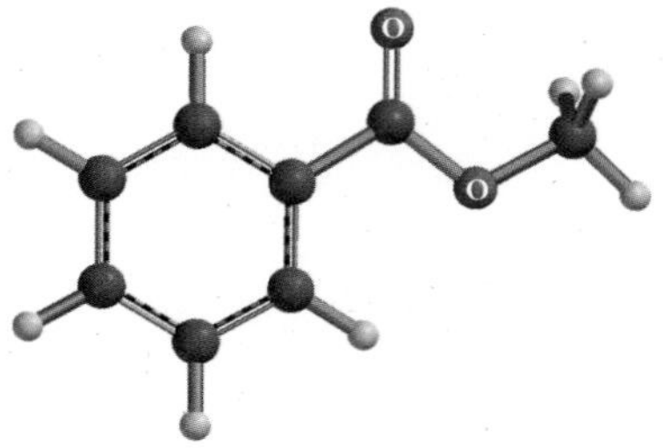

10.14 Ammonolysis of Esters

Ammonia converts esters to amides.

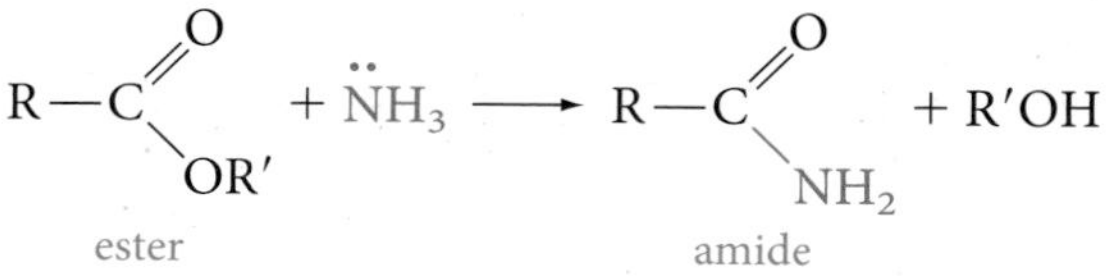

$$\underset{\text{ester}}{R{-}C(=O){-}OR'} + \ddot{N}H_3 \longrightarrow \underset{\text{amide}}{R{-}C(=O){-}NH_2} + R'OH \qquad (10.24)$$

For example,

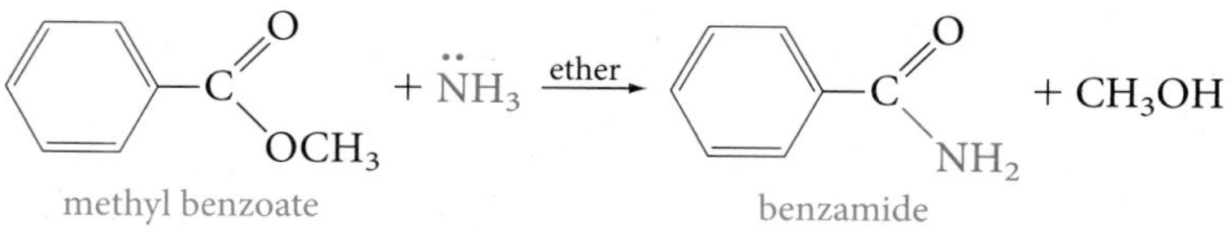

$$\underset{\text{methyl benzoate}}{C_6H_5{-}C(=O){-}OCH_3} + \ddot{N}H_3 \xrightarrow{\text{ether}} \underset{\text{benzamide}}{C_6H_5{-}C(=O){-}NH_2} + CH_3OH \qquad (10.25)$$

The reaction mechanism is very much like that of saponification. The unshared electron pair on the ammonia nitrogen initiates nucleophilic attack on the ester carbonyl group.

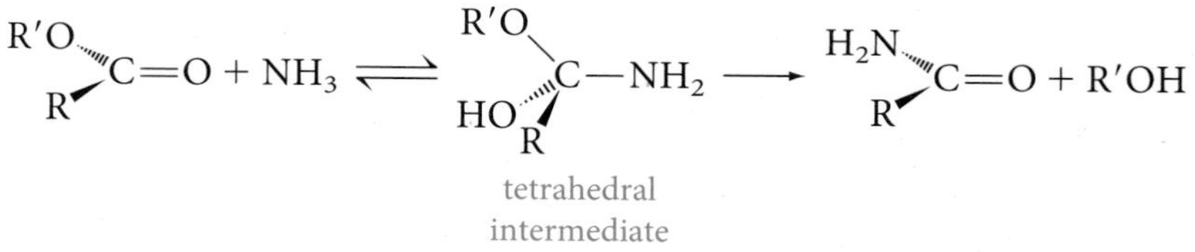

$$(R'O)(R)C{=}O + NH_3 \rightleftharpoons \underset{\text{tetrahedral intermediate}}{(R'O)(HO)(R)C{-}NH_2} \longrightarrow (H_2N)(R)C{=}O + R'OH \qquad (10.26)$$

PROBLEM 10.20 The first step in eq. 10.26 really involves two reactions, *addition* of ammonia to the carbonyl carbon to form an ammonium alkoxide followed by a *proton transfer* from the nitrogen to the alkoxide oxygen. Illustrate this process with equations using the arrow-pushing formalism. The second step in eq. 10.26 also involves two steps, *elimination* of an alkoxide ($R'O^-$) followed by deprotonation of the hydroxyl group. Write a detailed mechanism for these steps.

10.15 Reaction of Esters with Grignard Reagents

Esters react with two equivalents of a Grignard reagent to give tertiary alcohols. The reaction proceeds by *irreversible* nucleophilic attack of the Grignard reagent on the ester carbonyl group. The initial product, a ketone, reacts further in the usual way to give the tertiary alcohol.

$$\underset{\text{ester}}{R{-}\overset{O}{\overset{\|}{C}}{-}OR'} + 2\,R''MgBr \xrightarrow{\text{overall}} R{-}\underset{R''}{\overset{OMgBr}{C}}{-}R'' \xrightarrow[H^+]{H_2O} \underset{\text{tertiary alcohol}}{R{-}\underset{R''}{\overset{OH}{C}}{-}R''} \quad (10.27)$$

$$\text{ester} \xrightarrow{R''MgBr} R{-}\underset{R''}{\overset{BrMg{-}O}{C}}{-}OR' \xrightarrow{-R'OMgBr} \underset{\text{ketone}}{R{-}\overset{O}{\overset{\|}{C}}{-}R''} \xrightarrow{R''MgBr} R{-}\underset{R''}{\overset{OMgBr}{C}}{-}R''$$

This method is useful for making tertiary alcohols in which at least two of three alkyl groups attached to the hydroxyl-bearing carbon atom are identical.

PROBLEM 10.21 Using eq. 10.27 as a guide, write the structure of the tertiary alcohol that is obtained from

$$\text{cyclopropyl}{-}\overset{O}{\overset{\|}{C}}{-}OCH_3 + \text{excess } C_6H_5{-}MgBr$$

10.16 Reduction of Esters

Esters can be reduced to primary alcohols by lithium aluminum hydride ($LiAlH_4$).

$$\underset{\text{ester}}{R{-}\overset{O}{\overset{\|}{C}}{-}OR'} \xrightarrow[\text{ether}]{LiAlH_4} \underset{\text{primary alcohol}}{RCH_2OH} + R'OH \quad (10.28)$$

The mechanism is similar to the hydride reduction of aldehydes and ketones (eq. 9.33).

$$\underset{\text{ester}}{R{-}\overset{O}{\overset{\|}{C}}{-}OR'} \xrightarrow{H{-}\bar{A}lH_3} R{-}\underset{H}{\overset{O{-}\bar{A}lH_3}{C}}{-}OR' \xrightarrow{-\bar{A}lH_3(OR')} \underset{\text{aldehyde}}{R{-}\overset{O}{\overset{\|}{C}}{-}H} \xrightarrow{H{-}\bar{A}lH_2(OR')}$$

$$R{-}\underset{H}{\overset{O{-}\bar{A}lH_2(OR')}{C}}{-}H \xrightarrow[H^+]{H_2O} \underset{1^\circ\text{ alcohol}}{RCH_2OH} + R'OH \quad (10.29)$$

The intermediate aldehyde is not usually isolable and reacts rapidly with additional hydride to produce the alcohol.

Thus, with $LiAlH_4$, it is possible to reduce the carbonyl group of an ester without reducing a C═C bond in the same molecule. For example,

$$\underset{\text{ethyl 2-butenoate}}{CH_3CH{=}CH\overset{\overset{\displaystyle O}{\|}}{C}{-}OCH_2CH_3} \xrightarrow[2.\ H_2O,\ H^+]{1.\ LiAlH_4} \underset{\text{2-buten-1-ol}}{CH_3CH{=}CHCH_2OH} + CH_3CH_2OH \quad \textbf{(10.30)}$$

10.17 The Need for Activated Acyl Compounds

As we have seen, most reactions of carboxylic acids, esters, and related compounds involve, as the first step, nucleophilic attack on the carbonyl carbon atom. Examples are Fischer esterification, saponification and ammonolysis of esters, and the first stage of the reaction of esters with Grignard reagents or lithium aluminum hydride. All of these reactions can be summarized by a single mechanistic equation:

$$\underset{sp^2}{R(L)C{=}\ddot{O}\colon} + \colon Nu^- \underset{}{\overset{①}{\rightleftharpoons}} \underset{\text{tetrahedral intermediate}}{R(L)(Nu)C{-}\ddot{O}\colon^-} \overset{②}{\rightleftharpoons} \underset{sp^2}{R(Nu)C{=}\ddot{O}\colon} + \colon L^- \quad \textbf{(10.31)}$$

The carbonyl carbon, initially trigonal, is attacked by a nucleophile $Nu\colon^-$ to form a tetrahedral intermediate (step 1). Loss of a leaving group $\colon L^-$ (step 2) then regenerates the carbonyl group with its trigonal carbon atom. The net result is the replacement of L by Nu.

An **acyl transfer** is the transfer of an acyl group from a leaving group to a nucleophile.

Biochemists look at eq. 10.31 in a slightly different way. They refer to the overall reaction as an **acyl transfer**. The acyl group is transferred from L in the starting material to Nu in the product.

Regardless of how we consider the reaction, one important feature that can affect the rate of both steps is the nature of the leaving group. *The rates of both steps in a nucleophilic acyl substitution reaction are enhanced by increasing the electron-withdrawing properties of the leaving group.* Step 1 is favored because the more electronegative L is, the more positive the carbonyl carbon becomes, and therefore the more susceptible the carbonyl carbon is to nucleophilic attack. Step 2 is also facilitated because the more electronegative L is, the better leaving group it becomes.

In general, esters are *less* reactive toward nucleophiles than are aldehydes or ketones because the positive charge on the carbonyl carbon in esters can be delocalized to the oxygen atom. Consequently, the ester is more stable and less prone to attack.

$$\underset{\text{resonance in aldehydes and ketones}}{\left[RR'C{=}\ddot{O}\colon \longleftrightarrow RR'\overset{+}{C}{-}\ddot{O}\colon^- \right]} \qquad \underset{\text{resonance in esters}}{\left[R(R'\ddot{O})C{=}\ddot{O}\colon \longleftrightarrow R(R'\ddot{O})\overset{+}{C}{-}\ddot{O}\colon^- \longleftrightarrow R(R'\overset{+}{O}{=})C{-}\ddot{O}\colon^- \right]}$$

The carbonyl carbon has a partial positive charge.

The positive charge can be delocalized to the oxygen.

Now let us examine some of the ways in which the carboxyl group can be modified to *increase* its reactivity toward nucleophiles.

10.18 Acyl Halides

An **acyl halide** is a carboxylic acid derivative in which the —OH group is replaced by a halogen atom.

Acyl halides are among the most reactive of carboxylic acid derivatives. *Acyl chlorides* are more common and less expensive than bromides or iodides. They can be prepared from acids by reaction with thionyl chloride.

$$R-\overset{\overset{\large O}{\|}}{C}-OH + SOCl_2 \longrightarrow R-\overset{\overset{\large O}{\|}}{C}-Cl + HCl + SO_2 \qquad (10.32)$$

The mechanism is similar to that for the formation of chlorides from alcohols and thionyl chloride. The hydroxyl group is converted to a good leaving group by thionyl chloride, followed by a nucleophilic acyl substitution in which chloride is the nucleophile (compare with Sec. 7.10). Phosphorus pentachloride and other reagents can also be used to prepare acyl chlorides from carboxylic acids.

$$R-\overset{\overset{\large O}{\|}}{C}-OH + PCl_5 \longrightarrow R-\overset{\overset{\large O}{\|}}{C}-Cl + HCl + POCl_3 \qquad (10.33)$$

Acyl halides react rapidly with most nucleophiles. For example, they are rapidly hydrolyzed by water.

$$\underset{\text{acetyl chloride}}{CH_3-\overset{\overset{\large O}{\|}}{C}-Cl} + HOH \xrightarrow{\text{rapid}} \underset{\text{acetic acid}}{CH_3-\overset{\overset{\large O}{\|}}{C}-OH} + \underset{\text{(fumes)}}{HCl} \qquad (10.34)$$

For this reason, acyl halides have irritating odors. Benzoyl chloride (eq. 10.35), for example, is a lachrymator (tear gas).

EXAMPLE 10.7

Write a mechanism for the reaction shown in eq. 10.34.

Solution Nucleophilic addition of water to the carbonyl group, followed by proton transfer and elimination of HCl from the tetrahedral intermediate, gives the observed products.

$$CH_3-\overset{\overset{\large O}{\|}}{C}-Cl + H-\ddot{O}-H \underset{}{\overset{\text{addition}}{\rightleftharpoons}} CH_3-\underset{\underset{H-\overset{+}{O}-H}{|}}{\overset{\overset{:\ddot{O}:^-}{|}}{C}}-Cl \underset{\text{transfer}}{\overset{\text{proton}}{\rightleftharpoons}} CH_3-\underset{\underset{H-\ddot{O}}{|}}{\overset{\overset{:\ddot{O}-H}{|}}{C}}-Cl \xrightarrow{\text{elimination}} CH_3-\overset{\overset{\large O}{\|}}{C}-OH + HCl\,(\uparrow)$$

tetrahedral intermediates

Acyl halides react rapidly with alcohols to form esters.

$$\underset{\text{benzoyl chloride}}{C_6H_5-\overset{\overset{\large O}{\|}}{C}-Cl} + CH_3OH \xrightarrow[\text{temp.}]{\text{room}} \underset{\text{methyl benzoate}}{C_6H_5-\overset{\overset{\large O}{\|}}{C}-OCH_3} + HCl \qquad (10.35)$$

Indeed, the most common way to prepare an ester *in the laboratory* is to convert an acid to its acid chloride, then react the latter with an alcohol. Even though two steps are necessary (compared with one step for Fischer esterification), the method may be preferable, especially if either the acid or the alcohol is expensive. (Recall that Fischer esterification is an equilibrium reaction and must often be carried out with a large excess of one of the reactants.)

PROBLEM 10.22 Rewrite eq. 10.32 to show the preparation of benzoyl chloride (see eq. 10.35).

PROBLEM 10.23 Explain why acyl halides may be irritating to the nose.

PROBLEM 10.24 Write a mechanism for the reaction shown in eq. 10.35.

Acyl halides react rapidly with ammonia to form amides.

$$\underset{\text{acetyl chloride}}{CH_3\overset{\overset{O}{\|}}{C}-Cl} + 2\,NH_3 \longrightarrow \underset{\text{acetamide}}{CH_3\overset{\overset{O}{\|}}{C}-NH_2} + NH_4^+\;Cl^- \qquad (10.36)$$

The reaction is much more rapid than the ammonolysis of esters. Two equivalents of ammonia are required, however—one to form the amide and one to neutralize the hydrogen chloride.

Acyl halides are used to synthesize aromatic ketones, through Friedel–Crafts acylation of aromatic rings (review Sec. 4.9.d).

PROBLEM 10.25 Devise a synthesis of 4-methylphenyl propyl ketone from toluene and butanoic acid as starting materials.

10.19 Acid Anhydrides

Acid anhydrides are carboxylic acid derivatives formed by condensing two carboxylic acid molecules.

Acid anhydrides are derived from acids by removing water from two carboxyl groups and connecting the fragments.

$$\underset{\text{two acid molecules}}{R-\overset{\overset{O}{\|}}{C}-OH \quad HO-\overset{\overset{O}{\|}}{C}-R} \qquad \underset{\text{an acid anhydride}}{R-\overset{\overset{O}{\|}}{C}-O-\overset{\overset{O}{\|}}{C}-R}$$

The most important commercial aliphatic anhydride is acetic anhydride ($R{=}CH_3$). About 1 million tons are manufactured annually, mainly to react with alcohols to form acetates. The two most common uses are in making cellulose acetate (rayon) and aspirin (acetylsalicylic acid).

The name of an anhydride is obtained by naming the acid from which it is derived and replacing the word *acid* with *anhydride.*

$$\underset{\text{ethanoic anhydride or acetic anhydride}}{CH_3-\overset{\overset{O}{\|}}{C}-O-\overset{\overset{O}{\|}}{C}-CH_3}$$

PROBLEM 10.26 Write the structural formula for

a. butanoic anhydride

b. benzoic anhydride

Anhydrides are prepared by dehydration of acids. Dicarboxylic acids with appropriately spaced carboxyl groups lose water on heating to form cyclic anhydrides with five- and six-membered rings. For example,

$$\text{maleic acid} \xrightarrow{135^\circ C} \text{maleic anhydride} + H_2O \qquad (10.37)$$

maleic acid → maleic anhydride

PROBLEM 10.27 Predict and name the product of the following reaction:

$$C_6H_4(COOH)_2 \text{ (ortho)} \xrightarrow{\text{heat}}$$

PROBLEM 10.28 Do you expect fumaric acid (page 290) to form a cyclic anhydride on heating? Explain.

Anhydrides can also be prepared from acid chlorides and carboxylate salts in a reaction that occurs by a nucleophilic acyl substitution mechanism. This is a good method for preparing anhydrides derived from two different carboxylic acids, called **mixed anhydrides**.

Mixed anhydrides are prepared from two different carboxylic acids.

$$CH_3CH_2CH_2-\overset{O}{\overset{\|}{C}}-Cl + Na^{+}\,{}^{-}O-\overset{O}{\overset{\|}{C}}-CH_3 \longrightarrow CH_3CH_2CH_2-\overset{O}{\overset{\|}{C}}-O-\overset{O}{\overset{\|}{C}}-CH_3 + NaCl \qquad (10.38)$$

butanoic ethanoic anhydride

Anhydrides undergo nucleophilic acyl substitution reactions. They are more reactive than esters, but less reactive than acyl halides, toward nucleophiles. Some typical reactions of acetic anhydride follow:

$$CH_3-\overset{O}{\overset{\|}{C}}-O-\overset{O}{\overset{\|}{C}}-CH_3 \xrightarrow{HO-H} CH_3\overset{O}{\overset{\|}{C}}-OH + CH_3\overset{O}{\overset{\|}{C}}-OH \text{ (acid)}$$

$$\xrightarrow{RO-H} CH_3\overset{O}{\overset{\|}{C}}-OR + CH_3\overset{O}{\overset{\|}{C}}-OH \text{ (ester)} \qquad (10.39)$$

$$\xrightarrow{H_2N-H} CH_3\overset{O}{\overset{\|}{C}}-NH_2 + CH_3\overset{O}{\overset{\|}{C}}-OH \text{ (amide)}$$

acetic anhydride
bp 139.5°C

A WORD ABOUT... Thioesters, Nature's Acyl-Activating Groups

Acyl transfer plays an important role in many biochemical processes. However, acyl halides and anhydrides are far too corrosive to be cell constituents—they are hydrolyzed quite rapidly by water and are therefore incompatible with cellular fluid. Most ordinary esters, on the other hand, react too slowly with nucleophiles for acyl transfer to be carried out efficiently at body temperatures. Consequently, other functional groups have evolved to activate acyl groups in the cell. The most important of these is **coenzyme A** (the A stands for acetylation, one of the functions of this enzyme). Coenzyme A is a complex *thiol* (Figure 10.2). It is usually abbreviated by the symbol **CoA—SH**. Though its structure is made up of three parts—adenosine diphosphate (ADP), pantothenic acid (a vitamin), and 2-aminoethanethiol—it is the thiol group that gives coenzyme A its most important functions.

Coenzyme A can be converted to **thioesters**, the active acyl-transfer agents in the cell. Of the thioesters that coenzyme A forms, the acetyl ester, called **acetyl-coenzyme A** and abbreviated as

$$CH_3\overset{\overset{O}{\|}}{C}{-}S{-}CoA$$

is the most important. Acetyl-CoA reacts with many nucleophiles to transfer the acetyl group.

$$\underset{\text{acetyl-CoA}}{CH_3\overset{\overset{O}{\|}}{C}{-}S{-}CoA} + Nu{:} \xrightarrow[\text{enzyme}]{H_2O} CH_3\overset{\overset{O}{\|}}{C}{-}Nu + CoA{-}SH$$

The reactions are usually enzyme-mediated and occur rapidly at ordinary cell temperatures.

Why are thioesters superior to ordinary esters as acyl-transfer agents? Part of the answer lies in the acidity difference between alcohols and thiols (Sec. 7.17). Since thiols are much stronger acids than are alcohols, their conjugate bases, ^{-}SR, are much weaker bases than ^{-}OR. Thus, the —SR group of thioesters is a much better leaving group, in nucleophilic substitution reactions, than is the —OR group of ordinary esters. Thioesters are not so reactive that they hydrolyze in cellular fluid, but they are appreciably more reactive than simple esters. Nature makes use of this feature.

See Problem 10.64.

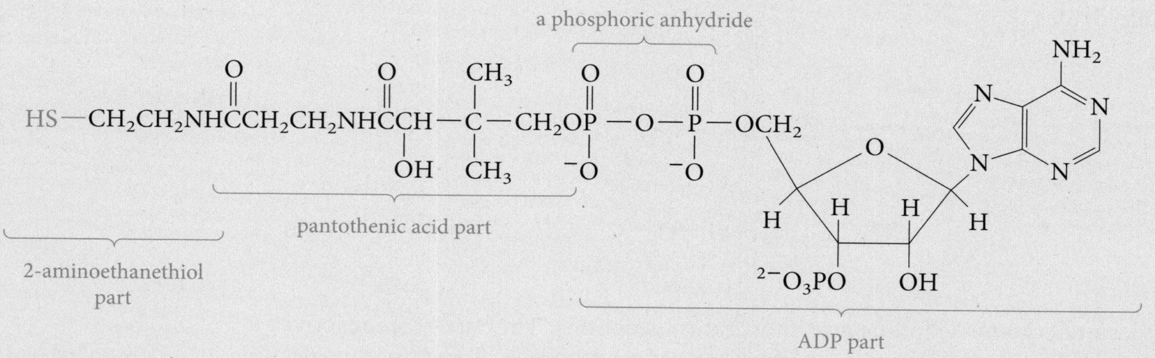

$$HS{-}CH_2CH_2NH\overset{\overset{O}{\|}}{C}CH_2CH_2NH\overset{\overset{O}{\|}}{C}\underset{OH}{\underset{|}{C}H}{-}\underset{CH_3}{\underset{|}{\overset{CH_3}{\overset{|}{C}}}}{-}CH_2O\underset{^{-}O}{\underset{|}{\overset{O}{\overset{\|}{P}}}}{-}O{-}\underset{^{-}O}{\underset{|}{\overset{O}{\overset{\|}{P}}}}{-}OCH_2\text{(ribose: H, H, H, H; }{}^{2-}O_3PO\text{, OH) – adenine (}NH_2\text{)}$$

a phosphoric anhydride

2-aminoethanethiol part

pantothenic acid part

ADP part

■ **Figure 10.2**
Coenzyme A.

Water hydrolyzes an anhydride to the corresponding acid. Alcohols give esters, and ammonia gives amides. In each case, one equivalent of acid is also produced.

PROBLEM 10.29 Write an equation for the reaction of acetic anhydride with 1-pentanol ($CH_3CH_2CH_2CH_2CH_2OH$).

PROBLEM 10.30 Write equations for the reactions of maleic anhydride (see eq. 10.37) with

a. water b. 1-propanol c. ammonia

The reaction of acetic anhydride with salicylic acid (*o*-hydroxybenzoic acid) is used to synthesize aspirin. In this reaction, the phenolic hydroxyl group is **acetylated** (converted to its acetate ester).

An alcohol is said to be **acetylated** when converted to its acetate ester.

Annual aspirin production in the United States is more than 24 million pounds, enough to produce over 30 billion standard 5-grain (325 mg) tablets. Aspirin is widely used, either by itself or mixed with other drugs, as an analgesic and antipyretic. It is not without dangers, however. Repeated use may cause gastrointestinal bleeding, and a large single dose (10 to 20 g) can cause death.

$$o\text{-}C_6H_4(OH)(CO_2H) + CH_3\overset{O}{\overset{\|}{C}}-O-\overset{O}{\overset{\|}{C}}CH_3 \longrightarrow o\text{-}C_6H_4(O\overset{O}{\overset{\|}{C}}CH_3)(CO_2H) + CH_3CO_2H \quad (10.40)$$

salicylic acid; acetic anhydride; acetylsalicylic acid (aspirin)

10.20 Amides

Amides are the least reactive of the common carboxylic acid derivatives. They occur widely in nature. The most important amides are the proteins, whose chemistry we will discuss in Chapter 17. Here we will concentrate on just a few properties of simple amides.

Amides are carboxylic acid derivatives in which the —OH group is replaced by $—NH_2$, —NHR, or $—NR_2$.

Primary amides have the general formula $RCONH_2$. They can be prepared by the reaction of ammonia with esters (eq. 10.24), with acyl halides (eq. 10.36), or with acid anhydrides (eq. 10.39). Amides can also be prepared by heating the ammonium salts of acids.

$$R-\overset{O}{\overset{\|}{C}}-OH + NH_3 \longrightarrow \underset{\text{ammonium salt}}{R-\overset{O}{\overset{\|}{C}}-O^-NH_4^+} \xrightarrow{\text{heat}} \underset{\text{amide}}{R-\overset{O}{\overset{\|}{C}}-NH_2} + H_2O \quad (10.41)$$

Amides are named by replacing the *-ic* or *-oic* ending of the acid name, either the common or the IUPAC name, with the *-amide* ending.

$$H-\overset{O}{\overset{\|}{C}}-NH_2 \qquad CH_3-\overset{O}{\overset{\|}{C}}-NH_2 \qquad CH_3CH_2CH_2\overset{O}{\overset{\|}{C}}-NH_2 \qquad C_6H_5-\overset{O}{\overset{\|}{C}}-NH_2$$

formamide (methanamide); acetamide (ethanamide); butanamide; benzamide (benzenecarboxamide)

PROBLEM 10.31

a. Name $(CH_3)_2CHCH_2CONH_2$
b. Write the structure of 1-phenylcyclopentanecarboxamide

The above examples are all primary amides. Secondary and tertiary amides, in which one or both of the hydrogens on the nitrogen atom are replaced by organic groups, are described in the next chapter.

Amides have a planar geometry. Even though the carbon–nitrogen bond is normally written as a single bond, rotation around that bond is restricted because of resonance.

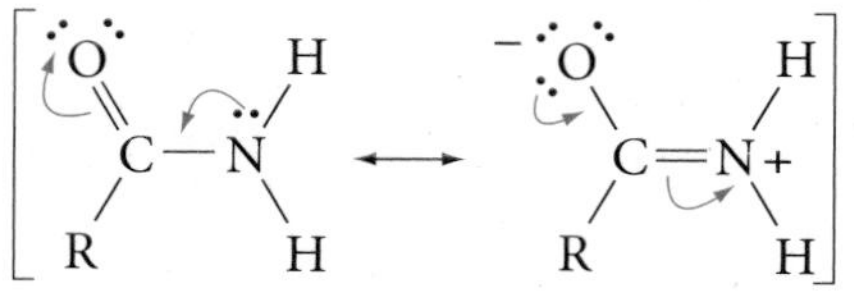

$$\left[\text{R}-\overset{\overset{\displaystyle :\ddot{O}:}{\|}}{C}-\ddot{N}H_2 \longleftrightarrow \text{R}-\overset{\overset{\displaystyle ^{-}:\ddot{O}:}{|}}{C}=\overset{+}{N}H_2 \right]$$

amide resonance

The dipolar contributor is so important that the carbon–nitrogen bond behaves much like a double bond. Consequently, the nitrogen and the carbonyl carbon, and the two atoms attached to each of them, lie in the same plane, and rotation at the C—N bond is restricted. Indeed, the C—N bond in amides is only 1.32 Å long—much shorter than the usual carbon–nitrogen single bond length (which is about 1.47 Å).

As the dipolar resonance contributor suggests, amides are highly polar and form strong hydrogen bonds.

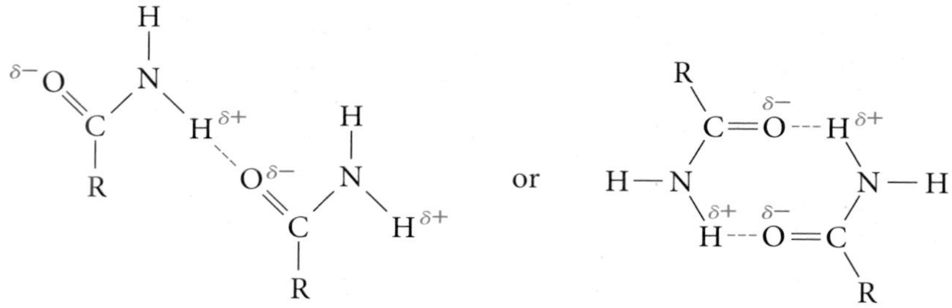

$$\text{R}-\overset{\overset{\delta-}{O}}{\|}{C}-\text{N(H)}-H^{\delta+}\cdots O^{\delta-}{=}\text{C(R)}-\text{NH}-H^{\delta+} \quad \text{or} \quad \text{cyclic dimer: } \text{R}-\text{C}{=}O^{\delta-}\cdots H^{\delta+}-\text{N(H)}-\text{C(R)}{=}O^{\delta-}\cdots H^{\delta+}-\text{N(H)}-$$

Amides have exceptionally high boiling points for their molecular weights, although alkyl substitution on the nitrogen lowers the boiling and melting points by decreasing the hydrogen-bonding possibilities, as shown in the following two pairs of compounds:

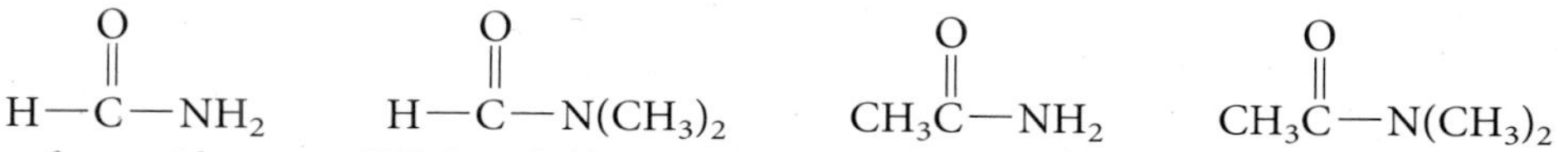

$H-\overset{O}{\overset{\|}{C}}-NH_2$	$H-\overset{O}{\overset{\|}{C}}-N(CH_3)_2$	$CH_3\overset{O}{\overset{\|}{C}}-NH_2$	$CH_3\overset{O}{\overset{\|}{C}}-N(CH_3)_2$
formamide	*N,N*-dimethylformamide	acetamide	*N,N*-dimethylacetamide
bp 210°C	153°C	222°C	165°C
mp 2.5°C	−60.5°C	81°C	−20°C

PROBLEM 10.32 Show that hydrogen bonding is possible for acetamide, but not for *N,N*-dimethylacetamide.

Like other acid derivatives, amides react with nucleophiles. For example, they can be hydrolyzed by water.

$$\underset{\text{amide}}{R-\overset{O}{\overset{\|}{C}}-NH_2} + H-OH \xrightarrow[HO^-]{H^+ \text{ or}} \underset{\text{acid}}{R-\overset{O}{\overset{\|}{C}}-OH} + NH_3 \qquad \textbf{(10.42)}$$

The reactions are slow, and prolonged heating or acid or base catalysis is usually necessary.

PROBLEM 10.33 Using eq. 10.42 as a model, write an equation for the hydrolysis of acetamide.

Amides can be reduced by lithium aluminum hydride to give amines.

$$\underset{\text{amide}}{R-\overset{\overset{\large O}{\|}}{C}-NH_2} \xrightarrow[\text{ether}]{LiAlH_4} \underset{\text{amine}}{RCH_2NH_2} \quad \textbf{(10.43)}$$

This is an excellent way to make primary amines, whose chemistry is discussed in the next chapter.

PROBLEM 10.34 Using eq. 10.43 as a model, write an equation for the reduction of acetamide with $LiAlH_4$.

Urea is a special amide, a diamide of carbonic acid. A colorless, water-soluble, crystalline solid, urea is the normal end product of protein metabolism. An average adult excretes approximately 30 g of urea in his or her urine daily. Urea is produced commercially from carbon dioxide and ammonia, mainly for use as a fertilizer.

$$\underset{\text{carbonic acid}}{HO-\overset{\overset{\large O}{\|}}{C}-OH} \qquad \underset{\substack{\text{urea} \\ \text{mp } 133^\circ\text{C}}}{H_2N-\overset{\overset{\large O}{\|}}{C}-NH_2}$$

10.21 A Summary of Carboxylic Acid Derivatives

We have studied a rather large number of reactions in this chapter. However, most of them can be summarized in a single chart, shown in Table 10.5, and visually presented in Figure 10.3.

Table 10.5 ■ Reactions of Acid Derivatives with Certain Nucleophiles

Acid derivative (decreasing reactivity ↓)	Nucleophile: HOH (hydrolysis)	Nucleophile: R′OH (alcoholysis)	Nucleophile: NH_3 (ammonolysis)
$R-\overset{\overset{O}{\|}}{C}-Cl$ acyl halide	$R-\overset{\overset{O}{\|}}{C}-OH + HCl$	$R-\overset{\overset{O}{\|}}{C}-OR' + HCl$	$R-\overset{\overset{O}{\|}}{C}-NH_2 + NH_4^+Cl^-$
$R-\overset{\overset{O}{\|}}{C}-O-\overset{\overset{O}{\|}}{C}-R$ acid anhydride	$2\,R-\overset{\overset{O}{\|}}{C}-OH$	$R-\overset{\overset{O}{\|}}{C}-OR' + RCO_2H$	$R-\overset{\overset{O}{\|}}{C}-NH_2 + RCO_2H$
$R-\overset{\overset{O}{\|}}{C}-O-R''$ ester	$R-\overset{\overset{O}{\|}}{C}-OH + R''OH$	$R-\overset{\overset{O}{\|}}{C}-OR' + R''OH$ (ester interchange)	$R-\overset{\overset{O}{\|}}{C}-NH_2 + R''OH$
$R-\overset{\overset{O}{\|}}{C}-NH_2$ amide	$R-\overset{\overset{O}{\|}}{C}-OH + NH_3$	—	—
Main organic product	acid	ester	amide

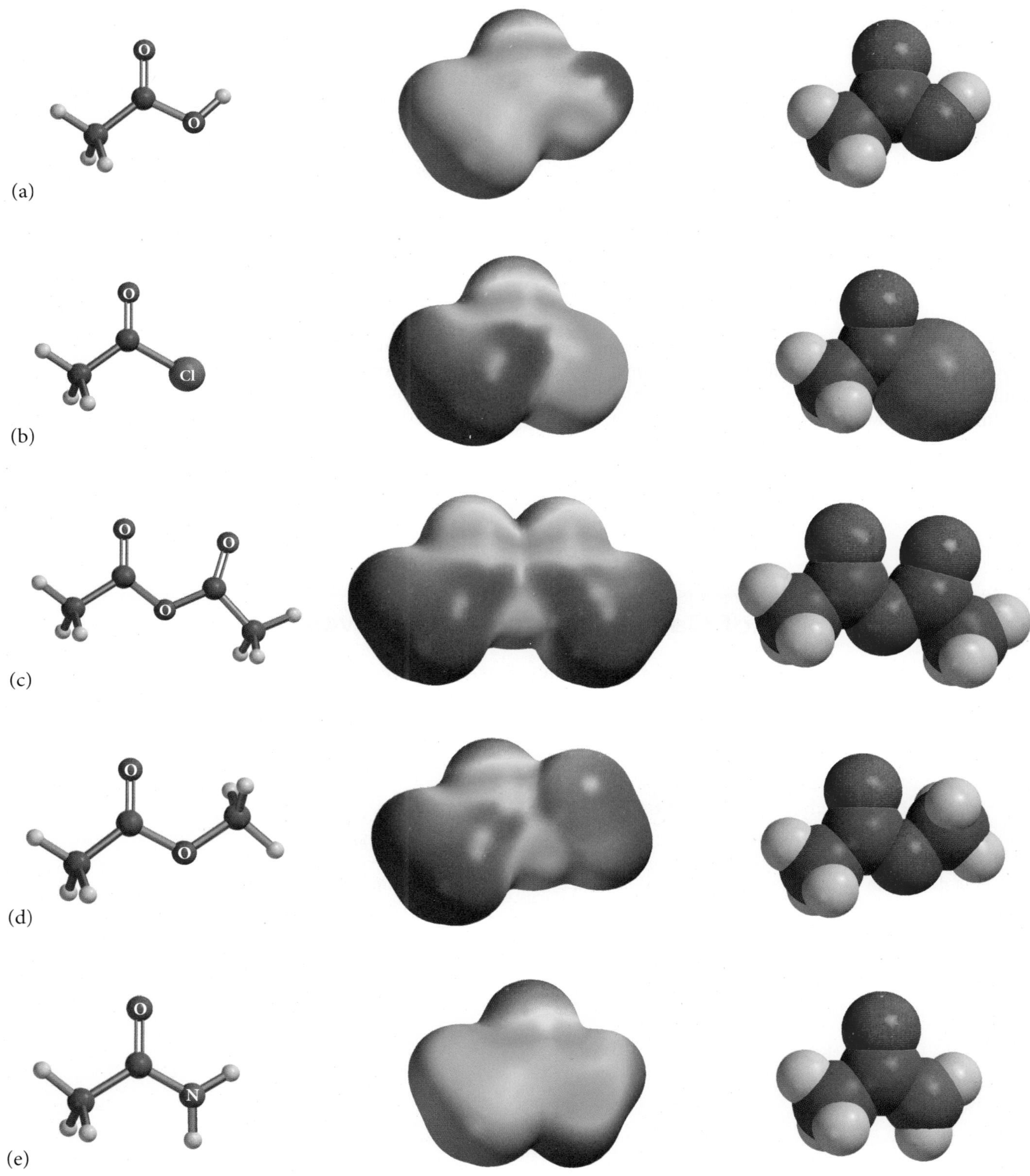

■ Figure 10.3

Three-dimensional representations and electrostatic potential maps of representative carboxylic acid derivatives. Ball-and-stick structures are provided on the left, electrostatic potential maps in the center, and space-filling models on the right for (a) acetic acid (CH_3CO_2H); (b) acetyl chloride ($CH_3C(=O)Cl$); (c) acetic anhydride ($CH_3C(=O)—O—C(=O)CH_3$); (d) methyl acetate ($CH_3C(=O)OCH_3$); and (e) acetamide ($CH_3C(=O)NH_2$).

The four types of acid derivatives are listed at the left of the chart in order of decreasing reactivity toward nucleophiles. Three common nucleophiles are listed across the top. Note that the main organic product in each column is the same, regardless of which type of carboxylic acid derivative we start with. For example, hydrolysis gives the corresponding organic acid, whether we start with an acyl halide, acid anhydride, ester, or amide. Similarly, alcoholysis gives an ester, and ammonolysis gives an amide. Note also that the *other* reaction product is generally the same from a given carboxylic acid derivative (horizontally across the table), regardless of the nucleophile. For example, starting with an ester, RCO_2R'', we obtain as the second product the alcohol R″OH, regardless of whether the reaction type is hydrolysis, alcoholysis, or ammonolysis.

All of the reactions in Table 10.5 take place via attack of the nucleophile on the carbonyl carbon of the carboxylic acid derivative, as described in eq. 10.31. Indeed, most of the reactions from Sections 10.10 through 10.19 occur by that same mechanism. We can sometimes use this idea to predict new reactions.

For example, the reaction of esters with Grignard reagents (eq. 10.27) involves nucleophilic attack of the Grignard reagent on the ester's carbonyl group. Keeping in mind that all carboxylic acid derivatives are susceptible to nucleophilic attack, it is understandable that acyl halides also react with Grignard reagents to give tertiary alcohols. The first steps involve ketone formation as follows:

$$R-\overset{O}{\overset{\|}{C}}-Cl + R'MgX \longrightarrow R-\overset{O^{-}\,\overset{+}{M}gX}{\overset{|}{\underset{R'}{\underset{|}{C}}}}-Cl \longrightarrow R-\overset{O}{\overset{\|}{C}}-R' + MgXCl \quad (10.44)$$

The ketone can sometimes be isolated, but usually it reacts with a second mole of Grignard reagent to give a tertiary alcohol.

$$R-\overset{O}{\overset{\|}{C}}-R' + R'MgX \longrightarrow R-\overset{O^{-}\,\overset{+}{M}gX}{\overset{|}{\underset{R'}{\underset{|}{C}}}}-R' \xrightarrow{H_3O^+} R-\overset{OH}{\overset{|}{\underset{R'}{\underset{|}{C}}}}-R' \quad (10.45)$$

PROBLEM 10.35 Predict the product from the reaction of phenylmagnesium bromide (C_6H_5MgBr) with benzoyl chloride (C_6H_5COCl).

10.22 The α-Hydrogen of Esters; the Claisen Condensation

In this final section, we describe an important reaction of esters that resembles the aldol condensation of aldehydes and ketones (Sec. 9.17). It makes use of the α-hydrogen (see pages 272–275) of an ester.

An **ester enolate** is the anion formed by removing the α-hydrogen of an ester.

Being adjacent to a carbonyl group, the α-hydrogens of an ester are weakly acidic ($pK_a \sim 23$) and can be removed by a *strong base*. The product is an **ester enolate**.

$$-\overset{H}{\underset{|}{\overset{|}{C}}}-\overset{O}{\overset{\|}{C}}-OR \xrightleftharpoons{\text{base}} \left[-\overset{..}{\underset{|}{C}}{}^{-}-\overset{O}{\overset{\|}{C}}-OR \longleftrightarrow -\underset{|}{C}=\overset{O^{-}}{\overset{|}{C}}-OR \right] \quad (10.46)$$

resonance contributors to an ester enolate

Common bases used for this purpose are sodium alkoxides or sodium hydride. The ester enolate, once formed, can act as a carbon nucleophile and add to the carbonyl group of another ester molecule. This reaction is called the Claisen condensation. It is a way of making β-keto esters. We will use ethyl acetate as an example to see how the reaction works.

Treatment of ethyl acetate with sodium ethoxide in ethanol produces the β-keto ester, ethyl acetoacetate:

$$\underset{\text{ethyl acetate}}{CH_3\overset{O}{\overset{\|}{C}}-OCH_2CH_3} + \underset{\text{ethyl acetate}}{H-\overset{\alpha}{C}H_2-\overset{O}{\overset{\|}{C}}-OCH_2CH_3} \xrightarrow[\text{2. } H_3O^+]{\text{1. } NaOCH_2CH_3 \text{ in ethanol}}$$

$$\underset{\substack{\text{ethyl acetoacetate} \\ \text{(ethyl 3-oxobutanoate)}}}{CH_3\overset{O}{\overset{\|}{C}}-CH_2-\overset{O}{\overset{\|}{C}}-OCH_2CH_3} + CH_3CH_2OH \quad (10.47)$$

The Claisen condensation takes place in three steps.

Step 1.
$$CH_3\overset{O}{\overset{\|}{C}}-OCH_2CH_3 + \underset{\text{sodium ethoxide}}{Na^{+}\,{}^{-}OCH_2CH_3} \rightleftharpoons \underset{\text{ester enolate}}{Na^{+}\,{}^{-}CH_2\overset{O}{\overset{\|}{C}}OCH_2CH_3} + CH_3CH_2OH \quad (10.48)$$

Step 2.
$$CH_3\overset{O}{\overset{\|}{C}}-OCH_2CH_3 + {}^{-}CH_2\overset{O}{\overset{\|}{C}}OCH_2CH_3 \rightleftharpoons$$

$$CH_3\overset{{}^{-}O}{\overset{|}{C}}(-OCH_2CH_3)-CH_2\overset{O}{\overset{\|}{C}}-OCH_2CH_3 \rightleftharpoons \underset{\beta\text{-keto ester}}{CH_3\overset{O}{\overset{\|}{C}}CH_2\overset{O}{\overset{\|}{C}}OCH_2CH_3} + {}^{-}OCH_2CH_3 \quad (10.49)$$

Step 3.
$$CH_3\overset{O}{\overset{\|}{C}}CH_2\overset{O}{\overset{\|}{C}}OCH_2CH_3 + {}^{-}OCH_2CH_3 \longrightarrow$$

$$\underset{\text{enolate ion of a }\beta\text{-keto ester}}{CH_3\overset{O}{\overset{\|}{C}}-\overset{-}{C}H-\overset{O}{\overset{\|}{C}}OCH_2CH_3} + CH_3CH_2OH \quad (10.50)$$

In step 1, the base (sodium ethoxide) removes an α-hydrogen from the ester to form an ester enolate. In step 2, this ester enolate, acting as a nucleophile, adds to the carbonyl group of a second ester molecule, displacing ethoxide ion. This step follows the mechanism in eq. 10.31 and proceeds through a tetrahedral intermediate. These first two steps of the reaction are completely reversible.

Step 3 drives the equilibrium forward. In this step, the β-keto ester is converted to *its* enolate anion. The methylene (CH_2) hydrogens in ethyl acetoacetate are *α to two carbonyl groups* and hence are appreciably more acidic than ordinary α-hydrogens. They have a pK_a of 12 and are easily removed by the base (ethoxide ion) to form a resonance-stabilized β-keto enolate ion, *with the negative charge delocalized to both carbonyl oxygen atoms.*

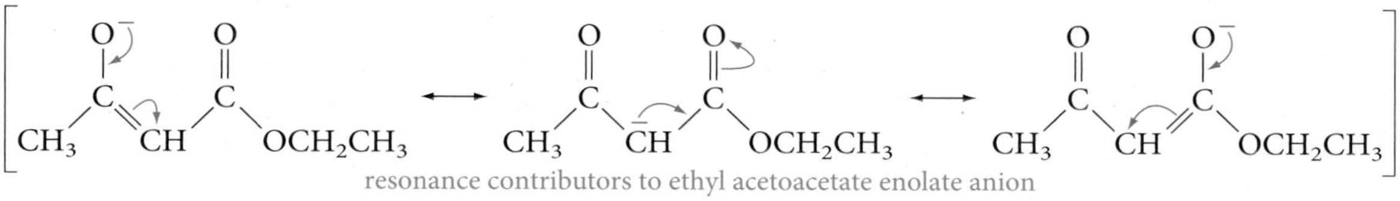

$$\left[CH_3C(-O^-)=CH-C(=O)OCH_2CH_3 \longleftrightarrow CH_3C(=O)-\overset{-}{C}H-C(=O)OCH_2CH_3 \longleftrightarrow CH_3C(=O)-CH=C(-O^-)OCH_2CH_3 \right]$$

resonance contributors to ethyl acetoacetate enolate anion

To complete the Claisen condensation, the solution is acidified, to regenerate the β-keto ester from its enolate anion.

EXAMPLE 10.8

Identify the product of the Claisen condensation of ethyl propanoate:

$$CH_3CH_2\overset{O}{\overset{\|}{C}}-OCH_2CH_3$$

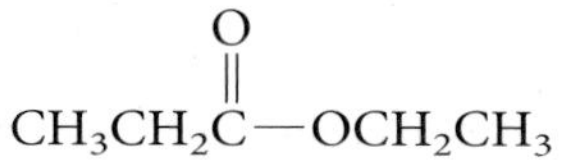

Solution The product is

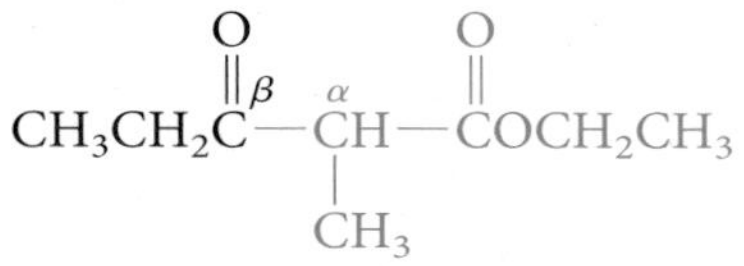

$$CH_3CH_2\overset{O}{\overset{\|}{C}}\,{}^{\beta}-\overset{\alpha}{C}H(CH_3)-\overset{O}{\overset{\|}{C}}OCH_2CH_3$$

The α-carbon of one ester molecule displaces the —OR group and becomes joined to the carbonyl carbon of the other ester. The product is always a β-keto ester.

PROBLEM 10.36 Using eqs. 10.48 through 10.50 as a model, write out the steps in the mechanism for the Claisen condensation of ethyl propanoate.

The Claisen condensation, like the aldol condensation (Sec. 9.17), is useful for making new carbon–carbon bonds. The resulting β-keto esters can be converted to a variety of useful products. For example, ethyl acetate can be converted to ethyl butanoate by the following sequence.

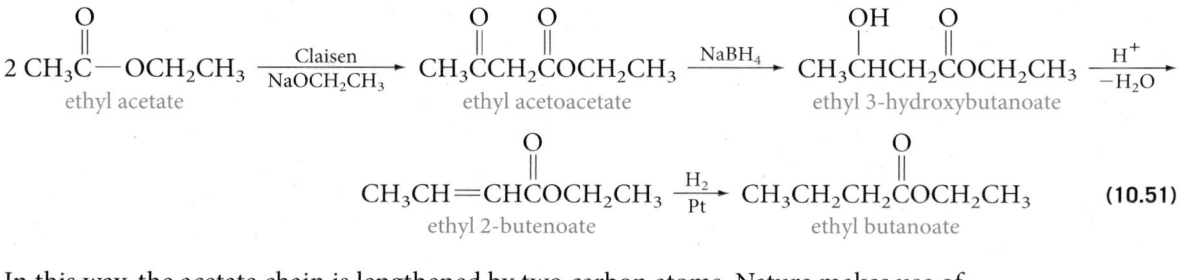

$$\underset{\text{ethyl acetate}}{2\,CH_3\overset{O}{\overset{\|}{C}}-OCH_2CH_3} \xrightarrow[\text{NaOCH}_2\text{CH}_3]{\text{Claisen}} \underset{\text{ethyl acetoacetate}}{CH_3\overset{O}{\overset{\|}{C}}CH_2\overset{O}{\overset{\|}{C}}OCH_2CH_3} \xrightarrow{NaBH_4} \underset{\text{ethyl 3-hydroxybutanoate}}{CH_3\overset{OH}{\overset{|}{C}}HCH_2\overset{O}{\overset{\|}{C}}OCH_2CH_3} \xrightarrow[-H_2O]{H^+}$$

$$\underset{\text{ethyl 2-butenoate}}{CH_3CH=CH\overset{O}{\overset{\|}{C}}OCH_2CH_3} \xrightarrow[Pt]{H_2} \underset{\text{ethyl butanoate}}{CH_3CH_2CH_2\overset{O}{\overset{\|}{C}}OCH_2CH_3} \qquad (10.51)$$

In this way, the acetate chain is lengthened by two carbon atoms. Nature makes use of a similar process, catalyzed by various enzymes, to construct the long-chain carboxylic acids that are components of fats and oils (Chapter 15).

REACTION SUMMARY

1. Preparation of Acids

a. From Alcohols or Aldehydes (Sec. 10.7)

$$RCH_2OH \xrightarrow{CrO_3,\ H_2SO_4,\ H_2O} RCO_2H \xleftarrow[\text{or } O_2 \text{ or } Ag^+]{CrO_3,\ H_2SO_4,\ H_2O} RCH{=}O$$

b. From Alkylbenzenes (Sec. 10.7)

$$ArCH_3 \xrightarrow[\text{or } O_2,\ Co^{+3}]{KMnO_4} ArCO_2H$$

c. From Grignard Reagents (Sec. 10.7)

$$RMgX + CO_2 \longrightarrow RCO_2MgX \xrightarrow{H_3O^+} RCO_2H$$

d. From Nitriles (Sec. 10.7)

$$RC{\equiv}N + 2\,H_2O \xrightarrow{H^+ \text{ or } HO^-} RCO_2H + NH_3$$

2. Reactions of Acids

a. Acid–Base (Secs. 10.4 and 10.6)

$$RCO_2H \rightleftharpoons RCO_2^- + H^+ \quad \text{(ionization)}$$

$$RCO_2H + NaOH \longrightarrow RCO_2^-Na^+ + H_2O \quad \text{(salt formation)}$$

b. Preparation of Esters (Secs. 10.10 and 10.12)

$$RCO_2H + R'OH \xrightarrow{H^+} RCO_2R' + H_2O$$

c. Preparation of Acid Chlorides (Sec. 10.18)

$$RCO_2H + SOCl_2 \longrightarrow RCOCl + HCl + SO_2$$

$$RCO_2H + PCl_5 \longrightarrow RCOCl + HCl + POCl_3$$

d. Preparation of Anhydrides (Sec. 10.19)

$$R{-}\overset{\overset{\displaystyle O}{\|}}{C}{-}Cl + Na^{+}\,{}^{-}O{-}\overset{\overset{\displaystyle O}{\|}}{C}{-}R' \longrightarrow R{-}\overset{\overset{\displaystyle O}{\|}}{C}{-}O{-}\overset{\overset{\displaystyle O}{\|}}{C}{-}R' + NaCl$$

e. Preparation of Amides (Sec. 10.20)

$$RCO_2^-NH_4^+ \xrightarrow{\text{heat}} RCONH_2 + H_2O$$

Also see reactions of esters, acid chlorides, and anhydrides in Section 10.21.

3. Reactions of Carboxylic Acid Derivatives

a. Saponification of Esters (Sec. 10.13)

$$RCO_2R' + NaOH \longrightarrow RCO_2^-Na^+ + R'OH$$

b. Ammonolysis of Esters (Sec. 10.14)

$$RCO_2R' + NH_3 \longrightarrow RCONH_2 + R'OH$$

c. Esters with Grignard Reagents (Sec. 10.15)

$$RCO_2R' \xrightarrow{2\,R''MgX} R{-}\underset{\underset{\displaystyle R''}{|}}{\overset{\overset{\displaystyle R''}{|}}{C}}{-}OMgX \xrightarrow{H_3O^+} R{-}\underset{\underset{\displaystyle R''}{|}}{\overset{\overset{\displaystyle R''}{|}}{C}}{-}OH$$

$$+\ R'OH$$

d. Reduction of Esters (Sec. 10.16)

$$RCO_2R' + LiAlH_4 \longrightarrow RCH_2OH + R'OH$$

e. Nucleophilic Acyl Substitution Reactions of Acid Chlorides and Anhydrides (Secs. 10.18 and 10.19)

$$R{-}\overset{O}{\overset{\|}{C}}{-}Cl \text{ or } R{-}\overset{O}{\overset{\|}{C}}{-}O{-}\overset{O}{\overset{\|}{C}}{-}R \;\begin{cases} \xrightarrow{H_2O} RCO_2H + HCl\ (\text{or } RCO_2H) \\ \xrightarrow{R'OH} RCO_2R' + HCl\ (\text{or } RCO_2H) \\ \xrightarrow{NH_3} RCONH_2 + NH_4Cl\ (\text{or } RCO_2H) \end{cases}$$

f. Hydrolysis of Amides (Sec. 10.20)

$$RCONH_2 + H_2O \xrightarrow{H^+ \text{ or } HO^-} RCO_2H + NH_3$$

g. Reduction of Amides (Sec. 10.20)

$$RCONH_2 \xrightarrow{LiAlH_4} RCH_2NH_2$$

h. Claisen Condensation (Sec. 10.22)

$$2\ RCH_2CO_2R' \xrightarrow[2.\ H_3O^+]{1.\ R'O^-Na^+} RCH_2\overset{O}{\overset{\|}{C}}\underset{R}{\underset{|}{C}}HCO_2R' + R'OH$$

MECHANISM SUMMARY

Nucleophilic Acyl Substitution (Secs. 10.11 and 10.17)

$$\bar{N}u{:} + \overset{R}{\underset{L}{>}}C{=}O \longrightarrow \left[\overset{Nu}{\underset{R\ \ L}{\diagdown}}C{-}\ddot{\underset{..}{O}}{:}^- \right] \longrightarrow \overset{R}{\underset{Nu}{>}}C{=}O + L{:}^-$$

ADDITIONAL PROBLEMS

OWL Interactive versions of these problems are assignable in OWL.

Nomenclature and Structure of Carboxylic Acids

10.37 Write a structural formula for each of the following acids:

a. 4-ethylhexanoic acid
b. 2-bromobutanoic acid
c. 3-chlorohexanoic acid
d. cyclopentanecarboxylic acid
e. 2-isopropylbenzoic acid
f. 3-oxooctanoic acid
g. *p*-toluic acid
h. 2-ethylbutanedioic acid
i. *p*-methoxyphenylacetic acid
j. 1-naphthoic acid
k. 2,3-dimethyl-3-butenoic acid

10.38 Name each of the following acids:

a. $(CH_3)_2C(Br)CH_2CH_2COOH$

b. $CH_3CH(OCH_3)CH(CH_3)COOH$

c. 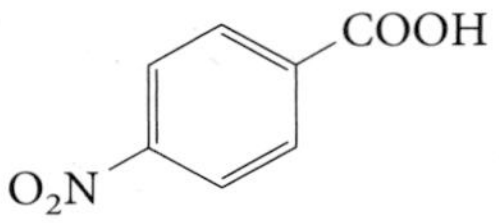 COOH, O_2N

d. 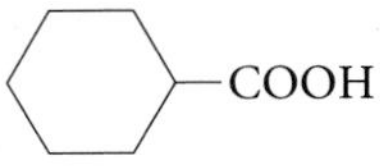 — COOH

e. $CH_2{=}CHCOOH$

f.

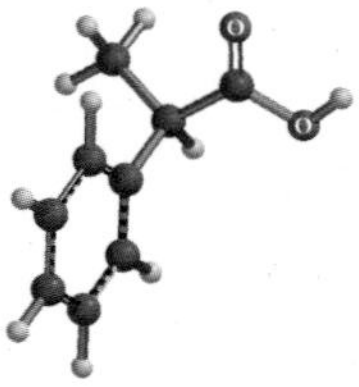

g. CH_3CF_2COOH

h. $HC{\equiv}CCH_2CO_2H$

Synthesis and Properties of Carboxylic Acids

10.39 Which will have the higher boiling point? Explain your reasoning.

a. CH_3CH_2COOH or $CH_3CH_2CH_2CH_2OH$

b. $CH_3CH_2CH_2CH_2COOH$ or $(CH_3)_3CCOOH$

10.40 In each of the following pairs of acids, which would be expected to be the stronger acid, and why?

a. $ClCH_2CO_2H$ and $BrCH_2CO_2H$

b. *o*-$BrC_6H_4CO_2H$ and *m*-$BrC_6H_4CO_2H$

c. Cl_3CCO_2H and F_3CCO_2H

d. $C_6H_5CO_2H$ and *p*-$CH_3OC_6H_4CO_2H$

e. $ClCH_2CH_2CO_2H$ and $CH_3CHClCO_2H$

10.41 Write a balanced equation for the reaction of

a. $ClCH_2CH_2CO_2H$ with KOH

b. $CH_3(CH_2)_4CO_2H$ with $Ca(OH)_2$

10.42 Give equations for the synthesis of

a. $CH_3CH_2CH_2CO_2H$ from $CH_3CH_2CH_2CH_2OH$

b. $CH_3CH_2CH_2CO_2H$ from $CH_3CH_2CH_2OH$ (two ways)

c. 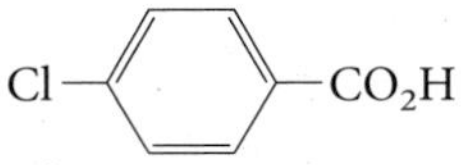 Cl — CO_2H from Cl — CH_3

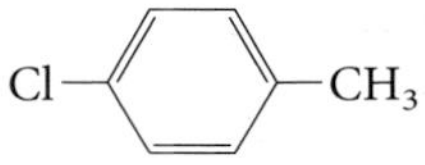

d. 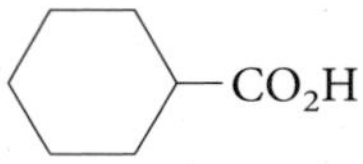 — CO_2H from — CN

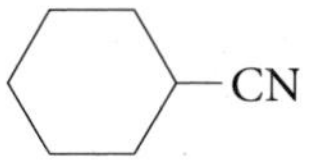

e. $CH_3OCH_2CO_2H$ from CH_2—CH_2 (O) (two steps)

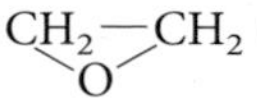

f. 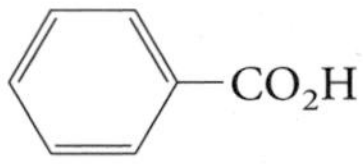 — CO_2H from — Br

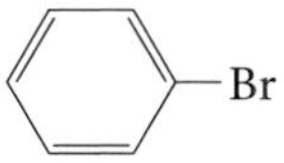

10.43 The Grignard route for the synthesis of $(CH_3)_3CCO_2H$ from $(CH_3)_3CBr$ (Example 10.5) is far superior to the nitrile route. Explain why.

Nomenclature and Structure of Carboxylic Acid Derivatives

10.44 Write a structure for each of the following compounds:

a. isobutyl acetate
b. isopropyl formate
c. sodium 2-chlorobutanoate
d. calcium acetate
e. phenyl benzoate
f. *o*-toluamide
g. 2-methoxybutanoyl chloride
h. benzonitrile
i. propanoic anhydride
j. 2-acetylcyclohexanecarboxylic acid
k. α-methyl-γ-butyrolactone

10.45 Name each of the following compounds:

a. Br—C_6H_4—$COO^- NH_4^+$ (para-substituted benzene ring)

b. $[CH_3(CH_2)_2CO_2^-]_2Ca^{2+}$

c. $(CH_3)_2CHCH_2CH_2COOC_6H_5$

d. $CF_3CO_2CH_3$

e. $HCONH_2$

f. $CH_3(CH_2)_2—\overset{O}{\overset{\|}{C}}—O—\overset{O}{\overset{\|}{C}}—(CH_2)_2CH_3$

10.46 Draw the structure of the mating pheromone of the female elephant, (*Z*)-7-dodecen-1-yl acetate (see page 302).

10.47 Organic emissions from mobile sources (cars, trucks, planes, and so on) become oxidized in the troposphere and can then assist the formation of particulate secondary organic aerosols. Such small particulate matter can penetrate deep into our lungs and cause acute irritations. It has been reported that carboxylic acids, such as benzoic acid, can form stable complexes with sulfuric acid (H_2SO_4) in a similar manner that carboxylic acids can form dimers (Sec. 10.2). Suggest a structure for a stable complex between benzoic acid and sulfuric acid.

Synthesis and Reactions of Esters

10.48 Write an equation for the Fischer esterification of butanoic acid ($CH_3CH_2CH_2CO_2H$) with ethanol.

10.49 Write out each step in the Fischer esterification of benzoic acid with methanol. (You may wish to use eq. 10.19 as a model.)

10.50 Starting from bromobenzene, provide a short synthesis of methyl benzoate ($C_6H_5CO_2CH_3$).

10.51 Write an equation for the reaction of propyl benzoate with

a. hot aqueous sodium hydroxide
b. ammonia (heat)
c. phenylmagnesium iodide (two equivalents), then H_3O^+
d. lithium aluminum hydride (two equivalents), then H_3O^+

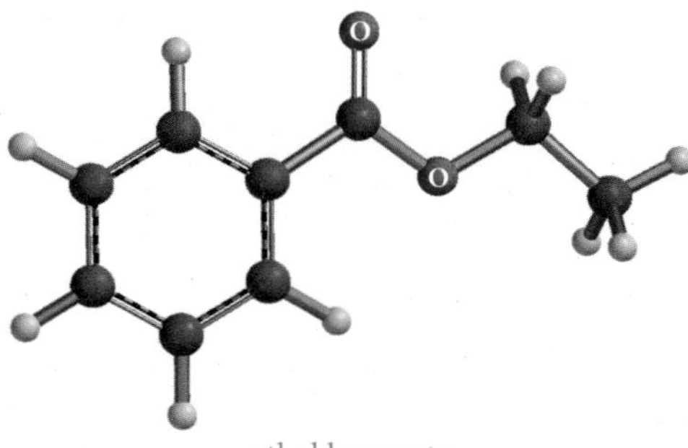

ethyl benzoate

10.52 Write out all of the steps in the mechanism for

a. saponification of $CH_3CH_2CO_2CH_2CH_3$

b. ammonolysis of $CH_3CH_2CO_2CH_2CH_3$

10.53 Identify the Grignard reagent and the ester that would be used to prepare

a. $CH_3CH_2{-}C(OH)(C_6H_5){-}CH_2CH_3$

b. $CH_3CH_2CH_2C(C_6H_5)_2OH$

Reactions of Carboxylic Acid Derivatives

10.54 Explain each difference in reactivity toward nucleophiles.

a. Esters are less reactive than ketones.
b. Benzoyl chloride is less reactive than cyclohexanecarbonyl chloride.

10.55 Write an equation for

a. hydrolysis of butanoyl chloride
b. ammonolysis of butanoyl bromide
c. 2-methylpropanoyl chloride + ethylbenzene + $AlCl_3$
d. succinic acid + heat (235°C)
e. benzoyl chloride with ethanol
f. esterification of 1-pentanol with acetic anhydride
g. phthalic anhydride + ethanol (1 equiv.) + H^+
h. phthalic anhydride + ethanol (excess) + H^+
i. adipoyl chloride + ammonia (excess)

10.56 Complete the equation for each of the following reactions:

a. $CH_3CH_2CH_2CO_2H + PCl_5 \longrightarrow$

b. $CH_3(CH_2)_6CO_2H + SOCl_2 \longrightarrow$

c. 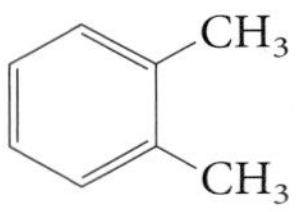 o-xylene (benzene ring with two adjacent CH_3 groups) $+ KMnO_4 \longrightarrow$

d.  $C_6H_5{-}CO_2^-NH_4^+$ + heat $\longrightarrow$

e. $CH_3(CH_2)_5CONH_2 + LiAlH_4 \longrightarrow$

f. 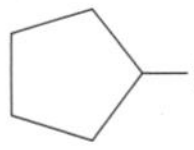 cyclopentyl$-CO_2CH_2CH_3 + LiAlH_4 \longrightarrow$

10.57 Considering the relative reactivities of ketones and esters toward nucleophiles, which of the following products seems the more likely?

$$CH_3\overset{O}{\overset{\|}{C}}CH_2CH_2CO_2CH_3 \xrightarrow{NaBH_4} CH_3\overset{O}{\overset{\|}{C}}CH_2CH_2CH_2OH \text{ or } CH_3\overset{OH}{\overset{|}{C}}HCH_2CH_2CO_2CH_3$$

10.58 Mandelic acid, which has the formula $C_6H_5CH(OH)COOH$, can be isolated from bitter almonds (called *mandel* in German). It is sometimes used in medicine to treat urinary infections. Devise a two-step synthesis of mandelic acid from benzaldehyde, using the latter's cyanohydrin (see Sec. 9.10) as an intermediate.

The Claisen Condensation

10.59 Write the structure of the Claisen condensation product of methyl 3-phenylpropanoate ($C_6H_5CH_2CH_2CO_2CH_3$), and show the steps in its formation.

10.60 Diethyl adipate, when heated with sodium ethoxide, gives the product shown, by an *intra*molecular Claisen condensation:

$$CH_3CH_2O\overset{O}{\overset{\|}{C}}{-}(CH_2)_4{-}\overset{O}{\overset{\|}{C}}OCH_2CH_3 \xrightarrow[2.\ H_3O^+]{1.\ NaOCH_2CH_3} \text{2-oxocyclopentyl}{-}\overset{O}{\overset{\|}{C}}OCH_2CH_3$$

diethyl adipate → ethyl 2-oxocyclopentanecarboxylate

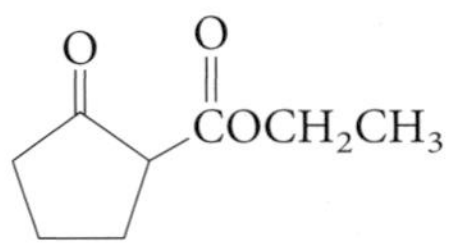

Write out the steps in a plausible mechanism for the reaction.

10.61 Analogous to the mixed aldol condensation (Sec. 9.18), mixed Claisen condensations are possible. Predict the structure of the product obtained when a mixture of ethyl benzoate and ethyl acetate is heated with sodium ethoxide in ethanol.

Miscellaneous Problems

10.62 Write the important resonance contributors to the structure of acetamide and tell which atoms lie in a single plane.

10.63 On page 299, "A Word About . . . Green Chemistry and Ibuprofen: A Case Study" showed how efficiently ibuprofen can be prepared on a large scale. Another nonsteroidal, anti-inflammatory drug (NSAID) is acetaminophen. Suggest a short synthesis of acetaminophen from *p*-hydroxyaniline.

H N CH_3 O HO

10.64 As noted in the "A Word About . . . Thioesters, Nature's Acyl-Activating Groups" on page 312, biological systems rely on thioesters for acyl transfer, but thioesters can be exploited in the laboratory as well. Provide the product for the treatment of

O HO S

with catalytic NaOH.

10.65 Consider the structure of the catnip ingredient nepetalactone (page 305).

a. Show with dotted lines that the structure is composed of two isoprene units.
b. Circle the stereogenic centers and determine their configurations (*R* or *S*).

10.66 The lactone shown below, known as *wine lactone*, is a sweet and coconut-like smelling odorant isolated recently from white wines such as Gewürztraminer.

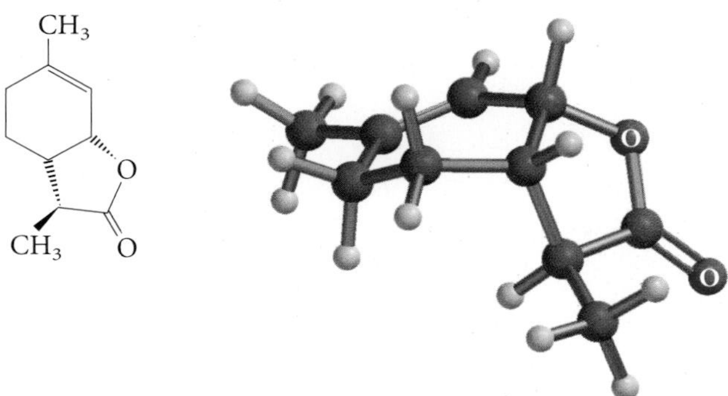

How many stereocenters are present, and what is the configuration (*R* or *S*) at each?

10.67 Provide a short synthesis of the ester-acid below, starting from maleic acid; see Section 10.1 for a structure of maleic acid. (Note, an ideal synthesis would avoid creating a mixture.)

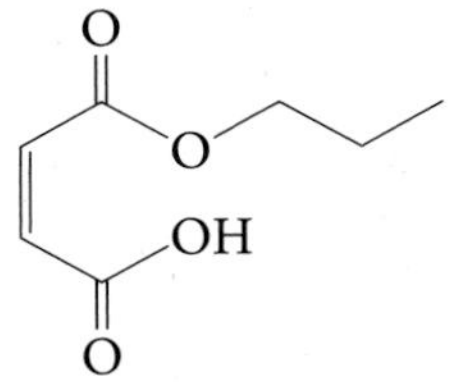

10.68 (5*R*,6*S*)-6-Acetoxy-5-hexadecanolide is a pheromone that attracts certain disease-carrying mosquitoes to sites where they like to lay their eggs. Such compounds might be used to lure these insects away from populated areas to locations where they can be destroyed. The last two steps in a recent synthesis of this compound are shown below. Provide reagents that would accomplish these transformations.

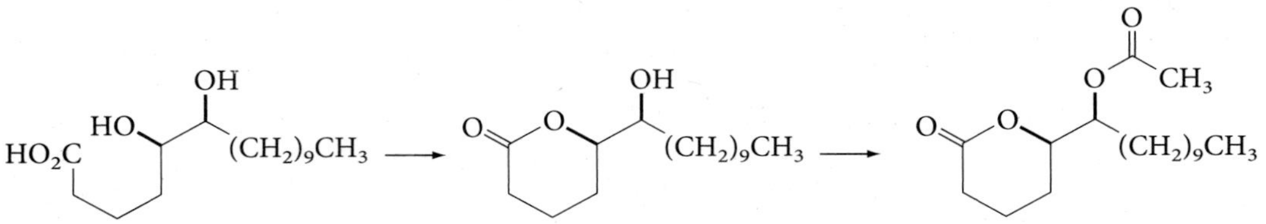

(5*R*, 6*S*)-6-Acetoxy-5-hexadecanolide

The painkiller morphine is obtained from opium, the dried sap of the unripe seed of the poppy *Papaver somniferum.*

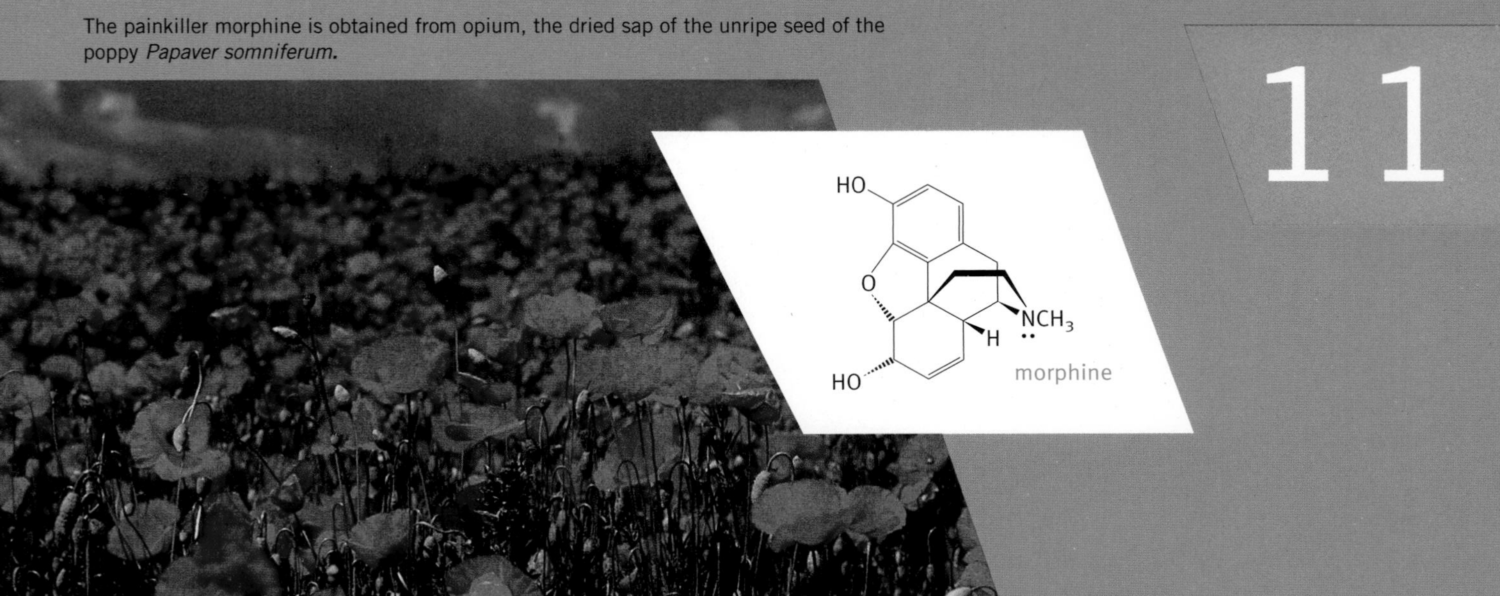

© Darek Karp / Animals Animals

Amines and Related Nitrogen Compounds

In this chapter, we will discuss the last of the major families of simple organic compounds—the amines. **Amines** are relatives of ammonia that abound in nature and play an important role in many modern technologies. Examples of important amines are the painkiller morphine, found in poppy seeds, and putrescine, one of several polyamines responsible for the unpleasant odor of decaying flesh. A diamine that is largely the creation of humans is 1,6-diaminohexane, used in the synthesis of nylon. Amine derivatives, known as quaternary ammonium salts, also touch our daily lives in the form of synthetic detergents. Several neurotoxins also belong to this family of compounds. They are toxic because they interfere with the key role that acetylcholine, also a quaternary ammonium salt, plays in the transmission of nerve impulses.

$$H_2\ddot{N}(CH_2)_4\ddot{N}H_2 \qquad H_2\ddot{N}(CH_2)_6\ddot{N}H_2 \qquad CH_3-\overset{CH_3}{\underset{CH_3}{\overset{|}{\underset{|}{N^+}}}}-CH_2CH_2-O-\overset{O}{\overset{\|}{C}}-CH_3$$

putrescine — 1,6-diaminohexane — acetylcholine

In this chapter, we will first describe the structure, preparation, chemical properties, and uses of some simple amines. Later in the chapter, we will discuss a few natural and synthetic amines with important biological properties.

OWL

Online homework for this chapter can be assigned in OWL, an online homework assessment tool.

11.1 Classification and Structure of Amines

Amines are organic bases derived from ammonia.

The relation between ammonia and amines is illustrated by the following structures:

$H{-}\ddot{N}{-}H$ with H on N	$R{-}\ddot{N}{-}H$ with H on N	$R{-}\ddot{N}{-}R$ with H on N	$R{-}\ddot{N}{-}R$ with R on N
ammonia	primary amine	secondary amine	tertiary amine

Primary amines have one organic group attached to nitrogen, **secondary** amines have two, and **tertiary** amines have three.

For convenience, amines are classified as **primary**, **secondary**, or **tertiary**, depending on whether one, two, or three organic groups are attached to the nitrogen. The *R* groups in these structures may be alkyl or aryl, and when two or more *R* groups are present, they may be identical to or different from one another. In some secondary and tertiary amines, the nitrogen may be part of a ring.

PROBLEM 11.1 Classify each of the following amines as primary, secondary, or tertiary:

a. $(CH_3)_3CCH_2NH_2$

b. (five-membered ring with N–H)

c. CH_3–(benzene ring)–NH_2

d. $(CH_3)_2N$–(benzene ring)

The nitrogen atom in amines is trivalent. Moreover, the nitrogen carries an unshared electron pair. Therefore, the nitrogen orbitals are sp^3-hybridized, and the overall geometry is pyramidal (nearly tetrahedral), as shown for trimethylamine in Figure 11.1. From this geometry, one might think that an amine with three different groups attached to the nitrogen would be chiral, with the unshared electron pair acting as the fourth group. This is true in principle, but in practice, the two enantiomers usually interconvert rapidly through inversion, via an "umbrella-in-the-wind" type of process, and are not resolvable.

$$R^1R^2R^3N \longrightarrow [R^1{-}N(R^2)(R^3)]^{TS} \longrightarrow R^1R^2R^3N \qquad (11.1)$$

planar transition state

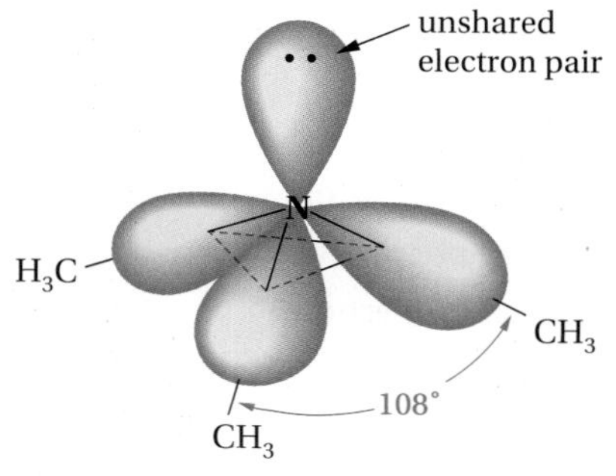

(a)

(b)

▰ **Figure 11.1**
(a) An orbital view of the pyramidal bonding in trimethylamine.
(b) Top view of a space-filling model of trimethylamine. The center ball represents the orbital with the unshared electron pair.

11.2 Nomenclature of Amines

Amines can be named in several different ways. Commonly, simple amines are named by specifying the alkyl groups attached to the nitrogen and adding the suffix *-amine.*

$CH_3CH_2NH_2$	$(CH_3CH_2)_2NH$	$(CH_3CH_2)_3N$
ethylamine (primary)	diethylamine (secondary)	triethylamine (tertiary)

In the IUPAC system, the amino group, $—NH_2$, is named as a substituent, as in the following examples:

$CH_3CH_2NH_2$ — aminoethane

$\overset{1}{C}H_3\overset{2}{C}H(NH_2)\overset{3}{C}H_2\overset{4}{C}H_2\overset{5}{C}H_3$ — 2-aminopentane

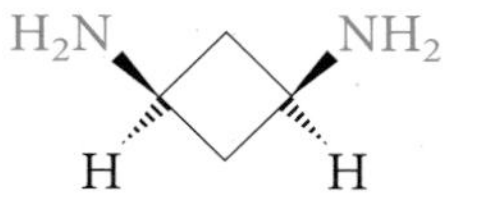

H_2N, NH_2, H, H (cyclobutane ring) — *cis*-1,3-diaminocyclobutane

In this system, secondary or tertiary amines are named by using a prefix that includes all but the longest carbon chain, as in

$CH_3NH\overset{1}{C}H_2\overset{2}{C}H_2\overset{3}{C}H_3$ — 1-methylaminopropane

$CH_3N(CH_2CH_3)—\overset{1}{C}H_2\overset{2}{C}H_2\overset{3}{C}H_3$ — 1-(ethylmethylamino)propane

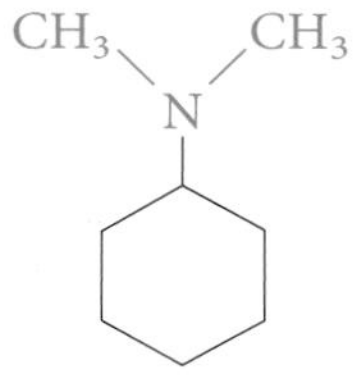

$(CH_3)_2N$–cyclohexane — dimethylaminocyclohexane

Recently, *Chemical Abstracts* (CA) introduced a system for naming amines that is rational and easy to use. In this system, amines are named as **alkanamines**. For example,

Amines are named as **alkanamines** in the CA system.

$CH_3CH_2CH_2NH_2$ — propanamine

$CH_3CH(NH_2)CH_3$ — 2-propanamine

$CH_3CH(NHCH_3)CH_2CH_2CH_3$ — *N*-methyl-2-pentanamine

EXAMPLE 11.1

Name cyclohexyl—$N(CH_3)_2$ by the CA system.

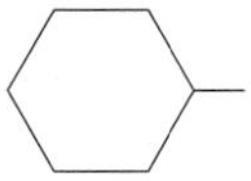

Solution The largest alkyl group attached to nitrogen is used as the root of the name. The compound is *N,N*-dimethylcyclohexanamine.

PROBLEM 11.2 Name $CH_3CH_2CH(N(CH_3)_2)CH_2CH_3$ by the CA system.

When other functional groups are present, the amino group is named as a substituent:

$\overset{4}{C}H_3\overset{3}{C}H(NH_2)\overset{2}{C}H_2\overset{1}{C}O_2H$ — 3-aminobutanoic acid

$H_2N\overset{1}{C}H_2\overset{2}{C}H_2\overset{3}{C}(=O)\overset{4}{C}H_2\overset{5}{C}H_3$ — 1-amino-3-pentanone

$CH_3NH\overset{2}{C}H_2\overset{1}{C}H_2OH$ — 2-methylaminoethanol

Aromatic amines are named as derivatives of aniline. In the CA system, aniline is called benzenamine; these CA names are shown in parentheses.

NH_2

aniline
(benzenamine)

NH_2 / Br

p-bromoaniline
(4-bromobenzenamine)

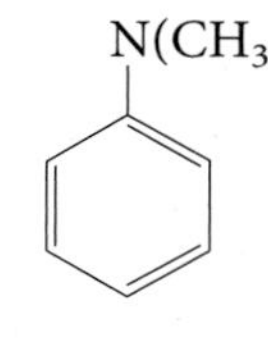

N,N-dimethylaniline
(*N,N*-dimethylbenzenamine)

$NHCH_3$ / CH_3

m-methyl-*N*-methylaniline, or
N-methyl-*m*-toluidine
(*N*-methyl-3-methylbenzenamine)

EXAMPLE 11.2

Give an acceptable name for the following compounds:

a. $(CH_3)_2CHCH_2NH_2$

b. $CH_3NHCH_2CH_3$

c. (benzene ring with Br, Br, NH_2)

d. (cyclopentane ring with OH, NH_2)

Solution

a. isobutylamine (common); 1-amino-2-methylpropane (IUPAC); 2-methylpropanamine (CA)
b. ethylmethylamine (common); methylaminoethane (IUPAC); *N*-methylethanamine (CA)
c. 3,5-dibromoaniline (common, IUPAC); 3,5-dibromobenzenamine (CA)
d. *trans*-2-aminocyclopentanol (only name)

PROBLEM 11.3 Give an acceptable name for the following compounds:

a. $(CH_3)_3CCH_2NH_2$ b. $H_2NCH_2CH_2CH_2OH$ c. O_2N–(benzene ring)–NH_2

PROBLEM 11.4 Write the structure for

a. dibutylamine
b. 2,4,6-trimethylaniline
c. 3-aminopentane
d. *N,N*-dimethyl-2-hexanamine

11.3 Physical Properties and Intermolecular Interactions of Amines

Table 11.1 lists the boiling points of some common amines. Methylamine and ethylamine are gases, but primary amines with three or more carbons are liquids. Primary amines boil well above alkanes with comparable molecular weights, but below comparable alcohols, as shown in Table 11.2. Intermolecular N—H···N hydrogen bonds are important and raise the boiling points of primary and secondary amines but are not as strong as the O—H···O bonds of alcohols (see Sec. 7.4). The reason for this is that nitrogen is not as electronegative as oxygen (see Table 1.4).

Table 11.1 The Boiling Points of Some Simple Amines

Name	Formula	bp, °C
ammonia	NH_3	−33.4
methylamine	CH_3NH_2	−6.3
dimethylamine	$(CH_3)_2NH$	7.4
trimethylamine	$(CH_3)_3N$	2.9
ethylamine	$CH_3CH_2NH_2$	16.6
propylamine	$CH_3CH_2CH_2NH_2$	48.7
butylamine	$CH_3CH_2CH_2CH_2NH_2$	77.8
aniline	$C_6H_5NH_2$	184.0

Table 11.2 A Comparison of Alkane, Amine, and Alcohol Boiling Points*

alkane	CH_3CH_3 (30) bp −88.6°C	$CH_3CH_2CH_3$ (44) bp −42.1°C
amine	CH_3NH_2 (31) bp −6.3°C	$CH_3CH_2NH_2$ (45) bp +16.6°C
alcohol	CH_3OH (32) bp +65.0°C	CH_3CH_2OH (46) bp +78.5°C

*Molecular weights (in g/mol) are given in parentheses.

PROBLEM 11.5 Explain why the tertiary amine $(CH_3)_3N$ boils so much lower than its primary isomer $CH_3CH_2CH_2NH_2$.

All three classes of amines can form hydrogen bonds with the —OH group of water (that is, O—H· · ·N). Primary and secondary amines can also form hydrogen bonds with the oxygen atom in water: N—H· · ·O. Thus, most simple amines with up to five or six carbon atoms are either completely or appreciably soluble in water.

Now we will describe some ways in which amines can be prepared.

11.4 Preparation of Amines; Alkylation of Ammonia and Amines

Ammonia reacts with alkyl halides to give amines via a two-step process. The first step is a nucleophilic substitution reaction (S_N2).

$$H_3N\!: + R{-}X \longrightarrow R{-}\overset{+}{N}H_3\ X^- \quad \textbf{(11.2)}$$

ammonia → alkylammonium halide

The free amine can then be obtained from its salt by treatment with a strong base.

$$R{-}\overset{+}{N}H_3\,X^- + NaOH \longrightarrow \underset{\text{primary amine}}{RNH_2} + H_2O + Na^+X^- \tag{11.3}$$

Primary, secondary, and tertiary amines can be similarly alkylated.

$$\underset{\text{primary amine}}{R\ddot{N}H_2} + R{-}X \longrightarrow R_2\overset{+}{N}H_2\,X^- \xrightarrow{NaOH} \underset{\text{secondary amine}}{R_2NH} \tag{11.4}$$

$$\underset{\text{secondary amine}}{R_2\ddot{N}H} + R{-}X \longrightarrow R_3\overset{+}{N}H\,X^- \xrightarrow{NaOH} \underset{\text{tertiary amine}}{R_3N} \tag{11.5}$$

$$\underset{\text{tertiary amine}}{R_3\ddot{N}} + R{-}X \longrightarrow \underset{\text{quaternary ammonium salt}}{R_4N^+\,X^-} \tag{11.6}$$

Unfortunately, mixtures of products are often obtained in these reactions because the starting ammonia or amine and the alkylammonium ion formed in the S_N2 step can equilibrate, as in the following equation:

$$NH_3 + R\overset{+}{N}H_3\,X^- \rightleftharpoons NH_4{}^+X^- + RNH_2 \tag{11.7}$$

So, in the reaction of ammonia with an alkyl halide (eq. 11.2), some primary amine is formed (eq. 11.7), and it may be further alkylated (eq. 11.4) to give a secondary amine, and so on. By adjusting the ratio of the reactants, however, a good yield of one desired amine may be obtained. For example, with a large excess of ammonia, the primary amine is the major product.

Aromatic amines can often be alkylated selectively.

$$\underset{\text{aniline}}{C_6H_5NH_2} \xrightarrow{CH_3I} \underset{N\text{-methylaniline}}{C_6H_5NHCH_3} \xrightarrow{CH_3I} \underset{N,N\text{-dimethylaniline}}{C_6H_5N(CH_3)_2} \tag{11.8}$$

Tobacco plant.

The alkylation can be intramolecular, as in the following final step in a laboratory synthesis of nicotine:

$$\text{3-pyridyl–CH(Br)–CH}_2\text{CH}_2\text{CH}_2\text{–}\ddot{N}HCH_3 \xrightarrow[-HBr]{\text{Intramolecular } S_N2} \underset{\text{nicotine}}{\text{3-pyridyl–(pyrrolidine)–}\ddot{N}CH_3} \tag{11.9}$$

EXAMPLE 11.3

Write an equation for the synthesis of benzylamine, $C_6H_5-CH_2NH_2$.

Solution

$$C_6H_5-CH_2X + 2\ddot{N}H_3 \longrightarrow C_6H_5-CH_2\ddot{N}H_2 + NH_4{}^+X^-$$

(X = Cl, Br, or I)

Use of excess ammonia helps prevent further substitution.

PROBLEM 11.6 Complete equations for the following reactions:

a. $CH_3CH_2CH_2CH_2Br + 2\ NH_3 \longrightarrow$

b. $CH_3CH_2I + 2(CH_3CH_2)_2NH \longrightarrow$

c. $(CH_3)_3N + CH_3I \longrightarrow$

d. $CH_3CH_2CH_2NH_2 + C_6H_5-CH_2Br \longrightarrow$

PROBLEM 11.7 Give a synthesis of $C_6H_5-NHCH_2CH_3$ from aniline.

11.5 Preparation of Amines; Reduction of Nitrogen Compounds

All bonds to the nitrogen atom in amines are either N—H or N—C bonds. Nitrogen in ammonia or amines is therefore in a reduced form. It is not surprising, then, that organic compounds in which a nitrogen atom is present in a more oxidized form can be reduced to amines by appropriate reducing agents. Several examples of this useful synthetic approach to amines are described here.

The best route to *aromatic primary amines* is by *reduction of the corresponding nitro compounds,* which are in turn prepared by electrophilic aromatic nitration. The nitro group is easily reduced, either catalytically with hydrogen or by chemical-reducing agents.

$$\underset{p\text{-nitrotoluene}}{CH_3-C_6H_4-NO_2} \xrightarrow[\text{1. SnCl}_2\text{,HCl} \quad \text{2. NaOH,H}_2\text{O}]{3\ H_2\text{, Ni catalyst} \quad \text{or}} \underset{p\text{-toluidine}}{CH_3-C_6H_4-NH_2} + 2\ H_2O \qquad (11.10)$$

EXAMPLE 11.4

Devise a synthesis of *p*-chloroaniline, $Cl-C_6H_4-NH_2$, from chlorobenzene.

Solution Chlorobenzene is first nitrated; —Cl is an *o,p*-directing group, so the major product is *p*-chloronitrobenzene. This product is then reduced.

$$\text{C}_6\text{H}_5\text{Cl} \xrightarrow[\text{H}_2\text{SO}_4]{\text{HONO}_2} p\text{-ClC}_6\text{H}_4\text{NO}_2 \xrightarrow[\text{Ni}]{\text{H}_2} p\text{-ClC}_6\text{H}_4\text{NH}_2$$

PROBLEM 11.8 Give a synthesis for H_2N–$C_6H_3(NH_2)$–CH_3 (2,4-diaminotoluene; NH_2 groups para and ortho to CH_3) from toluene.

As described in the previous chapter (eq. 10.43), *amides* can be *reduced to amines* with lithium aluminum hydride.

$$\text{R—}\overset{\text{O}}{\overset{\|}{\text{C}}}\text{—N}\begin{matrix}\text{R}'\\\text{R}''\end{matrix} \xrightarrow{\text{LiAlH}_4} \text{RCH}_2\text{N}\begin{matrix}\text{R}'\\\text{R}''\end{matrix} \quad \text{(R}' \text{ and R}'' \text{ may be H or organic groups.)} \tag{11.11}$$

Depending on the structures of R′ and R″, we can obtain primary, secondary, or tertiary amines in this way.

EXAMPLE 11.5

Complete the equation $CH_3\overset{O}{\overset{\|}{C}}NHCH_2CH_3 \xrightarrow{LiAlH_4}$.

Solution The C=O group is reduced to CH_2. The product is the secondary amine $CH_3CH_2NHCH_2CH_3$.

PROBLEM 11.9 Show how $CH_3CH_2N(CH_3)_2$ can be synthesized from an amide.

Reduction of nitriles (cyanides) gives *primary amines.*

$$\text{R—C}\equiv\text{N} \xrightarrow[\text{or H}_2\text{, Ni}]{\text{LiAlH}_4} \text{RCH}_2\text{NH}_2 \tag{11.12}$$

EXAMPLE 11.6

Complete the equation $NCCH_2CH_2CH_2CH_2CN \xrightarrow[\text{Ni catalyst}]{\text{excess } H_2}$.

Solution Both CN groups are reduced. The product $H_2N—(CH_2)_6—NH_2$, or 1,6-diaminohexane, is one of two raw materials for the manufacture of nylon (page 412).

PROBLEM 11.10 Devise a synthesis of

$$C_6H_5—CH_2CH_2NH_2 \text{ from } C_6H_5—CH_2Br$$

Aldehydes and ketones undergo **reductive amination** when treated with ammonia, primary, or secondary amines, to give primary, secondary, or tertiary amines, respectively. The most commonly used laboratory reducing agent for this purpose is the metal hydride sodium cyanoborohydride, $NaBH_3CN$.

Aldehydes and ketones undergo **reductive amination** when treated with amines in the presence of $NaBH_3CN$.

$$\underset{\text{aldehyde or ketone}}{>C=\ddot{O}:} + \underset{\text{primary amine}}{R\ddot{N}H_2} \underset{}{\overset{-H_2O}{\rightleftharpoons}} \underset{\text{imine}}{\left[>C=\ddot{N}R\right]} \xrightarrow{NaBH_3CN} \underset{\text{secondary amine}}{H—\overset{|}{\underset{|}{C}}—\ddot{N}HR} \quad (11.13)$$

The reaction involves nucleophilic attack on the carbonyl group, leading to an imine (in the case of ammonia or primary amines; compare with eq. 9.31) or an iminium ion with secondary amines. The reducing agent then reduces the C=N bond.

PROBLEM 11.11 Using eq. 11.13 as a guide, devise a synthesis of 2-aminopentane, shown below, from 2-pentanone.

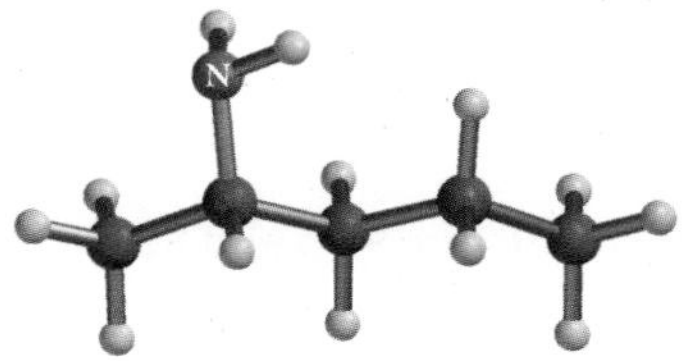

Now that we know several ways to make amines, let us examine some of their properties.

11.6 The Basicity of Amines

The unshared pair of electrons on the nitrogen atom dominates the chemistry of amines. Because of this electron pair, amines are both basic and nucleophilic.

Aqueous solutions of amines are basic because of the following equilibrium:

$$\underset{\text{amine}}{>N:} + H—\ddot{O}H \rightleftharpoons \underset{\text{ammonium ion}}{>N^+—H} + \underset{\text{hydroxide ion}}{^-:\ddot{O}H} \quad (11.14)$$

EXAMPLE 11.7

Write an equation that shows why aqueous solutions of ethylamine are basic.

Solution $CH_3CH_2\ddot{N}H_2 + H_2O \rightleftharpoons CH_3CH_2\overset{+}{N}H_3 + HO^-$

ethylamine → ethylammonium ion

Amines are more basic than water. They accept a proton from water, producing hydroxide ion, so their solutions are basic.

PROBLEM 11.12 Write an equation representing the equilibrium in an aqueous solution of trimethylamine, $(CH_3)_3N$.

An *amine* and its *ammonium ion* (eq. 11.14) are related as a *base* and its *conjugate acid*. For example, RNH_3^+ is the conjugate acid of the primary amine RNH_2. It is convenient, when comparing basicities of different amines, to compare instead the acidity constants (pK_a's) of their conjugate acids in water. Equation 11.15 expresses this acidity for a primary alkylammonium ion.

$$\underset{\text{conjugate acid}}{R\overset{+}{N}H_3} + H_2O \rightleftharpoons \underset{\text{base}}{RNH_2} + H_3O^+ \quad \textbf{(11.15)}$$

$$K_a = \frac{[RNH_2][H_3O^+]}{[RNH_3^+]}$$

The larger the K_a (or the smaller the pK_a), the stronger $R\overset{+}{N}H_3$ is as an acid, or the weaker RNH_2 is as a base.

EXAMPLE 11.8

The pK_a's of NH_4^+ and $CH_3\overset{+}{N}H_3$ are 9.30 and 10.64, respectively. Which is the stronger base, NH_3 or CH_3NH_2?

Solution NH_4^+ is the stronger acid (lower pK_a). Therefore, NH_3 is the *weaker* base, and CH_3NH_2 is the *stronger* base.

Table 11.3 lists some amine basicities. Alkylamines are approximately 10 times as basic as ammonia. Recall that alkyl groups are electron-donating relative to hydrogen. This electron-donating effect stabilizes the ammonium ion (positive charge) relative to the free amine (eq. 11.14). Hence it decreases the acidity of the ammonium ion, or increases the basicity of the amine. In general, *electron-donating groups increase the basicity of amines, and electron-withdrawing groups decrease their basicity.*

PROBLEM 11.13 Do you expect $ClCH_2CH_2NH_2$ to be a stronger or weaker base than $CH_3CH_2NH_2$? Explain.

Table 11.3 Basicities of Some Common Amines, Expressed as pK_a of the Corresponding Ammonium Ions

Name	Formula: Amine	Formula: Ammonium ion	pK_a of the ammonium ion
ammonia	NH_3	NH_4^+	9.30
methylamine	CH_3NH_2	$CH_3NH_3^+$	10.64
dimethylamine	$(CH_3)_2NH$	$(CH_3)_2NH_2^+$	10.71
trimethylamine	$(CH_3)_3N$	$(CH_3)_3NH^+$	9.77
ethylamine	$CH_3CH_2NH_2$	$CH_3CH_2NH_3^+$	10.67
propylamine	$CH_3CH_2CH_2NH_2$	$CH_3CH_2CH_2NH_3^+$	10.58
aniline	$C_6H_5NH_2$	$C_6H_5NH_3^+$	4.62
N-methylaniline	$C_6H_5NHCH_3$	$C_6H_5NH_2(CH_3)^+$	4.85
N,N-dimethylaniline	$C_6H_5N(CH_3)_2$	$C_6H_5NH(CH_3)_2^+$	5.04
p-chloroaniline	*p*-$ClC_6H_4NH_2$	*p*-$ClC_6H_4NH_3^+$	3.98

Aromatic amines are much weaker bases than aliphatic amines or ammonia. For example, aniline is less basic than cyclohexylamine by nearly a million times.

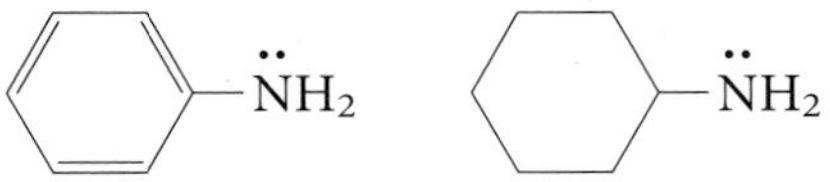

	aniline	cyclohexylamine
pK_a of ammonium ion	4.62	9.8

The reason for this huge difference is the resonance delocalization of the unshared electron pair that is possible in aniline, but not in cyclohexylamine (Figure 11.2).

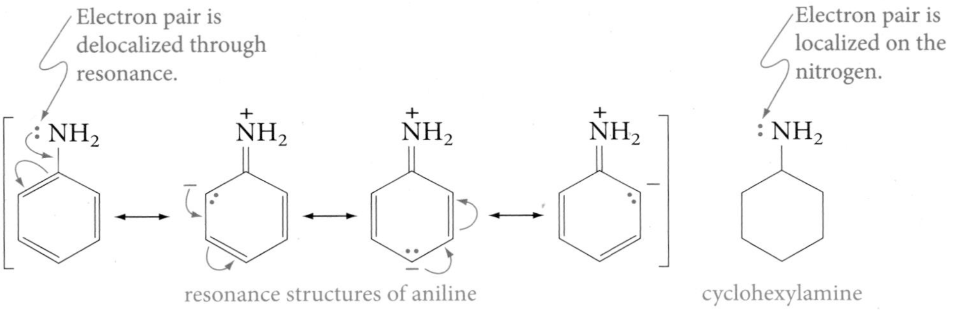

Electron pair is delocalized through resonance.

Electron pair is localized on the nitrogen.

resonance structures of aniline

cyclohexylamine

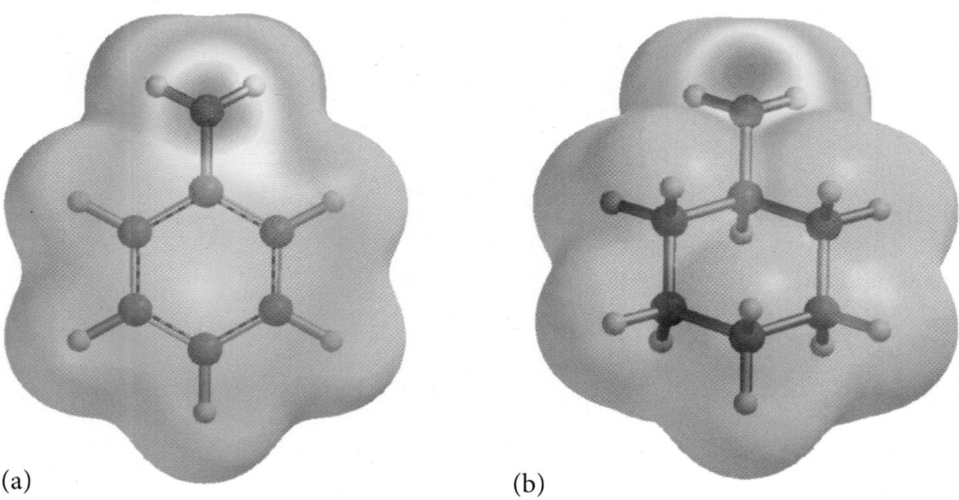

Figure 11.2
Electrostatic potential maps of (a) aniline and (b) cyclohexylamine, showing the enhanced electron-rich character of the N in (b).

Resonance stabilizes the unprotonated form of aniline. This shifts the equilibrium in eq. 11.15 to the right, increasing the acidity of the anilinium ion or decreasing the basicity of aniline. Another way to describe the situation is to say that the unshared electron pair in aniline is delocalized and therefore less available for donation to a proton than is the electron pair in cyclohexylamine.

PROBLEM 11.14 Compare the basicities of the last four amines in Table 11.3, and explain the reasons for the observed basicity order.

PROBLEM 11.15 Place aniline, *p*-toluidine, and *p*-nitroaniline in order of increasing basicity.

X—C_6H_4—NH_2

X = H aniline
X = CH_3 *p*-toluidine
X = NO_2 *p*-nitroaniline

11.7 Comparison of the Basicity and Acidity of Amines and Amides

Amines and amides each have nitrogens with an unshared electron pair. There is a huge difference, however, in their basicities. Aqueous solutions of *amines* are *basic;* aqueous solutions of *amides* are essentially *neutral.* Why this striking difference?

The answer lies in their structures, as illustrated in the following comparison of a primary amine with a primary amide:

localized; available for protonation

$R-\ddot{N}H_2$
amine

delocalized; less available for protonation

$\left[R-\overset{\overset{\displaystyle :\ddot{O}:}{\|}}{C}-\ddot{N}H_2 \longleftrightarrow R-\overset{\overset{\displaystyle :\ddot{O}:^-}{|}}{C}=\overset{+}{N}H_2 \right]$
amide

In the amine, the electron pair is mainly localized on the nitrogen. In the amide, the electron pair is delocalized to the carbonyl oxygen. The effect of this delocalization is seen in the low pK_a values for the conjugate acids of amides, compared with those for the conjugate acids of amines, for example:

	$CH_3CH_2\overset{+}{N}H_3$	$CH_3C(=\overset{+}{O}H)\ddot{N}H_2$
conjugate acid of:	ethylamine	acetamide
pK_a:	10.67	−0.6

Notice that amides are not protonated on nitrogen, but instead on the carbonyl oxygen. This is because protonation on oxygen gives a resonance-stabilized cation, while protonation on nitrogen does not.

Primary and secondary amines and amides have N—H bonds, and one might expect that they could on occasion behave as acids (proton donors).

$$R-\ddot{N}H_2 \rightleftharpoons R-\ddot{N}H^- + H^+ \qquad K_a \cong 10^{-40} \tag{11.16}$$

Primary amines are exceedingly weak acids, much weaker than alcohols. Their pK_a is about 40, compared with about 16 for alcohols. The main reason for the difference is that nitrogen is much less electronegative than oxygen and thus cannot stabilize a negative charge nearly as well.

Amides, on the other hand, are *much stronger acids than amines;* in fact, their pK_a (about 15) is comparable to that of alcohols:

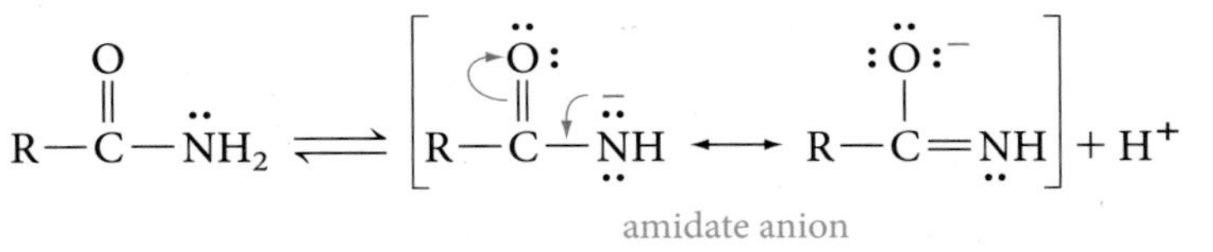

$$R-\overset{O}{\overset{\|}{C}}-\ddot{N}H_2 \rightleftharpoons \left[R-\overset{:\ddot{O}:}{\overset{\|}{C}}-\ddot{N}H^- \longleftrightarrow R-\overset{:\ddot{O}:^-}{\overset{|}{C}}=\ddot{N}H \right] + H^+ \qquad K_a \cong 10^{-15} \tag{11.17}$$

amidate anion

One reason is that the negative charge of the **amidate anion** can be delocalized through resonance. Another reason is that the nitrogen in an amide carries a partial positive charge (see page 338), making it easy to lose the attached proton, which is also positive.

The **amidate anion** is formed by removal of a proton from the amide nitrogen.

It is important to understand these differences between amines and amides, not only because they involve important chemical principles, but also because they help us understand the chemistry of certain natural products, such as peptides and proteins.

PROBLEM 11.16 Place the following compounds (a) in order of increasing basicity and (b) in order of increasing acidity.

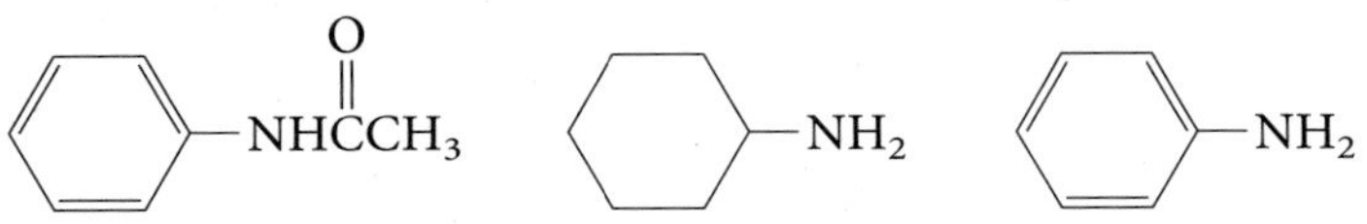

$C_6H_5-NHCCH_3$ (C=O) — acetanilide; $C_6H_{11}-NH_2$ — cyclohexylamine; $C_6H_5-NH_2$ — aniline

acetanilide cyclohexylamine aniline

11.8 Reaction of Amines with Strong Acids; Amine Salts

Alkylamines react with strong acids to form **alkylammonium salts**.

Amines react with strong acids to form **alkylammonium salts**. An example of this reaction for a primary amine and HCl is as follows:

$$\underset{\text{primary amine}}{R-\ddot{N}H_2} + HCl \longrightarrow \underset{\text{an alkylammonium chloride}}{R\overset{+}{N}H_3\ Cl^-} \quad (11.18)$$

EXAMPLE 11.9

Complete the following acid–base reaction, and name the product.

$$CH_3CH_2NH_2 + HI \longrightarrow$$

Solution

$$\underset{\text{ethylamine}}{CH_3CH_2\ddot{N}H_2} + HI \longrightarrow \underset{\text{ethylammonium iodide}}{CH_3CH_2N^+H_3\ I^-}$$

PROBLEM 11.17 Complete the following equation, and name the product:

$$C_6H_5-NH_2 + HCl \longrightarrow$$

This type of reaction is used to separate or extract amines from neutral or acidic water-insoluble substances. Consider, for example, a mixture of *p*-toluidine and *p*-nitrotoluene, which might arise from a preparation of the amine that for some reason does not go to completion (eq. 11.10). The amine can be separated from the unreduced nitro compound by the following scheme:

$$\underset{\substack{p\text{-toluidine}\\ \text{bp }200^\circ\text{C}}}{p\text{-}CH_3C_6H_4NH_2} + \underset{\substack{p\text{-nitrotoluene}\\ \text{bp }238^\circ\text{C}}}{p\text{-}CH_3C_6H_4NO_2} \xrightarrow{\text{shake with aqueous HCl and ether}} \begin{cases} \text{ether layer} \rightarrow p\text{-}CH_3C_6H_4NO_2 \quad \text{evaporate the ether} \\ \text{water layer} \rightarrow \underset{\text{amine salt}}{p\text{-}CH_3C_6H_4\overset{+}{N}H_3\ Cl^-} \xrightarrow[H_2O]{NaOH} \underset{\text{free amine}}{p\text{-}CH_3C_6H_4NH_2} + Na^+Cl^- + H_2O \end{cases} \quad (11.19)$$

A WORD ABOUT... Alkaloids and the Dart-Poison Frogs

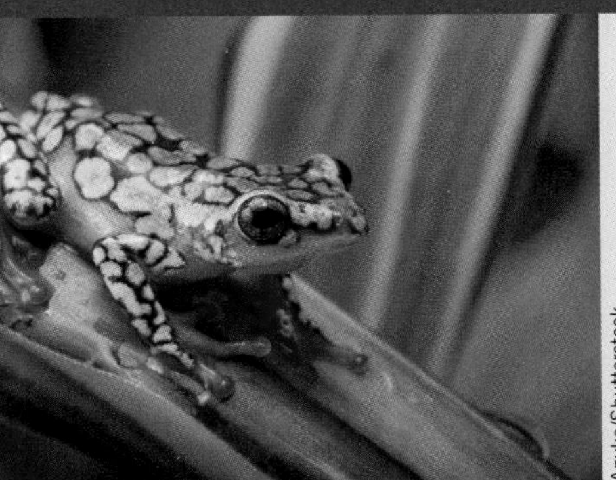
Anyka/Shutterstock

A dart-poison frog.

Alkaloids are basic, nitrogen-containing compounds of plant or animal origin, often having complex structures and significant pharmacological properties. Nicotine (from tobacco) and the amine obtained by deprotonation of squalamine (from sharks) are examples of alkaloids already mentioned in this chapter. Several more will appear in Chapter 13.

One prolific producer of alkaloids is the *dart-poison frogs*, a family of colorful and petite amphibians native to Costa Rica, Panama, Ecuador, and Colombia. The intense coloration of these frogs serves as a warning to potential predators who would have an unpleasant experience if they tried to make a meal of one of these tiny creatures. This is because the frogs secrete toxic alkaloids from glands on the surface of their skin. These secretions are so toxic that they have been used by locals to poison blowgun darts used in hunting; hence the name *dart-poison frogs*.

Scientists from the National Institutes of Health have isolated and determined the structures of many of the compounds present in these skin secretions using the techniques of mass spectrometry and NMR spectroscopy (Chapter 12). Thus far, the structures of close to 200 different alkaloids have been determined. The most potent of these toxins is batrachotoxin (from *Phyllobates terribilis*), an example of a steroidal alkaloid. Other toxins include

batrachotoxin

histrionicotoxin

pumiliotoxin B

histrionicotoxin (from *Dendrobates histrionicus*) and pumiliotoxin B (from *Dendrobates pumilio*).

Because many of these compounds are produced in only microgram quantities by the frogs, laboratory syntheses have often been developed to confirm structures and also to provide a supply of material for pharmacological studies. It turns out that most of these toxins act on the nervous system by affecting the manner in which ions are transported across cell membranes. As a result of this property, several of these alkaloids are now used as research tools in the field of neuroscience.

In 1992, an alkaloid with painkilling properties far greater than that of morphine was isolated from the Ecuadorian frog *Epipedobates tricolor*. Named epibatidine, it was hoped that this simple alkaloid would be a lead compound for the development of a new family of painkillers. Whether or not epibatidine or related synthetic materials will eventually result in new painkillers remains to be determined, but epibatidine is already being marketed in the form of its tartrate salt for use in biomedical research. Research in this interesting area of natural product chemistry once again confirms that big things can come in small packages.

epibatidine

See Problems 11.37 and 11.38.

The mixture, neither component of which is water soluble, is dissolved in an inert, low-boiling solvent such as diethyl ether and is shaken with aqueous hydrochloric acid. The amine reacts to form a salt, which is ionic and dissolves in the water layer. The nitro compound does not react and remains in the ether layer. The two layers are then separated. The nitro compound can be recovered by evaporating the ether. The amine can be recovered from its salt by making the aqueous layer alkaline with a strong base such as NaOH.

There are many natural and synthetic amine salts of biological interest. Two examples are squalamine, an antimicrobial steroid recently isolated from the dogfish shark (see Chapter 15 for more about steroids), and (+)-methamphetamine hydrochloride, the addictive and toxic stimulant commonly known as "ice" or "meth."

Dogfish shark.

squalamine

methamphetamine hydrochloride

11.9 Chiral Amines as Resolving Agents

Amines also form salts with organic acids. This reaction is used to resolve enantiomeric acids (Sec. 5.12). For example, (*R*)- and (*S*)-lactic acids can be resolved by reaction with a chiral amine such as (*S*)-1-phenylethylamine:

[(*R*)-lactic acid + (*S*)-lactic acid] + (*S*)-1-phenylethylamine ⟶ [(*R,S*) salt + (*S,S*) salt] (11.20)

The salts are *diastereomers*, not enantiomers, and can be separated by ordinary methods, such as fractional crystallization. Once separated, each salt can be treated with a strong acid, such as HCl, to liberate one enantiomer of lactic acid. For example,

(*R,S*) salt + HCl ⟶ (*R*)-lactic acid + (*S*)-1-phenylethylammonium chloride (11.21)

The chiral amine can be recovered for reuse by treating its salt with sodium hydroxide (as in the last step of eq. 11.19).

Numerous chiral amines are available from natural products and can be used to resolve acids. Conversely, some chiral acids are available to resolve amine enantiomers.

So far, we have considered reactions in which amines act as bases. Now we will examine some reactions in which they act as nucleophiles.

11.10 Acylation of Amines with Acid Derivatives

Amines are nitrogen nucleophiles. They react with the carbonyl group of carboxylic acid derivatives (acyl halides, anhydrides, and esters) by nucleophilic acyl substitution (Sec. 10.11).

Looked at from the viewpoint of the amine, we can say that the N—H bond in primary and secondary amines can be *acylated* by acid derivatives. For example, primary and secondary amines react with acyl halides to form amides (compare with eq. 10.36).

$$\underset{\text{acyl halide}}{R{-}\overset{\overset{\Large O}{\|}}{C}{-}Cl} + \underset{\text{primary amine}}{H_2\ddot{N}{-}R'} \longrightarrow \underset{\text{secondary amide}}{R{-}\overset{\overset{\Large O}{\|}}{C}{-}NHR'} + HCl \qquad (11.22)$$

$$\underset{\text{acyl halide}}{R{-}\overset{\overset{\Large O}{\|}}{C}{-}Cl} + \underset{\text{secondary amine}}{H\ddot{N}R'R''} \longrightarrow \underset{\text{tertiary amide}}{R{-}\overset{\overset{\Large O}{\|}}{C}{-}NR'R''} + HCl \qquad (11.23)$$

If the amine is inexpensive, two equivalents are used—one to form the amide and the second to neutralize the HCl. Alternatively, an inexpensive base may be added for the latter purpose. This can be sodium hydroxide (especially if R is *aromatic*) or a tertiary amine; having no N—H bonds, tertiary amines cannot be acylated, but they can neutralize the HCl.

EXAMPLE 11.10

Using eq. 10.31 as a guide, write out the steps in the mechanism for eq. 11.22.

Solution

$$R{-}\overset{\overset{\Large \ddot{O}:}{\|}}{C}{-}Cl + H_2\ddot{N}R' \longrightarrow R{-}\overset{\overset{:\ddot{O}:^-}{|}}{\underset{\underset{H}{|}}{\underset{H-N^+-R'}{|}}{C}}{-}Cl$$

$$\xrightarrow{-Cl^-} R{-}\overset{\overset{\Large O}{\|}}{C}{-}\underset{\underset{H}{|}}{\overset{\overset{H}{|}}{N^+}}{-}R' \quad (+\ H_2\ddot{N}R')$$

$$\xrightarrow{-H^+} \underset{\text{amide}}{R{-}\overset{\overset{\Large O}{\|}}{C}{-}\underset{\underset{H}{|}}{\ddot{N}}{-}R'}$$

The first step involves nucleophilic addition to the carbonyl group. Elimination of HCl completes the substitution reaction.

Acylation of amines is often put to practical use. For example, the insect repellent OFF® is the amide formed in the reaction of *m*-toluyl chloride and diethylamine.

$$m\text{-}CH_3C_6H_4\overset{O}{\overset{\|}{C}}{-}Cl + (CH_3CH_2)_2NH \xrightarrow{NaOH} m\text{-}CH_3C_6H_4\overset{O}{\overset{\|}{C}}{-}N(CH_2CH_3)_2 + Na^+Cl^- + H_2O \quad (11.24)$$

m-toluyl chloride; diethylamine; *N,N*-diethyl-*m*-toluamide (the insect repellent OFF®)

PROBLEM 11.18 Write out the steps in the mechanism for the synthesis of OFF® (eq. 11.24).

The antipyretic (fever-reducing substance) acetanilide is an amide prepared from aniline and acetic anhydride.

$$CH_3\overset{O}{\overset{\|}{C}}O\overset{O}{\overset{\|}{C}}CH_3 + H_2N{-}C_6H_5 \longrightarrow CH_3\overset{O}{\overset{\|}{C}}{-}NH{-}C_6H_5 + CH_3CO_2H \quad (11.25)$$

acetic anhydride; aniline; acetanilide

PROBLEM 11.19 Provide the structures of the amides obtained from reaction of acetic anhydride with

a. $(CH_3CH_2)_2NH$ b. $CH_3CH_2NH_2$

11.11 Quaternary Ammonium Compounds

In **quaternary ammonium salts**, all four hydrogens of the ammonium ion are replaced by organic groups.

Tertiary amines react with primary or secondary alkyl halides by an S_N2 mechanism (eq. 11.6). The products are **quaternary ammonium salts**, in which all four hydrogens of the ammonium ion are replaced by organic groups. For example,

$$(CH_3CH_2)_3N\!: + C_6H_5CH_2{-}Cl \longrightarrow (CH_3CH_2)_3\overset{+}{N}CH_2{-}C_6H_5 + Cl^- \quad (11.26)$$

triethylamine; benzyl chloride; benzyltriethylammonium chloride

Quaternary ammonium compounds are important in biological processes. One of the most common natural quaternary ammonium ions is **choline**, which is present in phospholipids (Sec. 15.6).

Choline and **acetylcholine** are important quaternary ammonium ions in biological processes.

$$CH_3-\overset{+}{N}(CH_3)_2-CH_2CH_2OH$$

choline

$$CH_3-\overset{+}{N}(CH_3)_2-CH_2\text{(tetrahydrofuran ring with OH and }CH_3)$$

muscarine

Choline is not only involved in various metabolic processes, but is also the precursor of **acetylcholine** (page 327), a compound that plays a key role in the transmission of nerve impulses. The mushroom *Amanita muscaria* contains the deadly neurotoxin muscarine, which structurally resembles acetylcholine and probably interferes with the function of this neurotransmitter.

© Okapia/Photo Researchers

***Amanita muscaria*, a source of muscarine.**

11.12 Aromatic Diazonium Compounds

Primary aromatic amines react with nitrous acid at 0°C to yield **aryldiazonium ions**. The process is called **diazotization**.

In **diazotization**, primary aromatic amines react with nitrous acid to form **aryldiazonium ions**.

$$C_6H_5-NH_2 + HONO + H^+Cl^- \xrightarrow[\text{aqueous solution}]{0\text{–}5^\circ C} C_6H_5-N_2^+Cl^- + 2\,H_2O \quad (11.27)$$

aniline; nitrous acid; benzenediazonium chloride

Diazonium compounds are extremely useful synthetic intermediates. Before we describe their chemistry, let us try to understand the steps in eq. 11.27. First, we need to examine the structure of nitrous acid.

Nitrous acid decomposes rather rapidly at room temperature. It is therefore prepared as needed by treating an aqueous solution of sodium nitrite with a strong acid at ice temperature. At that temperature, nitrous acid solutions are reasonably stable.

$$Na^+NO_2^- + H^+Cl^- \xrightarrow{0\text{–}5^\circ C} H-\ddot{O}-\ddot{N}=\ddot{O}: + Na^+Cl^- \quad (11.28)$$

sodium nitrite; nitrous acid

The reactive species in reactions of nitrous acid is the nitrosonium ion (NO^+). It is formed by protonation of the nitrous acid, followed by loss of water (compare with eq. 4.18):

$$H\ddot{O}-\ddot{N}=\ddot{O}: + H^+ \rightleftharpoons H\overset{+}{O}(H)-\ddot{N}=\ddot{O}: \rightleftharpoons H_2O + :\overset{+}{N}=\ddot{O}: \quad (11.29)$$

nitrosonium ion

How do the two nitrogens, one from the amine and one from the nitrous acid, become bonded to one another, as they appear in diazonium ions? This happens in the first step of diazotization (eq. 11.30), which involves nucleophilic attack of the primary amine on the nitrosonium ion, followed by proton loss.

$$\mathrm{Ar\ddot{N}H_2 + :\overset{+}{N}{=}\ddot{O}: \longrightarrow ArN^{+}H_2{-}\ddot{N}{=}\ddot{O}: \rightleftharpoons \underset{\text{a primary nitrosamine}}{Ar\ddot{N}H{-}\ddot{N}{=}\ddot{O}:} + H^+} \quad (11.30)$$

Protonation of the oxygen in the resulting nitrosamine, followed by elimination of water, then gives the aromatic diazonium ion.

$$\mathrm{Ar\ddot{N}H{-}\ddot{N}{=}\ddot{O}: + H^+ \longrightarrow Ar\overset{+}{N}H{=}\ddot{N}{-}\ddot{O}H \xrightarrow{-H_2O} \underset{\text{aryldiazonium ion}}{Ar\overset{+}{N}{\equiv}N:}} \quad (11.31)$$

Notice that in the final product, there are no N—H bonds; both hydrogens of the amino group are lost, the first in eq. 11.30 and the second in eq. 11.31. Therefore, *only primary amines can be diazotized.* (Secondary and tertiary amines do react with nitrous acid, but their reactions are different and less important in synthesis.)

Solutions of aryldiazonium ions are moderately stable and can be kept at 0°C for several hours. They are useful in synthesis because the diazonio group ($—N_2^+$) can be replaced by nucleophiles; the other product is nitrogen gas.

$$\mathrm{Ar{-}\overset{+}{N}{\equiv}N: + Nu:^- \longrightarrow Ar{-}Nu + N_2} \quad (11.32)$$

Specific useful examples are shown in eq. 11.33. The nucleophile always takes the position on the benzene ring that was occupied by the diazonio group.

(11.33) Benzenediazonium ($C_6H_5N_2^+$) reactions:
- KCN, $Cu_2(CN)_2$ → benzonitrile (C_6H_5CN)
- HCl, Cu_2Cl_2 → chlorobenzene (C_6H_5Cl)
- H_3PO_2 → benzene (C_6H_5H)
- H_2O, H^+ → phenol (C_6H_5OH)
- KI → iodobenzene (C_6H_5I)
- HBr, Cu_2Br_2 → bromobenzene (C_6H_5Br)
- HBF_4, heat → fluorobenzene (C_6H_5F)

Conversion of diazonium compounds to aryl chlorides, bromides, or cyanides is usually accomplished using cuprous salts, and is known as the **Sandmeyer reaction**. Since a CN group is easily converted to a CO_2H group (eq. 10.13), this provides another route to aromatic carboxylic acids. The reaction with KI gives aryl iodides, usually not easily accessible by direct electrophilic iodination. Similarly, direct aromatic fluorination is difficult, but aromatic fluorides can be prepared from diazonium compounds and tetrafluoroboric acid, HBF_4.

In the **Sandmeyer reaction**, diazonium ions react with cuprous salts to form aryl chlorides, bromides, or cyanides.

Phenols can be prepared by adding diazonium compounds to hot aqueous acid. This reaction is important because there are not many ways to introduce an —OH group directly on an aromatic ring.

Finally, we sometimes use the orienting effect of a nitro or amino group and afterwards remove this substituent from the aromatic ring. This can be done by diazotization followed by reduction. A common reducing agent for this purpose is **hypophosphorous acid**, H_3PO_2.

Hypophosphorous acid is used to reduce the diazonio group to H.

Here are some examples of ways that diazonium compounds can be used in synthesis:

EXAMPLE 11.11

How can *m*-dibromobenzene be prepared?

Solution It *cannot* be prepared by direct electrophilic bromination of bromobenzene, because the Br group is *o,p*-directing (Sec. 4.11). But we can take advantage of the *m*-directing effect of a nitro group and then convert the nitro group to a bromine atom, as follows:

NO_2 (benzene) $\xrightarrow[FeBr_3]{Br_2}$ NO_2, Br (meta) $\xrightarrow[HCl]{SnCl_2}$ NH_2, Br (meta) $\xrightarrow[0°]{HONO}$ N_2^+, Br (meta) $\xrightarrow[Cu_2Br_2]{HBr}$ Br, Br (meta)

EXAMPLE 11.12

How can *o*-toluic (*o*-methylbenzoic) acid be prepared from *o*-toluidine (*o*-methylaniline)?

Solution

NH_2, CH_3 (*o*-toluidine) $\xrightarrow[\text{2. } Cu_2(CN)_2,\ KCN]{\text{1. HONO}}$ CN, CH_3 $\xrightarrow{H_3O^+}$ CO_2H, CH_3 (*o*-toluic acid)

EXAMPLE 11.13

Design a route to 1,3,5-tribromobenzene from aniline.

Solution First brominate; the amino group is *o,p*-directing and ring-activating. Then remove the amino group by diazotization and reduction.

$$C_6H_5NH_2 \xrightarrow{3Br_2} \text{2,4,6-}Br_3C_6H_2NH_2 \xrightarrow[\text{2. } H_3PO_2]{\text{1. HONO}} \text{1,3,5-}C_6H_3Br_3$$

PROBLEM 11.20 Design a synthesis of each of the following compounds, using a diazonium ion intermediate.

a. *m*-chlorofluorobenzene from benzene
b. *m*-nitrophenol from *m*-nitroaniline
c. 3,5-dibromotoluene from *p*-toluidine
d. 2,4-difluorotoluene from toluene

11.13 Diazo Coupling; Azo Dyes

Being positively charged, aryldiazonium ions are electrophiles. They are *weak* electrophiles, however, because the positive charge can be delocalized through resonance.

EXAMPLE 11.14

Write the resonance contributors for the benzenediazonium ion that show how the nitrogen farthest from the benzene ring can become electrophilic.

Solution

$$\left[C_6H_5-\overset{+}{N}\equiv N: \longleftrightarrow C_6H_5-\ddot{N}=\overset{+}{N}: \right]$$

In the second contributor, the nitrogen at the right has only six electrons; it can react as an electrophile.

PROBLEM 11.21 Draw resonance contributors that show that the positive charge in benzenediazonium ion can also be delocalized to the *ortho* and *para* carbons of the benzene ring. (CAREFUL! These contributors have two positive charges and one negative charge.)

Aryldiazonium ions react with strongly activated aromatic rings (phenols and aromatic amines) to give **azo compounds**. For example,

Azo compounds contain the azo group, —N=N—.

$$C_6H_5\overset{+}{N}\equiv N: + C_6H_5OH \xrightarrow{HO^-} C_6H_5N=N\text{-}C_6H_4\text{-}OH + H_2O \quad (11.34)$$

benzenediazonium ion; phenol; *p*-hydroxyazobenzene, yellow leaflets, mp 155–157°C

The nitrogen atoms are retained in the product. This electrophilic aromatic substitution reaction is called **diazo coupling**, because in the product, two aromatic rings are coupled by the azo, or —N=N—, group. *Para* coupling is preferred, as in eq. 11.34, but if the *para* position is blocked by another substituent, *ortho* coupling can occur. *All azo compounds are colored,* and many are used commercially as dyes for cloth and in (film-based) color photography.*

Diazo coupling is an electrophilic aromatic substitution reaction in which phenols and aromatic amines react with aryldiazonium electrophiles to give azo compounds.

PROBLEM 11.22 Methyl orange is an azo dye used as an indicator in acid–base titrations. (It is yellow–orange above pH 4.5 and red below pH 3.) Show how it can be synthesized from *p*-aminobenzenesulfonic acid (sulfanilic acid) and *N,N*-dimethylaniline.

$$(CH_3)_2N\text{-}C_6H_4\text{-}N=N\text{-}C_6H_4\text{-}SO_3^-$$

methyl orange

At this point, we have completed a survey of the main functional groups in organic chemistry. By now, all of the structures in the table inside the front cover of this book should seem familiar to you. In the next chapter, we will describe some modern techniques that help us to assign a structure to a particular molecule. After that, we will conclude with a series of chapters on important commercial and biological applications of organic chemistry.

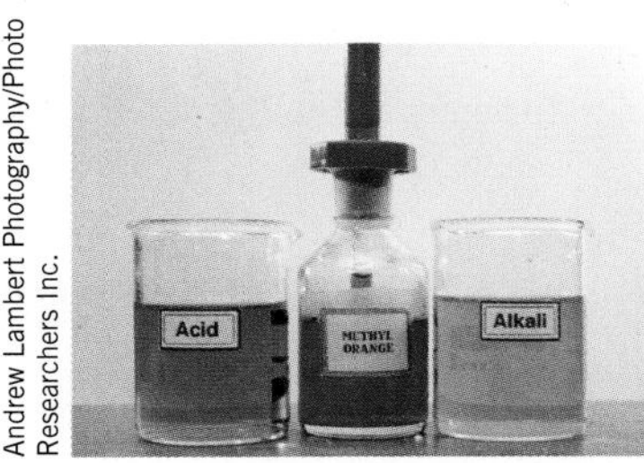

Andrew Lambert Photography/Photo Researchers Inc.

Methyl orange indicator in basic solution (right) and in acidic solution (left).

*For an interesting discussion of the diazo copying process, see the article by B. Osterby, *J. Chem. Educ.* **1989**, *66*, 1206–1208.

REACTION SUMMARY

1. Alkylation of Ammonia and Amines to Form Amines (Sec. 11.4)

$$R{-}X + 2\,NH_3 \longrightarrow R{-}NH_2 + NH_4^+\,X^-$$

$$C_6H_5{-}NH_2 \xrightarrow{R-X} C_6H_5{-}NHR \xrightarrow{R-X} C_6H_5{-}NR_2$$

2. Reduction Routes to Amines (Sec. 11.5)

a. Catalytic or Chemical Reduction of the Nitro Group

$$C_6H_5{-}NO_2 \xrightarrow[\text{1. SnCl}_2\text{, HCl; 2. NaOH, H}_2\text{O}]{\text{H}_2\text{, Ni catalyst or}} C_6H_5{-}NH_2$$

b. Hydride Reduction of Amides and Nitriles

$$R{-}\overset{O}{\overset{\|}{C}}{-}NR'R'' \xrightarrow{LiAlH_4} R{-}CH_2{-}NR'R''$$

$$R{-}C{\equiv}N \xrightarrow{LiAlH_4} R{-}CH_2{-}NH_2$$

c. Reductive Amination of Aldehydes and Ketones

$$R{-}\overset{O}{\overset{\|}{C}}{-}R' \xrightarrow[NaBH_3CN]{R''NH_2} R{-}\underset{H}{\underset{|}{\overset{NHR''}{\overset{|}{C}}}}{-}R'$$

3. Amines as Bases (Secs. 11.6 and 11.8)

$$R{-}NH_2 + H{-}OH \longrightarrow R{-}\overset{+}{N}H_3 + {}^-OH \qquad R{-}NH_2 + H{-}Cl \longrightarrow R{-}\overset{+}{N}H_3 + Cl^-$$

4. Amines as Nucleophiles

a. Acylation of Amines (Sec. 11.10)

Secondary and Tertiary Amides from Primary and Secondary Amines

$$\underset{\text{(primary amine)}}{R{-}NH_2} \xrightarrow[\text{or } R'COCl]{(R'CO)_2O} \underset{\text{(secondary amide)}}{R'{-}\overset{O}{\overset{\|}{C}}{-}NHR}$$

$$\underset{\text{(secondary amine)}}{R_2NH} \xrightarrow[\text{or } R'COCl]{(R'CO)_2O} \underset{\text{(tertiary amide)}}{R'{-}\overset{O}{\overset{\|}{C}}{-}NR_2}$$

b. Alkylation of Amines: Quaternary Ammonium Salts (Sec. 11.11)

$$R_3N + R'X \longrightarrow R_3\overset{+}{N}{-}R'\;X^-$$

5. **Aryldiazonium Salts: Formation and Reactions (Secs. 11.12 and 11.13)**

a. **Formation from Aniline and Nitrous Acid (Sec. 11.12)**

$$ArNH_2 + HONO \xrightarrow{HX} ArN_2^+ X^-$$

(aryldiazonium salt)

b. **Reactions to Form Substituted Benzenes (Sec. 11.12)**

$$ArN_2^+ + H_2O \xrightarrow{heat} ArOH + N_2 + H^+$$

(phenols)

$$ArN_2^+ + HX \xrightarrow{Cu_2X_2} ArX \quad (X{=}Cl, Br)$$

$$ArN_2^+ + KI \longrightarrow ArI$$

$$ArN_2^+ + KCN \xrightarrow{Cu_2(CN)_2} ArCN$$

$$ArN_2^+ + HBF_4 \longrightarrow ArF$$

$$ArN_2^+ + H_3PO_2 \longrightarrow ArH$$

c. **Diazo Coupling (Sec. 11.13)**

$$ArN_2^+ X^- + C_6H_5OH \longrightarrow p\text{-}HOC_6H_4{-}N{=}N{-}Ar + HX$$

(azo compound)

MECHANISM SUMMARY

Diazotization (Sec. 11.12)

$$\left[\begin{matrix}Ar \\ \;\;\ddot{N}{-}\ddot{N}{=}\ddot{O}: \\ H\end{matrix}\right] \rightleftharpoons Ar{-}\ddot{N}{=}\ddot{N}{-}\ddot{O}{-}H \xrightarrow{H^+} Ar{-}\ddot{N}{=}\ddot{N}{-}\overset{+}{\ddot{O}}(H){-}H \longrightarrow Ar{-}\overset{+}{N}{\equiv}N: + H_2O$$

ADDITIONAL PROBLEMS

OWL Interactive versions of these problems are assignable in OWL.

Nomenclature and Structure of Amines

11.23 Give an example of each of the following:

a. a primary amine
b. a cyclic tertiary amine
c. a secondary aromatic amine
d. a quaternary ammonium salt
e. an aryldiazonium salt
f. an azo compound
g. a primary amide

11.24 Write a structural formula for each of the following compounds:

a. diethylpropylamine
b. *N*-methylbenzylamine
c. *sec*-butylamine
d. 1,5-diaminopentane
e. *N*,*N*-dimethylaminocyclopentane
f. *N*,*N*-dimethyl-3-hexanamine
g. tetraethylammonium chloride
h. *p*-nitroaniline
i. 2-aminohexane
j. diphenylamine
k. *o*-toluidine
l. 3-methyl-2-pentanamine

11.25 Write a correct name for each of the following compounds:

a. Cl–C_6H_4–NH_2 (para)

b. $CH_3NHCH_2CH_2CH_3$

c. $(CH_3CH_2)_2NCH_3$

d. $(CH_3)_4N^+\ Cl^-$

e. $CH_3CH(OH)CH(NH_2)CH_3$

f. 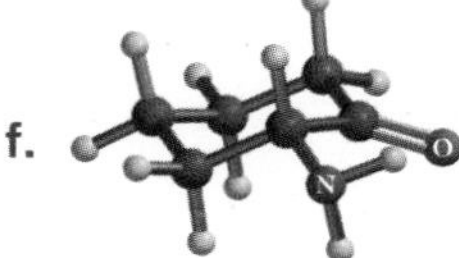

g. Cl–C_6H_4–$N_2^+Cl^-$ (para)

h. 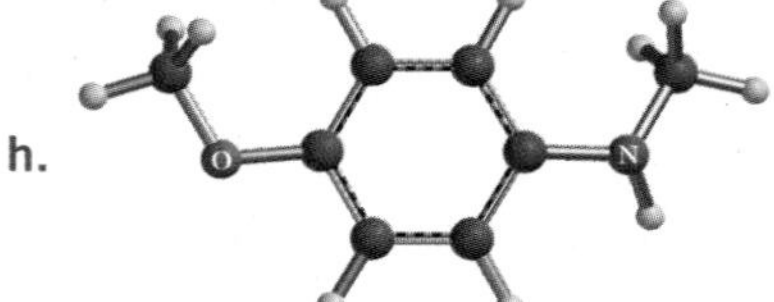

i.

j. $H_2N(CH_2)_6NH_2$

11.26 Draw the structures for, name, and classify as primary, secondary, or tertiary the eight isomeric amines with the molecular formula $C_4H_{11}N$.

Properties of Amines and Quaternary Ammonium Salts

11.27 Tell which is the stronger base and why.

a. aniline or *p*-cyanoaniline
b. aniline or diphenylamine

11.28 Write out a scheme similar to eq. 11.19 to show how you could separate a mixture of *p*-toluidine, *p*-methylphenol, and *p*-xylene.

CH_3–C_6H_4–NH_2 (*p*-toluidine) CH_3–C_6H_4–OH (*p*-methylphenol) CH_3–C_6H_4–CH_3 (*p*-xylene)

11.29 Draw the important contributors to the resonance hybrid structure of *p*-nitroaniline (page 338).

11.30 Place the following substances, which have nearly identical formula weights, in order of increasing boiling point: $CH_3CH_2CH_2CH_2NH_2$, $CH_3CH_2CH_2CH_2OH$, $CH_3CH_2CH_2OCH_3$, and $CH_3CH_2CH_2CH_2CH_3$.

11.31 Explain why compound A can be separated into its *R*- and *S*-enantiomers, but compound B cannot.

compound A: CH_3–$N^+(CH_2CH_3)(CH_2CH_2CH_3)$–$CH_2$–$C_6H_5$ Cl^-

compound B: CH_3–$\ddot{N}(CH_2CH_3)(CH_2CH_2CH_3)$

11.32 Give the priority order of groups in compound A (Problem 11.31), and draw a dash-wedge formula for its *R*-isomer.

Preparation and Reactions of Amines

11.33 Give equations for the preparation of the following amines from the indicated precursor:

a. *N,N*-diethylaniline from aniline
b. *m*-bromoaniline from benzene
c. *p*-bromoaniline from benzene
d. 1-aminohexane from 1-bromopentane

11.34 Complete the following equations:

a. $\text{cyclopentyl}{-}NH_2 + CH_2{=}CHCH_2Br \xrightarrow{\text{heat}}$

b. $CH_3\overset{O}{\overset{\|}{C}}Cl + H_2NCH_2CH_2CH(CH_3)_2 \longrightarrow A \xrightarrow{LiAlH_4} B$

c. $CH_3O\overset{O}{\overset{\|}{C}}{-}C_6H_5 \xrightarrow[H^+]{HONO_2} C \xrightarrow[\text{excess}]{LiAlH_4} D$

d. $C_6H_5{-}CH_2Br \xrightarrow{NaCN} E \xrightarrow{LiAlH_4} F \xrightarrow{(CH_3\overset{O}{\overset{\|}{C}})_2O} G$

e. $\text{cyclohexanone}\ (\text{C}_6\text{H}_{10}{=}O) + (CH_3)_2CHNH_2 \xrightarrow{NaBH_3CN} H$

11.35 Write an equation for the reaction of

a. *p*-toluidine (page 338) with hydrochloric acid
b. triethylamine, $(CH_3CH_2)_3N$, with sulfuric acid
c. dimethylammonium chloride, $(CH_3)_2\overset{+}{N}H_2Cl^-$, with sodium hydroxide
d. *N,N*-dimethylaniline (page 330) with methyl iodide
e. cyclohexylamine (page 337) with acetic anhydride

11.36 Write out the steps in the mechanism for the following reaction:

$$CH_3CH_2NH_2 + CH_3\overset{O}{\overset{\|}{C}}O\overset{O}{\overset{\|}{C}}CH_3 \longrightarrow CH_3CH_2NH\overset{O}{\overset{\|}{C}}CH_3 + CH_3COOH.$$

Explain why only one of the hydrogens of the amine is replaced by an acetyl group, even if a large excess of acetic anhydride is used.

Synthetic and Biological Applications of Amine Chemistry

11.37 As noted in the "A Word About... Alkaloids and the Dart-Poison Frogs" on page 341, many alkaloids have potent efficacy with regard to disrupting neurological function. Methamphetamine hydrochloride is an addictive and toxic stimulant.

a. Outline a synthesis of methamphetamine hydrochloride (shown below) that uses 1-phenyl-2-propanone (phenylacetone) and methylamine as carbon sources.

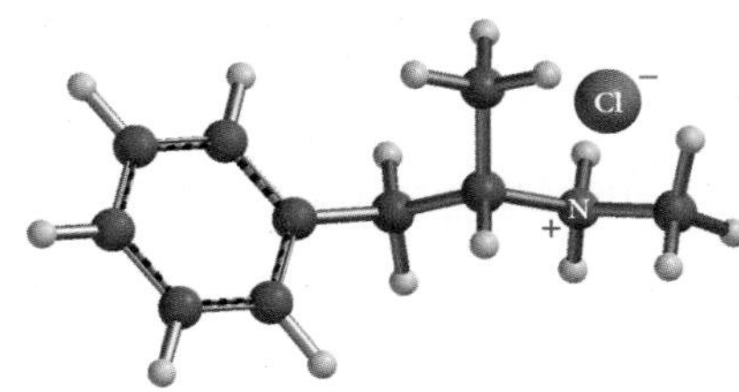

b. Will the product of your proposed synthesis be optically active or racemic? Explain why.

11.38 New evidence has been presented that dart-poison frogs partially obtain the bioactive alkaloids secreted on their skin from their diet. For example, ants found in the tropics generate alkaloid toxins that are very similar to those found in poison frogs.

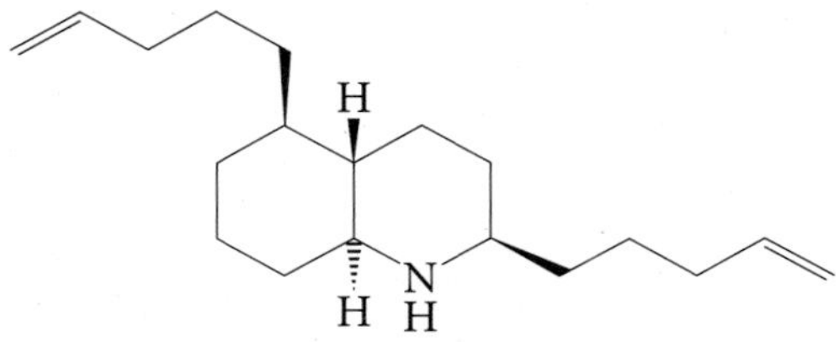

ant alkaloid

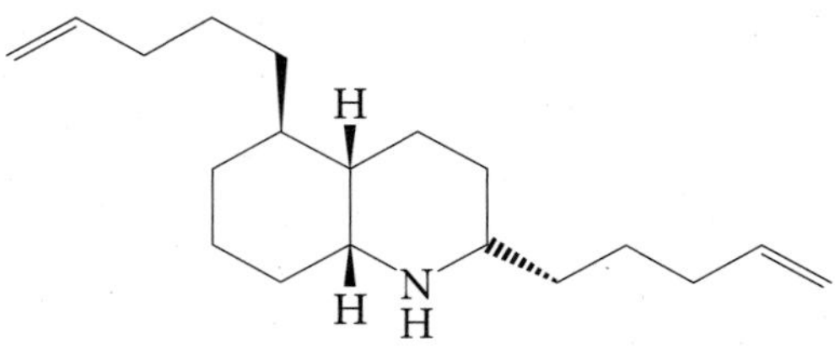

frog alkaloid

How many stereogenic centers are present in these alkaloids?

11.39 Decamethonium bromide is used in surgery as a muscle relaxant. It acts by preventing the enzyme acetylcholine esterase from destroying acetylcholine (see page 327), a necessary step in the transmission of nerve impulses. Show how decamethonium bromide can be synthesized from a diamine and an alkyl halide.

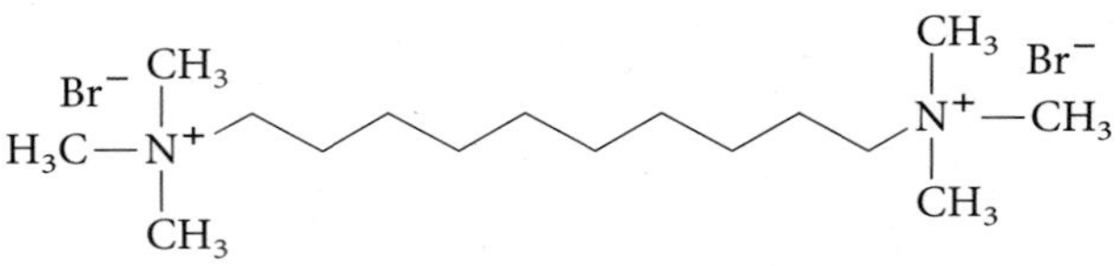

decamethonium bromide

11.40 Choline (Sec. 11.11) can be prepared by the reaction of trimethylamine with ethylene oxide. Write an equation for the reaction, and show its mechanism.

11.41 Acetylcholine (Sec. 11.11) is synthesized in the body's neurons. The enzyme choline acetyltransferase catalyzes its synthesis from acetyl-CoA (see "A Word About... Thioesters, Nature's Acyl-Activating Groups" on page 312) and choline. Write an equation for the reaction, using the formula $CH_3\underset{\underset{O}{\|}}{C}{-}S{-}CoA$ for acetyl-CoA.

Formation and Reactions of Aryldiazonium Ions

11.42 Primary aliphatic amines (RNH_2) react with nitrous acid in the same way that primary arylamines ($ArNH_2$) do, to form diazonium ions. But alkyldiazonium ions RN_2^+ are much less stable than aryldiazonium ions ArN_2^+ and readily lose nitrogen even at 0°C. Explain the difference.

11.43 Write an equation for the reaction of $CH_3{-}C_6H_4{-}N_2^+\ HSO_4^-$ with

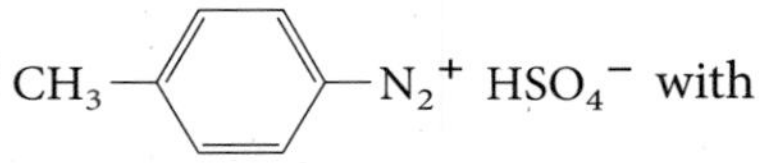

a. HBF_4, then heat
b. aqueous acid, heat
c. KCN and cuprous cyanide
d. *p*-methoxyphenol and HO^-
e. HCl and cuprous chloride
f. *N,N*-dimethylaniline and base
g. hypophosphorous acid
h. KI

11.44 Show how diazonium ions could be used to synthesize

a. *p*-chlorobenzoic acid from *p*-chloroaniline
b. *m*-iodochlorobenzene from benzene
c. *m*-iodoacetophenone from benzene
d. 3-cyano-4-methylbenzenesulfonic acid from toluene

11.45 Congo red is used as a direct dye for cotton. Write equations to show how it can be synthesized from benzidine and 1-aminonaphthalene-4-sulfonic acid.

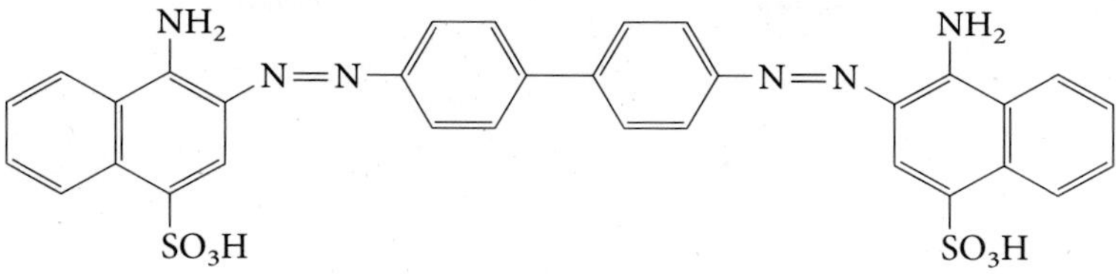

Congo red

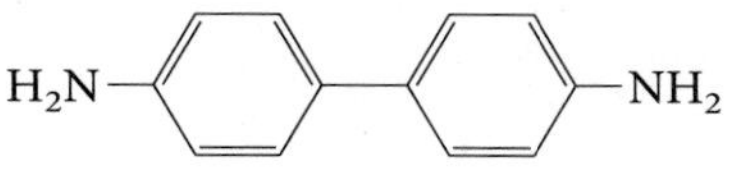

benzidine

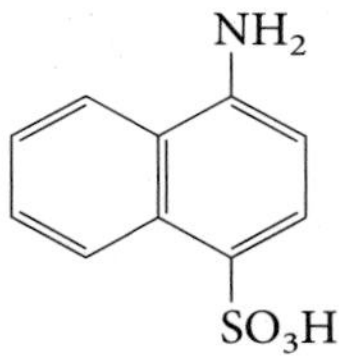

1-aminonaphthalene-4-sulfonic acid

11.46 Sunset yellow is a food dye that can be used to color Easter eggs. Write an equation for an azo coupling reaction that will give this dye.

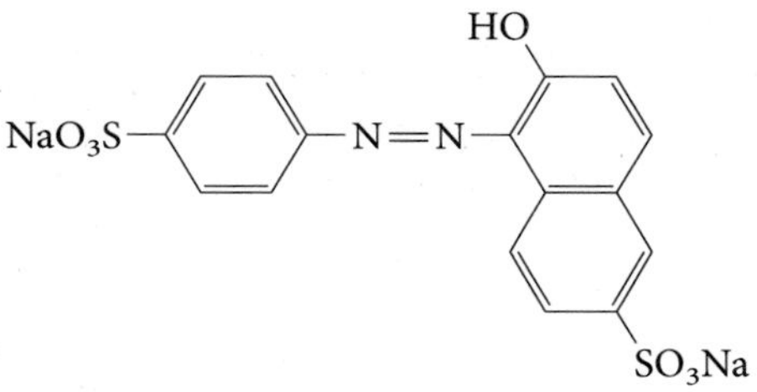

sunset yellow

Partial ^1H NMR spectrum of 2-phenylethanol: the two triplets reveal two different types of methylene protons in the molecule (a = left triplet, b = right triplet).

12

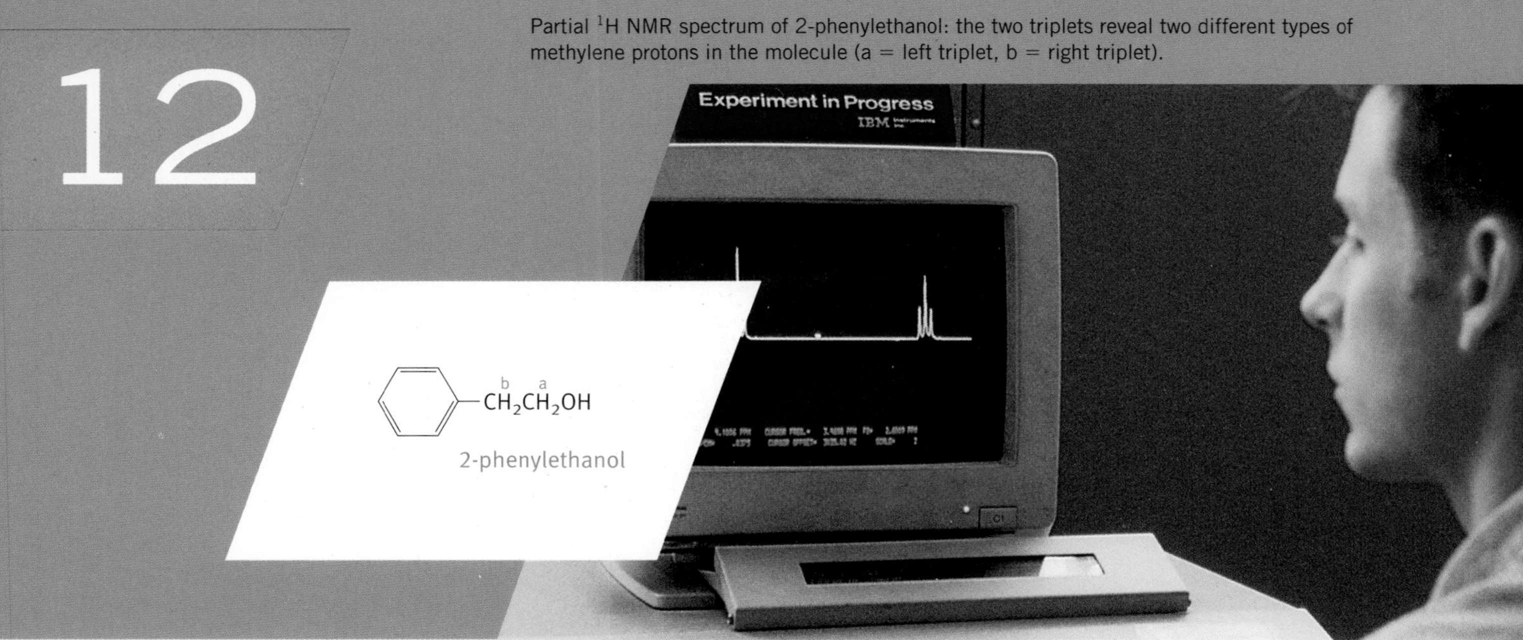

Spectroscopy and Structure Determination

An aspiring young chemist thinks he has just synthesized 2-phenylethanol, but how does he know for sure? In the early years of organic chemistry, determining the structure of a new compound was often a formidable task. The first step, of course, was an elemental analysis. Knowing the percentage of each element present allowed the empirical formula to be calculated; the molecular formula was then either the same as or a multiple of that formula. Elemental analysis is still an important criterion of the purity of a compound.

But how are the atoms arranged? What functional groups are present? And what about the carbon skeleton? Is it acyclic or cyclic? Are there branches and where are they located? Are benzene rings present? All of these questions and more had to be answered by chemical means. Reactions, such as ozonolysis (Sec. 3.17.b) or saponification (Sec. 10.13), could be used to convert complex molecules to simpler ones whose structures were easier to determine. To identify functional groups, various chemical tests could be applied (such as the bromine or permanganate tests for unsaturation or the Tollens' silver mirror test for an aldehyde group).

Once the functionality was known, reactions whose chemistry was well understood could be used to convert the unknown compound to a compound whose structure was already known. For example, if the compound was an aldehyde suspected to have the same R group as a known acid, it could be oxidized (eq. 12.1). If the physical properties (bp, mp, specific rotation if chiral, and so on)

Online homework for this chapter can be assigned in OWL, an online homework assessment tool.

and chemical reactions of the acid obtained from the aldehyde agreed with those of the known acid, it could safely be concluded that the two R groups *were* the same, and the structure of the aldehyde also became known. If they did *not* agree, one had to do some rethinking about the suspected structure. Ultimate structure proof came through synthesis of the unknown from compounds whose structures were already known, by reactions whose outcome was unambiguous. Gradually, over the years, a vast network of compounds with known structures was built up and catalogued in reference books.

$$\text{RCH}{=}\text{O} \xrightarrow{\text{KMnO}_4} \text{RCO}_2\text{H} \tag{12.1}$$

These methods—which often required weeks, months, even years—are still used in appropriate situations. But since the 1940s, various types of spectroscopy have simplified and speeded up the process of structure determination greatly. Automated instruments have been developed that permit us to determine and record spectroscopic properties often with little more effort than pushing a button. And these spectra, if properly interpreted, yield a great deal of structural information. For example, 2-phenylethanol can easily be identified from its ^{1}H NMR spectrum alone.

Spectroscopic methods have many advantages. Usually only a very small sample of material is required, and it can often be recovered if necessary. The methods are rapid, sometimes requiring only a few minutes. And usually we obtain more detailed structural information from spectra than from ordinary laboratory methods.

In this chapter, we will describe some of the more important spectroscopic techniques used today and how they can be applied to structural problems. But first, let us examine some general principles that form the basis of most of these techniques.

12.1 Principles of Spectroscopy

Equation 12.2 describes the relationship between the **energy** of light (or any other form of radiation), ***E***, and its **frequency**, ν (Greek nu, pronounced “new”).

E **=** ***h*****ν**. The **energy** of light is directly proportional to its **frequency**, ν.

$$E = h\nu \tag{12.2}$$

The equation says that there is a direct relationship between the frequency of light and its energy; the higher the frequency, the higher the energy. The proportionality constant between the two is known as Planck’s constant, *h*.* Because the frequency of light and its wavelength are *inversely* proportional, the equation can also be written

$$E = hc/\lambda, \text{ because } \nu = c/\lambda \tag{12.3}$$

where λ (Greek lambda) is the **wavelength** of light and *c* is the speed of light. In this form, the equation tells us that the shorter the wavelength of light, the higher its energy.

E **=** ***hc*****/λ**. The energy of light is inversely proportional to its **wavelength**, λ.

Molecules can exist at various energy levels. For example, the bonds in a given molecule may stretch, bend, or rotate; electrons may move from one orbital to another; and so on. These processes are quantized; that is, bonds may stretch,

*Named after the German physicist Max Planck, who introduced the term in 1900 in the context of his proposed quantum theory.

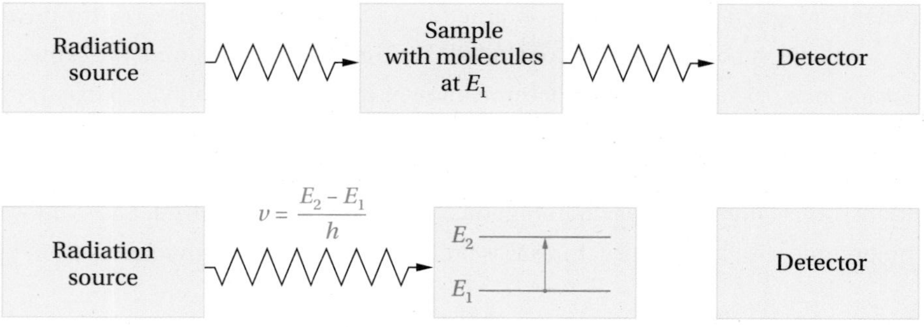

▪ Figure 12.1
Radiation passes through the sample unchanged (top), except when its frequency corresponds to the energy difference between two energy states of the molecule (bottom).

bend, or rotate only with certain frequencies (or energies; the two are proportional), and electrons may only jump between orbitals with well-defined energy differences. It is these energy (or frequency) differences that we measure by various types of spectra.

The idea behind most forms of spectroscopy is very simple and is expressed schematically in Figure 12.1. A molecule at some energy level, E_1, is exposed to radiation. The radiation passes through the molecule to a detector. As long as the molecule does not absorb the radiation, the amount of radiation detected will be equal to the amount of radiation emitted by the source (top part of Figure 12.1). At a frequency that corresponds to some molecular energy transition, from E_1 to E_2, the radiation will be absorbed by the molecule and will *not* appear at the detector (bottom part of Figure 12.1). The spectrum, then, consists of a record or plot of the amount of energy (radiation) received by the detector as the input energy is gradually varied.

Some transitions require more energy than others, so we must use radiation of the appropriate frequency to determine them. In this chapter, we will discuss three types of spectroscopy that depend on such transitions. They are nuclear magnetic resonance (NMR), infrared (IR), and ultraviolet-visible (UV-vis) spectroscopy. Table 12.1 summarizes the regions of the electromagnetic spectrum in which transitions for these three types of spectroscopy can be observed. We will begin with NMR spectroscopy and nuclear spin transitions, which require exceedingly small amounts of energy.

Table 12.1 ▪ Types of Spectroscopy and the Electromagnetic Spectrum

		Region of the spectrum			
Type of spectroscopy	**Radiation source**	**Frequency (hertz)**	**Wavelength (meters)**	**Energy (kcal/mol)**	**Type of transition**
nuclear magnetic resonance	radio waves	$60–600 \times 10^6$ (depends on magnet strength of the instrument)	5–0.5	$6–60 \times 10^{-6}$	nuclear spin
infrared	infrared light	$0.2–1.2 \times 10^{14}$	$15.0–2.5 \times 10^{-6}$	2–12	molecular vibrations
visible-ultraviolet (electronic)	visible or ultraviolet light	$0.375–1.5 \times 10^{15}$	$8–2 \times 10^{-7}$	37–150	electronic states

12.2 Nuclear Magnetic Resonance Spectroscopy

The kind of spectroscopy that has had by far the greatest impact on the determination of organic structures is nuclear magnetic resonance (NMR) spectroscopy. Commercial instruments became available in the late 1950s, and since then, NMR spectroscopy has become an indispensable tool for the organic chemist. Let us look briefly at the theory and then see what practical information we can obtain from an NMR spectrum.

Certain nuclei behave as though they are spinning. Because nuclei are charged and a spinning charge creates a magnetic field, these spinning nuclei behave like tiny magnets. The most important nuclei for organic structure determination are ^1H (ordinary hydrogen) and ^{13}C (a stable, nonradioactive isotope of ordinary carbon). Although ^{12}C and ^{16}O are present in most organic compounds, they do not possess a spin and do not give NMR spectra.

When nuclei with spin are placed between the poles of a powerful magnet, they align their magnetic fields *with* or *against* the field of the magnet. Nuclei aligned with the applied field have a slightly lower energy than those aligned against the field (Figure 12.2). By applying energy in the radio frequency (rf) range, it is possible to excite nuclei in the lower energy spin state to the higher energy spin state (we sometimes say that the spins "flip").

The energy gap between the two spin states depends on the strength of the applied magnetic field; the stronger the field, the larger the energy gap. Routine instruments currently in use have magnetic fields that range from about 1.4 to 14 tesla (T) (by comparison, the earth's magnetic field is only about 0.0007 T). At these field strengths, the energy gap corresponds to a rf of 60 MHz to 600 MHz (megahertz; 1 MHz $= 10^6$ Hz or 10^6 cycles per second). Translated to energy units to which chemists are more accustomed, the energy gap between the spin states is only $6–60 \times 10^{-6}$ kcal/mol. Even though this gap is exceedingly small, modern technology permits its detection with great accuracy.

12.2.a Measuring an NMR Spectrum

A ^1H NMR* spectrum is usually obtained in the following way. A sample of the compound being studied (usually only a few milligrams) is dissolved in some inert solvent that does not contain ^1H nuclei. Examples of such solvents are CCl_4, or solvents with the hydrogens replaced by deuterium, such as $CDCl_3$ (deuteriochloroform) and $CD_3C(=O)CD_3$ (hexa-deuterioacetone). A small amount of a reference compound is also added (we will say more about this in the next section). The solution, in a thin glass tube, is placed in the center of a rf coil, between the pole faces of a powerful magnet. The nuclei align themselves with or against the field. Continuously increasing amounts of energy can then be applied to the nuclei by the rf coil. When this energy corresponds exactly to the energy gap between the lower and higher energy spin states,

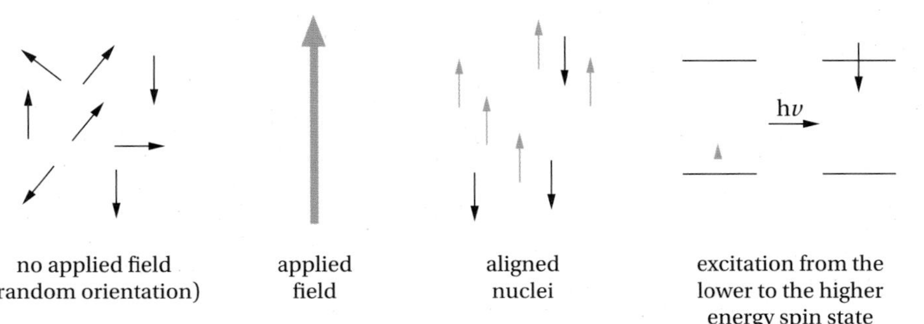

▪ **Figure 12.2**
Orientation of nuclei in an applied field, and excitation of nuclei from the lower to the higher energy spin state.

*The term *proton* is often used interchangeably with *hydrogen* or ^1H in discussing NMR spectra, even though the hydrogens are covalently bound (and not H^+). This is an inexact but common usage.

Nuclei in a magnetic field exhibit **nuclear magnetic resonance** when they absorb a specific rf of energy to go from a lower to a higher spin state.

it is absorbed by the nuclei. At this point, the nuclei are said to be in resonance with the applied frequency—hence, the term **nuclear magnetic resonance**. A plot of the energy absorbed by the sample against the applied frequency of the rf coil gives an NMR spectrum.

The **applied magnetic field** increases in strength from left to right on the recorded NMR spectrum.

In practice, there are two ways by which the resonance frequencies of ^{1}H nuclei can be determined. Because the magnetic field strength and the size of the energy gap between nuclear spin states are directly related, either the magnetic field strength or the rf can be varied. In earlier NMR spectrometers, a constant rf was applied, the strength of the **applied magnetic field** was varied, and different ^{1}H nuclei resonated at different magnetic field strengths. In modern Fourier transform (FT) NMR spectrometers, the applied magnetic field is held constant, and a short pulse of rf energy causes all of the ^{1}H nuclei to resonate simultaneously at their resonance rfs. The instrument computer uses a mathematical process called Fourier transformation to sort the signal that is produced into the resonance rfs of the different ^{1}H nuclei. Whether it is the magnetic field strength or the rf that is varied, this variable increases from left to right in the recorded spectra. All ^{1}H NMR spectra in this book, unless otherwise indicated, were obtained by FT-NMR techniques at a magnetic field strength of 4.7 T and an operating frequency of 200 MHz.

12.2.b Chemical Shifts and Peak Areas

Not all ^{1}H nuclei flip their spins at precisely the same rf because they may differ in chemical (and, more particularly, electronic) environment. We will return to this point, but first let us examine some spectra.

Tetramethylsilane (TMS) is the common reference compound for determining chemical shifts in ^{1}H NMR spectra.

Figure 12.3 shows the ^{1}H NMR spectrum of *p*-xylene. The spectrum is very simple and consists of two peaks. The positions of the peaks are measured in δ (delta) units from the peak of a reference compound, which is **tetramethylsilane (TMS)**, $(CH_3)_4Si$. The reasons for selecting TMS as a reference compound are (1) all 12 of its hydrogens are equivalent, so it shows only one sharp NMR signal, which serves as a reference point; (2) its ^{1}H signals appear at higher field than do most ^{1}H signals in other organic compounds, thus making it easy to identify the TMS peak; and (3) TMS is inert, so it does not react with most organic compounds, and it is low boiling and can be removed easily at the end of a measurement.

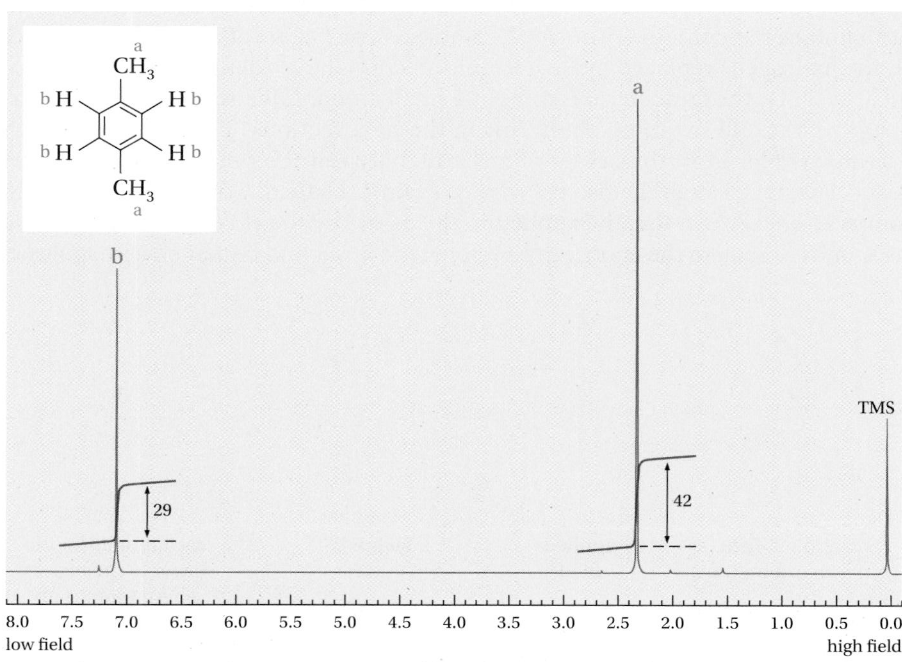

■ **Figure 12.3**
^{1}H NMR spectrum of *p*-xylene.

Most organic compounds have peaks *downfield* (at lower field) from TMS and are given positive δ values. A δ value of 1.00 means that a peak appears 1 part per million (ppm) downfield from the TMS peak. If the spectrum is measured at 60 MHz (60×10^6 Hz), then 1 ppm is 60 Hz (one-millionth of 60 MHz) downfield from TMS. If the spectrum is run at 100 MHz, a δ value of 1 ppm is 100 Hz downfield from TMS and so on. The **chemical shift** of a particular kind of ^{1}H signal is *its δ value with respect to TMS.* It is called a chemical shift because it depends on the chemical environment of the hydrogens. The chemical shift is independent of the instrument on which it is measured.

The **chemical shift** of a ^{1}H NMR signal is its δ (delta) value with respect to TMS.

$$\text{Chemical shift} = \delta = \frac{\text{distance of peak from TMS, in Hz}}{\text{spectrometer frequency in MHz}}\ \text{ppm} \qquad (12.4)$$

In the spectrum of *p*-xylene, we see a peak at δ 2.30 and another at δ 7.10. It seems reasonable that these peaks are caused by the two different "kinds" of ^{1}H nuclei in the molecule: the methyl hydrogens and the aromatic ring hydrogens. How can we tell which is which?

One way is to determine, by integration, the area under each peak. The **peak area** is directly proportional to the number of ^{1}H nuclei responsible for the particular peak. All commercial NMR spectrometers are equipped with electronic integrators that can print an **integration line** over the peaks (Figure 12.3). The ratio of heights of the vertical parts of this line is the ratio of peak areas. Thus, we find that the areas of the peaks at δ 2.30 and δ 7.10 in the *p*-xylene spectrum give a ratio of 3:2 (or 6:4).* These areas allow us to assign the peak at δ 2.30 to the six methyl hydrogens and the peak at δ 7.10 to the four aromatic ring hydrogens.

The **peak area** of an ^{1}H NMR signal is directly proportional to the number of ^{1}H nuclei responsible for that peak. An **integration line** is used to determine the ratio of peak areas.

EXAMPLE 12.1

How many peaks do you expect to see in the ^{1}H NMR spectrum of each of the following compounds? If you expect several peaks, what will their relative areas be?

a. $CH_3—C(CH_3)_2—CH_3$ (central C bearing CH$_3$ above and below)

b. benzene ring with CO_2CH_3 at top and CO_2CH_3 at bottom

c. $BrCH_2—C(CH_3)_2—CH_2Br$ (central C bearing CH$_3$ above and below)

Solution

a. All twelve ^{1}H nuclei are equivalent and appear as a single peak.
b. The four aromatic hydrogens are equivalent, and the six methyl hydrogens on the ester functions are equivalent. There will be two peaks in the spectrum, with an area ratio of 4:6 (or 2:3).
c. There are two kinds of hydrogens, $CH_3—C$ and $CH_2—Br$. There will be two peaks, with the area ratio 6:4 (or 3:2).

PROBLEM 12.1 Which of the following compounds show only a single peak in their ^{1}H NMR spectrum?

a. $CH_3CH_2OCH_2CH_3$ b. cyclopentane (pentagon) c. $CH_3CH_2\overset{\underset{\|}{O}}{C}Cl$

*The ratio of areas is often approximate rather than exact. For example, in this case, it is 42:29 or 1.45:1 or 2.9:2 and has been rounded up to 3:2.

PROBLEM 12.2 Each of the following compounds shows more than one peak in its ^{1}H NMR spectrum. What will the area ratios be?

a. CH_3CH_2OH b. $CH_3\overset{\overset{\displaystyle O}{\|}}{C}OCH_3$ c. $CH_3CH_2OCH_2CH_3$

PROBLEM 12.3 How could ^{1}H NMR spectroscopy be used to distinguish 1,1-dichloroethane from 1,2-dichloroethane?

A more general way to assign peaks is to compare chemical shifts with those of similar protons in a known reference compound. For example, benzene has six equivalent hydrogens and shows a single peak in its ^{1}H NMR spectrum, at δ 7.24. Other aromatic compounds also show a peak in this region. We can conclude that most aromatic ring hydrogens will have chemical shifts at about δ 7. Similarly, most CH_3—Ar hydrogens appear at δ 2.2–δ 2.5 (see Figure12.3).

The chemical shifts of ^{1}H nuclei in various chemical environments have been determined by measuring the ^{1}H NMR spectra of a large number of compounds with known, relatively simple structures. Table 12.2 gives the chemical shifts for several common types of ^{1}H nuclei.

Table 12.2 ■ Typical ^{1}H Chemical Shifts (Relative to Tetramethylsilane)

Type of ^{1}H	δ (ppm)	Type of ^{1}H	δ (ppm)
C—CH_3	0.85–0.95	—CH_2—F	4.3–4.4
C—CH_2—C	1.20–1.35	—CH_2—Br	3.4–3.6
		—CH_2—I	3.1–3.3
C—CH(—C)—C	1.40–1.65	CH_2=C	4.6–5.0
CH_3—C=C	1.6–1.9	—CH=C	5.2–5.7
CH_3—Ar	2.2–2.5	Ar—H	6.6–8.0
CH_3—C=O	2.1–2.6	—C≡C—H	2.4–2.7
CH_3—N<	2.1–3.0	—C(=O)—H	9.5–9.7
CH_3—O—	3.5–3.8	—C(=O)—OH	10–13
—CH_2—Cl	3.6–3.8	R—OH	0.5–5.5
—$CHCl_2$	5.8–5.9	Ar—OH	4–8

EXAMPLE 12.2

Using the data in Table 12.2, describe the expected ^{1}H NMR spectrum of

a. $CH_3C(=O)—OCH_3$ b. $Cl_2CH—C(CH_3)_2—CH_2Cl$

Solution

a. The spectrum will consist of two peaks, equal in area, at about δ 2.3 (for the $CH_3C(=O)—$ hydrogens) and δ 3.6 (for the $—OCH_3$ hydrogens).

b. The spectrum will consist of three peaks, with relative areas 6 : 2 : 1 at δ 0.9 (the two methyl groups), δ 3.5 (the $—CH_2—Cl$ hydrogens), and δ 5.8 (the $—CHCl_2$ hydrogen).

PROBLEM 12.4 Describe the expected ^{1}H NMR spectrum of

a. $CH_3C(=O)OH$ b. $CH_3—C{\equiv}C—H$

PROBLEM 12.5 An ester is suspected of being either $(CH_3)_3CC(=O)OCH_3$ or $CH_3C(=O)—OC(CH_3)_3$. Its ^{1}H NMR spectrum consists of two peaks at δ 0.9 and δ 3.6 (relative areas 3:1). Which compound is it? Describe the spectrum that would be expected if it had been the other ester.

Now let us return to the point mentioned at the beginning of this section, the factors that influence chemical shifts. One important factor is the electronegativity (page 9) of groups in the immediate environment of the ^{1}H nuclei. *Electron-withdrawing groups generally cause a downfield chemical shift.* Compare, for example, the following chemical shifts from Table 12.2:

$—CH_3$	$—CH_2Cl$	$—CHCl_2$
~ 0.9	~ 3.7	~ 5.8

Electrons in motion near a ^{1}H nucleus create a small magnetic field in its microenvironment that tends to shield the nucleus from the externally applied magnetic field. Chlorine is an electron-withdrawing group. Therefore, withdrawal of electron density by the chlorines "deshields" the nucleus, allowing it to flip its spin at a lower applied external field or lower frequency. The more chlorines, the larger the effect.

EXAMPLE 12.3

Predict the order of chemical shifts of the various ^{1}H signals for 1-bromopropane.

Solution

$\overset{3}{C}H_3\overset{2}{C}H_2\overset{1}{C}H_2Br$

The hydrogens at C-1 will be at lowest field because they are closest to the electron-withdrawing Br atom. The methyl hydrogens will be at highest field because they are farthest from the Br, and the peak for the C-2 hydrogens will appear between the other two. The inductive effect falls off rapidly with distance, as seen by the actual chemical shift values.

$$\overset{1.06}{CH_3}-\overset{1.81}{CH_2}-\overset{3.47}{CH_2}-Br$$

PROBLEM 12.6 Compare and explain the following four chemical shifts:

$$\overset{0.23}{CH_4} \quad \overset{3.05}{CH_3Cl} \quad \overset{2.68}{CH_3Br} \quad \overset{2.16}{CH_3I}$$

A second factor that influences chemical shifts is the presence of pi electrons. Hydrogens attached to a carbon that is part of a multiple bond or aromatic ring usually appear downfield from hydrogens attached to saturated carbons. Compare these values from Table 12.2:

$C-CH_2-C$	$CH_2{=}C$	$-CH{=}C$	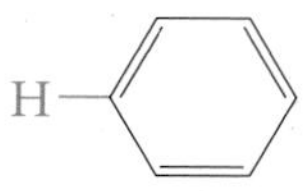 H–(benzene ring)
δ 1.2–1.35	4.6–5.0	5.2–5.7	6.6–8.0

The reasons for this effect are complex, but it is useful in assigning structures.

PROBLEM 12.7 Describe the ^1H NMR spectrum of *trans*-2,2,5,5-tetramethyl-3-hexene.

12.2.c Spin–Spin Splitting

A ^1H nucleus with no ^1H neighbors gives a **singlet** peak. Neighboring ^1H nuclei cause **spin–spin splitting** of the ^1H signal.

Many compounds give spectra that show more complex peaks than just single peaks (**singlets**) for each type of hydrogen. Let us examine some spectra of this type to see what additional structural information they convey.

Figure 12.4 shows the ^1H NMR spectrum of diethyl ether, $CH_3CH_2OCH_2CH_3$. From the information given in Table 12.2, we might have expected the ^1H NMR spectrum of diethyl ether to consist of two lines: one in the region of δ 0.9 for the six equivalent CH_3 hydrogens and one at about δ 3.5 for the four equivalent CH_2 hydrogens adjacent to the oxygen atom, with relative areas 6:4. Indeed, in Figure 12.4, we see absorptions in each of these regions, with the expected total area ratio. But we do not see singlets! Instead, the methyl signal is split into three peaks, a triplet, with relative areas 1:2:1; and the methylene signal is split into four peaks, a quartet, with relative areas 1:3:3:1. These **spin–spin splittings**, as they are called, tell us quite a bit about molecular structure, and they arise in the following way.

We know that each ^1H nucleus in the molecule acts as a tiny magnet. When we run a ^1H NMR spectrum, each hydrogen "feels" not only the very large applied magnetic field but also a tiny field due to its neighboring hydrogens. When we excite the ^1H nuclei on one carbon, the ^1H nuclei on neighboring carbons can be in either the lower or the higher energy spin state, with nearly equal probabilities (nearly equal because, as we have said, the energy difference between the two states is exceedingly small). So the magnetic field of the nuclei whose peak we are observing is perturbed slightly by the tiny fields of its neighboring ^1H nuclei.

The ***n* + 1 rule**: If a ^1H nucleus has n ^1H neighbors with a different chemical shift, its NMR signal is split into $n + 1$ peaks.

We can predict the splitting pattern by the ***n* + 1 rule**: If a ^1H nucleus or a set of equivalent ^1H nuclei has n ^1H neighbors with a substantially different chemical shift, its NMR signal will be split into $n + 1$ peaks. In diethyl ether, each CH_3 hydrogen has

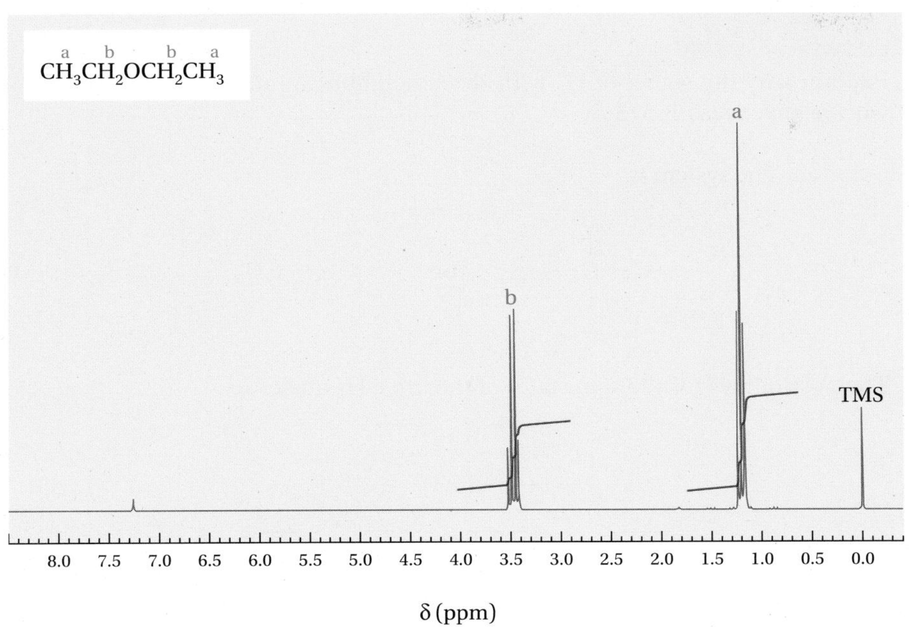

Figure 12.4
^1H NMR spectrum of diethyl ether, showing spin–spin splitting.

two ^1H neighbors (on the CH_2 group). Therefore, the CH_3 signal is split into 2 + 1 = 3 peaks. At the same time, each CH_2 hydrogen has three ^1H neighbors (on the CH_3 group). The CH_2 signal is therefore split into 3 + 1 = 4 peaks. Let us see why this rule works and why the split peaks have the area ratios they do.

Consider first the system

$$\begin{array}{cc} | & | \\ -\mathrm{C}- & \mathrm{C}- \\ | & | \\ \mathrm{H}_a & \mathrm{H}_b \end{array}$$

H_a has one nonequivalent neighbor, H_b. At the time we pass through the H_a signal, H_b can be in either the lower or the higher energy spin state. Because these two possibilities are nearly equal, the H_a signal will be split into two equal peaks: a doublet. The same is true for the peak due to H_b.

Now consider the system

$$\begin{array}{cc} & \mathrm{H}_b \\ | & | \\ -\mathrm{C}- & \mathrm{C}- \\ | & | \\ \mathrm{H}_a & \mathrm{H}_b \end{array}$$

H_a has two neighbors, H_b. At the time we pass through the H_a signal, there are three possibilities for the two H_b nuclei:

$\rightarrow$ $\rightarrow$	$\rightarrow$ $\leftarrow$ $\leftarrow$ $\rightarrow$	$\leftarrow$ $\leftarrow$
both aligned with the field	one with and one against the field	both aligned against the field

Both can be in the lower energy state, both can be in the higher energy state, or one can be in each state, with this latter arrangement being possible in two ways. Hence, the H_a signal will be a triplet with relative areas 1 : 2 : 1. The H_b signal, on the other hand, will be a doublet because of the two possible spin states for H_a.

EXAMPLE 12.4

Explain why the signal of H_a with three neighboring nuclei H_b is a quartet with relative areas 1 : 3 : 3 : 1.

Solution The system is

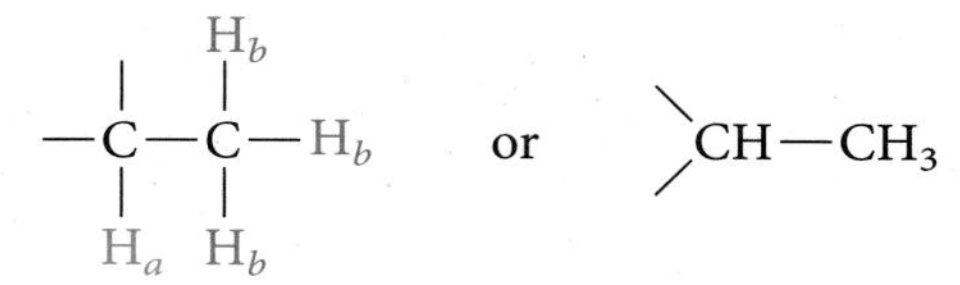

$$-\overset{|}{\underset{|}{\text{C}}}-\overset{H_b}{\underset{H_b}{\text{C}}}-H_b \quad \text{with } H_a \text{ on the first C} \qquad \text{or} \qquad >\text{CH}-\text{CH}_3$$

The possibilities for the spin states of the three H_b nuclei are

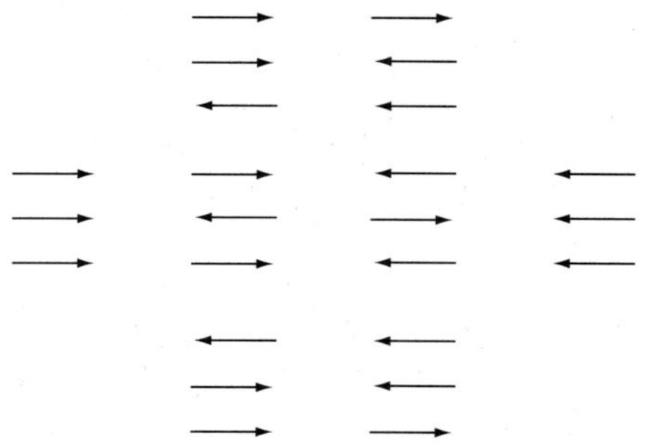

	→ → ←	→ ← ←	
→ → →	→ ← →	← → ←	← ← ←
	← → →	← ← →	

Hence the H_a signal will appear as four peaks (a quartet), with the area ratio 1 : 3 : 3 : 1. The H_b peak will appear as a doublet, because the possible alignments of H_a are ⟶ or ⟵.

PROBLEM 12.8 Use the data in Table 12.2 to predict the ^{1}H NMR spectrum of CH_3CHCl_2. Give the approximate chemical shifts and the splitting patterns of the various peaks.

^{1}H nuclei that split each other's signals are **coupled**. The **coupling constant, *J***, is the number of hertz by which the peaks are split.

^{1}H nuclei that split one another's signals are said to be **coupled**. The extent of the coupling, or the number of hertz by which the signals are split, is called the **coupling constant** (abbreviated ***J***). A few typical coupling constants are shown in Table 12.3. *Spin–spin splitting falls off rapidly with distance.* Whereas hydrogens on adjacent carbons may show appreciable splitting (J = 6 Hz–8 Hz), hydrogens farther apart hardly "feel" each other's presence (J = 0 Hz–1 Hz). As seen in Table 12.3, coupling constants can even be used at times to distinguish between *cis–trans* isomers or between positions of substituents on a benzene ring.

Chemically equivalent ^{1}H nuclei do not split each other. For example, $BrCH_2CH_2Br$ shows only a sharp singlet in its ^{1}H NMR spectrum for all four hydrogens. Even though they are on adjacent carbons, the hydrogens do not split each other because they have identical chemical shifts.

PROBLEM 12.9 Describe the ^{1}H NMR spectrum of

a. $BrCH_2CH_2Cl$ b. $ClCH_2CH_2Cl$

Not all ^{1}H NMR spectra are simple; they may sometimes be quite complex. This complexity can arise when adjacent hydrogens have nearly the same, but not identical, chemical shifts. An example is phenol (see Figure 12.5). We can easily distinguish the aromatic hydrogens (δ 6.8–δ 7.4) from the hydroxyl hydrogen (δ 4.95), but the splitting

Table 12.3 ▪ Some Typical Coupling Constants

Group	J (Hz)	Group	J (Hz)
$-\underset{H}{\overset{\mid}{C}}-\underset{H}{\overset{\mid}{C}}-$	6–8	H on benzene ring, with a second H on the ring	*ortho:* 6–10 *meta:* 1–3 *para:* 0–1
$-\underset{H}{\overset{\mid}{C}}-\overset{\mid}{\underset{\mid}{C}}-\underset{H}{\overset{\mid}{C}}-$	0–1	$(H)_2C{=}C(R_1)(R_2)$	0–3
$H(R_1)C{=}C(R_2)H$ (H's trans)	12–18	$H(R_1)C{=}C(R_2)H$ (H's cis)	6–12

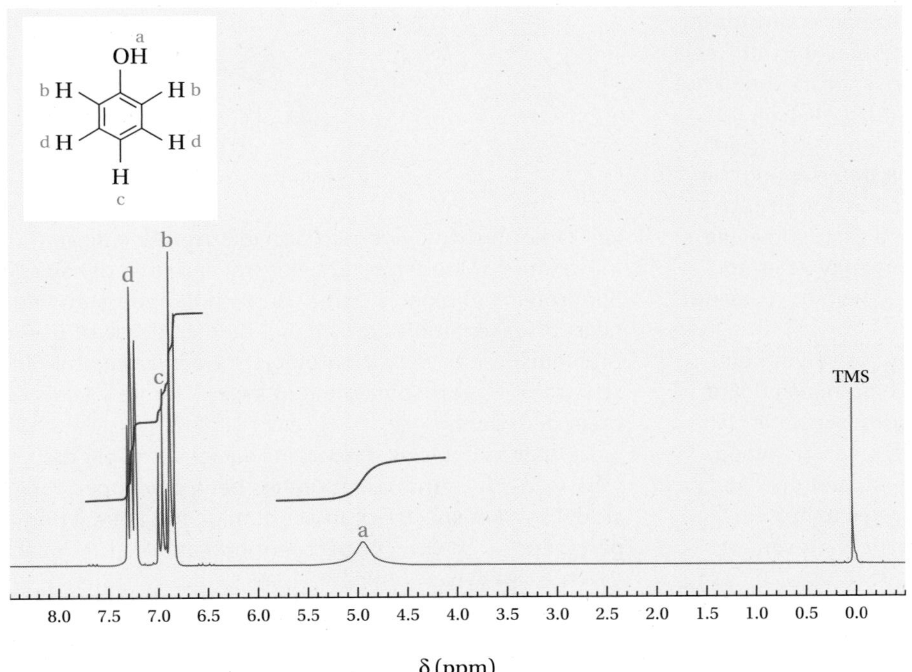

▪ Figure 12.5
The ^{1}H NMR spectrum of phenol. Note the complexity in the aromatic ^{1}H region (δ 6.8–δ 7.4). Reprinted with permission from University Science Books.

pattern of the complex **multiplet** seen for the aromatic hydrogens cannot be analyzed with the simple $n + 1$ rule. Such spectra can, however, be thoroughly analyzed by specially designed computer programs.

The splitting pattern of a proton is often called a **multiplet** when it is too complex to be analyzed by the $n + 1$ rule.

In summary, then, ^{1}H NMR spectroscopy can give us the following kinds of structural information:

1. **The number of signals and their chemical shifts can be used to identify the kinds of chemically different ^{1}H nuclei in the molecule.**
2. **The peak areas tell us how many ^{1}H nuclei of each kind are present.**
3. **The spin–spin splitting pattern gives us information about the number of nearest ^{1}H neighbors that a particular kind of ^{1}H nucleus may have.**

A WORD ABOUT... NMR in Biology and Medicine

© Jonathan Nourok / Photo Edit

Magnetic resonance imaging (MRI) is used to produce an image of a brain.

In this chapter, we have presented only the bare bones of NMR spectroscopy, with examples of organic structure determination. But with the application of modern computer technology to NMR instrumentation, the field has developed rapidly and added many sophisticated techniques.* It is possible, for example, to measure spectra at variable temperatures (−180°C to +200°C) and to measure the rate of processes such as rotations about single bonds, "flipping" of cyclohexane chair conformations, nitrogen inversion in amines, and reactions of free radicals. With other techniques, it is possible to obtain not only ^{13}C spectra of *all* of the carbons in a molecule, but separate spectra of *only* CH_3 groups (or only CH_2 or only CH groups), which helps enormously in determining the structures of complex biological molecules. Other techniques allow one to locate groups in the same molecule that are close in space even though they may seem to be far apart if we count by bonds, a real help in studying conformations of complex molecules.

The potential of NMR spectroscopy for solving biological and medical problems is now being realized. For example, instruments are now sufficiently sensitive that one can study intact bodily fluids, such as urine, blood plasma, seminal fluid, cerebrospinal fluid, and eye fluid. **Creatinine**, for example, is a normal organic protein metabolite excreted in the urine, where it is present at millimolar levels. The singlets due to its CH_3 (δ 3.1) and CH_2 (δ 4.2) groups are easily observed at these low concentrations with a high-resolution (500 MHz) spectrometer, even in the presence of all of the other constituents in urine. Its concentration, which can be determined in just a few minutes, provides a useful indicator of kidney function. NMR spectroscopy has been used to detect inherited metabolic diseases by studying the urine of newborn infants.

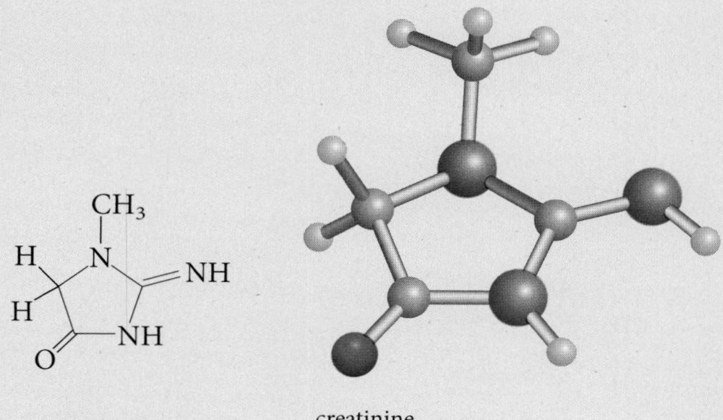

creatinine

Other bodily fluids also provide useful medical information. Monitoring low-density and high-density lipoproteins in blood plasma can help in understanding heart disease. Studies of seminal fluid could be of use for problems of infertility. Neurologists are examining NMR spectra of cerebrospinal fluid in infants for new leads on brain disorders.**

Other biologically important nuclei with spin are ^{31}P, ^{23}Na, and ^{19}F. Instrumentation has been developed for studying NMR spectra of intact human and animal body parts. For example, ^{31}P spectra of human forearm muscle

12.3 ^{13}C NMR Spectroscopy

Whereas ^{1}H NMR spectroscopy gives information about the arrangement of hydrogens in a molecule, ^{13}C NMR spectroscopy gives information about the carbon skeleton. The ordinary isotope of carbon, carbon-12, does not have a nuclear spin, but carbon-13 does. Carbon-13 constitutes only 1.1% of naturally occurring carbon atoms. Also, the energy gap between the higher and lower spin states of ^{13}C is very small. For these two reasons, ^{13}C NMR spectrometers must be exceedingly sensitive. Nevertheless, modern high-field FT-NMR spectrometers are quite sensitive, and ^{13}C NMR spectroscopy has become routine.

Carbon-13 spectra differ from ^{1}H spectra in several ways. Carbon-13 chemical shifts occur over a wider range than those of ^{1}H nuclei. They are measured against the same reference compound, TMS, whose methyl carbons are all equivalent and give a sharp signal. Chemical shifts for ^{13}C are reported in δ units, but the usual range is (the forearm is placed directly in the magnetic field) taken before, during, and after exercise have allowed the monitoring of several phosphorus-containing components of muscle tissue—for example, adenosine triphosphate (ATP), phosphocreatine, and inorganic phosphate. By comparing concentration changes of these components in normal people with those in patients with muscle disorders, the nature of the disorders and methods of treatment can be devised.[†]

In **topical NMR**, the NMR magnet is brought to the object being studied, instead of vice versa. A magnetic probe of some sort placed on the surface induces resonances in molecules close to the surface, and useful ^{1}H, ^{13}C, or ^{31}P spectra of molecules within living bodies can be obtained in this way. The method has been used, for example, to monitor effects of various drugs on metabolic processes.

Magnetic resonance imaging (MRI) is a technique that has been used clinically in hospitals since the mid-1980s. The method allows one to obtain internal images of whole-body parts and has several advantages over x-rays. For one, it is far less hazardous, since it does not cause radiation damage. For another, it gives good images of soft tissues, which are more difficult to obtain with x-rays. How does it work?

A large percentage of the human body is water. Most MRI uses the ^{1}H spectra of water in various tissues (although ^{31}P has also been used) to create images. The magnet must be very large, with about a 25-cm-diameter gap to accommodate a human head, and an even larger one for whole-body work. Fortunately, the magnetic field need not be as uniform as in high-resolution NMR for structure determination, so that construction of such large magnets becomes feasible. A gradient field, instead of a uniform field, is essential for imaging. The field strength can also be appreciably lower than for laboratory NMR work. This is fortunate as well because, as you can imagine, the large magnets required for imaging are quite expensive.

The time required for MRI varies from about 30 seconds to 10 minutes, with about 2 minutes being fairly typical. Imaging can be done at preselected planar sections through the body or in three dimensions. The picture at the beginning of this "A Word About . . ." shows the imaging of a human brain by MRI. Functional MRI (fMRI) is a recent variant in which changes in blood flow in the brain can be directly monitored as a function of neural activity, thereby expanding opportunities for studying neuroscience and neural cognition.

The principal use of MRI is, of course, in *diagnostic medicine,* but it can also be used in *food science* to assess correct conditions for picking, storing, or marketing food; in *agriculture* to study seed germination; in the *building industry* to study water concentration and distribution in timbers; and in many other areas.[‡]

Nuclear magnetic resonance was discovered by physicists and developed by chemists and is now being applied by biologists and others. The benefits to humanity could not have been foreseen in the beginning—another example of the wisdom of investing in fundamental research without immediate regard for practical applications.

See Problem 12.50.

*For excellent descriptions of these methods, see Lambert, J. B., Shurvell, H. F., Lightner, D. A., and Cooks, R. G., *Organic Structural Spectroscopy,* Prentice Hall, New York, 1998.

For further information, see Bell, J. D., Brown, J. C. C., and Sadler, P. J., "NMR Spectroscopy of Body Fluids," *Chemistry in Britain,* **1988, 1021–1024.

†For more details, see Radda, G. K., "The Use of NMR Spectroscopy for the Understanding of Disease," *Science* **1986**, *233,* 640–645.

‡For more about MRI, see the book by Morris, P. G., *Nuclear Magnetic Resonance in Medicine and Biology,* Clarendon Press, Oxford, England, 1986.

about 0 ppm to 200 ppm downfield from TMS (instead of the smaller range of 0 ppm to 10 ppm observed for ^{1}H). This wide range of chemical shifts tends to simplify ^{13}C spectra relative to ^{1}H spectra.

Because of the low natural abundance of ^{13}C, the chance of finding two adjacent ^{13}C atoms in the same molecule is small. Hence ^{13}C–^{13}C spin–spin splitting is ordinarily not seen. This feature simplifies ^{13}C spectra. However, ^{13}C–^{1}H spin–spin splitting can occur. A spectrum can be run in such a way as to show this splitting or not, as desired. Figure 12.6 shows the ^{13}C NMR spectrum of 2-butanol measured with and without ^{13}C–^{1}H splitting. The spectrum without ^{1}H splitting (called a **proton-decoupled spectrum**) shows four sharp singlets, one for each type of carbon atom. The hydroxyl-bearing carbon occurs at the lowest field (δ 69.3), and the two methyl carbons are well separated (δ 10.8 and δ 22.9). In the spectrum *with* ^{13}C–^{1}H splitting, the $n + 1$ rule applies. The signal for each type of ^{13}C nucleus is split by the ^{1}H nuclei bonded directly to it. Both CH_3 signals are quartets (three hydrogens; therefore, $n + 1 = 4$), the CH_2 carbon is a triplet, and the CH carbon is a doublet.

A ^{13}C NMR spectrum that is **proton-decoupled** shows one singlet for each different type of carbon atom.

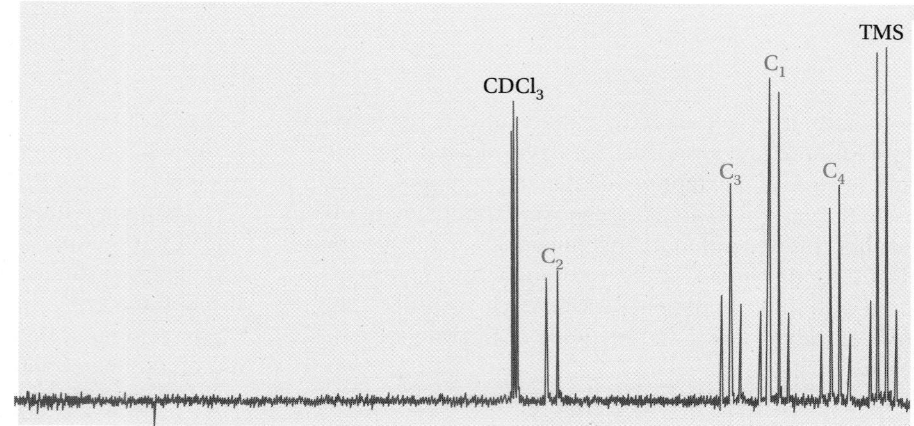

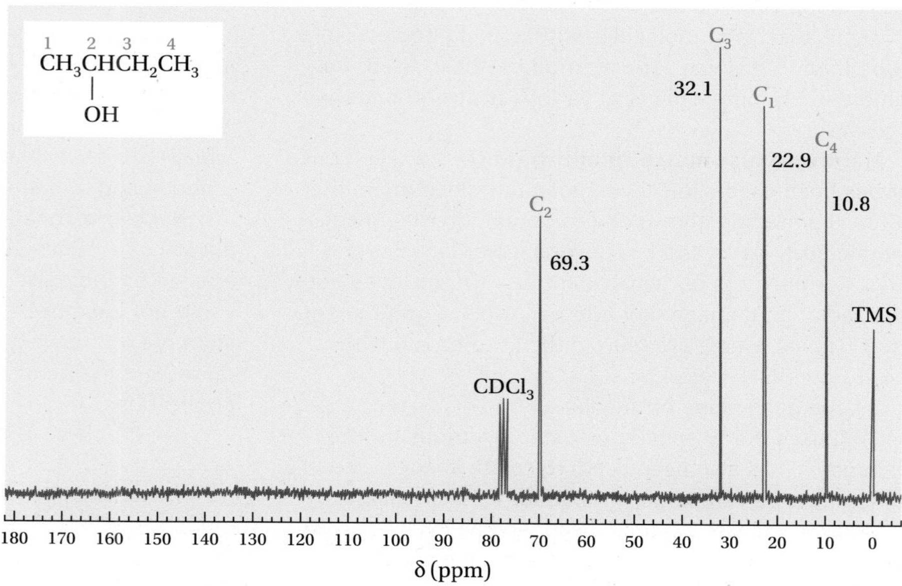

Figure 12.6
The ^{13}C NMR spectrum of 2-butanol without (bottom) and with (top) $^{13}C–^{1}H$ coupling. δ values are shown in the lower spectrum. The solvent is $CDCl_3$. Reprinted with permission from University Science Books.

EXAMPLE 12.5

Describe the ^{13}C spectrum of CH_3CH_2OH.

Solution The spectrum without $^{13}C–^{1}H$ splitting consists of two lines because there are two nonequivalent carbons (their signals come at δ 18.2 and δ 57.8). With $^{13}C–^{1}H$ splitting, the signal at δ 18.2 is a quartet, and the one at δ 57.8 is a triplet.

PROBLEM 12.10 Describe the main features of the ^{13}C spectrum of $CH_3CH_2CH_2OH$.

PROBLEM 12.11 How many peaks would you expect to see in the ^{1}H-decoupled ^{13}C NMR spectrum of

a. 2-methyl-2-propanol?
b. 4-methylcyclohexanone?
c. 2-methyl-1-propanol?
d. *cis*-1,3-dimethylcyclopentane?

12.4 Infrared Spectroscopy

Even though NMR spectroscopy is a powerful tool for deducing structures, it is usually supplemented by other spectroscopic methods that provide additional structural information. One of the more important of these is **infrared (IR) spectroscopy**.

Infrared (IR) spectroscopy is used to determine the types of bonds present in a molecule.

Infrared frequency is usually expressed in units of **wavenumber**, defined as the number of waves per centimeter. Ordinary instruments scan the range of about 700 cm^{-1} to 5000 cm^{-1}. This frequency range corresponds to energies of about 2 kcal/mol to 12 kcal/mol (Table 12.1). This amount of energy is sufficient to affect bond vibrations (motions such as bond stretching or bond bending) but is appreciably less than would be needed to break bonds. These motions are exemplified for a CH_2 group in Figure 12.7.

The wavenumber of an IR frequency is defined as the number of waves per centimeter.

Particular types of bonds usually stretch within certain rather narrow frequency ranges. *IR spectroscopy is particularly useful for determining the types of bonds that are present in a molecule.* Table 12.4 gives the ranges of stretching frequencies for some bonds commonly found in organic molecules.

The IR spectrum of a compound can easily be obtained in a few minutes. A small sample of the compound is placed in an instrument with an IR radiation source. The spectrometer automatically scans the amount of radiation that passes through the sample over a given frequency range and records on a chart the percentage of radiation that is transmitted. Radiation absorbed by the molecule appears as a band in the spectrum.

Figure 12.8 on pages 373–375 shows two typical IR spectra. Both spectra show C—H stretching bands near 3000 cm^{-1} and a C═O stretching band near 1700 cm^{-1}. **Functional group bands** will appear in the same range regardless of the details of the molecular structure. Both cyclopentanone and cyclohexanone have C—H and C═O bonds and therefore have similar spectra in the functional group region of the spectrum (1500 cm^{-1} to 4000 cm^{-1}).

The functional group bands for a particular functional group appear in the same region of the IR spectrum.

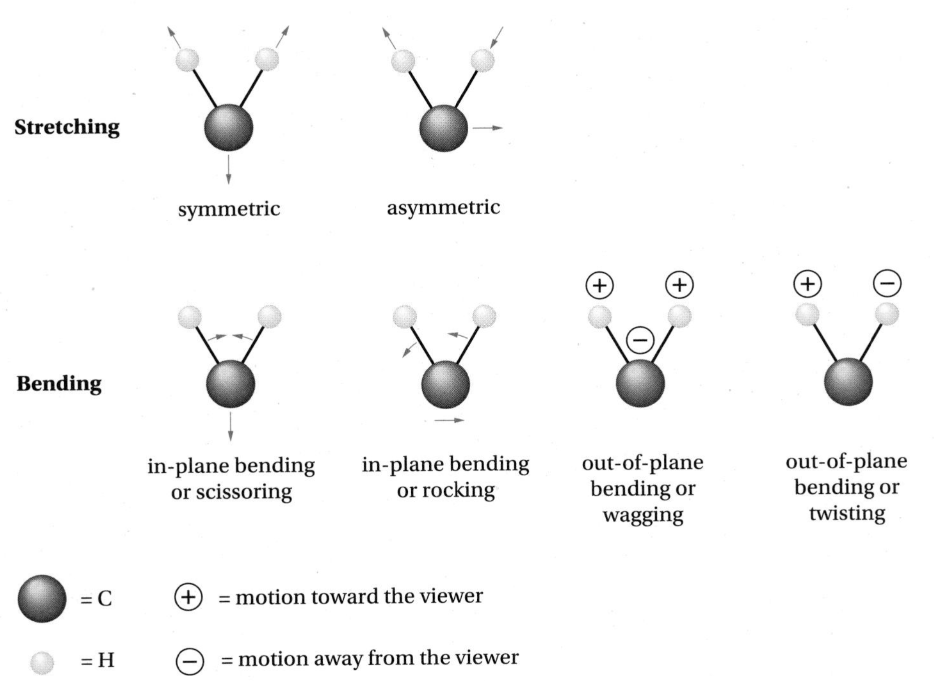

■ **Figure 12.7**
Stretching and bending motions of a CH_2 group that require energies in the IR region.

Table 12.4 ■ Infrared Stretching Frequencies of Some Typical Bonds

Bond type	Group	Class of compound	Frequency range (cm^{-1})
single bonds to hydrogen	C—H	alkanes	2850–3000
	=C—H	alkenes and aromatic compounds	3030–3140
	≡C—H	alkynes	3300
	O—H	alcohols and phenols	3500–3700 (free) 3200–3500 (hydrogen-bonded)
	O—H	carboxylic acids	2500–3000
	N—H	amines	3200–3600
	S—H	thiols	2550–2600
double bonds	C=C	alkenes	1600–1680
	C=N	imines, oximes	1500–1650
	C=O	aldehydes, ketones, esters, acids	1650–1780
triple bonds	C≡C	alkynes	2100–2260
	C≡N	nitriles	2200–2400

Bands in the **fingerprint region** of an IR spectrum are unique for each particular compound.

The spectra of cyclopentanone and cyclohexanone differ in the low frequency or **fingerprint region**, from 700 cm^{-1} to 1500 cm^{-1}. Bands in this region result from combined bending and stretching motions of the atoms and are unique for each particular compound.

Spectra of 1-butanol, butanoic acid, and 1-butanamine illustrate typical functional group bands for alcohols, carboxylic acids, and amines, respectively, in Figure 12.9 (pages 374–375). Again, the fingerprint region is unique for each compound, although all are derived from the parent hydrocarbon butane.

EXAMPLE 12.6

Using IR spectroscopy, how could you quickly distinguish between the structural isomers benzyl alcohol and anisole?

C_6H_5—CH_2OH (benzyl alcohol) C_6H_5—OCH_3 (anisole)

Solution Benzyl alcohol's IR spectrum will show a band in the O—H stretching frequency region (3200 cm^{-1} to 3700 cm^{-1}); the spectrum of anisole will not have a band in that region.

PROBLEM 12.12 How could you use IR spectroscopy to distinguish between the isomers 1-hexyne and 1,3-hexadiene?

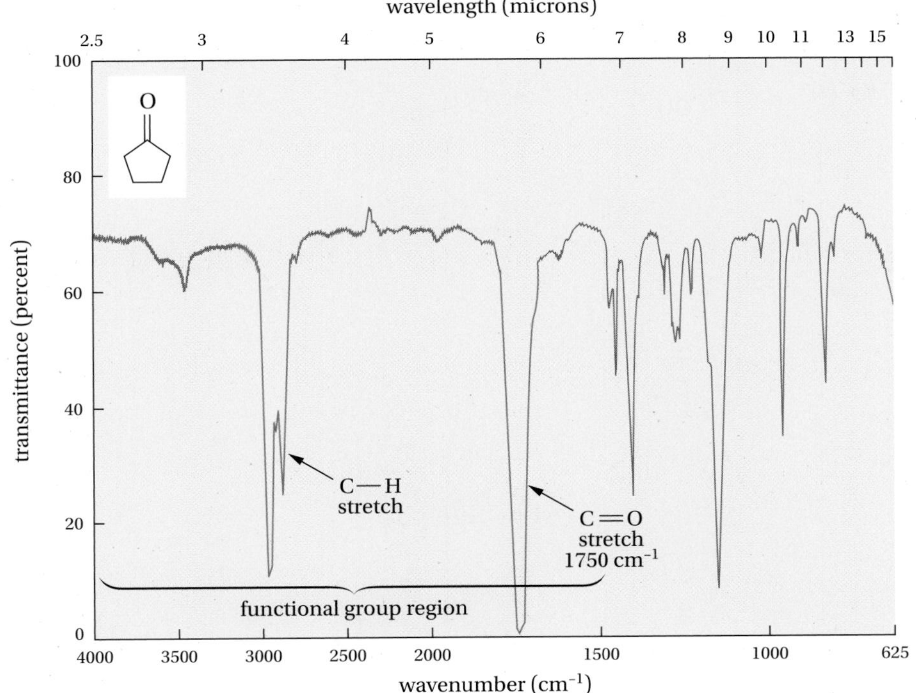

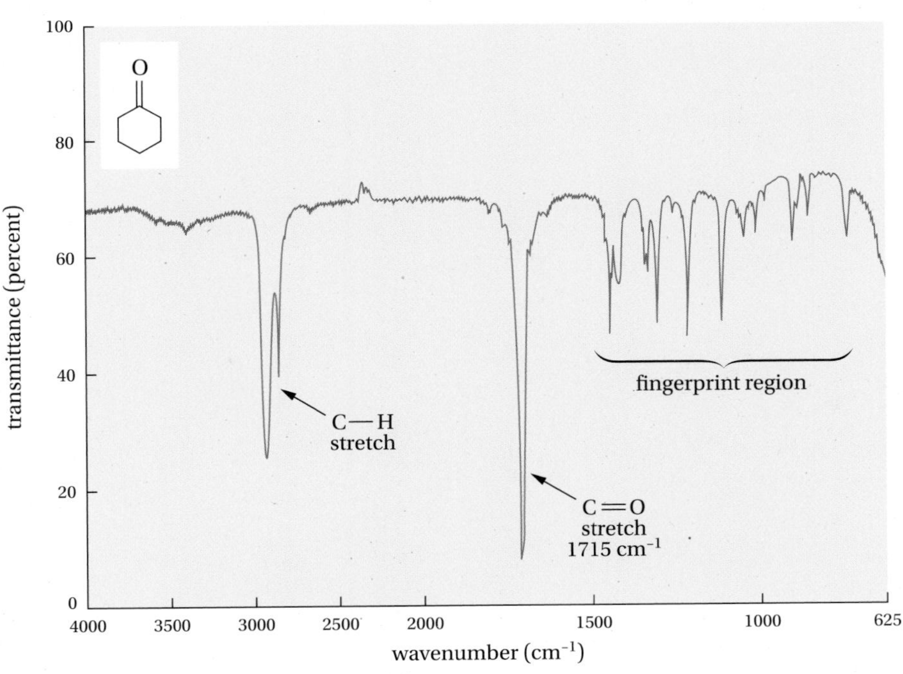

▰ Figure 12.8

The IR spectra of two similar ketones. The spectra have similar functional group bands but differ in the fingerprint region.

▪ Figure 12.9
The IR spectra of (a) 1-butanol, (b) butanoic acid, and (c) 1-butanamine illustrate the typical functional group bands for alcohols, carboxylic acids, and amines, respectively.

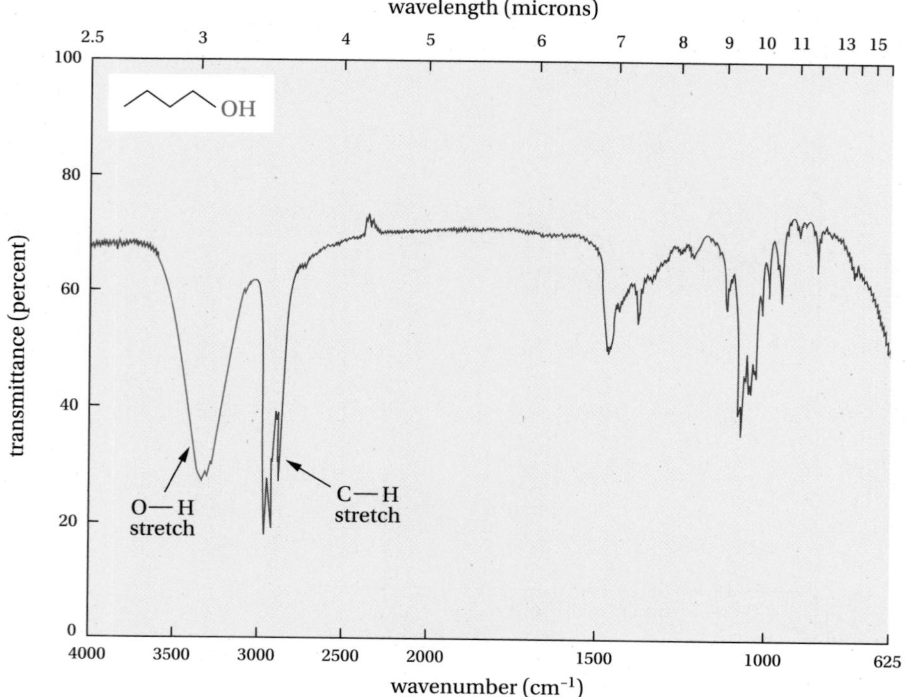

(a)

transmittance (percent)
100 80 60 40 20 0
O OH
O—H stretch
C—H stretch
C=O stretch 1710 cm^{-1}
4000 3500 3000 2500 2000 1500 1000 625
wavenumber (cm^{-1})

(b)

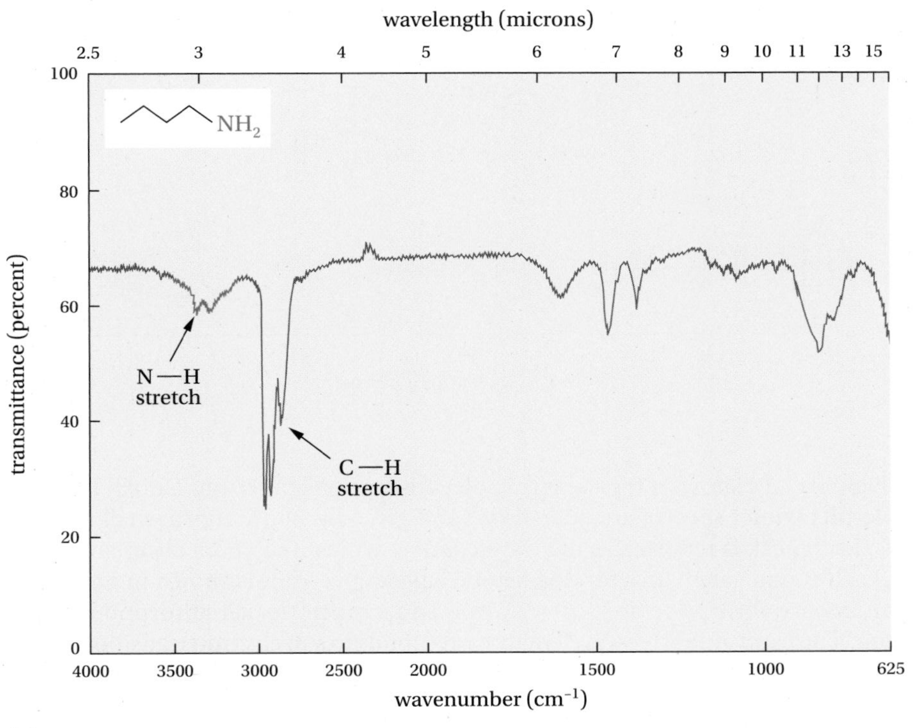

Figure 12.9 (continued)

To summarize, IR spectra can be used to tell what types of bonds might be present in a molecule (by using the functional group region) and to tell whether two substances are identical or different (by using the fingerprint region).

12.5 Visible and Ultraviolet Spectroscopy

The visible region of the spectrum (visible to the human eye, that is) corresponds to light with wavelengths of 400 to 800 **nanometers** (nm; 1 nm $= 10^{-9}$ meters). Ultraviolet light has a shorter wavelength, about 200 nm to 400 nm, whereas IR light has a longer wavelength, about 2500 nm. Older units for reporting these spectra are **millimicrons** (mμ; 1 mμ = 1 nm) or **angstrom units** (Å; 10 Å = 1 nm). Table 12.5 summarizes these units.

1 **nanometer** (nm) = 10^{-9} meters

1 **millimicron** (mμ) = 1 nm

10 **angstroms** (Å) = 1 nm

The amounts of energy associated with light are 37 kcal/mol to 75 kcal/mol for the visible region and 75 kcal/mol to 150 kcal/mol for the ultraviolet region (Table 12.1). These energies are much larger than those involved in IR spectroscopy (2 to 12 kcal/mol). They correspond to the amounts of energy needed to cause an electron to jump from a filled molecular orbital to a higher-energy, vacant molecular orbital. Such electron jumps are called **electronic transitions**.

A **visible-ultraviolet spectrum** records **electronic transitions**, electron jumps from filled molecular orbitals to higher energy vacant molecular orbitals.

Table 12.5 ■ Units for Visible-Ultraviolet Spectra

visible (vis)	400–800 nm (or mμ)	4000–8000 Å
ultraviolet (uv)	200–400 nm (or mμ)	2000–4000 Å

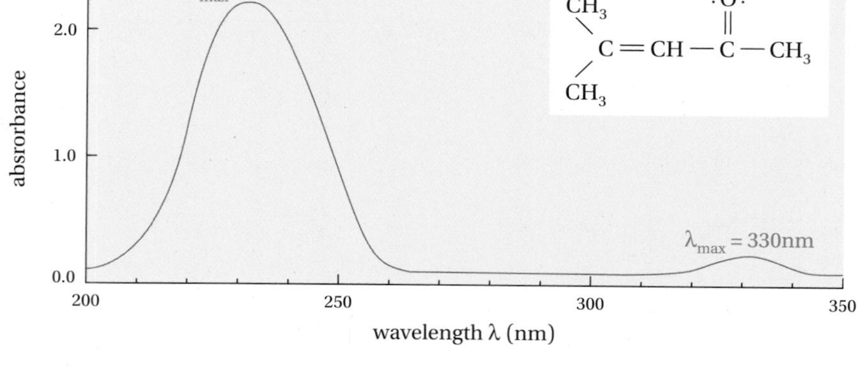

Figure 12.10
The absorption spectrum of 4-methyl-3-penten-2-one.

Figure 12.10 shows a typical ultraviolet absorption spectrum. Unlike IR spectra, **visible-ultraviolet spectra** are quite broad and generally show only a small number of peaks. Each peak is reported as the wavelength at which maximum absorbance occurs (λ_{max}). The conjugated, unsaturated ketone whose spectrum is shown in Figure 12.10 has an intense absorption at λ_{max} = 232 nm and a much weaker absorption at λ_{max} = 330 nm. The band at shorter wavelength corresponds to a pi electron transition, whereas the longer-wavelength, weaker-intensity band corresponds to a transition of the nonbonding electrons on the carbonyl oxygen atom.

The intensity of an absorption band can be expressed quantitatively. Band intensity depends on the particular molecular structure and also on the number of absorbing molecules in the light path. **Absorbance**, which is the log of the ratio of light intensities entering and leaving the sample, is given by the equation

The **absorbance** of a sample of molecules depends on the particular molecular structure and the number of absorbing molecules.

$$A = \epsilon c l \text{ (Beer's law)} \quad (12.5)$$

where ϵ is the **molar absorptivity** (sometimes called the **extinction coefficient**), c is the concentration of the solution in moles per liter, and l is the length in centimeters of the sample through which the light passes. The value of ϵ for any peak in the spectrum of a compound is a constant characteristic of that particular molecular structure. For example, the values of ϵ for the peaks in the spectrum of the unsaturated ketone shown in Figure 12.10 are λ_{max} = 232 nm (ϵ = 12,600) and λ_{max} = 330 nm (ϵ = 78).

EXAMPLE 12.7

What is the effect of doubling the concentration of a particular absorbing sample on A? On ϵ?

Solution The observed absorbance A will be doubled since A is directly proportional to c. The value of ϵ, however, is a function of molecular structure and is a constant, independent of the concentration.

PROBLEM 12.13 A particular solution of $(CH_3)_2C{=}CH{-}C({=}O){-}CH_3$, the ketone whose spectrum is shown in Figure 12.10, placed in a 1-cm absorption cell shows a peak at λ_{max} = 232 nm with an observed absorbance A = 2.2. Calculate the concentration of the solution, using the value of ϵ given in the text.

Visible-ultraviolet spectra are most commonly used to detect conjugation (page 69). In general, molecules with no double bonds or with only one double bond do not absorb in the visible-ultraviolet region (200 to 800 nm). Conjugated systems do absorb there, however, and the greater the conjugation, the longer the wavelength of maximum absorption, as seen in the following examples:

$CH_2{=}CH{-}CH{=}CH_2$
$\lambda_{max} = 220$ nm
($\epsilon = 20{,}900$)

$CH_2{=}CH{-}CH{=}CH{-}CH{=}CH_2$
$\lambda_{max} = 257$ nm
($\epsilon = 35{,}000$)

$CH_2{=}CH{-}CH{=}CH{-}CH{=}CH{-}CH{=}CH_2$
$\lambda_{max} = 287$ nm
($\epsilon = 52{,}000$)

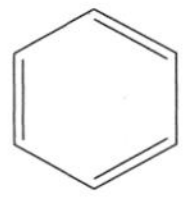

$\lambda_{max} = 255$ nm
($\epsilon = 215$)

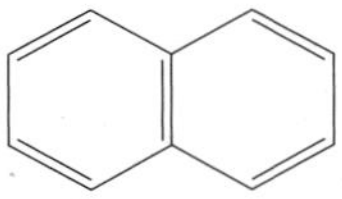

$\lambda_{max} = 314$ nm
($\epsilon = 289$)

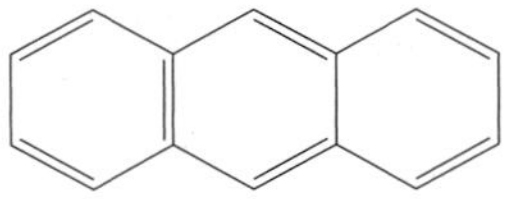

$\lambda_{max} = 380$ nm
($\epsilon = 9000$)

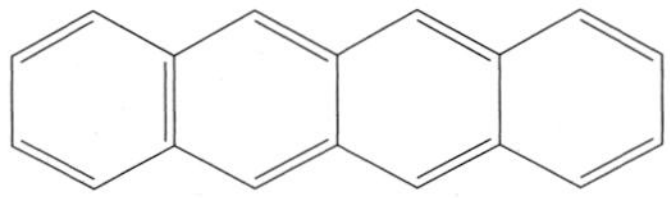

$\lambda_{max} = 480$ nm: a yellow compound
($\epsilon = 12{,}500$)

PROBLEM 12.14 Which of the following aromatic compounds do you expect to absorb at the longer wavelength?

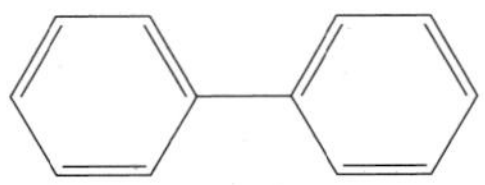

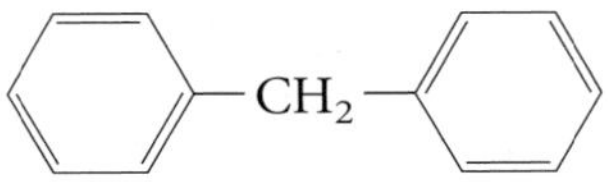

CH_2

PROBLEM 12.15 Naphthalene is colorless, but its isomer azulene is blue. Which compound has the lower-energy pi electronic transition?

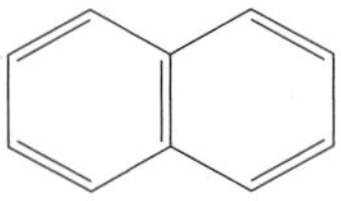

naphthalene

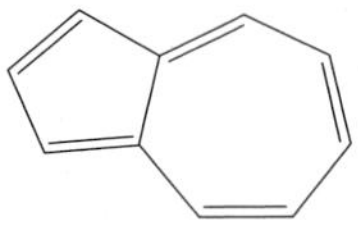

azulene

12.6 Mass Spectrometry

Mass spectrometry (MS) differs from the other types of spectroscopy discussed in this chapter, in that it does not depend on transitions between energy states. Instead, a mass spectrometer converts molecules to ions, sorts them according to their mass-to-charge (m/z) ratio, and determines the relative amounts of each ion present. A small sample of the substance is introduced into a high-vacuum chamber, where it is vaporized and bombarded with high-energy electrons. These bombarding electrons eject an electron from the molecule M, to give a **cation radical** called the **molecular ion** $M^{+\cdot}$ (sometimes referred to as the **parent ion**).

The **molecular ion**, or **parent ion**, in a mass spectrum is a **cation radical**, with the same mass but one less electron than the neutral molecule.

$$M + e^- \longrightarrow \underset{\text{molecular ion}}{M^{+\cdot}} + 2\,e^- \quad (12.6)$$

Methanol, for example, forms a molecular ion in the following way:

$$e^- + CH_3\ddot{\underset{\cdot\cdot}{O}}H \longrightarrow \underset{\text{methanol molecular ion } (m/z = 32)}{[CH_3\dot{\underset{\cdot\cdot}{O}}H]^+} + 2\,e^- \quad (12.7)$$

The beam of these parent ions then passes between the poles of a powerful magnet, which deflects the beam. The extent of the deflection depends on the mass of the ion. Since $M^{+\cdot}$ has a mass that is essentially identical to the mass of the molecule M (the mass of the ejected electron is trivial compared to the mass of the rest of the molecule), *mass spectrometers can be used to determine molecular weights.*

Frequently, mass spectra show a peak one or two mass units *higher* than the molecular weights. How can this be? Recall that the isotope ^{13}C (one mass unit higher than ordinary ^{12}C) has a natural abundance of about 1.1%. This gives rise to an $(M + 1)^{+\cdot}$ peak in carbon compounds. The intensity of this peak relative to the $M^{+\cdot}$ peak is approximately 1.1% times the number of carbons in the compound (because the chance of finding a ^{13}C atom in a compound is proportional to the number of carbon atoms present).

PROBLEM 12.16 An alkane shows an $M^{+\cdot}$ peak at m/z 114. What is its molecular formula? What will be the relative intensities of the 115/114 peaks?

Other isotopic peaks can also be useful. For example, chlorine consists of a mixture of ^{35}Cl (75%) and ^{37}Cl (25%), and bromine consists of a 50 : 50 mixture of ^{79}Br and ^{81}Br. A monochloroalkane will therefore show two parent ion peaks, two mass units apart and in an intensity ratio of 3 : 1. Monobromoalkanes also show two parent ion peaks two mass units apart, but in a 1 : 1 intensity ratio. These isotopic peaks can be used to obtain structural information, as in the following example.

EXAMPLE 12.8

A bromoalkane shows two equal-intensity parent ion peaks at m/z 136 and m/z 138. Deduce its molecular formula.

Solution Only one Br can be present (the molecular weight is not high enough for two bromines). Subtract 136 − 79 (or 138 − 81) to get 57 for the mass of the carbons and hydrogens. Dividing 57 by 12 (the mass of carbon), we get 4, with 9 mass units left over for the hydrogens. The formula is C_4H_9Br.

PROBLEM 12.17 A compound containing only C, H, and Cl shows parent ion peaks at m/z 74 and m/z 76 in a ratio of 3:1. Suggest possible structures for the compound.

If bombarding electrons have enough energy, they produce not only parent ions but also fragments called **daughter ions**. That is, the original molecular ion breaks into smaller fragments, some of which are ionized and get sorted on an m/z basis by the spectrometer. A prominent peak in the mass spectrum of methanol, for example, is the $M^+ - 1$ peak at $m/z = 31$. This peak arises through loss of a hydrogen atom from the molecular ion.

Daughter ions, the smaller fragments produced when a parent ion breaks up, occur in patterns typical for certain molecular structures.

$$\underset{m/z\,=\,32}{H_3C\!-\!\ddot{O}^{\cdot +}\!-\!H} \longrightarrow \underset{m/z\,=\,31}{H_2C\!=\!\ddot{O}^{+}\!-\!H} + H\cdot \qquad (12.8)$$

You will recognize this daughter ion as protonated formaldehyde, a resonance-stabilized carbocation.

A mass spectrum consists, then, of a series of signals of varying intensities at different m/z ratios. In practice, most of the ions are singly charged ($z = 1$) so that we can readily obtain their masses, m. Figure 12.11 shows a mass spectrum printed as the computer output of a mass spectrometer. It is the mass spectrum of a typical ketone, 4-octanone. The peak at $m/z = 128$ is the most intense high mass peak in the spectrum and corresponds to the molecular weight of the ketone. In addition, we see certain prominent daughter ion peaks. For example, the peaks at $m/z = 85$ and $m/z = 71$ correspond in mass to $C_4H_9CO^+$ and $C_3H_7CO^+$, respectively. This suggests that one easy fragmentation path for the parent ion is to break the carbon–carbon bond adjacent to the carbonyl group. Ion fragmentation paths depend on ion structure, and the interpretation of mass spectral fragmentation patterns can give significant information about molecular structure.

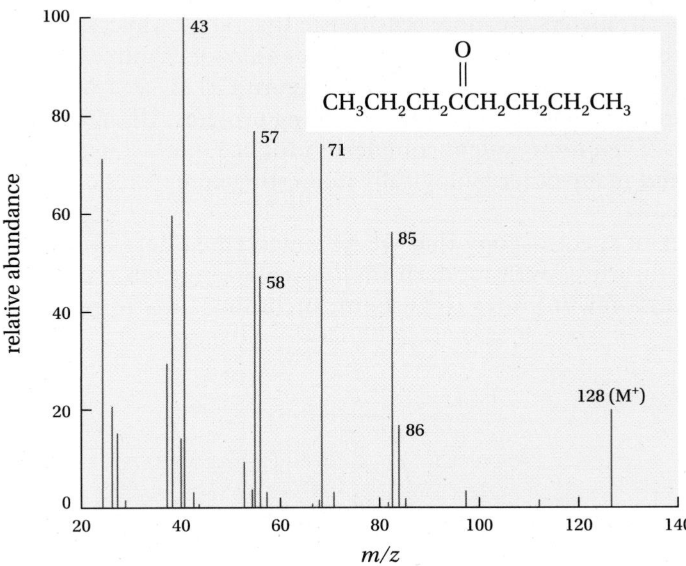

▰ **Figure 12.11**
The mass spectrum of 4-octanone.

EXAMPLE 12.9

The most intense peak (called the *base peak*) in Figure 12.11 occurs at $m/z = 43$. Suggest how it might arise.

Solution This peak corresponds to the m/z for $C_3H_7^+$, suggesting that the daughter ion $C_3H_7CO^+$ loses carbon monoxide to give $C_3H_7^+$. This explanation becomes more plausible when we consider that the spectrum also contains an intense peak at $m/z = 57$, corresponding to the analogous process for $C_4H_9CO^+$. We can summarize these conclusions in the following "family tree" of ions:

$$C_8H_{16}O^{+\cdot}\ M^+\ (128) \xrightarrow{-C_3H_7\cdot} C_4H_9CO^+\ (85) \xrightarrow{-CO} C_4H_9^+\ (57)$$

$$C_8H_{16}O^{+\cdot}\ M^+\ (128) \xrightarrow{-C_4H_9\cdot} C_3H_7CO^+\ (71) \xrightarrow{-CO} C_3H_7^+\ (43)$$

PROBLEM 12.18 In what ways will the mass spectrum of 4-heptanone be similar to and in what ways will it differ from the mass spectrum in Figure 12.11?

MALDI (matrix-assisted laser desorption ionization) and **ESI** (electrospray ionization) are two modern mass spectrometric techniques for analyzing large biomolecules.

Until recently, large biomolecules such as proteins could not be studied by MS because these large polar molecules are nonvolatile and difficult to convert to gaseous molecular ions. Newer techniques such as matrix-assisted laser desorption ionization (**MALDI**) and electrospray ionization (**ESI**) have solved this problem. Often, MALDI-MS produces a molecular ion with one positive charge. Thus, the m/z ratio for the molecular ion gives the molecular weight of the molecule.

The molecular ions produced by ESI-MS are typically formed by the addition of protons (H^+) to the molecule, and typical ions have more than one positive charge. The ions, therefore, have a small m/z ratio, typical of small molecules, an advantage since mass spectrometers are more sensitive in this range. Molecular ions with different charges are formed for the same molecule, as seen for example in the ESI-MS spectrum of the α-chain of human hemoglobin (Figure 12.12).* ESI-MS has been used to study a wide range of macromolecules including proteins, DNA, synthetic polymers, and fullerenes. Even noncovalent complexes produce intact ions, so that DNA–drug interactions and many other biologically interesting complexes can be studied using this technique.

The types of spectroscopy that we have described here are routinely used in research laboratories. With modern instrumentation, each type of spectrum can be obtained in a few minutes to an hour, including time to prepare the sample.

*The molecular weight is determined from the mathematical relationship of the ion peaks. For more details about ESI-MS, see Hofstadler, S. A., Bakhtiar, R., and Smith, R. D., "Electrospray Ionization Mass Spectrometry," *J. Chem. Educ.* **1996**, 73, A82–A88.

A CLOSER LOOK AT... Mass Spectrometry and Carbon Dating

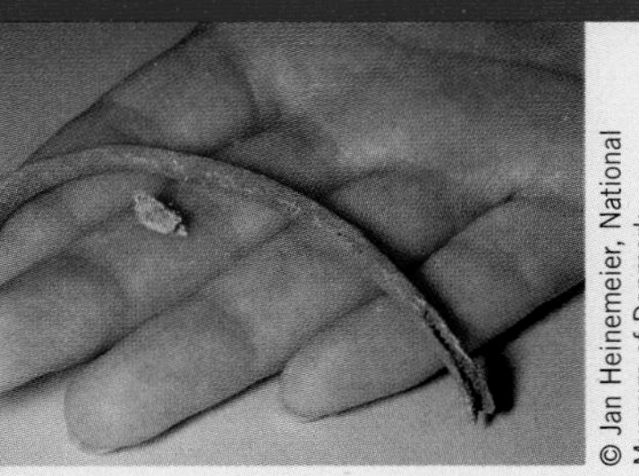
© Jan Heinemeier, National Museum of Denmark

Conduct research on the Internet to find more information on MS and carbon dating and to answer the following questions.

Radiocarbon Dating

1. What is radiocarbon dating?
2. What materials can be dated using this method?
3. How old can a sample be for its age to be accurately determined by this method?

Accelerator Mass Spectrometry and Radiocarbon Dating

1. What is accelerator mass spectrometry (AMS) and how does it work? What information can be obtained from the mass spectrum that is produced?
2. What are the advantages of using AMS for radiocarbon dating rather than the traditional beta-counting method? What are some disadvantages?
3. In what fields of study is AMS used? Use the link to a field that interests you to find out the kinds of samples that can be studied using AMS.
4. Research C-14 and C-13 dating using AMS to resolve a debate about the cause of the disappearance of Norsemen from Greenland. Based on recent radiocarbon dating, what happened to the diet of the Norsemen during the 500 years they were present in Greenland? How does understanding the diet of the Norsemen help resolve the debate about their disappearance from Greenland?

Interpretation of the spectra may take longer, but investigators with experience can often deduce the structure of even complex molecules from their spectra alone in a relatively short time.

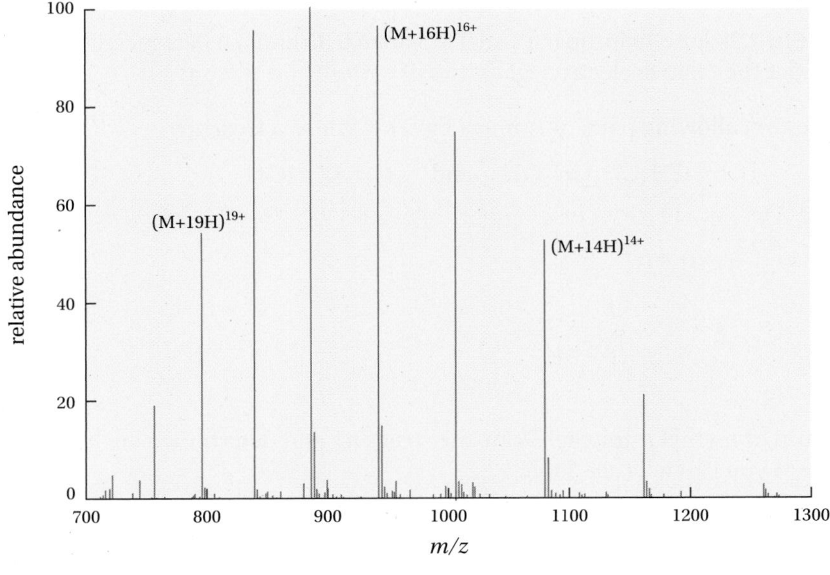

■ **Figure 12.12**
ESI-MS spectrum of the α-chain of human hemoglobin. Reprinted with permission from the *Journal of Chemical Education,* **1996**, Vol. 73, No. 4, pp. A82–A88; copyright © 1996, Division of Chemical Education, Inc.

ADDITIONAL PROBLEMS

OWL Interactive versions of these problems are assignable in OWL.

^{1}H NMR Spectroscopy

12.19 Draw the structure of a compound with each of the following molecular formulas that will show only one peak in its ^{1}H NMR spectrum.

a. C_6H_{12} b. $C_3H_6Cl_2$ c. $C_{12}H_{18}$ d. C_6H_6 e. C_5H_{12} f. C_2H_6S

12.20 Tell how many chemically different types of ^{1}H nuclei are present in

a. $(CH_3)_3C{-}CH_2CH_2CH_3$

b. $CH_3CH_2NHCH(CH_3)_2$

c. 1,2-dimethylcyclopentene (cyclopentene ring with CH_3 on each carbon of the double bond)

d. $H_3C{-}CH(OH)CH_2CH_3$

12.21 The ^{1}H NMR spectrum of a compound, C_4H_9Br, consists of a single sharp peak. What is its structure? The spectrum of an isomer of this compound consists of a doublet at δ 3.2, a complex pattern at δ 1.9, and a doublet at δ 0.9, with relative areas 2 : 1 : 6. What is its structure?

12.22 Predict the ^{1}H NMR spectrum of isobutyl acetate (ball-and-stick model shown below).

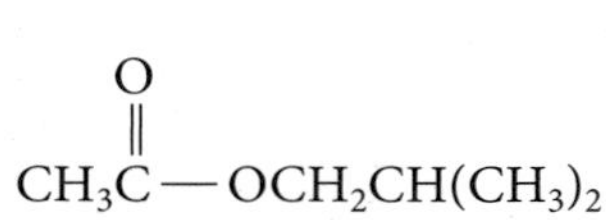
$CH_3C(=O){-}OCH_2CH(CH_3)_2$

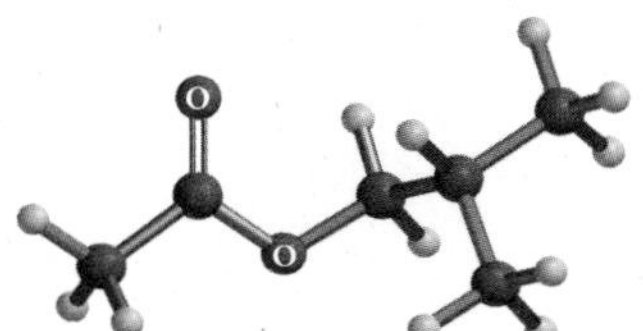

12.23 The chemical shifts of the ^{1}H nuclei in 2,2-dimethylpropane and TMS are δ 0.95 and δ 0.0, respectively. From these data, what can you deduce about the relative electronegativities of carbon and silicon?

12.24 How could you distinguish between the following pairs of isomers by ^{1}H NMR spectroscopy?

a. CH_3CCl_3 and $CH_2ClCHCl_2$

b. $CH_3CH_2CH_2OH$ and $(CH_3)_2CHOH$

c. $CH_3{-}C(=O){-}OCH_3$ and $H{-}C(=O){-}OCH_2CH_3$

d. $C_6H_5{-}CH_2{-}CH{=}O$ and $C_6H_5{-}C(=O){-}CH_3$

12.25 Figure 12.13 is the ^{1}H NMR spectrum of methyl *p*-toluate. Draw the structure and determine which hydrogens are responsible for each peak, as far as you can tell (use Table 12.2).

12.26 Using the information in Table 12.2, sketch the ^{1}H NMR spectrum of each of the following compounds. Be sure to show all splitting patterns.

a. $(CH_3)_2CH{-}C(=O){-}H$

b. $CH_3CH_2OCH(CH_3)_2$

c.

d.

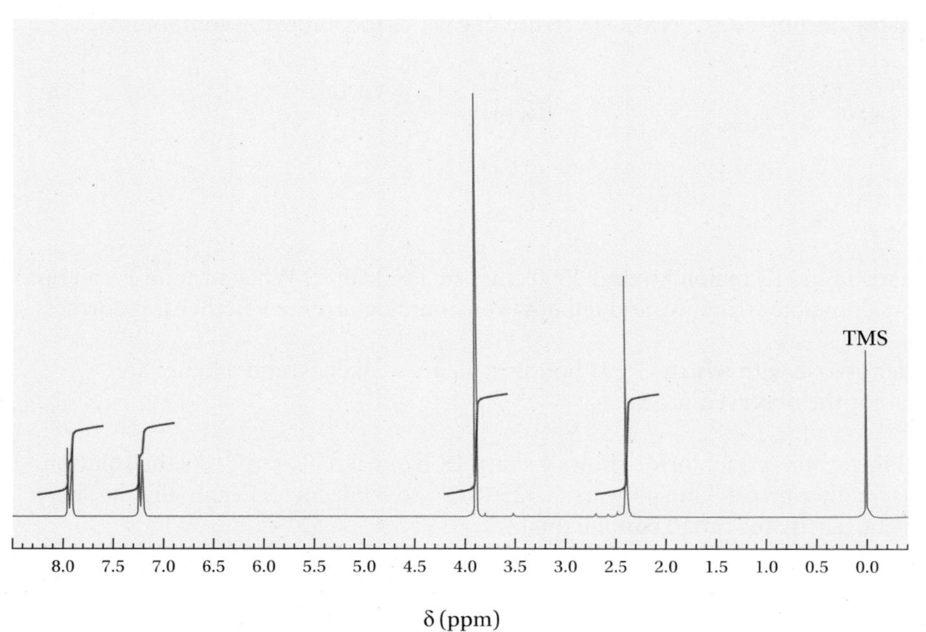

Figure 12.13
^{1}H NMR spectrum of methyl *p*-toluate.

^{13}C NMR Spectroscopy

12.27 A compound is known to be a methyl anisole, but the orientation of the two substituents (—CH_3 and —OCH_3) on the aromatic ring is not known. The ^{13}C NMR spectrum shows six peaks. Which isomer is it? What will the ^{1}H NMR spectrum look like?

12.28 Draw the structures of the four isomeric alcohols with molecular formula C_4H_9OH. (a) Which isomer has the ^{13}C NMR spectrum shown in Figure 12.14? Explain why. (b) Two of the four isomers will have the same number of peaks (4) in their decoupled ^{13}C NMR spectra. What are their structures? How could a ^{13}C–^{1}H coupled spectrum be used to distinguish between them?

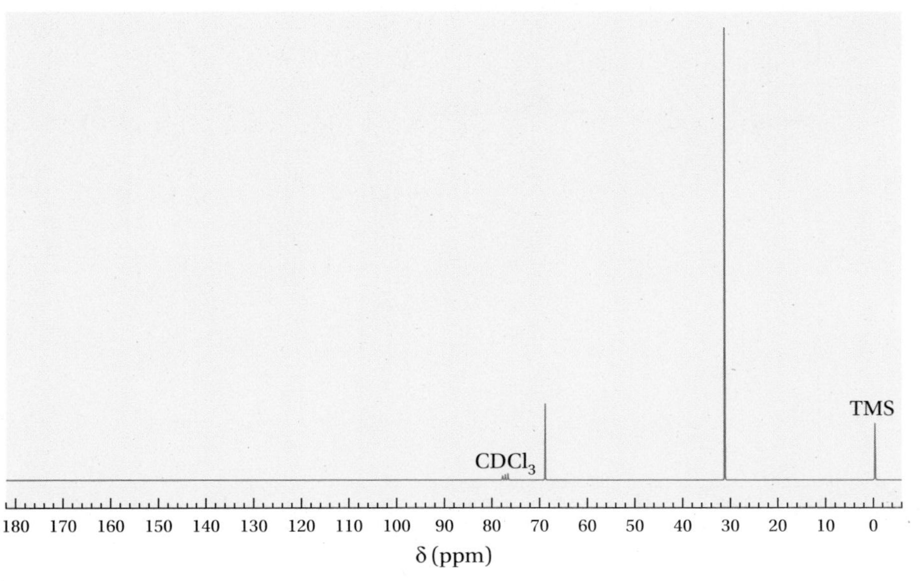

Figure 12.14
^{13}C NMR spectrum.

12.29 How many peaks will there be in the decoupled ^{13}C NMR spectrum of each of the following compounds?

a. $CH_3CH_2-\overset{\overset{O}{\|}}{C}-OH$ b. (benzene ring)$-NH_2$ c. (cyclohexene ring with two CH_3 groups on the double-bond carbons)

Infrared Spectroscopy

12.30 A compound, C_3H_6O, has no bands in the IR region around 3500 cm^{-1} or 1720 cm^{-1}. What structures can be eliminated by these data? Suggest a possible structure, and tell how you could determine whether it is correct.

12.31 From Table 12.4, deduce the order of ease with which C—H bonds in alkanes, alkenes, and alkynes are stretched. Suggest an explanation for the observed order.

12.32 A very dilute solution of ethanol in carbon tetrachloride shows a sharp IR band at 3580 cm^{-1}. As the solution is made more concentrated, a new, rather broad band appears at 3250 cm^{-1} to 3350 cm^{-1}. Eventually the sharp band disappears and is replaced entirely by the broad band. Explain.

12.33 Figures 12.15, 12.16, 12.17, and 12.18 are the IR spectra of four compounds: hexanoic acid, 1-pentanol, cyclohexane, and 3-pentanone, respectively. Match each compound with the correct spectrum, indicating which IR bands you used to make your assignments.

12.34 What features will be similar in the IR spectra of the following compounds, and how will their IR spectra differ?

$$H_2C{=}CH\underset{\underset{OH}{|}}{C}HCH_3 \quad \text{and} \quad CH_3CH{=}CHCH_2\underset{\underset{OH}{|}}{C}HCH_3$$

■ **Figure 12.15**

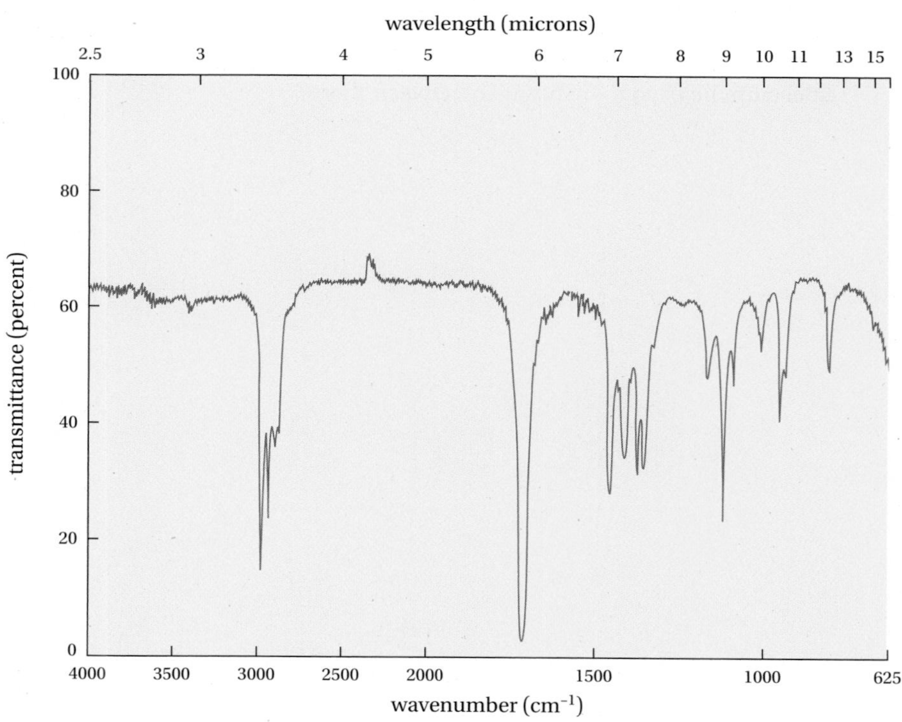

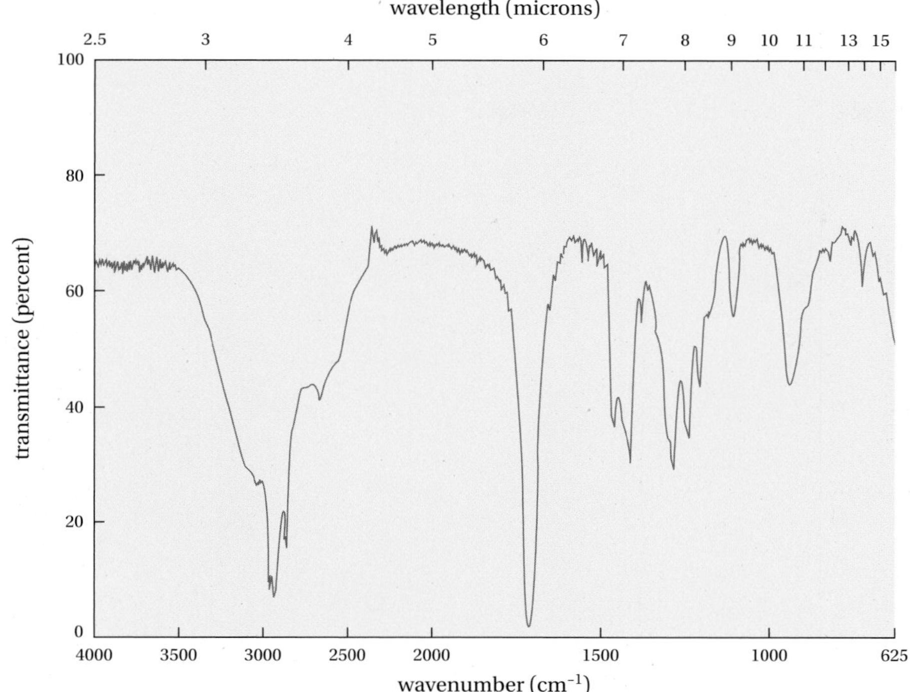

Figure 12.16

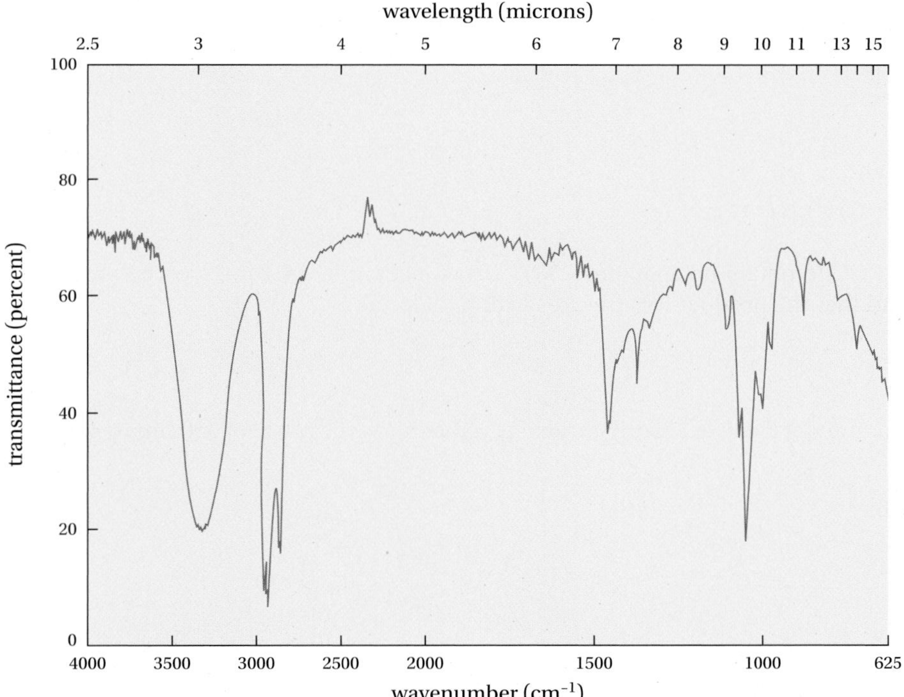

Figure 12.17

■ **Figure 12.18**

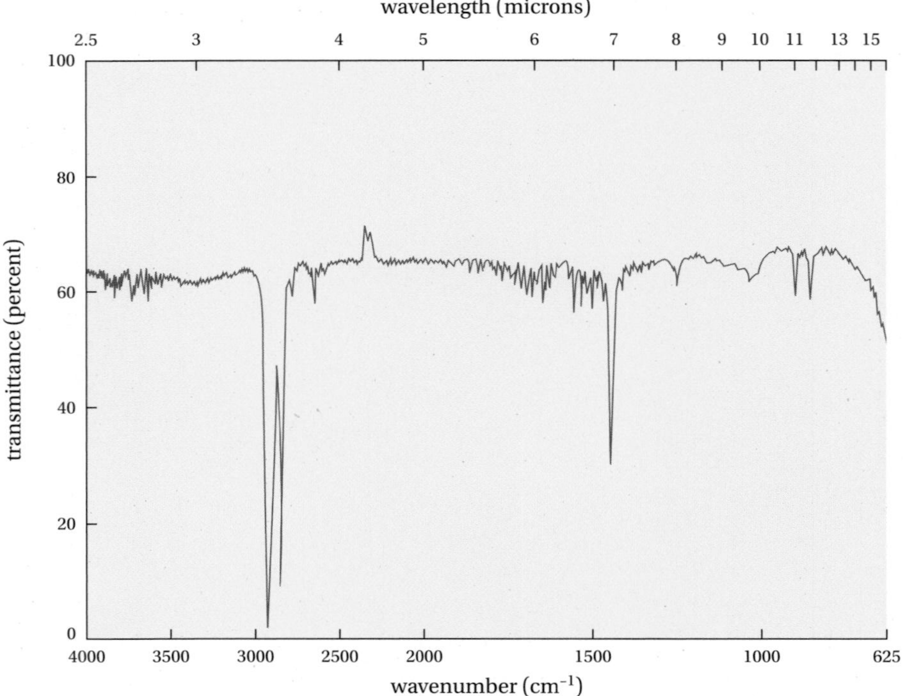

12.35 How could you use IR spectroscopy to distinguish between the following pairs of isomers?

a. $CH_3\overset{\overset{\displaystyle O}{\|}}{C}CH_2CH_3$ and $CH_3\overset{\overset{\displaystyle OH}{|}}{C}HCH{=}CH_2$

b. $C_6H_5-\overset{\overset{\displaystyle CH_3}{|}}{C}HCHO$ and $C_6H_5-CH{=}CHOCH_3$

c. $(CH_3CH_2)_3N$ and $(CH_3CH_2CH_2)_2NH$

12.36 You are oxidizing 1-octanol with PCC to obtain octanal (eq. 7.38). How could you use IR spectroscopy to tell that the reaction was complete and that the product was free of starting material?

Visible-Ultraviolet Spectroscopy

12.37 Which of the following compounds are not likely to absorb ultraviolet radiation in the range of 200 nm to 400 nm?

a. $CH_3CH_2CH_2CH_2OH$ b. (benzene) c. (cyclohexane)

d. $CH_3OCH_2CH_2CH_3$ e. (methylenecyclopentadiene) f. $H_2C{=}CHCH{=}CHCH_3$

12.38 In Chapter 9 ("A Word About . . . Tautomerism and Photochromism," page 274), we discussed the difference in color between the phenol-imine (yellow) and the keto-enamine (red). Rationalize the difference in color on the basis of the differences in aromaticity and their UV-vis spectra.

12.39 The unsaturated aldehydes $CH_3(CH{=}CH)_nCH{=}O$ have ultraviolet absorption spectra that depend on the value of n; the λ_{max} values are 220 nm, 270 nm, 312 nm, and 343 nm as n changes from 1 to 4. Explain.

12.40 The λ_{max} for *cis*-1,2-diphenylethene is at shorter wavelength (280 nm) than for *trans*-1,2-diphenylethene (295 nm). Suggest an explanation.

12.41 A sample of methylcyclohexane is suspected of being contaminated with toluene, from which it had been prepared by hydrogenation. At 261 nm, toluene has a molar absorptivity $\epsilon = 224$, whereas methylcyclohexane does not absorb at that wavelength ($\epsilon = 0$). An ultraviolet spectrum of the contaminated methylcyclohexane (obtained in a 1.0 cm cell) shows an absorbance $A = 0.65$ at 261 nm. Calculate the concentration of toluene in the methylcyclohexane.

Mass Spectrometry

12.42 Write a formula for the molecular ion of 1-hexanol.

12.43 The mass spectrum of 1-butanol shows an intense daughter ion peak at $m/z = 31$. Explain how this peak might arise.

Problem Solving with More Than One Spectroscopic Method

12.44 A compound, $C_5H_{10}O$, has an intense IR band at 1725 cm^{-1}. Its ^1H NMR spectrum consists of a quartet at δ 2.7 and a triplet at δ 0.9, with relative areas 2:3. What is its structure?

12.45 A compound, $C_5H_{10}O_3$, has a strong IR band at 1745 cm^{-1}. Its ^1H NMR spectrum consists of a quartet at δ 4.15 and a triplet at δ 1.20; relative areas are 2:3. What is the correct structure?

12.46 Compound A, $C_4H_{10}O$, is oxidized with PCC (see Sec. 7.12) to Compound B, C_4H_8O, which gives a positive Tollens' test and has a strong IR band at 1725 cm^{-1}. The ^1H NMR spectrum of Compound A is given in Figure 12.19. What are the structures of Compound A and Compound B?

12.47 An alcohol, $C_5H_{12}O$, shows a daughter ion peak at $m/z = 59$ in its mass spectrum. An isomeric alcohol shows no daughter ion at $m/z = 59$ but does have a peak at $m/z = 45$. Suggest possible structures for each isomer. How could you confirm these structures by ^1H NMR spectroscopy? by ^{13}C NMR spectroscopy?

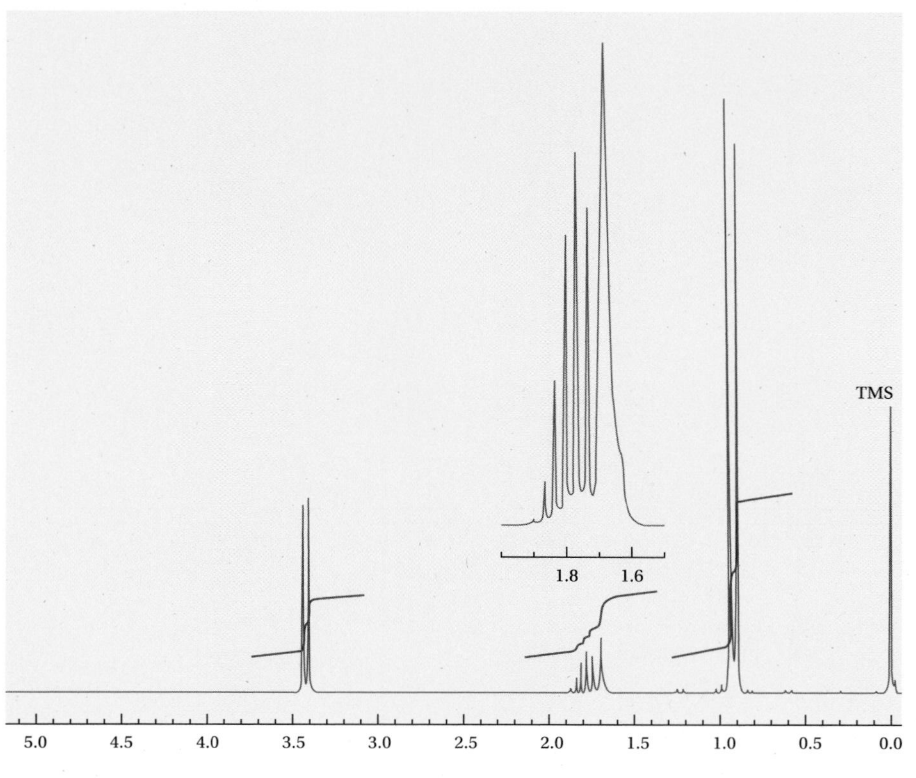

Figure 12.19
^1H NMR spectrum of Compound A (Problem 12.46).

12.48 A hydrocarbon shows a parent ion peak in its mass spectrum at $m/z = 102$. Its ^{1}H NMR spectrum shows peaks at δ 2.7 and δ 7.4, with relative areas 1 : 5. What is the correct structure?

12.49 Below (Figure 12.20) are the ^{1}H NMR, ^{13}C NMR, IR, and mass spectra of the compound C_8H_8O. What is the structure of the compound?

12.50 As discussed in the "A Word About . . . NMR in Biology and Medicine" on page 368, MRI typically uses ^{1}H NMR spectra of water in various tissues. What is the disadvantage of using ^{13}C NMR spectra? Suggest a method that could be used with MRI to enhance the signal strength (actually the signal-to-noise ratio) for a nucleus besides [1]H.

■ **Figure 12.20**

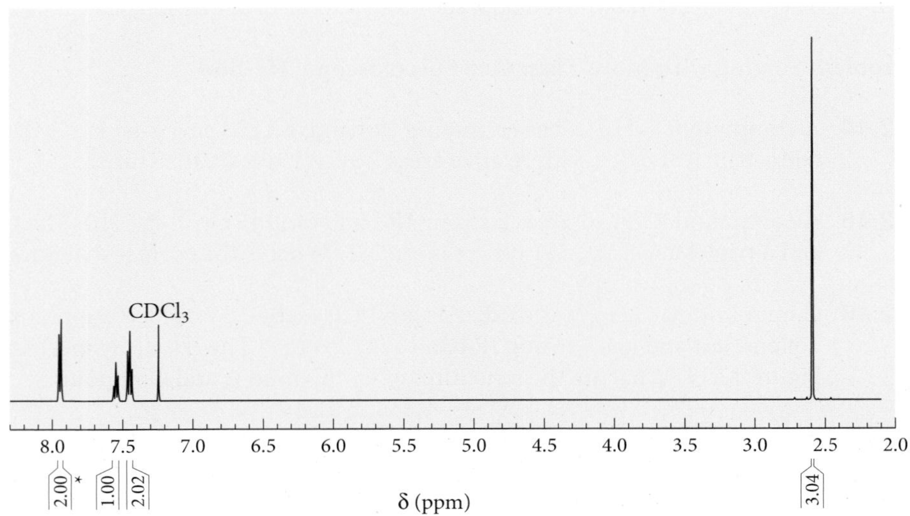

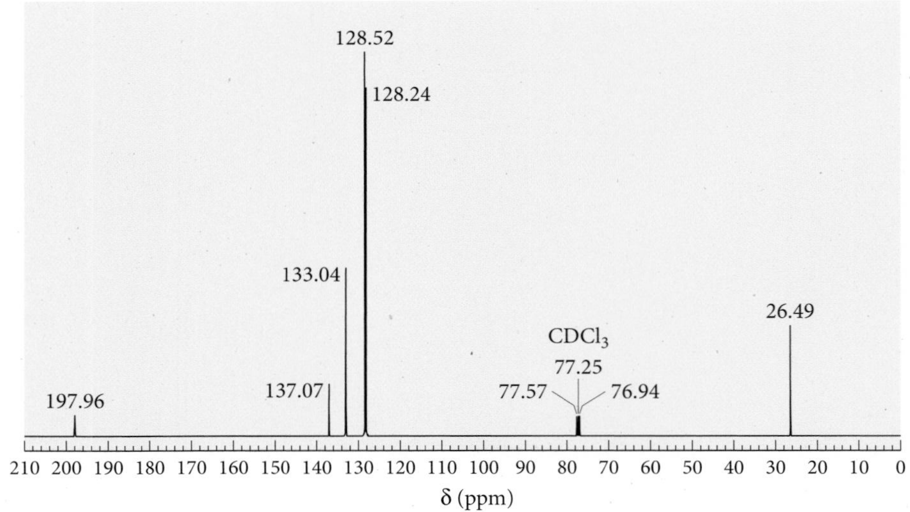

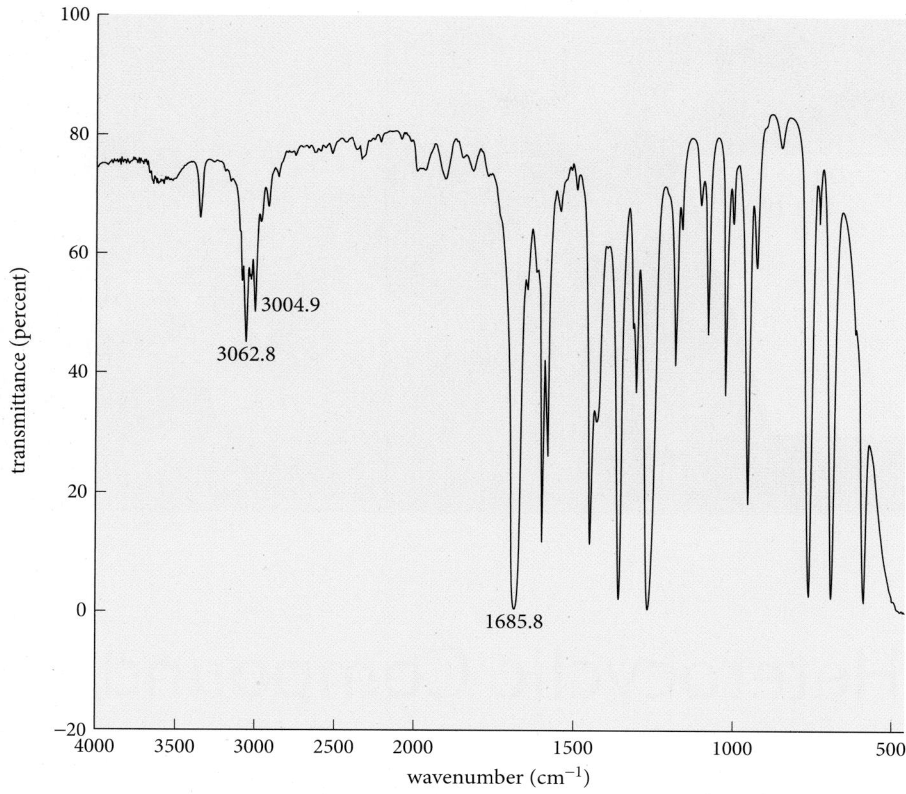
transmittance (percent)
100
80
60
40
20
0
−20
3004.9
3062.8
1685.8
4000 3500 3000 2500 2000 1500 1000 500
wavenumber (cm⁻¹)

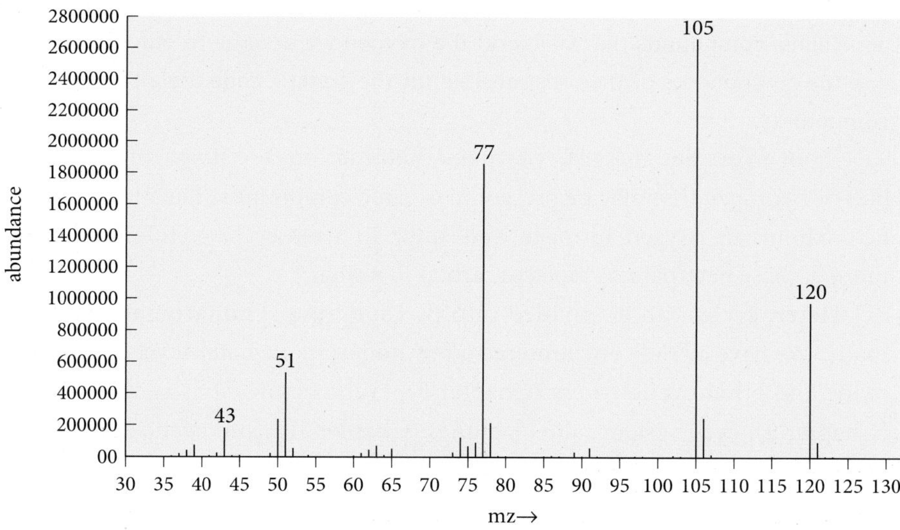
abundance
2800000
2600000
2400000
2200000
2000000
1800000
1600000
1400000
1200000
1000000
800000
600000
400000
200000
00
105
77
120
51
43
30 35 40 45 50 55 60 65 70 75 80 85 90 95 100 105 110 115 120 125 130
mz→

13

The red color of roses is produced by an anthocyanin (page 397).

Heterocyclic Compounds

Heterocycles form the largest class of organic compounds. In fact, many natural products and most drugs contain heterocyclic rings. The colors of flowers and plants, antibiotics known to all as penicillins, compounds that transport the oxygen we breathe to our vital organs, and the components of DNA responsible for the genetic code are all heterocyclic compounds.

From an organic chemist's viewpoint, **heteroatoms** are atoms other than carbon or hydrogen that may be present in organic compounds. The most common heteroatoms are oxygen, nitrogen, and sulfur. In heterocyclic compounds, one or more of these heteroatoms replaces carbon in a ring.

Heterocycles can be divided into two subgroups: nonaromatic and aromatic. We have already encountered a few nonaromatic heterocycles—ethylene oxide and other cyclic ethers (Chapter 8), cyclic hemiacetals such as glucose (Chapter 9), cyclic esters called lactones (Chapter 10), and cyclic amines such as morphine (Chapter 11). In general, these nonaromatic heterocycles behave a great deal like their acyclic analogs and do not require special discussion.

Aromatic heterocycles are extremely important, and that is where we will focus most of our attention. We begin with an important six-membered ring aromatic nitrogen heterocycle, pyridine.

Online homework for this chapter can be assigned in OWL, an online homework assessment tool.

13.1 Pyridine: Bonding and Basicity

Heterocycles are cyclic organic compounds in which one or more carbon atoms is replaced by **heteroatoms**, atoms other than C or H.

Pyridine is a benzene ring in which one CH unit has been replaced by an N atom.

Pyridine has a structure similar to that of benzene, except that one CH unit is replaced by a nitrogen atom. As with benzene, pyridine is a resonance hybrid of Kekulé-type structures (see Sec. 4.2).

Pyridine is a resonance hybrid of these two contributing structures

The orbital pictures for benzene (Sec. 4.4 and Figure 4.2) and pyridine are similar. The nitrogen atom, as with the carbons, is sp^2-hybridized, with one electron in a *p* orbital perpendicular to the ring plane. Thus, the nitrogen contributes one electron to the six electrons that form the aromatic pi cloud above and below the ring plane. On the other hand, the unshared electron pair on nitrogen lies in the ring plane (as with the C—H bonds) in an sp^2 orbital.

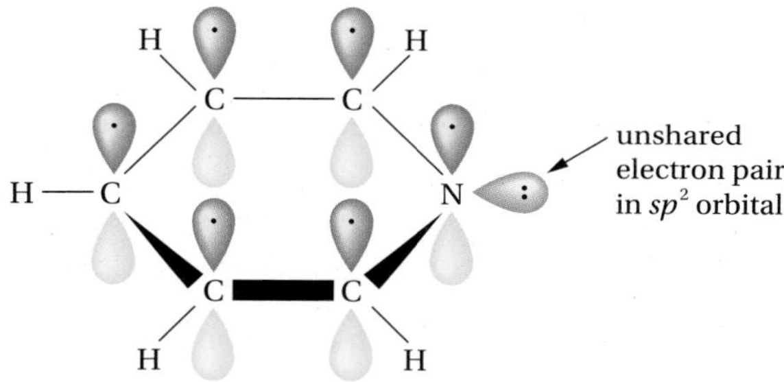

bonding in pyridine

Because of the similarities in bonding, pyridine resembles benzene in shape. It is planar, with nearly perfect hexagonal geometry. It is aromatic and tends to undergo substitution rather than addition reactions.

But the substitution of nitrogen for carbon changes many of the properties. Like benzene, pyridine is miscible with most organic solvents, but unlike benzene, pyridine is also completely miscible with water! One explanation lies in its hydrogen-bonding capability (Sec. 2.7).

Another reason is that pyridine is much more polar than benzene. The nitrogen atom is electron-withdrawing compared to carbon; hence, there is a shift of electrons away from the ring carbons and toward the nitrogen, making it partially negative and the ring carbons partially positive (Figure 13.1). This polarity enhances the solubility of pyridine in polar solvents like water, and also increases the boiling point of pyridine (115°C) relative to benzene (80°C).

Pyridine is weakly basic. It is a much weaker base than aliphatic amines, mainly because of the different hybridization of the nitrogen (sp^2 in pyridine and sp^3 in aliphatic amines). The greater *s*-character of the orbital containing the basic nonbonded lone pair (one-third *s* in pyridine and one-fourth *s* in aliphatic amines) means that the unshared electron pair is held closer to the nitrogen nucleus in pyridine, decreasing its basicity.

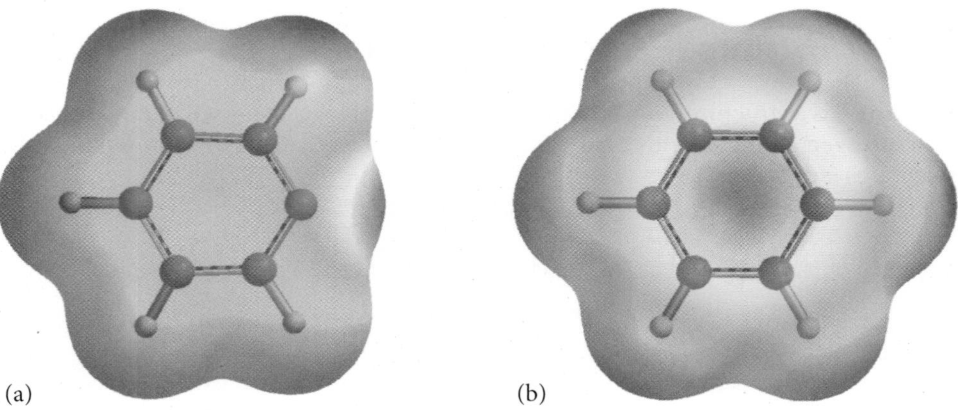

■ **Figure 13.1**
Electrostatic potential maps of (a) pyridine and (b) benzene.

Pyridinium salts are formed when pyridine reacts with strong acids.

Pyridine does react with strong acids to form **pyridinium salts**. For this reason, pyridine is often used as a scavenger in acid-producing reactions; for example, in the reaction of thionyl chloride with alcohols (page 219).

$$\text{C}_5\text{H}_5\text{N:} + \text{H}^+\text{Cl}^- \longrightarrow \text{C}_5\text{H}_5\overset{+}{\text{N}}\text{—H Cl}^- \qquad (13.1)$$

pyridinium chloride
$pK_a = 5.29$

The weak basicity of pyridine is also reflected in the pK_a of its conjugate acid (pyridinium chloride) when compared to the pK_a of ammonium ions, the conjugate acids of aliphatic amines (see Table 11.3). The weaker base (pyridine) has the stronger conjugate acid.

PROBLEM 13.1 Write an equation for the reaction of pyridine with

a. cold sulfuric acid (H_2SO_4) b. cold nitric acid (HNO_3)

13.2 Substitution in Pyridine

Though aromatic, *pyridine is very resistant to electrophilic aromatic substitution* and undergoes reaction only under drastic conditions. For example, nitration or bromination requires high temperatures and strong acid catalysis.

Pyridine (positions numbered 1 (N), 2, 3, 4) →

- KNO_3, HNO_3, fuming H_2SO_4; 300°C, Fe → 3-nitropyridine
- Br_2, H_2SO_4, SO_3; 300°C → 3-bromopyridine

(13.2)

One reason for this sluggishness is that electron withdrawal by the nitrogen makes the ring partially positive and therefore not receptive to attack by electrophiles, which are also positive (see Figure 13.1). A second reason is that, under the acidic conditions for these reactions, most of the pyridine is protonated and present as the positively charged pyridinium ion, which is even more unlikely to be attacked by electrophiles than is neutral pyridine.

When substitution does occur, electrophiles attack pyridine mainly at C-3. The cationic intermediate (review Sec. 4.9) is *least unfavorable* in this case, because it does not put a positive charge on the electron-deficient nitrogen (especially bad if the nitrogen is protonated).

EXAMPLE 13.1

Draw all contributors to the resonance hybrid for electrophilic attack at C-3 of pyridine.

Solution

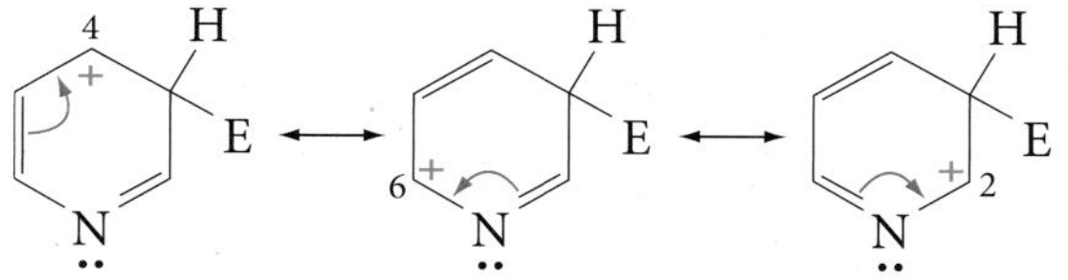

4 H E ⟷ H E ⟷ H E
6 N N N 2

For substitution at C-3, the "pyridinonium ion" positive charge is delocalized to C-2, C-4, and C-6, but not to the nitrogen.

PROBLEM 13.2 Repeat Example 13.1, but for electrophilic substitution at C-2 or C-4 of pyridine. Explain why substitution at C-3 (eq. 13.2) is preferred.

Although resistant to electrophilic substitution, pyridine undergoes **nucleophilic aromatic substitution**. The pyridine ring is partially positive (due to electron withdrawal by the nitrogen) and is therefore susceptible to attack by nucleophiles. Here are two examples:

In **nucleophilic aromatic substitution** reactions of pyridine, a nucleophile displaces a hydride or halide ion from the aromatic ring.

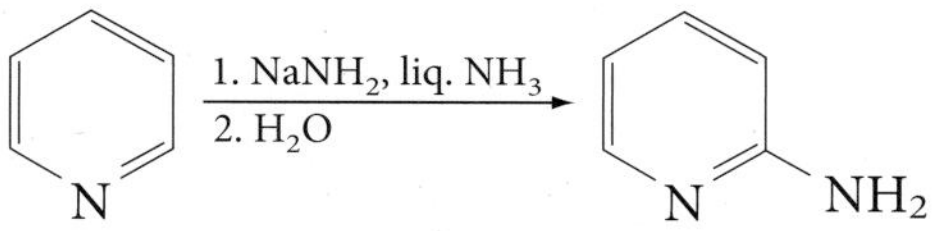

N —(1. $NaNH_2$, liq. NH_3; 2. H_2O)→ N NH_2 **(13.3)**

2-aminopyridine

Cl N —($NaOCH_3$, CH_3OH)→ OCH_3 N **(13.4)**

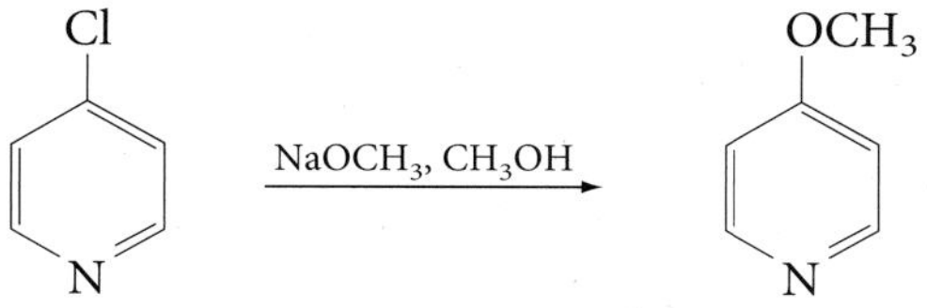

4-chloropyridine 4-methoxypyridine

EXAMPLE 13.2

Write a mechanism for eq. 13.3.

Solution Attack of amide ion at C-2 gives an anionic intermediate with the negative charge mainly on nitrogen.

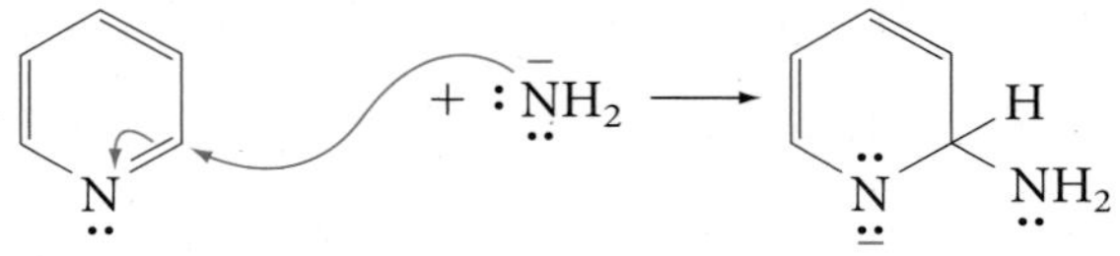

You can think of this as nucleophilic addition to a C═N bond, analogous to addition of nucleophiles to a C═O bond (Sec. 9.6).

To restore aromaticity, hydride ion is displaced. The hydride then attacks the amino group to give hydrogen gas and an amide-type anion (see eq. 11.17).

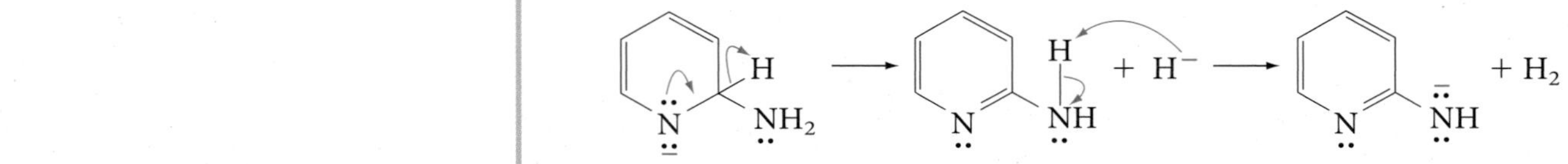

In the final step, this ion is protonated by water.

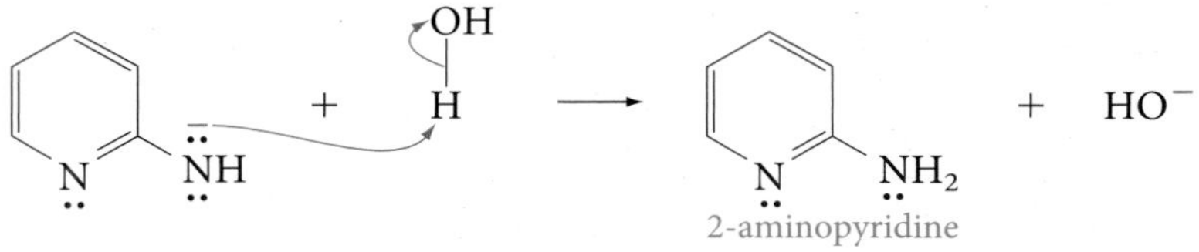

2-aminopyridine

PROBLEM 13.3 Write a mechanism for eq. 13.4.

Pyridine and alkylpyridines are found in coal tar. The monomethyl pyridines (called picolines) undergo side-chain oxidation to carboxylic acids (review Sec. 10.7.b). For example, 3-picoline gives nicotinic acid (or niacin), a vitamin essential in the human diet to prevent the disease *pellagra*.

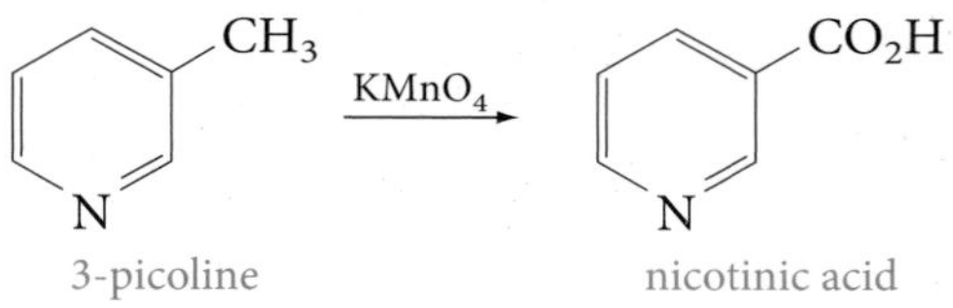

3-picoline $\xrightarrow{KMnO_4}$ nicotinic acid **(13.5)**

Pyridine can be reduced by catalytic hydrogenation to the fully saturated secondary amine piperidine.

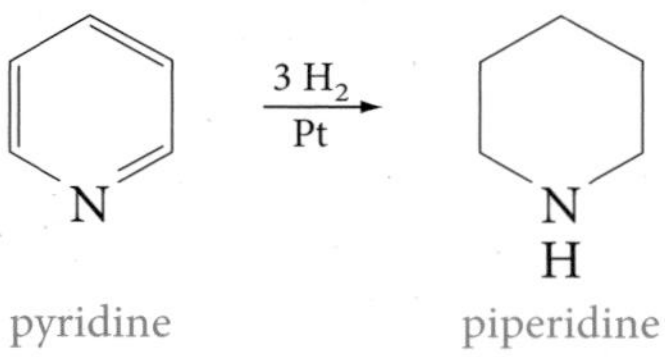

pyridine $\xrightarrow[Pt]{3\ H_2}$ piperidine **(13.6)**

The pyridine and piperidine rings are found in many natural products. Examples are **nicotine** (the major alkaloid in tobacco, used as an agricultural insecticide and highly toxic to humans), pyridoxine (vitamin B_6, a coenzyme), and coniine (the toxic principle of poison hemlock, taken by Socrates).

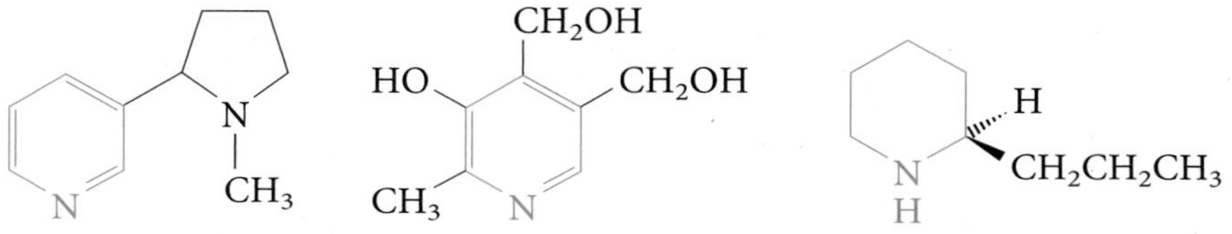

nicotine   pyridoxine   (+)-coniine

PROBLEM 13.4 Naturally occurring coniine is the (+)-isomer shown. What is its configuration, *R* or *S*?

PROBLEM 13.5 Naturally occurring nicotine is the (*S*)-(−) isomer. Locate the stereogenic center, and draw the three-dimensional structure.

PROBLEM 13.6 Nicotine contains two nitrogens, one in a pyridine ring and one in a pyrrolidine ring. It reacts with *one* equivalent of HCl to form a crystalline salt, $C_{10}H_{15}N_2Cl$. Draw its structure. Nicotine also reacts with *two* equivalents of HCl to form another crystalline salt, $C_{10}H_{16}N_2Cl_2$. Draw its structure.

13.3 Other Six-Membered Heterocycles

The pyridine ring can be fused with benzene rings to produce polycyclic aromatic heterocycles. The most important examples are **quinoline** and **isoquinoline**, analogs of naphthalene (Sec. 4.13) but with N in place of CH at C-1 or C-2.

Quinoline and **isoquinoline** are naphthalene analogs in which an N atom replaces a CH unit at C-1 or C-2.

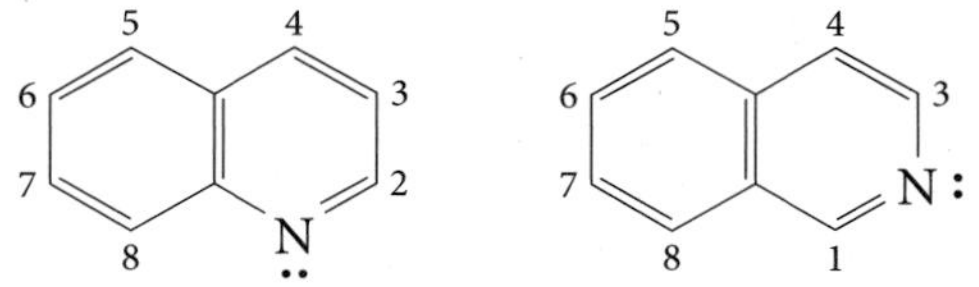

quinoline
bp 237°C

isoquinoline
bp 243°C, mp 26.5°C

Electrophilic substitution in these amines occurs in the carbocyclic ring, illustrating the inactivity toward electrophiles of the pyridine ring relative to the benzene ring.

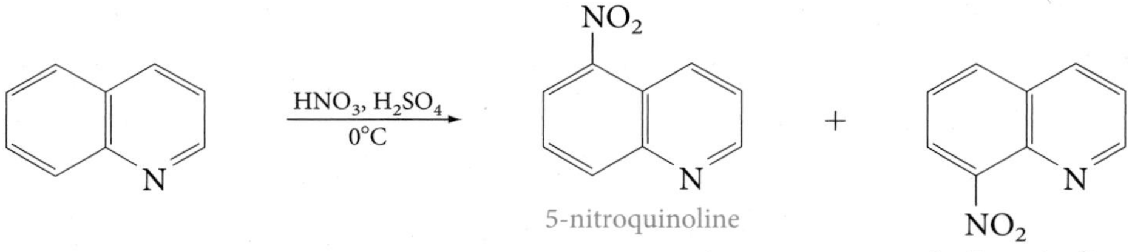

HNO_3, H_2SO_4, 0°C → 5-nitroquinoline + 8-nitroquinoline **(13.7)**

The stability of the pyridine ring is also illustrated by its resistance to oxidation. Thus, when quinoline is treated with potassium permanganate, the benzene ring is oxidized.

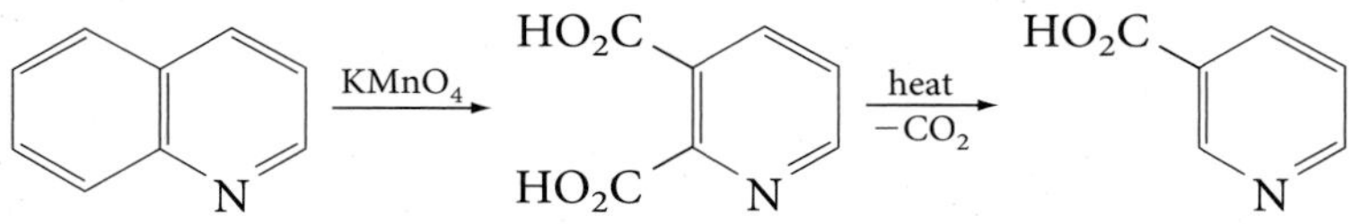

quinoline —$KMnO_4$→ quinolinic acid —heat, $-CO_2$→ nicotinic acid **(13.8)**

The quinoline and isoquinoline rings occur in many natural products. Good examples are quinine (which occurs in cinchona bark and is used to treat malaria) and papaverine (present in opium and used as a muscle relaxant).

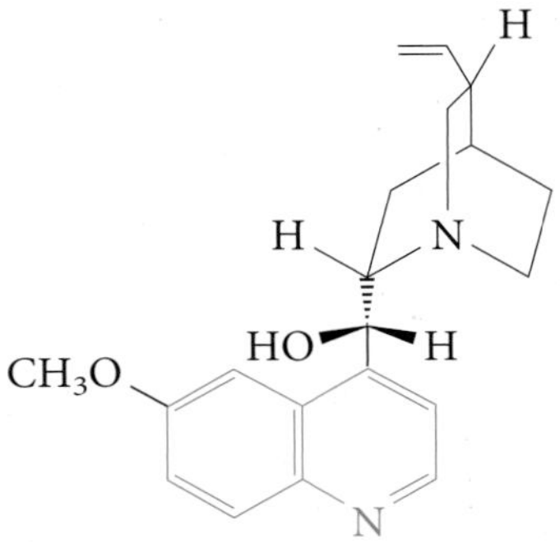

quinine

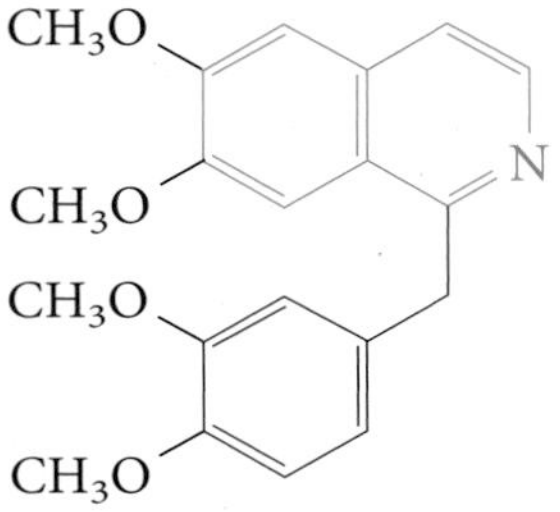

papaverine

It is logical to ask: If we can replace one CH with N in a benzene ring to create pyridine, can we replace more than one of the benzene CH groups with nitrogens? The answer is yes. For example, there are *three* **diazines**.

Diazines are six-membered ring heterocycles containing two N atoms. In **pyrimidines**, the N atoms are located at ring positions 1 and 3.

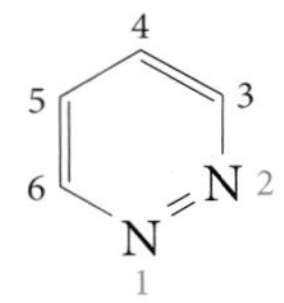

pyridazine (bp 208°C)

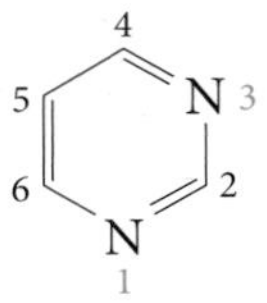

pyrimidine (bp 134°C)

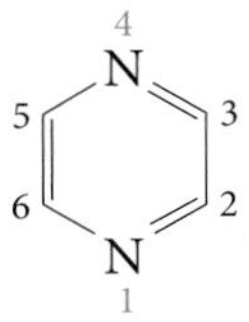

pyrazine (bp 118°C)

Of these, the most important are the **pyrimidines**, derivatives of which (**cytosine**, **thymine**, and **uracil**) are important bases in nucleic acids DNA and RNA (see Chapter 18).

Cytosine, **thymine**, and **uracil** are pyrimidines, and they are important bases in nucleic acids DNA and RNA.

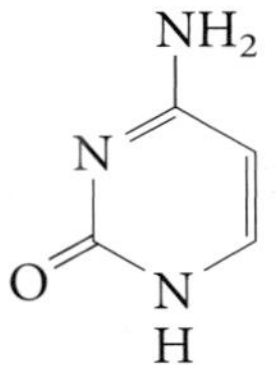

cytosine

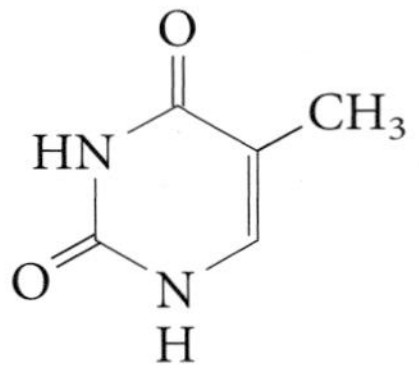

thymine

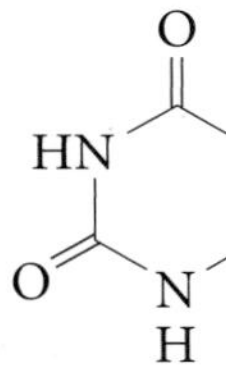

uracil

Triazines and tetrazines are also known, but neither pentazine nor hexazine (which would really not be a heterocycle, since it would contain only one element and be an allotrope of nitrogen) is known.

Similar analogs of naphthalene with more than one nitrogen are also known. The pteridine ring, with nitrogens replacing C-1, C-3, C-5, and C-8 in naphthalene, is present in many natural products, such as the butterfly wing pigment xanthopterin and the blood-forming vitamin B9, which is also called folic acid. The analog of folic acid, but with an NH_2 group in place of the OH group on the pteridine ring and a methyl group on the first nitrogen in the side chain, is useful in cancer chemotherapy (it is called methotrexate).

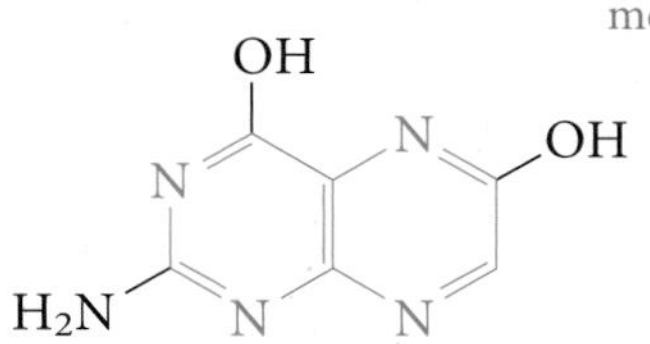

xanthopterin

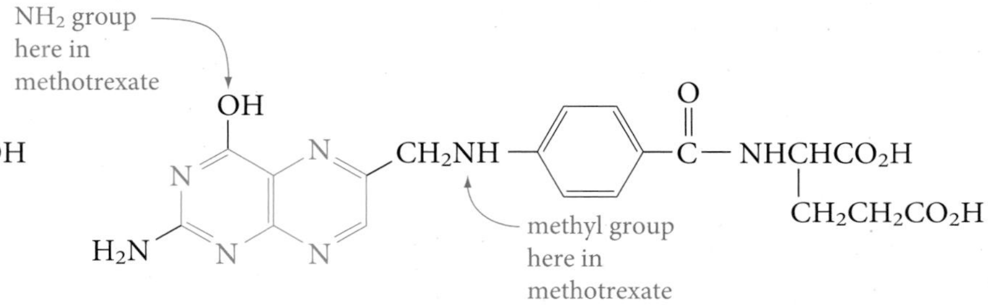

NH_2 group here in methotrexate

methyl group here in methotrexate

folic acid

To summarize, six-membered heterocycles with nitrogen (one or more) in place of the CH groups in benzenoid aromatics are also aromatic. Each nitrogen atom contributes one electron to the 6π aromatic system. The nitrogens are also basic, because of unshared electron pairs located in the ring plane. The nitrogen, being electron-withdrawing, deactivates the ring toward electrophilic aromatic substitution but activates it toward nucleophilic aromatic substitution. These heterocyclic aromatic rings are present in many natural products.

If we replace a benzene CH group with oxygen (instead of nitrogen), we obtain an aromatic cation called a **pyrylium ion**.

Replacing a CH unit in benzene with an O atom produces the **pyrylium ion**, an aromatic cation.

$\overset{+}{O}$

pyrylium ion

EXAMPLE 13.3

Describe the bonding in a pyrylium ion.

Solution The oxygen is sp^2-hybridized, with only five electrons (hence the $+1$ formal charge). Two of these, the unshared electron pair, are in an sp^2 orbital that lies in the ring plane. Two others are in sp^2 orbitals that form the sigma bonds with the adjacent carbon atoms, also sp^2-hybridized. The fifth electron is in a p orbital perpendicular to the ring plane; it overlaps with similar orbitals on each ring carbon atom (as in benzene) to form the aromatic six-electron pi cloud above and below the ring plane.

© Patricia J. Bruno/Positive Images

Blue cornflowers.

The red and blue colors of many flowers are due to anthocyanins, compounds in which a pyrylium ring is present. For example, the pigment responsible for the color of red roses is

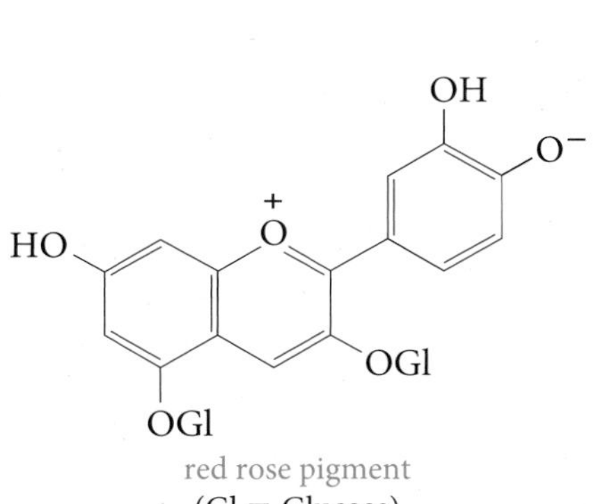

red rose pigment
(Gl = Glucose)

The glucose units (page 53) solubilize the pigment in the aqueous cellular fluid. The color of blue cornflowers is due to the same pigment, but complexed with metal ions, such as Fe^{3+} or Al^{3+}. Other flower pigments have the same basic structures, but with fewer, more, or differently located hydroxyl groups. Anthocyanins are also responsible for the color of certain vegetables, including the red onion.

13.4 Five-Membered Heterocycles: Furan, Pyrrole, and Thiophene

Furan, pyrrole, and **thiophene** are five-membered ring aromatic heterocycles containing one heteroatom (O, N, and S, respectively) in the ring.

Now let us examine rather different types of heteroaromatic compounds: those with five-membered rings. **Furan, pyrrole,** and **thiophene** are important five-membered ring heterocycles with one heteroatom.

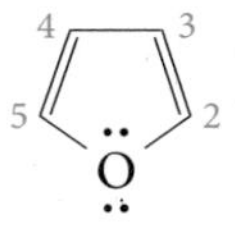

furan (bp 32°C) | pyrrole (bp 131°C) | thiophene (bp 84°C)

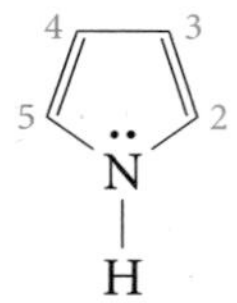

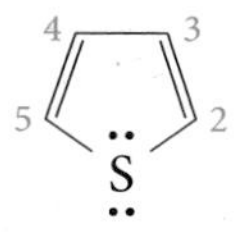

Numbering begins with the heteroatom and proceeds around the ring.

The most important commercial source of furans is furfural (2-furaldehyde), obtained by heating oat hulls, corn cobs, or straw with strong acid. These naturally occurring materials are polymers of a five-carbon sugar, which is dehydrated by the acid to furfural.

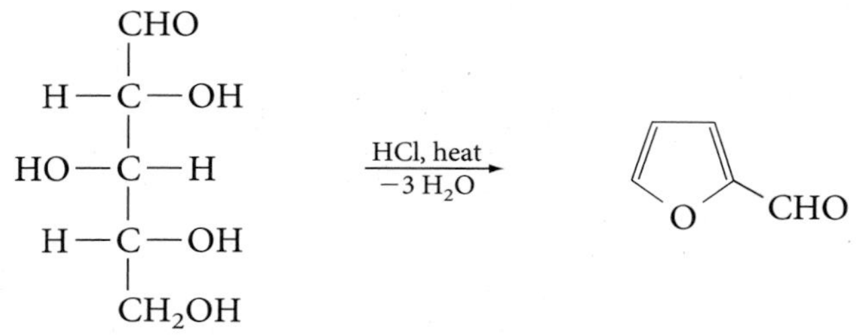

$$\text{a pentose (5-carbon sugar)} \xrightarrow[-3\,H_2O]{\text{HCl, heat}} \text{furfural (2-furaldehyde)} \quad (13.9)$$

Pyrrole is obtained commercially by distillation of coal tar or from furan, ammonia, and a catalyst. Thiophene is obtained by heating a mixture of butanes and butenes with sulfur.

As drawn, the structures of these heterocycles look as if they ought to be dienes, but in fact, these ring systems are aromatic; they behave like benzene in many ways, particularly in their tendency to undergo electrophilic aromatic substitution. The reasons for this behavior will become clear if we examine the bonding in these molecules.

Furan has a planar, pentagonal structure in which each ring atom is sp^2-hybridized (Figure 13.2). Each ring atom uses two of these orbitals to form sigma bonds with its neighbors. Each carbon also uses one sp^2 orbital to form a sigma bond in the ring plane with a hydrogen atom and has one electron in a p orbital perpendicular to the ring plane, exactly analogous to the carbons in benzene. Now look at the oxygen. It has an unshared electron pair in an sp^2 orbital in the ring plane and *two electrons in a p orbital perpendicular to the ring plane.* These two electrons overlap with those in the p orbitals on the carbons to form a 6π electron cloud above and below the ring plane, just as in benzene. The bonding in pyrrole and thiophene is similar to that in furan.

Some striking differences are seen between five-membered ring heterocyclic compounds and their six-membered ring counterparts. For example, pyrrole (Figure 13.3) is an exceedingly weak base compared to pyridine (Figure 13.1). Its conjugate acid has a pK_a of −4.4 compared with a pK_a of 5.29 for pyridinium (see eq. 13.1). How can this be explained? In pyrrole, the unshared electron pair on nitrogen is part of the aromatic 6π system.

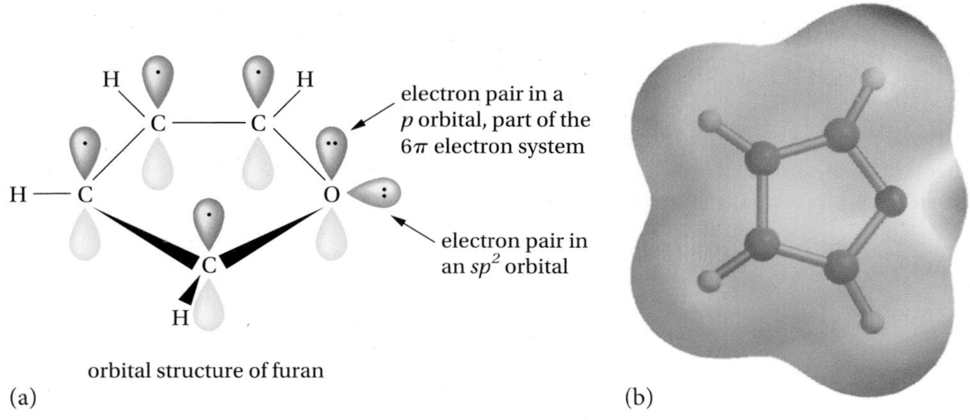

Figure 13.2
(a) Bonding in furan and
(b) electrostatic potential map.

Protonation of the nitrogen of pyrrole destroys the aromatic system, thus forfeiting its resonance energy. Hence pyrrole is a very weak base. In fact, pyrrole can be protonated by very strong acids, but it is *protonated on the carbon* rather than on the nitrogen! In pyridine, the electron pair on nitrogen is not part of the aromatic π system and is available for protonation.

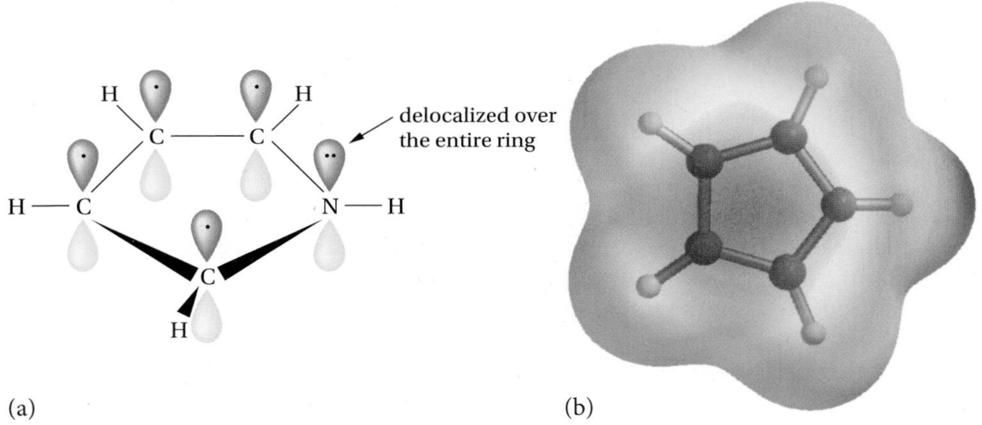

Figure 13.3
(a) Bonding in pyrrole and
(b) electrostatic potential map that illustrates how pyrrole is an electron-rich relative to furan.

PROBLEM 13.7 What is the structure of the cation derived from protonation of pyrrole at C-2?

Another important difference between five- and six-membered aromatic heterocycles is that, in the five-membered heterocycles, the heteroatom contributes *two* electrons to the aromatic 6π systems, whereas in six-membered heterocycles, the heteroatom contributes only *one* electron to that system. This difference has important consequences for the chemical behavior of the two types of heterocycles.

Before we consider those consequences, let us examine the bonding in furan in another way. We can write these heterocycles as a resonance hybrid in which an electron pair from the heteroatom is delocalized to all ring atoms.

contributors to the resonance hybrid structure of furan

Notice that four of these structures are dipolar and place a *negative* charge on the ring carbons. As might be expected, this enhances their susceptibility to attack by electrophiles.

13.5 Electrophilic Substitution in Furan, Pyrrole, and Thiophene

Furan, pyrrole, and thiophene are all much more reactive than benzene toward electrophilic substitution. Each reacts predominantly at the 2-position (and, if that position is already substituted, at the 5-position). Here are typical examples:

pyrrole + HNO_3 $\xrightarrow{0°C}$ 2-nitropyrrole + H_2O **(13.10)**

furan + Br_2 $\xrightarrow[0°C]{ether}$ 2-bromofuran + HBr **(13.11)**

2-methylthiophene + CH_3CCl (O) $\xrightarrow{SnCl_4}$ 2-acetyl-5-methylthiophene + HCl **(13.12)**

The reason for predominant attack at C-2 (instead of the other possibility, C-3) becomes clear if we examine the carbocation intermediate in each case:

Attack of electrophile at C-2 (X = NH, O, or S):

X + E^+ ⟶ [three resonance structures] ⟶ 2-E product + H^+ **(13.13)**

Attack of electrophile at C-3:

X + E^+ ⟶ [two resonance structures] ⟶ 3-E product + H^+ **(13.14)**

Attack at C-2 is preferred because, in the carbocation intermediate, the positive charge can be delocalized over *three* atoms, whereas attack at C-3 allows delocalization of the charge over only two positions.

PROBLEM 13.8 Write out the steps in the mechanism for bromination of furan (eq. 13.11).

13.6 Other Five-Membered Heterocycles: Azoles

It is possible to introduce a second heteroatom (and even a third and fourth) into five-membered heterocycles. The most important of these are the **azoles**, in which the second heteroatom, located at position 3, is nitrogen.

Azoles are five-membered heterocycles with an O, N, or S atom at position 1 and an N atom at position 3.

oxazole imidazole thiazole

Analogs in which the two heteroatoms are adjacent are also known.

As with pyridine, these heterocycles can be thought of as derived by replacing an aromatic CH group by N. The unshared electron pair on this nitrogen (at position 3) is therefore *not* part of the 6π aromatic system, as seen in the orbital picture for imidazole in Figure 13.4.

Consequently, the N-3 nitrogen is basic and can be protonated. The imidazolium ion is very stable because the positive charge can be delocalized equally over both nitrogens. Imidazole is even more basic than pyridine, and as a consequence, the pK_a of the imidazolium ion (7.0) is greater than the pK_a of the pyridinium ion (5.3).

resonance in the imidazolium ion (13.15)

N³ is basic because of this unshared pair

N¹ is not basic; it is like the nitrogen in pyrrole

bonding in imidazole

(a)

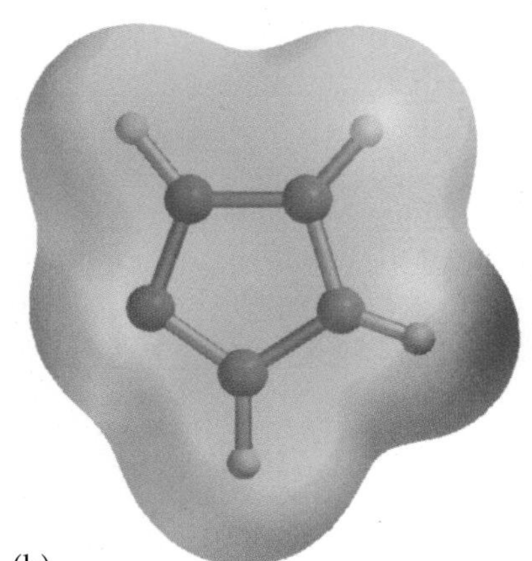

(b)

■ **Figure 13.4**
(a) Bonding in imidazole and (b) electrostatic potential map illustrating the difference in properties of the two nitrogen atoms.

A WORD ABOUT... Porphyrins: What Makes Blood Red and Grass Green?

When green chlorophyll decays, the colors of autumn leaves are revealed.

Pyrrole rings form the building blocks of several biologically important pigments. The **porphyrins** are macrocyclic compounds that contain four pyrrole rings linked by one-carbon bridges. The molecules are flat and have a conjugated system of 18π electrons, shown in color in the parent molecule, **porphine**.

Porphyrins are exceptionally stable and highly colored. They form complexes with metallic ions. In these complexes, the hydrogens bonded to two of the nitrogens are absent, and each of the four nitrogens donates an electron pair to the metal, which sits in the middle of the structure.

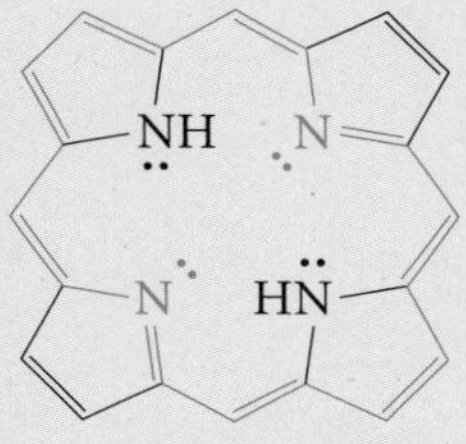

porphine
red crystals

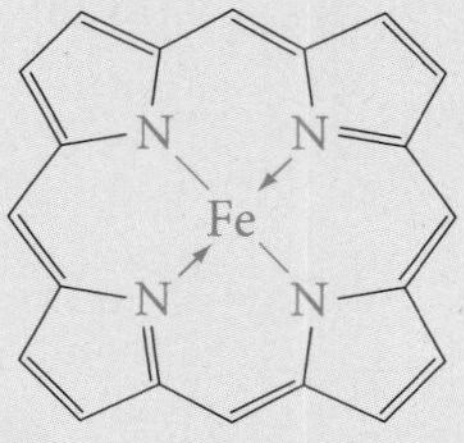

iron-porphine complex
brown cubic crystals

Porphine itself does not occur in nature, but analogous compounds with various side chains on the pyrrole rings are some of the most important life-sustaining compounds of nature. One example is **heme**, the iron-porphyrin complex responsible for the red color of arterial blood.

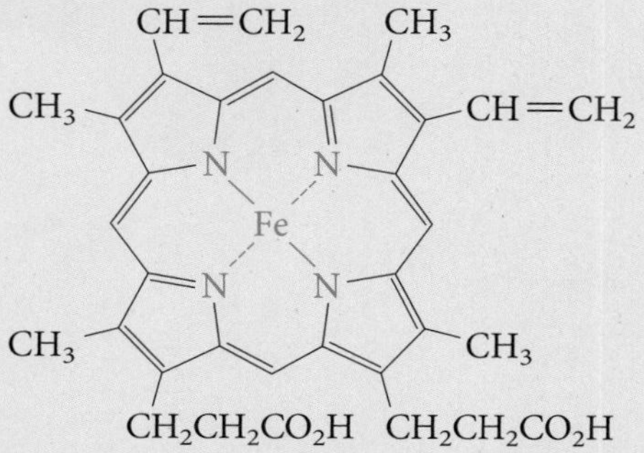

heme (brown needles with violet sheen)

Heme is present in red blood cells as a complex with a protein called **globin**. The complex is called **hemoglobin**. This complex is responsible for binding molecular oxygen and transporting it to sites where it is needed. The iron atom is complexed with the four porphyrin nitrogens but also has two additional coordination sites, above and below the plane of the porphyrin ring. One of these sites is occupied by an imidazole ring from a histidine unit in the protein, but the second site is available for **reversible binding to oxygen**. Carbon monoxide is toxic because it, too, can bind to this site, preventing oxygen transport to the lungs and elsewhere and ultimately leading to death by suffocation. Fortunately, CO binding can be reversed if oxygen is given quickly to a patient poisoned by the gas.

Heme is also associated with **myoglobin**, the oxygen-transport protein of muscle, where it performs a similar function.

The green color of plants is due to **chlorophyll**, a magnesium complex of a modified porphyrin. This pigment is present in chloroplasts, one or more of which can be present in cells, depending on the plant. Chlorophyll-*a* is an ester of the long-chain alcohol **phytol**, which helps solubilize the pigment in the chloroplasts.

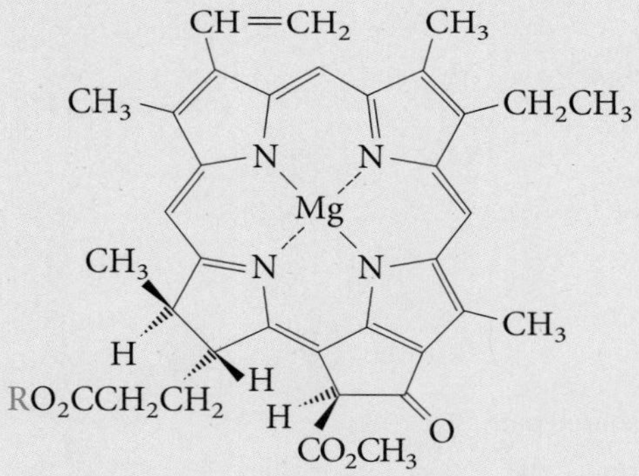

chlorophyll-*a*
blue-black crystals (mp 117–120°C)

R =

Chlorophyll-*b*, another plant pigment, differs from chlorophyll-*a* only in the replacement of one ring methyl group by an aldehyde group. The mechanism of photosynthesis (the conversion of CO_2 and H_2O by plants to carbohydrates) is quite complex and involves many steps. The first of these is the absorption of sunlight by chlorophyll and its conversion into chemical energy. Thus, chlorophyll is, in a sense, a solar energy converter.

See Problem 13.32.

These ring systems occur in nature. For example, the imidazole skeleton is present in the amino acid histidine, where it plays an important role in the reactions of many enzymes. Decarboxylation of histidine gives **histamine**, a toxic substance present in combination with proteins in body tissues. It is released as a consequence of allergic hypersensitivity or inflammation (for example, in hay fever sufferers). Many **antihistamines**, compounds that counteract the effects of histamine, have been developed. One of the better known of these is the drug benadryl (diphenylhydramine).

Antihistamines are compounds that counteract the effects of the toxin **histamine**, which contains an imidazole ring.

CH_2CHCO_2H / NH_2 / N / NH — histidine

$CH_2CH_2NH_2$ / N / NH — histamine

$(C_6H_5)_2CHOCH_2CH_2N(CH_3)_2$ — benadryl (an antihistamine)

The thiazole ring occurs in thiamin (vitamin B_1), a coenzyme required for certain metabolic processes and hence essential to life (thiamin also contains a pyrimidine ring). In its reduced form, the tetrahydrothiazole ring appears in penicillins, which are important antibiotics.

NH_2, Cl^-, CH_3, CH_2, N^+, N, CH_2CH_2OH, CH_3, N, S — thiamin (vitamin B_1)

O, H H, R—C—NH, S, CH_3, N, CH_3, O, H, CO_2H — penicillin

13.7 Fused-Ring Five-Membered Heterocycles: Indoles and Purines

Another aromatic or heteroaromatic ring can be fused to the double bonds of five-membered heterocycles. For example, **indole** has a benzene ring fused to the C-2–C-3 bond of pyrrole.

The **indole** ring system, consisting of a benzene ring fused to the C-2–C-3 bond of pyrrole, occurs in many natural products.

N / H — indole

The indole ring system, which occurs in several important natural products, is usually biosynthesized from the amino acid tryptophan, one of the protein building blocks. Indole itself and its 3-methyl derivative, skatole, are formed during protein decay.

Decarboxylation of tryptophan gives tryptamine. Several compounds with this skeleton have a profound effect on the brain and nervous system. One example is serotonin (5-hydroxytryptamine), a neurotransmitter and vasoconstrictor active in the central nervous system.

CH_2CHCO_2H / NH_2 / N / H — tryptophan

$CH_2CH_2NH_2$ / N / H — tryptamine

HO / $CH_2CH_2NH_2$ / N / H — serotonin

The tryptamine skeleton is disguised but present (shown in color) in more complex molecules. Reserpine, present in Indian snake root (*Rauwolfia serpentina*) that grows wild on the foothills of the Himalayas, has been used medically for centuries. It lowers blood pressure and is used to calm schizophrenics and improve their accessibility to psychiatric treatment. Lysergic acid is present in the fungus *ergot,* which grows on rye and other grains. Conversion of the carboxyl group to its diethylamide gives the extremely potent hallucinogen LSD.

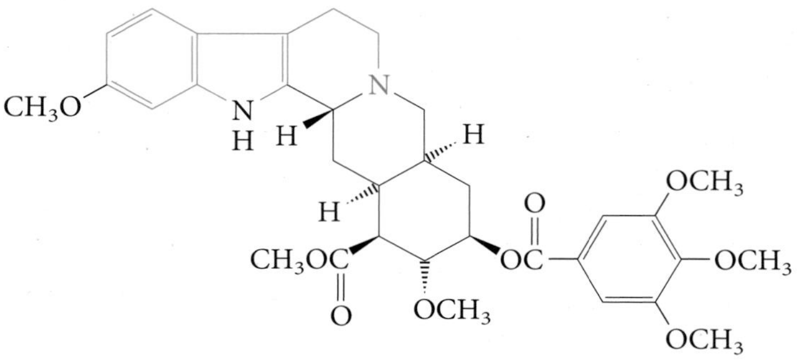

CH_3O N H H N H H CH_3OC O OCH_3 O OC OCH_3 OCH_3 OCH_3

reserpine

H HO_2C NCH_3 H N H

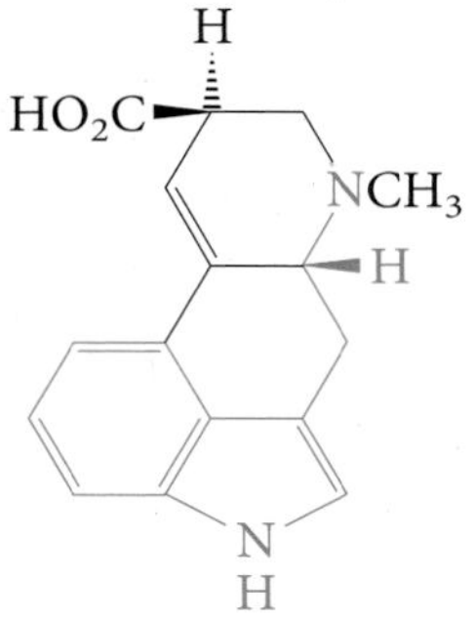

lysergic acid

Purines contain a pyrimidine ring fused to an imidazole ring.

The **purines** are another biologically important class of fused-ring heterocycles. They contain a pyrimidine ring fused to an imidazole ring.

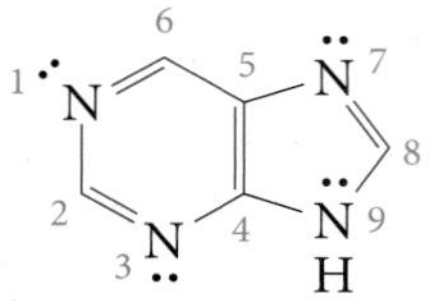

purine
(mp 217°C)

Uric acid is present in the urine of all carnivores and is the main product of nitrogen metabolism in the excrement of birds and reptiles. The disease gout results from deposition of sodium urate (the salt of uric acid) in joints and tendons. Caffeine, present in coffee, tea, and cola beverages, and theobromine (in cocoa) are also purines.

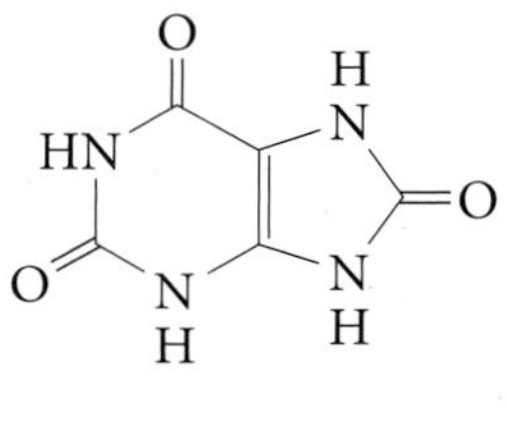

uric acid

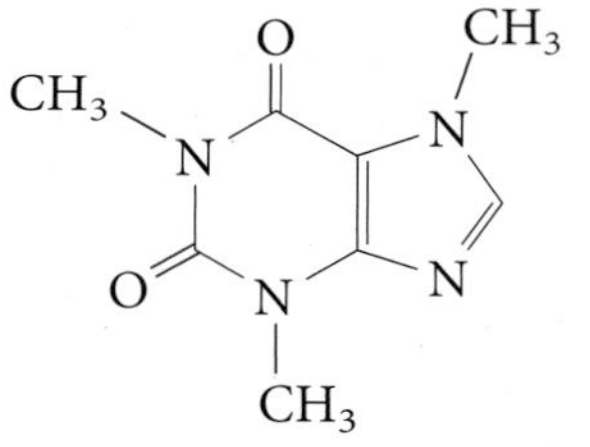

caffeine

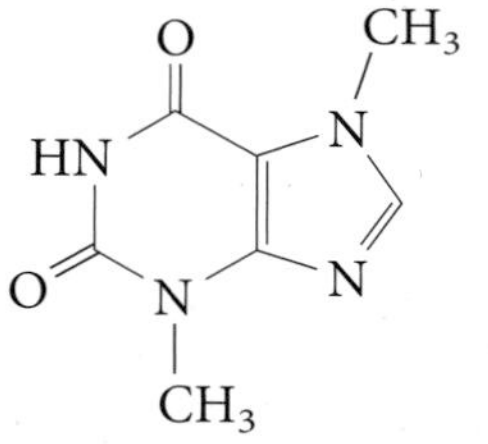

theobromine

Adenine and **guanine** are purines, and they are present in nucleic acids DNA and RNA.

Perhaps the most important purines in nature are **adenine** and **guanine**, two of the nitrogen bases that are present in nucleic acids (DNA and RNA; for further details, see Chapter 18).

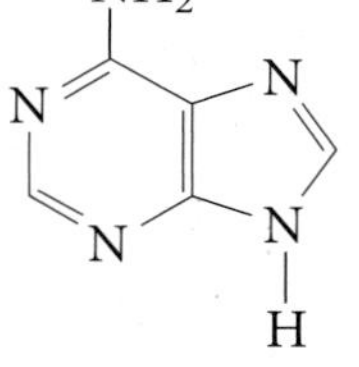

adenine

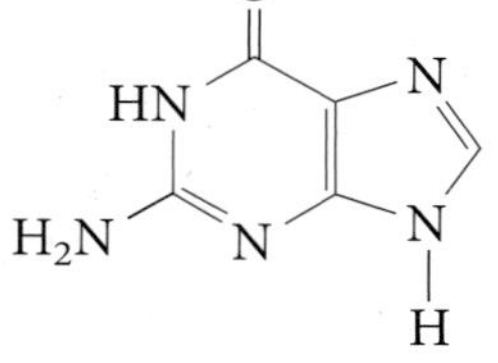

guanine

Many nitrogen heterocycles play a role in medicine. One leading actor in this field is morphine.

A WORD ABOUT... Morphine and Other Nitrogen-Containing Drugs

Morphine (named after Morpheus, the Greek god of dreams) is the major alkaloid present in opium. Opium is the dried sap of the unripe seed capsule of the poppy *Papaver somniferum;* its medical properties have been known since ancient times. Morphine was not isolated in pure form until 1805; its correct structure was not established until 1925, and it was not synthesized in the laboratory until 1952.

Pain is a major problem in medicine, and relief of pain has long been a medical goal. Morphine is an **analgesic**, a substance that relieves pain without causing unconsciousness. It was used for the large-scale relief of pain from battle wounds during the American Civil War (largely as a consequence of the invention, at about that time, of the hypodermic syringe). But morphine has serious side effects. It is addictive and also can cause nausea, a decrease in blood pressure, and a depressed breathing rate that can be fatal to the very young or the severely debilitated.

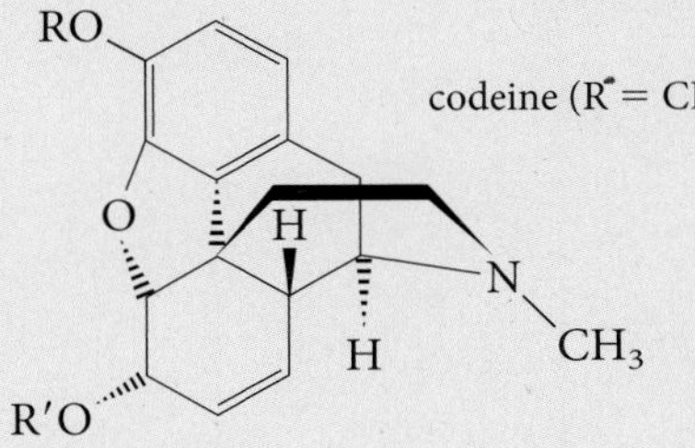

morphine (R = R′ = H)
heroin (R = R′ = $-\underset{\displaystyle O}{\overset{\|}{C}}CH_3$)
codeine (R = CH_3, R′ = H)

RO, R′O, O, H, H, N, CH_3

The first attempts to find a substance with morphine's benefits but without its side effects involved minor modification of its structure. Acetylation with acetic anhydride gave its diacetyl derivative **heroin**, which is a good analgesic with less of a respiratory depressant effect than morphine. But heroin is severely addictive, and its abuse has become a serious problem. Partial methylation of morphine gave **codeine**, which is useful as an antitussive (anticough) agent. Unfortunately, it is less than one-tenth as effective as morphine as an analgesic.

Many compounds similar to various parts of the morphine structure have been synthesized and tested for their analgesic properties. Two of these are shown here. Their structural similarity to morphine is evident in the colored parts of the structures.

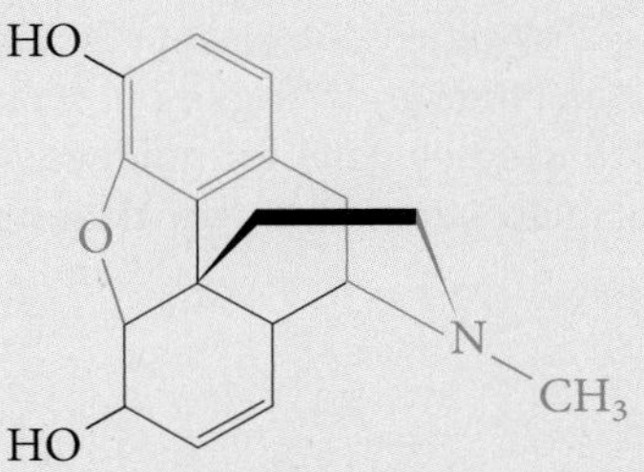

HO, HO, O, N, CH_3

morphine

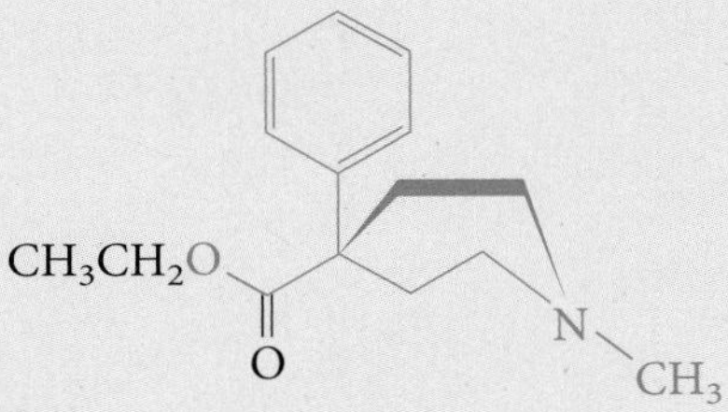

CH_3CH_2O, O, N, CH_3

demerol
(meperidine)

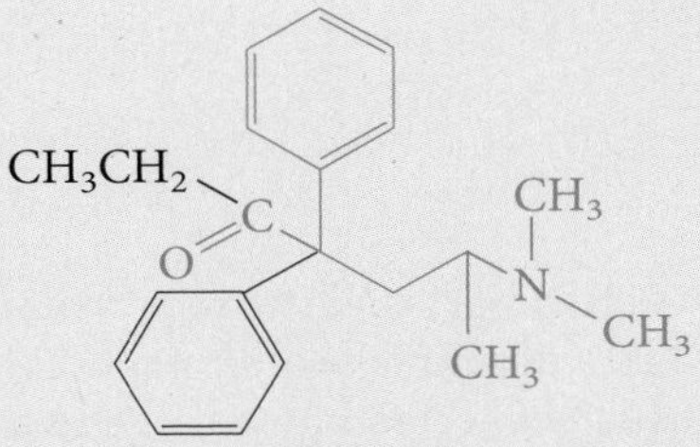

CH_3CH_2, C, O, CH_3, N, CH_3, CH_3

methadone

Demerol is an effective analgesic with a relatively simple structure compared to that of morphine. Notice that it still retains the piperidine ring present in morphine. **Methadone**, which retains the nitrogen of morphine but is no longer heterocyclic, was synthesized and used as an analgesic by the Germans during World War II, when natural sources of morphine were in short supply. Later it was used in substitution therapy for heroin addiction, but it, too, is addictive. The search for a perfect analgesic continues.

The pain associated with surgery or injury can sometimes be treated with local or regional anesthetics, many of which are nitrogen-containing drugs. **Cocaine**, an alkaloid found in the plant *Erythroxylum coca,* was one of the first anesthetics of this type. It constricts blood vessels, thus producing a bloodless surgical area. But cocaine is addictive and has other undesirable properties. Its medical use (it is also used illegally for nonmedical purposes) has been supplanted largely by **procaine hydrochloride** (Novocain).

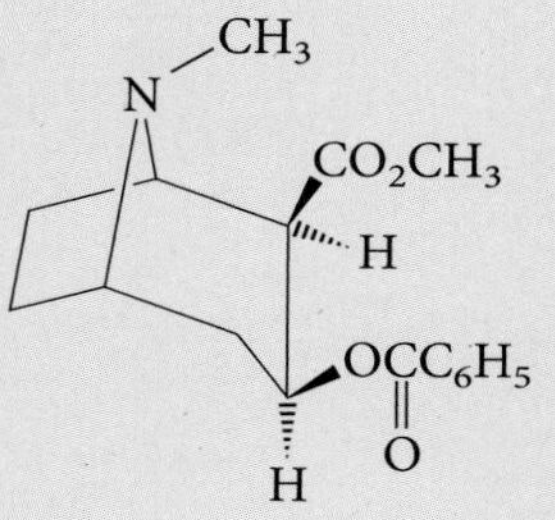

N–CH_3, CO_2CH_3, H, OCC_6H_5 (C=O), H

cocaine
mp 98°C

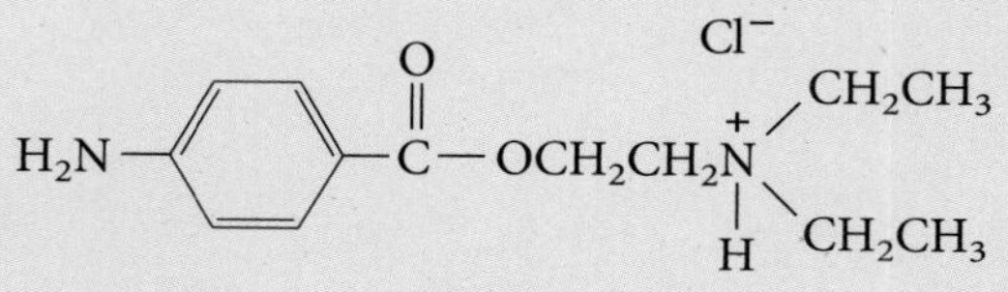

$H_2N-C_6H_4-\overset{O}{\overset{\|}{C}}-OCH_2CH_2\overset{+}{N}H(CH_2CH_3)_2\ Cl^-$

procaine hydrochloride
(mp 153–156°C)

Procaine is less toxic, is easier to synthesize and sterilize, and has a desirably shorter period of action than cocaine. It is usually injected into a nerve to anesthetize a small region of the body. It acts by inhibiting nerve impulse transmission by acetylcholine. Procaine hydrochloride is a widely used drug, sold to dentists and to doctors of human and veterinary medicine under many different trade names.

Benzocaine (ethyl *p*-aminobenzoate), which has a very simple structure, is used as a mild topical anesthetic in ointments for burns, insect bites, and open wounds. Note that procaine and benzocaine are different esters of the same acid, *p*-aminobenzoic acid.

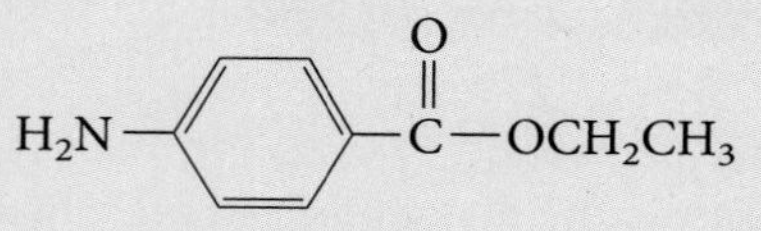

$H_2N-C_6H_4-\overset{O}{\overset{\|}{C}}-OCH_2CH_3$

benzocaine
(mp 88–90°C)

Fentanyl, a very short-acting drug about 100 times as potent as morphine, is extremely important in medicine today. It is used as an anesthetic in many surgical procedures.

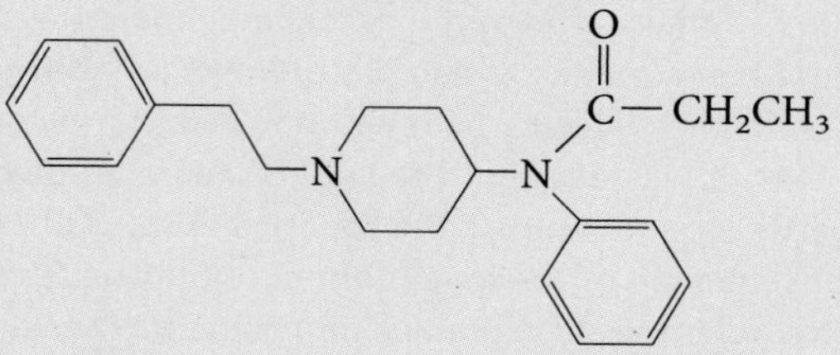

$\overset{O}{\overset{\|}{C}}-CH_2CH_3$, N, N

fentanyl

There are, of course, many types of pain. Sometimes a mild tranquilizer can be medically useful. Two of the most commonly prescribed contain seven-membered heterocyclic rings; they are the well-known twins of modern psychiatry: **librium** and **valium**.

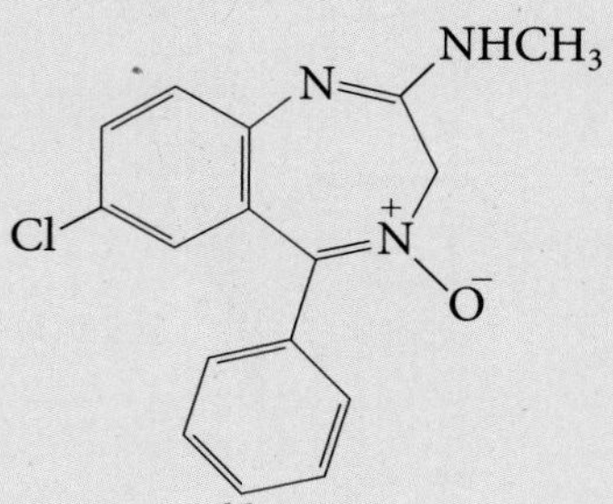

$NHCH_3$, N, Cl, $\overset{+}{N}$, O^-

librium

H_3C, N, O, Cl, N

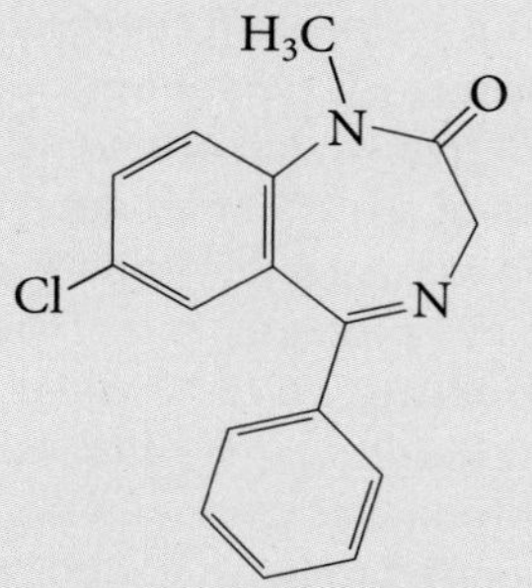

valium

See Problems 13.25 and 13.27.

REACTION SUMMARY

1. Reactions of Pyridine and Related Six-Membered Ring Aromatic Heterocycles

a. Protonation (Sec. 13.1)

pyridine + HX ⟶ pyridinium (N^+–H) X^-

$X = Cl, Br, I, HSO_4$

b. Electrophilic Aromatic Substitution (Sec. 13.2)

pyridine + E^+ ⟶ 3-E-pyridine + H^+

c. Nucleophilic Aromatic Substitution (Sec. 13.2)

pyridine + Nu–metal ⟶ 2-Nu-pyridine + metal–H

Examples: Nu—metal = H_2N—Na, Ph—Li

4-Cl-pyridine + Nu–metal ⟶ 4-Nu-pyridine + metal–Cl

Example: Nu—metal = CH_3O—Na

d. Alkyl Side Chain Oxidation (Sec. 13.2)

3-CH_3-pyridine $\xrightarrow{KMnO_4}$ 3-CO_2H-pyridine

e. Ring Reduction (Sec. 13.2)

pyridine $\xrightarrow{H_2}$ piperidine (N–H)

2. Electrophilic Aromatic Substitution Reactions of Five-Membered Ring Aromatic Heterocycles (Sec. 13.5)

five-membered ring (X with lone pair) + E^+ ⟶ 2-E-substituted ring + H^+

X = O, S, N—H

MECHANISM SUMMARY

1. Electrophilic Aromatic Substitution of Pyridine (Sec. 13.2)

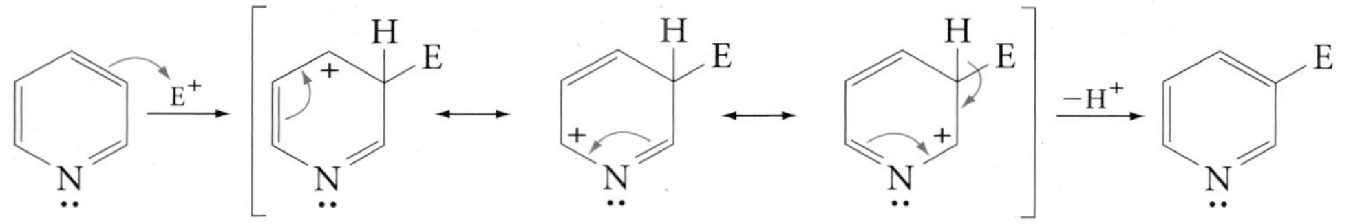

E^+ $-H^+$

2. Nucleophilic Aromatic Substitutions of Pyridines (Sec. 13.2)

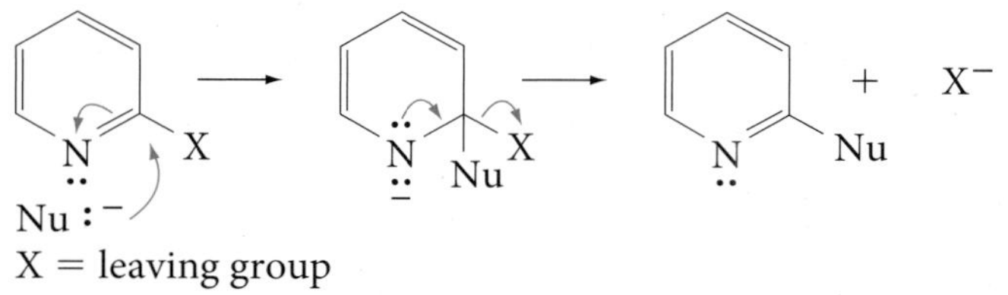

+ X^-

Nu :$^-$
X = leaving group

3. Electrophilic Aromatic Substitution of Five-Membered Ring Heterocycles (Sec. 13.5)

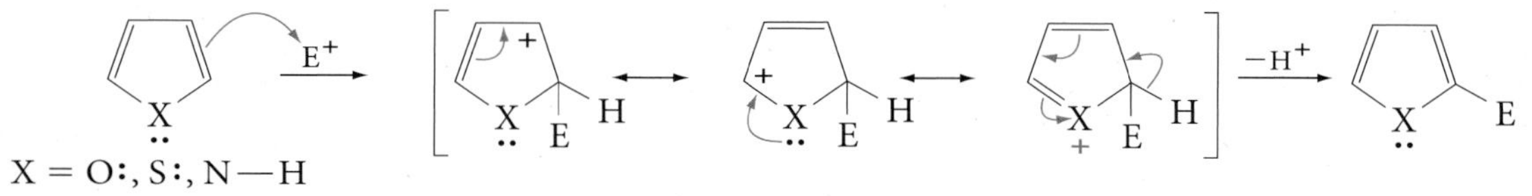

E^+ $-H^+$

X = O:, S:, N—H

ADDITIONAL PROBLEMS

OWL Interactive versions of these problems are assignable in OWL.

Reactions of Pyridine and Related Six-Membered Ring Heterocycles

13.9 In addition to the Kekulé-type contributors to the pyridine resonance hybrid shown on page 391, there are three minor dipolar contributors. Draw their structures. Do they suggest a reason why pyridine is deactivated (relative to benzene) toward reaction with electrophiles and a reason why substitution, when it does occur, takes place at the 3-position?

13.10 Although nitration of pyridine requires a temperature of 300°C (eq. 13.2), 2,6-dimethylpyridine is readily nitrated at 100°C. Write an equation for the reaction, and explain why milder conditions suffice.

13.11 Pyridine reacts with phenyllithium to give a good yield of 2-phenylpyridine. Write an equation and a mechanism for the reaction.

13.12 Oxidation of nicotine (page 395) with $KMnO_4$ gives nicotinic acid (page 394). Write an equation for the reaction.

13.13 Draw the product of the reaction of nicotinic acid with

a. catalytic sulfuric acid and excess methanol **b.** $LiAlH_4$, then H_3O^+ **c.** H_2 (excess) with Pt

13.14 Write equations for the reaction of coniine (page 395) with

a. hydrochloric acid
b. benzyl iodide (1 equivalent)
c. allyl iodide (2 equivalents)
d. acetic anhydride

13.15 Explain why nitration of quinoline (eq. 13.7) occurs mainly at C5 and C8.

13.16 Write an equation for each of the following reactions:

a. quinoline (page 395) + HCl
b. nitration of quinoline
c. quinoline + $NaNH_2$
d. quinoline + phenyllithium
e. quinoline + CH_3I

Properties and Reactions of Five-Membered Ring Heterocycles

13.17 Write equations for the reactions of furan with

a. Br_2
b. HNO_3
c. CH_3COCl (acetyl chloride), $SnCl_4$

13.18 Write equations for the reactions 2-methylfuran with the reagents shown in Problem 13.16.

13.19 Although electrophilic substitution occurs at C2 in pyrrole, it occurs predominantly at C3 in indole. Suggest an explanation.

13.20 As we saw in Chapter 10 in "A Word about . . . Green Chemistry" (page 299), ionic liquids are used as alternatives for other organic solvents. Suggest a synthetic scheme to prepare 1-butyl-3-methylimidazolium chloride from imidazole (Sec. 13.6).

$[H_3C-N\oplus N-CH_2CH_2CH_2CH_3]\,Cl^-$

13.21 Draw a molecular orbital picture of the bonding in oxazole (Sec. 13.6), using the bonding in imidazole as an example. Do you expect oxazole to be more or less basic than pyrrole? Explain.

Structure, Nomenclature, and Properties of Heterocycles

13.22 Using structures for the parent compounds given in the text, write the structural formula for

a. 2,4-difluorofuran
b. 2-bromothiophene
c. 4-methoxyquinoline
d. 5-hydroxyisoquinoline
e. 2-isopropylimidazole
f. 3-pyridinecarboxylic acid
g. 3,4-diethylpyrrole
h. 5-hydroxyindole
i. 4-chloropyridinium bromide
j. 4-aminopyrimidine

13.23 What is the configuration (*R* or *S*) of the hydroxyl-bearing carbon atom in codeine (page 405)?

13.24 The water solubility of morphine (page 405) is only 0.2 g/L, but morphine hydrochloride has a water solubility of 57 g/L. Write equations to show how morphine could be extracted and isolated from opium.

13.25 Nitrogen is present in many pharmaceutical drugs, as described in "A Word About . . . Morphine and Other Nitrogen-Containing Drugs" (pages 405–406). Varenicline is a drug inspired by morphine and is a prescription medication used to aid in smoking cessation.

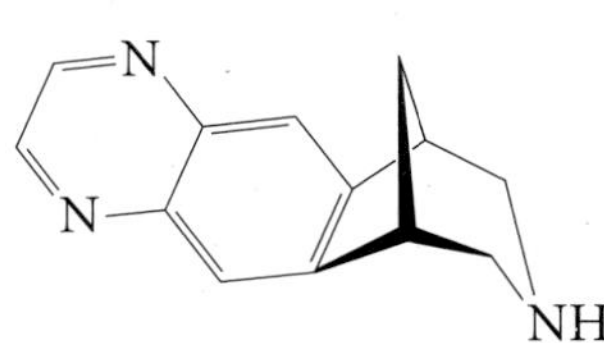

What is your prediction of the pK_a values for the conjugate acids of each of the nitrogens in varenicline?

13.26 Identify the stereogenic centers in nicotine and cocaine shown below, and assign an absolute configuration (*R* or *S*) for each center.

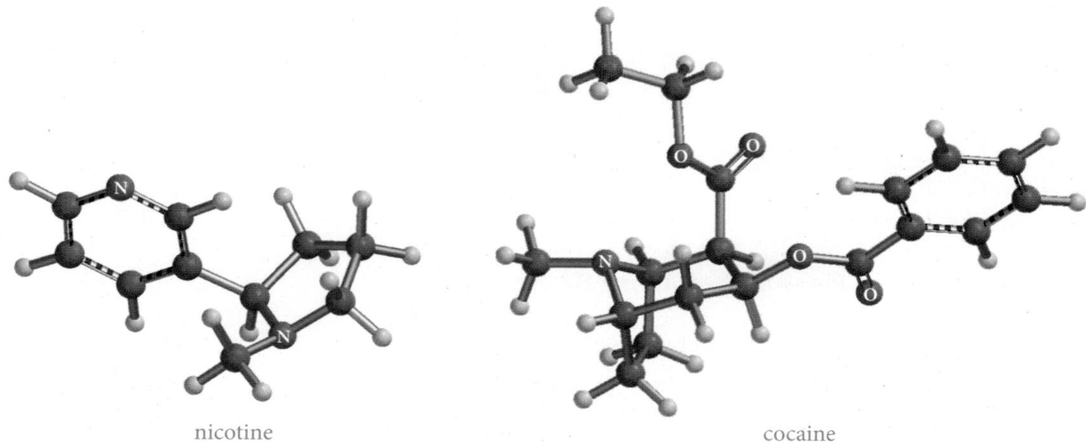

nicotine cocaine

13.27 Lysergic acid (page 404) has two nitrogens. Which do you expect is the more basic, and why?

Hydroxypyridines, Hydroxypyrimidines, and Related Heterocycles

13.28 In contrast with phenol, which exists almost entirely in the enol form, 2-hydroxypyridine exists mainly in the keto form. Draw the structures, and suggest a reason for the difference.

13.29 The RNA base **uracil** is a pyrimidine derivative, usually depicted in its keto form (page 396). The enol tautomer readily undergoes electrophilic substitution, unlike most pyrimidines.

a. Draw the structure of the enol tautomer of uracil.
b. Draw the product of the nitration of uracil with HNO_3. Where on the ring is the nitronium (NO_2^+) ion most likely to attack (see Chapter 4)?

13.30 Uric acid (page 404) has four hydrogens. Which do you expect to be the most acidic, and why?

Puzzle Problem

13.31 Benadryl can be synthesized in two steps from diphenylmethane, according to the following sequence:

$$C_6H_5{-}CH_2{-}C_6H_5 \xrightarrow[\text{light}]{Br_2} \text{A} \xrightarrow{(CH_3)_2NCH_2CH_2OH} \text{benadryl}$$

a. What is the structure of A?
b. How can the aminoalcohol needed for the second step be synthesized from ethylene oxide?
c. Write a mechanism for each step.

13.32 The structure of porphine is shown on page 402 as part of the "A Word About . . . Porphyrins: What Makes Blood Red and Grass Green?" There are two different kinds of nitrogens in porphine.

a. Which one is more basic? Explain.
b. If porphine is treated with two equivalents of a very strong base, what would be the product?

The Velcro fastener, the modern equivalent of zippers and shoelaces, is made of nylon "hooks" and "loops."

$$\left[NH(CH_2)_6NH - \overset{\overset{\large O}{\|}}{C} - (CH_2)_4 - \overset{\overset{\large O}{\|}}{C} \right]$$

nylon

Eye of Science/Photo Researchers

14

Synthetic Polymers

Polymers, or macromolecules as they are sometimes called, are large molecules that are built by repetitive linking of many smaller units called monomers. Polymers can be natural or synthetic. The most important natural polymers are carbohydrates (starch and cellulose), proteins, and nucleic acids (DNA and RNA). We will study these biopolymers in the last three chapters of this book. In this chapter, we will focus on some of the most important *synthetic* polymers.

Synthetic polymers whose names may already be familiar to you include polyethylene, Teflon, Styrofoam, nylon, Dacron, saran, and polyurethanes. There are many others. In the United States alone, annual synthetic polymer production exceeds 87 billion pounds. Synthetic polymers touch our daily lives perhaps to a greater extent than any other synthetic organic chemicals. They make up a substantial percentage of our clothing, appliances, vehicles, homes, packaging, toys, paints, plywood, fiberboard, and tires. It is virtually impossible to live in the modern world without, dozens of times each day, making use of these materials—all of which were totally unknown less than a century ago. Our standard of living could not possibly approach what it is today without products of the synthetic organic chemical industry.

Online homework for this chapter can be assigned in OWL, an online homework assessment tool.

Polymers are large molecules **(macromolecules)** built from repeating units called **monomers**.

14.1 Classification of Polymers

Chain-growth (or **addition**) **polymers** are made by adding **monomers** together.

Synthetic polymers can be classified into two main types, depending on how they are made. **Chain-growth polymers** (also called **addition polymers**) are made by the addition of one monomer unit to another in a repetitive manner. Alkenes serve as monomers for the preparation of many important chain-growth polymers; a catalyst is required to initiate their polymerization.* The catalyst adds to a carbon–carbon double bond to form a reactive intermediate and this intermediate then adds to the double bond of a second monomer unit to yield a new intermediate. The process continues until the polymer chain is built. Eventually the process is terminated in some way. *Chain-growth polymers retain all the atoms of the monomer units in the polymer.* A good example of a chain-growth polymer is polyethylene (Sec. 3.16).

$$\underset{\text{ethylene}}{n\,CH_2{=}CH_2} \xrightarrow[\text{initiator}]{\text{polymerization}} \underset{\text{polyethylene}}{\left(CH_2{-}CH_2\right)_n} \tag{14.1}$$

A **step-growth (condensation) polymer** is formed by reaction of two different functional groups with loss of a small molecule.

Step-growth polymers (also called **condensation polymers**) are usually formed by a reaction between two different functional groups with the loss of some small molecule, such as water. Thus, *a step-growth polymer does not contain all the atoms initially present in the monomers;* some atoms are lost in the small molecule that is eliminated. The monomer units are usually di- or polyfunctional, and the monomers usually appear in alternating order in the polymer chain. Perhaps the best-known example of a step-growth polymer is the polyamide nylon, prepared from 1,6-diaminohexane (hexamethylenediamine) and hexanedioic acid (adipic acid).

$$\underset{\substack{\text{1,6-diaminohexane}\\ \text{(hexamethylenediamine)}}}{H_2N(CH_2)_6NH_2} + \underset{\substack{\text{hexanedioic acid}\\ \text{(adipic acid)}}}{HO\overset{O}{\overset{\|}{C}}(CH_2)_4\overset{O}{\overset{\|}{C}}OH} \xrightarrow{200\text{–}300^\circ C} \underset{\text{nylon, a polyamide}}{\left[NH(CH_2)_6NH\overset{O}{\overset{\|}{C}}(CH_2)_4\overset{O}{\overset{\|}{C}}\right]_n} + \underset{(H_2O)}{2n\,H{-}OH} \tag{14.2}$$

Let us consider each of these ways of making polymers in greater detail.

14.2 Free-Radical Chain-Growth Polymerization

The free-radical chain mechanism described earlier for polyethylene (Sec. 3.16) is typical of chain-growth polymers. The overall reaction is

$$\underset{\text{vinyl monomer}}{CH_2{=}\underset{L}{\underset{|}{CH}}} \xrightarrow[\text{initiator}]{\text{radical}} \underset{\text{vinyl (or chain-growth) polymer}}{{-}CH_2\underset{L}{\underset{|}{CH}}{-}\left(CH_2\underset{L}{\underset{|}{CH}}\right)_n{-}CH_2\underset{L}{\underset{|}{CH}}{-}} \tag{14.3}$$

where L is some substituent. Table 14.1 lists some common commercial chain-growth polymers and their uses.

*Other compounds, such as formaldehyde and epoxides, also serve as monomers for the preparation of important chain-growth polymers.

Table 14.1 ■ Some Commercial Chain-Growth (Vinyl) Polymers Prepared by Free-Radical Polymerization

Monomer name	Formula	Polymer	Uses
ethylene (ethene)	$CH_2{=}CH_2$	polyethylene	sheets and films, blow-molded bottles, injection-molded toys and housewares, wire and cable coverings, shipping containers
propylene (propene)	$CH_2{=}CHCH_3$	polypropylene	fiber products such as indoor–outdoor carpeting, car and truck parts, packaging, toys, housewares
styrene	$CH_2{=}CH{-}C_6H_5$	polystyrene	packaging and containers (Styrofoam), toys, recreational equipment, appliance parts, disposable food containers and utensils, insulation
acrylonitrile (propenenitrile)	$CH_2{=}CHCN$	polyacrylonitrile (Orlon, Acrilan)	sweaters and other clothing
vinyl acetate (ethenyl ethanoate)	$CH_2{=}CH{-}OC(=O)CH_3$	polyvinyl acetate	adhesives, latex paints
methyl methacrylate (methyl 2-methylpropenoate)	$CH_2{=}C(CH_3){-}C(=O)OCH_3$	polymethyl methacrylate (Plexiglas, Lucite)	objects that must be clear, transparent, and tough
vinyl chloride (chloroethene)	$CH_2{=}CHCl$	polyvinyl chloride (PVC)	plastic pipe and pipe fittings, films and sheets, floor tile, coatings
tetrafluoroethylene (tetrafluoroethene)	$CF_2{=}CF_2$	polytetrafluoroethylene (Teflon)	coatings for utensils, electric insulators

Free-radical chain-growth polymerization requires a radical initiator, of which benzoyl peroxide is an example. It decomposes at about 80°C to give benzoyloxy radicals. These radicals may initiate chains or may lose carbon dioxide to give phenyl radicals that can also initiate chains.

$$C_6H_5{-}\overset{O}{\overset{\|}{C}}{-}O{-}O{-}\overset{O}{\overset{\|}{C}}{-}C_6H_5 \xrightarrow{80^\circ C} 2\, C_6H_5{-}\overset{O}{\overset{\|}{C}}{-}O\cdot \xrightarrow{-CO_2} 2\, C_6H_5\cdot \quad (14.4)$$

weak bond (O–O)

benzoyl peroxide → benzoyloxy radicals → phenyl radicals

For simplicity, we can represent the initiator radicals by the In· symbol.

Initiator radicals add to the carbon–carbon double bond of the vinyl monomer to produce a carbon radical.

Initiation step

$$\text{In}\cdot + \underset{\text{monomer}}{\text{CH}_2{=}\underset{\substack{|\\ \text{L}}}{\text{CH}}} \longrightarrow \underset{\text{carbon radical}}{\text{In}{-}\text{CH}_2{-}\underset{\substack{|\\ \text{L}}}{\dot{\text{C}}\text{H}}} \qquad (14.5)$$

Experience shows that the initiator usually adds to the *least* substituted carbon of the monomer; that is, to the CH_2 group. This gives a carbon radical adjacent to the substituent. There are two reasons for this preference: First, the terminal vinylic carbon is less hindered and therefore more easily attacked, and second, the substituent L usually can stabilize an adjacent radical.

The carbon radical formed in the initiation step then adds to another monomer molecule, and the adduct adds to another, and so on.

Propagation steps

$$\text{InCH}_2\underset{\substack{|\\ \text{L}}}{\dot{\text{C}}\text{H}} \xrightarrow{\text{CH}_2=\text{CHL}} \text{InCH}_2\underset{\substack{|\\ \text{L}}}{\text{CH}}\text{CH}_2\underset{\substack{|\\ \text{L}}}{\dot{\text{C}}\text{H}} \xrightarrow{\text{CH}_2=\text{CHL}} \underset{\text{growing polymer chain}}{\text{InCH}_2\underset{\substack{|\\ \text{L}}}{\text{CH}}\text{CH}_2\underset{\substack{|\\ \text{L}}}{\text{CH}}\text{CH}_2\underset{\substack{|\\ \text{L}}}{\dot{\text{C}}\text{H}}} \longrightarrow \text{and so on} \qquad (14.6)$$

Chain propagation (eq. 14.6) occurs in the same sense as initiation (eq. 14.5) so that the monomer units are linked in a head-to-tail manner, with the substituent on alternate carbon atoms.

PROBLEM 14.1 In polystyrene, the chain grows in a strictly head-to-tail arrangement because the intermediate radical is stabilized through resonance. Draw the intermediate radical, and show how it is stabilized through resonance. (The structure of styrene is shown in Table 14.1.)

Two common chain-terminating reactions are **radical coupling** and **radical disproportionation**.

Chain propagation may continue until anywhere from a few hundred to several thousand monomer units are linked. The extent of reaction depends on several factors, some of which are the reaction conditions (temperature, pressure, solvent, concentration of monomer, and catalyst, for example); the nature of the monomer, especially the substituent L; and the rates of competing reactions, which may terminate the chain. Two common chain-terminating reactions are **radical coupling** and **radical disproportionation**.

Termination Steps

$$\sim\text{CH}_2\underset{\substack{|\\ \text{L}}}{\dot{\text{C}}\text{H}} \quad \underset{\substack{|\\ \text{L}}}{\dot{\text{C}}\text{H}}\text{CH}_2\sim \xrightarrow[\text{coupling}]{\text{radical}} \overbrace{\sim\text{CH}_2\underset{\substack{|\\ \text{L}}}{\text{CH}}{-}\underset{\substack{|\\ \text{L}}}{\text{CH}}\text{CH}_2\sim}^{\text{head-to-head}} \qquad (14.7)$$

$$\sim\text{CH}_2\underset{\substack{|\\ \text{L}}}{\dot{\text{C}}\text{H}} \quad \sim\overset{\text{H}}{\overset{|}{\text{CH}}}{-}\underset{\substack{|\\ \text{L}}}{\dot{\text{C}}\text{H}} \xrightarrow[\text{disproportionation}]{\text{radical}} \underset{\text{alkane}}{\sim\text{CH}_2\underset{\substack{|\\ \text{L}}}{\text{CH}_2}} + \underset{\text{alkene}}{\sim\text{CH}{=}\text{CHL}} \qquad (14.8)$$

The ~ bonds at the ends of structures in eqs. 14.7–14.11 indicate the continuation of the polymer chain.

Radical coupling (eq. 14.7) gives rise to a head-to-head arrangement of two monomers. In radical disproportionation (eq. 14.8), one radical abstracts a hydrogen atom from the carbon adjacent to another radical site, producing a saturated and an unsaturated polymer.

EXAMPLE 14.1

What feature distinguishes propagation steps from termination steps?

Solution In propagation steps, one radical is destroyed, but another radical is created. In other words, the number of radicals on the left and right sides of the equation for a propagation step is always the same. In termination steps, however, radicals are destroyed, and no new radicals are generated. Therefore, chain growth terminates.

If eqs. 14.5 through 14.8 told the whole story, all polymers formed by free-radical chain growth would be linear. But we know from physical measurements that many such polymers have branched chains, so something is missing from the scheme.

What is missing is a **chain-transfer reaction**. A growing polymer radical may abstract a hydrogen atom from another polymer chain.

A **chain-transfer reaction** terminates one chain but initiates a branch on another chain.

Chain-transfer step

$$\sim CH_2\dot{C}H(L) + \sim CH_2CH(L)\sim \longrightarrow \sim CH_2CH_2(L) + \sim CH_2\dot{C}(L)\sim \quad (14.9)$$

This step terminates one chain but initiates another chain somewhere along the length of the polymer, not at its end, so that, when the polymerization continues, a branch is produced.

$$\underset{\text{midchain radical from eq. 14.9}}{\sim CH_2\dot{C}(L)\sim} \xrightarrow{CH_2{=}CHL} \underset{\text{branch point}}{\sim CH_2C(L)(CH_2\dot{C}HL)\sim} \quad (14.10)$$

Chain transfer and radical disproportionation are similar in that both involve a hydrogen abstraction reaction.

The extent of chain branching in a particular polymer depends on the relative rates of the chain-propagating and chain-transfer steps. If radical addition is very fast compared with hydrogen abstraction, the polymer will be mainly linear. On the other hand, if the chain-transfer rate were, for example, one-tenth of the addition rate, we might expect an average of one branch for every ten linearly linked monomers.

Chain-transfer reactions can be used to control the molecular weight of a polymer. Certain reagents, such as thiols, have a hydrogen that is easily abstracted. The resulting RS· radical is not reactive enough to add to double bonds. Instead, it dimerizes to form a disulfide.

$$\sim CH_2\dot{C}H(L) + \underset{\text{thiol}}{RSH} \xrightarrow[\text{abstraction}]{\text{hydrogen}} \sim CH_2CH_2(L) + RS\cdot \rightarrow RSSR \quad (14.11)$$

RS·: Does not add to double bonds; dimerizes instead.

Therefore, thiols are chain terminators, and when added to a polymerization reaction mixture in small amounts, they limit the polymer chain length.

Free-radical chain-growth polymerization is a very fast reaction. A chain may grow to 1000 monomer units or more in less than a second! The polymer chains contain one or two groups derived from the initiator radical, but those groups make up only a small fraction of the polymer molecule so that the polymer properties are determined largely by the particular monomer used.

David Buffington/PhotoDisc/ Getty Images

Polystyrene popcorn used in packaging.

Now let us consider two typical free-radical growth polymers: polystyrene and polyvinyl chloride.

Styrene is easily polymerized by benzoyl peroxide, and the product polystyrene has a molecular weight in the range of 1 to 3 million (that is, n = 10,000–30,000).

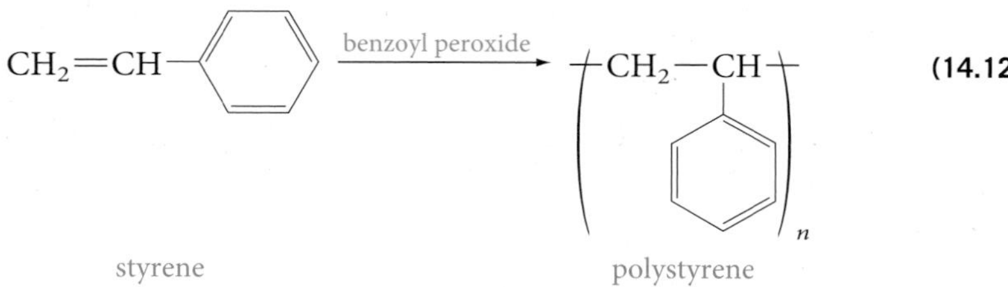

(14.12)

Amorphous polymer chains are randomly arranged.

Thermoplastic polymers can be reshaped because they melt on heating and harden again on cooling. Polymers can be made more rigid by **cross-linking** polymer chains.

Polystyrene is an amorphous, thermoplastic polymer. By **amorphous**, we mean that the polymer chains are irregularly arranged in a random manner; they are not regularly aligned, as in a crystal. By **thermoplastic**, we mean that the polymer melts or softens on heating and hardens again on cooling. Polystyrene can be molded or extruded to produce parts for housewares, toys, radio and television chassis, and bottles, jars, and containers of all kinds. Styrofoam is produced by including a low-boiling hydrocarbon (such as pentane) in the processing. When the polymer is heated, the pentane volatilizes, producing bubbles that expand the polymer into a foam. These foams are used for insulation, packaging, cups for hot drinks, egg cartons, and many other purposes. Annual U.S. production of polystyrene is more than 6 billion pounds.

Polystyrene can be modified in various ways. For example, it can be rigidified through **cross-linking**, by including small amounts of *p*-divinylbenzene with the monomer.

$CH{=}CH_2$ (on benzene ring) + $CH{=}CH_2$ / $CH{=}CH_2$ (para on benzene ring) —benzoyl peroxide→ ~$CH-CH_2-CH-CH_2-CH-CH_2$~ / ~$CH-CH_2-CH-CH_2-CH-CH_2$~ (14.13)

styrene (mostly)

p-divinylbenzene (0.1–1%)

cross-linked polystyrene

The resulting polymer is more rigid and less soluble in organic solvents than ordinary polystyrene. By sulfonation, this cross-linked polymer can be converted to an ion-exchange resin used in water softeners. As hard water percolates through the resin, the Ca^{2+} and Mg^{2+} ions that are present are exchanged for Na^+ ions.

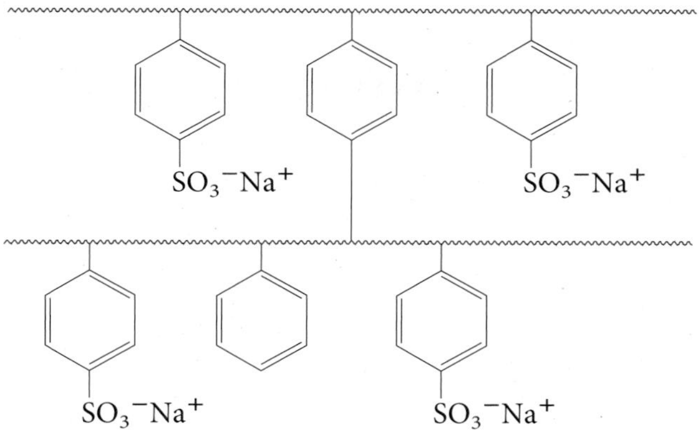

$SO_3^-Na^+$ $SO_3^-Na^+$

$SO_3^-Na^+$ $SO_3^-Na^+$

schematic structure of an ion-exchange resin

PROBLEM 14.2 Explain why sulfonation of polystyrene occurs in the positions shown in the ion-exchange resin formula. (*Hint:* Review Sec. 4.11.)

Polyvinyl chloride (PVC) can be represented by the general formula

$$\text{+}CH_2\underset{\underset{Cl}{|}}{CH}\text{+}_n$$

polyvinyl chloride (PVC)

It has a head-to-tail structure. PVC is a hard polymer but can be softened by adding **plasticizers**, usually low-molecular-weight esters that act as lubricants between the polymer chains. A good example is bis-2-ethylhexyl phthalate.

Plasticizers such as low-molecular-weight esters act as lubricants between polymer chains to soften hard polymers.

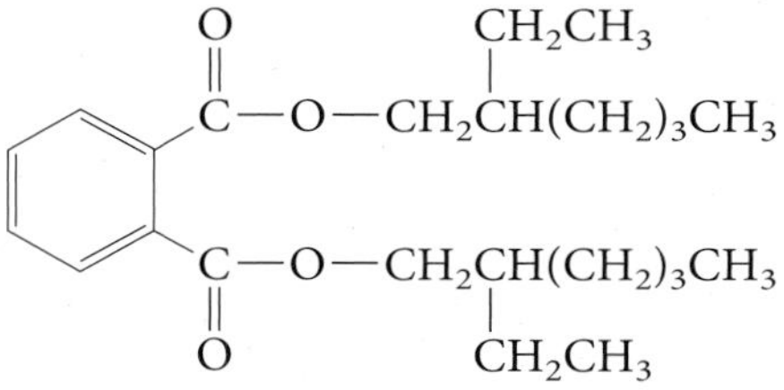

$$C_6H_4[C(=O)-O-CH_2CH(CH_2CH_3)(CH_2)_3CH_3]_2$$

bis-2-ethylhexyl phthalate
(a plasticizer)

PROBLEM 14.3 Bis-2-ethylhexyl phthalate can be prepared from 2-ethylhexanol (review Problem 9.29) and phthalic anhydride (review Problem 10.27). Write an equation for the reaction.

Polyvinyl chloride plastic pipes.

PVC is used to make floor tiles, vinyl upholstery (imitation leather), plastic pipes, plastic squeeze bottles, and so on. Annual U.S. production exceeds 15 billion pounds.

PROBLEM 14.4 Write the structural formula for a three-monomer segment of

a. polypropylene
b. polyvinyl acetate
c. poly(methyl methacrylate)
d. polyacrylonitrile

See Table 14.1 for structures of these monomers.

PROBLEM 14.5 Using a three-monomer segment, write an equation for

a. the reaction of polystyrene with $Cl_2 + FeCl_3$
b. the reaction of polystyrene with $HNO_3 + H_2SO_4$

14.3 Cationic Chain-Growth Polymerization

Certain vinyl compounds are best polymerized via cationic rather than free-radical intermediates. The most common commercial example is isobutylene (2-methylpropene), which can be polymerized with Friedel–Crafts catalysts in a reaction that involves tertiary carbocation intermediates.

Initiation

$$\underset{\text{isobutylene}}{CH_2{=}C(CH_3)_2} \xrightarrow[BF_3;\ H^+]{AlCl_3\ \text{or}} \underset{\text{tertiary carbocation}}{CH_3{-}\overset{+}{C}(CH_3)_2} \tag{14.14}$$

Propagation

$$CH_3{-}\overset{+}{C}(CH_3)_2 + CH_2{=}C(CH_3)_2 \longrightarrow CH_3C(CH_3)_2{-}CH_2{-}\overset{+}{C}(CH_3)_2 \longrightarrow CH_3{-}C(CH_3)_2{+}CH_2{-}C(CH_3)_2{+}_n CH_2{-}\overset{+}{C}(CH_3)_2 \tag{14.15}$$

Termination

$$(CH_3)_3C{+}CH_2{-}C(CH_3)_2{+}_n CH_2{-}\overset{+}{C}(CH_3){-}CH_2{-}H \xrightarrow{-H^+} \underset{\text{polyisobutylene}}{(CH_3)_3C{+}CH_2{-}C(CH_3)_2{+}_n CH_2{-}C(CH_3){=}CH_2} \tag{14.16}$$

Initiation gives the *tert*-butyl cation (eq. 14.14), which, in the propagation step, adds to the CH_2 carbon of the double bond in a Markovnikov manner to produce another tertiary carbocation and so on (eq. 14.15). The chain is terminated by loss of a proton from a carbon atom adjacent to the positive carbon (eq. 14.16).

Polyisobutylenes prepared this way (n = about 50) are used as additives in lubricating oil and as adhesives in pressure-sensitive tape and removable paper labels. Higher-molecular-weight polymers are used in the manufacture of inner tubes for truck and bicycle tires.

14.4 Anionic Chain-Growth Polymerization

Alkenes with electron-withdrawing substituents can be polymerized via carbanionic intermediates. The catalyst may be an organometallic compound; for example, an alkyllithium.

Initiation $R^- \quad Li^+ \quad CH_2{=}CH(L) \longrightarrow R{-}CH_2{-}\ddot{C}H(L) \quad Li^+$ (14.17)

Propagation

$RCH_2\ddot{C}H(L) \quad Li^+ \xrightarrow{CH_2=CHL} RCH_2CH(L)CH_2\ddot{C}H(L) \quad Li^+$, and so on (14.18)

Addition of the catalyst to the double bond gives a carbanion intermediate (eq. 14.17) in which the substituent L usually delocalizes the negative charge through resonance. Common L groups of this type are cyano (CN), carbomethoxy (CO_2CH_3), phenyl, and vinyl ($CH{=}CH_2$).

EXAMPLE 14.2

Draw the carbanion intermediate for anionic polymerization of acrylonitrile ($CH_2{=}CHCN$), and show how it is stabilized by resonance.

Solution

$$\left[\sim CH_2{-}\ddot{C}^-(H){-}C{\equiv}N: \longleftrightarrow \sim CH_2{-}C(H){=}C{=}\ddot{N}:^- \right]$$

PROBLEM 14.6 Methyl methacrylate (Table 14.1) can be polymerized by catalytic amounts of n-butyllithium at −78°C. Using eqs. 14.17 and 14.18 as a model, write a mechanism for the reaction. Show how the intermediate carbanion is resonance-stabilized.

Anionic polymerizations are terminated by quenching the reaction mixture with a proton source (water or alcohol).

PROBLEM 14.7 Ethylene oxide can be polymerized by base to give carbowax, a water-soluble wax. Suggest a mechanism for the reaction.

$$\underset{\text{ethylene oxide}}{CH_2{-}CH_2\ (\text{-O-})} \xrightarrow{HO^-} \underset{\text{carbowax}}{HOCH_2CH_2{+}OCH_2CH_2{+}_n OCH_2CH_2OH}$$

14.5 Stereoregular Polymers; Ziegler–Natta Polymerization

When a monosubstituted vinyl compound is polymerized, every other carbon atom in the chain becomes a stereogenic center:

$$\underset{\mathrm{L}}{\overset{}{CH_2{=}\underset{|}{CH}}} \longrightarrow -CH_2-\underset{\mathrm{L}}{\underset{|}{\overset{*}{C}H}}-CH_2-\underset{\mathrm{L}}{\underset{|}{\overset{*}{C}H}}-CH_2-\underset{\mathrm{L}}{\underset{|}{\overset{*}{C}H}}- \qquad (14.19)$$

The carbons marked with an asterisk have four different groups attached and are therefore stereogenic centers. Three classes of such polymers are recognized (Fig. 14.1):

An **atactic** (random configurations) polymer is **stereorandom**, but an **isotactic** (same configurations) or **syndiotactic** (alternate configurations) polymer is **stereoregular**.

atactic: stereocenters have random configurations.
isotactic: all stereocenters have the same configuration.
syndiotactic: stereocenters alternate in configuration.

An atactic polymer is **stereorandom**, but an isotactic or syndiotactic polymer is **stereoregular**. These three classes of polymers, *even if derived from the same monomer,* will have different physical properties.

EXAMPLE 14.3

Draw a chain segment of isotactic polypropylene.

Solution For polypropylene, the group L in eq. 14.19 is $—CH_3$. With the chain extended in zigzag fashion, all methyl substituents occupy identical positions.

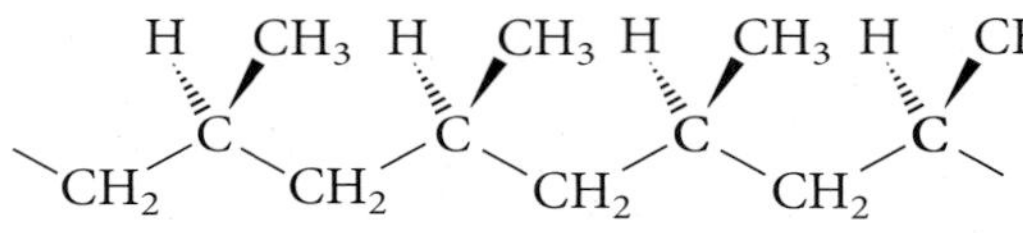

H CH_3 H CH_3 H CH_3 H CH_3
C C C C
CH_2 CH_2 CH_2 CH_2

PROBLEM 14.8 Using the definitions just given, draw a chain segment of

a. syndiotactic polypropylene b. atactic polypropylene

(a)

(b)

(c)

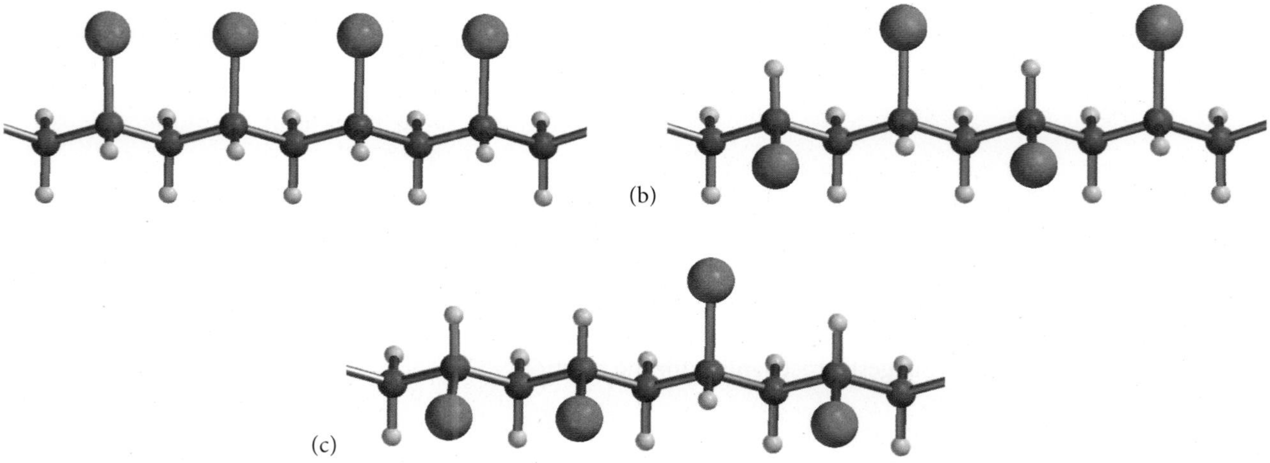

Figure 14.1
Ball-and-stick structures representing units of (a) isotactic, (b) syndiotactic, and (c) atactic vinyl polymers. The red balls represent "L" in polymers of the type shown in eq. 14.19.

A WORD ABOUT... Polyacetylene and Conducting Polymers

Materials that conduct electricity are an integral part of our everyday lives. The sizes of these materials range from enormous to miniscule. For example, huge metallic power lines are used to carry electricity hundreds of miles to light our cities, and tiny silicon wafers are used in the circuitry of computers. Most conducting materials used today are metallic, but the possibility of developing conducting organic polymers is being studied in many laboratories. Polyacetylene is an example of one such polymer.

While studying the polymerization of acetylene using various Ziegler–Natta catalysts, Hideki Shirakawa (University of Tsukuba) and his students accidentally discovered a way to prepare thin films of materials that were coppery or silvery in their appearance. These materials were the all-*cis* and all-*trans* forms of polyacetylene, respectively. Whereas these materials were not good conductors of electricity, in collaboration with Alan MacDiarmid (University of Pennsylvania) and Alan Heeger (University of California, Santa Barbara), it was discovered that treating the polyacetylene with iodine increases its conductivity by a factor of several million. This is because the iodine oxidizes the polyacetylene to radical cations (the iodine is reduced to triiodide anions) in which the charge can move along the polyacetylene chain and also be transferred from one polyacetylene chain to another.

Because organic polymers are generally lightweight and are strong yet flexible, it is hoped that conducting organic polymers will be ideal for constructing useful miniature devices that conduct electricity. Although polyacetylene itself has not yet proved useful for such purposes, this discovery stimulated research that has led to the development of related organic polymers that are now being used commercially; for example, in video matrix displays. Karl Ziegler (Germany) and Giulio Natta (Italy) shared the 1963 Nobel Prize in chemistry for their development of polymerization catalysts. Shirakawa, MacDiarmid, and Heeger shared the 2000 Nobel Prize in chemistry for their discovery of electrically conducting polymers.

acetylene molecules —polymerization→ polyacetylene —oxidation→ radical cation —charge transport→

See Problem 14.41.

Stereoregularity imparts certain favorable properties to polymers. Since free-radical polymerization usually results in an atactic polymer, the discovery in the 1950s by Ziegler and Natta of mixed organometallic catalysts that produce stereoregular polymers was a landmark in polymer chemistry. One such catalyst system is a mixture of triethylaluminum (or other trialkylaluminums) and titanium tetrachloride. With this catalyst, for example, propylene gives a polymer that is more than 98% isotactic.

The mechanism of Ziegler–Natta catalysis is quite complex. A key step in the chain growth involves an alkyl–titanium bond and coordination of the monomer to the metal. The coordinated monomer then inserts into the carbon–titanium bond, and the process continues.

$$\sim\text{CH(R)CH}_2\text{—Ti} \xrightarrow[\text{coordination step}]{\text{RCH=CH}_2} \sim\text{CH(R)CH}_2\text{—Ti}\cdots(\text{CH}_2\text{=CHR}) \xrightarrow{\text{insertion step}} \sim\text{CH(R)CH}_2\text{CH(R)CH}_2\text{—Ti} \quad \text{and so on} \tag{14.20}$$

Because of the various ligands attached to the titanium atom, coordination and insertion occur in a stereoregular manner and can be controlled to give either an isotactic or a syndiotactic polymer.

Commercial production of polypropylene is performed exclusively with Ziegler–Natta catalysts. A stereoregular, isotactic polymer that is highly crystalline is obtained. It is used for interior trim and battery cases in automobiles, for packaging (for example, containers for nested potato chips), and for furniture (such as plastic stacking chairs). It is also spun into fibers for ropes that float (an advantage for sailors and dockers), synthetic grass, carpet backings, and related materials.

Polyethylene obtained through Ziegler–Natta catalysis is linear, in contrast with the highly branched polyethylene obtained through free-radical processes. Linear polyethylene has a more crystalline structure, a higher density, and greater tensile strength and hardness than the branched polymer. It is used in thin-wall containers like those used to hold laundry bleach and detergents; in molded housewares such as mixing bowls, refrigerator containers, and toys; and in extruded plastic pipes and conduits.

14.6 Diene Polymers: Natural and Synthetic Rubber

Emma Lee/Life File/Getty Images

Latex (natural rubber) is obtained from the rubber tree.

Natural rubber is an unsaturated hydrocarbon polymer. It is obtained commercially from the milky sap (latex) of the rubber tree. Its chemical structure was deduced in part from the observation that, when latex is heated *in the absence of air,* it breaks down to give mainly a single unsaturated hydrocarbon product, **isoprene.**

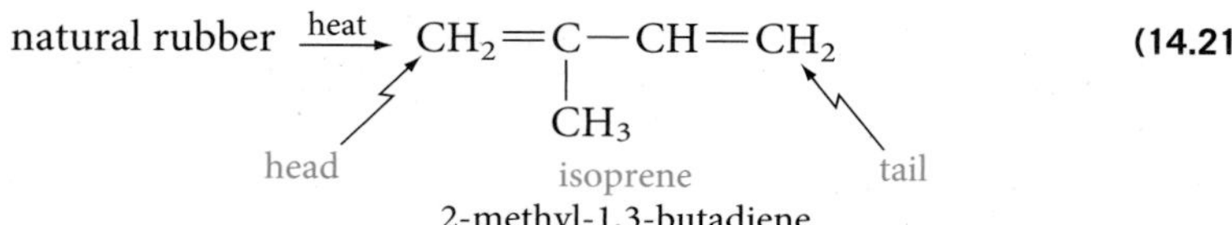

(14.21)

It is possible to synthesize a material that is nearly identical to natural rubber by treating isoprene with a Ziegler–Natta catalyst, such as a mixture of triethylaluminum, $(CH_3CH_2)_3Al$, and titanium tetrachloride, $TiCl_4$. The isoprene molecules add to one another by a head-to-tail, 1,4-addition.

head tail head tail head tail head tail

isoprene molecules

Ziegler–Natta catalyst ($R_3Al—TiCl_4$)

natural rubber segment (all Z)

(14.22)

The dashed lines in eq. 14.22 mark individual **isoprene** units.

The double bonds in natural rubber are *isolated;* that is, they are separated from one another by three single bonds. Moreover, these double bonds have a *Z* geometry.

PROBLEM 14.9 Gutta-percha, a less common form of natural rubber, is also a 1,4-polymer of isoprene, but with *E* double bonds. Draw the structural formula for a three-monomer segment of gutta-percha.

Most rubber has a molecular weight in excess of one million, though the value varies with the source and method of processing. This corresponds to about 15,000 isoprene monomers per rubber molecule. Crude plantation rubber contains, in addition to polyisoprene, about 2.5% to 3.5% protein, 2.5% to 3.2% fats, 0.1% to 1.2% water, and traces of inorganic matter.

Although natural rubber has many useful properties, it also has some undesirable ones. Early manufactured rubber goods were often sticky and smelly, and they softened in warm weather and hardened in cold. Some of these weaknesses were overcome when Charles Goodyear invented **vulcanization**, a process of cross-linking polymer chains by heating rubber with sulfur. The cross-links add strength to the rubber and act as a kind of "memory" that helps the polymer recover its original shape after stretching.

Vulcanization is a process of cross-linking rubber by heating it with sulfur.

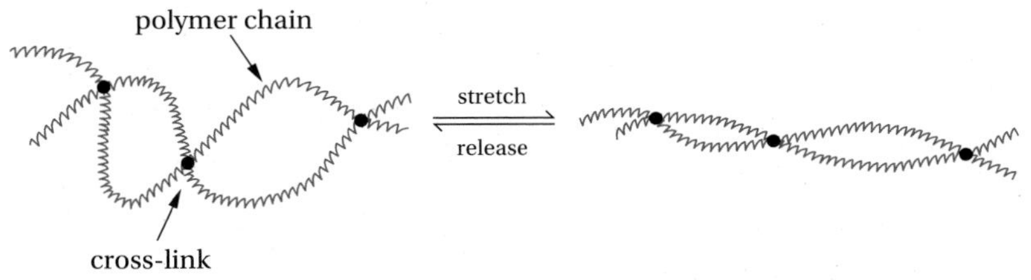

polymer chain

stretch ⇌ release

cross-link

In spite of such improvements, there were still problems. For example, it was not uncommon years ago to have to check the air pressure in tires almost every time one purchased gasoline, because the rubber inner tubes were somewhat porous. Therefore, there was a need to develop synthetic rubber, a name given to polymers with properties similar to those of natural rubber but superior to and somewhat chemically different from it.

Many monomers or mixtures of monomers form **elastomers** (rubber-like substances) when they are polymerized. The largest scale commercial synthetic rubber is a copolymer of 25% styrene and 75% 1,3-butadiene, called SBR (styrene-butadiene rubber).

Elastomers are rubber-like polymers.

$$n\text{CH}_2{=}\text{CHC}_6\text{H}_5 + 3n\,\text{CH}_2{=}\text{CH}{-}\text{CH}{=}\text{CH}_2 \xrightarrow[\text{initiator}]{\text{free radical}}$$

styrene

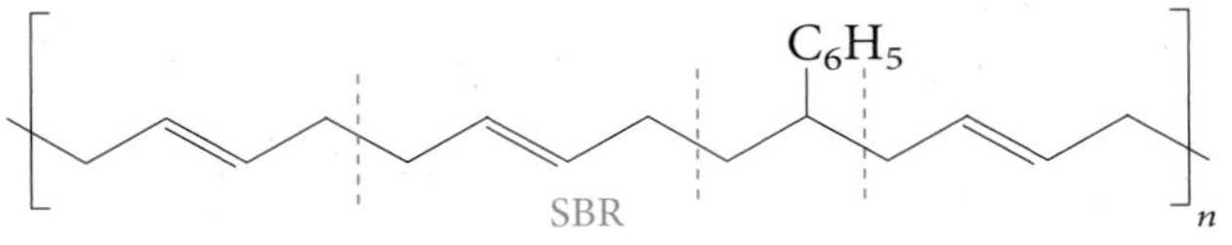

C_6H_5

SBR

n

(14.23)

The structure is approximately as shown, although about 20% of the butadiene adds in a 1,2-fashion, instead of 1,4-. Unlike those in natural rubber, the double bonds in this polymer have an *E* geometry. The dashed lines in the structure show the units from which the polymer is constructed. About two-thirds of SBR goes into tires. Its annual production exceeds that of natural rubber by a factor of more than two.

PROBLEM 14.10 Draw the structural formula for a three-monomer segment of poly(1,3-butadiene) in which

a. addition is 1,4 and double bonds are *Z*
b. addition is 1,4 and double bonds are *E*
c. addition is 1,2 for the middle unit and 1,4 for the outer units, with double bonds *Z*

14.7 Copolymers

A **homopolymer** is formed from one kind of monomer; a **copolymer** is formed from two different monomers.

Most of the polymers we have described so far are **homopolymers**, polymers made from a single monomer. But the variety and utility of chain-growth polymerization can sometimes be enhanced (as we have just seen with SBR synthetic rubber) by using mixtures of monomers to give **copolymers**. Figure 14.2 summarizes some of the ways in which monomers can be arranged in homo- and copolymers. The copolymers depicted are limited to two different monomers (A and B); in principle, of course, the possibilities are unlimited.

The exact arrangement of monomers along a copolymer chain will depend on a number of factors. One of these is the relative reactivity of the two monomers. Let us assume that we polymerize a 1 : 1 mixture of A and B by a free-radical chain-growth process. Here are some of the possibilities:

1. Radical A· reacts rapidly with B but slowly with A, and B· reacts rapidly with A but slowly with B. The polymer will then be alternating: —ABABAB—. Many copolymers tend toward this arrangement, though not always perfectly.
2. Monomers A and B are equally reactive toward radicals, and each reacts readily with radical A· or B· . The polymer will then be random: —AABABBA—.
3. Monomer A is much more reactive than B toward all radicals. In this case, A will be consumed first, followed more slowly by B. We will obtain a mixture of two homopolymers, $—(A)_n—$ and $—(B)_m—$.

PROBLEM 14.11 1,1-Dichloroethene and vinyl chloride form a copolymer called *saran,* used in food packaging. The monomer units tend to alternate in the chain. Draw the structural formula for a 4-monomer segment of the chain.

Block copolymers contain alternating blocks of monomers.

Graft copolymers are made by adding a second monomer to a homopolymer chain that contains double bonds.

Block and graft copolymers are made by special methods. If we first initiate polymerization of monomer A, then add some B, then add A again, and so on, we can obtain a block polymer with alternating segments of blocks of A units, then B units, and so on. This is particularly easy with anionic polymerizations, where there are no significant termination steps.

Graft polymers are made by taking advantage of functionality present in a homopolymer. For example, if a polymer contains double bonds (as in poly-1,3-butadiene),

▰ Figure 14.2
Arrangements of monomers in polymers.

Homopolymers

```
                    AA—
                    |
—AAAAA—        —AAAAA—        —AAAAA—
                 |                |
                 AA—           —AAAAA—
 linear        branched       cross-linked
```

Copolymers

```
—ABABAB—     —AABABBA—     —AAAAABBBB—     —AAAAAAA—
                                               |    |
                                            —BBB   BBB—
alternating    random          block          graft
```

addition of a free-radical initiator R· and second monomer (such as styrene) will "graft" polystyrene chains onto the polybutadiene backbone.

$$\begin{array}{c}\text{R} \qquad\qquad\qquad\qquad\qquad\qquad \text{R}\\ -CH_2CHCHCH_2CH_2CH{=}CHCH_2CH_2CHCHCH_2-\\ CH_2CHCH_2CH- \qquad\qquad CH_2CHCH_2CH-\\ \text{Ph}\quad\ \text{Ph} \qquad\qquad\qquad\qquad \text{Ph}\quad\ \text{Ph}\end{array}$$

poly-1,3-butadiene with polystyrene grafts

This particular graft polymer is used to make rubber soles for shoes.

14.8 Step-Growth Polymerization: Dacron and Nylon

Step-growth polymers are usually produced by a reaction between two monomers, each of which is at least difunctional. Many of them can be represented by the following overall equation:

$$A{\sim}A + B{\sim}B \longrightarrow -A{\sim}A-B{\sim}B-A{\sim}A-B{\sim}B- \qquad (14.24)$$

where A ~ A and B ~ B are difunctional molecules with groups A and B that can react with one another. For example, A might be an OH group, and B might be a CO_2H group, in which case A ~ A would be a diol, B ~ B would be a dicarboxylic acid, and ~ A—B ~ would be an ester. The polymer would be a **polyester**.

A **polyester** is a step-growth polymer made from diacid and diol monomers.

Unlike chain-growth polymers, which grow by one monomer unit at a time, step-growth polymers are formed in steps (or leaps), often by reaction of one polymer molecule with another. The way this works is best illustrated with a specific example.

Consider the formation of a polyester from a diol and a diacid. In the first step, the product will be an ester, with an alcohol group at one end and an acid at the other (eq. 14.25).

$$\underset{\text{diol}}{HO{\sim}OH} + \underset{\text{diacid}}{HO_2C{\sim}CO_2H} \xrightarrow{-H_2O} \underset{\text{alcohol}}{HO{\sim}}\underset{\text{ester}}{O-\overset{O}{\overset{\|}{C}}}\underset{\text{acid}}{{\sim}CO_2H} \qquad (14.25)$$

At the next stage, the alcohol-ester-acid can react with another diol, with another diacid, *or with another trifunctional molecule like itself.*

$$\underset{\text{alcohol-ester-acid}}{HO{\sim}O-\overset{O}{\overset{\|}{C}}{\sim}CO_2H} \begin{cases} \xrightarrow[-H_2O]{HO{\sim}OH} \underset{\text{diester-diol}}{HO{\sim}O-\overset{O}{\overset{\|}{C}}{\sim}\overset{O}{\overset{\|}{C}}-O{\sim}OH} \\ \xrightarrow[-H_2O]{HO_2C{\sim}CO_2H} \underset{\text{diester-diacid}}{HO_2C{\sim}\overset{O}{\overset{\|}{C}}-O{\sim}O-\overset{O}{\overset{\|}{C}}{\sim}CO_2H} \\ \xrightarrow[-H_2O]{HO{\sim}O-\overset{O}{\overset{\|}{C}}{\sim}CO_2H} \underset{\text{alcohol-triester-acid}}{HO{\sim}O-\overset{O}{\overset{\|}{C}}{\sim}\overset{O}{\overset{\|}{C}}-O{\sim}O-\overset{O}{\overset{\|}{C}}{\sim}CO_2H} \end{cases} \qquad (14.26)$$

A WORD ABOUT... Degradable Polymers

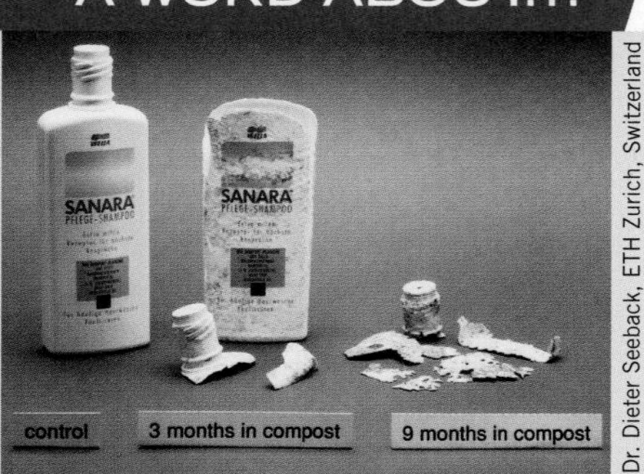

Shampoo bottles made of BIOPOL, after 0, 3, and 9 months in compost. The slower degradation of the printed areas is visible.

Widespread use of polymeric plastic materials that contribute so much to our standard of living has brought with it a pollution problem. What is to be done with these materials once they are no longer useful? Incineration represents one solution to the problem, but it can be accompanied by release of toxic substances into the environment. Another solution is recycling. For example, poly(ethylene terephthalate) (PETE), which is widely used in the production of plastic soft drink bottles (look for the symbol PETE on the bottom of your next bottle of soda), can be recycled as polyester fiber for use in carpets or furniture stuffing. Also, milk jugs made from high-density poly(ethylene) (HDPE) can be recycled in a number of ways; for example, as plastic lumber.

A third solution to the pollution problem is to develop polymers that are degraded in the environment once they are no longer useful. This approach is quite difficult because it is often the durability of plastics that makes them so useful. It would be useless to produce a product that began to disintegrate as it came off the production line. Nonetheless, several commercially useful **degradable polymers** have been developed.

R = Me poly(3-hydroxybutyrate)
R = Et poly(3-hydroxyvalerate)

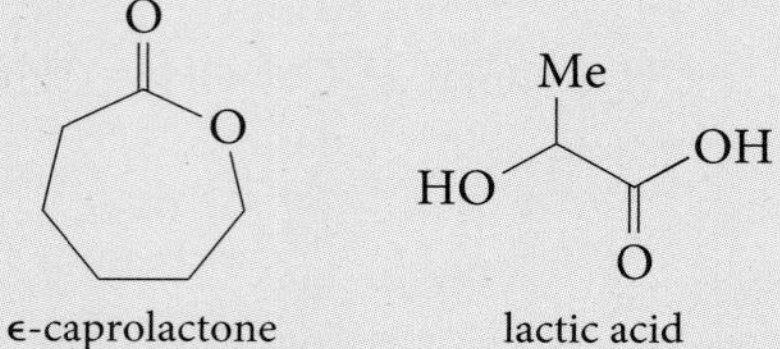

For example, natural biopolymers have been developed for use in the manufacturing of plastic containers. **Poly(hydroxyalkanoates)** such as poly(3-hydroxybutyrate) (PHB) and poly(3-hydroxyvalerate) (PHV) are naturally occurring polyesters produced by certain bacteria. PHB has many physical properties in common with poly(propylene), and a PHB–PHV copolymer (BIOPOL) has recently been used to manufacture plastic shampoo bottles. PHB–PHV is of special interest because it is biodegradable. Since it is a naturally occurring polymer, it is easily degraded by enzymes produced by soil microorganisms and therefore does not persist in the environment after disposal. Other biodegradable polymers, such as polyesters derived from ϵ-caprolactone and lactic acid, are also known and have been commercialized. Although it remains to be seen how widespread the use of biodegradable plastics will become, research and development of these materials is sure to continue as we try to deal with contemporary environmental issues.*

See Problem 14.39.

*For additional reading, see articles by J. D. Evans and S. K. Sikdar in *Chemtech* **1990**, *38*, and H. Muller and D. Seebach in *Angew. Chem., Int. Ed. Engl.* **1993**, *32*, 477.

Polyethylene terephthalic (PETE), a recyclable polyester.

The consequences of these alternatives are different; each of the first two products contains three monomer units, but in the third alternative, we go directly from two-monomer fragments to a product with four monomer units. Since the reactivity of the —OH group or of the —CO_2H group in all these reactants is quite similar, there is no particular preference among them, and the rates of the various reactions will depend mainly on the concentrations of the particular reactants.

PROBLEM 14.12 How many monomer units will be present in the next product if the diester-diol and diester-diacid in eq. 14.26 react? Draw the structure of the product.

A WORD ABOUT... Aramids, the Latest in Polyamides

© Jennifer Meyer Dare

Flame-resistant Nomex is used in firefighters' coats.

Aromatic polyamides (called aramids) are being produced at a rapidly growing rate because of their special properties of heat resistance, low flammability, and exceptional strength. The best known is **Kevlar** (see Figure 14.3). Because of the aromatic rings, this type of polyamide has a much stiffer structure than do the nylons. It is being used in place of steel to make tire cord for radial tires, because a Kevlar fiber is five times stronger than an equal-weight steel fiber. Kevlar is used in lightweight personal body armor (such as bullet-resistant vests), which offers protection against handgun fire, shotgun pellets, and knife slashes.

Nomex has a structure similar to that of Kevlar but uses *meta-* instead of *para*-oriented monomers. It is used in flame-resistant clothing because its fibers char rather than melt when exposed to flame. It has wide applications, ranging from firefighters' coats to racing drivers' uniforms. It has also been used in flame-resistant building materials. Honeycomb core constructions of Nomex are used in many internal and external parts of military and commercial jets, helicopters, and space vehicles. For example, close to 25,000 square feet of a Boeing 747's exterior structures include Nomex construction. The combination of strength and light weight has also made both Nomex and Kevlar popular in boat construction. For an account of the Kevlar story, see Tanner, D., Fitzgerald, J. A., and Phillips, B. R., *Angewandte Chemie International Edition in English, Advanced Materials* **1989**, *28,* 649–654.

See Problem 14.42.

$$H_2N-C_6H_4-NH_2 + Cl-\overset{O}{\overset{\|}{C}}-C_6H_4-\overset{O}{\overset{\|}{C}}-Cl \xrightarrow{\text{base}} \left(HN-C_6H_4-NH-\overset{O}{\overset{\|}{C}}-C_6H_4-\overset{O}{\overset{\|}{C}}\right)_n$$

p-phenylenediamine terephthaloyl chloride Kevlar

Figure 14.3
Reaction used to synthesize Kevlar.

If we start with exactly one equivalent each of a diol and a diacid, we should, in principle, be able to form one giant polyester molecule. In practice, this does not happen. In fact, to form a polymer with an average of 100 monomer units or more, the reaction must go to at least 99% completion. Consequently, the starting materials for this type of polymerization must be exceedingly pure, their mole ratio must be controlled precisely, and the reaction must be forced to completion, usually by distilling or otherwise removing, as it is formed, the small molecule that is eliminated.

PROBLEM 14.13 What product will mainly be formed if a diacid is treated with a *large excess* of a diol? If a diol is treated with a *large excess* of a diacid? These reactions represent two extremes of what can happen as the ratio of reactant concentrations deviates from 1:1 in a step-growth polymerization.

Although many polyesters are known, the most common example is **Dacron**, the polyester of terephthalic acid and ethylene glycol.

$$\left[\overset{O}{\overset{\|}{C}} - C_6H_4 - \overset{O}{\overset{\|}{C}} - OCH_2CH_2O \right]_n$$

the polyester Dacron, poly(ethylene terephthalate)

Dacron, the polyester of terephthalic acid and ethylene glycol, is used in fibers. **Mylar** is a strong film made of the same polyester.

The value of n is about 100 ± 20. The crude polyester can be spun into fibers for use in textiles. The fibers are highly resistant to wrinkling.

The same polyester can be fabricated into a particularly strong film called **Mylar**. Mylar polyester film is used for the long-term protection of artwork and historical documents because of its transparency, strength, and inertness. Mylar is popular for its use in the manufacture of balloons for festive occasions. Mylar was also used to cover the 94-foot wingspan of the Gossamer Albatross, a human-powered aircraft used to fly the English Channel in 1979. Only two pounds of the polyester was needed to cover the entire wingspan, and the polyamide Kevlar was the material used for construction of the rudder of this aircraft! In the United States, production of polyester fibers exceeds 3 billion pounds per year.

Gossamer Albatross.

PROBLEM 14.14 *Kodel* is a polyester with the following structure:

$$\left[\overset{O}{\overset{\|}{C}} - C_6H_4 - \overset{O}{\overset{\|}{C}} - O - CH_2 - C_6H_{10} - CH_2 - O \right]_n$$

From what two monomers is it made?

Nylons are polyamides: step-growth polymers made from diacids and diamines.

Nylons are polyamide step-growth polymers. The formula for **nylon-6,6**, so called because each monomer (diamine and diacid) has six carbon atoms, is shown in eq. 14.2. This polymer was first made by W. H. Carothers at the DuPont Company in 1933 and was commercialized five years later.* When mixed, the two monomers form a polysalt, which, on heating, loses water to form a polyamide. The molten polymer can be molded or spun into fibers.

The second most important polyamide is nylon-6, made from caprolactam.

$$\underset{\text{caprolactam}}{\text{cyclo-}[\overset{O}{\overset{\|}{C}} - NH - CH_2 - CH_2 - CH_2 - CH_2 - CH_2]} \xrightarrow{250\text{–}270°C} \underset{\text{nylon-6}}{\left[NHCH_2CH_2CH_2CH_2CH_2\overset{O}{\overset{\|}{C}} \right]_n} \quad (14.27)$$

*For an account of its discovery, the strange origin of its name, and a description of its properties and many uses, see the article by G. B. Kauffman in *J. Chem. Educ.* **1988**, *65*, 803–808. Also see Matthew E. Hermes's biography *Enough for One Lifetime: Wallace Carothers, Inventor of Nylon*, 1996, ACS/Chemical Heritage Foundation.

A CLOSER LOOK AT... Nylon

Conduct research on the Internet to find more information on nylon and synthetic fibers and to answer the following questions.

The History of Nylon

1. When was the first viable synthetic fiber drawn?
2. What was the first product to be marketed as nylon?

Wallace Hume Carothers: The Inventor of Nylon

1. Before making nylon, Carothers and his group tried to use polyesters to make commercially viable fibers. What were the problems with the first polyesters?
2. What advantages did polyamides have over polyesters?

The Nylon Legacy

1. Why was nylon important to the United States during World War II?
2. What are some post–World War II uses of nylon?
3. What are Kevlar and Nomex? What are they used for?
4. What is the process that is used to make nylon so that it has all the properties for which we value it?

Lactams are cyclic amides (compare with lactones, Sec. 10.12). On heating, the seven-membered ring of caprolactam opens as the amino group of one molecule reacts with the carbonyl group of the next, and so on, to produce the polyamide.

A **lactam** is a cyclic amide.

Nylons are extremely versatile polymers that can be processed to give materials as delicate as sheer fabrics, as long-wearing as carpets, as tough as molded automobile parts, or as useful as Velcro fasteners. In the United States, annual production of nylon fibers exceeds 3.7 billion pounds.

14.9 Polyurethanes and Other Step-Growth Polymers

A **urethane** (also called a **carbamate**) is a functional group that is simultaneously an ester and an amide at a single carbonyl group. Urethanes are commonly prepared from isocyanates and alcohols.

A **urethane** (or **carbamate**) contains the ester and amide functional groups on the same carbonyl. A **polyurethane** is a step-growth polymer prepared from diisocyanates and diols.

$$\underset{\text{isocyanate}}{R{-}N{=}C{=}O} + \underset{\text{alcohol}}{R'OH} \longrightarrow \underset{\text{urethane}}{RNH\overset{\overset{\displaystyle O}{\|}}{C}OR'} \quad (14.28)$$

The reaction is an example of a nucleophilic addition to the carbonyl group:

$$R{-}\ddot{N}{=}C{=}\ddot{O}{:} + R'\ddot{O}H \longrightarrow R{-}\ddot{N}{=}\overset{\overset{\displaystyle :\ddot{O}:^-}{|}}{C}{-}\overset{+}{\ddot{O}}{-}R' \;(H^+,\ \text{on O: } {-}H) \longrightarrow R{-}\underset{H}{\ddot{N}}{-}\overset{\overset{\displaystyle :\ddot{O}}{\|}}{C}{-}\ddot{O}R' \quad (14.29)$$

PROBLEM 14.15 The highly effective, biodegradable insecticide Sevin is a urethane called 1-naphthyl-*N*-methylcarbamate. It is made from methyl isocyanate and 1-naphthol. Using eq. 14.28 as a guide, write an equation for its preparation.

(In the urethane, R = CH_3— and R′ = .)

Polyurethanes are made from *di*isocyanates and *di*ols. The most important commercial example is prepared from 2,4-tolylenediisocyanate (TDI) and ethylene glycol.

$$\text{tolylene diisocyanate (TDI)} + \underset{\text{ethylene glycol}}{HOCH_2CH_2OH} \longrightarrow \text{a polyurethane (urethane groups } -NH-\overset{O}{\overset{\|}{C}}-O-) \tag{14.30}$$

This reaction is a little different from most step-growth polymerizations in that no small molecule is eliminated. Like other step-growth polymerizations, however, it does involve a reaction of two different difunctional monomers.

PROBLEM 14.16 Rewrite the partial polyurethane structure shown in eq. 14.30, but with one more monomer unit attached to each wavy bond.

© Richard Pasley/Stock Boston

Polyurethane varnish.

The reaction in eq. 14.30 produces a polyurethane but does not produce a foam. To obtain a foam as the product, the polymerization is carried out with a little water present. The water reacts with isocyanate groups in the starting material or in the growing polymer to produce a carbamic acid. This acid spontaneously loses carbon dioxide, which creates bubbles as the polymer forms.

$$\underset{\text{isocyanate}}{\sim N{=}C{=}O} + \underset{\text{water}}{HOH} \longrightarrow \underset{\text{carbamic acid}}{\sim NH\overset{O}{\overset{\|}{C}}OH} \longrightarrow \underset{\text{amine}}{\sim NH_2} + CO_2\uparrow \tag{14.31}$$

The amount of carbon dioxide formed, which determines the density of the foam, can be controlled by the amount of water used. The resulting amine can also react with isocyanate groups to form a urea, which may serve as a cross-link between polymer chains.

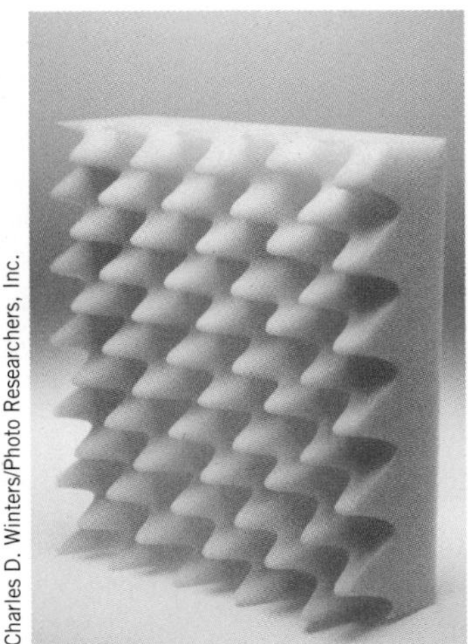

Polyurethane foam.

$$\sim N{=}C{=}O + H_2N\sim \longrightarrow \sim NH\overset{O}{\overset{\|}{C}}NH\sim \quad (14.32)$$

isocyanate, amine, urea link

Polyurethanes have a wide range of applications. Polymers with relatively little cross-linking produce stretchable fibers (Spandex and Lycra) used for bathing suits. Polyurethane foams are used in furniture, mattresses, and car seats and as insulation in portable ice chests. Cross-linked polyurethanes form very tough surface coatings in paints and varnishes. The major component of the Jarvik-7 artificial heart is also a polyurethane.

Several commercially important step-growth polymers are based on reactions of formaldehyde. Bakelite, the oldest totally synthetic polymer, was invented by Leo Baekeland in 1907. It is prepared from phenol and formaldehyde. The polymer is highly cross-linked, with methylene groups *ortho* and/or *para* to the phenolic hydroxyl group.

$$\text{phenol} + CH_2{=}O \text{ (formaldehyde)} \xrightarrow[-H_2O]{H^+,\ \text{heat}} \text{segment of Bakelite} \quad (14.33)$$

segment of Bakelite

Bakelite is a **thermosetting polymer**. Heating leads to further cross-linking, producing a hard, infusible material. This process cannot be reversed, and once setting has occurred, the polymer cannot be melted. Bakelite is used for molded plastic parts, such as appliance handles, and for applications requiring light materials that can withstand high temperatures, such as missile nose cones.

A **thermosetting polymer**, when heated, forms additional cross-links and hardens irreversibly.

Urea and formaldehyde also form an important commercial polymer.

$$H_2N{-}\overset{O}{\overset{\|}{C}}{-}NH_2 + CH_2{=}O \xrightarrow[-H_2O]{\text{base}} \sim HN{-}\overset{O}{\overset{\|}{C}}{-}N(CH_2\sim){-}CH_2{-}NH{-}\overset{O}{\overset{\|}{C}}{-}NH{-}CH_2\sim \quad (14.34)$$

$$\sim CH_2{-}N{-}\underset{O}{\underset{\|}{C}}{-}NH{-}CH_2{-}NH{-}\overset{O}{\overset{\|}{C}}{-}NH\sim$$

urea, formaldehyde, urea-formaldehyde polymer

© David Young-Wolff/Photo Edit

Urea-formaldehyde polymer used in Formica countertops.

This kind of polymer is used in molded materials (electrical fittings and kitchenware), in laminates such as Formica, in plywood and particle board as adhesive, and in foams.

Epoxy resins are an important class of step-growth polymers used as adhesives for bonding to metal, glass, and ceramics. They are also used in paints because of their exceptional inertness, hardness, and flexibility. Two raw materials for the manufacture of epoxy resins are epichlorohydrin and bisphenol-A. Reaction of a mixture of these compounds with a base gives the "linear" epoxy resin shown in eq. 14.35. The remaining epoxides and hydroxyl groups can be used to form cross-links between the polymer chains, thus substantially increasing the molecular weight of the polymer. This is especially important when the end use is as a surface coating.

$$\underset{\text{epichlorohydrin}}{\overset{\text{O}}{H_2C{-}CH}{-}CH_2{-}Cl} + \underset{\text{bisphenol-A}}{HO{-}C_6H_4{-}C(CH_3)_2{-}C_6H_4{-}OH} \xrightarrow{\text{base}}$$

$$\overset{\text{O}}{H_2C{-}CH}{-}CH_2{-}\left[O{-}C_6H_4{-}C(CH_3)_2{-}C_6H_4{-}O{-}CH_2{-}CH(OH){-}CH_2\right]_n{-}O{-}C_6H_4{-}C(CH_3)_2{-}C_6H_4{-}O{-}CH_2{-}\overset{\text{O}}{CH{-}CH_2} \qquad \textbf{(14.35)}$$

We have barely touched on polymer chemistry. The field is vast and constantly developing. No doubt you will see many new types of polymers reach the marketplace during your lifetime.*

*For further reading, try Alper, J. and Nelson, G. L., *Polymeric Materials: Chemistry for the Future* (American Chemical Society, Washington, DC, 1989) or Munk, P., *Introduction to Macromolecular Science* (John Wiley, New York, 1989).

REACTION SUMMARY

1. Chain-Growth Polymerization (Secs. 14.2–14.7)

a. Free Radical Chain-Growth Polymerization (Sec. 14.2)

$$CH_2{=}CH{-}L \xrightarrow[\text{initiator}]{\text{free radical}} {-}\!\!\left(CH_2{-}\overset{\overset{\large L}{|}}{C}H\right)\!\!{-}_n$$

L = H, alkyl, aryl, electron-donating group (OAc), electron-withdrawing group (CN)

b. Cationic Chain-Growth Polymerization (Sec. 14.3)

$$CH_2{=}CH{-}L \xrightarrow[\text{initiator}]{\text{cationic}} {-}\!\!\left(CH_2{-}\overset{\overset{\large L}{|}}{C}H\right)\!\!{-}_n$$

L = carbocation stabilizing group (alkyl, Ph)

c. Anionic Chain-Growth Polymerization (Sec. 14.4)

$$CH_2{=}CH{-}L \xrightarrow[\text{initiator}]{\text{anionic}} {-}\!\!\left(CH_2{-}\overset{\overset{\large L}{|}}{C}H\right)\!\!{-}_n$$

L = carbanion stabilizing group (CN, Ph)

d. Ziegler–Natta Polymerization (Secs. 14.5 and 14.6)

(1) $$\underset{\text{propene}}{CH_2{=}CH{-}CH_3} \xrightarrow[TiCl_4]{Et_3Al} \underset{\text{isotactic polypropylene}}{{-}\!\!\left(CH_2{-}\overset{\overset{\large CH_3}{|}}{C}H\right)\!\!{-}_n}$$

(2) isoprene (CH_3) $\xrightarrow[TiCl_4]{Et_3Al}$ all Z polyisobutylene (CH_3), n

2. Step-Growth Polymerization (Secs. 14.8 and 14.9)

a. Preparation of Dacron (Sec. 14.8)

$$HO_2C{-}C_6H_4{-}CO_2H + HOCH_2CH_2OH \xrightarrow{\Delta} \underset{\text{Dacron}}{{-}\!\!\left(\overset{O}{\overset{\|}{C}}{-}C_6H_4{-}\overset{O}{\overset{\|}{C}}{-}OCH_2CH_2O\right)\!\!{-}_n} + H_2O$$

b. Preparation of Nylon (Secs. 14.1 and 14.8)

$$HO\overset{O}{\overset{\|}{C}}(CH_2)_4\overset{O}{\overset{\|}{C}}OH + H_2N(CH_2)_6NH_2 \xrightarrow{\Delta} \underset{\text{nylon-6,6}}{{-}\!\!\left(\overset{O}{\overset{\|}{C}}(CH_2)_4\overset{O}{\overset{\|}{C}}NH(CH_2)_6NH\right)\!\!{-}_n} + H_2O$$

MECHANISM SUMMARY

1. Free-Radical Chain-Growth Polymerization (Sec. 14.2)

Initiation: Initiator $\xrightarrow{\text{heat or light}}$ 2 In·

Propagation: In· + $H_2C{=}CH(L)$ ⟶ In—CH_2—$\dot{C}H(L)$ $\xrightarrow{H_2C=CH(L)}$ In—CH_2—CH(L)—CH_2—$\dot{C}H(L)$ and so on

Termination: ~~$\dot{C}H(L)$ + $\dot{C}H(L)$~~ $\xrightarrow[\text{coupling}]{\text{radical}}$ ~~HC(L)—CH(L)~~

~~$CH_2\dot{C}H(L)$ + ~CH(H)—$\dot{C}H(L)$ $\xrightarrow{\text{disproportionation}}$ ~~$CH_2CH_2(L)$ (alkane) + ~~CH=CH(L) (alkene)

2. Chain Transfer (Hydrogen Abstraction) (Sec. 14.2)

~~$\dot{C}H(L)$ + ~~C(H)(L)~~ ⟶ ~~$CH_2(L)$ + ~~$\dot{C}(L)$~~

3. Cationic Chain-Growth Polymerization (Sec. 14.3)

R^+ + $H_2C{=}CH(L)$ ⟶ RCH_2—$\overset{+}{C}H(L)$ $\xrightarrow{H_2C=CH(L)}$ RCH_2—CH(L)—CH_2—$\overset{+}{C}H(L)$ and so on

4. Anionic Chain-Growth Polymerization (Sec. 14.4)

R^- + $H_2C{=}CH(L)$ ⟶ RCH_2—$\overset{\cdot\cdot -}{C}H(L)$ $\xrightarrow{H_2C=CH(L)}$ RCH_2—CH(L)—CH_2—$\overset{\cdot\cdot -}{C}H(L)$ and so on

5. Step-Growth Polymerization (Secs. 14.8 and 14.9)

(illustrated for polyesters)

HO—C(=O)~~C(=O)—OH + HO~~OH $\xrightarrow{-H_2O}$ HO~~O (from diol)—C(=O)~~C(=O)—OH (from diacid) (two units) $\xrightarrow{\text{HO~~O—C(=O)~~C(=O)—OH}}$ four units

(two units) $\xrightarrow{\text{HO—C(=O)~~C(=O)—OH or HO~~OH}}$ three units

ADDITIONAL PROBLEMS

OWL Interactive versions of these problems are assignable in OWL.

Definitions

14.17 Define and give an example of the following terms:

a. homopolymer
b. copolymer
c. chain-growth polymerization
d. cross-linked polymer
e. thermoplastic
f. thermosetting
g. isotactic
h. atactic
i. chain transfer
j. step-growth polymer

Chain-Growth Polymers

14.18 Write out all the steps in the free-radical chain-growth polymerization of vinyl acetate ($CH_2{=}CH{-}O\overset{\overset{O}{\|}}{C}CH_3$).

14.19 The structure of poly(vinyl alcohol) is shown below. This polymer cannot be made by polymerizing its monomer. Why not? How could poly(vinyl alcohol) be prepared from poly(vinyl acetate)?

$$-(CH_2\overset{\overset{OH}{|}}{C}H)_n-$$

poly(vinyl alcohol)

14.20 Although propylene can be polymerized with a free-radical initiator, the molecular weight of the polymer is never very high by this method because chain transfer from the methyl group of propylene monomer keeps the chains short. Draw the structures of the products of this chain transfer. Explain why this chain transfer occurs so easily.

14.21 Draw the expected structure of "propylene tetramer," made by the acid-catalyzed polymerization of propene.

14.22 Propylene oxide can be converted to a polyether by anionic chain-growth polymerization. What is the structure of the polymer, and how is it formed?

14.23 Superglue is made by the on-the-spot polymerization of methyl α-cyanoacrylate (methyl 2-cyanopropenoate) with a little water or base. Draw the structure of the repeating unit. Why is this monomer so susceptible to anionic polymerization?

14.24 Draw the structural formula for a segment of alternating copolymer made from styrene and methyl methacrylate.

14.25 Provide the structures of the isomeric monomers from which poly(acrylamide) and poly(*N*-vinylformamide) are prepared.

$$-(CH_2\overset{\overset{CONH_2}{|}}{C}H)_n- \qquad -(CH_2\overset{\overset{NHCHO}{|}}{C}H)_n-$$

poly(acrylamide) poly(*N*-vinylformamide)

Natural and Synthetic Rubber

14.26 Draw six-unit chain segments of isotactic and syndiotactic polystyrene.

14.27 Can isobutylene polymerize in isotactic, syndiotactic, or atactic forms? Explain.

14.28 Explain how polyethylenes obtained by free-radical and by Ziegler–Natta polymerization differ in structure.

14.29 The ozonolysis of natural rubber gives levulinic aldehyde, $CH_3\overset{\overset{\displaystyle O}{\|}}{C}CH_2CH_2CH{=}O$. Explain how this result is consistent with natural rubber's formula (eq. 14.22).

14.30 Write out the steps in a mechanism that explains the free-radical copolymerization of 1,3-butadiene and styrene to give the synthetic rubber SBR (eq. 14.23).

14.31 Neoprene is a synthetic rubber that was invented more than 50 years ago at the DuPont Company. It is used to manufacture industrial hoses, drive belts, window gaskets, shoe soles, and packaging materials. Neoprene is a polymer of 2-chloro-1,3-butadiene. Assuming mainly 1,4-addition, draw the structure of the repeating unit in neoprene.

Step-Growth Polymers

14.32 Draw the repeating unit in the polymer that results from each of the following step-growth polymerizations:

a. $Cl{-}\overset{\overset{\displaystyle O}{\|}}{C}(CH_2)_8\overset{\overset{\displaystyle O}{\|}}{C}{-}Cl + H_2N(CH_2)_6NH_2 \longrightarrow$

b. $O{=}C{=}N{-}C_6H_4{-}CH_2{-}C_6H_4{-}N{=}C{=}O + HOCH_2CH_2OH \longrightarrow$ (para-substituted benzene rings)

c. $CH_3O\overset{\overset{\displaystyle O}{\|}}{C}(CH_2)_4\overset{\overset{\displaystyle O}{\|}}{C}OCH_3 + HOCH_2CH_2OH \xrightarrow{H^+}$

14.33 Lexan is a tough polycarbonate used to make molded articles and also for impact-resistant shields, including bullet-proof "glass." Lexan is made from diphenyl carbonate and bisphenol A. Draw the structure of its repeating unit.

$C_6H_5{-}O{-}\overset{\overset{\displaystyle O}{\|}}{C}{-}O{-}C_6H_5$

diphenyl carbonate

$HO{-}C_6H_4{-}\underset{\underset{\displaystyle CH_3}{|}}{\overset{\overset{\displaystyle CH_3}{|}}{C}}{-}C_6H_4{-}OH$

bisphenol A

14.34 Formaldehyde polymerizes in aqueous solution to give paraformaldehyde, $HO{-}(CH_2O)_n{-}H$. Although high molecular weights are achieved, the polymer "unzips" readily. However, if the polymer is treated with acetic anhydride, the resulting material (an important commercial polymer called Delrin) no longer "unzips." Explain the chemistry described.

14.35 Reaction of phthalic anhydride with glycerol gives a cross-linked polyester called a *glyptal resin.* Draw a partial structure, showing the cross-linking.

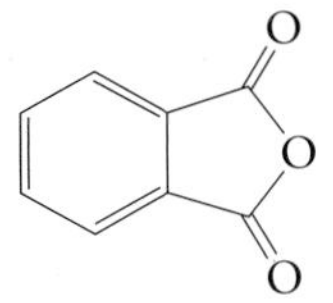

phthalic anhydride

14.36 Using eq. 14.33 as a model, draw a segment of the polymer that will be formed from formaldehyde and *p*-methylphenol. Will the polymer be cross-linked? Explain.

Puzzle Problems

14.37 Poly(ethylene naphthalate) (PEN) has a more rigid structure than PETE (page 428). This rigidity gives PEN increased strength and heat stability, features that are useful for food packaging. PEN is made from ethylene glycol and dimethyl naphthalene-2,6-dicarboxylate. Draw the structure of the repeating unit of PEN. Suggest a reason for the greater rigidity of PEN than PETE.

14.38 To make polymers commercially, it is necessary to make the monomers inexpensively and on a large scale. One commercial method for making hexamethylenediamine (for nylon-6,6) starts with the 1,4-addition of chlorine to 1,3-butadiene. Suggest a possibility for the remaining steps. (*Hint:* See Table 6.1 and eq. 11.12.)

14.39 As discussed in "A Word About . . . Degradable Polymers" (p. 426), some polymers can be naturally recycled in the environment. One can also imagine an application in drug delivery by the slow hydrolysis of a polymer to generate *in vivo* a pharmaceutical drug. Suggest the biologically relevant drug that is liberated from the aqueous hydrolysis of this polyester.

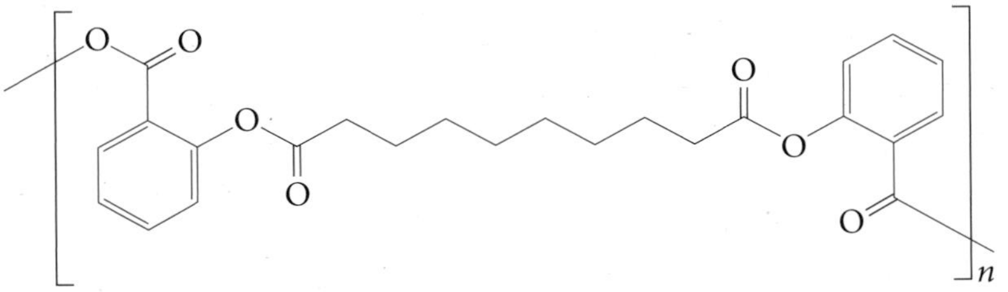

14.40 Water solutions of aqueous poly(vinyl alcohol) react with aqueous sodium borate to give a cross-linked polymer known as "slime." Given that alcohols react with sodium borate to give borate esters as shown below, propose a structure for the cross-link in "slime."

$$HO{-}\overset{\displaystyle OH}{\underset{\displaystyle OH}{B^-}}{-}OH\ Na^+ \xrightarrow{2\ ROH} RO{-}\overset{\displaystyle OH}{\underset{\displaystyle OH}{B^-}}{-}OR\ Na^+ + 2\ H_2O$$

14.41 As discussed in "A Word About . . . Polyacetylene and Conducting Polymers" (p. 421), thin films of all-*cis* or all-*trans* polyacetylene have different appearances, being coppery or silvery, respectively. Draw structures of the all-*cis* or all-*trans* isomers of polyacetylene as well as the alternating *cis–trans* isomer.

14.42 As discussed in "A Word About . . . Aramids, the Latest in Polyamides" (p. 427), Nomex is similar in structure to Kevlar (Figure 14.3), except that Nomex is prepared from *meta*- rather than *para*-oriented monomers. Draw a structure for Nomex.

15

Cocoa butter, which comes from cocoa beans, contains a large amount of glyceryl tristearate (tristearin), the same saturated fat found in beef.

© David Cain/Photo Researchers

Lipids and Detergents

Lipids (from the Greek *lipos,* fat) are constituents of plants or animals that are characterized by their solubility properties. In particular, *lipids are insoluble in water but are soluble in nonpolar organic solvents,* such as diethyl ether. Lipids can be extracted from cells and tissues by organic solvents. This solubility property distinguishes lipids from three other major classes of natural products—carbohydrates, proteins, and nucleic acids—which in general are *not* soluble in organic solvents.

Lipids may vary considerably in chemical structure, even though they have similar solubility properties. Some are esters and some are hydrocarbons; some are acyclic and others are cyclic, even polycyclic. We will take up each structural type separately.

OWL
Online homework for this chapter can be assigned in OWL, an online homework assessment tool.

15.1 Fats and Oils; Triesters of Glycerol

Fats and oils are familiar parts of daily life. Common fats include butter, lard, and the fatty portions of meat. Oils come mainly from plants and include corn, cottonseed, olive, peanut, and soybean oils. Although fats are solids and oils are liquids, they have the same basic organic structure. **Fats** and **oils** are triesters of glycerol and are called **triglycerides**. When we boil a fat or oil with alkali and acidify the resulting solution, we obtain glycerol and a mixture of **fatty acids**. The reaction is called saponification (Sec. 10.13).

Lipids are substances that are insoluble in water but soluble in nonpolar organic solvents.

Fats and **oils** are **triglycerides**, triesters of glycerol. **Fatty acids** are the acids obtained from saponification of fats and oils.

$$\begin{array}{l} CH_2-O-\overset{O}{\overset{\|}{C}}-R \\ | \\ CH-O-\overset{O}{\overset{\|}{C}}-R' \\ | \\ CH_2-O-\overset{O}{\overset{\|}{C}}-R'' \end{array} \xrightarrow[2.\ H^+]{1.\ NaOH,\ H_2O,\ heat} \begin{array}{l} CH_2OH \\ | \\ CHOH \\ | \\ CH_2OH \end{array} + \begin{array}{l} HO\overset{O}{\overset{\|}{C}}R \\ HO\overset{O}{\overset{\|}{C}}R' \\ HO\overset{O}{\overset{\|}{C}}R'' \end{array} \quad (15.1)$$

a triglyceride (fat or oil) — glycerol — three equivalents of fatty acids

The most common saturated and unsaturated fatty acids obtained in this way are listed in Table 15.1. Although exceptions are known, *most fatty acids are unbranched and contain an even number of carbon atoms.* If double bonds are present, they usually have the *cis* (or *Z*) configuration and are not conjugated.

EXAMPLE 15.1

Draw the structure of linoleic acid, showing the geometry at each double bond.

Solution Both double bonds have the *Z* configuration.

COOH

The preferred conformation is fully extended, with a staggered arrangement at each C—C single bond.

PROBLEM 15.1 Draw the structure of linolenic acid.

Table 15.1 ■ Common Acids Obtained from Fats

	Common name	Number of carbons	Structural formula	mp, °C
Saturated	lauric	12	$CH_3(CH_2)_{10}COOH$	44
	myristic	14	$CH_3(CH_2)_{12}COOH$	58
	palmitic	16	$CH_3(CH_2)_{14}COOH$	63
	stearic	18	$CH_3(CH_2)_{16}COOH$	70
	arachidic	20	$CH_3(CH_2)_{18}COOH$	77
Unsaturated	oleic	18	$CH_3(CH_2)_7CH{=}CH(CH_2)_7COOH$ (*cis*)	13
	linoleic	18	$CH_3(CH_2)_4CH{=}CHCH_2CH{=}CH(CH_2)_7COOH$ (*cis, cis*)	−5
	linolenic	18	$CH_3CH_2CH{=}CHCH_2CH{=}CHCH_2CH{=}CH(CH_2)_7COOH$ (all *cis*)	−11

There are two types of triglycerides: **simple triglycerides**, in which all three fatty acids are identical, and **mixed triglycerides**.

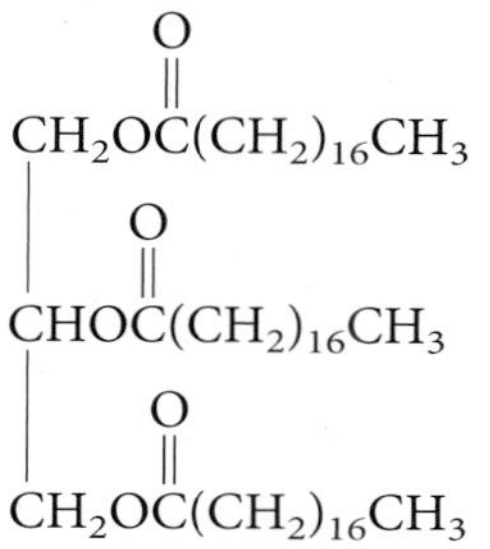

$$\begin{array}{l} CH_2OC(=O)(CH_2)_{16}CH_3 \\ | \\ CHOC(=O)(CH_2)_{16}CH_3 \\ | \\ CH_2OC(=O)(CH_2)_{16}CH_3 \end{array}$$

a simple triglyceride
(glyceryl tristearate or tristearin)

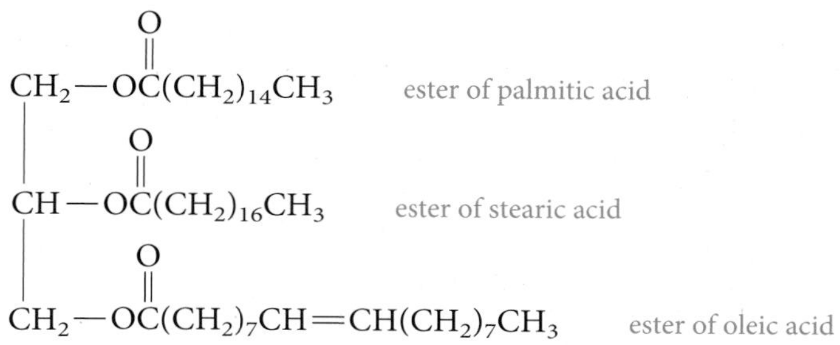

$$\begin{array}{ll} CH_2{-}OC(=O)(CH_2)_{14}CH_3 & \text{ester of palmitic acid} \\ | & \\ CH{-}OC(=O)(CH_2)_{16}CH_3 & \text{ester of stearic acid} \\ | & \\ CH_2{-}OC(=O)(CH_2)_7CH{=}CH(CH_2)_7CH_3 & \text{ester of oleic acid} \end{array}$$

a mixed triglyceride
(glyceryl palmitostearoöleate)

EXAMPLE 15.2

Draw the structure of glyceryl stearopalmitoöleate, an isomer of the mixed triglyceride shown above.

Solution

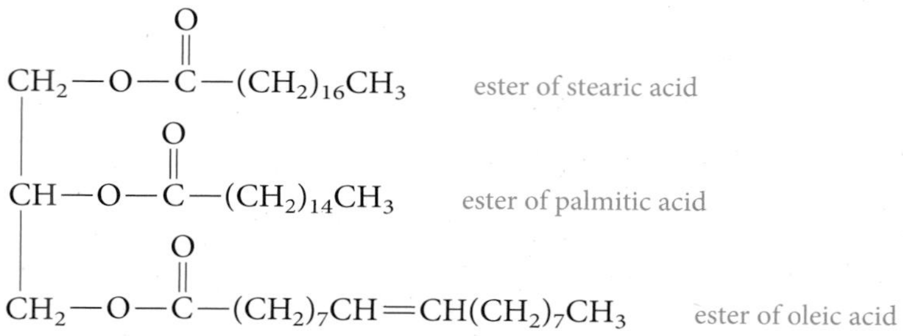

$$\begin{array}{ll} CH_2{-}O{-}C(=O){-}(CH_2)_{16}CH_3 & \text{ester of stearic acid} \\ | & \\ CH{-}O{-}C(=O){-}(CH_2)_{14}CH_3 & \text{ester of palmitic acid} \\ | & \\ CH_2{-}O{-}C(=O){-}(CH_2)_7CH{=}CH(CH_2)_7CH_3 & \text{ester of oleic acid} \end{array}$$

Notice that glyceryl palmitostearoöleate and glyceryl stearopalmitoöleate give identical saponification products.

PROBLEM 15.2 Draw the structure for

a. glyceryl trilaurate

b. glyceryl palmitoöleostearate

PROBLEM 15.3 What saponification products would be obtained from each of the triglycerides in Problem 15.2?

In general, a particular fat or oil consists not of a single triglyceride but of a complex mixture of triglycerides. For this reason, the composition of a fat or oil is usually expressed in terms of the percentages of the various acids obtained from it by saponification (Table 15.2). Some fats and oils give mainly one or two acids, with only minor amounts of others. Olive oil, for example, gives 83% oleic acid. Palm oil gives 43% palmitic acid and 43% oleic acid, with lesser amounts of stearic and linoleic acids. Butterfat, on the other hand, gives at least 14 different fatty acids on hydrolysis and is somewhat exceptional in that about 9% of these acids have fewer than 10 carbon atoms.

PROBLEM 15.4 From the data in Table 15.2, what can you say in general about the ratio of saturated to unsaturated acids in fats and oils?

Table 15.2 Fatty Acid Composition of Some Fats and Oils (Approximate)

	Saturated acids (%)					Unsaturated acids (%)	
Source	C_{10} and less	C_{12} lauric	C_{14} myristic	C_{16} palmitic	C_{18} stearic	C_{18} oleic	C_{18} linoleic
Animal Fats:							
Butter	12	3	12	28	10	26	2
Lard	—	—	1	28	14	46	5
Beef tallow	—	0.2	3	28	24	40	2
Human	—	1	3	25	8	46	10
Vegetable Oils:							
Olive	—	—	1	5	2	83	7
Palm	—	—	2	43	2	43	8
Corn	—	—	1	10	2	40	40
Peanut	—	—	—	8	4	60	25

What is it that makes some triglycerides solids (fats) and others liquids (oils)? The distinction is clear from their composition. *Oils contain a much higher percentage of unsaturated fatty acids than do fats.* For example, most vegetable oils (such as corn oil or soybean oil) give about 80% unsaturated acids on hydrolysis. For fats (such as beef tallow), the figure is much lower, just a little over 50%.

Although, in general, fats come from animal sources and oils come from vegetable sources, this is not always so. For example, fish oils are high in unsaturated fatty acids. The unsaturated fatty acid contents of sardine oil and cod-liver oil are 77% and 84%, respectively. Likewise, not all fats come from animal sources. Cocoa butter (see opening photo), with a fatty acid composition of 24% palmitic acid and 35% stearic acid, is a solid at room temperature.

Table 15.1 shows that the melting points of unsaturated fatty acids are appreciably lower than those of saturated acids. Compare, for example, the melting points of stearic and oleic acids, which differ structurally by only one double bond. The same difference applies to triglycerides: *The more double bonds in the fatty acid portion of the triester, the lower its melting point.*

The reason for the effect of saturation or unsaturation on the melting point becomes apparent when we examine space-filling models. Figure 15.1 shows a model

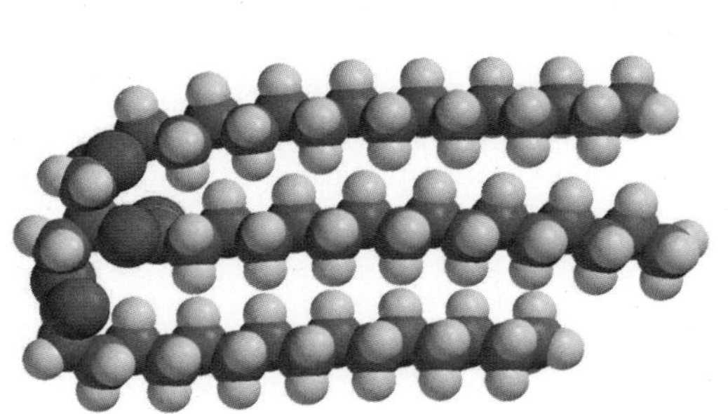

$CH_2OC(=O)$
$CHOC(=O)$
$CH_2OC(=O)$

■ Figure 15.1
Space-filling and schematic models of glyceryl tripalmitate.

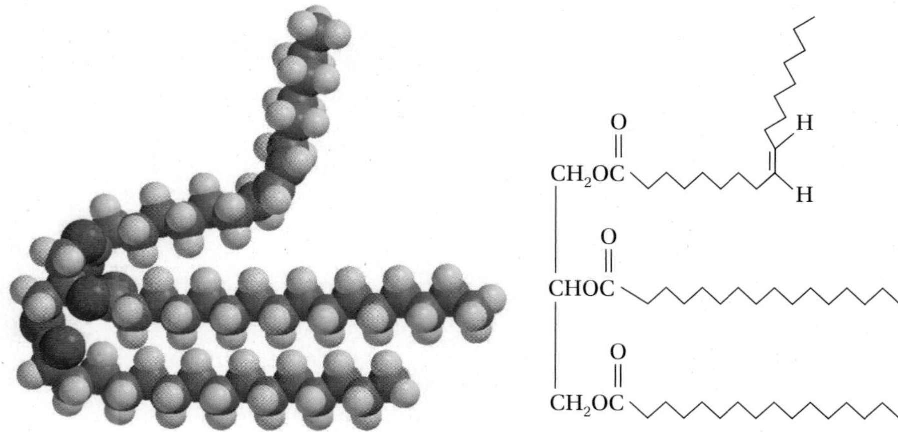

Figure 15.2
Space-filling and schematic models of glyceryl dipalmitoöleate.

of a fully saturated triglyceride. The long, saturated chains are able to adopt fully extended, staggered conformations. Therefore, they can pack together fairly regularly, as in a crystal. Consequently, saturated triglycerides are usually solids at room temperature.

The result of introducing just one *cis* double bond in one of the chains is shown in Figure 15.2. Clearly, the chains in this kind of molecule (and the molecules themselves, when many are close to one another) cannot align nicely in a crystalline array. Thus, the substance remains a liquid. The more double bonds in the fatty acid chain, the more disordered the structure and the lower the melting point.

15.2 Hydrogenation of Vegetable Oils

Vegetable oils, which are highly unsaturated, are converted into solid vegetable fats, such as Crisco, by catalytically hydrogenating some or all of the double bonds. This process, called **hardening**, is illustrated by the hydrogenation of glyceryl trioleate to glyceryl tristearate.

Hardening is the process of converting oils to fats by catalytic hydrogenation of double bonds. **Margarine** is made by hydrogenating oils.

$$\begin{array}{l} CH_2OC(=O)(CH_2)_7CH{=}CH(CH_2)_7CH_3 \\ | \\ CHOC(=O)(CH_2)_7CH{=}CH(CH_2)_7CH_3 \\ | \\ CH_2OC(=O)(CH_2)_7CH{=}CH(CH_2)_7CH_3 \end{array} \xrightarrow[\text{Ni catalyst, heat}]{3\,H_2} \begin{array}{l} CH_2OC(=O)(CH_2)_{16}CH_3 \\ | \\ CHOC(=O)(CH_2)_{16}CH_3 \\ | \\ CH_2OC(=O)(CH_2)_{16}CH_3 \end{array} \quad (15.2)$$

glyceryl trioleate (triolein) (mp −17°C) → glyceryl tristearate (tristearin) (mp 55°C)

Hydrogenation of corn oil, and other vegetable oils, provides margarine.

Margarine is made by hydrogenating cottonseed, soybean, peanut, or corn oil until the desired butter-like consistency is obtained. The product may be churned with milk and artificially colored to mimic butter's flavor and appearance.

15.3 Saponification of Fats and Oils; Soap

When a fat or oil is heated with alkali (e.g., NaOH), the ester is converted to glycerol and the salts of fatty acids. The reaction is illustrated here with the saponification of glyceryl tripalmitate.

$$\begin{array}{l} \overset{\displaystyle O}{\|} \\ CH_2OC(CH_2)_{14}CH_3 \\ |\quad \overset{\displaystyle O}{\|} \\ CHOC(CH_2)_{14}CH_3 \\ |\quad \overset{\displaystyle O}{\|} \\ CH_2OC(CH_2)_{14}CH_3 \end{array} + 3\,Na^{+}\,{}^{-}OH \xrightarrow{\text{heat}} \begin{array}{l} CH_2OH \\ | \\ CHOH \\ | \\ CH_2OH \end{array} + 3\,CH_3(CH_2)_{14}CO_2{}^{-}Na^{+} \tag{15.3}$$

glyceryl tripalmitate (tripalmitin) (from palm oil) — glycerol — sodium palmitate (a soap)

The salts (usually sodium) of long-chain fatty acids are **soaps**.

Soaps are the salts of long-chain fatty acids.

The conversion of animal fats (e.g., goat tallow) into soap by heating with wood ashes (which are alkaline) is one of the oldest of chemical processes. Soap has been produced for at least 2,300 years, having been known to the ancient Celts and Romans. Yet, as recently as the sixteenth and seventeenth centuries, soap was still a rather rare substance, used mainly in medicine. But by the nineteenth century, soap had come into such widespread use that the German organic chemist Justus von Liebig was led to remark that the quantity of soap consumed by a nation was an accurate measure of its wealth and civilization. At present, annual world production of ordinary soaps (not including synthetic detergents) is well over 12 billion pounds.

Soaps are made by either a batch process or a continuous process. In the batch process, the fat or oil is heated with a slight excess of alkali (e.g., NaOH) in an open kettle. When saponification is complete, salt is added to precipitate the soap as thick curds. The water layer, which contains salt, glycerol, and excess alkali, is drawn off, and the glycerol is recovered by distillation. The crude soap curds, which contain some salt, alkali, and glycerol as impurities, are purified by boiling with water and reprecipitating with salt several times. Finally, the curds are boiled with enough water to form a smooth mixture that, on standing, gives a homogeneous upper layer of soap. This soap may be sold without further processing, as a cheap industrial soap. Various fillers, such as sand or pumice, may be added, to make scouring soaps. Other treatments transform the crude soap to toilet soaps, powdered or flaked soaps, medicated or perfumed soaps, laundry soaps, liquid soaps, or (by blowing air in) floating soaps.

In the continuous process, which is more common today, the fat or oil is hydrolyzed by water at high temperatures and pressures in the presence of a catalyst, usually a zinc soap. The fat or oil and the water are introduced continuously into opposite ends of a larger reactor, and the fatty acids and glycerol are removed as formed, by distillation. The acids are then carefully neutralized with an appropriate amount of alkali to make soap.

■ **Figure 15.3**

The structure of (a) sodium stearate and (b) an electrostatic potential map of the stearate anion that illustrates the polar nature of the carboxylate ion and nonpolar nature of the hydrocarbon chain.

$$\underbrace{CH_3CH_2CH_2CH_2CH_2CH_2CH_2CH_2CH_2CH_2CH_2CH_2CH_2CH_2CH_2CH_2CH_2}_{\text{nonpolar, lipophilic}}C\begin{matrix} \diagup\!\!\diagup O \\ \diagdown \underbrace{O^-Na^+}_{\text{polar, hydrophilic}} \end{matrix}$$

(a)

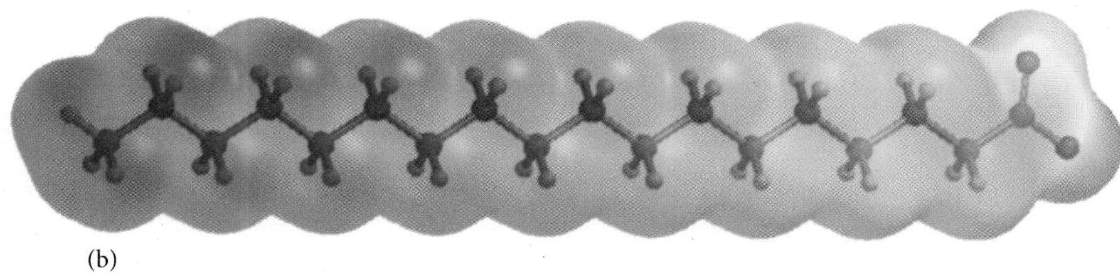

(b)

15.4 How Do Soaps Work?

Most dirt on clothing or skin adheres to a thin film of oil. If the oil film can be removed, the dirt particles can be washed away. A soap molecule consists of a long, hydrocarbon-like chain of carbon atoms with a highly polar or ionic group at one end (Figure 15.3). The carbon chain is **lipophilic** (attracted to or soluble in fats and oils), and the polar end is **hydrophilic** (attracted to or soluble in water). In a sense, soap molecules are "schizophrenic," having two different "personalities." Let us see what happens when we add soap to water.

Soap molecules form globular aggregates in water called **micelles**, with their polar **hydrophilic** heads facing the water and their nonpolar **lipophilic** tails in the center.

When soap is shaken with water, it forms a colloidal dispersion—not a true solution. These soap solutions contain aggregates of soap molecules called **micelles**. The nonpolar, or lipophilic, carbon chains are directed toward the center of the micelle. The polar, or hydrophilic, ends of the molecule form the "surface" of the micelle that is presented to the water (Figure 15.4). In ordinary soaps, the outer part of each micelle is negatively charged, and the positively charged sodium ions congregate near the periphery of each micelle.

In acting to remove dirt, soap molecules surround and emulsify the droplets of oil or grease. The lipophilic "tails" of the soap molecules dissolve in the oil. The hydrophilic ends extend out of the oil droplet toward the water. In this way, the oil droplets are stabilized in the water solution because the negative surface charge of the droplets prevents their coalescence (as shown in Figure 15.5).

Another striking property of soap solutions is their unusually low surface tension, which gives a soap solution more "wetting" power than plain water. As a consequence,

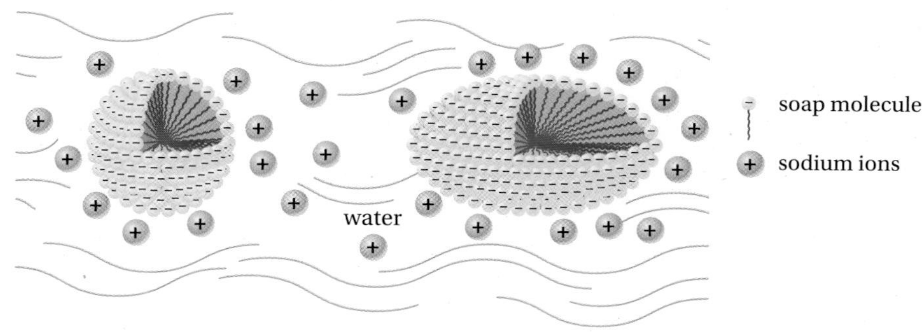

■ **Figure 15.4**

Soap molecules form micelles when "dissolved" in water.

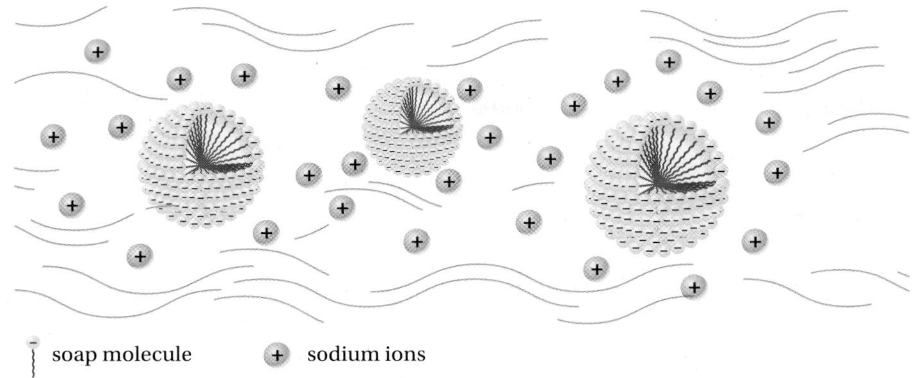

▰ **Figure 15.5**
Oil droplets (shown in orange) become emulsified by soap molecules.

soaps belong to a class of substances called **surfactants**. A combination of the emulsifying power and the surface action of soap solutions enables them to detach dirt, grease, and oil particles from the surface being cleaned and to emulsify them so that they can be washed away. These same principles of cleansing action apply to synthetic detergents.

Surfactants contain polar and nonpolar parts and therefore act at the surfaces where different substances meet.

15.5 Synthetic Detergents (Syndets)

Annual world production of **synthetic detergents** (sometimes called **syndets**) now exceeds that of ordinary soaps. Syndets evolved in response to two problems with ordinary soaps. First, being salts of weak acids, *soaps give somewhat alkaline solutions in water.* This is due to partial hydrolysis of the sodium salts.

Syndets or **synthetic detergents** are soap-like molecules designed to work well in hard water and produce neutral solutions.

$$\underset{\text{soap}}{R-\overset{\overset{\displaystyle O}{\|}}{C}-O^-Na^+} + H-OH \rightleftharpoons R-\overset{\overset{\displaystyle O}{\|}}{C}-OH + \underset{\text{alkali}}{Na^{+-}OH} \quad \textbf{(15.4)}$$

Alkali can be harmful to certain fabrics. Yet ordinary soaps cannot function well in acid because the long-chain fatty acid will precipitate from the solution as a scum. For example, sodium stearate, a typical soap, is destroyed by conversion to stearic acid on acidification.

$$\underset{\substack{\text{sodium stearate}\\ \text{(soluble)}}}{C_{17}H_{35}C(=O)O^-Na^+} + H^+Cl^- \longrightarrow \underset{\substack{\text{stearic acid}\\ \text{(insoluble)}}}{C_{17}H_{35}C(=O)OH}\downarrow + Na^+Cl^- \quad \textbf{(15.5)}$$

The second problem with ordinary soaps is that *they form insoluble salts with the calcium, magnesium, or ferric ions that may be present in hard water.*

$$2\,\underset{\substack{\text{sodium stearate}\\ \text{(soluble)}}}{C_{17}H_{35}C(=O)O^-Na^+} + Ca^{2+} \longrightarrow \underset{\substack{\text{calcium stearate}\\ \text{(insoluble)}}}{(C_{17}H_{35}COO^-)_2Ca^{2+}}\downarrow + 2\,Na^+ \quad \textbf{(15.6)}$$

The insoluble salts are responsible for the "rings" around bathtubs or collars and for films that dull the look of clothing and hair.

These problems with ordinary soaps have been solved or diminished in several ways. For example, water can be "softened," either municipally or in individual households, to remove the offending calcium or magnesium ions.* In softened water, those ions are replaced by sodium ions. If this water is also used for drinking, however, it may cause health problems, especially for people who have to limit their sodium ion intake.

Phosphates can also be added to soaps. Phosphates form soluble complexes with metal ions, thus keeping these ions from forming insoluble salts with the soap. But widespread use of phosphates in the past has created environmental problems. Because of their use in detergents, tremendous quantities of phosphates eventually found their way into lakes, rivers, and streams. Since they are fertilizers, these phosphates stimulated plant growth to such an extent that the plants exhausted the dissolved oxygen in the water, in turn causing fish to die. Phosphates are still used in detergents, but their use is now limited by law to levels that are unlikely to be harmful.

Another way to eliminate the problems associated with ordinary soaps is to design more effective detergents. These syndets must have several features. Like ordinary soaps, they must have a long lipophilic chain and a polar, or ionic, hydrophilic end. However, the polar end should not form insoluble salts with metal ions present in hard water and should not affect the acidity of water.

Long-chain alcohols are produced by reductive hydrolysis or **hydrogenolysis** of fats.

The first syndets were sodium salts of alkyl hydrogen sulfates. A long-chain alcohol was prepared by the **hydrogenolysis** of a fat or oil. For example, glyceryl trilaurate can be reduced to a mixture of 1-dodecanol (from the acid part of the fat) and glycerol. Because glycerol is water soluble, whereas the long-chain alcohol is not, the two hydrogenolysis products can be easily separated. The long-chain alcohol is next treated with sulfuric acid to make the alkyl hydrogen sulfate ($R{-}OSO_2OH$), which is then neutralized with base (eq. 15.8).

$$\begin{array}{l} CH_3(CH_2)_{10}\overset{O}{\overset{\|}{C}}{-}OCH_2 \\ CH_3(CH_2)_{10}\overset{O}{\overset{\|}{C}}{-}OCH \\ CH_3(CH_2)_{10}\overset{O}{\overset{\|}{C}}{-}OCH_2 \\ \text{glyceryl trilaurate} \end{array} + 6\,H_2 \xrightarrow[\text{heat, pressure}]{\text{copper chromite}} \underset{\substack{\text{1-dodecanol} \\ \text{(lauryl alcohol)}}}{3\,CH_3(CH_2)_{10}CH_2OH} + \begin{array}{c} HOCH_2 \\ | \\ HOCH \\ | \\ HOCH_2 \\ \text{glycerol} \end{array} \qquad (15.7)$$

$$\underset{\text{lauryl alcohol}}{CH_3(CH_2)_{10}CH_2OH} + \underset{\text{sulfuric acid}}{HOSO_2OH} \longrightarrow \underset{\text{lauryl hydrogen sulfate}}{CH_3(CH_2)_{10}CH_2OSO_2OH} + H_2O$$

$$\downarrow \text{NaOH}$$

$$\underbrace{\overbrace{CH_3CH_2CH_2CH_2CH_2CH_2CH_2CH_2CH_2CH_2CH_2CH_2}^{\text{lipophilic chain}}{-}O{-}\underbrace{\overset{O}{\overset{\|}{\underset{\underset{O}{\|}}{S}}}{-}O^-Na^+}_{\text{polar, hydrophilic end}}}_{\text{sodium lauryl sulfate}} + H_2O \qquad (15.8)$$

*See pages 417 and 447 for the chemistry of water softeners.

A WORD ABOUT... Commercial Detergents

© Tom Pantages

Some commercial detergents.

The design and manufacture of a successful commercial detergent is a highly sophisticated process because, more and more, detergents are developed for a specific purpose and market. No one formulation fits every use. An inventory of the cleaning materials in a typical household might disclose half a dozen or more products designed to be most suitable for a specific job—to clean clothes, dishes, floors, automobiles, the human body (special hand and bath soaps and hair shampoo), and so on.

We have described the structures of common surface active agents (*surfactants*). As we saw, these are organic molecules with a lipophilic portion and a polar portion, whose function is to emulsify and disperse oil and grease and to lower the surface tension of water to aid in wetting clothes, dishes, or other surfaces. Surfactants form an important part of all commercial detergents, but depending on the end use, these detergents will also contain a variety of other components, such as builders, bleaches, fabric softeners, enzymes, antiredeposition agents, and optical brighteners.

Builders are perhaps the second most important component of commercial detergents. They are added to remove calcium and magnesium ions ("hardness") from wash water. They may do this by chelation (forming a complex) or by exchanging these ions for sodium ions. Builders also raise the pH to aid oil emulsification and to buffer against pH changes. The most common builder is **sodium tripolyphosphate** ($5Na^+ \; P_3O_{10}{}^{5-}$), but because waste phosphates can be environmental pollutants, the amount that may be used is restricted by law. Recently, sodium citrate, sodium carbonate, and sodium silicate have begun to replace sodium tripolyphosphate as builders. Zeolites (sodium aluminosilicates) are used as ion exchangers, especially for calcium ions.

Bleaches that contain chlorine (hypochlorites) have been used for years, sometimes separately and sometimes incorporated into the detergent. The chlorine acts as an oxidizing agent. Problems with residual chlorine and its odor have led more recently to the use of peroxides, especially sodium perborate ($NaBO_3$). Hydrolysis of the perborate produces hydrogen peroxide (H_2O_2), which does the bleaching.

Fabric softeners are cationic surfactants that give clothes a soft feel. They may be incorporated in a laundry detergent or added to the wash separately.

Enzymes are added to detergents to deal with specific types of clothing stains. *Proteases* hydrolyze protein-based stains, and *amylases* convert starch-based stains to soluble materials that are easily washed away.

Antiredeposition agents are compounds added to laundry detergents to prevent the redeposition of soil on clothes. The most common examples are cellulose ethers or esters.

Optical brighteners are usually organic dyes that absorb ultraviolet light and fluoresce blue. In this way, a yellowed appearance of some "white" clothing can be avoided. In the mid-nineteenth century, "bluing" (a blue dye, such as ultramarine) was used for this purpose and is sometimes still used separately from the detergent. Brighteners that are incorporated in the detergent formulation are usually aromatic or heteroaromatic amines.

In addition to all of these components, a modern commercial detergent may also contain *antistatic agents* (cationic surfactants added to reduce static cling); *hydrotropes* (compounds added to liquid detergents to hold less-soluble surfactants and other compounds in solution); and, of course, *fragrances* and *perfumes* and inert fillers and formulation aids that keep powdered detergents free flowing.

Next time you do the laundry, bathe, or wash your hair, think about the complexity and diversity of the product you are using. It is not just soap—there is a lot of science, and especially chemistry, in each product.

See Problems 15.16, 15.17, 15.18, and 15.19.

Sodium lauryl sulfate is an excellent detergent. Because it is a salt of a *strong* acid, its solutions are nearly neutral. Its calcium and magnesium salts do not precipitate from solution, so it is effective in hard as well as soft water. Unfortunately, its supply is too limited to meet the demand, so the need for other syndets persists.

At present, the most widely used syndets are straight-chain alkylbenzenesulfonates. They are made in three steps, shown in eq. 15.9. A straight-chain alkene with 10 to 14 carbons is treated with benzene and a Friedel–Crafts catalyst ($AlCl_3$ or HF) to form an alkylbenzene. Sulfonation and neutralization of the sulfonic acid with base complete the process.

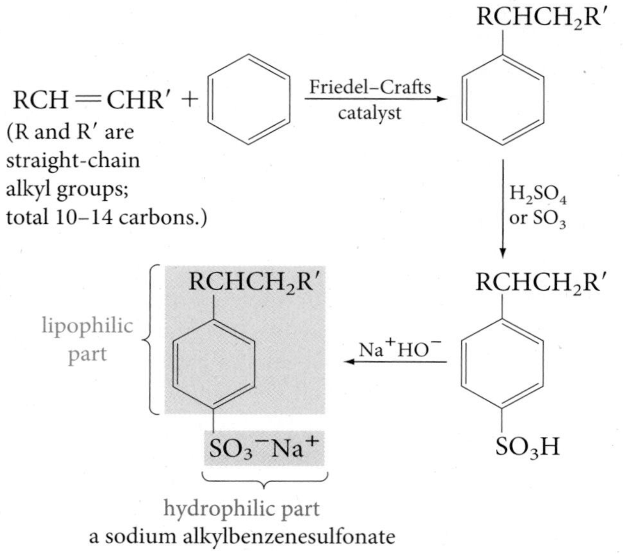

$$RCH{=}CHR' + C_6H_6 \xrightarrow[\text{catalyst}]{\text{Friedel–Crafts}} RCHCH_2R'\text{-}C_6H_5 \xrightarrow{H_2SO_4 \text{ or } SO_3} RCHCH_2R'\text{-}C_6H_4\text{-}SO_3H \xrightarrow{Na^+HO^-} RCHCH_2R'\text{-}C_6H_4\text{-}SO_3^-Na^+ \quad (15.9)$$

(R and R′ are straight-chain alkyl groups; total 10–14 carbons.)

lipophilic part: $RCHCH_2R'\text{-}C_6H_4$; hydrophilic part: $SO_3^-Na^+$

a sodium alkylbenzenesulfonate

It is important that the alkyl chain in these detergents have no branches so that they are fully biodegradable and do not accumulate in the environment.

The soaps and syndets we have mentioned so far are *anionic* detergents; they have a lipophilic chain with a negatively charged polar end. But there are also *cationic, neutral,* and even *amphoteric* detergents, in which the polar portion of the molecule is positive, neutral, or dipolar, respectively. Here are some examples:

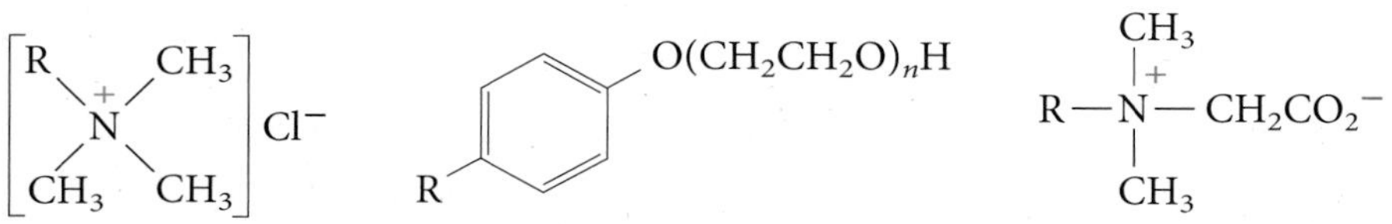

$$[R\overset{+}{N}(CH_3)_3]\,Cl^- \qquad R\text{-}C_6H_4\text{-}O(CH_2CH_2O)_nH \qquad R\text{-}\overset{+}{N}(CH_3)_2\text{-}CH_2CO_2^-$$

cationic detergent ($R = C_{16-18}$)

neutral detergent ($R = C_{8-12}$; $n = 5–10$)

amphoteric detergent ($R = C_{12-18}$)

The essential features of all of these detergents are a lipophilic portion with a hydrocarbon chain of appropriate length to dissolve in oil or grease droplets and a polar portion to create a micelle surface that is attractive to water.

EXAMPLE 15.3

Design a synthesis for the cationic detergent shown above ($R = C_{16}$).

Solution

$$CH_3(CH_2)_{14}CH_2Cl + (CH_3)_3N\!: \longrightarrow \left[CH_3(CH_2)_{14}CH_2\text{—}\overset{+}{N}(CH_3)_2\text{—}CH_3\ \ Cl^-\right]$$

The reaction is an S_N2 displacement (see entry 9 of Table 6.1 and eqs. 11.6 and 11.26).

PROBLEM 15.5 Design a synthesis for the neutral detergent shown on the preceding page ($R = C_8$, $n = 5$), starting with *p*-octylphenol and ethylene oxide (review Sec. 8.8).

PROBLEM 15.6 Design a synthesis for the amphoteric detergent shown on the preceding page, using an S_N2 displacement with $CH_3(CH_2)_{14}CH_2\ddot{N}(CH_3)_2$ as the nucleophile and an appropriate halide.

15.6 Phospholipids

Phospholipids constitute about 40% of cell membranes, the remaining 60% being proteins. Phospholipids are related structurally to fats and oils, except that one of the three ester groups is replaced by a phosphatidylamine.

Phospholipids are like fats except that one ester group is replaced by a phosphatidylamine. Phospholipids form **bilayers** in membranes.

$$CH_3CH_2CH_2CH_2CH_2CH_2CH_2CH_2CH_2CH_2CH_2CH_2CH_2CH_2CH_2\overset{O}{\overset{\|}{C}}-O-CH_2$$

$$\underbrace{CH_3CH_2CH_2CH_2CH_2CH_2CH_2CH_2CH_2CH_2CH_2CH_2CH_2CH_2CH_2\overset{O}{\overset{\|}{C}}}_{\text{nonpolar tail}}-O-CH$$

$$CH_2-O-\underbrace{\overset{O^-}{\overset{|}{\underset{\underset{O}{\|}}{P}}}-OCH_2CH_2\overset{+}{N}H_3}_{\text{polar head}}$$

a phospholipid

The fatty acid portions are usually palmityl, stearyl, or oleyl. The structure shown is a cephalin; the three protons on the nitrogen are replaced by methyl groups in the lecithins. Both types of phospholipids are widely distributed in the body, especially in the brain and nerve tissues.

Phospholipids arrange themselves in **bilayers** in membranes, with the two hydrocarbon "tails" pointing in and the polar phosphatidylamine ends constituting the membrane surface, as shown in Figure 15.6. Membranes play a key role in biology, controlling diffusion of substances into and out of cells.

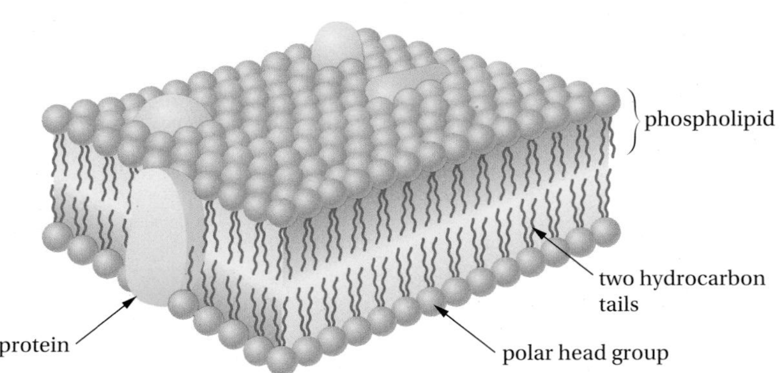

▪ **Figure 15.6** Schematic diagram of a bilayered cell membrane.

A WORD ABOUT... Prostaglandins, Aspirin, and Pain

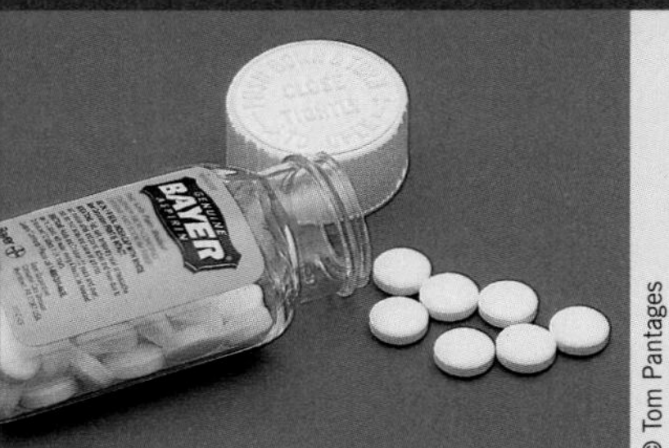

© Tom Pantages

Aspirin, the world's most widely used painkiller.

Aspirin is the most widely used painkiller in the world.* It has been used to treat toothaches, headaches, arthritis, and other painful maladies for 100 years. Only recently, however, have we come to understand how aspirin works. It is an interesting story with a direct connection to the biosynthesis of **prostaglandins**.

CO_2H O O CH_3

aspirin
(acetylsalicylic acid)

Prostaglandins are made in cells by a series of enzyme-catalyzed reactions. A critical step is the reaction of **arachidonic acid** (eq. 15.10) with oxygen to produce PGG_2, a highly reactive peroxide-containing prostaglandin. In this reaction, promoted by an enzyme called **cyclooxygenase**, carbon-9 and carbon-11 of arachidonic acid bond to oxygen while carbon-8 and carbon-12 bond to one another to form the cyclopentane ring that is characteristic of all prostaglandins. Once PGG_2 is formed, it is reduced to PGH_2, which is subsequently converted to PGE_2 (eq. 15.10) and a host of other prostaglandins.

Many of these prostaglandins play important regulatory roles in life processes such as digestion, blood circulation, and reproduction. It has also been established that injured cells in the body produce and release prostaglandins that, in this situation, give rise to the phenomena of inflammation and pain. This is where the connection with aspirin begins.

In 1969, Dr. John Vane and his collaborators at the Royal College of Surgeons in London discovered that aspirin inhibited production of prostaglandins by injured tissue. Eventually it was shown that this happens because aspirin binds to cyclooxygenase, inhibiting the conversion of arachidonic acid to PGG_2. This stops the production of prostaglandins, resulting in a reduction of inflammation and pain. Compounds that keep enzymes from playing their roles as catalysts are known as **enzyme inhibitors**, and the synthesis of inhibitors of specific enzymes now plays an important role in the discovery of new drugs.

arachidonic acid

O_2 | cyclooxygenase

9 8 O O CO_2H 11 12 15 OR

PGG_2 R = OOH
PGH_2 R = H

PGE_2 and other prostaglandins

One side effect of taking aspirin is stomach irritation, and this is also a result of inhibition of prostaglandin synthesis. It turns out that PGE_2 protects the cells of the stomach wall by stimulating formation of a protective layer of mucous. PGE_2 also helps regulate acid levels in the stomach, and in its absence, hydrochloric acid production rises. Thus, it is understandable how inhibition of PGE_2 release in the stomach by orally ingested aspirin could lead to an upset stomach. Of course, this does not keep aspirin from being useful. As is the case with many elixirs, when taken in moderation, aspirin is very useful, but when taken in excess, it can cause problems.**

The discovery of how aspirin works has inspired medicinal chemists to search for new cyclooxygenase inhibitors that might behave as painkillers.† Celebrex is a recently developed prescription drug now sold for this purpose.

See Problems 15.21 and 15.22.

*For more information on painkillers (analgesics), see "A Word About . . . Morphine and Other Nitrogen-Containing Drugs" in Chapter 13.

**In 1982, John Vane, Bengt Samuelsson, and Sune Bergstrom received the Nobel Prize in physiology or medicine for their research on the mechanism of action of aspirin and the biosynthesis of prostaglandins.

†For a discussion of the issues involved in developing cyclooxygenase inhibitors, see *Science* **1998**, *280*, 1191–1192.

15.7 Prostaglandins, Leukotrienes, and Lipoxins

Prostaglandins are a group of compounds related to the unsaturated fatty acids. They were discovered in the 1930s, when it was found that human semen contained substances that could stimulate smooth muscle tissue, such as uterine muscle, to contract. On the assumption that these substances came from the prostate gland, they were named prostaglandins. We now know that prostaglandins are widely distributed in almost all human tissues, that they are biologically active in minute concentration, and that they have various effects on fat metabolism, heart rate, and blood pressure.

Prostaglandins are biosynthesized from the 20-carbon unsaturated fatty acid, **arachidonic acid.** **Leukotrienes** and **lipoxins** are acyclic derivatives of prostaglandins.

Prostaglandins have 20 carbon atoms. They are synthesized in the body by oxidation and cyclization of the 20-carbon unsaturated fatty acid **arachidonic acid**. Carbon-8 through carbon-12 of the chain are looped to form a cyclopentane ring, and an oxygen function (carbonyl or hydroxyl group) is always present at carbon-9. Various numbers of double bonds or hydroxyl groups may also be present elsewhere in the structure.

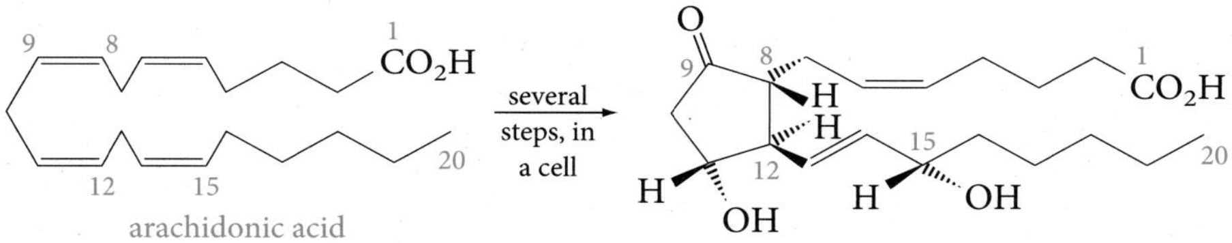

9 8 1 CO_2H 12 15 20 arachidonic acid —several steps, in a cell→ O 9 8 H H 12 15 H OH H OH 1 CO_2H 20 prostaglandin E_2 (PGE_2) **(15.10)**

Prostaglandins have excited much interest in the medical community, where they are used in the treatment of inflammatory diseases, such as asthma and rheumatoid arthritis; treatment of peptic ulcers; control of hypertension; regulation of blood pressure and metabolism; and inducing of labor and therapeutic abortions.

Enzymatic oxidation of arachidonic acid also leads to two important classes of *acyclic* products, **leukotrienes** and **lipoxins**, that arise from oxidation at C-5 and/or C-15.

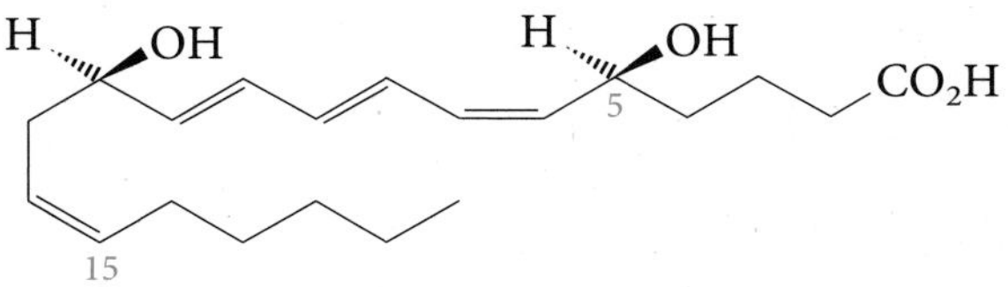

H OH H OH 5 CO_2H 15 leukotriene B_4

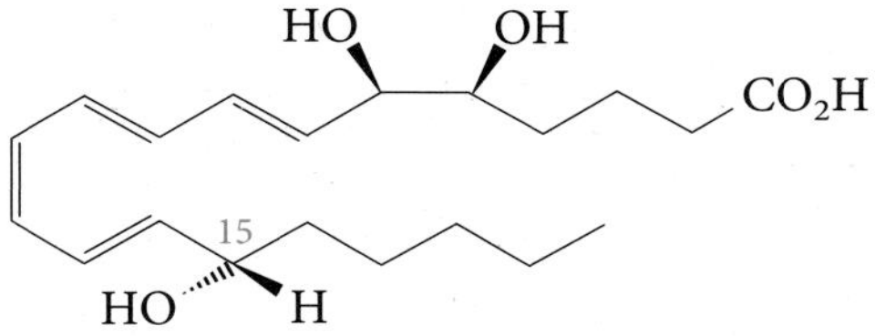

HO OH CO_2H 15 HO H lipoxin A

These classes of compounds can regulate specific cellular responses important in inflammation and immune reactions and are the subject of intense current research.

15.8 Waxes

Waxes are monoesters of fatty acids.

Waxes differ from fats and oils in that they are simple monoesters. The acid and alcohol portions of a wax molecule both have long saturated carbon chains.

$$CH_3(CH_2)_{13}CH_2\overset{\overset{\displaystyle O}{\|}}{C}{-}O(CH_2)_{15}CH_3 \qquad C_{25\text{-}27}H_{51\text{-}55}\overset{\overset{\displaystyle O}{\|}}{C}{-}OC_{30\text{-}32}H_{61\text{-}65}$$

cetyl palmitate (component of spermaceti, a wax in sperm whale oil)

components of beeswax

Some plant waxes are simply long-chain saturated hydrocarbons (page 45).

Waxes are more brittle, harder, and less greasy than fats. They are used to make polishes, cosmetics, ointments, and other pharmaceutical preparations, as well as candles and phonograph records. In nature, waxes coat the leaves and stems of plants that grow in arid regions, thus reducing evaporation. Similarly, insects with a high surface-area-to-volume ratio often have a coating of a natural protective wax.

15.9 Terpenes and Steroids

Terpenes are compounds containing multiples of **isoprene**, a five-carbon unit.

Essential oils of many plants and flowers are obtained by distilling the plant with water. The water-insoluble oil that separates usually has an odor characteristic of the particular plant (rose oil, geranium oil, and others). Compounds isolated from these oils contain multiples of five carbon atoms (that is, 5, 10, 15, and so on) and are called **terpenes** (some compounds of this type were described on page 222). They are synthesized in the plant from acetate by way of an important biochemical intermediate, isopentenyl pyrophosphate. The five-carbon unit with a four-carbon chain and a one-carbon branch at C-2 is called an **isoprene unit** (see "A Word About . . . Biologically Important Alcohols and Phenols," in Chapter 7).

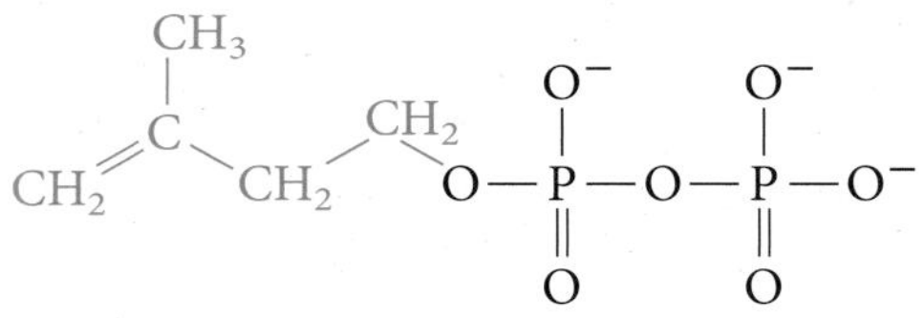

$$CH_2{=}C(CH_3){-}CH_2{-}CH_2{-}O{-}\underset{\underset{\displaystyle O}{\|}}{\overset{\overset{\displaystyle O^-}{|}}{P}}{-}O{-}\underset{\underset{\displaystyle O}{\|}}{\overset{\overset{\displaystyle O^-}{|}}{P}}{-}O^- \qquad C{-}\overset{\overset{\displaystyle C}{|}}{C}{-}C{-}C$$

isopentenyl pyrophosphate

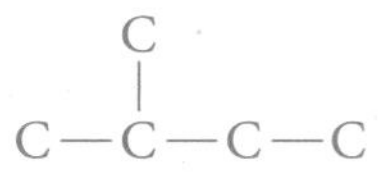

isoprene unit

Monoterpenes contain 2 isoprene units. **Sesquiterpenes**, **diterpenes**, **triterpenes**, and **tetraterpenes** contain 3, 4, 6, and 8 isoprene units, respectively.

Most terpene structures can be broken down into multiples of isoprene units. Terpenes contain various functional groups (C=C, OH, C=O) as part of their structures and may be acyclic or cyclic.

Compounds with a single isoprene unit (C_5) are relatively rare in nature, but compounds with two such units (C_{10}), called **monoterpenes**, are common. Examples are geraniol (page 222), citronellal, and myrcene (all acyclic) as well as menthol and β-pinene (both cyclic).

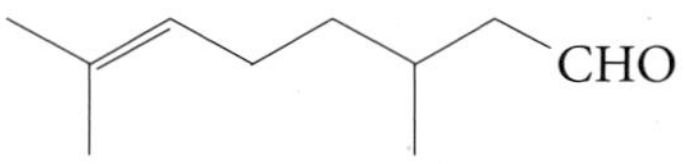

citronellal
(lemon oil)

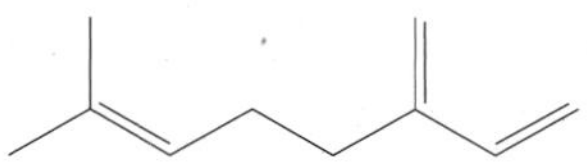

myrcene
(bay leaves)

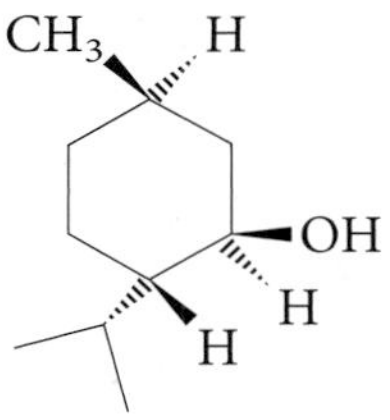

menthol
(peppermint oil)

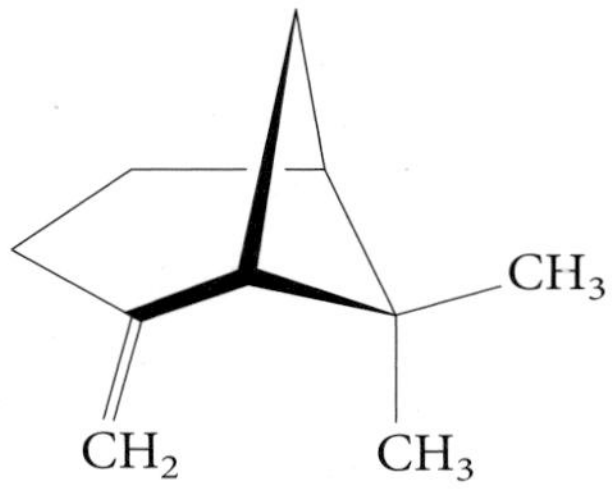

β-pinene
(turpentine)

EXAMPLE 15.4

Mark off the isoprene units in citronellal and menthol.

Solution

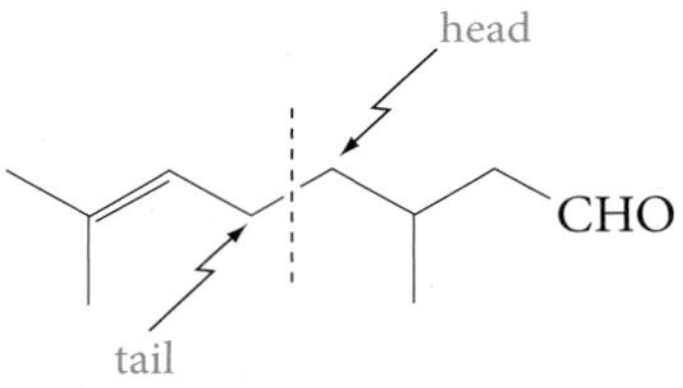

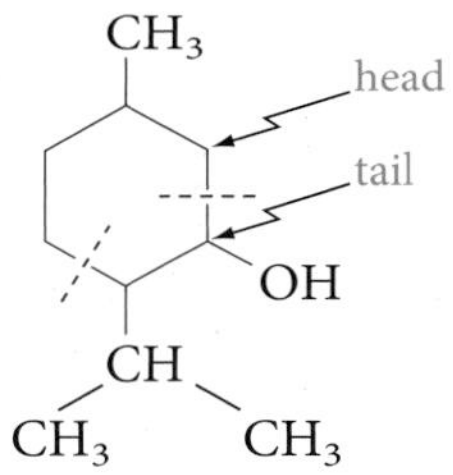

The dashed lines divide the molecules into two isoprene units.

PROBLEM 15.7 There is another way to divide menthol into two isoprene units. Can you find it? Notice that both ways of dividing the structure join the isoprene units in a head-to-tail manner.

We have already seen examples of a **sesquiterpene** (C_{15}, farnesol, page 222), a **diterpene** (C_{20}, retinal, page 76), a **triterpene** (C_{30}, squalene, page 222), and even a **tetraterpene** (C_{40}, β-carotene, page 77).

PROBLEM 15.8 Draw the structures of farnesol, retinal, squalene, and β-carotene, and mark off the component isoprene units. Identify the unusual tail-to-tail linkage in squalene.

Steroids are tetracyclic lipids derived from the acyclic triterpene squalene.

Steroids constitute a major class of lipids. They are related to terpenes in the sense that they are biosynthesized by a similar route. Through a truly remarkable reaction sequence, the acyclic triterpene squalene is converted *stereospecifically* to the tetracyclic steroid lanosterol, from which other steroids are subsequently synthesized.

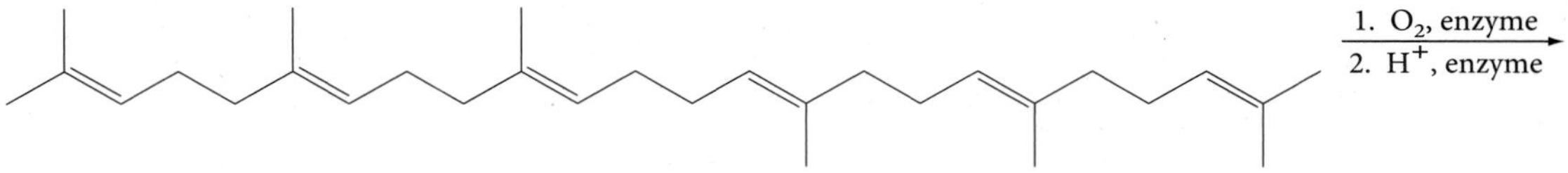

1. O_2, enzyme
2. H^+, enzyme

squalene (C_{30})

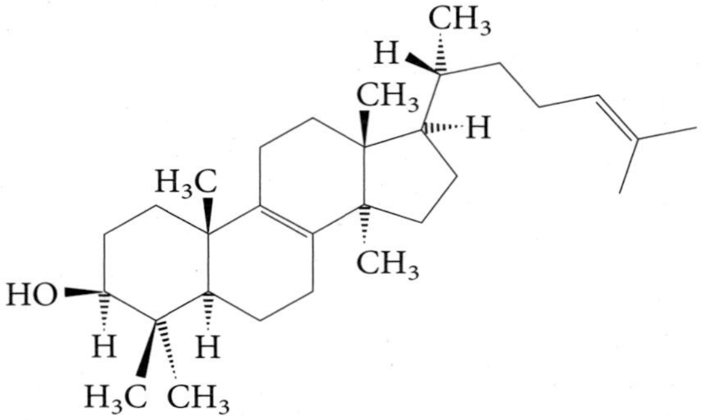

lanosterol (C_{30})

(15.11)

PROBLEM 15.9 How many stereogenic centers are present in squalene? In lanosterol?

The common structural feature of steroids is a system of four fused rings. The A, B, and C rings are six membered, and the D ring is five membered, usually all fused in a *trans* manner.

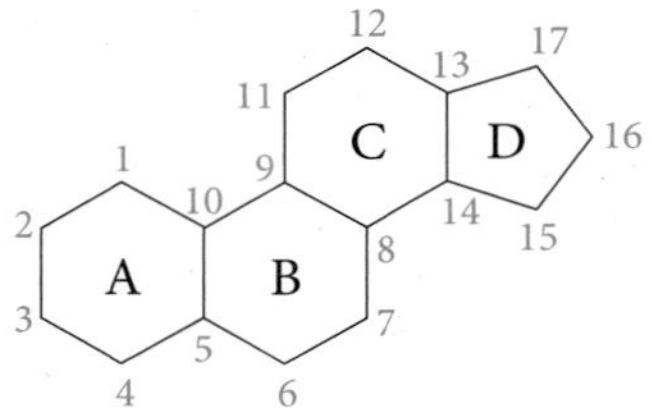

the steroid ring system, showing the numbering

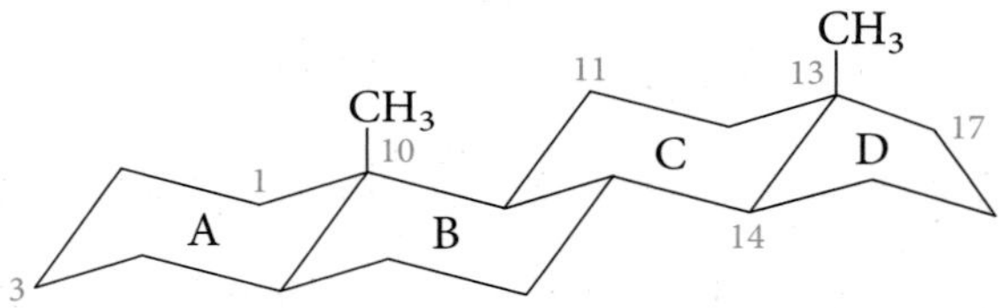

steroid shape, with chair cyclohexanes

In most steroids, the six-membered rings are not aromatic, although there are exceptions. Usually there are methyl substituents attached to C-10 and C-13 (called "angular" methyl groups) and some sort of side chain attached to C-17.

Perhaps the best known steroid is cholesterol. With 27 carbons, it is biosynthesized from lanosterol by a sequence of reactions that includes removing three carbon atoms.

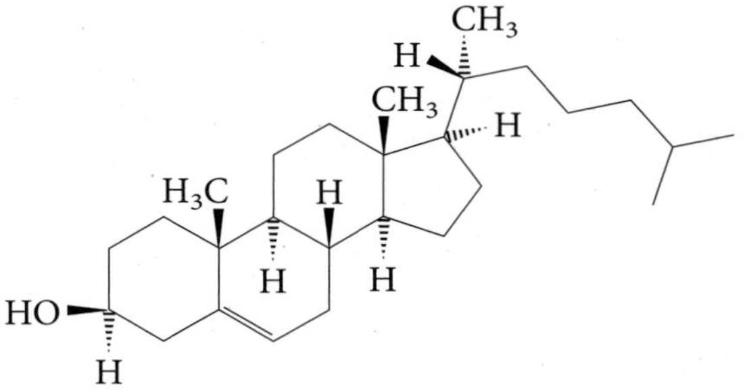

cholesterol

Cholesterol is present in all animal cells but is mainly concentrated in the brain and spinal cord. It is also the chief constituent of gallstones. The total amount of cholesterol present in an average human is about half a pound! There appears to be some connection between the concentration of blood serum cholesterol and coronary heart disease. Levels below 200 mg/dL seem desirable, whereas levels above 280 mg/dL may constitute a high risk.

Southern Illinois University/Photo Researchers, Inc

Cholesterol is the chief constituent of gallstones.

Other steroids are common in animal tissues and play important biological roles. Cholic acid, for example, occurs in the bile duct, where it is present mainly in the form of various amide salts. These salts have a polar region (hydrophilic) and a largely hydrocarbon region (lipophilic) and function as emulsifying agents to facilitate the absorption of fats in the intestinal tract. They are, in a sense, biological soaps.

Z = OH cholic acid

$Z = NHCH_2CH_2S(=O)_2-O^-Na^+$ a bile salt

The sex hormones are compounds, produced in the ovaries and testes, that control reproductive physiology and secondary sex characteristics. Those sex hormones that predominate in females are of two types. The estrogens, of which the most plentiful is estradiol, are essential for initiating changes during the menstrual cycle and for the development of female secondary sex characteristics. Progesterone, which prepares the uterus for implantation of the fertilized egg, also maintains a pregnancy and prevents further ovulation during that time. Progesterone is administered clinically to prevent abortion in difficult pregnancies. It differs structurally from estrogens, such as estradiol, in that the A ring is not aromatic.

estradiol progesterone

Oral contraceptives, sometimes called "the pill," have structures similar to that of progesterone. An example is the acetylenic alcohol norethindrone, which prevents conception. On the other hand, RU 486, which also resembles progesterone, is a contragestive. It interferes with gestation of a fertilized ovum and, if taken in conjunction with prostaglandins, terminates a pregnancy within the first nine weeks of gestation more effectively and safely than surgical methods. It was discovered in France and its use in the United States has only recently been approved.

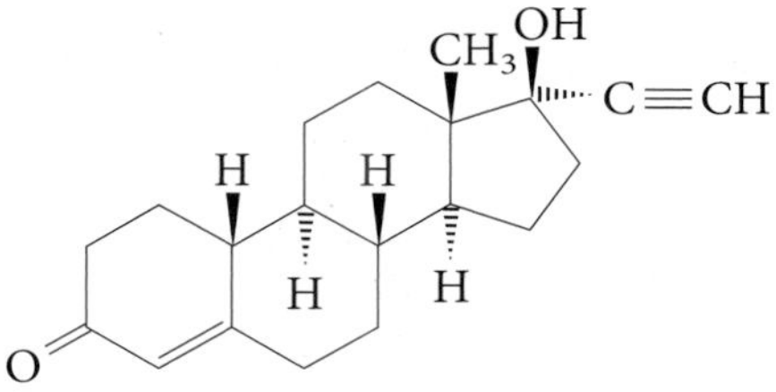

norethindrone (Norlutin)

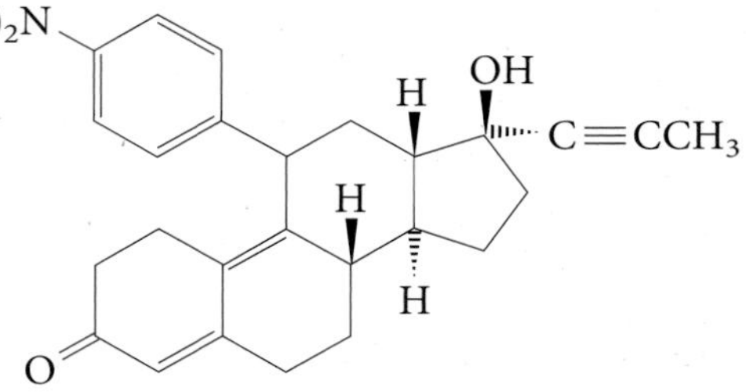

RU 486 (mifepristone)

Sex hormones that predominate in males are called androgens. They regulate the development of male reproductive organs and secondary sex characteristics, such as facial and body hair, deep voice, and male musculature. Two important androgens are testosterone and androsterone.

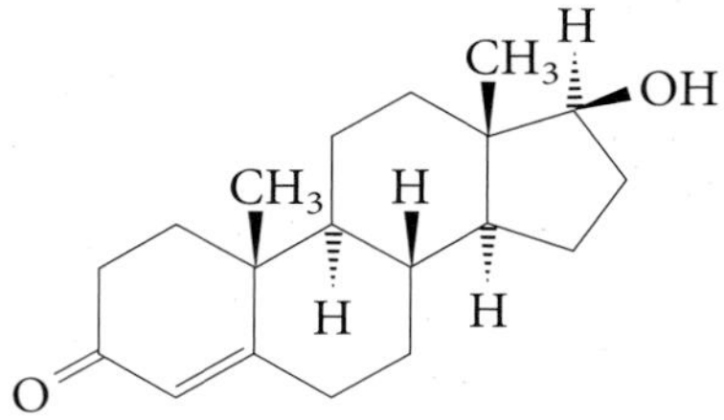

testosterone

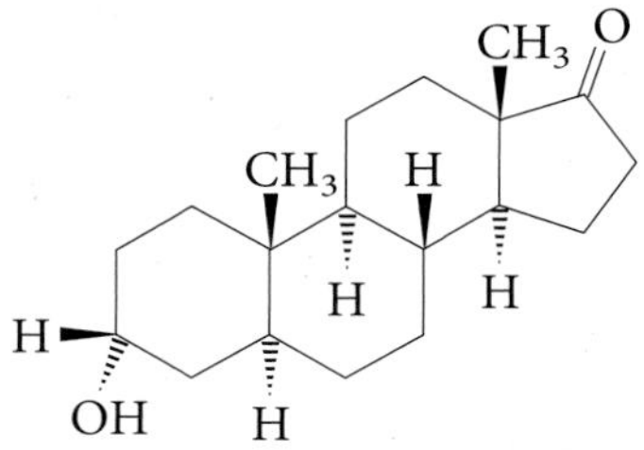

androsterone

Testosterone is an anabolic (muscle-building) steroid. Drugs based on its structure are sometimes administered to prevent withering of muscle in people recovering from surgery, starvation, or similar trauma. These same drugs, however, are sometimes illegally administered to healthy athletes and race horses to increase muscle mass and endurance. If taken in high doses, they can have serious side effects, including sexual malfunctions and liver tumors.

The only structural difference between testosterone and progesterone is the replacement of a hydroxyl group by an acetyl group at C-17 in the D ring. This great change in bioactivity, due to a seemingly minor change in structure, exemplifies the extreme specificity of biochemical reactions.

Another steroid bearing a resemblance to testosterone and progesterone is cortisone, a drug used in the treatment of arthritis.*

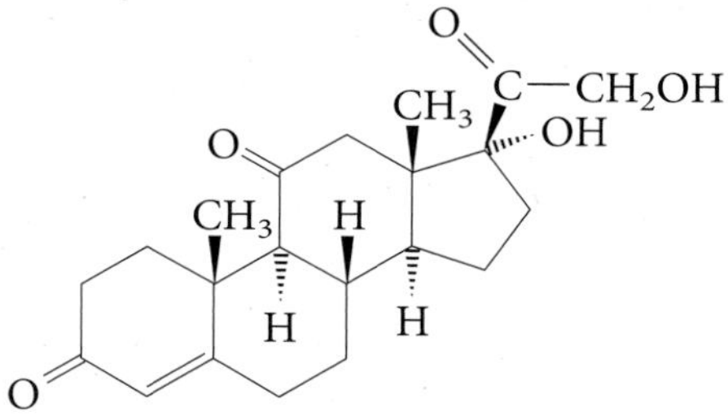

cortisone

*A low-cost method for synthesizing cortisone was developed by the American chemist Percy Julian (1899–1975). Julian also discovered an economical way to produce sex hormones from soybean oil and completed the first total synthesis of the glaucoma drug physostigmine (a nonaromatic nitrogen heterocycle). For a brief biography of Julian and his struggle with racial prejudice in the United States, see *Chemical and Engineering News,* Feb. 1, **1993**, p. 9.

REACTION SUMMARY

1. Saponification of a Triglyceride (Secs. 15.1 and 15.3)

$$\begin{array}{l} H_2C-O-\overset{\displaystyle O}{\overset{\|}{C}}-R \\ HC-O-\overset{\displaystyle O}{\overset{\|}{C}}-R \\ H_2C-O-\overset{\displaystyle O}{\overset{\|}{C}}-R \end{array} + 3\,NaOH \longrightarrow \begin{array}{l} H_2C-OH \\ HC-OH \\ H_2C-OH \end{array} + 3\,Na^{+}\,{}^{-}O-\overset{\displaystyle O}{\overset{\|}{C}}-R$$

2. Hydrogenation of a Triglyceride (Hardening) (Sec. 15.2)

$$\begin{array}{l} H_2C-O-\overset{\displaystyle O}{\overset{\|}{C}}-(CH_2)_nCH=CH(CH_2)_mCH_3 \\ HC-O-\overset{\displaystyle O}{\overset{\|}{C}}-(CH_2)_nCH=CH(CH_2)_mCH_3 \\ H_2C-O-\overset{\displaystyle O}{\overset{\|}{C}}-(CH_2)_nCH=CH(CH_2)_mCH_3 \end{array} \xrightarrow[\text{Ni, heat}]{3\,H_2} \begin{array}{l} H_2C-O-\overset{\displaystyle O}{\overset{\|}{C}}-(CH_2)_nCH_2CH_2(CH_2)_mCH_3 \\ HC-O-\overset{\displaystyle O}{\overset{\|}{C}}-(CH_2)_nCH_2CH_2(CH_2)_mCH_3 \\ H_2C-O-\overset{\displaystyle O}{\overset{\|}{C}}-(CH_2)_nCH_2CH_2(CH_2)_mCH_3 \end{array}$$

3. Hydrogenolysis of a Triglyceride (Sec. 15.5)

$$\begin{array}{l} H_2C-O-\overset{\displaystyle O}{\overset{\|}{C}}-R \\ HC-O-\overset{\displaystyle O}{\overset{\|}{C}}-R \\ H_2C-O-\overset{\displaystyle O}{\overset{\|}{C}}-R \end{array} \xrightarrow[\text{copper chromite}]{6\,H_2} \begin{array}{l} H_2C-OH \\ HC-OH \\ H_2C-OH \end{array} + 3\,HOH_2C-R$$

ADDITIONAL PROBLEMS

OWL Interactive versions of these problems are assignable in OWL.

Nomenclature and Structure

15.10 Using Table 15.1 as a guide, write the structural formula for

a. sodium laurate
b. magnesium oleate
c. glyceryl trimyristate
d. glyceryl palmitolauroöleate
e. linoleyl stearate
f. methyl arachidate

15.11 Write the general structure for

a. a fat
b. a vegetable oil
c. a wax
d. an ordinary soap
e. a synthetic detergent
f. a steroid
g. a phospholipid
h. a terpene
i. an isoprene unit

Fats, Soaps, and Fatty Acids

15.12 Write equations for the (a) saponification, (b) hydrogenation, and (c) hydrogenolysis of glyceryl trilaurate.

15.13 Saponification of castor oil gives glycerol and mainly (80 to 90%) ricinoleic acid, also called 12-hydroxyoleic acid. Draw the structure of the main component of castor oil.

15.14 Complete the equation for each of the following reactions:

a. $C_{15}H_{31}\overset{\overset{O}{\|}}{C}O^-K^+ + HCl \longrightarrow$ **b.** $C_{15}H_{31}\overset{\overset{O}{\|}}{C}O^-Na^+ + Ca^{2+} \longrightarrow$

Synthetic Detergents

15.15 In 1789, the year of the French Revolution, the young chemist Nicolas Leblanc developed the first method for the inexpensive mass production of sodium carbonate, Na_2CO_3. His discovery ultimately made soap available to the masses. For what purpose can sodium carbonate be used, in the synthesis of soap?*

15.16 Using eq. 15.9 as a model, write equations for the preparation of an alkylbenzenesulfonate synthetic detergent, starting with 1-dodecene and benzene.

15.17 A synthetic detergent widely used in dishwashing liquids has the structure $CH_3(CH_2)_{11}(OCH_2CH_2)_3OSO_3^-Na^+$. Write a series of equations showing how this detergent can be synthesized from $CH_3(CH_2)_{10}CH_2OH$ and ethylene oxide.

15.18 What disadvantages of a soap are syndets designed to overcome?

15.19 Why are synthetic detergents made with unbranched alkyl chains?

Stereochemistry of Lipids

15.20 When a beet armyworm caterpillar eats corn seedlings, it secretes volicitin onto the injured leaves. The leaves respond by releasing a volatile cocktail of terpenes and indoles that attract parasitic wasps, the natural enemies of the caterpillar. (For more details, read "New Fatty Acid-Based Signals: A Lesson from the Plant World" by E. E. Farmer in *Science,* May 9, **1997**, p. 912.)

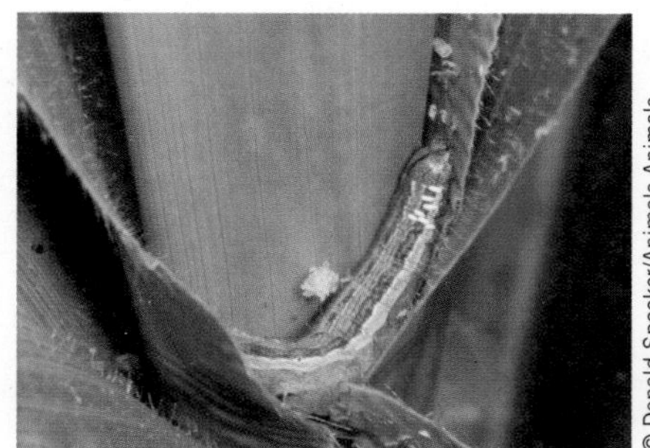

© Donald Specker/Animals Animals

The beet armyworm caterpillar eats corn seedlings and excretes the fatty acid derivative volicitin.

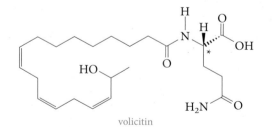

volicitin

a. The amide volicitin is derived from a fatty acid and the amino acid glutamine (see Table 17.1). Which fatty acid is present in volicitin? What is the configuration (*Z* or *E*) of each double bond?

b. What is the configuration (*R* or *S*) of the stereogenic center marked with an asterisk?

15.21 Consider leukotriene B_4 (Sec. 15.7), which is formed from the enzymatic oxidation of arachidonic acid, as are the prostaglandins described in the "A Word About . . . Prostaglandins, Aspirin, and Pain" (page 450). Provide a complete IUPAC name for leukotriene B_4, including stereogenic centers and double bonds.

---

*For an account of this revolutionary discovery and its positive and negative effects, read Chapter 1 of *Prometheans in the Laboratory* by Sharon Bertsch McGrayne, McGraw-Hill, New York, 2001.

15.22 Consider the structure of prostaglandin E_2 shown on page 451.

a. How many stereogenic centers are present?
b. What is the configuration (*R* or *S*) of each?
c. What is the configuration of the double bonds (*Z* or *E*) in the two side chains?
d. Are the two side chains *cis* or *trans* to one another?

Waxes

15.23 Write an equation for the saponification of cetyl palmitate, the main component of spermaceti, a wax found in the head cavities of sperm whales.

15.24 When boiled with concentrated aqueous alkali, fats and oils dissolve, but waxes do not. Explain the difference.

Terpenes

15.25 Divide the structures of (a) myrcene and (b) β-pinene (page 453) into their component isoprene units.

15.26 Notice (from your answer to Problem 15.8) that all of the isoprene units in farnesol and retinal are joined in a head-to-tail manner, but that they are not all arranged that way in squalene and β-carotene. Where does the head-to-tail arrangement break down in these two terpenes? (*Hint:* Count carbons to locate the terpene units.) What does this suggest about the way these two terpenes might be biosynthesized?

15.27 Ball-and-stick structures of the odiferous components of (a) spearmint and (b) caraway seed are shown below. What are the stereochemical relationships between these compounds?

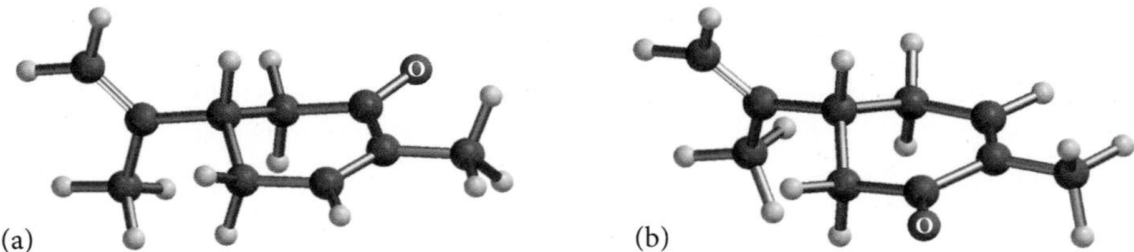

Steroids

15.28 Predict the products of the following reactions. (Consult the text for the structure of the starting materials. Look up the reagents in the index if you cannot remember what they do.)

a. testosterone + $LiAlH_4$
b. androsterone + methylmagnesium bromide
c. lanosterol + acetic anhydride
d. androsterone + Jones' reagent

15.29 Cholesterol has a rigid fused-ring structure (page 454) and is found in phospholipid bilayers (Figure 15.6), where it adds rigidity to cell membranes. Cholesterol contains both polar and nonpolar groups. Taking into account nonbonding intermolecular interactions (Sec. 2.7), draw a schematic picture showing the probable orientation of a cholesterol molecule in the lipid bilayer.

15.30 Androstenedione is a steroid touted for its muscle-building ability through metabolic conversion to testosterone (page 456). What reaction must occur in the body to convert androstenedione to testosterone? Why is the stereochemistry of this reaction important?

androstenedione

Fructose, also known as fruit sugar, is 50% sweeter than sucrose (table sugar) and is the major sugar in honey.

16

© Dan McCoy/Rainbow

Online homework for this chapter can be assigned in OWL, an online homework assessment tool.

Carbohydrates

Carbohydrates occur in all plants and animals and are essential to life. Through photosynthesis, plants convert atmospheric carbon dioxide to carbohydrates, mainly cellulose, starch, and sugars. Cellulose is the building block of rigid cell walls and woody tissues in plants, whereas starch is the chief storage form of carbohydrates for later use as a food or energy source. Some plants (cane and sugar beets) produce sucrose, ordinary table sugar. Another sugar, glucose, is an essential component of blood. Two other sugars, ribose and 2-deoxyribose, are components of the genetic materials RNA and DNA. Other carbohydrates are important components of coenzymes, antibiotics, cartilage, the shells of crustaceans, bacterial cell walls, and mammalian cell membranes.

In this chapter, we will describe the structures and a few reactions of the more important carbohydrates.

16.1 Definitions and Classification

The word *carbohydrate* arose because molecular formulas of these compounds can be expressed as *hydrates* of *carbon*. Glucose, for example, has the molecular formula $C_6H_{12}O_6$, which might be written as $C_6(H_2O)_6$. Although this type of formula is useless in studying the chemistry of carbohydrates, the old name persists.

We can now define carbohydrates more precisely in terms of their organic structures. **Carbohydrates** are polyhydroxyaldehydes, polyhydroxyketones, or substances that give such compounds on hydrolysis. The chemistry of carbohydrates is mainly the combined chemistry of two functional groups: the hydroxyl group and the carbonyl group.

Carbohydrates are polyhydroxyaldehydes, polyhydroxyketones, or substances that give such compounds on hydrolysis. The hydroxyl group and the carbonyl group are the major functional groups in carbohydrates.

Carbohydrates are usually classified according to their structure as monosaccharides, oligosaccharides, or polysaccharides. The term *saccharide* comes from Latin (*saccharum*, sugar) and refers to the sweet taste of some simple carbohydrates. The three classes of carbohydrates are related to each other through hydrolysis.

$$\text{polysaccharide} \xrightarrow[H^+]{H_2O} \text{oligosaccharides} \xrightarrow[H^+]{H_2O} \text{monosaccharides} \quad \textbf{(16.1)}$$

For example, hydrolysis of starch, a polysaccharide, gives first maltose and then glucose.

$$\underset{\substack{\text{starch}\\ \text{(a polysaccharide)}}}{[C_{12}H_{20}O_{10}]_n} \xrightarrow[H^+]{n\,H_2O} \underset{\substack{\text{maltose}\\ \text{(a disaccharide)}}}{n\,C_{12}H_{22}O_{11}} \xrightarrow[H^+]{n\,H_2O} \underset{\substack{\text{glucose}\\ \text{(a monosaccharide)}}}{2n\,C_6H_{12}O_6} \quad \textbf{(16.2)}$$

Monosaccharides (or simple sugars, as they are sometimes called) are carbohydrates that cannot be hydrolyzed to simpler compounds. **Polysaccharides** contain many monosaccharide units—sometimes hundreds or even thousands. Usually, but not always, the units are identical. Two of the most important polysaccharides, starch and cellulose, contain linked units of the same monosaccharide, glucose. **Oligosaccharides** (from the Greek *oligos*, few) contain at least two and generally no more than a few linked monosaccharide units. They may be called **disaccharides**, **trisaccharides**, and so on, depending on the number of units, which may be the same or different. Maltose, for example, is a disaccharide made of two glucose units, but sucrose, another disaccharide, is made of two different monosaccharide units: glucose and fructose.

Monosaccharides (simple sugars) cannot be hydrolyzed to simpler compounds. **Oligosaccharides** contain a few linked monosaccharide units, whereas **polysaccharides** contain many monosaccharide units.

In the next section, we will describe the structures of monosaccharides. Later, we will see how these units are linked to form oligosaccharides and polysaccharides.

16.2 Monosaccharides

Monosaccharides are classified according to the number of carbon atoms present (**triose**, **tetrose**, **pentose**, **hexose**, and so on) and according to whether the carbonyl group is present as an aldehyde (**aldose**) or as a ketone (**ketose**).

An **aldose** and a **ketose** contain the aldehyde and ketone functional groups, respectively. A **triose** has three carbon atoms, a **tetrose** has four, and so on.

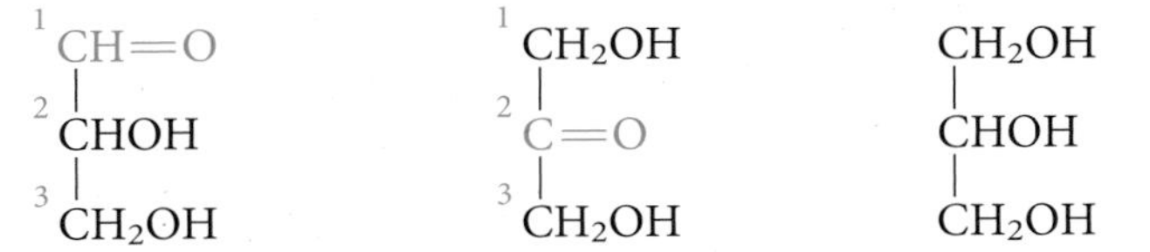

There are only two trioses: glyceraldehyde and dihydroxyacetone. Each has two hydroxyl groups, attached to different carbon atoms, and one carbonyl group.

glyceraldehyde (an aldose)	dihydroxyacetone (a ketose)	glycerol
1 CH=O	1 CH_2OH	CH_2OH
2 CHOH	2 C=O	CHOH
3 CH_2OH	3 CH_2OH	CH_2OH

Glyceraldehyde is the simplest aldose, and dihydroxyacetone is the simplest ketose. Each is related to glycerol in that each has a carbonyl group in place of one of the hydroxyl groups.

Other aldoses or ketoses can be derived from glyceraldehyde or dihydroxyacetone by adding carbon atoms, each with a hydroxyl group. In aldoses, the chain is numbered from the aldehyde carbon. In most ketoses, the carbonyl group is located at C-2.

aldoses: tetrose	pentose	hexose	ketoses: tetrose	pentose	hexose
1 CH=O	1 CH=O	1 CH=O	1 CH_2OH	1 CH_2OH	1 CH_2OH
2 CHOH	2 CHOH	2 CHOH	2 C=O	2 C=O	2 C=O
3 CHOH	3 CHOH	3 CHOH	3 CHOH	3 CHOH	3 CHOH
4 CH_2OH	4 CHOH	4 CHOH	4 CH_2OH	4 CHOH	4 CHOH
	5 CH_2OH	5 CHOH		5 CH_2OH	5 CHOH
		6 CH_2OH			6 CH_2OH

16.3 Chirality in Monosaccharides; Fischer Projection Formulas and D,L-Sugars

You will notice that glyceraldehyde, the simplest aldose, has one stereogenic carbon atom (C-2) and hence can exist in two enantiomeric forms.

CH=O
H—C—OH
CH_2OH
R-(+)-glyceraldehyde
$[\alpha]_D^{25}$ +8.7(c = 2,H_2O)

CH=O
HO—C—H
CH_2OH
S-(−)-glyceraldehyde
$[\alpha]_D^{25}$ −8.7(c = 2,H_2O)

The dextrorotatory form has the *R* configuration.

It was in connection with his studies on carbohydrate stereochemistry that Emil Fischer invented his system of projection formulas. Because we will be using these formulas here, it might be wise for you to review Sections 5.7 through 5.9. Recall that, in a Fischer projection formula, *horizontal* lines show groups that project *above* the plane of the paper *toward* the viewer; *vertical* lines show groups that project *below* the plane of the paper *away* from the viewer. Thus, *R*-(+)-glyceraldehyde can be represented as

CH=O
H—C—OH
CH_2OH
R-(+)-glyceraldehyde

≡

CH=O
H—+—OH
CH_2OH
Fischer projection formula for *R*-(+)-glyceraldehyde

with the stereogenic center represented by the intersection of two crossed lines.

Fischer also introduced a stereochemical nomenclature that preceded the *R,S* system and that is still in common use for sugars and amino acids. He used a small capital D to represent the configuration of (+)-glyceraldehyde, with the hydroxyl group on the *right;* its enantiomer, with the hydroxyl group on the *left,* was designated L-(−)-glyceraldehyde. The most oxidized carbon (CHO) was placed at the top.

```
     CHO              CHO
 H ─┼─ OH        HO ─┼─ H
    CH2OH            CH2OH
```

D-(+)-glyceraldehyde   L-(−)-glyceraldehyde

Fischer extended his system to other monosaccharides in the following way. If the stereogenic carbon *farthest* from the aldehyde or ketone group had the same configuration as D-glyceraldehyde (hydroxyl on the right), the compound was called a D-sugar. If the configuration at the remote carbon had the same configuration as L-glyceraldehyde (hydroxyl on the left), the compound was an L-sugar.

```
                                       CH2OH            CH2OH
                                         |                |
    CH=O            CH=O                C=O              C=O
     |               |                   |                |
   (CHOH)n         (CHOH)n             (CHOH)n          (CHOH)n
 H ─┼─ OH       HO ─┼─ H            H ─┼─ OH         HO ─┼─ H
    CH2OH          CH2OH               CH2OH            CH2OH
```

a D-aldose   an L-aldose   a D-ketose   an L-ketose

Figure 16.1 shows the Fischer projection formulas for all of the D-aldoses through the hexoses. Starting with D-glyceraldehyde, one CHOH unit at a time is inserted in the chain. This carbon, which adds a new stereogenic center to the structure, is shown in black. In each case, the new stereogenic center can have the hydroxyl group at the right or at the left in the Fischer projection formula (*R* or *S* absolute configuration).

EXAMPLE 16.1

Using Figure 16.1, write the Fischer projection formula for L-erythrose.

Solution L-Erythrose is the enantiomer of D-erythrose. Since both —OH groups are on the right in D-erythrose, they will both be on the left in its mirror image.

```
      CH=O
 HO ─┼─ H
 HO ─┼─ H
      CH2OH
```

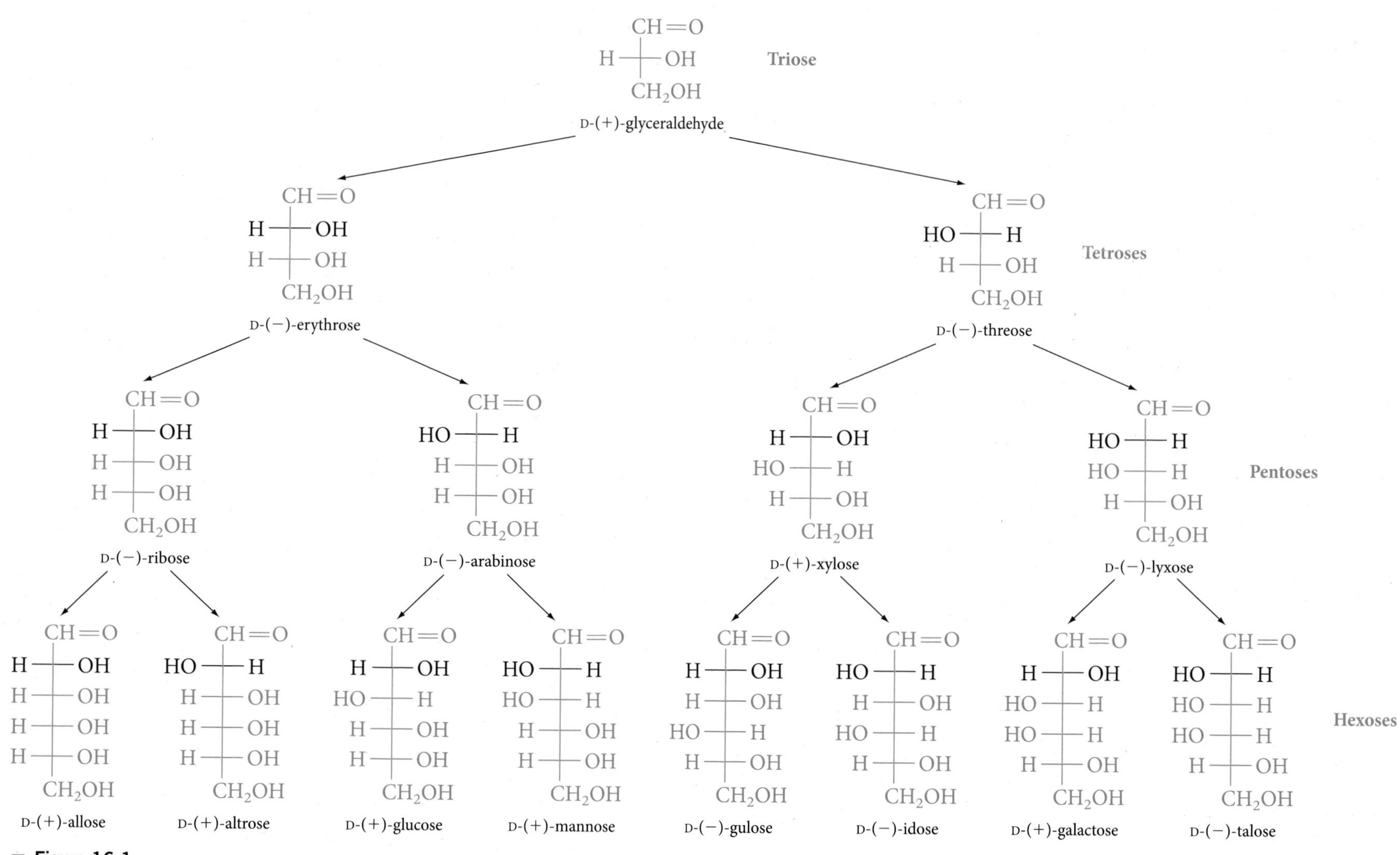

■ **Figure 16.1**
Fischer projection formulas and genealogy chart for the D-aldoses with up to six carbon atoms.

EXAMPLE 16.2

Convert the Fischer projection formula for D-erythrose to a three-dimensional structural formula.

Solution

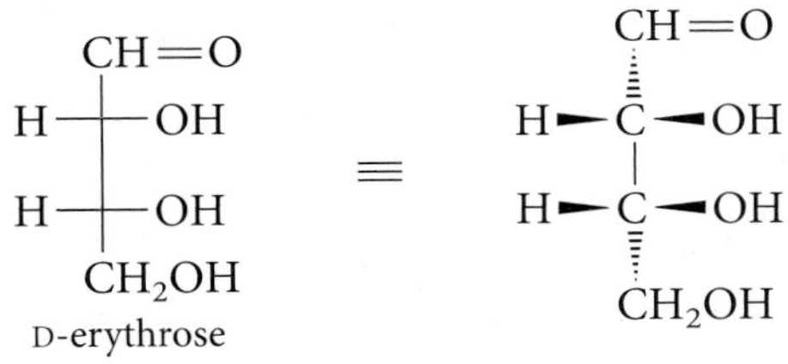
CH=O
H OH
H OH
CH2OH
CH=O
H C OH
H C OH
CH2OH

D-erythrose

We can also write the structure as

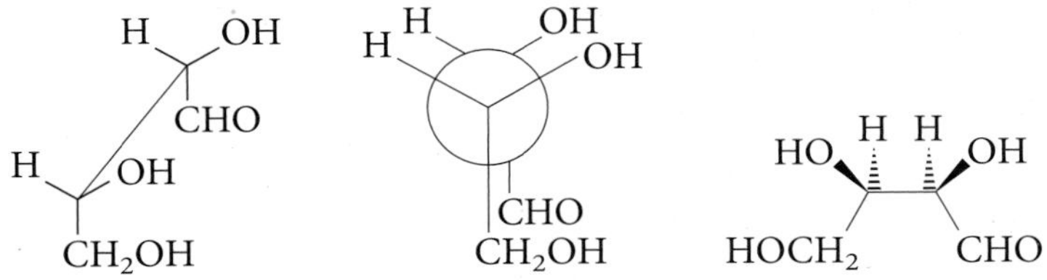
H OH
CHO
H OH
CH2OH
H H OH OH
CHO
CH2OH
HO H H OH
HOCH2 CHO

sawhorse Newman dash-wedge

and we can then rotate around the central C—C bond to more favorable staggered (instead of eclipsed) conformations, such as

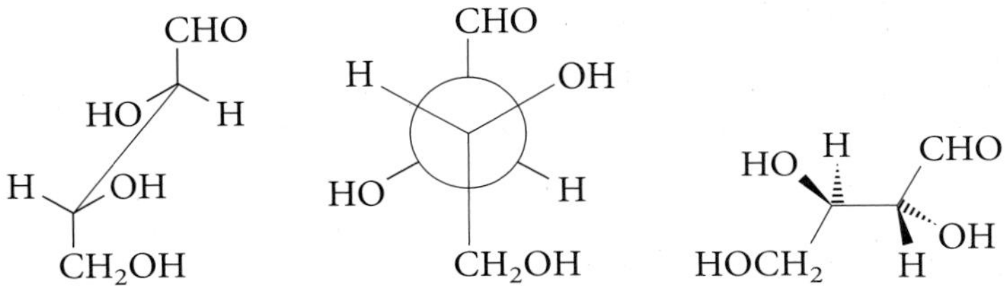
CHO
HO H
H OH
CH2OH
CHO
H OH
HO H
CH2OH
HO H CHO
OH
HOCH2 H

Molecular models may help you to follow these interconversions.

PROBLEM 16.1 Using Figure 16.1, write the Fischer projection formula for

a. L-arabinose b. L-mannose

PROBLEM 16.2 Convert the Fischer projection formula for D-threose to three-dimensional representations.

PROBLEM 16.3 How many D-aldoheptoses are possible?

How are the sugars with identical numbers of carbon atoms, shown horizontally across Figure 16.1, related to one another? Compare, for example, D-(−)-erythrose and D-(−)-threose. They have the same configuration at C-3 (D, with the OH on the right), but opposite configurations at C-2. These sugars are stereoisomers, but *not* mirror images (*not* enantiomers). In other words, they are *diastereomers* (review Sec. 5.8). Similarly, there are four diastereomeric D-pentoses and eight diastereomeric D-hexoses.

A special name is given to diastereomers that differ in configuration *at only one stereogenic center;* they are called **epimers.** D-(−)-Erythrose and D-(−)-threose are not only diastereomers, they are epimers. Similarly, D-glucose and D-mannose are epimers (at C-2), and D-glucose and D-galactose are epimers (at C-4). Each pair has the same configurations at all stereogenic centers except one.

Epimers are diastereomers that differ in configuration at only one stereogenic center.

PROBLEM 16.4 What pairs of D-pentoses are epimeric at C-3?

Notice that there is no direct relationship between configuration and the sign of optical rotation. Although all of the sugars in Figure 16.1 are D-sugars, some are dextrorotatory (+) and others are levorotatory (−).

16.4 The Cyclic Hemiacetal Structures of Monosaccharides

The monosaccharide structures described so far are consistent with much of the known chemistry of these compounds, but they are oversimplified. We now examine the true structures of these compounds.

We learned earlier that alcohols undergo rapid and reversible addition to the carbonyl group of aldehydes and ketones, to form hemiacetals (review Sec. 9.7). This can happen *intramolecularly* when the hydroxyl and carbonyl groups are properly located in the same molecule (eqs. 9.14 and 9.15), which is the situation in many monosaccharides. *Monosaccharides exist mainly in cyclic, hemiacetal forms* and not in the acyclic aldo- or keto-forms we have depicted so far.

As an example, consider D-glucose. First, let us rewrite its Fischer projection formula in a way that brings the OH group at C-5 within bonding distance of the carbonyl group (as in eq. 9.14). This is shown in Figure 16.2. The Fischer projection is first converted to its three-dimensional (dash-wedge) structure, which is then turned on its side and bent around so that C-1 and C-6 are close to one another. Finally, rotation about the C-4—C-5 bond brings the hydroxyl oxygen at C-5 close enough for nucleophilic addition to the carbonyl carbon (C-1). Reaction then leads to the cyclic, hemiacetal structure shown at the bottom left of the figure.

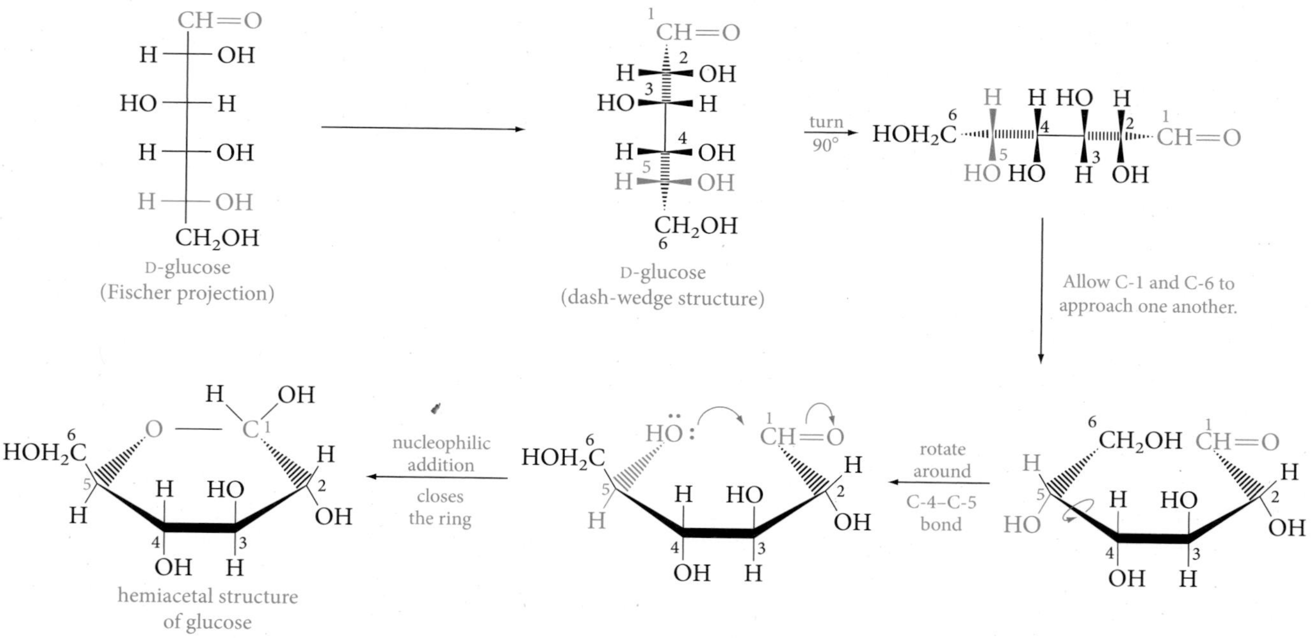

D-glucose (Fischer projection)

D-glucose (dash-wedge structure)

turn 90°

Allow C-1 and C-6 to approach one another.

rotate around C-4–C-5 bond

nucleophilic addition closes the ring

hemiacetal structure of glucose

■ **Figure 16.2**

Manipulation of the Fischer projection formula of D-glucose to bring the C-5 hydroxyl group in position for cyclization to the hemiacetal form.

The British carbohydrate chemist W. N. Haworth (Nobel Prize, 1937) introduced a useful way of representing the cyclic forms of sugars. In a **Haworth projection**, the ring is represented as if it were planar and viewed edge on, with the oxygen at the upper right. The carbons are arranged clockwise numerically, with C-1 at the right. Substituents attached to the ring lie above or below the plane. For example, the Haworth formula for D-glucose (Figure 16.2) is written as

In **Haworth projections**, the carbohydrate ring is represented as planar and viewed edge on. The carbons are arranged clockwise numerically, with C-1 at the right.

Haworth projection formulas for D-glucose

Sometimes, as in the structure at the right, the ring hydrogens are omitted so that attention can be focused on the hydroxyl groups.

In converting from one type of projection formula to another, notice that hydroxyl groups on the *right* in the Fischer projection are *down* in the Haworth projection (and conversely, hydroxyl groups on the *left* in the Fischer projection are *up* in the Haworth projection). For D-sugars, the terminal —CH_2OH group is *up* in the Haworth projection; for L-sugars, it is down.

© Hulton Archive/Getty Images

W. N. Haworth was a pioneer in the field of carbohydrate chemistry. For his 1937 Nobel Prize address and other information on Nobel Prizes in chemistry, see http://nobelprize.org/chemistry/laureates/1937/haworth-bio.html

EXAMPLE 16.3

Draw the Haworth projection for the six-membered cyclic structure of D-mannose.

Solution Notice from Figure 16.1 that D-mannose differs from D-glucose *only* in the configuration at C-2. In the Fischer projection formula, the C-2 hydroxyl is on the *left;* therefore, it will be *up* in the Haworth projection. Otherwise, the structure is identical to that of D-glucose.

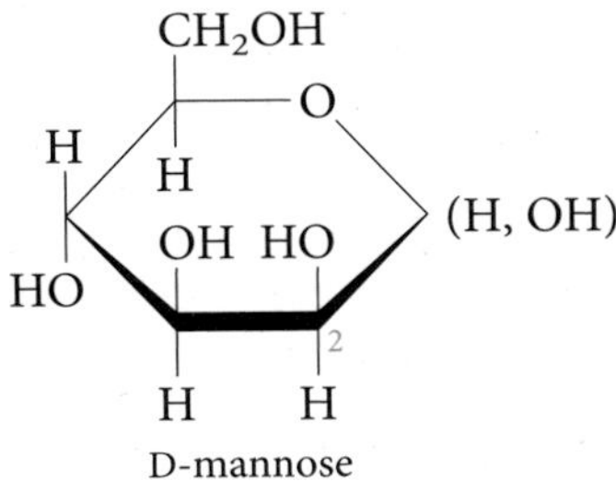

D-mannose

PROBLEM 16.5 Draw the Haworth projection formula for the six-membered cyclic structure of D-galactose.

Now notice three important features of the hemiacetal structure of D-glucose. First, *the ring is heterocyclic,* with five carbons and an oxygen. Carbons 1 through 5 are part of the ring structure, but carbon 6 (the —CH_2OH group) is a substituent on the ring. Next, C-1 is special. *C-1 is the hemiacetal carbon,* simultaneously an alcohol and an ether carbon (it carries a hydroxyl group, and it is also connected to C-5 by an ether linkage). In contrast, all of the other carbons are monofunctional. C-2, C-3, and C-4 are secondary alcohol carbons; C-6 is a primary alcohol carbon; and C-5 is an ether carbon.

These differences show up in the different chemical reactions of D-glucose. Finally, *C-1 in the cyclic, hemiacetal structure is a stereogenic center.* It has four different groups attached to it (H, OH, OC-5, and C-2) and can therefore exist in two configurations, *R* or *S*. Let us consider this last feature in greater detail.

16.5 Anomeric Carbons; Mutarotation

The hemiacetal carbon in a cyclic monosaccharide is the **anomeric carbon**. Two monosaccharides that differ in configuration only at the anomeric carbon are **anomers**.

In the acyclic, aldehyde form of glucose, C-1 is achiral, but in the cyclic structures, this carbon becomes chiral. Consequently, *two* hemiacetal structures are possible, depending on the configuration at the new chiral center. The hemiacetal carbon, the carbon that forms the new stereogenic center, is called the **anomeric carbon**. Two monosaccharides that differ only in configuration at the anomeric center are **anomers** (a special kind of epimers). Anomers are called α or β, depending on the position of the hydroxyl group. For monosaccharides in the D-series, the hydroxyl group is "down" in the α anomer and "up" in the β anomer, when the structure is written in the usual way (eq. 16.3).

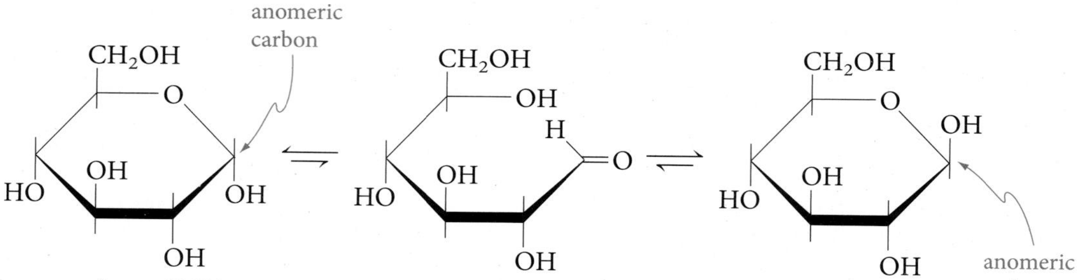

anomeric carbon

CH_2OH ... $\rightleftharpoons$ ... $\rightleftharpoons$ ... (16.3)

anomeric carbon

α-D-glucose (36%)
(mp 146°C)
[α]+112°

D-glucose
(acyclic, aldehyde form)

β-D-glucose (64%)
(mp 150°C)
[α]+19°

The α and β forms of D-glucose have identical configurations at every stereogenic center *except at C-1, the anomeric carbon.*

How do we know that monosaccharides exist mainly as cyclic hemiacetals? There is direct physical evidence. For example, if D-glucose is crystallized from methanol, the pure α form is obtained. On the other hand, crystallization from acetic acid gives the β form. The α and β forms of D-glucose are *diastereomers.* Being diastereomers, they have different physical properties, as shown under their structures in eq. 16.3; note that they have different melting points and different specific optical rotations.

Changes in optical rotation due to interconversion of anomers in solution are called **mutarotation**.

The α and β forms of D-glucose interconvert in aqueous solution. For example, if crystalline α-D-glucose is dissolved in water, the specific rotation drops gradually from an initial value of +112° to an equilibrium value of +52°. Starting with the pure crystalline β form results in a gradual rise in specific rotation from an initial +19° to the same equilibrium value of +52°. These changes in optical rotation are called **mutarotation**. They can be explained by the equilibria shown in eq. 16.3. Recall that hemiacetal formation is a *reversible equilibrium process* (Sec. 9.7). Starting with either pure hemiacetal form, the ring can open to the acyclic aldehyde, which can then recyclize to give either the α or the β form. Eventually, an equilibrium mixture is obtained.

At equilibrium, an aqueous solution of D-glucose contains 35.5% of the α form and 64.5% of the β form. There is only about 0.003% of the open-chain aldehyde form present.

EXAMPLE 16.4

Show that the percentages of α- and β-D-glucose in aqueous solution at equilibrium can be calculated from the specific rotations of the pure α and β forms and the specific rotation of the solution at equilibrium.

Solution The equilibrium rotation is $+52°$, and the rotations of pure α and β forms are $+112°$ and $+19°$, respectively. Assuming that no other forms are present, we can express these values graphically as follows:

+112°	+52°	+19°
100% α	equilibrium	100% β

The percentage of the β form at equilibrium is then

$$\frac{112 - 52}{112 - 19} \times 100 = \frac{60}{93} \times 100 = 64.5\%$$

The percentage of the α form at equilibrium is $100 - 64.5 = 35.5\%$.

16.6 Pyranose and Furanose Structures

The six-membered cyclic form of most monosaccharides is the preferred structure. These structures are called **pyranose** forms after the six-membered oxygen heterocycle pyran. The formula at the extreme left of eq. 16.3 is more completely named α-D-glucopyranose, with the last part of the name showing the ring size.

Cyclic monosaccharides with six-membered and five-membered rings are called **pyranoses** and **furanoses**, respectively.

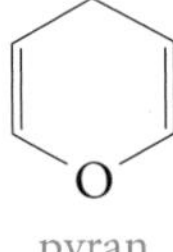

pyran

Pyranoses are formed by reaction of the hydroxyl group at C-5, with the carbonyl group. With some sugars, however, the hydroxyl group at C-4 reacts instead. In these cases, the cyclic hemiacetal that is formed has a five-membered ring. This type of cyclic monosaccharide is called a **furanose**, after the parent five-membered oxygen heterocycle furan.

furan

For example, D-glucose could, in principle, exist in two furanose forms (α and β at C-1) through attack of the C-4 hydroxyl on the aldehyde carbon.

D-glucose ⇌ α- and β-D-glucofuranose (16.4)

In practice, these forms are present to less than 1% in glucose solutions, but they are important with other monosaccharides. The ketose D-fructose, for example, exists in solution mainly in two furanose forms. The carbonyl carbon at C-2 and hydroxyl group at C-5 cyclize to give the furanose ring.

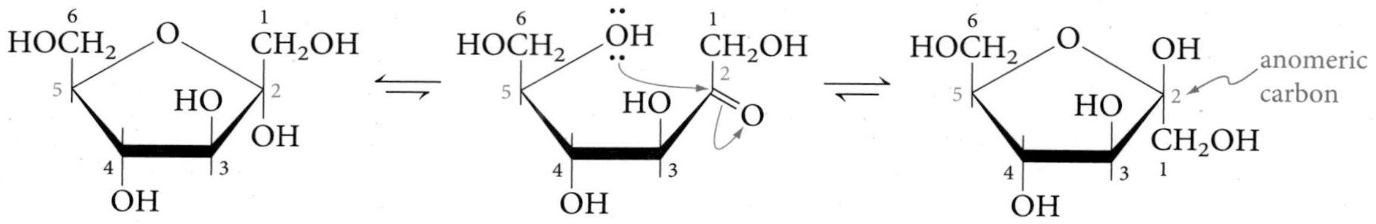

α-D-fructofuranose
(—OH at C-2 is "down")

D-fructose
(acyclic keto form)

β-D-fructofuranose
(—OH at C-2 is "up")

anomeric carbon

(16.5)

PROBLEM 16.6 Draw Haworth projections of the α and β forms of D-glucofuranose (eq. 16.4).

PROBLEM 16.7 D-Erythrose cannot exist in pyranose forms, but furanose cyclic forms are possible. Explain. Draw the structure for α-D-erythrofuranose.

16.7 Conformations of Pyranoses

Haworth projections depict pyranose rings as planar. However, as with cyclohexane, the rings generally prefer a chair conformation (review Sec. 2.9). Consequently, we can rewrite eq. 16.3 more accurately as eq. 16.6.

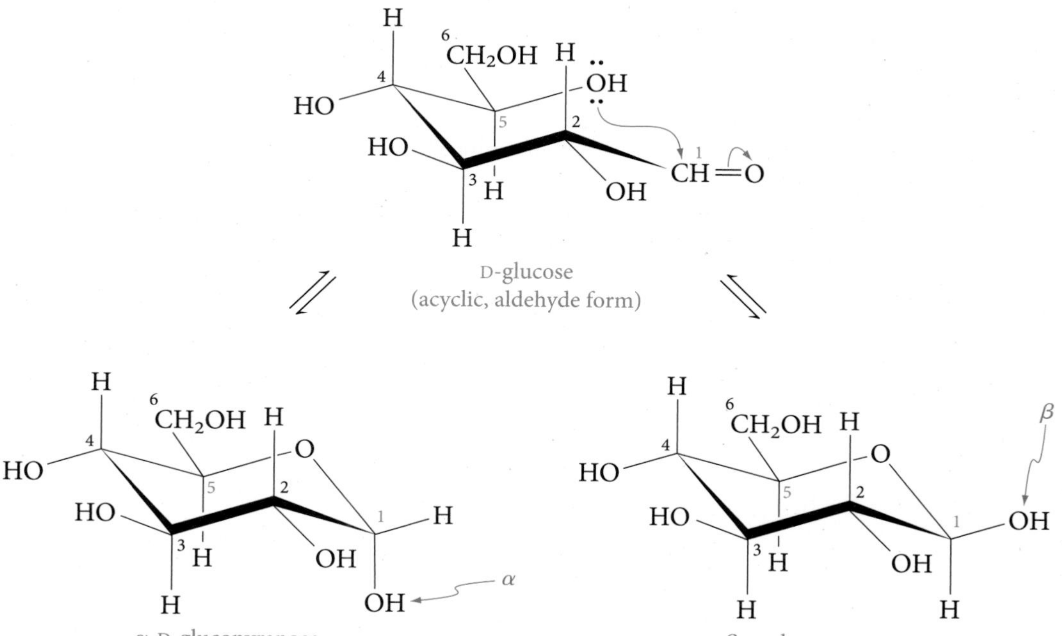

D-glucose
(acyclic, aldehyde form)

(16.6)

α-D-glucopyranose

β-D-glucopyranose

It is probably no accident that glucose is the most abundant natural monosaccharide because in D-glucose, the larger substituent at each ring carbon is equatorial. The only exception occurs at the anomeric carbon (C-1), where the hydroxyl group may be axial (in the α anomer) or equatorial (in the β anomer). This difference provides one reason why the β form is preferred at equilibrium (eq. 16.3). For a ball-and-stick model of β-D-glucopyranose, see Section 2.9.

EXAMPLE 16.5

Draw the most stable chair conformation of α-D-mannopyranose.

Solution Recall from Example 16.3 that D-mannose differs from D-glucose only at C-2. Using the cyclic structure at the left of eq. 16.6 as a guide, we can write

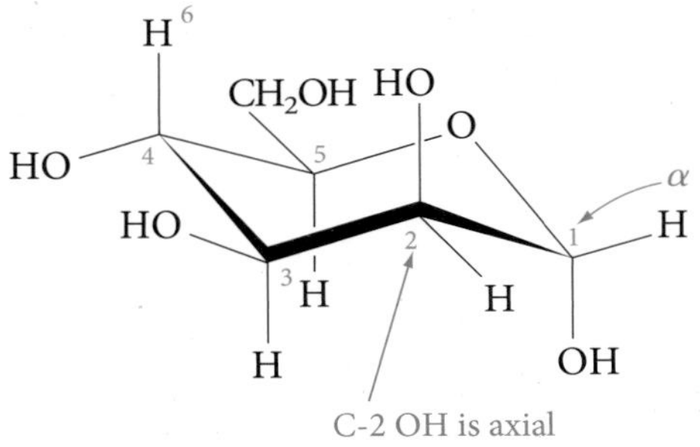

C-2 OH is axial

α-D-mannopyranose

PROBLEM 16.8 D-Galactose differs from D-glucose only in the configuration at C-4. Draw the most stable chair conformation of β-D-galactopyranose.

Now that we have described the structures of monosaccharides, let us examine some of their common reactions.

16.8 Esters and Ethers from Monosaccharides

Monosaccharides contain hydroxyl groups. It is not surprising, then, that they undergo reactions typical of alcohols. For example, they can be converted to esters by reaction with acid halides or anhydrides. The conversion of β-D-glucose to its pentaacetate by reaction with excess acetic anhydride is typical; all five hydroxyl groups, including the hydroxyl at the anomeric C-1, are esterified. (To clarify the structure, the ring H's are omitted.)

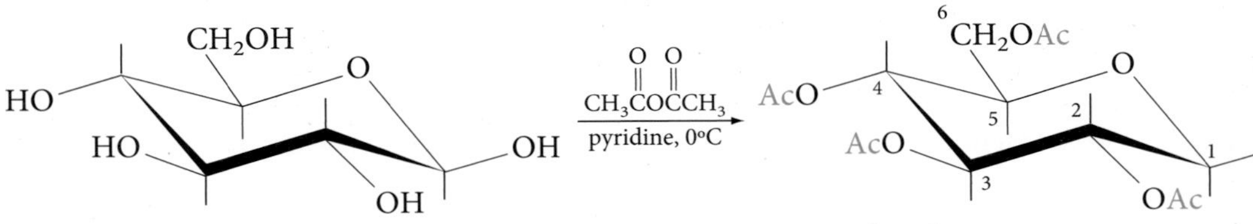

β-D-glucopyranose $\xrightarrow[\text{pyridine, 0°C}]{CH_3COCCH_3}$ β-D-glucopyranose pentaacetate **(16.7)**

$Ac = CH_3C(=O)—$

The hydroxyl groups can also be converted to ethers by treatment with an alkyl halide and a base (the Williamson synthesis, Sec. 8.5). Because sugars are sensitive to strong bases, the mild base silver oxide is preferred.

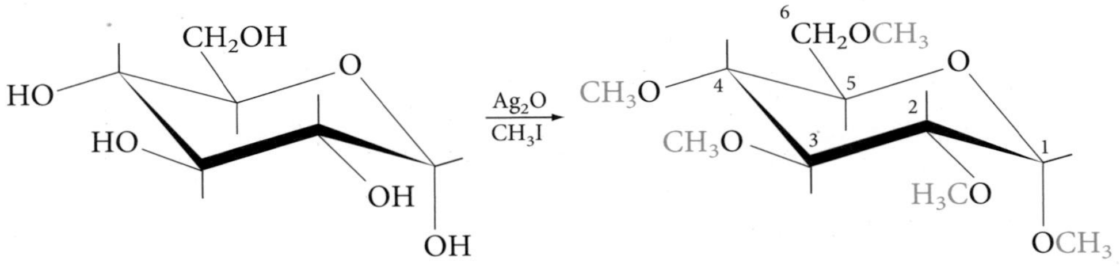

α-D-glucopyranose $\xrightarrow[CH_3I]{Ag_2O}$ α-D-glucopyranose pentamethyl ether **(16.8)**

Whereas sugars tend to be soluble in water and insoluble in organic solvents, the reverse is true for their esters and ethers. This often facilitates their purification and manipulation with organic reagents.

16.9 Reduction of Monosaccharides

When the carbonyl group of an aldose or ketose is reduced, the product is an **alditol**, an acyclic **polyol**.

The carbonyl group of aldoses and ketoses can be reduced by various reagents. The products are **polyols**, called **alditols**. For example, catalytic hydrogenation or reduction with sodium borohydride ($NaBH_4$) converts D-glucose to D-glucitol (also called sorbitol; review Sec. 9.12).

CH₂OH, HO, HO, O, OH, OH — D-glucose (cyclic) ⇌ CH=O; H–OH; HO–H; H–OH; H–OH; CH_2OH — D-glucose (acyclic) $\xrightarrow[\text{or } NaBH_4]{H_2,\ \text{catalyst}}$ CH_2OH; H–OH; HO–H; H–OH; H–OH; CH_2OH — D-glucitol (sorbitol) **(16.9)**

Reaction occurs by reduction of the small amount of aldehyde in equilibrium with the cyclic hemiacetal. As that aldehyde is reduced, the equilibrium shifts to the right, so that eventually all of the sugar is converted. Sorbitol is used commercially as a sweetener and sugar substitute.

When the aldehyde group of an aldose is oxidized, the product is an **aldonic acid**.

PROBLEM 16.9 D-Mannitol, which occurs naturally in olives, onions, and mushrooms, can be made by $NaBH_4$ reduction of D-mannose. Draw its structure.

16.10 Oxidation of Monosaccharides

Benedict's reagent (blue) reacts with an aldose to give Cu_2O (red precipitate).

Although aldoses exist primarily in cyclic hemiacetal forms, these structures are in equilibrium with a small but finite amount of the open-chain aldehyde. These aldehyde groups can be easily oxidized to acids (review Sec. 9.13). The products are called **aldonic acids**. For example, D-glucose is easily oxidized to D-gluconic acid.

CH=O; H–OH; HO–H; H–OH; H–OH; CH_2OH — D-glucose $\xrightarrow[\text{or } Ag^+ \text{ or } Cu^{2+}]{Br_2,\ H_2O}$ COOH; H–OH; HO–H; H–OH; H–OH; CH_2OH — D-gluconic acid **(16.10)**

The oxidation of aldoses is so easy that they react with such mild oxidizing agents as Tollens' reagent (Ag^+ in aqueous ammonia), Fehling's reagent (Cu^{2+} complexed with tartrate ion), or Benedict's reagent (Cu^{2+} complexed with citrate ion). With Tollens' reagent, they give a silver mirror test (Sec. 9.13), and with the copper reagents, the blue solution gives a red precipitate of cuprous oxide, Cu_2O. A carbohydrate that reacts with Ag^+ or Cu^{2+} is called a **reducing sugar** because reduction of the metal accompanies oxidation of the aldehyde group. These reagents are used in laboratory tests for this property.

An aldose that reduces Ag^+ or Cu^{2+} and is itself oxidized is called a **reducing sugar**.

$$\underset{\text{blue solution}}{RCH{=}O + 2\,Cu^{2+}} + 5\,OH^- \longrightarrow RC(=O)O^- + \underset{\text{red precipitate}}{Cu_2O} + 3\,H_2O \quad (16.11)$$

PROBLEM 16.10 Write an equation for the reaction of D-mannose with Fehling's reagent (Cu^{2+}) to give D-mannonic acid.

Stronger oxidizing agents, such as aqueous nitric acid, oxidize the aldehyde group *and* the primary alcohol group, producing dicarboxylic acids called **aldaric acids**. For example, D-glucose gives D-glucaric acid.

Aldoses are oxidized by aqueous nitric acid to dicarboxylic acids called **aldaric acids**.

CH=O; H–C–OH; HO–C–H; H–C–OH; H–C–OH; CH₂OH (D-glucose) $\xrightarrow{HNO_3}$ COOH; H–C–OH; HO–C–H; H–C–OH; H–C–OH; COOH (D-glucaric acid) (16.12)

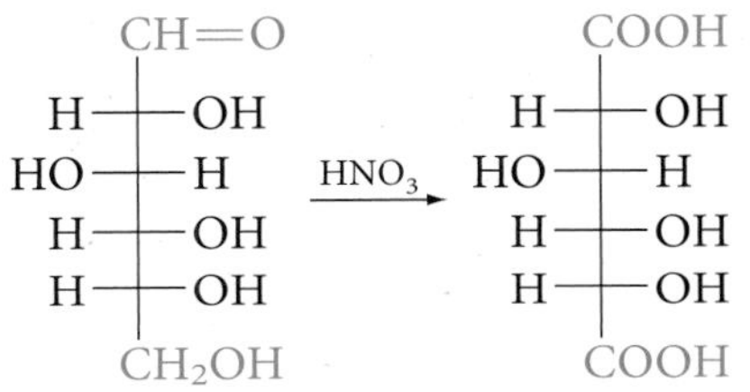

D-glucose

D-glucaric acid

PROBLEM 16.11 Write the structure of D-mannaric acid.

16.11 Formation of Glycosides from Monosaccharides

Because monosaccharides exist as cyclic hemiacetals, they can react with one equivalent of an alcohol to form acetals. An example is the reaction of β-D-glucose with methanol.

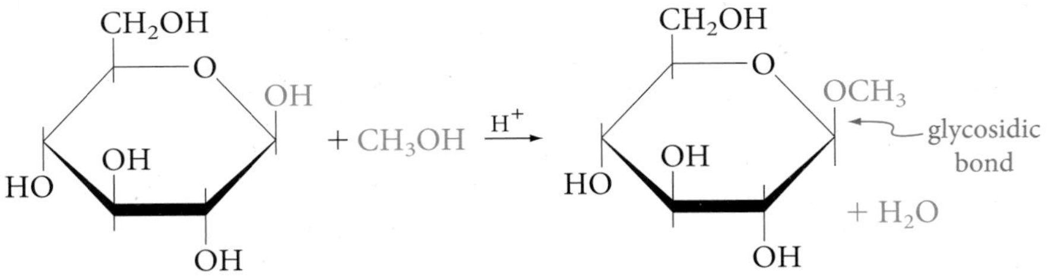

β-D-glucopyranose $+ CH_3OH \xrightarrow{H^+}$ methyl β-D-glucopyranoside (anomeric OCH_3, glycosidic bond) $+ H_2O$ (16.13)

β-D-glucopyranose

methyl β-D-glucopyranoside (mp 115–116°C)

Note that *only the —OH on the anomeric carbon is replaced by an —OR group.* Such acetals are called **glycosides**, and the bond from the anomeric carbon to the OR group is called the **glycosidic bond**. Glycosides are named from the corresponding

In a **glycoside**, the anomeric —OH group is replaced by an —OR group. The bond from the anomeric carbon to the —OR group is called the **glycosidic bond**.

monosaccharide by changing the -*e* ending to -*ide*. Thus, glucose gives glucosides, mannose gives mannosides, and so on.

EXAMPLE 16.6

Write a Haworth formula for ethyl α-D-mannoside.

Solution

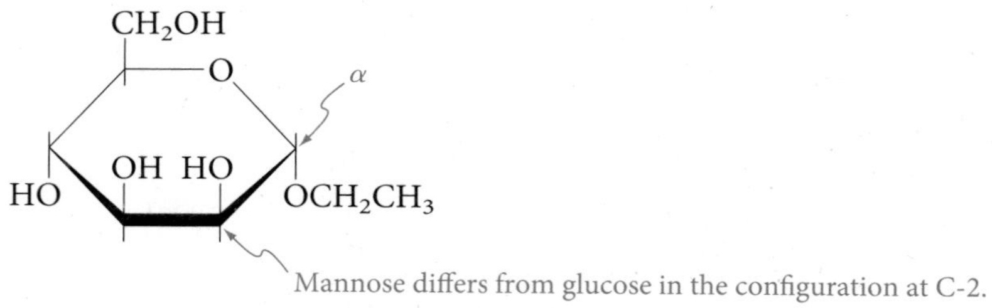

CH2OH
O
α
OH HO
HO
OCH2CH3
Mannose differs from glucose in the configuration at C-2.

PROBLEM 16.12 Write an equation for the acid-catalyzed reaction of β-D-galactose with methanol.

The mechanism of glycoside formation is the same as that described in eq. 9.13 of Section 9.7. The acid catalyst can protonate any of the six oxygen atoms, since each has unshared electron pairs and is basic. However, *only protonation of the hydroxyl oxygen at C-1 leads, after water loss, to a resonance-stabilized carbocation.* In the final step, methanol can attack from either "face" of the six-membered ring, to give either the β-glycoside as shown or the α-glycoside.

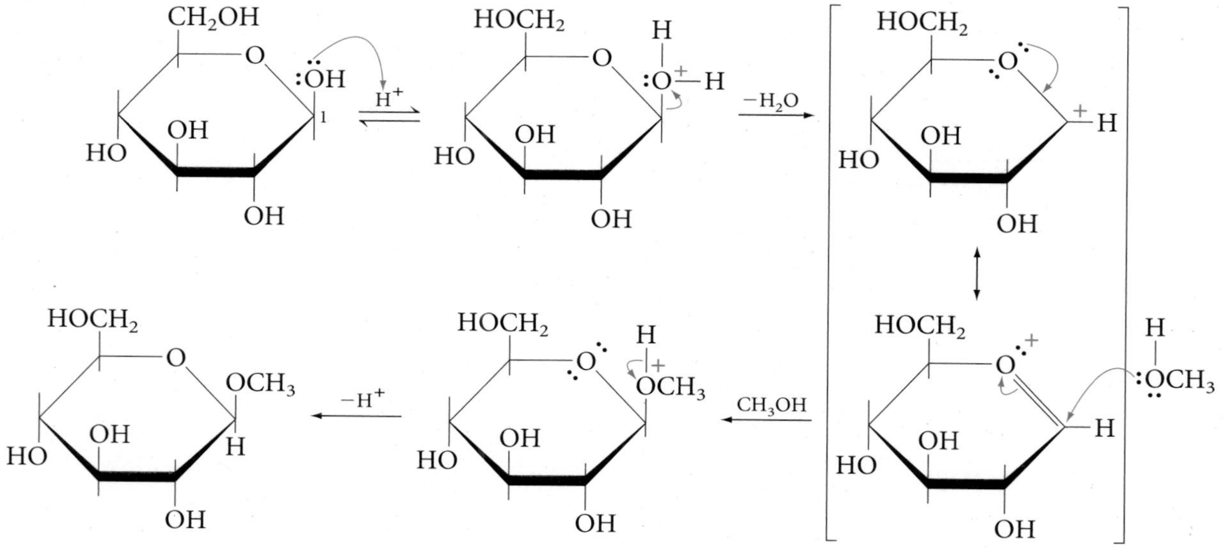

CH2OH
:ÖH
H+
−H2O
CH3OH
−H+
HOCH2
OCH3
:OCH3

(16.14)

Naturally occurring alcohols or phenols often occur in cells combined as a glycoside with some sugar—most commonly, glucose. In this way, the many hydroxyl groups of the sugar portion of the glycoside solubilize compounds that would otherwise be insoluble in cellular protoplasm. An example is the bitter-tasting glucoside salicin, which occurs in willow bark and whose fever-reducing power was known to the ancients.

salicin
(the β-D-glucoside of salicyl alcohol)

Extracts of willow bark were used for medicinal purposes as early as the time of Hippocrates.

The glycosidic bond is the key to understanding the structure of oligosaccharides and polysaccharides, as we will see in the following sections.

16.12 Disaccharides

The most common oligosaccharides are **disaccharides.** *In a disaccharide, two monosaccharides are linked by a glycosidic bond between the anomeric carbon of one monosaccharide unit and a hydroxyl group on the other unit.* In this section, we will describe the structure and properties of four important disaccharides.

A **disaccharide** consists of two monosaccharides linked by a glycosidic bond between the anomeric carbon of one unit and a hydroxyl group on the other unit.

16.12.a Maltose

Maltose is the disaccharide obtained by the partial hydrolysis of starch. Further hydrolysis of maltose gives only D-glucose (eq. 16.2). Maltose must, therefore, consist of two linked glucose units. It turns out that the anomeric carbon of the left unit is linked to the C-4 hydroxyl group of the unit at the right as an acetal (glycoside). The configuration at the anomeric carbon of the left unit is α. In the crystalline form, the anomeric carbon of the right unit has the β configuration. Both units are pyranoses, and the right-hand unit fills the same role as the methanol in eq. 16.13.

Some common disaccharides are **maltose, cellobiose, lactose,** and **sucrose.**

acetal (glycosidic) carbon

maltose
4-O-(α-D-glucopyranosyl)-β-D-glucopyranose

The systematic name for maltose, shown beneath the common name, describes the structure fully, including the name of each unit (D-glucose), the ring sizes (pyranose), the configuration at each anomeric carbon (α or β), and the location of the hydroxyl group involved in the glycosidic link (4-O).

The anomeric carbon of the right glucose unit in maltose is a hemiacetal. Naturally, when maltose is in solution, this hemiacetal function will be in equilibrium with the open-chain aldehyde form. Maltose, therefore, gives a positive Tollens' test and other reactions similar to those of the anomeric carbon in glucose.

PROBLEM 16.13 When crystalline maltose is dissolved in water, the initial specific rotation changes and gradually reaches an equilibrium value. Explain.

16.12.b Cellobiose

Cellobiose is the disaccharide obtained by the partial hydrolysis of cellulose. Further hydrolysis of cellobiose gives only D-glucose. Cellobiose must therefore be an isomer of maltose. In fact, *cellobiose differs from maltose* only *in having the β configuration at C-1 of the left glucose unit.* Otherwise, all other structural features are identical, including a link from C-1 of the left unit to the hydroxyl group at C-4 in the right unit.

cellobiose
4-O-(β-D-glucopyranosyl)-β-D-glucopyranose

Note that, in the conformational formula for cellobiose, one ring oxygen is drawn to the "rear" and one to the "front" of the molecule. This is the way the rings exist in the cellulose chain.

Cheese whey is a source of lactose, also known as milk sugar.

16.12.c Lactose

Lactose is the major sugar in human and cow's milk (4% to 8% lactose). Hydrolysis of lactose gives equimolar amounts of D-galactose and D-glucose. The anomeric carbon of the galactose unit has the β configuration at C-1 and is linked to the hydroxyl group at C-4 of the glucose unit. The crystalline anomer, with the α configuration at the glucose unit, is made commercially from cheese whey.

galactose unit glucose unit

lactose
4-O-(β-D-galactopyranosyl)-α-D-glucopyranose

PROBLEM 16.14 Will lactose give a positive Fehling's test? Will it mutarotate?

Norbert Rillieux, inventor of the "triple-effect evaporator" used in sugar processing.

Some human infants are born with a disease called *galactosemia*. They lack the enzyme that isomerizes galactose to glucose and therefore cannot digest milk. If milk is excluded from such infants' diets, the disease symptoms caused by accumulation of galactose can be avoided.

16.12.d Sucrose

The most important commercial disaccharide is **sucrose**, ordinary table sugar. More than 130 million tons are produced annually worldwide. Sucrose occurs in all photosynthetic plants, where it functions as an energy source. It is obtained commercially from sugar cane and sugar beets, in which it constitutes 14% to 20% of the plant juices.

Sugar beets are a source of table sugar, sucrose.

One of the major engineering advances of the industrial revolution was developed to reduce the cost and labor associated with isolating sucrose from sugar cane and sugar beets. Norbert Rillieux, a free African American living in Louisiana in pre–Civil War days, invented the "triple-effect evaporator" to remove water from the juices of sugar cane and sugar beets in 1844. His invention modernized the sugar industry, and versions of his equipment are still used today wherever large amounts of liquid must be quickly evaporated.*

Sucrose is very water soluble (2 grams per milliliter at room temperature) because it is polar due to the presence of eight hydroxyl groups on its surface (Fig. 16.3). Hydrolysis of sucrose gives equimolar amounts of D-glucose and the ketose D-fructose. *Sucrose differs from the other disaccharides we have discussed in that the anomeric carbons of both units are involved in the glycosidic link;* that is, C-1 of the glucose unit is

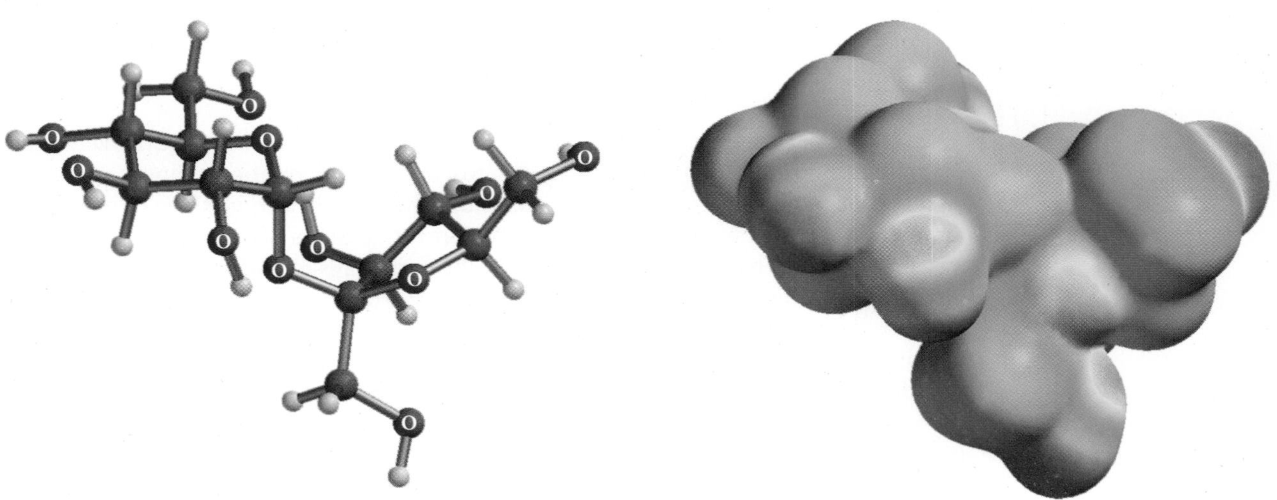

Figure 16.3
The polar nature of the surface of sucrose is illustrated by a ball-and-stick structure on the left (oxygen atoms are red) and electrostatic potential map (hydroxyl groups are in orange-red).

*A marvelous account of this invention can be found in *Prometheans in the Lab* by Sharon B. McGrayne (McGraw-Hill, 2001).

linked via oxygen to C-2 of the fructose unit. An additional difference is that the fructose unit is in the furanose form.

sucrose
α-D-glucopyranosyl-β-D-fructofuranoside
(or β-D-fructofuranosyl-α-D-glucopyranoside)

Since both anomeric carbons are linked in the glycosidic bond, neither monosaccharide unit has a hemiacetal group. Therefore, neither unit is in equilibrium with an acyclic form. Sucrose cannot mutarotate. And, because there is no free or potentially free aldehyde group, sucrose cannot reduce Tollens', Fehling's, or Benedict's reagent. Sucrose is therefore referred to as a *nonreducing sugar,* in contrast with the other disaccharides and monosaccharides we have discussed, all of which are reducing sugars.

© Philip Gould/CORBIS

Sugar cane field in Louisiana.

PROBLEM 16.15 Although β-D-glucose is a reducing sugar, methyl β-D-glucopyranoside (eq. 16.13) is not. Explain.

Hydrolysis of sucrose ($[\alpha] = +66°$), catalyzed by enzymes called **invertases**, produces a mixture of glucose and fructose called **invert sugar** ($[\alpha] = +20°$).

Sucrose has an optical rotation of $[\alpha] = +66°$. When sucrose is hydrolyzed to an equimolar mixture of D-glucose and D-fructose, the optical rotation changes value and sign and becomes $[\alpha] = -20°$. This is because the equilibrium mixture of D-glucose anomers (α and β) has a rotation of +52°, but the mixture of fructose anomers has a strong negative rotation, $[\alpha] = -92°$. In the early days of carbohydrate chemistry, glucose was called dextrose (because it was dextrorotatory), and fructose was called levulose (because it was levorotatory). Because hydrolysis of sucrose inverts the sign of optical rotation (from + to −), enzymes that bring about sucrose hydrolysis are called **invertases**, and the resulting equimolar mixture of glucose and fructose is called **invert sugar**. A number of insects, including the honeybee, possess invertases. Honey is largely a mixture of D-glucose, D-fructose, and some unhydrolyzed sucrose. It also contains flavors from the particular flowers whose nectars are collected.

A WORD ABOUT... Sweetness and Sweeteners

© 2005 Tom Pantages

Several products containing saccharin and aspartame.

Sweetness is literally a matter of taste. Although individuals vary greatly in their sensory perceptions, it is possible to make some quantitative comparisons of sweetness. For example, we can take some standard sugar solution (say 10% sucrose in water) and compare its sweetness with that of solutions containing other sugars or sweetening agents. If a 1% solution of some compound tastes as sweet as the 10% sucrose solution, we can say that the compound is 10 times sweeter than sucrose.

D-Fructose is the sweetest of the simple sugars—almost twice as sweet as sucrose. D-Glucose is almost as sweet as sucrose. On the other hand, sugars like lactose and galactose have less than 1% of the sweetness of sucrose.

Many synthetic sweeteners are known, perhaps the most familiar being **saccharin**. It was discovered in 1879 in the laboratory of Professor Ira Remsen at Johns Hopkins University. Its structure has no relation whatsoever to that of the saccharides, but saccharin is about 300 times sweeter than sucrose. For most tastes, 0.5 grain (0.03 g) of saccharin is equivalent in sweetness to a heaping teaspoon (10 g) of sucrose. Saccharin is made commercially from toluene, as shown in Figure 16.4.

Saccharin is very sweet yet has virtually no caloric content. It is useful as a sugar substitute for those who must restrict their sugar intake and also for those who wish to control their weight but still have a desire for sweets.

In 1981, **aspartame** became the first new sweetener to be approved by the U.S. Food and Drug Administration (FDA) in nearly 25 years. It is about 160 times sweeter than sucrose. Structurally, aspartame is the methyl ester of a dipeptide of two amino acids that occur naturally in proteins—aspartic acid and phenylalanine—and is sold under the trade name NutraSweet®. (Amino acids and peptides will be discussed in the next chapter.)

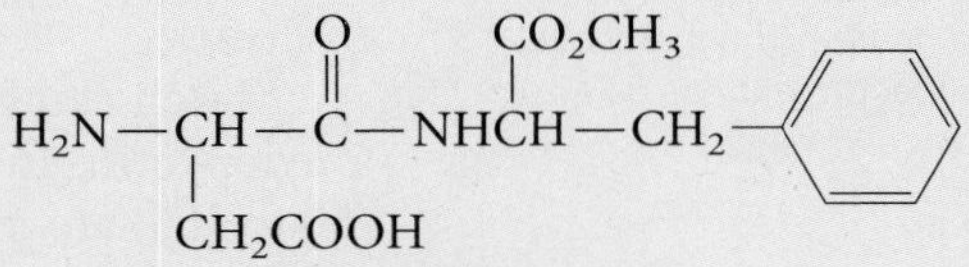

$$H_2N-\underset{\displaystyle CH_2COOH}{CH}-\overset{\displaystyle O}{\overset{\|}{C}}-NH\overset{\displaystyle CO_2CH_3}{CH}-CH_2-C_6H_5$$

the methyl ester of *N*-L-α-aspartyl-L-phenylalanine (aspartame)

Aspartame has about the same caloric content as sucrose, but because of its intense sweetness, much less is used, so its energy value becomes insignificant.

D-Sucralose is a trichloro derivative of sucrose whose structure is shown on page 490. This disaccharide is 600 times sweeter than sucrose and is sold under the brand name of Splenda®. For a nice review of the synthesis and properties of commercial, synthetic nonnutritive sweeteners, see *Angew. Chem. Int. Ed.* **1998**, *37*, 1802–1817 by David J. Ager and colleagues.

See Problems 16.44, 16.53, and 16.54.

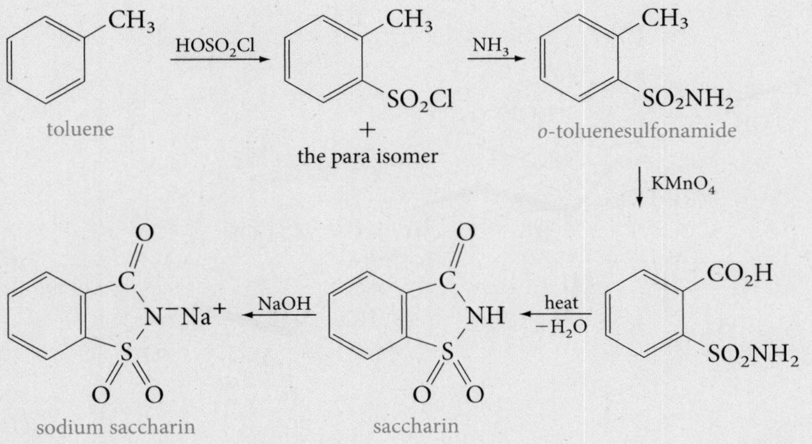

toluene ($C_6H_5CH_3$) $\xrightarrow{HOSO_2Cl}$ o-$CH_3C_6H_4SO_2Cl$ + the para isomer $\xrightarrow{NH_3}$ o-$CH_3C_6H_4SO_2NH_2$ (*o*-toluenesulfonamide) $\xrightarrow{KMnO_4}$ o-$HO_2CC_6H_4SO_2NH_2$ $\xrightarrow[-H_2O]{heat}$ saccharin $\xrightarrow{NaOH}$ sodium saccharin (N^-Na^+)

■ **Figure 16.4**
Synthesis of saccharin.

16.13 Polysaccharides

Polysaccharides contain many linked monosaccharides and vary in chain length and molecular weight. Most polysaccharides give a single monosaccharide on complete hydrolysis. The monosaccharide units may be linked linearly, or the chains may be branched. In this section, we will describe a few of the more important polysaccharides.

16.13.a Starch and Glycogen

The polysaccharide **starch** contains glucose units joined by 1,4-α-glycosidic bonds. **Amylose** is an unbranched form of starch, whereas **amylopectin** is highly branched. **Glycogen,** found in animals, is more highly branched than amylopectin and has a higher molecular weight.

Starch is the energy-storing carbohydrate of plants. It is a major component of cereals, potatoes, corn, and rice. It is the form in which glucose is stored by plants for later use.

Starch is made up of glucose units joined mainly by 1,4-α-glycosidic bonds, although the chains may have a number of branches attached through 1,6-α-glycosidic bonds. Partial hydrolysis of starch gives maltose, and complete hydrolysis gives only D-glucose.

Starch can be separated by various techniques into two fractions: amylose and amylopectin. In **amylose**, which constitutes about 20% of starch, the glucose units (50 to 300) are in a continuous chain, with 1,4 linkages (Figure 16.5).

Amylopectin (Figure 16.6) is highly branched. Although each molecule may contain 300 to 5000 glucose units, chains with consecutive 1,4 links average only 25 to 30 units in length. These chains are connected at branch points by 1,6 linkages. Because of this highly branched structure, starch granules swell and eventually form colloidal systems in water.

Glycogen is the energy-storing carbohydrate of animals. Like starch, it is made of 1,4- and 1,6-linked glucose units. Glycogen has a higher molecular weight than starch (perhaps 100,000 glucose units), and its structure is even more branched than that of amylopectin, with a branch every 8 to 12 glucose units. Glycogen is produced from glucose that is absorbed from the intestines into the blood; transported to the liver, muscles, and elsewhere; and then polymerized enzymatically. Glycogen helps maintain the glucose balance in the body by removing and storing excess glucose from ingested food and later supplying it to the blood when various cells need it for energy.

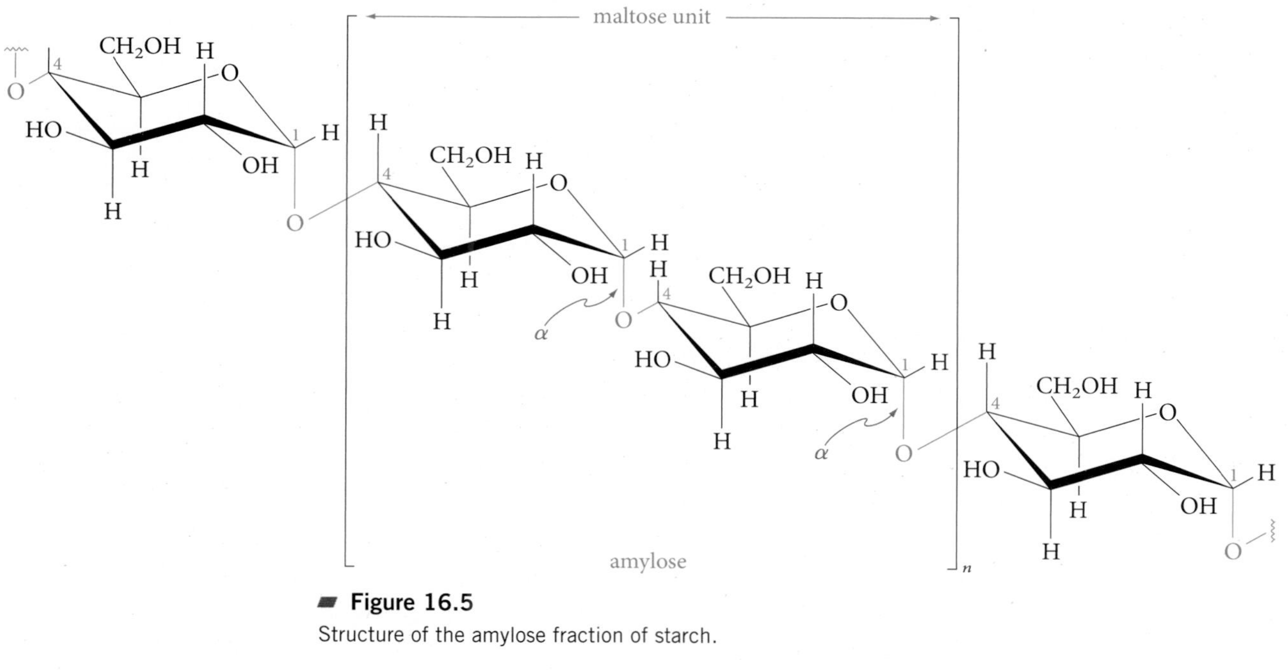

■ **Figure 16.5**
Structure of the amylose fraction of starch.

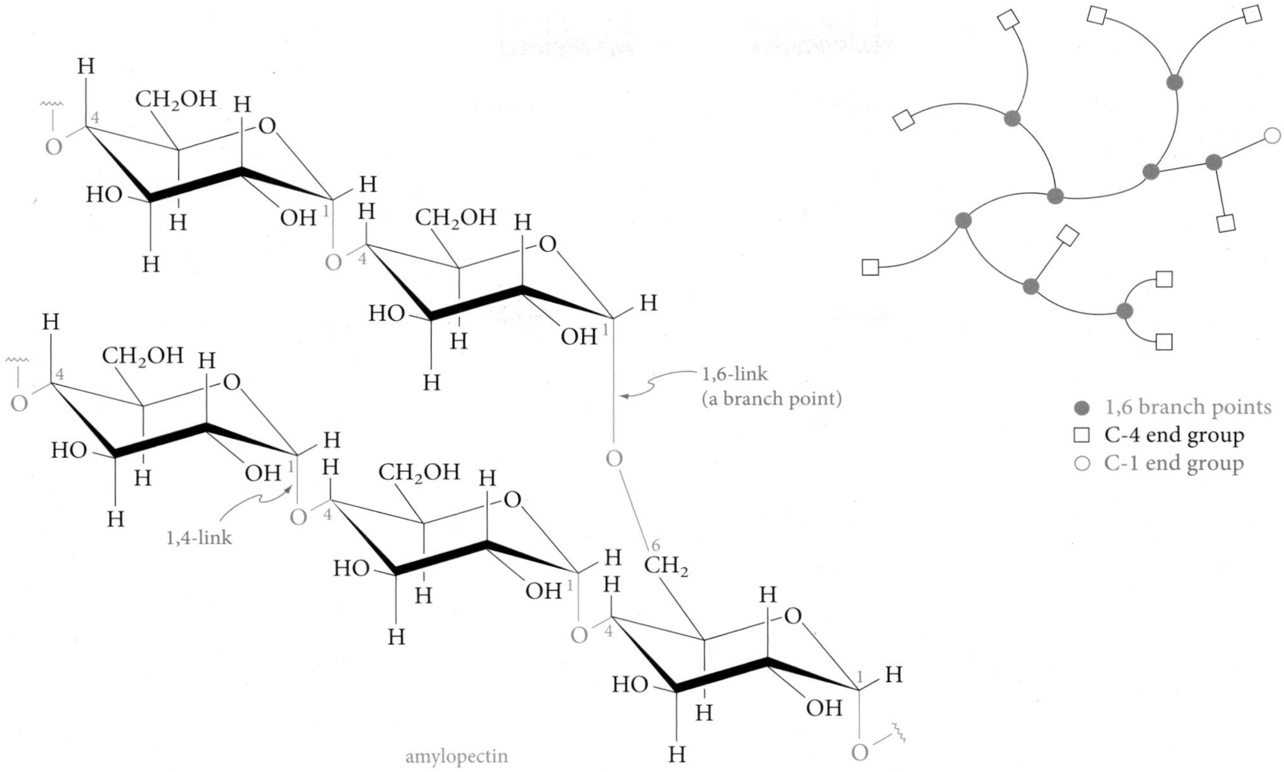

amylopectin

● 1,6 branch points
□ C-4 end group
○ C-1 end group

■ **Figure 16.6**
Structure of the amylopectin fraction of starch. Adapted from Peter M. Collins and Robert J. Ferrier, *Monosaccharides: Their Chemistry and Their Roles in Natural Products,* p. 491. Copyright © 1995 by John Wiley & Sons, Ltd. Reprinted with permission.

16.13.b Cellulose

Cellulose is an *unbranched* polymer of glucose joined by 1,4-β-glycosidic bonds. X-ray examination of cellulose shows that it consists of linear chains of cellobiose units in which the ring oxygens alternate in "forward" and "backward" positions (Figure 16.7). These linear molecules, containing an average of 5000 glucose units, aggregate to give fibrils bound together by hydrogen bonds between hydroxyls on adjacent chains. Cellulose fibers having considerable physical strength are built up from these fibrils, wound spirally in opposite directions around a central axis. Wood, cotton, hemp, linen, straw, and corncobs are mainly cellulose.

Cellulose is an unbranched polymer of glucose joined by 1,4-β-glycosidic bonds. In **cellulose acetate**, the free hydroxyl groups are acetylated; in **cellulose nitrate**, they are nitrated. **Guncotton**, a highly nitrated cellulose, is an explosive.

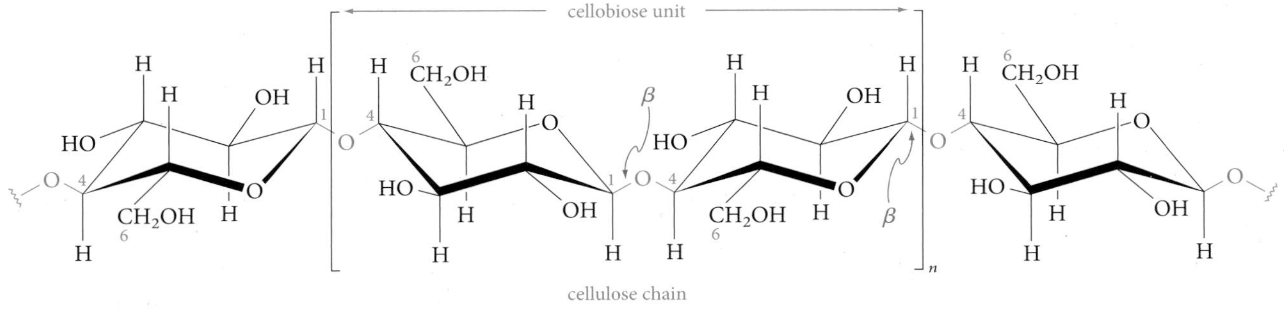

cellobiose unit

cellulose chain

■ **Figure 16.7**
Partial structure of a cellulose molecule showing the β linkages of each glucose unit.

A WORD ABOUT... Fat Substitutes from Carbohydrates

© Tom Pantages

Several products containing Olestra, Oatrim, and Z-Trim.

Like it or not, many of us eat numerous food additives as a consequence of our diets. We are all familiar with artificial sweeteners (see "A Word about Sweetness and Sweeteners" earlier in this chapter), which satisfy our desire for the taste of sugar without providing the caloric intake and weight problems associated with sweets. Fats also taste good and are an even better source of energy than carbohydrates. Every gram of fat we eat provides us with twice the energy (in calories) of a gram of carbohydrate. Therefore, it is not surprising that considerable effort has gone into developing artificial fats.

ester linkages

triglycerides
R = long chain saturated and unsaturated alkyl

Fats are triesters of the polyol glycerol. They are good at delivering flavors and provide food with a pleasing texture. When we eat fats, they are metabolized when needed as an energy source, or stored for later use when carbohydrates are no longer available as an energy source. Both metabolism and storage of fats involve hydrolysis of the ester linkages by esterases (see "A Word about Prostaglandins, Aspirin, and Pain" in Chapter 15) in the gut to give glycerol and fatty acids.

One family of compounds that has been used as artificial fats is polyesters of sucrose. They are prepared by esterifying sucrose (table sugar) with derivatives of fatty acids obtained from vegetable oils such as cottonseed or soybean oil. Esters of this type are marketed under the name **Olestra**. Olestra has been used to replace fat in a number of snacks and can replace cooking oil in preparing fried foods, such as potato chips. Olestra has a structure similar to that of a normal fat, being a fatty acid ester of a polyol. It has the physical and textural properties of fats, but because it is so large, it is not hydrolyzed by the esterases that normally begin the process of fat storage and metabolism. It simply passes through the digestive tract and is excreted without being used as an energy source.

$R = C(=O)(CH_2)_nCH_3$

Olestra

Other carbohydrate-derived fat substitutes have been developed from oats. Slightly less than 50% of the weight of oats consists of fiber, cellulose-like polysaccharides not easily digested by humans. Some of this fiber is water soluble, and some of it is not. Both kinds of fibers have been processed to give materials (Oatrim and Z-Trim) that mimic the texture of fat but have very little caloric value. These products are being used in the preparation of some low-fat food products, such as cheese and baked goods. For an interesting overview of fat substitutes, see "Fake Fats in Real Food" by Elisabeth M. Kirschner in *Chemical and Engineering News*, **1997**, *75*, 19.

See Problems 16.40 and 16.45.

Although humans and other animals can digest starch and glycogen, they cannot digest cellulose. This is a truly striking example of the specificity of biochemical reactions. *The only chemical difference between starch and cellulose is the stereochemistry of the glucosidic link*—more precisely, the stereochemistry at C-1 of each glucose unit. The human digestive system contains enzymes that can catalyze the hydrolysis of α-glucosidic bonds, but it lacks the enzymes necessary to hydrolyze β-glucosidic bonds. Many bacteria, however, do contain β-glucosidases and can hydrolyze cellulose. Termites, for example, have such bacteria in their intestines and thrive on wood (cellulose) as their main food. Ruminants (cud-chewing animals such as cows) can digest grasses and other forms of cellulose because they harbor the necessary microorganisms in their rumen.

Cellulose is the raw material for several commercially important derivatives. *Each glucose unit in cellulose contains three hydroxyl groups.* These hydroxyl groups can be modified by the usual reagents that react with alcohols. For example, cellulose reacts with acetic anhydride to give **cellulose acetate**.

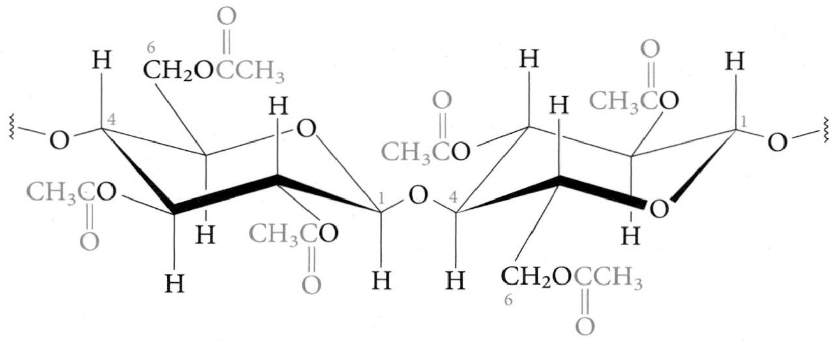

segment of a cellulose acetate molecule

Cellulose with about 97% of the hydroxyl groups acetylated is used to make acetate rayon.

Cellulose nitrate is another useful cellulose derivative. Like glycerol, cellulose can be converted with nitric acid to a nitrate ester (compare eq. 7.41). The number of hydroxyl groups nitrated per glucose unit determines the properties of the product. **Guncotton**, a highly nitrated cellulose, is an efficient explosive used in smokeless powders.

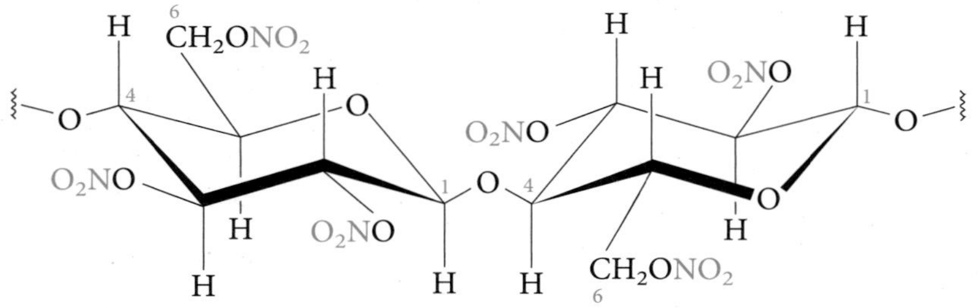

segment of a cellulose nitrate molecule

16.13.c Other Polysaccharides

Chitin is a nitrogen-containing polysaccharide that forms the shells of crustaceans and the exoskeletons of insects. It is similar to cellulose, except that the hydroxyl group at C-2 of each glucose unit is replaced by an acetylamino group, CH_3CONH—.

Chitin is found in crustacean shells and insect exoskeletons; **pectins**, found in fruits, are used in making jellies.

Pectins, which are obtained from fruits and berries, are polysaccharides used in making jellies. They are linear polymers of D-galacturonic acid, linked with 1,4-α-glycosidic bonds. D-Galacturonic acid has the same structure as D-galactose, except that the C-6 primary alcohol group is replaced by a carboxyl group.

Numerous other polysaccharides are known, such as gum arabic and other gums and mucilages, chondroitin sulfate (found in cartilage), the blood anticoagulant heparin (found in the liver and heart), and the dextrans (used as blood plasma substitutes).

Some saccharides have structures that differ somewhat from the usual polyhydroxyaldehyde or polyhydroxyketone pattern. In the final sections of this chapter, we will describe a few such modified saccharides that are important in nature.

A WORD ABOUT... Green Chemistry: Alternative Uses for Carbohydrates

A sugar cane plantation.

As we've discussed, there is currently a great push to find renewable sources for energy. Carbohydrates have received the greatest attention in the recent quest for renewable energy sources, primarily due to their direct formation from photosynthesis. With the environmental switch from MTBE (Chapter 8) to ethanol as an additive in gasoline, ethanol was an immediate target for preparation from renewable resources such as biomass.

Extensive research has been extended toward the fermentation of sucrose and glucose into ethanol. As we saw in Section 16.12.d, invertase enzymes in yeast can convert sucrose into a mixture of glucose and fructose. An enzyme called zymase, also available in yeast, can then convert glucose into ethanol as part of the fermentation process, creating CO_2 as a by-product. A variety of carbohydrate sources can be used in these fermentation steps, including corn, sugar beets, and sugar cane.

Conversely, there are many plant components, specifically cellulose, that are not used for human consumption. Hence, there is growing interest in ethanol fermentation from cellulose (Sec. 16.13.b). However, only recently has there been commercial production of cellulase enzymes that can hydrolyze cellulose into separate glucose units. Indeed, the first cellulose-based facility for producing ethanol was constructed in 2004. If such efforts are successful, there are many biomass sources of cellulosic materials, such as corn stover, switchgrass, sawdust, and straw, each of which can be used as a raw material for fermentation to make ethanol from cellulose.

Moreover, there are other uses for renewable carbohydrates. Many chemical feedstocks that are used for making polymers (Chapter 14) are derived from petroleum sources. New efforts are aimed at making some of those important chemicals from renewable biomass. The U.S. Department of Energy released two reports, one in 2004 and another in 2007, that outline the "Top Value Added Chemicals from Biomass."* The first report explored carbohydrates as the biomass feedstock and the second report focused on lignins. Technology currently exists for the biological conversion of glucose to succinic acid (Table 10.2) in modified forms of the bacterial strain *Escherichia coli*. The four-carbon succinic acid can be used to prepare a variety of other small organic chemicals, many of which are important solvents, water-soluble polymers (2-pyrrolidinone), or straight-chain and branched polymers used in fabrics, such as Spandex or Lycra.

However, currently, the cost associated with the fermentation route is not very competitive relative to petroleum-based sources. Further research is needed to develop renewable sources of energy and value-added chemicals. So while carbohydrates provide a diverse source of these renewable chemicals, success in this area will come from chemical, biological, and engineering efforts going forward.

See Problem 16.55.

glucose $\xrightarrow[\text{fermentation}]{\text{anaerobic}}$ HO_2C–CH_2CH_2–CO_2H (succinic acid)

succinic acid → tetrahydrofuran; 2-pyrrolidone; succindiamide (H_2N–CO–CH_2CH_2–CO–NH_2); 1,4-diaminobutane (H_2N–$(CH_2)_4$–NH_2); 1,4-butanediol (HO–$(CH_2)_4$–OH); γ-butyrolactone

*These reports are available electronically from the DOE at http://www.osti.gov/bridge—search for "Top Value Added Chemicals from Biomass."

16.14 Sugar Phosphates

Phosphate esters of monosaccharides are found in all living cells, where they are intermediates in carbohydrate metabolism. Some common sugar phosphates are the following:

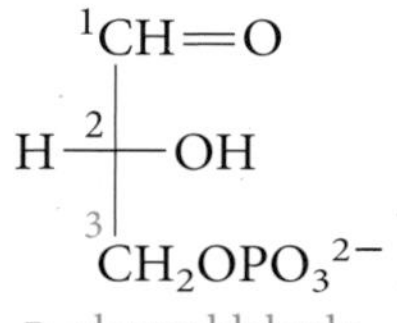

D-glyceraldehyde-3-phosphate

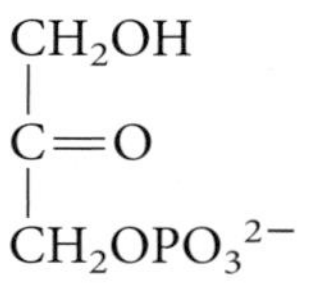

dihydroxyacetone phosphate

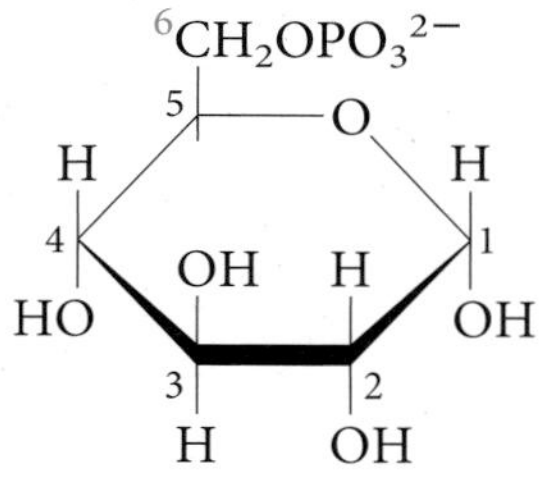

α-D-glucose-6-phosphate

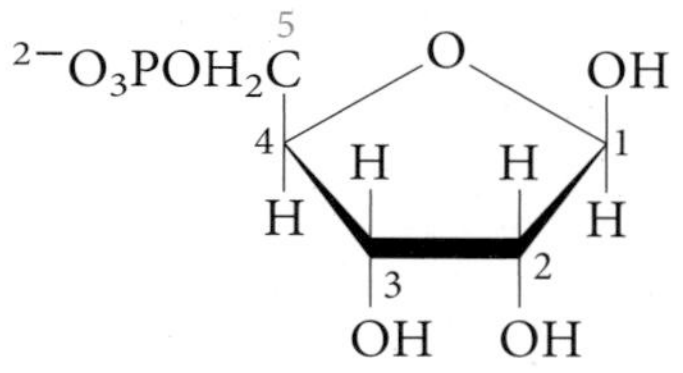

β-D-ribose-5-phosphate

Phosphates of the five-carbon sugar ribose and its 2-deoxy analog are important in nucleic acid structures (DNA and RNA) and in other key biological compounds (Sec. 18.12).

16.15 Deoxy Sugars

In **deoxy sugars**, one or more of the hydroxyl groups is replaced by a hydrogen atom. The most important example is 2-deoxyribose, the sugar component of DNA. It lacks the hydroxyl group at C-2 and occurs in DNA in the furanose form.

In **deoxy sugars**, one or more of the hydroxyl groups of the parent carbohydrate is replaced by a hydrogen atom.

no OH group here

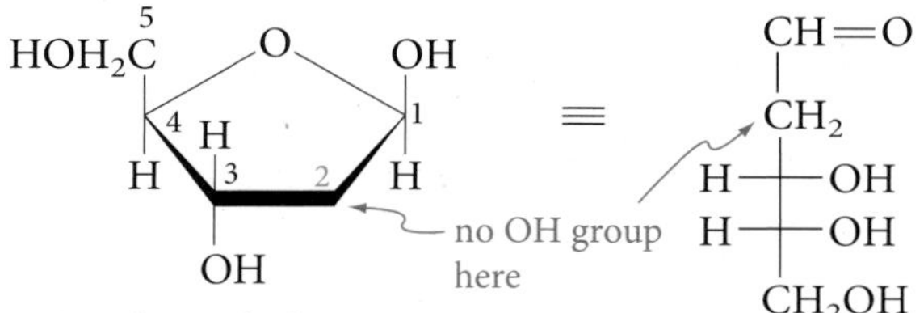

β-D-deoxyribofuranose (the sugar of DNA)

16.16 Amino Sugars

In **amino sugars**, one of the sugar hydroxyl groups is replaced by an amino group. Usually the —NH_2 group is also acetylated. **D-Glucosamine** is one of the more abundant amino sugars.

In **amino sugars**, an —OH group is replaced by an —NH_2 group. **D-Glucosamine** is an abundant amino sugar.

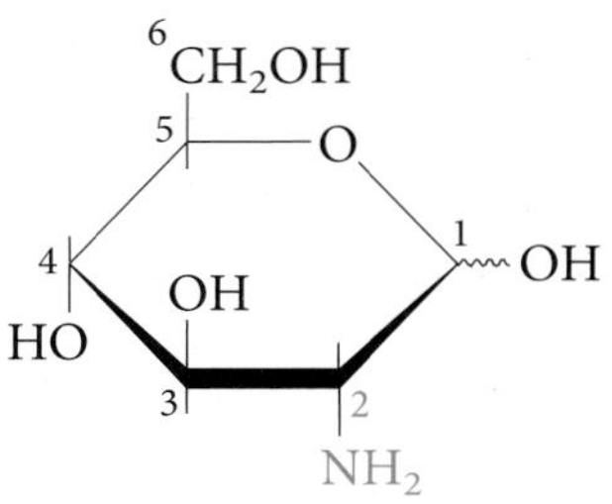

D-glucosamine
α mp 88°C
β [mp 110°C (decomposes)]

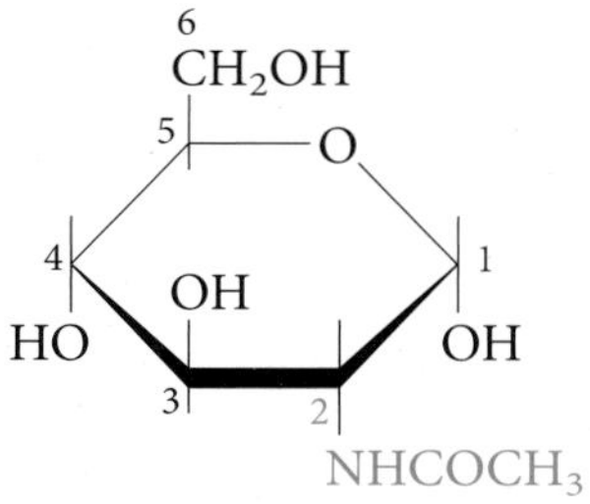

N-acetyl-α-D-glucosamine
[mp 211°C (decomposes)]

In its *N*-acetyl form, β-D-glucosamine is the monosaccharide unit of chitin, which forms the shells of lobsters, crabs, shrimp, and other shellfish.

16.17 Ascorbic Acid (Vitamin C)

L-Ascorbic acid (vitamin C) has a five-membered lactone ring containing an **enediol**. The hydroxyl proton at C-3 is acidic.

L-Ascorbic acid (vitamin C) resembles a monosaccharide, but its structure has several unusual features. The compound has a five-membered unsaturated lactone ring (review Sec. 10.12) with two hydroxyl groups attached to the doubly bonded carbons. This **enediol** structure is relatively uncommon.

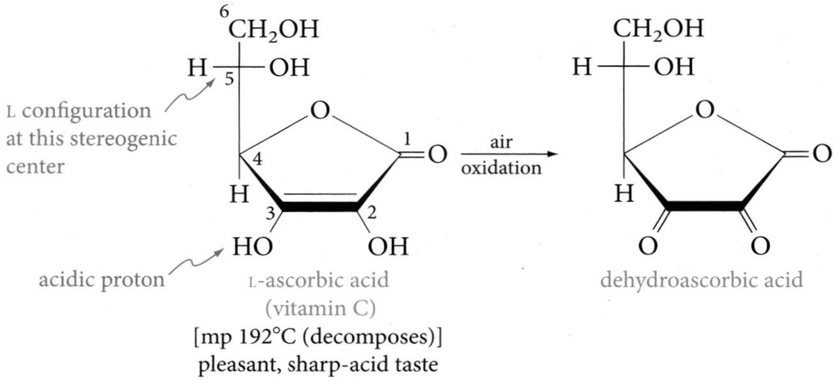

L configuration at this stereogenic center

acidic proton

L-ascorbic acid (vitamin C)
[mp 192°C (decomposes)]
pleasant, sharp-acid taste

air oxidation

dehydroascorbic acid

As a consequence of this structural feature, ascorbic acid is easily oxidized to dehydroascorbic acid. Both forms are biologically effective as a vitamin.

There is no carboxyl group in ascorbic acid, but it is nevertheless an acid with a pK_a of 4.17. The proton of the hydroxyl group at C-3 is acidic, because the anion that results from its loss is resonance stabilized and similar to a carboxylate anion.

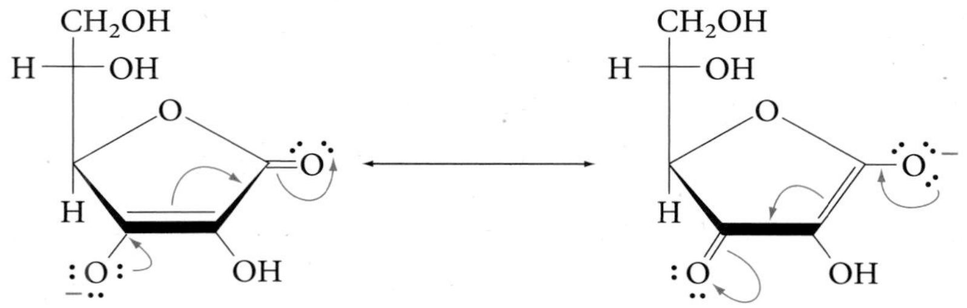

resonance stabilized ascorbate anion

Limes, used by eighteenth-century sailors as a source of vitamin C to prevent scurvy.

Humans, monkeys, guinea pigs, and a few other vertebrates lack an enzyme that is essential for the biosynthesis of ascorbic acid from D-glucose. Hence ascorbic acid must be included in the diet of humans and these other species. Ascorbic acid is abundant in citrus fruits and tomatoes. Its lack in the diet causes scurvy, a disease that results in weak blood vessels, hemorrhaging, loosening of teeth, lack of ability to heal wounds, and eventual death. Ascorbic acid is needed for collagen synthesis (collagen is the structural protein of skin, connective tissue, tendon, cartilage, and bone). In the eighteenth century, British sailors were required to eat fresh limes (a vitamin C source) to prevent outbreaks of the dreaded scurvy; hence their nickname "limeys."

REACTION SUMMARY

1. Reactions of Monosaccharides

a. Mutarotation (Sec. 16.5)

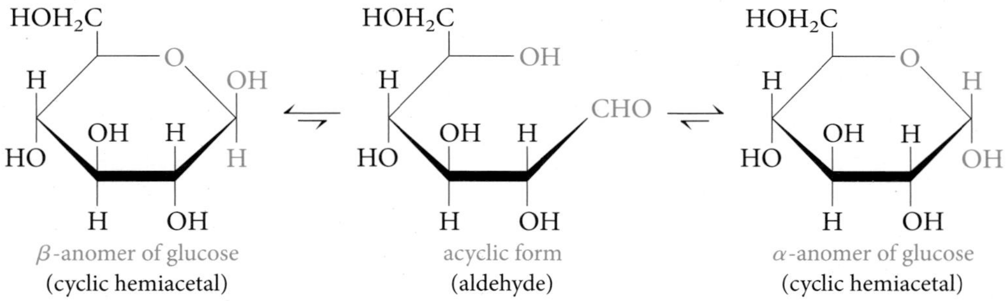

β-anomer of glucose (cyclic hemiacetal) ⇌ acyclic form (aldehyde) ⇌ α-anomer of glucose (cyclic hemiacetal)

b. Esterification (Sec. 16.8)

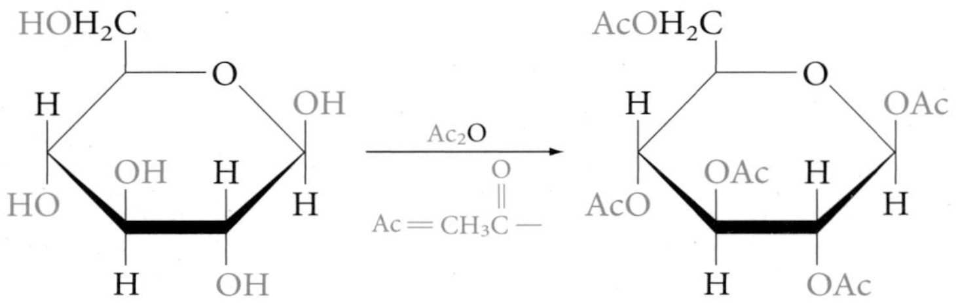

Glucose (pyranose, all OH) $\xrightarrow{Ac_2O}$ pentaacetate (AcO at all positions, $AcOH_2C$)

$Ac = CH_3C(=O)-$

c. Etherification (Sec. 16.8)

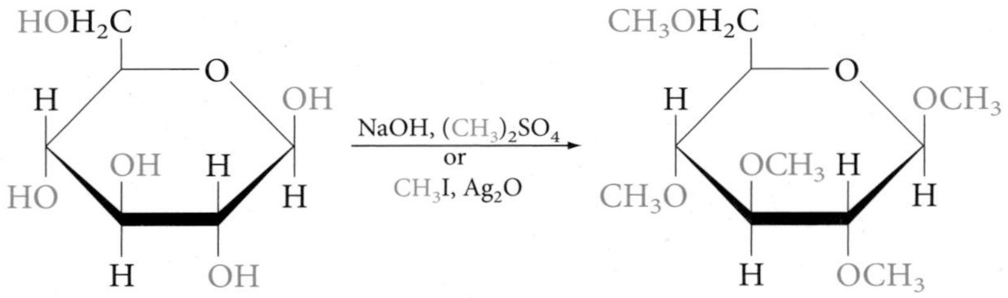

Glucose (pyranose, all OH) $\xrightarrow[\text{or } CH_3I,\ Ag_2O]{NaOH,\ (CH_3)_2SO_4}$ pentamethyl ether (OCH_3 at all positions, CH_3OH_2C)

d. Reduction (Sec. 16.9)

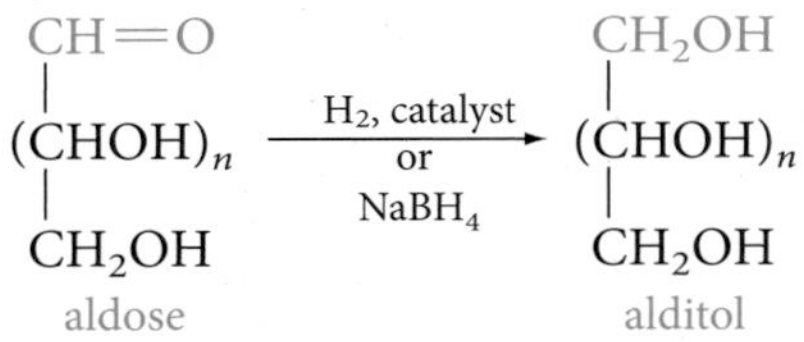

$$\underset{\text{aldose}}{CH{=}O-(CHOH)_n-CH_2OH} \xrightarrow[\text{or } NaBH_4]{H_2,\ \text{catalyst}} \underset{\text{alditol}}{CH_2OH-(CHOH)_n-CH_2OH}$$

e. Oxidation (Sec. 16.10)

$$\underset{\text{aldaric acid}}{CO_2H-(CHOH)_n-CO_2H} \xleftarrow{HNO_3} \underset{\text{aldose}}{CH{=}O-(CHOH)_n-CH_2OH} \xrightarrow[\text{or } Ag^+ \text{ or } Cu^{2+}]{Br_2,\ H_2O} \underset{\text{aldonic acid}}{CO_2H-(CHOH)_n-CH_2OH}$$

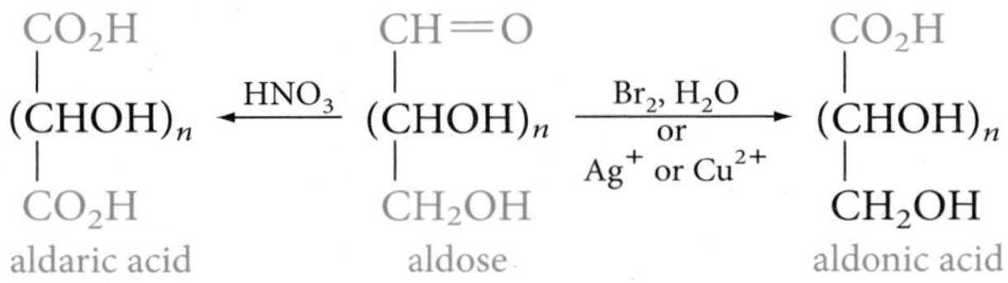

f. Preparation of Glycosides (Sec. 16.11)

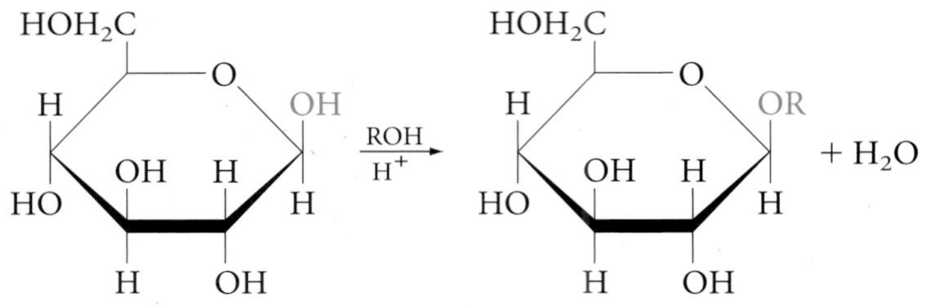

$$\text{glucopyranose (anomeric OH)} \xrightarrow[H^+]{ROH} \text{glycoside (anomeric OR)} + H_2O$$

glycoside formation

2. Hydrolysis of Polysaccharides (Sec. 16.1)

$$\text{polysaccharide} \xrightarrow{H_3O^+} \text{oligosaccharide} \xrightarrow{H_3O^+} \text{monosaccharide}$$

ADDITIONAL PROBLEMS

OWL Interactive versions of these problems are assignable in OWL.

Nomenclature and Structure of Carbohydrates

16.16 Define each of the following, and give the structural formula of one example:

a. aldohexose
b. ketopentose
c. monosaccharide
d. disaccharide
e. polysaccharide
f. furanose
g. pyranose
h. glycoside
i. anomeric carbon

16.17 Explain, using formulas, the difference between a D-sugar and an L-sugar.

16.18 Three of the four hydroxyl groups at the stereogenic centers in the Fischer projection of D-talose (Figure 16.1) are on the left, yet it is called a D-sugar. Explain.

16.19 What term would you use to describe the stereochemical relationship between D-xylose and D-lyxose?

Monosaccharides: Fischer and Haworth Projections

16.20 Construct an array analogous to Figure 16.1 for the D-ketoses as high as the hexoses. Dihydroxyacetone should be at the head, in place of glyceraldehyde.

16.21 Using Figure 16.1 if necessary, write a Fischer projection formula and a Haworth projection formula for

a. methyl β-D-glucopyranoside
b. α-D-mannopyranose
c. β-D-arabinofuranose
d. methyl α-L-glucopyranoside

16.22 Draw the Fischer projection formula for

a. L-(−)-mannose
b. L-(+)-fructose

16.23 At equilibrium in aqueous solution, D-ribose exists as a mixture containing 20% α-pyranose, 56% β-pyranose, 6% α-furanose, and 18% β-furanose forms. Draw Haworth formulas for each of these forms.

16.24 Which monosaccharide in Figure 16.1 is represented by the ball-and-stick structure shown below?

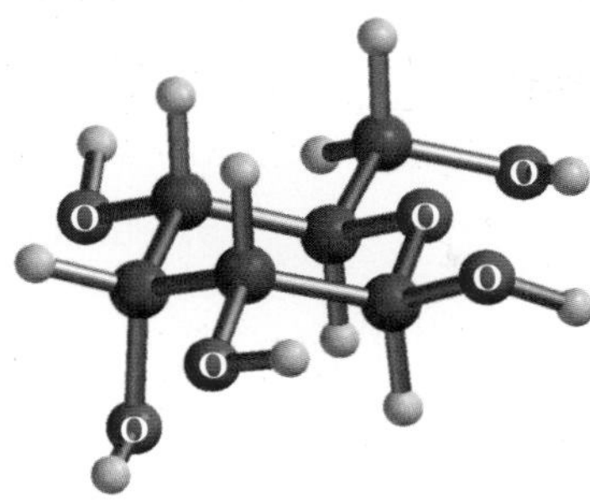

16.25 D-Threose can exist in a furanose form but *not* in a pyranose form. Explain. Draw the β-furanose structure.

16.26 Draw the Fischer and Newman projection formulas for L-threose.

16.27 L-Fucose is a component of bacterial cell walls. It is also called 6-deoxy-L-galactose. Using the description in Section 16.15, write its Fischer projection formula.

Anomers and Mutarotation

16.28 The solubilities of α- and β-D-glucose in water at 25°C are 82 g/100 mL and 178 g/100 mL, respectively. Why are their solubilities not identical?

16.29 The specific rotations of pure α- and β-D-fructofuranose are $+21°$ and $-133°$, respectively. Solutions of each isomer mutarotate to an equilibrium specific rotation of $-92°$. Assuming that no other forms are present, calculate the equilibrium concentrations of the two forms.

16.30 Starting with β-D-glucose and using acid (H^+) as a catalyst, write out all of the steps in the mechanism for the mutarotation process.

16.31 Lactose exists in α and β forms, with specific rotations of $+92.6°$ and $+34°$, respectively.

a. Draw their structures.
b. Solutions of each isomer mutarotate to an equilibrium value of $+52°$. What is the percentage of each isomer at equilibrium?

Reactions of Monosaccharides

16.32 Oxidation of either D-erythrose or D-threose with nitric acid gives tartaric acid. In one case, the tartaric acid is optically active; in the other, it is optically inactive. How can these facts be used to assign stereochemical structures to erythrose and threose?

16.33 Write a structure for

a. D-gluconic acid
b. L-galactaric acid

16.34 Using complete structures, write out the reaction of D-galactose with

a. bromine water
b. nitric acid
c. sodium borohydride
d. acetic anhydride

16.35 Reduction of D-fructose with $NaBH_4$ gives a mixture of D-glucitol and D-mannitol. What does this result prove about the configurations of D-fructose, D-mannose, and D-glucose?

16.36 Although D-galactose contains five stereogenic centers (in its cyclic form) and is optically active, its oxidation with nitric acid gives an optically inactive dicarboxylic acid (called galactaric or mucic acid). What is the structure of this acid, and why is it optically inactive?

16.37 Write an equation for the acid-catalyzed hydrolysis of salicin (page 475). Notice that one of the products is structurally similar to aspirin (eq. 10.40), perhaps accounting for salicin's fever-reducing property.

16.38 Write a balanced equation for the reaction of D-(+)-galactose (use either an acyclic or a cyclic structure, whichever seems more appropriate) with each of the following:

a. hydroxylamine (to form an oxime)
b. ethanol, H^+
c. bromine water
d. sodium borohydride
e. hydrogen cyanide (to form a cyanohydrin)
f. acetic anhydride (excess)
g. Tollens' reagent

Disaccharides

16.39 Write equations that clearly show the mechanism for the acid-catalyzed hydrolysis of

a. maltose to glucose
b. lactose to galactose and glucose
c. sucrose to fructose and glucose

16.40 What would be the structures of the monosaccharide esters obtained from hydrolysis of Olestra ("A Word about Fat Substitutes from Carbohydrates," page 482) with an aqueous acid?

16.41 Write equations for the reaction of maltose with

a. Tollens' reagent
b. bromine water
c. ethanol and H^+
d. acetic anhydride

16.42 Trehalose is a disaccharide that is the main carbohydrate component in the blood of insects. Its Haworth structure is

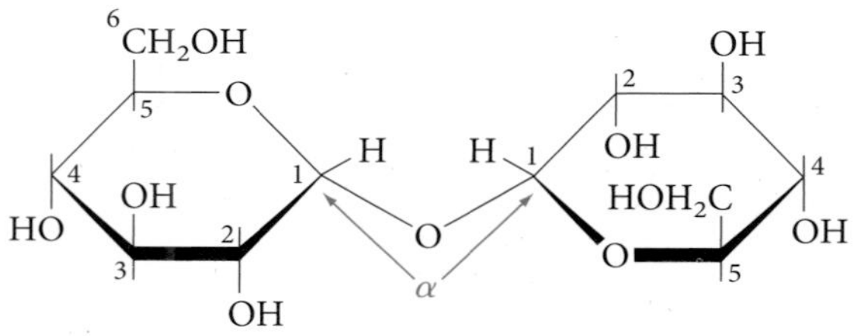

trehalose

a. What are its hydrolysis products?
b. Will trehalose give a positive or a negative test with Fehling's reagent? Explain.

16.43 Explain why sucrose is a nonreducing sugar but maltose is a reducing sugar.

16.44 Sucralose is a chlorinated derivative of sucrose that is about 600 times sweeter than its parent. It is hydrolyzed only slowly at the mildly acidic pH of soft drinks. Draw the structure of the hydrolysis products.

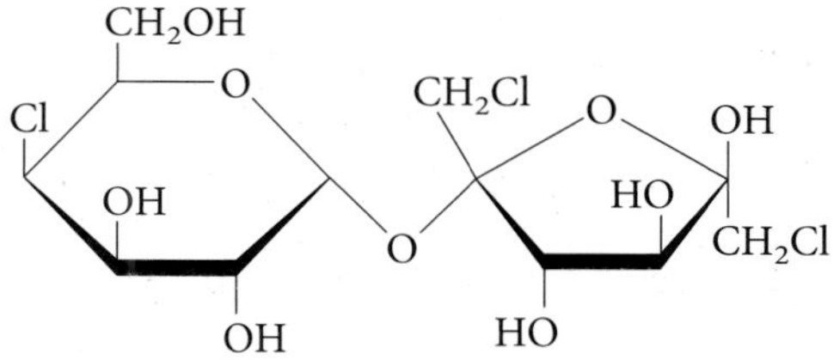

sucralose

16.45 Suggest a synthesis of Olestra (see "A Word About . . . Fat Substitutes from Carbohydrates," on page 482) starting from sucrose and lauric acid.

16.46 An important portion of bacterial cell walls is a disaccharide in which C-1 of *N*-acetylglucosamine (Sec. 16.16) is linked to the C-4 oxygen of *N*-acetylmuramic acid by a β-glycosidic bond. Draw the structure of this disaccharide.

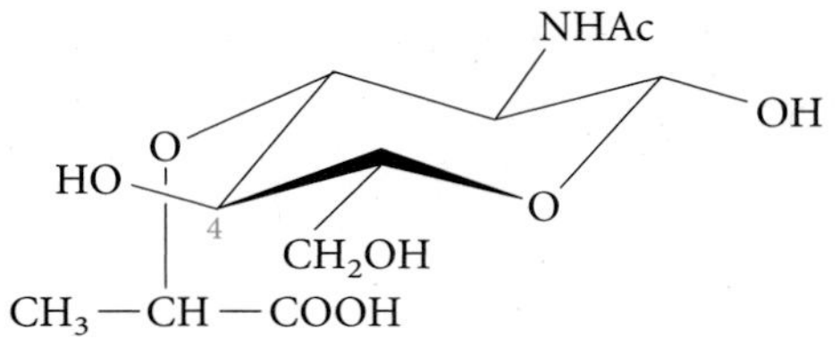

NHAc, OH, HO, O, 4, CH_2OH, $CH_3-CH-COOH$

N-acetylmuramic acid

Polysaccharides

16.47 Using the descriptions in Section 16.13.c, write a formula for

a. chitin b. pectin

16.48 Hemicelluloses are noncellulose materials produced by plants and found in straw, wood, and other fibrous tissues. Xylans are the most abundant hemicelluloses. They consist of 1,4-β-linked D-xylopyranoses. Draw the structure for the repeating unit in xylans.

16.49 What are the products expected from sequential treatment of maltose (page 475) with Br_2 in water followed by hydrolysis with aqueous acid?

16.50 Hydrolysis of disaccharides **A** and **B** gives the same products D-glucose and D-galactose. Describe a reaction sequence that would clearly distinguish between these isomers. Use Problem 16.49 for one idea about how to proceed.

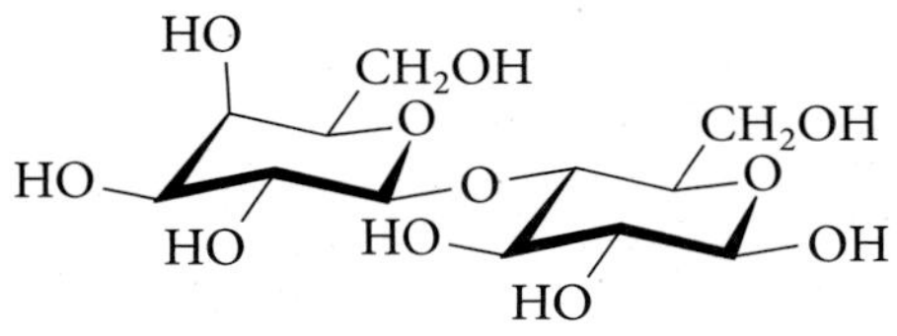

A 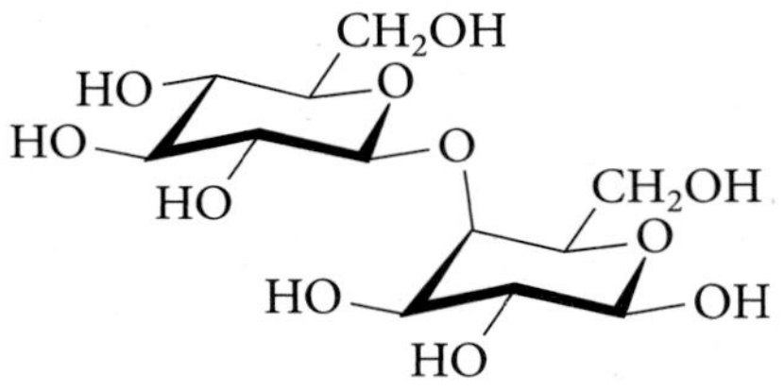 **B**

Other Sugars and Sweeteners

16.51 Write the main contributors to the resonance hybrid anion formed when ascorbic acid acts as an acid (loses the proton from the —OH group on C-3).

16.52 Inositols are hexahydroxycyclohexanes, with one hydroxyl group on each carbon atom of the ring. Although not strictly carbohydrates, they are obviously similar to pyranose sugars and do occur in nature. There are nine possible stereoisomers. Draw Haworth formulas for all possibilities (all are known), and tell which are chiral.

16.53 Aspartame has the *S* configuration at each of its two stereogenic centers (see "A Word About… Sweetness and Sweeteners," page 479). Draw a structure of aspartame that shows its stereochemistry.

16.54 Sorbitol (D-glucitol) is used as a sugar substitute. Starting with maltose, suggest a synthesis of sorbitol.

16.55 As described in "A Word About… Green Chemistry: Alternative Uses for Carbohydrates" (page 484), succinic acid can be prepared enzymatically from renewable resources, such as glucose. Suggest a chemical synthesis to convert succinic acid to

a. 1,4-butanediol b. succindiamide c. 1,4-diaminobutane

Silk, produced by numerous insects and spiders, is the common name for β-keratin (Fig. 17.11), a fibrous protein composed largely of the amino acids glycine and alanine.

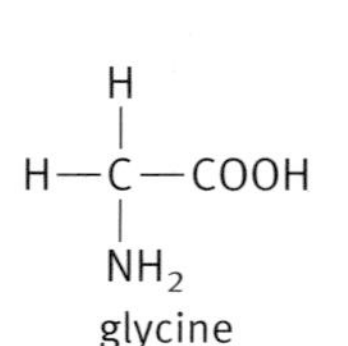

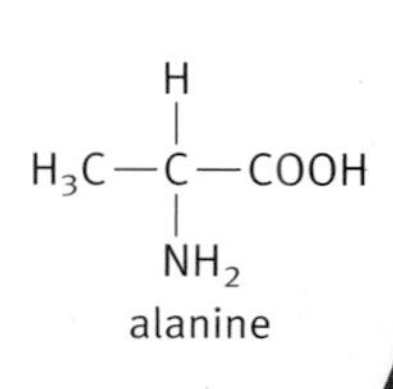

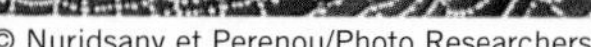

17 Amino Acids, Peptides, and Proteins

Proteins are naturally occurring polymers composed of amino acid units joined one to another by amide (or peptide) bonds. Spider webs, animal hair and muscle, egg whites, and hemoglobin (the molecule that transports oxygen in the body to where it is needed) are all proteins. **Peptides** are oligomers of amino acids that play important roles in many biological processes. For example, the peptide hormone insulin controls our blood sugar levels, bradykinin controls our blood pressure, and oxytocin regulates uterine contraction and lactation. Thus, proteins, peptides, and amino acids are essential to the structure, function, and reproduction of living matter. In this chapter, we will first discuss the structure and properties of amino acids, then the properties of peptides, and finally the structures of proteins.

Online homework for this chapter can be assigned in OWL, an online homework assessment tool.

17.1 Naturally Occurring Amino Acids

The amino acids obtained from protein hydrolysis are **α-amino acids**. That is, the amino group is on the α-carbon atom, the one adjacent to the carboxyl group.

Proteins are composed of **α-amino acids**, carboxylic acids with an amino group on the α-carbon atom. **Peptides** consist of a few linked amino acids.

$$\mathrm{R{-}\overset{\alpha}{C}H(NH_2){-}C({=}O){-}OH}$$

an α-amino acid

With the exception of glycine, where R = H, α-amino acids have a stereogenic center at the α-carbon. All except glycine are therefore optically active. They have the L configuration relative to glyceraldehyde (Figure 17.1, page 495). Note that the Fischer convention, used with carbohydrates, is also applied to amino acids.

Table 17.1 lists the 20 α-amino acids commonly found in proteins. The amino acids are known by common names. Each also has a three-letter abbreviation based on this name, which is used when writing the formulas of peptides, and a one-letter

Table 17.1 ■ Names and Formulas of the Common Amino Acids

Name	Three-letter abbreviation (isoelectric point) one-letter abbreviation	Formula	R
A. One amino group and one carboxyl group			
1. glycine	Gly (6.0) G	$H{-}CH(NH_2){-}CO_2H$	
2. alanine	Ala (6.0) A	$CH_3{-}CH(NH_2){-}CO_2H$	
3. valine	Val (6.0) V	$CH_3CH(CH_3){-}CH(NH_2){-}CO_2H$	R is hydrogen or an alkyl group.
4. leucine	Leu (6.0) L	$CH_3CH(CH_3)CH_2{-}CH(NH_2){-}CO_2H$	
5. isoleucine	Ile (6.0) I	$CH_3CH_2CH(CH_3){-}CH(NH_2){-}CO_2H$	
6. serine	Ser (5.7) S	$CH_2(OH){-}CH(NH_2){-}CO_2H$	
7. threonine	Thr (5.6) T	$CH_3CH(OH){-}CH(NH_2){-}CO_2H$	R contains an alcohol function.
8. cysteine	Cys (5.0) C	$CH_2(SH){-}CH(NH_2){-}CO_2H$	
9. methionine	Met (5.7) M	$CH_3S{-}CH_2CH_2{-}CH(NH_2){-}CO_2H$	R contains sulfur.

(continued)

Table 17.1 ■ (continued)

Name	Three-letter abbreviation (isoelectric point) one-letter abbreviation	Formula	R
10. proline	Pro (6.3) P	$CH_2-CH-CO_2H$, ring: CH_2, CH_2, NH (five-membered ring $CH_2CH_2CH_2$–NH–CH)	The amino group is secondary and part of a ring.
11. phenylalanine	Phe (5.5) F	$C_6H_5-CH_2-CH(NH_2)-CO_2H$	One hydrogen in alanine is replaced by an aromatic or heteroaromatic (indole) ring.
12. tyrosine	Tyr (5.7) Y	$HO-C_6H_4-CH_2-CH(NH_2)-CO_2H$	
13. tryptophan	Trp (5.9) W	indolyl(N–H)$-CH_2-CH(NH_2)-CO_2H$	
B. One amino group and two carboxyl groups			
14. aspartic acid	Asp (3.0) D	$HOOC-CH_2-CH(NH_2)-CO_2H$	
15. glutamic acid	Glu (3.2) E	$HOOC-CH_2CH_2-CH(NH_2)-CO_2H$	
16. asparagine	Asn (5.4) N	$H_2N-C(=O)-CH_2-CH(NH_2)-CO_2H$	
17. glutamine	Gln (5.7) Q	$H_2N-C(=O)-CH_2CH_2-CH(NH_2)-COOH$	
C. One carboxyl group and two basic groups			
18. lysine	Lys (9.7) K	$CH_2(NH_2)CH_2CH_2CH_2-CH(NH_2)-CO_2H$	The second basic group is a primary amine, a guanidine, or an imidazole.
19. arginine	Arg (10.8) R	$NH_2-C(=NH)-NH-CH_2CH_2CH_2-CH(NH_2)-CO_2H$	
20. histidine	His (7.6) H	$CH{=}C-CH_2-CH(NH_2)-CO_2H$, ring: N, NH, CH (imidazole)	

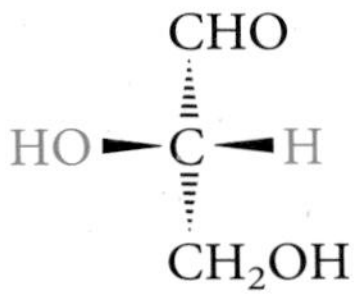

$$\begin{array}{c} CHO \\ HO \blacktriangleright C \blacktriangleleft H \\ CH_2OH \end{array}$$

L-(−)-glyceraldehyde

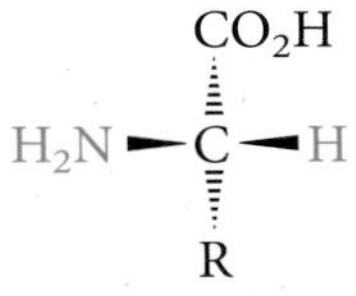

$$\begin{array}{c} CO_2H \\ H_2N \blacktriangleright C \blacktriangleleft H \\ R \end{array}$$

a naturally occurring L-amino acid

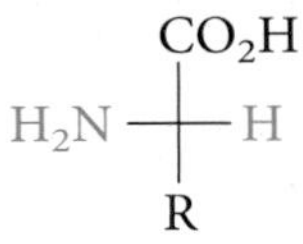

$$\begin{array}{c} CO_2H \\ H_2N - \!\!\!+\!\!\! - H \\ R \end{array}$$

Fischer projection formula of an L-amino acid

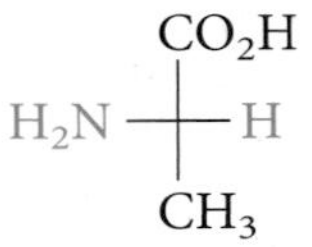

$$\begin{array}{c} CO_2H \\ H_2N - \!\!\!+\!\!\! - H \\ CH_3 \end{array}$$

L-(+)-alanine

▪ **Figure 17.1**
Naturally occurring α-amino acids have the L configuration.

abbreviation used to describe the amino acid sequence in a protein. The amino acids in Table 17.1 are grouped to emphasize structural similarities. Of the 20 amino acids listed in the table, 12 can be synthesized in the body from other foods. The other 8, those with names shown in color and referred to as essential amino acids, cannot be synthesized by adult humans and therefore must be included in the diet in the form of proteins.

17.2 The Acid–Base Properties of Amino Acids

The carboxylic acid and amine functional groups are *simultaneously* present in amino acids, and we might ask whether they are mutually compatible since one group is acidic and the other is basic. Although we have represented the amino acids in Table 17.1 as having amino and carboxyl groups, these structures are oversimplified.

Amino acids with one amino group and one carboxyl group are better represented by a **dipolar ion structure**.*

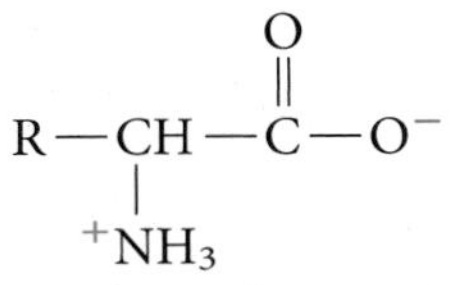

$$R-\underset{\underset{^{+}NH_3}{|}}{CH}-\overset{\overset{O}{\|}}{C}-O^-$$

dipolar structure of an α-amino acid

The amino group is protonated and present as an ammonium ion, whereas the carboxyl group has lost its proton and is present as a carboxylate anion. This dipolar structure is consistent with the salt-like properties of amino acids, which have rather high melting points (even the simplest, glycine, melts at 233°C) and relatively low solubilities in organic solvents.

Amino acids are *amphoteric* (page 212). They can behave as acids and donate a proton to a strong base, or they can behave as bases and accept a proton from a strong acid. These behaviors are expressed in the following equilibria for an amino acid with one amino group and one carboxyl group:

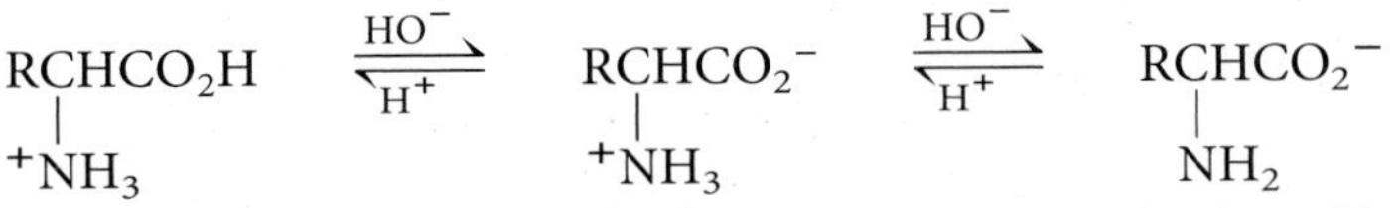

$$\underset{\underset{^{+}NH_3}{|}}{RCHCO_2H} \underset{H^+}{\overset{HO^-}{\rightleftharpoons}} \underset{\underset{^{+}NH_3}{|}}{RCHCO_2^-} \underset{H^+}{\overset{HO^-}{\rightleftharpoons}} \underset{\underset{NH_2}{|}}{RCHCO_2^-} \quad (17.1)$$

amino acid at low pH (acid) — dipolar ion form (neutral) — amino acid at high pH (base)

*Such structures are sometimes called *zwitterions* (from a German word for hybrid ions).

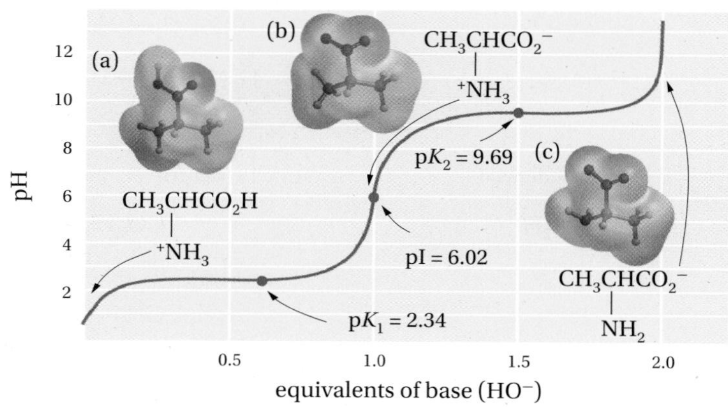

Figure 17.2
Titration curve for alanine, showing how its structure varies with pH. Electrostatic potential maps of alanine at low pH (a), at the isoelectric point (b), and at high pH (c) illustrate the difference in charge distribution as a function of pH.

Figure 17.2 shows a titration curve for alanine, a typical amino acid of this kind. At low pH (acidic solution), the amino acid is in the form of a substituted ammonium ion. At high pH (basic solution), it is present as a substituted carboxylate ion. At some intermediate pH (for alanine, pH 6.02), the amino acid is present as the dipolar ion with an ammonium ($—NH_3^+$) and a carboxylate ($—CO_2^-$) unit. A simple rule to remember for any acidic site is that *if the pH of the solution is* less than *the* pK_a, *the proton is on; if the pH of the solution is* greater than *the* pK_a, *the proton is off.*

EXAMPLE 17.1

Starting with alanine hydrochloride (its structure at low pH in hydrochloric acid is shown in Figure 17.2), write equations for its reaction with one equivalent of sodium hydroxide and then with a second equivalent of sodium hydroxide.

Solution

$$\underset{\text{ammonium salt}}{\underset{^+NH_3\ Cl^-}{CH_3\underset{|}{C}HCO_2H}} + Na^+\ HO^- \longrightarrow \underset{\text{dipolar ion}}{\underset{^+NH_3}{CH_3\underset{|}{C}HCO_2^-}} + Na^+\ Cl^- + H_2O \quad \textbf{(17.2)}$$

$$\underset{\text{dipolar ion}}{\underset{^+NH_3}{CH_3\underset{|}{C}HCO_2^-}} + Na^+\ HO^- \longrightarrow \underset{\text{carboxylate salt}}{\underset{NH_2}{CH_3\underset{|}{C}HCO_2^-Na^+}} + H_2O \quad \textbf{(17.3)}$$

The first equivalent of base removes a proton from the carboxyl group to give the dipolar ion, and the second equivalent of base removes a proton from the ammonium ion to give the sodium carboxylate.

PROBLEM 17.1 Starting with the sodium carboxylate salt of alanine, write equations for its reaction with one equivalent of hydrochloric acid and then with a second equivalent, and explain what each equivalent of acid does.

PROBLEM 17.2 Which group in the ammonium salt form of alanine is more acidic, the $—\overset{+}{N}H_3$ group or the $—CO_2H$ group?

PROBLEM 17.3 Which group in the carboxylate salt form of alanine is more basic, the $—NH_2$ group or the $—CO_2^-$ group?

Note from Figure 17.2 and from eq. 17.1 that the charge on an amino acid changes as the pH changes. At low pH, for example, the charge on alanine is positive; at high pH, it is negative; and near neutrality, the ion is dipolar. If placed in an electric field, the amino acid will therefore migrate toward the cathode (negative electrode) at low pH and toward the anode (positive electrode) at high pH (Figure 17.3). At some intermediate pH, called the **isoelectric point (pI)**, the amino acid will be dipolar and have a net charge of zero. It will be unable to move toward either electrode. The isoelectric points of the various amino acids are listed in Table 17.1.

The **isoelectric point (pI)** of an amino acid is the pH at which the amino acid is mainly in its dipolar form and has no net charge.

In general, amino acids with one amino group and one carboxyl group, and no other acidic or basic groups in their structure, have two pK_a values: one around 2 to 3 for proton loss from the carboxyl group and the other around 9 to 10 for proton loss from the ammonium ion. The isoelectric point is about halfway between the two pK_a values, near pH 6.

R is neutral

$$\underset{\substack{\text{low pH}\\ +1}}{\underset{^{+}NH_3}{\underset{|}{RCHCO_2H}}} \underset{}{\overset{pK_a = 2\text{–}3}{\rightleftharpoons}} \underset{\substack{\\ 0}}{\underset{^{+}NH_3}{\underset{|}{RCHCO_2^-}}} \underset{}{\overset{pK_a = 9\text{–}10}{\rightleftharpoons}} \underset{\substack{\text{high pH}\\ -1}}{\underset{NH_2}{\underset{|}{RCHCO_2^-}}} \quad (17.4)$$

net charge: +1 (low pH), 0, −1 (high pH)

EXAMPLE 17.2

Write the structure of leucine

a. at pI b. at high pH c. at low pH

Solution

a. $(CH_3)_2CHCH_2\underset{^{+}NH_3}{\underset{|}{C}}HCO_2^-$
dipolar and neutral

b. $(CH_3)_2CHCH_2\underset{NH_2}{\underset{|}{C}}HCO_2^-$
negative

c. $(CH_3)_2CHCH_2\underset{^{+}NH_3}{\underset{|}{C}}HCO_2H$
positive

PROBLEM 17.4 Write the structure for the predominant form of each of the following amino acids at the indicated pH. If placed in an electric field, toward which electrode (+ or −) will each amino acid migrate?

a. serine at its pI b. methionine at low pH c. tyrosine at high pH

The situation is more complex with amino acids containing two acidic or two basic groups.

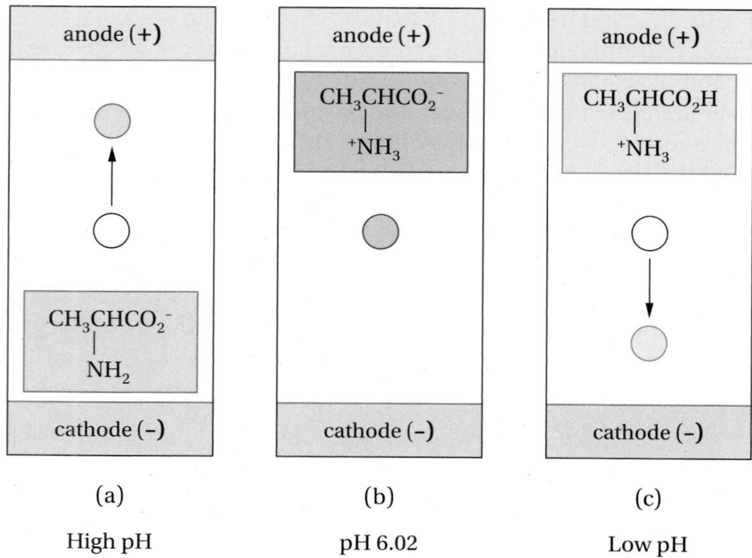

Figure 17.3
The migration of an amino acid (such as alanine) in an electric field depends on pH: (a) At high pH, the alanine is negatively charged and migrates toward the positive anode. (b) At the isoelectric point (pH 6.02), the alanine is neutral and does not migrate. (c) At low pH, the alanine is positively charged and migrates toward the negatively charged cathode.

17.3 The Acid–Base Properties of Amino Acids with More Than One Acidic or Basic Group

Aspartic and glutamic acids (entries 14 and 15 in Table 17.1) have two carboxyl groups and one amino group. In strong acid (low pH), all three of these groups are in their acidic form (protonated). As the pH is raised and the solution becomes more basic, each group in succession gives up a proton. The equilibria are shown for aspartic acid, with the three pK_a values over the equilibrium arrows:

$$\underset{\substack{+\mathrm{NH_3}}}{\mathrm{HO_2CCH_2CHCO_2H}} \xrightleftharpoons{\mathrm{p}K_a = 2.09} \underset{\substack{+\mathrm{NH_3}}}{\mathrm{HO_2CCH_2CHCO_2^-}} \xrightleftharpoons{\mathrm{p}K_a = 3.86} \underset{\substack{+\mathrm{NH_3}}}{{}^-\mathrm{O_2CCH_2CHCO_2^-}} \xrightleftharpoons{\mathrm{p}K_a = 9.82} \underset{\substack{\mathrm{NH_2}}}{{}^-\mathrm{O_2CCH_2CHCO_2^-}} \quad (17.5)$$

low pH → high pH

net charge: +1, 0, −1, −2

The isoelectric point for aspartic acid, the pH at which it is mainly in the neutral dipolar form, is 2.87 (in general, the pI is close to the average of the two pK_a's on either side of the neutral, dipolar species).

EXAMPLE 17.3

Which carboxyl group is the stronger acid in the most acidic form of aspartic acid?

Solution As shown at the extreme left of eq. 17.5, the first proton to be removed from the most acidic form of aspartic acid is the proton on the carboxyl group nearest the $^+NH_3$ substituent. The $^+NH_3$ group is electron-withdrawing due to its positive charge and enhances the acidity of the carboxyl group closest to it.

PROBLEM 17.5 Use eq. 17.5 to tell which is the least acidic group in aspartic acid and why.

The situation differs for amino acids with two basic groups and only one carboxyl group (entries 18, 19, and 20 in Table 17.1). With lysine, for example, the equilibria are

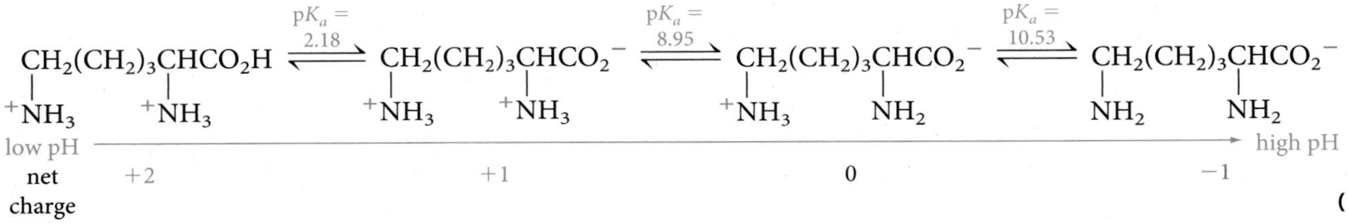

$$\underset{\text{net charge } +2}{\underset{{}^{+}NH_3\qquad {}^{+}NH_3}{CH_2(CH_2)_3CHCO_2H}} \underset{}{\overset{pK_a = 2.18}{\rightleftharpoons}} \underset{+1}{\underset{{}^{+}NH_3\qquad {}^{+}NH_3}{CH_2(CH_2)_3CHCO_2^-}} \overset{pK_a = 8.95}{\rightleftharpoons} \underset{0}{\underset{{}^{+}NH_3\qquad NH_2}{CH_2(CH_2)_3CHCO_2^-}} \overset{pK_a = 10.53}{\rightleftharpoons} \underset{-1}{\underset{NH_2\qquad NH_2}{CH_2(CH_2)_3CHCO_2^-}} \tag{17.6}$$

low pH → high pH

The pI for lysine comes in the basic region, at 9.74.

The second basic groups in arginine and histidine are not simple amino groups. They are a guanidine group and an imidazole ring, respectively, shown in color. The most protonated forms of these two amino acids are

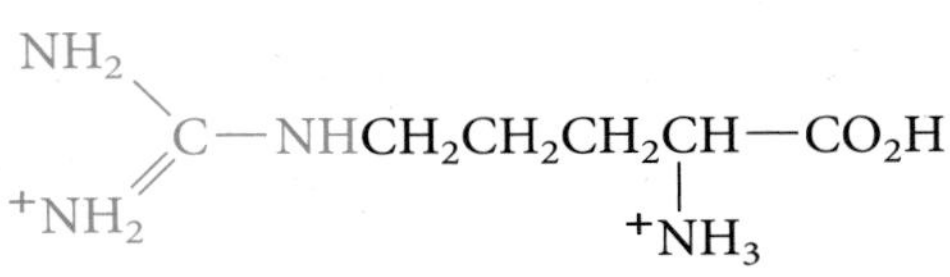

$$(NH_2)({}^{+}NH_2){=}C{-}NHCH_2CH_2CH_2\underset{{}^{+}NH_3}{CH}{-}CO_2H$$

arginine at pH 1

$$\text{(imidazolium: } H\overset{+}{N}{-}CH,\ HC,\ N{-}H\text{ ring)}\ C{-}CH_2\underset{{}^{+}NH_3}{CH}CO_2H$$

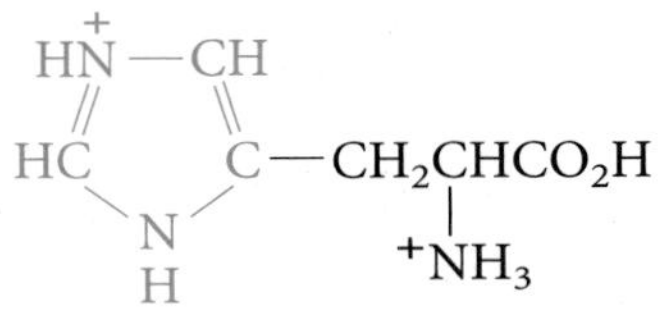

histidine at pH 1

PROBLEM 17.6 Arginine shows three pK_a's: at 2.17 (the —COOH group), at 9.04 (the $-\overset{+}{N}H_3$ group), and at 12.48 (the guanidinium ion). Write equilibria (similar to eq. 17.6) for its dissociation. At approximately what pH will the isoelectric point come, and what is the structure of the dipolar ion?

Table 17.2 summarizes the approximate pK_a values and isoelectric points for the three types of amino acids.

Table 17.2 Approximate Acidity Constants and Isoelectric Points (pI) for the Three Types of Amino Acids

Type		pK_a 1	2	3	pI
1 acidic and 1 basic group		2.3	9.4	—	6.0
2 acidic and 1 basic group		2.2	4.1	9.8	3.0
1 acidic and 2 basic groups	(Lys, Arg)	2.2	9.0	11.5	10.0
	(His)	1.8	6.0	9.2	7.6

17.4 Electrophoresis

Electrophoresis is a method for separating amino acids and proteins, based on their charge differences.

As we have seen, the charge on an amino acid depends on the pH of its environment. **Electrophoresis** is an important method for separating amino acids that takes advantage of these charge differences. In a typical electrophoresis experiment, a mixture of amino acids is placed on a solid support (e.g., paper), and the support is bathed in an aqueous solution at a controlled pH. An electrical field is then applied across the paper. Amino acids that are positively charged at that pH migrate toward the negatively charged cathode. Amino acids that are negatively charged migrate toward the positively charged anode. Migration ceases when the electrical field is turned off.

An example of an electrophoresis experiment is illustrated in Figure 17.3. In this case, the α-amino acid alanine migrates toward the anode at high pH, toward the cathode at low pH, and does not migrate when the pH is adjusted to the isoelectric point of the amino acid. Because most amino acids do not have exactly the same charge at a given pH, mixtures of amino acids can be separated by electrophoresis because the components migrate toward the anode or cathode at different rates. Electrophoresis is also used to separate peptides and proteins because they behave much like their building blocks, the amino acids, at a given pH.

EXAMPLE 17.4

Predict the direction of migration (toward the positive or negative electrode) of alanine in an electrophoresis apparatus at pH 5. Do the same for aspartic acid.

Solution A pH of 5 is *less* than the pI of alanine (~6). Therefore, the dipolar ions will be protonated (positive) and migrate toward the negative electrode. But pH 5 is *greater* than the pI of aspartic acid (~3). Therefore aspartic acid will exist mainly as the −1 ion (eq. 17.5) and migrate toward the positive electrode. A mixture of the two amino acids could therefore easily be separated in this way.

PROBLEM 17.7 Predict the direction of migration in an electrophoresis apparatus (toward the positive or negative electrode) of each component of the following amino acid mixtures:

a. glycine and lysine at pH 7
b. phenylalanine, leucine, and proline at pH 6

17.5 Reactions of Amino Acids

In addition to their acidic and basic behavior, amino acids undergo other reactions typical of carboxylic acids or amines. For example, the carboxyl group can be esterified:

$$\underset{\substack{|\\ {}^{+}NH_3}}{R-CH}-CO_2^- + R'OH + H^+ \xrightarrow{\text{heat}} \underset{\substack{|\\ {}^{+}NH_3}}{R-CH}-CO_2R' + H_2O \quad (17.7)$$

The amino group can be acylated to an amide:

$$\underset{\underset{^{+}NH_3}{|}}{R{-}CH}{-}CO_2^- + R'{-}\overset{\overset{O}{\|}}{C}{-}Cl \xrightarrow{2\,HO^-} \underset{\underset{\underset{\underset{O}{\|}}{R'C}{-}NH}{|}}{R{-}CH}{-}CO_2^- + 2\,H_2O + Cl^- \qquad (17.8)$$

These types of reactions are useful in temporarily modifying or protecting either of the two functional groups, especially during the controlled linking of amino acids to form peptides or proteins.

PROBLEM 17.8 Using eqs. 17.7 and 17.8 as models, write equations for the following reactions:

a. glutamic acid + CH_3OH + HCl $\longrightarrow$
b. proline + benzoyl chloride + NaOH $\longrightarrow$
c. phenylalanine + acetic anhydride $\xrightarrow{\text{heat}}$

17.6 The Ninhydrin Reaction

Ninhydrin is a useful reagent for detecting amino acids and determining the concentrations of their solutions. It is the hydrate of a cyclic triketone, and when it reacts with an amino acid, a violet dye is produced. The overall reaction, whose mechanism is complex and need not concern us in detail here, is as follows:

$$2\ \text{ninhydrin} + \underset{\underset{^{+}NH_3}{|}}{RCH}CO_2^- \longrightarrow \text{violet anion} + RCHO + CO_2 + 3\,H_2O + H^+ \qquad (17.9)$$

ninhydrin

violet anion

Only the nitrogen atom of the violet dye comes from the amino acid; the rest of the amino acid is converted to an aldehyde and carbon dioxide. Therefore, *the same violet dye is produced from all α-amino acids with a primary amino group,* and the intensity of its color is directly proportional to the concentration of the amino acid present. Only proline, which has a secondary amino group, reacts differently to give a yellow dye, but this, too, can be used for analysis.

PROBLEM 17.9 Write an equation for the reaction of alanine with ninhydrin.

17.7 Peptides

An amide bond linking two amino acids is called a **peptide bond**. A peptide has an **N-terminal amino acid** with a free $^{+}NH_3$ group and a **C-terminal amino acid** with a free CO_2^- group.

Amino acids are linked in peptides and proteins by an amide bond between the carboxyl group of one amino acid and the α-amino group of another amino acid. Emil Fischer, who first proposed this structure, called this amide bond a **peptide bond**. A molecule containing only *two* amino acids (the shorthand aa is used for amino acid) joined in this way is a **dipeptide**:

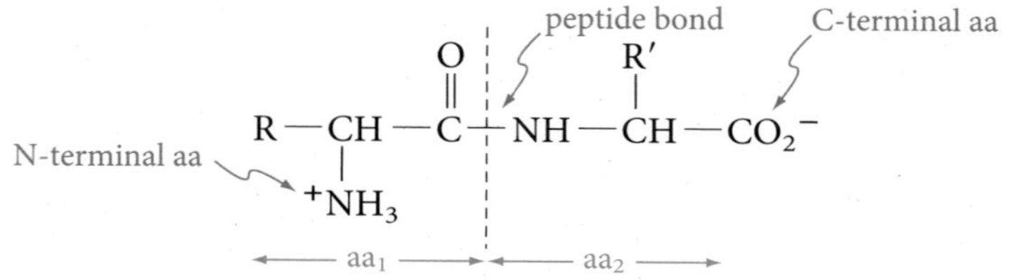

$$\underbrace{R{-}\underset{\displaystyle ^{+}NH_3}{\underset{|}{CH}}{-}\overset{\displaystyle O}{\overset{\|}{C}}}_{aa_1}\;\vdots\;\underbrace{NH{-}\overset{\displaystyle R'}{\overset{|}{CH}}{-}CO_2^-}_{aa_2}$$

N-terminal aa → $^{+}NH_3$; peptide bond → C–NH; C-terminal aa → CO_2^-

By convention, the peptide bond is written with the amino acid having a free $^{+}NH_3$ group at the left and the amino acid with a free CO_2^- group at the right. These amino acids are called, respectively, the **N-terminal amino acid** and the **C-terminal amino acid**.

EXAMPLE 17.5

Write the dipeptide structures that can be made by linking alanine and glycine with a peptide bond.

Solution There are two possibilities:

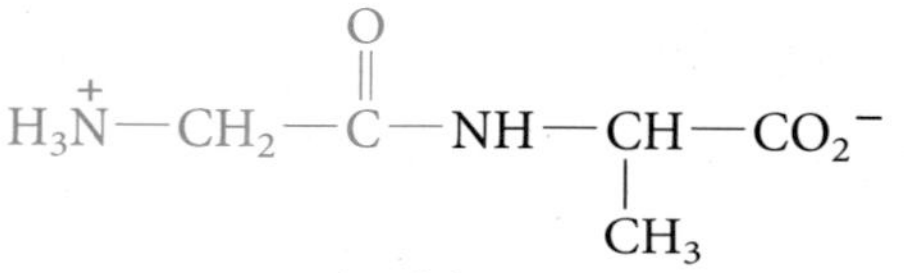

$$H_3\overset{+}{N}{-}CH_2{-}\overset{\displaystyle O}{\overset{\|}{C}}{-}NH{-}\underset{\displaystyle CH_3}{\underset{|}{CH}}{-}CO_2^- \qquad H_3\overset{+}{N}{-}\underset{\displaystyle CH_3}{\underset{|}{CH}}{-}\overset{\displaystyle O}{\overset{\|}{C}}{-}NH{-}CH_2{-}CO_2^-$$

glycylalanine 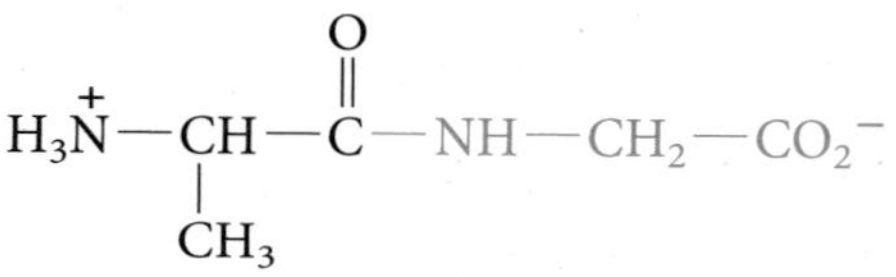   alanylglycine

In glycylalanine, glycine is the N-terminal amino acid, and alanine is the C-terminal amino acid. In alanylglycine, these roles are reversed. The two dipeptides are structural isomers.

We often write the formulas for peptides in a kind of shorthand by simply linking the three-letter abbreviations for each amino acid, *starting with the N-terminal one at the left.* For example, glycylalanine is Gly—Ala, and alanylglycine is Ala—Gly.

PROBLEM 17.10 In Example 17.5, the formulas for Gly—Ala and Ala—Gly are written in their dipolar forms. At what pH do you expect these structures to predominate? Draw the expected structure of Gly—Ala in solution at pH 3 and at pH 9.

PROBLEM 17.11 Write the dipolar structural formula for

a. leucylalanine   b. alanylisoleucine

A WORD ABOUT... Some Naturally Occurring Peptides

Peptides with just a few linked amino acids per molecule have been isolated from living matter, where they often perform important roles in biology. Here are a few examples.

Bradykinin is a nonapeptide present in blood plasma and involved in regulating blood pressure. Several peptides found in the brain act as chemical transmitters of nerve impulses. One of these is the undecapeptide **substance P**, thought to be a transmitter of pain impulses. Notice that the C-terminal amino acid, methionine, is present as the primary amide. This is quite common in peptide chains and is symbolized by placing an NH_2 at the right-hand end of the formula.

Arg—Pro—Pro—Gly—Phe—Ser—
Pro—Phe—Arg

bradykinin

Arg—Pro—Lys—Pro—Gln—Gln—Phe—Phe—
Gly—Leu—Met—NH_2

substance P

Oxytocin and **vasopressin** are two cyclic nonapeptide hormones produced by the posterior pituitary gland. Oxytocin regulates uterine contraction and lactation and may be administered when it is necessary to induce labor at childbirth. Note that its structure includes two cysteine units joined by a disulfide bond, and, once again, the C-terminal amino acid is present as the amide. Vasopressin differs from oxytocin only in the substitution of Phe for Ile and Arg for Leu. Vasopressin regulates the excretion of water by the kidneys and also affects blood pressure. The disease *diabetes insipidus,* in which too much urine is excreted, is a consequence of vasopressin deficiency and can be treated by administering this hormone.

Ile — Gln
Tyr Asn NH_2
Cys Cys Gly
S —— S Pro — Leu
N terminus C terminus

oxytocin

Cyclosporin A is another cyclic peptide that was first isolated from the fungus *Trichoderma polysporum.* Cyclosporin A has immunosuppressive activity and is used to help prevent organ rejection after transplant operations. Notice that cyclosporin A contains several unusual amino acids and that a number of the amide nitrogens are methylated.

One area of current research involves modifying the structures of natural, biologically important peptides (by replacing one amino acid with another or by altering side-chain structures or by replacing portions of the peptide chain with nonpeptidic structures) with the intent of developing new and useful drugs.

See Problems 17.43, 17.44, and 17.50.

cyclosporin A

EXAMPLE 17.6

Consider the abbreviated formula Gly—Ala—Ser for a tripeptide. Which is the N-terminal amino acid, and which is the C-terminal amino acid?

Solution Such formulas always read from the N-terminal amino acid at the left to the C-terminal amino acid at the right. Glycine is the N-terminal amino acid, and serine is the C-terminal amino acid. Both the amino group *and* the carboxyl group of the middle amino acid, alanine, are tied up in peptide bonds.

PROBLEM 17.12 Write out the complete structural formula for Gly—Ala—Ser.

PROBLEM 17.13 Write out the *abbreviated* formulas for all possible tripeptide isomers of Leu—Ala—Met.

The complexity that is possible in peptide and protein structures is truly astounding. For example, Problem 17.13 shows that there are 6 possible arrangements of 3 different amino acids in a tripeptide. For a tetrapeptide, this number jumps to 24, and for an octapeptide (constructed from 8 different amino acids), there are 40,320 possible arrangements!

Now we must introduce one small additional complication before we consider the structures of particular peptides and proteins.

17.8 The Disulfide Bond

The **disulfide bond** is an S—S single bond. In proteins, it links two **cysteine** amino acid units.

Aside from the peptide bond, the only other type of covalent bond between amino acids in peptides and proteins is the **disulfide bond**. It links two **cysteine** units. Recall that thiols are easily oxidized to disulfides (eq. 7.49). Two cysteine units can be linked by a disulfide bond.

$$\begin{array}{c} -\text{NH}-\text{CH}-\overset{\text{O}}{\overset{\|}{\text{C}}}- \\ | \\ \text{CH}_2\text{SH} \\ \\ \text{CH}_2\text{SH} \\ | \\ -\text{NH}-\text{CH}-\underset{\text{O}}{\underset{\|}{\text{C}}}- \\ \text{two cysteine units} \end{array} \quad \underset{\text{reduction}}{\overset{\text{oxidation}}{\rightleftharpoons}} \quad \begin{array}{c} -\text{NH}-\text{CH}-\overset{\text{O}}{\overset{\|}{\text{C}}}- \\ | \\ \text{CH}_2-\text{S} \\ \quad | \leftarrow \text{disulfide bond} \\ \text{CH}_2-\text{S} \\ | \\ -\text{NH}-\text{CH}-\underset{\text{O}}{\underset{\|}{\text{C}}}- \\ -\text{Cys}-\text{S}-\text{S}-\text{Cys}- \end{array} \qquad (17.10)$$

If the two cysteine units are in different parts of the *same* chain of a peptide or protein, a disulfide bond between them will form a "loop," or large ring. If the two units are on different chains, the disulfide bond will cross-link the two chains. We will see examples of both arrangements. Disulfide bonds can easily be broken by mild reducing agents (see "A Word about Hair—Curly or Straight," page 228).

17.9 Proteins

Proteins are biopolymers composed of many amino acids connected to one another through amide (peptide) bonds. They play numerous roles in biological systems. Some proteins are major components of structural tissue (muscle, skin, nails, and hair). Others transport molecules from one part of a living system to another. Yet others serve as catalysts for the many biological reactions needed to sustain life.

In the remainder of this chapter, we will describe the main features of peptide and protein structure. We will first examine what is called the *primary structure* of peptides and proteins; that is, how many amino acids are present and what their sequence is in the peptide or protein chain. We will then examine three-dimensional aspects of peptide and protein structure, usually referred to as their *secondary, tertiary,* and *quaternary structures.*

17.10 The Primary Structure of Proteins

The backbone of proteins is a repeating sequence of one nitrogen and two carbon atoms.

```
    H   R   O   H   R   O   H   R   O
    |   |   ||  |   |   ||  |   |   ||
---N — C — C — N — C — C — N — C — C---
        |           |           |
        H           H           H
```

protein chain, showing amino acids linked by amide bonds

Things we must know about a peptide or protein, if we are to write down its structure, are (1) which amino acids are present and how many of each there are and (2) the sequence of the amino acids in the chain. In this section, we will briefly describe ways to obtain this kind of information.

17.10.a Amino Acid Analysis

Since peptides and proteins consist of amino acids held together by amide bonds, they can be hydrolyzed to their amino acid components. This hydrolysis is typically accomplished by heating the peptide or protein with 6 *M* HCl at 110°C for 24 hours. Analysis of the resulting amino acid mixture requires a procedure that separates the amino acids from one another, identifies each amino acid present, and determines its amount.

An instrument called an **amino acid analyzer** performs these tasks automatically in the following way. The amino acid mixture from the complete hydrolysis of a few milligrams of the peptide or protein is placed at the top of a column packed with material that selectively absorbs amino acids. The packing is an insoluble resin that contains strongly acidic groups. These groups protonate the amino acids. Next, a buffer solution of known pH is pumped through the column. The amino acids pass through the column at different rates, depending on their structure and basicity, and are thus separated.

Amino acid analyzer is used to determine the amino acid content of peptides or proteins.

The column effluent is met by a stream of ninhydrin reagent. Therefore, the effluent is alternately violet or colorless, depending on whether or not an amino acid is being eluted from the column. The intensity of the color is automatically recorded as a function of the volume of effluent. Calibration with known amino acid mixtures allows each amino acid to be identified by the appearance time of its peak. Furthermore, the intensity of each peak gives a quantitative measure of the amount of each amino acid that is present. Figure 17.4 shows a typical plot obtained from an automatic amino acid analyzer.

PROBLEM 17.14 Show the products expected from complete hydrolysis of Gly—Ala—Ser.

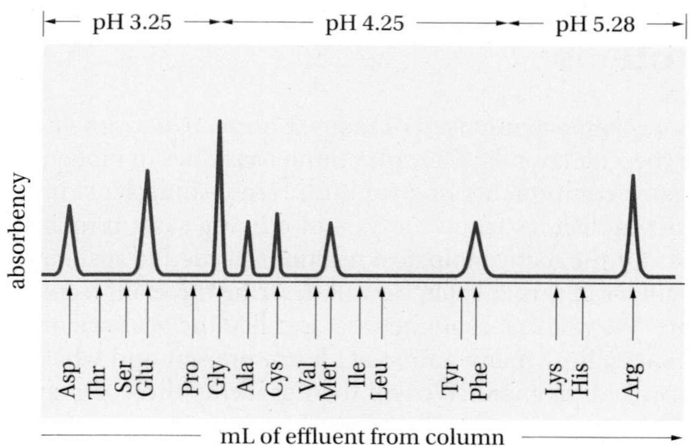

Figure 17.4
Different amino acids in a peptide hydrolysate are separated on an ion-exchange resin. Buffers with different pH's elute the amino acids from the column. Each amino acid is identified by comparing the peaks with the standard elution profile shown near the bottom on the figure. The amount of each amino acid is proportional to the area under its peak. This sample contains eight amino acids: aspartic acid, glutamic acid, glycine, alanine, cysteine, methionine, phenylalanine, and arginine.

17.10.b Sequence Determination

Frederick Sanger* devised a method for sequencing peptides based on the observation that the N-terminal amino acid differs from all others in the chain by having a free amino group. If that amino group were to react with some reagent prior to hydrolysis, then after hydrolysis, that amino acid would be labeled and could be identified. **Sanger's reagent** is 2,4-dinitrofluorobenzene, which reacts with the NH_2 group of amino acids and peptides to give yellow 2,4-dinitrophenyl (DNP) derivatives.

Sanger's reagent (2,4-dinitrofluorobenzene) is used to identify the N-terminal amino acid of a peptide.

$$O_2N-C_6H_3(NO_2)-F + H_2N-\underset{R}{CH}-\overset{O}{\overset{\|}{C}}--- \xrightarrow[\text{base}]{\text{mild}} O_2N-C_6H_3(NO_2)-NH-\underset{R}{CH}-\overset{O}{\overset{\|}{C}}--- + F^- \quad (17.11)$$

2,4-dinitrofluoro-benzene; N-terminal amino acid; DNP-peptide, labeled at the N terminus

Hydrolysis of a peptide treated this way (eq. 17.11) would give the DNP derivative of the N-terminal amino acid; other amino acids in the chain would be unlabeled. In this way, the N-terminal amino acid could be identified.

EXAMPLE 17.7

How might alanylglycine be distinguished from glycylalanine?

Solution Each dipeptide will give one equivalent each of alanine and glycine on hydrolysis. Therefore, we cannot distinguish between them without applying a sequencing method.

Treat the dipeptide with 2,4-dinitrofluorobenzene and *then* hydrolyze. If the dipeptide is alanylglycine, we will obtain DNP-alanine and glycine; if the dipeptide is glycylalanine, we will get DNP-glycine and alanine.

PROBLEM 17.15 Write out equations for the reactions described in Example 17.7.

*Frederick Sanger (Cambridge University, England) received *two* Nobel Prizes—the first in 1958 for his landmark work in amino acid sequencing and the second in 1980 for methodology in the base sequencing of RNA and DNA.

Sanger used his method with great ingenuity to deduce the complete sequence of insulin, a protein hormone with 51 amino acid units. But the method suffers in that it identifies only the N-terminal amino acid.

An ideal method for sequencing a peptide or protein would have a reagent that clips off just one amino acid at a time from the end of the chain, and identifies it. Such a method was devised by Pehr Edman (professor at the University of Lund in Sweden), and it is now widely used.

Edman's reagent is phenyl isothiocyanate, $C_6H_5N{=}C{=}S$. The steps in selectively labeling and releasing the N-terminal amino acid are shown in Figure 17.5. In the first step, the N-terminal amino acid acts as a nucleophile toward the C=S bond of the reagent to form a thiourea derivative. In the second step, the N-terminal amino acid is removed in the form of a heterocyclic compound, a phenylthiohydantoin. The specific phenylthiohydantoin that is formed can be identified by comparison with reference compounds separately prepared from the known amino acids. Then the two steps are repeated to identify the next amino acid, and so on. The method has been automated, so currently amino acid "sequenators" can easily determine, in a day, the sequence of the first 50 or so amino acids in a peptide, starting at the N-terminal end. But the Edman method cannot be used indefinitely, due to the gradual buildup of impurities.

Edman's reagent (phenylisothiocyanate, $C_6H_5N{=}C{=}S$) reacts with the N-terminal amino acid and is used to sequence small peptides.

PROBLEM 17.16 Write the structure of the phenylthiohydantoin derived from the first cycle of Edman degradation of Phe—Ala—Ser.

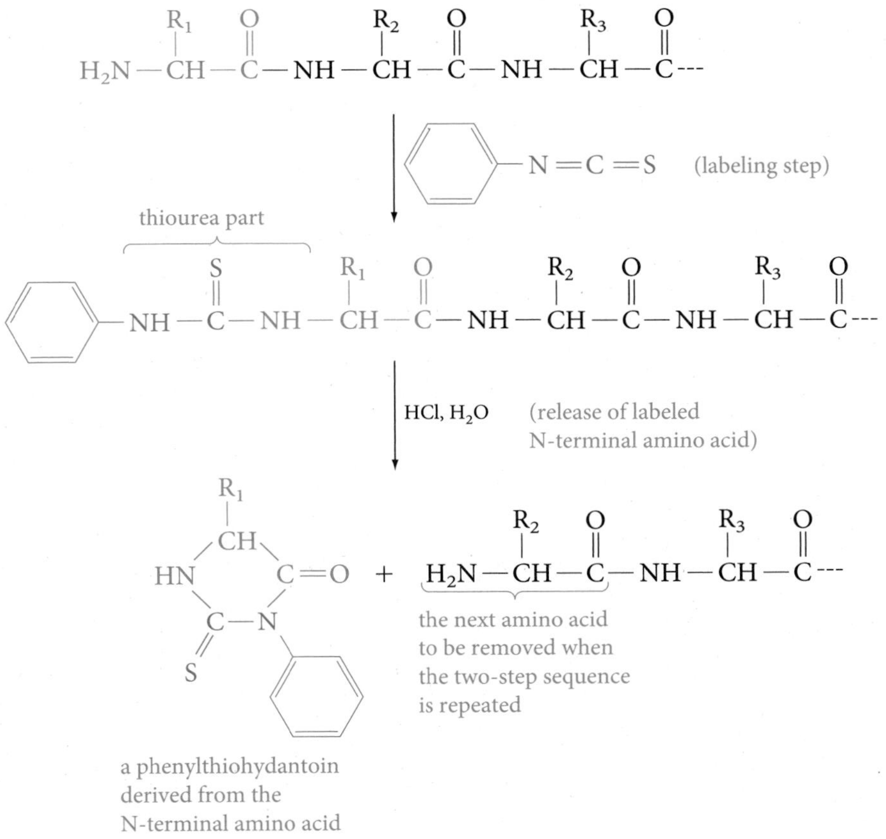

$H_2N-CH(R_1)-C(=O)-NH-CH(R_2)-C(=O)-NH-CH(R_3)-C(=O)$---

$\downarrow$ $C_6H_5-N{=}C{=}S$ (labeling step)

thiourea part

$C_6H_5-NH-C(=S)-NH-CH(R_1)-C(=O)-NH-CH(R_2)-C(=O)-NH-CH(R_3)-C(=O)$---

$\downarrow$ HCl, H_2O (release of labeled N-terminal amino acid)

[phenylthiohydantoin ring: HN, CH(R_1), C=O, C(=S), N–C_6H_5] + $H_2N-CH(R_2)-C(=O)-NH-CH(R_3)-C(=O)$---

a phenylthiohydantoin derived from the N-terminal amino acid

the next amino acid to be removed when the two-step sequence is repeated

■ **Figure 17.5** The Edman degradation of peptides.

17.10.c Cleavage of Selected Peptide Bonds

If a protein contains several hundred amino acid units, it is best to first partially hydrolyze the chain to smaller fragments that can be separated and subsequently sequenced by the Edman method. Certain chemicals or enzymes are used to cleave proteins at *particular* peptide bonds. For example, the enzyme *trypsin* (an intestinal digestive enzyme) specifically hydrolyzes polypeptides only at the carboxy end of arginine and lysine. A few of the many reagents of this type are listed in Table 17.3.

EXAMPLE 17.8

Consider the following peptide:

Ala—Gly—Tyr—Trp—Ser—Lys—Gly—Leu—Met—Gly

By referring to Table 17.3, determine what fragments will be obtained when this peptide is hydrolyzed with

a. trypsin   b. chymotrypsin   c. cyanogen bromide

Solution

a. The enzyme trypsin will split the peptide on the carboxyl side of lysine to give

Ala—Gly—Tyr—Trp—Ser—Lys and Gly—Leu—Met—Gly

b. The enzyme chymotrypsin will split the peptide on the carboxyl sides of tyrosine and tryptophan to give

Ala—Gly—Tyr and Trp and Ser—Lys—Gly—Leu—Met—Gly

c. Cyanogen bromide will split the peptide on the carboxyl side of methionine, thus splitting off the C-terminal glycine and leaving the rest of the peptide untouched. (Carboxypeptidase would do the same thing, confirming that the C-terminal amino acid is glycine.)

PROBLEM 17.17 Determine what fragments will be obtained if bradykinin (its abbreviated formula is given on page 503) is hydrolyzed enzymatically with

a. trypsin   b. chymotrypsin

Over the years, the methods discussed here have been improved and expanded; the separation and sequencing of peptides and proteins can now be accomplished even if only minute amounts are available.

Table 17.3 ■ Reagents for Specific Cleavage of Polypeptides

Reagent	Cleavage site
trypsin	carboxyl side of Lys, Arg
chymotrypsin	carboxyl side of Phe, Tyr, Trp
cyanogen bromide (CNBr)	carboxyl side of Met
carboxypeptidase	the C-terminal amino acid

17.11 The Logic of Sequence Determination

Frederick Sanger developed methods for determining the amino acid sequence of peptides and proteins.

A specific example will illustrate the reasoning that is used to fully determine the amino acid sequence in a particular peptide with 30 amino acid units.

First we hydrolyze the peptide completely, subject it to amino acid analysis, and find that it has the formula

$$\text{Ala}_2\text{ArgAsnCys}_2\text{GlnGlu}_2\text{Gly}_3\text{His}_2\text{Leu}_4\text{LysPhe}_3\text{ProSerThrTyr}_2\text{Val}_3$$

Using the Sanger method, we find that the N-terminal amino acid is Phe.

Since the chain is rather long, we decide to simplify the problem by digesting the peptide with chymotrypsin. (We select chymotrypsin because the peptide contains three Phe's and two Tyr's and will undoubtedly be cleaved by that reagent.) When we carry out this cleavage, we get three fragment peptides. In addition, we get two equivalents of Phe and one of Tyr. After separation, we subject the three fragment peptides to Edman degradation and obtain their structures.

A. Leu—Val—Cys—Gly—Glu—Arg—Gly—Phe

B. Val—Asn—Gln—His—Leu—Cys—Gly—Ser—His—Leu—Val—Glu—Ala—Leu—Tyr

27 28 29 30
C. Thr—Pro—Lys—Ala

We still cannot write a unique structure for the intact peptide, but we can say that the C-terminal amino acid must be Ala and that the last four amino acids must be in the sequence shown for fragment C. We deduce this because we know that Ala is *not* cleaved at its carboxyl end by chymotrypsin, yet it appears at the C-terminal end of one of the fragments. (Note that the C-terminal amino acids in fragments A and B are Phe and Tyr, both cleaved at the carboxyl ends by chymotrypsin.) That the C-terminal amino acid is Ala can be confirmed using carboxypeptidase. We can number the amino acids in fragment C as 27 through 30 in the chain.

What do we do next? Cyanogen bromide is no help, because the peptide does not contain Met. But the peptide does contain Lys and Arg, so we go back to the beginning and digest the intact peptide with trypsin, which cleaves peptides on the carboxy side of these amino acid units. We obtain (not surprisingly) some Ala (the C-terminal amino acid) because it comes right after a Lys. We also obtain two peptides. One of them is relatively short, so we determine its sequence by the Edman method and find it to be

23 24 25 26 27 28 29
D. Gly—Phe—Phe—Tyr—Thr—Pro—Lys

Because the last three amino acids in fragment D *overlap* with 27, 28, and 29 of fragment C, we can number the rest of that chain, back to 23. We now note that amino acids 23 and 24 appear at the end of fragment A, so originally A must have been connected to C. The only place left for fragment B is in front of A. This leaves only one of the Phe's unaccounted for, and it must occupy the N-terminal position (recall the Sanger result). We can now write out the complete sequence!

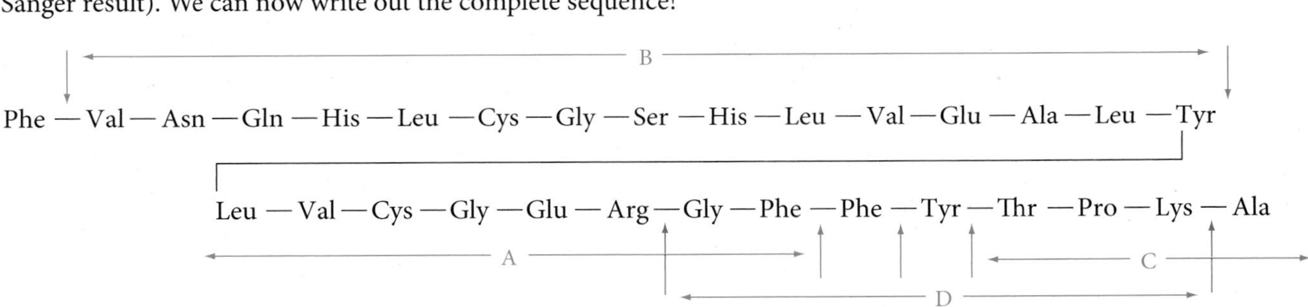

A WORD ABOUT... Protein Sequencing and Evolution

There are many reasons why it is important to determine the sequences of amino acids in proteins. First, we must know the detailed structures of proteins if we are to understand, at a molecular level, the way they function. The amino acid sequence is the link between the genetic message coded in DNA and the three-dimensional protein structure that forms the basis for biological function.

There are medical benefits to knowing amino acid sequences. Certain genetic diseases, such as sickle-cell anemia, can result from the change of a single amino acid unit in a protein. In this case, it is a replacement of glutamic acid at position 6 in the β chain of hemoglobin by a valine unit. Thus, sequence determination is an important part of medical pathology. One future possible application of genetic engineering is devising ways to correct these amino acid sequence errors.

Another important reason for determining amino acid sequences is that they provide a chemical tool for studying our evolutionary history. Proteins resemble one another in amino acid sequence if they have a common evolutionary ancestry. Let's look at a specific example.

Cytochrome c, an enzyme important in the respiration of most plants and animals, is an electron-transport protein with 104 linked amino acids. It is involved in oxidation–reduction processes. In these reactions, cytochrome *c* must react with and transfer an electron from one enzyme complex (cytochrome reductase) to another (cytochrome oxidase).

Cytochrome *c* probably evolved more than 1.5 billion years ago, even before the evolutionary divergence of plants and animals. The function of this protein has been preserved all that time! We know this because the cytochrome *c* isolated from any eukaryotic microorganism (one that contains a cell nucleus) reacts *in vitro* with the cytochrome oxidase of every other species tested so far. For example, wheat germ cytochrome *c*, a plant-derived enzyme, reacts with human cytochrome *c* oxidase. Also, the three-dimensional structures of cytochrome *c* isolated from such diverse species as tuna and photosynthetic bacteria are very similar.

Although *the shape and functions are similar* for cytochrome *c* samples isolated from different sources, *the amino acid sequence varies somewhat* from one species to another. The amino acid sequence of cytochrome *c* isolated from humans differs from that of monkeys in *just 1* (out of 104) amino acid residues! On the other hand, cytochrome *c* from dogs, a more distant evolutionary relative of humans, differs from the human protein by 11 amino acid residues.

See Problems 17.45, 17.49, 17.52, 17.53, and 17.54.

The blue vertical arrows show the cleavage points with chymotrypsin, and the red ones show the cleavage points with trypsin.

Insulin is a peptide hormone that regulates blood glucose levels.

The peptide just used for illustration is the B chain of the protein hormone **insulin**, whose structure was first determined by Sanger and is shown schematically in Figure 17.6. Insulin consists of an A chain with 21 amino acid units and a B chain with 30 amino acid units. The two chains are joined by two disulfide bonds, and the A chain also contains a small disulfide loop.

PROBLEM 17.18 How could the A and B chains of insulin be separated chemically? (*Hint:* See eq. 7.49.)

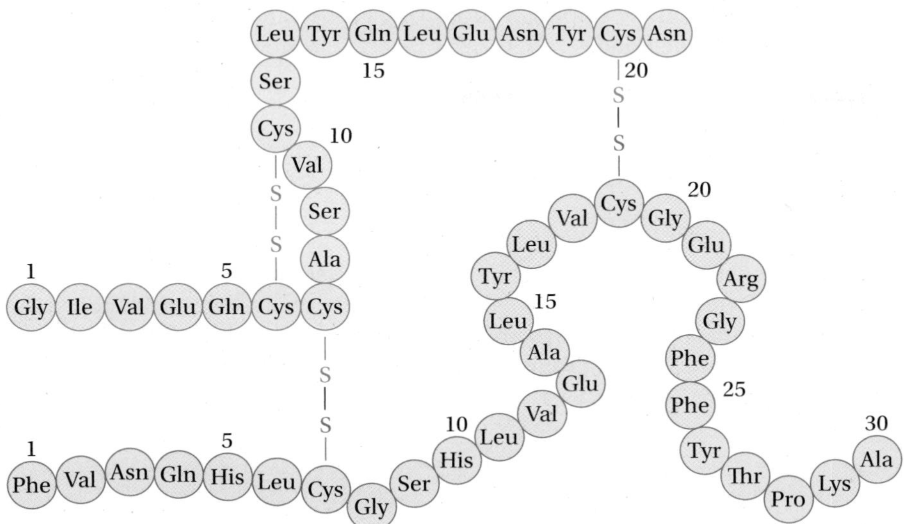

▪ **Figure 17.6**
Primary structure of beef insulin. The A chain is shown in blue, and the B chain, whose structure determination is described in the text, is shown in pink with black letters.

17.12 Peptide Synthesis

Once we know the amino acid sequence in a peptide or protein, we are in a position to synthesize it from its amino acid components. Why would we want to do this? There are several reasons. For example, we might wish to verify a particular peptide structure by comparing the properties of the synthetic and natural substances. Or we might wish to study the effect of substituting one amino acid for another on the biological properties of a peptide or protein. Such modified proteins could be valuable in treating disease or in understanding how the protein functions.

Many methods have been developed to link amino acids in a controlled manner. They require careful strategy. Amino acids are bifunctional. To link the carboxyl group of one amino acid to the amino group of a second amino acid, we must first prepare each compound by protecting the amino group of the first and the carboxyl group of the second.

$$\underset{aa_1}{H_2N-\overset{R_1}{\overset{|}{C}H}-CO_2H} \xrightarrow[\text{amino group}]{\text{protect the}} \boxed{P_1}-NH-\overset{R_1}{\overset{|}{C}H}-CO_2H \quad (17.12)$$

$$\underset{aa_2}{H_2N-\overset{R_2}{\overset{|}{C}H}-CO_2H} \xrightarrow[\text{carboxyl group}]{\text{protect the}} H_2N-\overset{R_2}{\overset{|}{C}H}-\overset{O}{\overset{\|}{C}}-\boxed{P_2} \quad (17.13)$$

In this way, we can control the linking of the two amino acids so that the carboxyl group of aa_1 combines with the amino group of aa_2.

$$\boxed{P_1}-NH\overset{R_1}{\overset{|}{C}}HCO_2H + H_2N-\overset{R_2}{\overset{|}{C}H}-\overset{O}{\overset{\|}{C}}-\boxed{P_2} \xrightarrow{-H_2O}$$

$$\underset{\text{doubly protected dipeptide}}{\boxed{P_1}-NH\overset{R_1}{\overset{|}{C}}H-\underbrace{\boxed{\overset{O}{\overset{\|}{C}}-NH}}_{\text{peptide bond}}-\overset{R_2}{\overset{|}{C}H}-\overset{O}{\overset{\|}{C}}-\boxed{P_2}} \quad (17.14)$$

Later we will give specific examples of protecting groups and of a reagent that can be used to form the peptide bond.

EXAMPLE 17.9

What would happen if we tried to combine aa_1 with aa_2 without using protecting groups?

Solution Since each amino acid could react either as an amine or as a carboxylic acid, we could get not only $aa_1—aa_2$ but also $aa_2—aa_1$, $aa_1—aa_1$, and $aa_2—aa_2$. Furthermore, since the resulting dipeptides would still have a free amino and a free carboxyl group, we could also get trimers, tetramers, and so on. In other words, a mess.

After the peptide bond is formed, we must be able to *remove the protecting groups under conditions that do not hydrolyze the peptide bond.* Or, if more amino acids are to be added to the chain, we must be able to *selectively* remove one of the two protecting groups from the doubly protected dipeptide before joining the next amino acid to it. All of this can be quite a tricky and tedious process. Yet these methods were used by Vincent du Vigneaud* and his colleagues to synthesize oxytocin and vasopressin (page 503), the first naturally occurring polypeptides to be synthesized in the laboratory.

Peptides are synthesized by Merrifield's **solid-phase technique**, in which one end of the growing peptide is chemically attached to an insoluble, inert solid.

In 1965, R. Bruce Merrifield** developed a technique that revolutionized peptide synthesis. This **solid-phase technique** avoids many of the tedious aspects of previous methods and is now universally used. The principle is to *assemble the peptide chain while one end of it is chemically anchored to an insoluble, inert solid.* In this way, excess reagents and by-products can be removed simply by washing and filtering the solid. The growing peptide chain does not need to be purified at any intermediate stage. When the peptide is fully constructed, it is cleaved chemically from the solid support.

Typically, the solid phase is a cross-linked polystyrene in which some (usually 1% to 10%) of the aromatic rings contain chloromethyl ($ClCH_2—$) groups.

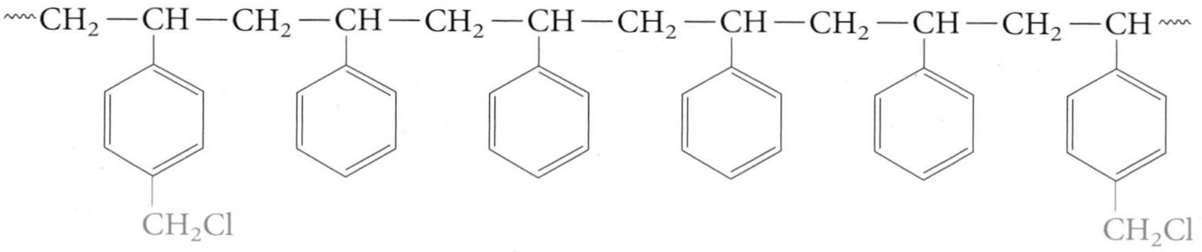

$$\sim CH_2—CH—CH_2—CH—CH_2—CH—CH_2—CH—CH_2—CH—CH_2—CH \sim$$

(each CH bears a benzene ring; the first and last rings carry CH_2Cl groups)

The polymer behaves chemically like benzyl chloride, an alkyl halide that is quite reactive in nucleophilic substitution reactions (S_N2).

The steps in a Merrifield synthesis are summarized for a dipeptide in Figure 17.7. In step 1, the polymer is first treated with an N-protected amino acid. The carboxylate ion acts as an oxygen nucleophile and displaces the chloride ion from the polymer, thus forming an ester link. *The first amino acid attached to the polymer will eventually become the C-terminal amino acid of the synthetic peptide.*

---

*Vincent du Vigneaud (Cornell University) was awarded the 1955 Nobel Prize in chemistry for this achievement.

R. Bruce Merrifield (Rockefeller University) received the 1984 Nobel Prize in chemistry for his contribution, which not only revolutionized peptide synthesis but also affected many other areas of chemistry through the use of polymer-bound reagents. For an account of the history of this discovery, see the article by Merrifield in *Science* **1986, 232, 341, and for a personal history, see *Chemistry in Britain* **1987**, 816.

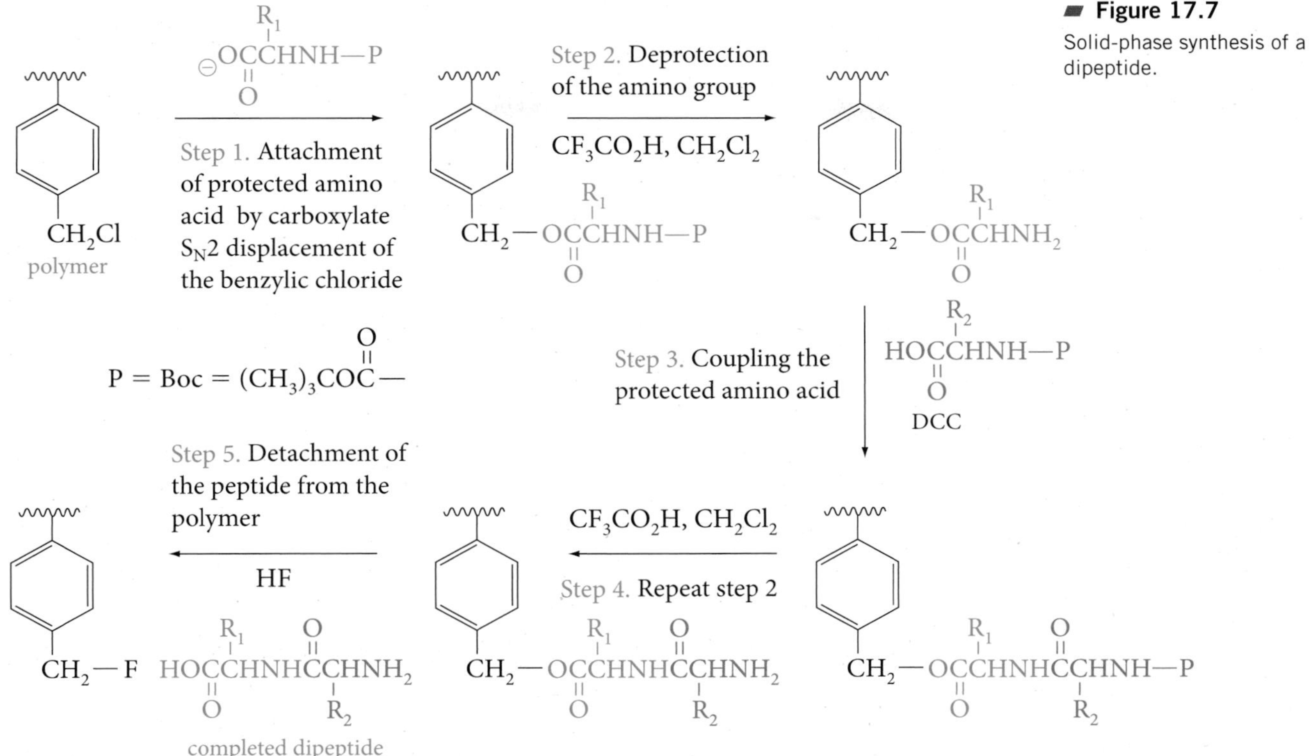

■ **Figure 17.7**
Solid-phase synthesis of a dipeptide.

Many protecting groups are known, but in solid-phase peptide synthesis, the most frequently used N-protecting group is the ***t*-butoxycarbonyl (Boc)** group. The amino acid is protected by reaction with di-*t*-butyl dicarbonate.

The most frequently used N-protecting group in solid-phase peptide synthesis is the ***t*-butoxycarbonyl (Boc)** group.

$$[(CH_3)_3CO{-}\overset{O}{\overset{\|}{C}}]_2O + H_3\overset{+}{N}{-}\underset{}{\overset{R}{\overset{|}{C}H}}{-}CO_2^- \xrightarrow{\text{base}} \underbrace{(CH_3)_3CO{-}\overset{O}{\overset{\|}{C}}}_{\boxed{P}}{-}NH{-}\overset{R}{\overset{|}{C}H}{-}CO_2H \qquad (17.15)$$

di-*t*-butyl dicarbonate

After the amino acid is attached to the polymer, the protecting group is removed. This is accomplished (step 2) by reaction with acid under mild conditions.

$$(CH_3)_2(H{-}CH_2)C{-}O{-}\overset{O}{\overset{\|}{C}}{-}NH\overset{R}{\overset{|}{C}H}{-}\underset{O}{\underset{\|}{C}}{-}O{-}CH_2{-}C_6H_4{-}\text{polymer} \xrightarrow[CH_2Cl_2]{CF_3CO_2H}$$

$$CH_2{=}C(CH_3)_2 + CO_2 + H_2N{-}\overset{R_1}{\overset{|}{C}H}{-}\underset{O}{\underset{\|}{C}}{-}O{-}CH_2{-}C_6H_4{-}\text{polymer} \qquad (17.16)$$

2-methylpropene; ester group still intact; deprotected amino acid

The by-products of deprotection are gaseous (2-methylpropene and carbon dioxide) and are thus easily removed from the reaction mixture. The ester group that links the first amino acid to the polymer is *not* hydrolyzed under these conditions.

Dicyclohexylcarbodiimide (DCC) is a reagent used to link carboxyl and amino groups in a peptide bond.

In step 3, the next N-protected amino acid is linked to the first one. This is accomplished with the aid of **dicyclohexylcarbodiimide (DCC)**. DCC is able to link carboxyl and amino groups in a peptide bond; in the process, the DCC is converted to dicyclohexylurea. The mechanism of this step is described in Figure 17.8.

Steps 2 and 3 may be repeated to add a third amino acid, a fourth, and so on. Finally, when the desired amino acids have been connected in the proper sequence and the N-terminal amino group has been deprotected (step 4 in Figure 17.7), the complete polypeptide chain is detached from the polymer. This can be accomplished by treatment with anhydrous hydrogen fluoride, which cleaves the benzyl ester without hydrolyzing the amide bonds in the polypeptide (step 5 in Figure 17.7).

All operations in solid-phase peptide synthesis have been automated. The reactions occur in a single reaction vessel, with reagents and wash solvents automatically added from reservoirs by mechanical pumps. Merrifield synthesized the nonapeptide bradykinin (page 503) in just 27 hours using this technique. And, in 1969, he used the automated synthesizer to prepare the enzyme ribonuclease (124 amino acid residues), the first enzyme to be prepared synthetically from its amino acid components. The synthesis, which required 369 chemical reactions and 11,391 steps, was completed in only six weeks. Automated computerized peptide synthesis, though still not without occasional problems, is now a fairly routine matter.

Automated polypeptide synthesizer.

PROBLEM 17.19 Write all of the equations for the synthesis of Gly—Ala—Phe using Merrifield solid-phase technology.

We have seen how the primary structure of peptides and proteins can be determined and how peptides can be synthesized in the laboratory. Now let us examine some further details of protein structure.

N-protected amino acid: H—O—C(=O)—CHNH—P (with R on CH) + DCC: Cy—N=C=N—Cy, DCC (Cy = cyclohexyl)

Step 1 → $^{\ominus}$O—C(=O)—CHNH—P (R) + Cy—N(H)$^{\oplus}$=C=N—Cy

Step 2 → polymer-bound amino acid: ■—NH$_2$ attacking C(=O)(—CHNH—P, R)—O—C(=N—Cy)—N(H)—Cy

Step 3 → ■—N$^{\oplus}$(H)(H)—C(O$^{\ominus}$)(—CHNH—P, R)—O—C(=N—Cy)—N(H)—Cy, H$^{\oplus}$

Step 4 → ■—N(H)—C(=O)—CHNH—P (R) + Cy—N(H)—C(=O)—N(H)—Cy

polymer-bound amino acid with new amino acid attached

dicyclohexylurea

■ **Figure 17.8**

The mechanism of DCC coupling involves conversion of the carboxyl group of the acid to a good acylating agent (steps 1 and 2) followed by acylation of the free amino group of the polymer-bound amino acid (steps 3 and 4).

17.13 Secondary Structure of Proteins

Because proteins consist of long chains of amino acids strung together, one might think that their shapes are rather amorphous, or "floppy" and ill-defined. This is incorrect. Many proteins have been isolated in pure crystalline form and are polymers with well-defined shapes. Indeed, even in solution, the shapes seem to be quite regular. Let us examine some of the structural features of peptide chains that are responsible for their definite shapes.

17.13.a Geometry of the Peptide Bond

We pointed out earlier that simple amides have a planar geometry, that the amide C—N bond is shorter than usual, and that rotation around that bond is restricted (Sec. 10.20). Bond planarity and restricted rotation, which are consequences of resonance, are also important in peptide bonds.

X-ray studies of crystalline peptides by Linus Pauling and his colleagues determined the precise geometry of peptide bonds. The characteristic dimensions, which are common to all peptides and proteins, are shown in Figure 17.9.

Things to notice about peptide geometry are as follows: (1) The amide group is flat; the carbonyl carbon, the nitrogen, and the four atoms connected to them all lie in a single plane. (2) The short amide C—N distance (1.32 Å, compared with 1.47 Å for the other C—N bond) and the 120° bond angles around that nitrogen show that it is essentially sp^2-hybridized and that the bond between it and the carbonyl carbon is like a double bond. (3) Although each amide group is planar, two adjacent amide groups need not be *co*planar because of rotation about the other single bonds in the chain; that is, rotation can occur around the two single bonds to the —CHR— group.

The rather rigid geometry and restricted rotation of the peptide bond help to impart a definite shape to proteins.

17.13.b Hydrogen Bonding

We pointed out earlier that amides readily form *inter*molecular hydrogen bonds between the carbonyl group and the N—H group, bonds of the type C═O···H—N. Such bonds are present and important in peptide chains. The chain may coil in such a way that the N—H of one peptide bond can hydrogen-bond with a carbonyl group of another peptide bond farther down the *same* chain, thus rigidifying the coiled structure. Alternatively, carbonyl groups and N—H groups on *different* peptide chains may hydrogen-bond, linking the two chains. Although a single hydrogen bond is relatively weak (perhaps only 5 kcal/mol of energy), the possibility of forming multiple intra-chain or inter-chain hydrogen bonds makes this a very important factor in protein structure, as we will now see.

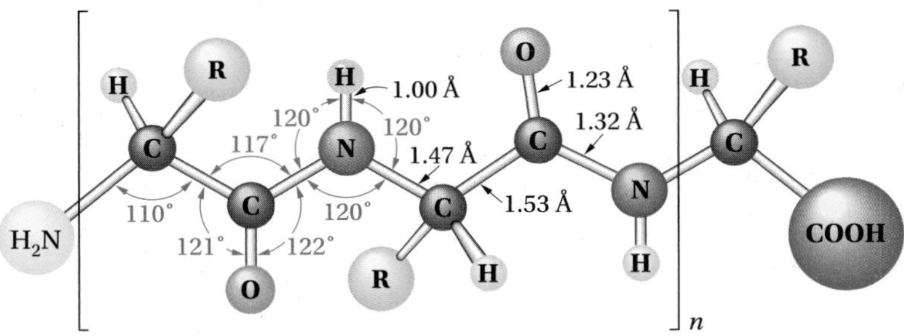

■ **Figure 17.9**
The characteristic bond angles and bond lengths in peptide bonds.

17.13.c The α Helix and the Pleated Sheet

The α **helix** and the **pleated sheet** are two common secondary structures of proteins or segments of proteins.

X-ray crystallographic studies of α-keratin, a structural protein present in hair, wool, horns, and nails, showed that some feature of the structure repeats itself every 5.4 Å. Using molecular models with the correct geometry of the peptide bond, Linus Pauling was able to suggest a structure that explains this and other features of the x-ray studies. Pauling proposed that the polypeptide chain coils about itself in a spiral manner to form a helix, held rigid by intra-chain hydrogen bonds. The ***α* helix**, as it is called, is right-handed and has a pitch of 5.4 Å, or 3.6 amino acid units (Figure 17.10).*

Note several features of the α helix. Proceeding from the N terminus (at the top of the structure as drawn in the figure), each carbonyl group points ahead or down toward the C terminus and is hydrogen-bonded to an N—H bond farther down the chain. The N—H bonds all point back toward the N terminus. All of the hydrogen bonds are roughly aligned with the long axis of the helix. The very large number of hydrogen bonds (one for each amino acid unit) strengthens the helical structure. The R groups of the individual amino acid units are directed *outward* and do not disrupt the central core of the helix. It turns out that the α helix is a natural pattern into which many proteins or segments of proteins fold.

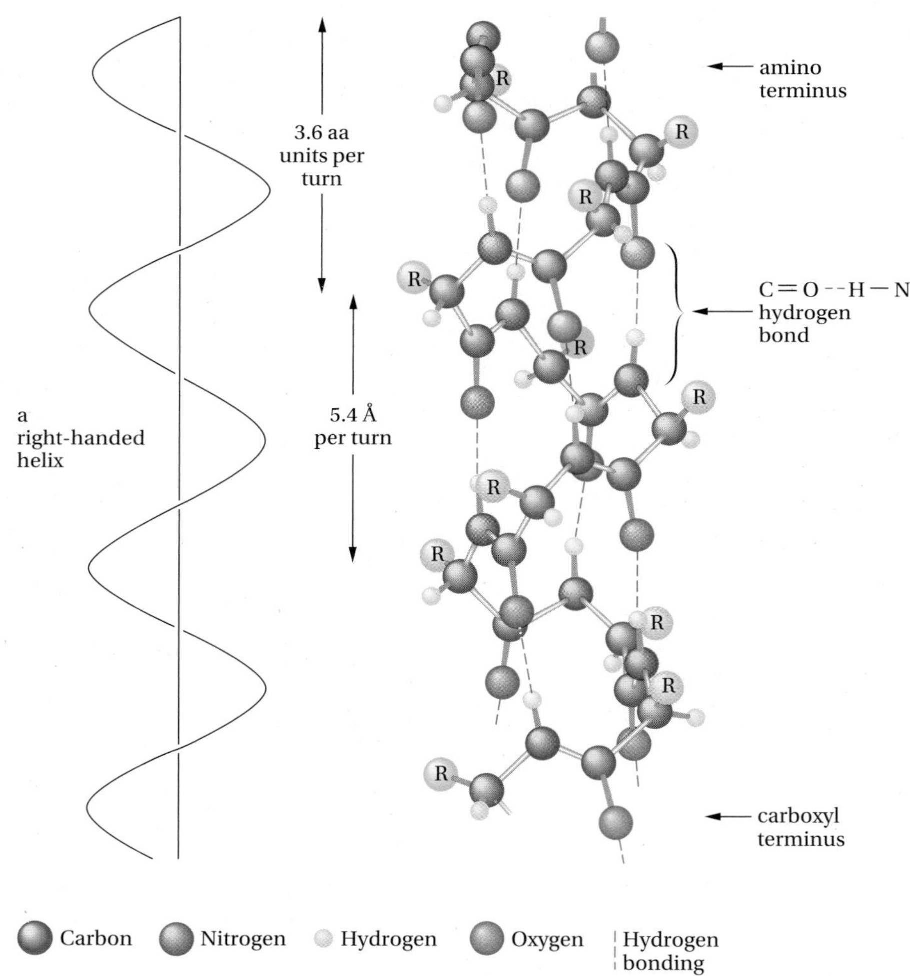

▪ **Figure 17.10**
Segment of an α helix, showing three turns of the helix, with 3.6 amino acid units per turn. Hydrogen bonds are shown as dashed colored lines.

*Linus Pauling (1901–1994) made many contributions to our knowledge of organic structures. He did fundamental work on the theory of resonance, on the measurement of bond lengths and energies, and on the structure of proteins and the mechanism of antibody action. He received the Nobel Prize in chemistry in 1954 and the Nobel Peace Prize in 1962.

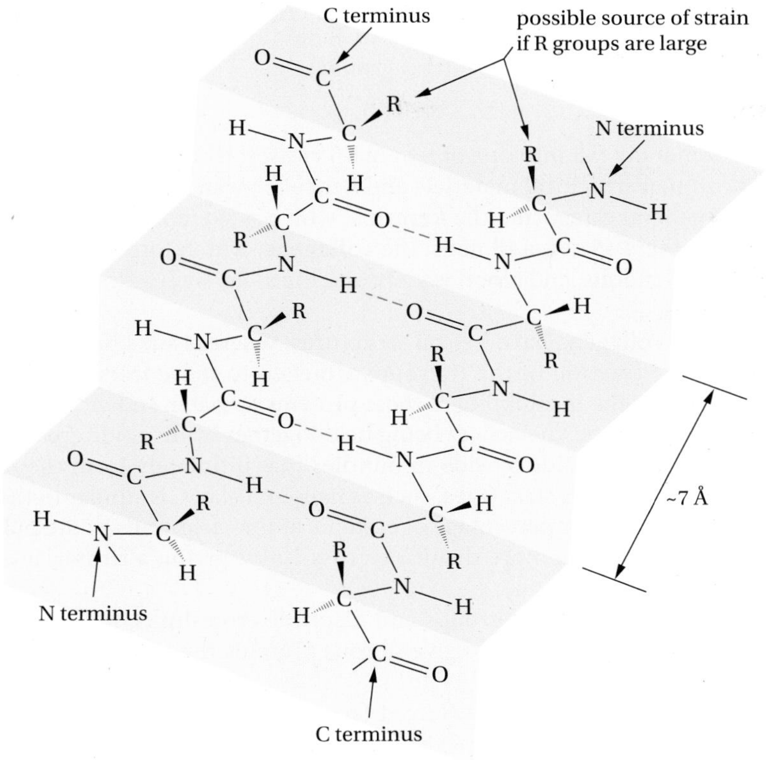

■ **Figure 17.11**
A segment of the pleated-sheet structure of β-keratin. Adjacent chains run in opposite directions and are held together by hydrogen bonds (shown in color). R groups project above or below the mean plane of the sheet.

The structural protein β-keratin, obtained from silk fibroin, shows a different repeating pattern (7 Å) in its x-ray crystal structure. To explain the data, Pauling suggested a **pleated-sheet** arrangement of the peptide chain (Figure 17.11). In the pleated sheet, peptide chains lie side by side and are held together by *inter-chain* hydrogen bonds. Adjacent chains run in opposite directions. The repeating unit in each chain, which is stretched out compared with the α helix, is about 7 Å. In the pleated-sheet structure, the R groups of amino acid units in any one chain alternate above and below the mean plane of the sheet. If the R groups are large, there will be appreciable steric repulsion between them on adjacent chains. For this reason, the pleated-sheet structure is important *only* in proteins that have a high percentage of amino acid units with *small* R groups. In the β-keratin of silk fibroin, for example, 36% of the amino acid units are glycine (R = H) and another 22% are alanine (R = CH_3).

17.14 Tertiary Structure: Fibrous and Globular Proteins

We may well ask how materials as rigid as horses' hoofs, as springy as hair, as soft as silk, as slippery and shapeless as egg white, as inert as cartilage, and as reactive as enzymes, can all be made of the same building blocks: amino acids and proteins. The key lies mainly in the amino acid makeup itself. So far we have focused on the protein backbone and its shape. But what about the diverse R groups of the various amino acids? How do they affect protein structure?

Some amino acids have nonpolar R groups, simple alkyl or aromatic groups. Others have highly polar R groups, with carboxylate or ammonium ions and hydroxyl or other polar groups. Still others have flat, rigid aromatic rings that may interact in specific ways. *Different R groups affect the gross properties of a protein.*

PROBLEM 17.20 Which amino acids in Table 17.1 have nonpolar R groups? Highly polar groups? Relatively flat R groups?

Fibrous proteins (including **keratins**, **collagens**, and **silks**) have rather rigid shapes and are not water soluble.

Proteins generally fall into one of two main classes: **fibrous** or **globular**. **Fibrous proteins** are animal structural materials and hence are water insoluble. They, in turn, fall into three general categories: the **keratins**, which make up protective tissue, such as skin, hair, feathers, claws, and nails; the **collagens**, which form connective tissue, such as cartilage, tendons, and blood vessels; and the **silks**, such as the fibroin of spider webs and cocoons.

Keratins and collagens have helical structures, whereas silks have pleated-sheet structures. A large fraction of the R groups attached to these frameworks are nonpolar, accounting for the insolubility of these proteins in water. In hair, three α helices are braided to form a rope, the helices being held together by disulfide cross-links. The ropes are further packed side by side in bundles that ultimately form the hair fiber. The α-keratin of more rigid structures, such as nails and claws, is similar to that of hair, except that there is a higher percentage of cysteine amino acid units in the polypeptide chain. Therefore, there are more disulfide cross-links, giving a firmer, less flexible overall structure.

To summarize, nonpolar R groups and disulfide cross-links, together with helical or sheet-like backbones, tend to give fibrous proteins their rather rigid, insoluble structures.

Globular proteins (including **enzymes**, **hormones**, **transport proteins**, and **storage proteins**) are spherical in shape and tend to be water soluble.

Globular proteins are very different from fibrous proteins. They tend to be water soluble and have roughly spherical shapes, as their name suggests. Instead of being structural, globular proteins perform various other biological functions. They may be **enzymes** (biological catalysts), **hormones** (chemical messengers that regulate biological processes), **transport proteins** (carriers of small molecules from one part of the body to another, such as hemoglobin, which transports oxygen in the blood), or **storage proteins** (which act as food stores; ovalbumin of egg white is an example).

Globular proteins have more amino acids with polar or ionic side chains than the water-insoluble fibrous proteins. An enzyme or other globular protein that carries out its function mainly in the aqueous medium of the cell will adopt a structure in which the nonpolar, hydrophobic R groups point in toward the center and the polar or ionic R groups point out toward the water.

Globular proteins are mainly helical, but they have folds that permit the overall shape to be globular. One of the 20 amino acids, proline has a secondary amino group. Wherever a proline unit occurs in the primary peptide structure, there will be no N—H group available for intra-chain hydrogen bonding.

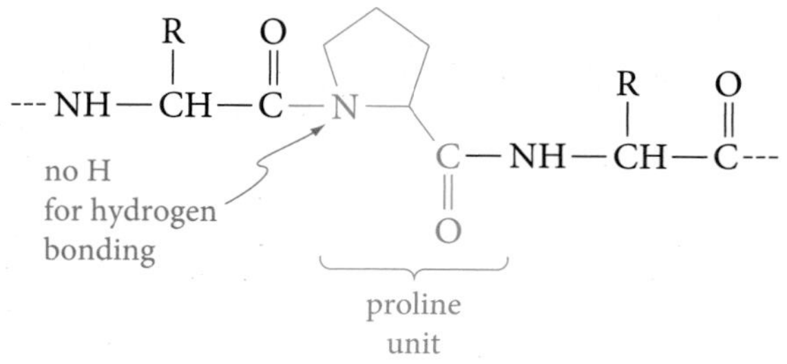

---NH—CH(R)—C(=O)—N(proline ring)—C(=O)—NH—CH(R)—C(=O)---

no H for hydrogen bonding

proline unit

Proline units tend therefore to disrupt an α helix, and we frequently find them at "turns" in a protein structure.

Myoglobin, the oxygen-transport protein of muscle, is a good example of a globular protein (Figure 17.12). It contains 153 amino acid units, yet is extremely compact,

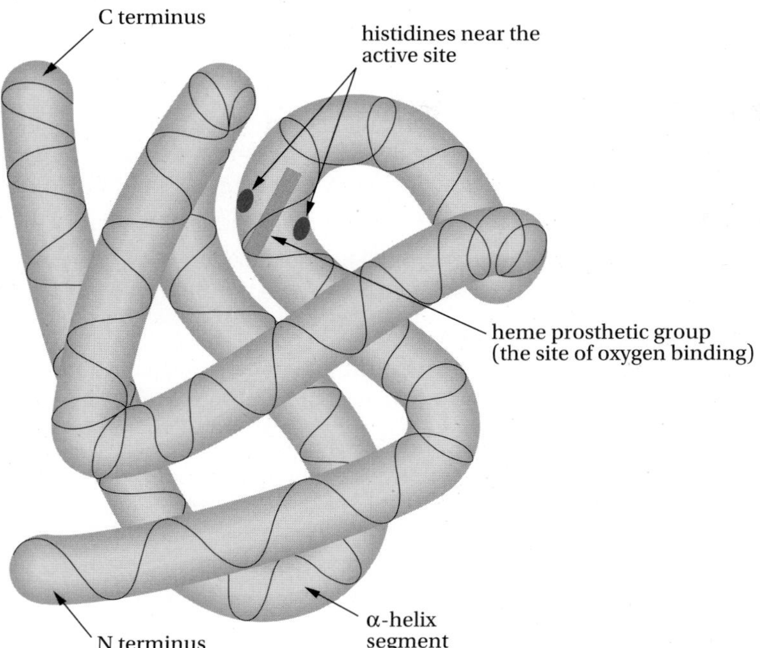

Figure 17.12
Schematic drawing of myoglobin. Each of the tubular sections is a segment of α helix, but the overall shape is globular.

with very little empty space in its interior. Approximately 75% of the amino acid units in myoglobin are part of eight right-handed α-helical sections. There are four proline units, and each occurs at or near "turns" in the structure. There are also three other "turns" caused by structural features of other R groups. The interior of myoglobin consists almost entirely of nonpolar R groups, such as those of leucine, valine, phenylalanine, and methionine. The only interior polar groups are two histidines. These perform a necessary function at the *active site* of the protein, where the nonprotein portion, a molecule of the porphyrin *heme* (page 402), binds the oxygen. The outer surface of the protein includes many highly polar amino acid residues (lysine, arginine, glutamic acid, etc.).

To summarize this section, we see that the particular amino acid content of a peptide or protein influences its shape. These interactions are mainly a consequence of disulfide bonds and of the polarity or nonpolarity of the R groups, their shape, and their ability to form hydrogen bonds. When we refer to the **tertiary structure** of a protein, we refer to all contributions of these factors to its three-dimensional structure.

The particular amino acid content of a protein influences its overall shape or **tertiary structure**.

The structure of the aggregate formed by subunits of a high-molecular-weight protein is called its **quaternary structure**.

17.15 Quaternary Protein Structure

Some high-molecular-weight proteins exist as aggregates of several subunits. These aggregates are referred to as the **quaternary structure** of the protein. Aggregation helps to keep nonpolar portions of the protein surface from being exposed to the aqueous cellular environment. **Hemoglobin**, the oxygen-transport protein of red cells, provides an example of such aggregation. It consists of four almost spherical units, two α units with 141 amino acids and two β units with 146 amino acids. The four units come together in a tetrahedral array, as shown in Figure 17.13.

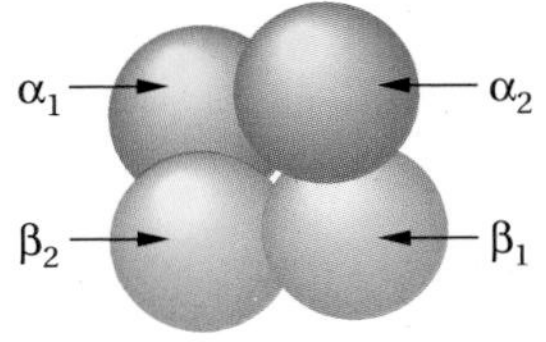

Figure 17.13
Schematic drawing of the four hemoglobin subunits.

A CLOSER LOOK AT... Nobel Laureates and Protein Chemistry

Conduct research on the Internet to find information on Nobel laureates in protein chemistry and to answer the following questions.

Nobel Prizes Related to Chemistry of Peptides and Proteins

1. The 2004 Nobel Prize in chemistry was the last in a series of prizes awarded for developments in the fields of amino acid, peptide, and protein chemistry. Who were the recipients of the 2004 award and for what contribution to protein chemistry were they recognized?
2. What were the names of the other scientists awarded the Nobel Prize in chemistry for contributions to the fields of amino acid, peptide, and protein chemistry? Briefly describe their contributions.
3. What are the nationalities of all of the Nobel laureates recognized for their contributions to peptide and protein chemistry?
4. Which of these Nobel laureates contributed to the development of solid-phase synthesis techniques?
5. What, if any, other Nobel prizes were awarded for contributions to the field of solid-phase synthesis?

Many other proteins form similar aggregates. Some are active only in their aggregate state, whereas others are active only when the aggregate dissociates into subunits. Aggregation in quaternary structures, then, provides an additional control mechanism over biological activity.

REACTION SUMMARY

1. Reactions of Amino Acids

a. Acid–Base Reactions (Secs. 17.2 and 17.3)

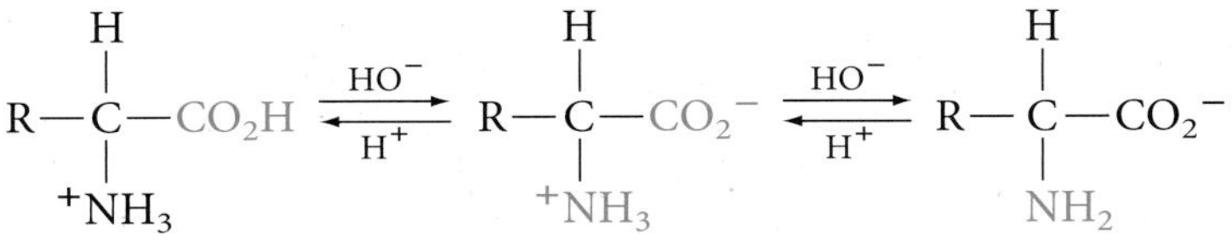

$$\mathrm{R{-}\underset{\underset{{}^{+}NH_3}{|}}{\overset{\overset{H}{|}}{C}}{-}CO_2H} \underset{H^+}{\overset{HO^-}{\rightleftharpoons}} \mathrm{R{-}\underset{\underset{{}^{+}NH_3}{|}}{\overset{\overset{H}{|}}{C}}{-}CO_2^-} \underset{H^+}{\overset{HO^-}{\rightleftharpoons}} \mathrm{R{-}\underset{\underset{NH_2}{|}}{\overset{\overset{H}{|}}{C}}{-}CO_2^-}$$

dipolar ion

b. Esterification (Sec. 17.5)

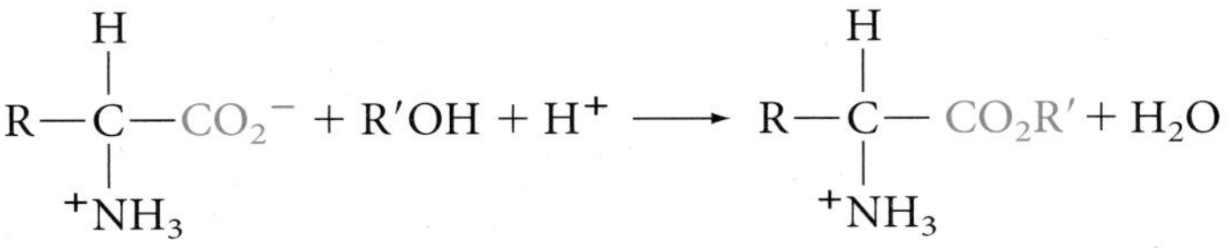

$$\mathrm{R{-}\underset{\underset{{}^{+}NH_3}{|}}{\overset{\overset{H}{|}}{C}}{-}CO_2^-} + \mathrm{R'OH} + \mathrm{H^+} \longrightarrow \mathrm{R{-}\underset{\underset{{}^{+}NH_3}{|}}{\overset{\overset{H}{|}}{C}}{-}CO_2R'} + \mathrm{H_2O}$$

c. Amide Formation (Sec. 17.5)

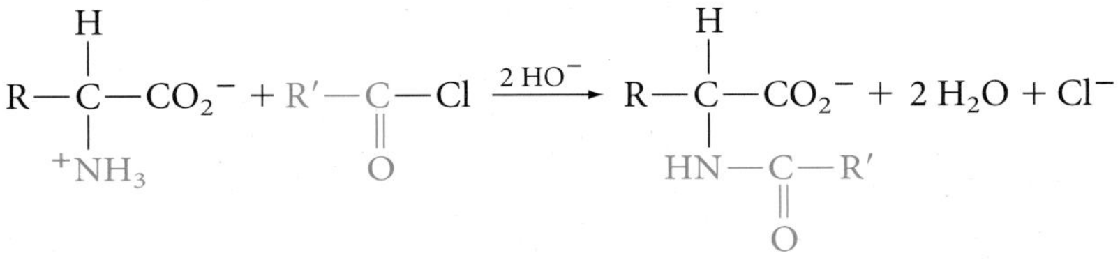

$$\mathrm{R{-}\underset{\underset{{}^{+}NH_3}{|}}{\overset{\overset{H}{|}}{C}}{-}CO_2^-} + \mathrm{R'{-}\underset{\underset{O}{\|}}{C}{-}Cl} \xrightarrow{2\,HO^-} \mathrm{R{-}\underset{\underset{HN{-}\underset{\underset{O}{\|}}{C}{-}R'}{|}}{\overset{\overset{H}{|}}{C}}{-}CO_2^-} + 2\,\mathrm{H_2O} + \mathrm{Cl^-}$$

d. Ninhydrin Reaction (Sec. 17.6)

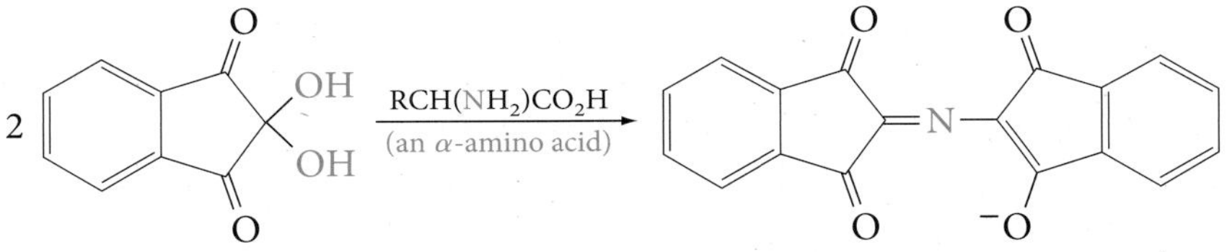

2 ninhydrin $\xrightarrow[\text{(an } \alpha\text{-amino acid)}]{\mathrm{RCH(NH_2)CO_2H}}$ (purple product: indanedione=N–enolate indanone anion) $+ \mathrm{RCHO} + \mathrm{CO_2} + 3\,\mathrm{H_2O} + \mathrm{H^+}$

ninhydrin

(purple)

2. Reactions of Proteins and Peptides

a. Hydrolysis (Sec. 17.10)

$$\text{proteins} \xrightarrow[\mathrm{H_2O},\ \Delta]{\mathrm{HCl}} \text{peptides} \xrightarrow[\mathrm{H_2O},\ \Delta]{\mathrm{HCl}} \alpha\text{-amino acids}$$

b. Sanger's Reagent (Sec. 17.10; used to identify the N-terminal amino acid of a peptide or protein)

$$O_2N-C_6H_3(NO_2)-F + H_2N-\underset{\displaystyle R}{\underset{|}{CH}}-\overset{\displaystyle O}{\overset{\|}{C}}-\underset{\displaystyle H}{\underset{|}{N}}-(P) \longrightarrow \text{Peptide with labeled N-terminal amino acid}$$

Sanger's reagent

(P) = peptide chain

hydrolysis

$$\text{amino acids from peptide} + O_2N-C_6H_3(NO_2)-\underset{\displaystyle H}{\underset{|}{N}}-\underset{\displaystyle R}{\underset{|}{CH}}-\overset{\displaystyle O}{\overset{\|}{C}}-OH \quad \text{(yellow)}$$

labeled amino acid from N terminus of peptide

c. Edman degradation (Sec. 17.10; used to determine the amino acid sequence of a peptide)

$$H_2N-\underset{\displaystyle R}{\underset{|}{CH}}-\overset{\displaystyle O}{\overset{\|}{C}}-\underset{\displaystyle H}{\underset{|}{N}}-(P) \xrightarrow[\text{(Ph = phenyl group)}]{Ph-N=C=S} \xrightarrow[H_2O]{HCl} \text{(cyclic HN–C(=S)–N(Ph)–C(=O)–CH(R))} + H_2N-(P)$$

(P) = peptide chain

d. Solid-Phase Peptide Synthesis (Sec. 17.12)

Details are given in Figure 17.7.

ADDITIONAL PROBLEMS (See pages 493–494 for structures of amino acids.)

OWL Interactive versions of these problems are assignable in OWL.

Amino Acids: Definitions, Formulas, and Properties

17.21 Give a definition or illustration of each of the following terms:

a. peptide bond
b. dipolar ion
c. dipeptide
d. L configuration of amino acids
e. essential amino acid
f. amino acid with a nonpolar *R* group
g. amino acid with a polar R group
h. amphoteric compound
i. isoelectric point
j. ninhydrin

17.22 Draw a Fischer projection for L-leucine. What is the priority order of the groups attached to the stereogenic center? What is the absolute configuration, *R* or *S*?

17.23 Write Fischer projection formulas for

a. L-tyrosine
b. L-isoleucine

17.24 Illustrate the amphoteric nature of amino acids by writing an equation for the reaction of valine in its dipolar ion form with one equivalent of

a. hydrochloric acid b. sodium hydroxide

17.25 Write the formula for each of the following in its dipolar ion form:

a. threonine b. cysteine
c. proline d. phenylalanine

17.26 Locate the most acidic proton in each of the following species, and draw the structure of the product formed by reaction with one equivalent of base (HO^-).

a. $HOOC{-}CH_2CH_2CH(\overset{+}{N}H_3)CO_2H$ b. $HOCH_2{-}CH(\overset{+}{N}H_3)CO_2^-$

c. $(CH_3)_2CHCH(\overset{+}{N}H_3)CO_2H$ d. $(NH_2)(H_2\overset{+}{N}{=})C{-}NHCH_2CH_2CH_2CH(NH_2)CO_2^-$

17.27 What species is obtained by adding a proton to each of the following?

a. $CH_3CH(OH){-}CH(\overset{+}{N}H_3)CO_2^-$ b. $^-O_2CCH_2CH(\overset{+}{N}H_3){-}CO_2^-$

17.28 Protonated alanine, $CH_3CH(\overset{+}{N}H_3)CO_2H$, has a p$K_a$ of 2.34, whereas propanoic acid, $CH_3CH_2CO_2H$, has a pK_a of 4.85. Explain the increase in acidity due to replacing an α-hydrogen with an $-\overset{+}{N}H_3$ substituent.

17.29 The pK_a's of glutamic acid are 2.19 (the α carboxyl group), 4.25 (the other carboxyl group), and 9.67 (the α ammonium ion). Write equations for the sequence of reactions that occurs when base is added to a strongly acidic (pH = 1) solution of glutamic acid.

17.30 The pK_a's of arginine are 2.17 for the carboxyl group, 9.04 for the ammonium ion, and 12.48 for the guanidinium ion. Write equations for the sequence of reactions that occurs when acid is gradually added to a strongly alkaline solution of arginine.

17.31 Draw the structure of histidine at pH 1, and show how the positive charge in the second basic group (the imidazole ring) can be delocalized.

17.32 Predict the direction of migration in an electrophoresis apparatus of each component in a mixture of glutamine, histidine, and glutamic acid at pH 6.

Reactions of Amino Acids

17.33 Write equations for the reaction of leucine with

a. $CH_3CH_2OH + HCl$ b. $C_6H_5C({=}O)Cl$ + base c. acetic anhydride

17.34 Write equations for the following reactions:

a. serine + excess acetic anhydride $\longrightarrow$
b. threonine + excess benzoyl chloride $\longrightarrow$
c. glutamic acid + excess methanol + HCl $\longrightarrow$
d. valine + di-*t*-butyl dicarbonate (p. 513) $\longrightarrow$

17.35 Write the equations that describe what occurs when serine is treated with ninhydrin.

Peptides

17.36 Write structural formulas for the following peptides:

a. serylalanine
b. valylphenylalanine
c. tryptophanylisoleucine
d. glycylalanylserine

17.37 Write an equation for the hydrolysis of

a. leucylthreonine
b. serylmethionine
c. cysteinyltyrosylmethionine

17.38 Write an equation for the acid-catalyzed hydrolysis of the artificial sweetener aspartame (page 479).

17.39 Write formulas that show how the structure of valylglycine changes as the pH of the solution changes from 1 to 10. Estimate the pI (isoelectric point) of this dipeptide.

17.40 Use the one-letter abbreviations to write out all possible tetrapeptides containing one unit each of cysteine, alanine, serine, and leucine. How many structures are possible?

17.41 Write the structure of the product expected from the reaction of alanylcysteine with a mild oxidizing agent, such as hydrogen peroxide (see Sec. 17.8).

17.42 Write equations for the following reactions of Sanger's reagent:

a. 2,4-dinitrofluorobenzene + glycine $\rightarrow$
b. excess 2,4-dinitrofluorobenzene + lysine $\rightarrow$

17.43 Examine the structure of cyclosporin A (page 503).

a. By drawing a dashed line at each peptide bond, deduce how many amino acid units are present in cyclosporin A.
b. Three of the units are identical. What are they?
c. Two of the units are unusual and are *not* listed in Table 17.1. What are their structures?
d. There are only three other types of amino acids present, though there may be more than one unit of each. What are they, and how many units are there of each?

17.44 Verify that the one-letter representation of bradykinin (page 503) is RPPGFSPFR. What is the one-letter representation of substance P (page 503)?

Primary Structure of Peptides and Proteins

17.45 A pentapeptide was converted to its DNP derivative, then completely hydrolyzed and analyzed quantitatively. It gave DNP-methionine, 2 moles of methionine, and 1 mole each of serine and glycine. The peptide was then partially hydrolyzed, the fragments were converted to their DNP derivatives, and each of them was hydrolyzed and analyzed quantitatively. Two tripeptides and two dipeptides isolated in this way gave the following products:

Tripeptide A: DNP-methionine and 1 mole each of methionine and glycine
Tripeptide B: DNP-methionine and 1 mole each of methionine and serine
Dipeptide C: DNP-methionine and 1 mole of methionine
Dipeptide D: DNP-serine and 1 mole of methionine

Deduce the structure of the original pentapeptide, and explain your reasoning.

17.46 Write the equations for the removal of one amino acid from the peptide alanylglycylvaline by the Edman method. What is the name of the remaining dipeptide?

17.47 Express the B chain of insulin (Figure 17.6) using one-letter abbreviations for the amino acids, and compare it with the space taken by the three-letter abbreviations on page 511.

17.48 Insulin (Figure 17.6), when subjected to the Edman degradation, gives *two* phenylthiohydantoins. From which amino acids are they derived? Draw their structures.

17.49 The following compounds are isolated as hydrolysis products of a peptide: Ala—Gly, Tyr—Cys—Phe, Phe—Leu—Trp, Cys—Phe—Leu, Val—Tyr—Cys, Gly—Val, and Gly—Val—Tyr. Complete hydrolysis of the peptide shows that it contains one unit of each amino acid. What is the structure of the peptide, and what are its N- and C-terminal amino acids?

17.50 Vasopressin regulates the excretion of water by the kidneys, as noted in the "A Word about Some Naturally Occurring Peptides" (on page 503).

a. Write the structure of the product expected from the reaction of vasopressin with a mild oxidizing agent, such as hydrogen peroxide.
b. What are your expectations of the three-dimensional structure for vasopressin as compared to the product from the H_2O_2 reaction?

17.51 Simple pentapeptides called *enkephalins* are abundant in certain nerve terminals. They have opiate-like activity and are probably involved in organizing sensory information pertaining to pain. An example is *methionine enkephalin,* Tyr—Gly—Gly—Phe—Met. Write out its complete structure, including all of the side chains.

17.52 Angiotensin II is an octapeptide with vasoconstrictor activity. Complete hydrolysis gives one equivalent each of Arg, Asp, His, Ile, Phe, Pro, Tyr, and Val. Reaction with Sanger's reagent gives, after hydrolysis,

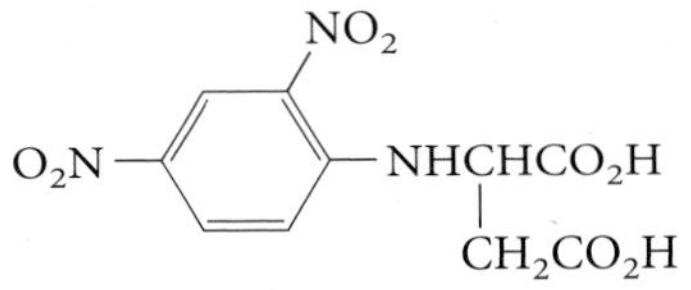

NO_2

O_2N—(benzene ring)—$NHCHCO_2H$

CH_2CO_2H

and seven amino acids. Treatment with carboxypeptidase gives Phe as the first released amino acid. Treatment with trypsin gives a dipeptide and a hexapeptide, whereas with chymotrypsin, two tetrapeptides are formed. One of these tetrapeptides, by Edman degradation, has the sequence Ile—His—Pro—Phe. From these data, deduce the complete sequence of angiotensin II.

17.53 *Endorphins* were isolated from the pituitary gland in 1976. They are potent pain relievers. β-Endorphin is a polypeptide containing 32 amino acid residues. Digestion of β-endorphin with trypsin gave the following fragments:

Lys
Gly—Gln
Asn—Al—His—Lys
Asn—Ala—Ile—Val—Lys
Tyr—Gly—Gly—Phe—Leu—Met—Thr—Ser—Glu—Lys
Ser—Gln—Thr—Pro—Leu—Val—Thr—Leu—Phe—Lys

Using this information only, what is the C-terminal amino acid of β-endorphin?

Treatment with cyanogen bromide gave the hexapeptide

Tyr—Gly—Gly—Phe—Leu—Met

and a 26 amino acid fragment. With this information, what is the N-terminal amino acid of β-endorphin? Digestion of β-endorphin with chymotrypsin gave, among other fragments, a 15-unit fragment identified as

Leu—Met—Thr—Ser—Glu—Lys—Ser—Gln—Thr—Pro—Leu—Val—Thr—Leu—Phe

You should now be able to locate 22 of the 32 amino acid units. Write out as much as you can of the sequence. What further information do you need to complete the sequence?

17.54 *Glucagon* is a polypeptide hormone secreted by the pancreas when the blood sugar level is low. It increases the blood sugar level by stimulating the breakdown of glycogen in the liver. The primary structure of glucagon is

His—Ser—Glu—Gly—Thr—Phe—Thr—Ser—Asp—Tyr—Ser—Lys—Tyr—Leu—Asp—Ser—Arg—Arg—Ala—Gln—Asp—Phe—Val—Gln—Trp—Leu—Met—Asn—Thr

What fragments would you expect to obtain from digestion of glucagon with

a. trypsin?
b. chymotrypsin?

Peptide Synthesis

17.55 The attachment of the N-protected C-terminal amino acid to the polymer in solid-phase peptide synthesis (Figure 17.7) is an S_N2 displacement reaction. What is the nucleophile? What is the leaving group? Write an equation that clearly shows the reaction mechanism.

17.56 The detachment of the peptide chain from the polymer in solid-phase peptide synthesis (Figure 17.7) occurs by an acid-catalyzed S_N2 mechanism. Write an equation to show this mechanism.

17.57 Write out all of the steps in a Merrifield solid-phase synthesis of Leu—Pro.

Protein Structure

17.58 Write the structure for alanylalanine, and show the resonance contributors to the peptide bond. At which bond is rotation restricted?

17.59 In a globular protein, which of the following amino acid's side chains are likely to point toward the center of the structure? Which will point toward the surface when the protein is dissolved in water?

a. leucine
b. glutamine
c. arginine
d. valine
e. lysine
f. phenylalanine

17.60 What secondary and tertiary structures are characteristic of fibrous proteins?

17.61 What secondary and tertiary structures are characteristic of globular proteins?

17.62 Enzymes are globular proteins that catalyze reactions in biological systems. What amino acid side chains might play the role of bases in enzyme-catalyzed reactions? What amino acid side chains might play the role of acids in enzyme-catalyzed reactions?

Every living organism possesses a unique genetic "blueprint" encoded in its DNA.

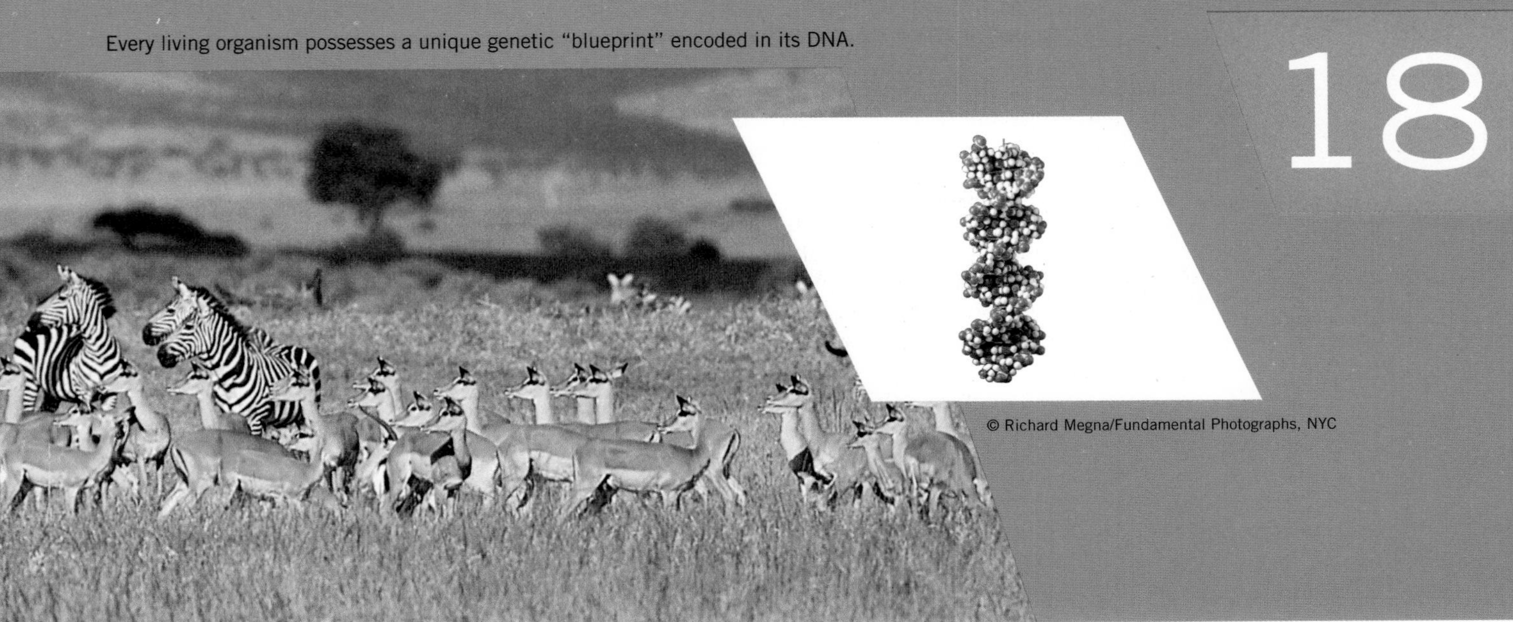

© Richard Megna/Fundamental Photographs, NYC

© Tim Davis/Photo Researchers

18

Nucleotides and Nucleic Acids

DNA, the double helix, and the genetic code—through the media's popularization of science—have become household words. And they represent one of the greatest triumphs ever for chemistry and biology.

In this chapter, we will describe the structure of the nucleic acids, DNA and RNA. We will first look at their building blocks, the nucleosides and nucleotides, and then describe how these building blocks are linked to form giant nucleic acid molecules. Later we will consider the three-dimensional structures of these vital biopolymers and how the information they contain (the genetic code) was unraveled.

Online homework for this chapter can be assigned in OWL, an online homework assessment tool.

18.1 The General Structure of Nucleic Acids

A **nucleic acid** is a macromolecule with a backbone of sugar molecules, each with a base attached, connected by phosphate linkages.

Nucleic acids are linear, chain-like macromolecules that were first isolated from cell nuclei. Hydrolysis of nucleic acids gives nucleotides, which are the building blocks of nucleic acids, just as amino acids are the building blocks of proteins. A complete description of the primary structure of a nucleic acid requires knowledge of its nucleotide sequence, which is comparable to knowing the amino acid sequence in a protein.

Hydrolysis of a nucleotide gives 1 mole each of phosphoric acid and a nucleoside. The nucleoside can be hydrolyzed further in aqueous acid, to one equivalent each of a sugar and a heterocyclic base.

$$\text{nucleic acid} \xrightarrow[\text{enzyme}]{H_2O} \text{nucleotide (phosphate-sugar-heterocyclic base)} \xrightarrow{H_2O,\ HO^-} \text{nucleoside (sugar-base)} + H_3PO_4 \xrightarrow[H^+]{H_2O} \text{heterocyclic base} + \text{sugar} \quad (18.1)$$

The overall structure of the nucleic acid itself is a macromolecule with a backbone of sugar molecules connected by phosphate links and with a base attached to each sugar unit.

```
—sugar—phosphate—sugar—phosphate—sugar—phosphate
   |                 |                 |
  base              base              base
 nucleotide1       nucleotide2       nucleotide3
```

schematic structure of a nucleic acid

18.2 Components of Deoxyribonucleic Acid (DNA)

DNA contains the sugar **2-deoxy-D-ribose**.

Complete hydrolysis of DNA gives phosphoric acid, a single sugar, and a mixture of four heterocyclic bases. The sugar is **2-deoxy-D-ribose**.

HOCH$_2$ (5), O, OH (1), H, H, H (4), H, OH (3), H (2)

Note that there is no hydroxyl group at C-2.

2-deoxy-D-ribose

The heterocyclic bases (Figure 18.1) fall into two categories, the pyrimidines (*cytosine* and *thymine*; review Sec. 13.3) and the purines (*adenine* and *guanine*; review

the pyrimidines

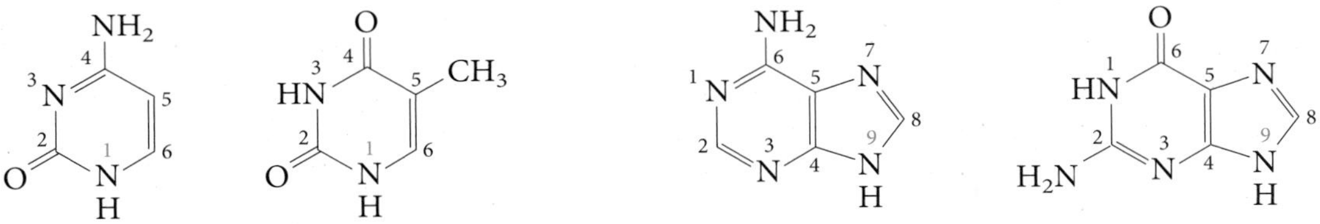

cytosine (C)

thymine (T)

the purines

adenine (A)

guanine (G)

■ **Figure 18.1**
The DNA bases.

Sec. 13.7). When we refer to these bases later, especially in connection with the genetic code, we will use the first letters of their names (capitalized) as abbreviations for their structures.

Now let us see how the sugar and bases are linked.

18.3 Nucleosides

A **nucleoside** is an *N-glycoside*. The pyrimidine or purine base is connected to the anomeric carbon (C-1) of the sugar. The pyrimidines are connected at N-1 and the purines at N-9 (Figure 18.2). Nucleoside structures are numbered in the same way as their component bases and sugars, except that primes are added to the numbers for the sugar part.

A **nucleoside** is an *N-glycoside;* a nitrogen atom of the heterocyclic base is connected to the anomeric carbon of the sugar.

N-glycosides have structures similar to those of *O*-glycosides (Sec. 16.11). In *O*-glycosides, the —OH group on the anomeric carbon is replaced by —OR; in *N*-glycosides, that group is replaced by $—NR_2$.

EXAMPLE 18.1

Draw the structure of

a. the *β*-*O*-glycoside of 2-deoxy-D-ribose and methanol
b. the *β*-*N*-glycoside of 2-deoxy-D-ribose and dimethylamine

Solution

a. 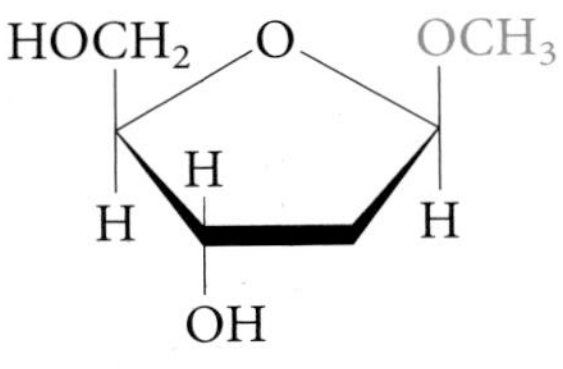 $HOCH_2$ O OCH_3 H H H OH

b. 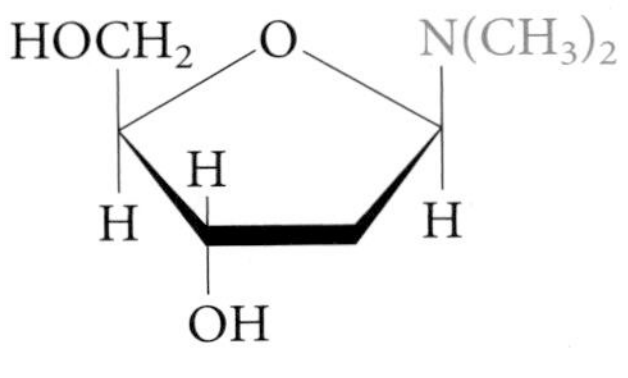 $HOCH_2$ O $N(CH_3)_2$ H H H OH

Note the similarity between *N*- and *O*-glycosides.

PROBLEM 18.1 Figure 18.2 shows the structures of two DNA nucleosides. Draw the structures for the remaining two nucleosides of DNA: 2′-deoxythymidine and 2′-deoxyguanosine.

Figure 18.2
Schematic formation of nucleosides.

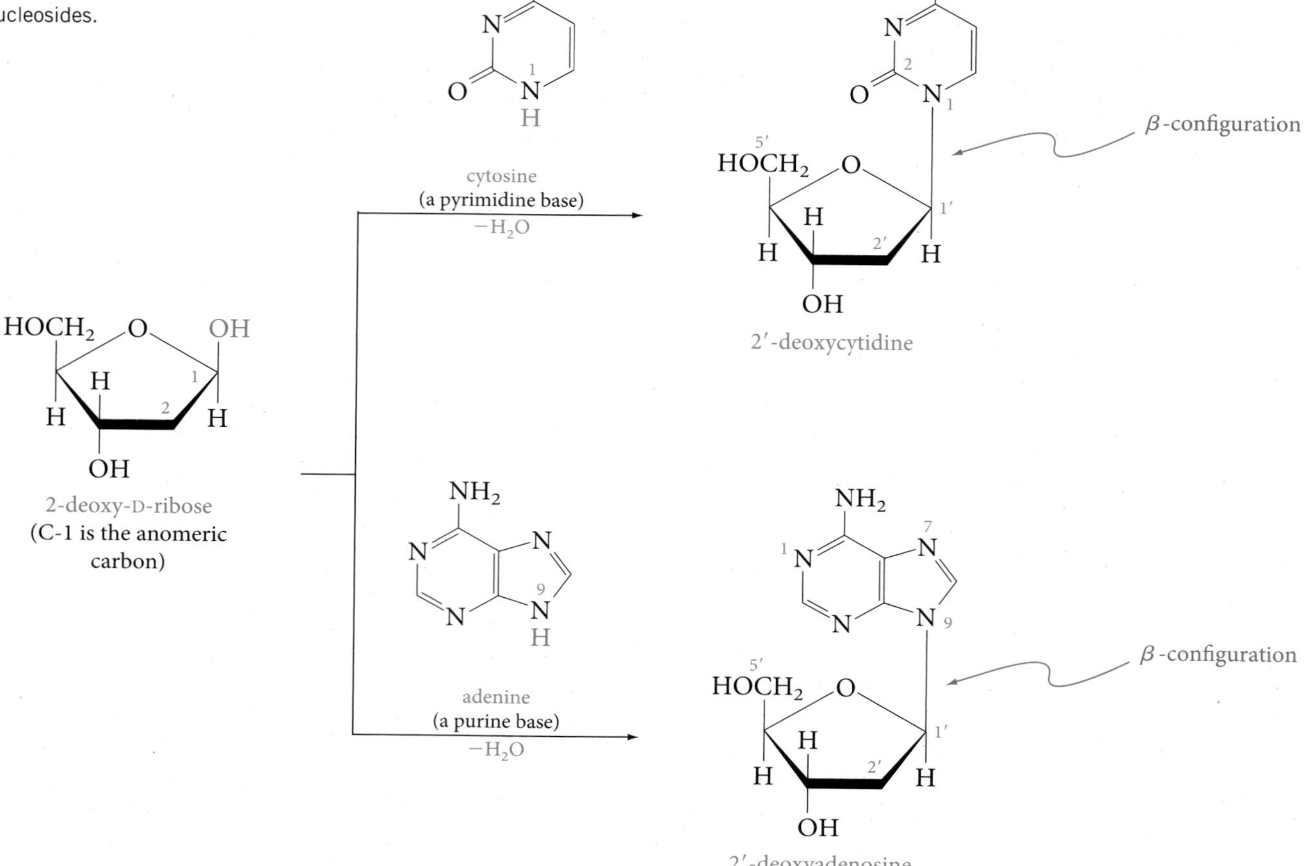

Because of their many polar groups, nucleosides are water soluble. Like other glycosides, they can be hydrolyzed readily by aqueous acid (or by enzymes) to the sugar and the heterocyclic base. For example,

2′-deoxyadenosine $\xrightarrow[H^+]{H_2O}$ 2-deoxy-D-ribose + adenine (18.2)

PROBLEM 18.2 Using eq. 18.2 as a guide, write an equation for hydrolysis of

a. 2′-deoxythymidine b. 2′-deoxyguanosine

18.4 Nucleotides

Nucleotides are phosphate esters of nucleosides. A hydroxyl group in the sugar part of a nucleoside is esterified with phosphoric acid. In DNA nucleotides, either the 5′ or the 3′ hydroxyl group of 2-deoxy-D-ribose is esterified.

Nucleotides are phosphate esters of nucleosides.

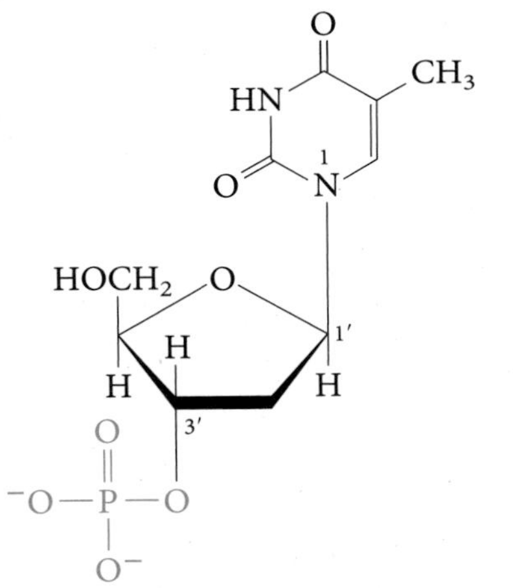

2′-deoxythymidine 3′-monophosphate

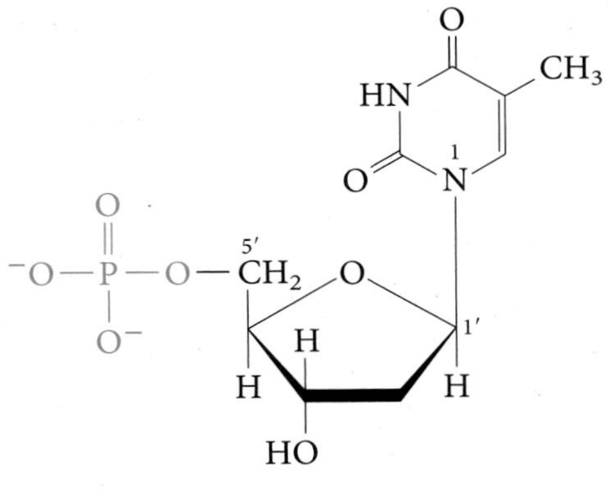

2′-deoxythymidine 5′-monophosphate

Nucleotides are named as the 3′- or 5′-monophosphate esters of a nucleoside, as shown above. These names are frequently abbreviated as in Table 18.1. In these abbreviations, letter d stands for 2-deoxy-D-ribose, the next letter refers to the heterocyclic base, and MP stands for monophosphate. (Later we will see that some nucleotides are diphosphates, abbreviated DP, or triphosphates, TP.) Unless otherwise stated, the abbreviations usually refer to the 5′-phosphates.

Table 18.1 ■ The Common 2-Deoxyribonucleotides

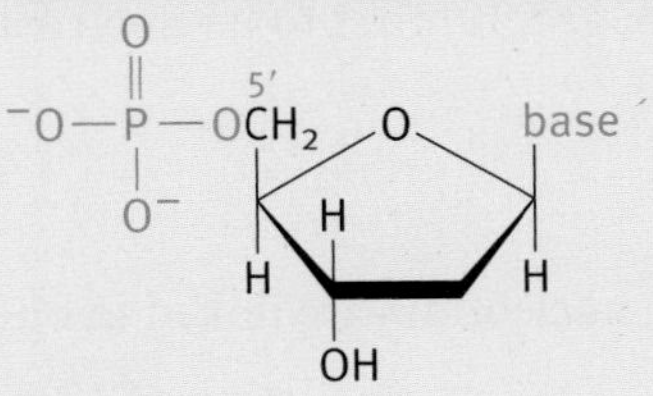

Base	Monophosphate name	Abbreviation
cytosine (C)	2′-deoxycytidine 5′-monophosphate	dCMP
thymine (T)	2′-deoxythymidine 5′-monophosphate	dTMP
adenine (A)	2′-deoxyadenosine 5′-monophosphate	dAMP
guanine (G)	2′-deoxyguanosine 5′-monophosphate	dGMP

EXAMPLE 18.2

Write the structure of dAMP.

Solution The letter d tells us that the sugar is 2-deoxy-D-ribose. Letter A stands for the base adenine, and the MP indicates a monophosphate. The structure is the same as that of the nucleotide shown in eq. 18.3.

PROBLEM 18.3 Write the structure for

a. dCMP b. dGMP

The phosphoric acid groups of nucleotides are acidic, and under physiological conditions (pH 7), these groups exist mainly as dianions, as shown in the structures.

Nucleotides can be hydrolyzed by aqueous base (or by enzymes) to nucleosides and phosphoric acid. Phosphoric acid is sometimes abbreviated P_i, meaning inorganic phosphate.

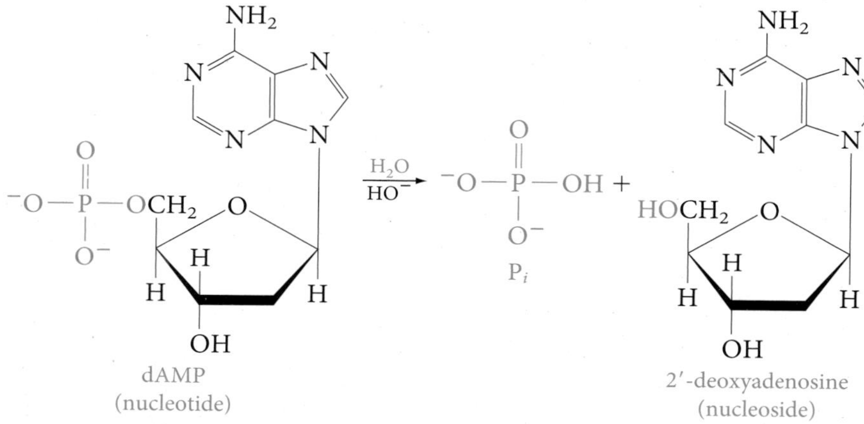

dAMP (nucleotide) $\xrightarrow[HO^-]{H_2O}$ $^-O-P(=O)(O^-)-OH$ (P_i) + 2′-deoxyadenosine (nucleoside) (18.3)

PROBLEM 18.4 Write a sequence of two equations that shows the stepwise hydrolysis of dTMP (page 531) first to its nucleoside, then to the sugar and free base.

Now let us see how the nucleotides are linked to one another in DNA.

18.5 The Primary Structure of DNA

In *deoxyribonucleic acid (DNA)*, 2-deoxy-D-ribose and phosphate units alternate in the backbone. The 3′ hydroxyl of one ribose unit is linked to the 5′ hydroxyl of the next ribose unit by a phosphodiester bond. The heterocyclic base is connected to the anomeric carbon of each deoxyribose unit by a β-*N*-glycosidic bond. Figure 18.3 shows a schematic drawing of a DNA segment.

In DNA, there are no remaining hydroxyl groups on any deoxyribose unit. Each phosphate, however, still has one acidic proton that is usually ionized at pH 7, leaving a negatively charged oxygen, as shown in Figure 18.3. If this proton were present,

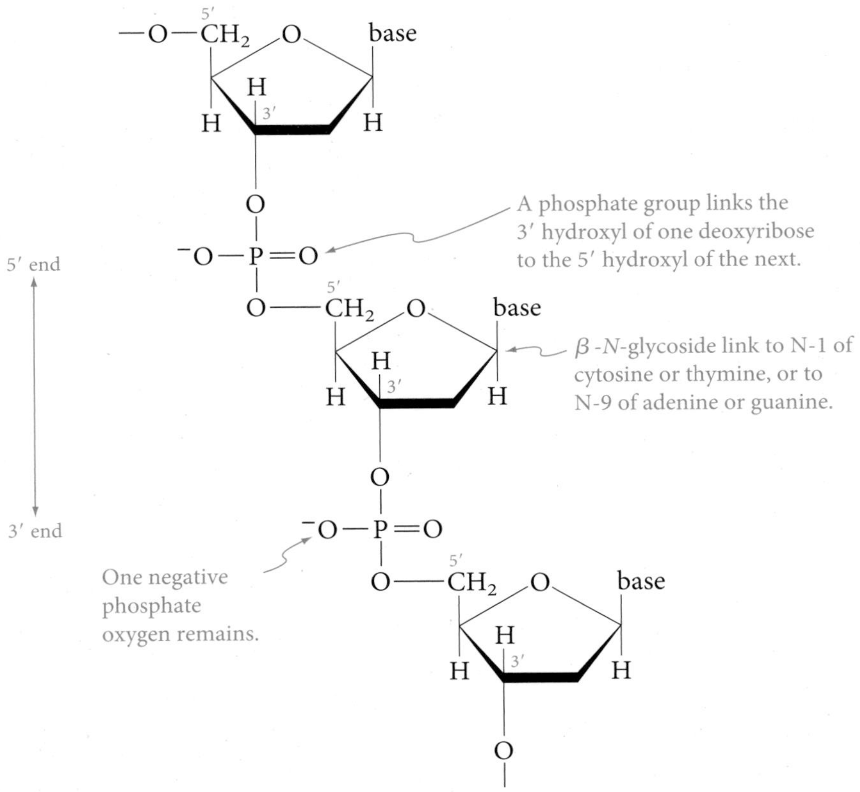

Figure 18.3
A segment of a DNA chain.

the substance would be an acid; hence the name nucleic *acid.* A complete description of any particular DNA molecule, which may contain thousands or even millions of nucleotide units, would have to include the exact sequence of heterocyclic bases (A, C, G, and T) along the chain.

18.6 Sequencing Nucleic Acids

The problem of sequencing nucleic acids is in principle similar to that of sequencing proteins. At first, the job might appear to be easier because there are only four bases compared to 20 common amino acids. In fact, it is much more difficult. Even the smallest DNA molecule contains at least 5,000 nucleotide units, and some DNA molecules contain 1 million or more nucleotide units. To determine the exact base sequence in such a molecule is a task of considerable magnitude.

Without trying to discuss nucleic acid sequencing in detail, we can describe the strategy. The strategy basically relies on breaking the DNA into small identifiable fragments using a combination of enzymatic and chemical reactions. First, enzymes called **restriction endonucleases**, which split the DNA chain at known four-base sequences, are used to break the huge DNA molecule into smaller fragments with perhaps 100 to 150 nucleotide units. The purified fragments are then further degraded using four different and carefully controlled reaction conditions, which selectively split the chains at a particular base, A, G, C, or T. Each set of conditions gives a different group of yet smaller nucleotides that are subjected to gel electrophoresis (a technique similar to that used in separating peptides), which separates them based on the number of nucleotide units they contain. These experiments provide enough information such that the DNA sequence can be deciphered.

Enzymes called **restriction endonucleases** are used to split the DNA chain at known four-base sequences.

A WORD ABOUT... DNA and Crime

© James Holmes/Cellmark Diagnostics/Science Photo Library/Photo Researchers

Scientist studying DNA autoradiograms.

DNA profiling, which is sometimes called **DNA fingerprinting**, is one of the most powerful new techniques of forensic science. Here's how it works.

A small quantity of DNA is obtained from some source associated with a crime—semen, blood, or hair roots, perhaps associated with a rape, murder, or other violent crime. The DNA is purified, cut with established restriction enzymes, and sequenced as described in Section 18.6. Functional genes—those that code for enzymes, hormones, and other peptides common to all humans—do not vary from person to person, but these genes account for only about 5% of human DNA. The remaining DNA varies enormously from person to person, and in fact (except for identical twins) is characteristic of the individual. *The DNA in these genes can identify a single person.* And one tremendous advantage of the method is that the DNA sample can be very small, as small as a few micrograms.

There are two main uses of DNA profiling in dealing with a crime. One is to compare a suspect's profile with a sample from the scene of the crime. Some years ago, in one of the first applications of the DNA technique, this kind of pairwise comparison cleared one individual suspected of two murders and led to the conviction of another.

But DNA profiling also has the potential for use in crime investigation, through the accumulation of databases similar to those used for fingerprinting. Once that is done, a search of the database should answer such questions as: Does the assailant in this crime match up with one from a previous crime? Is there a match with an individual already on file?

The use of DNA profiling in forensic science is not restricted to humans. In a recent murder case, it was applied to hair from a cat named Snowball. Snowball's hair was found on a jacket that linked her owner to the victim, ultimately leading to conviction of the cat's owner.

Despite the tremendous potential of DNA profiling, it is not without problems. As with all analytical methods, DNA typing must be done carefully and with proper controls; otherwise, errors are possible, especially with band shifting, when one lane in the gel electrophoresis step runs faster or slower than another. When comparing patterns from two DNA samples, it is essential to show that the bands match in order to conclude that the samples come from the same individual. When the job is done properly and a match is obtained, the odds for identity are very long indeed—usually better than one in a hundred million.

See Problem 18.30.

Progress in nucleic acid sequencing has been spectacular. In 1978, the longest known nucleic acid sequences (in RNA chains, which are shorter than DNA chains) were of about 200 nucleotide units. Later, the base sequence of DNA in a virus chromosome with 5,375 nucleotide units was worked out by Frederick Sanger, who earned his second Nobel Prize in chemistry in 1980 for this achievement. In 1977, the Maxam–Gilbert* method for sequencing was introduced; and by 1985, base sequences of more than 170,000 nucleotide units became known. With present instrumentation, several thousand nucleotide bases can be sequenced in a day. Indeed, DNA sequencing, together with a knowledge of the genetic code (Sec. 18.11), is now often used to sequence very large proteins.

18.7 Laboratory Synthesis of Nucleic Acids

It is important to be able to synthesize specific DNA segments in the laboratory, just as it is important to synthesize specific peptide segments. Short nucleotide chains of known sequence were needed, as we will see, to solve the genetic code. Longer

*Walter Gilbert (Harvard University) shared a Nobel Prize in 1980 with F. Sanger and P. Berg.

chains of known sequence could, through genetic engineering techniques, be used to induce microorganisms to synthesize useful proteins—insulin, for example. Synthetic **oligonucleotides** of known sequence can be inserted into the DNA of microorganisms, where they can serve as templates for biological DNA synthesis.

Oligonucleotides are chains of several nucleotides.

The problems of laboratory DNA synthesis are similar to, but even greater than, those of peptide synthesis, mainly because nucleotides have more complicated structures than do amino acids. Each nucleotide has several functional groups that must be protected, and later deprotected, during synthesis. Despite these difficult chemical problems, methods for oligonucleotide synthesis have been developed in various laboratories. Using these methods, Khorana* and coworkers were able to synthesize a gene for the first time, by a combination of chemical and enzymatic means.

More recently, automated gene synthesizers have been developed that operate on principles similar to the Merrifield solid-phase technique for peptides. A protected nucleotide is covalently bonded to a polymer. Other protected nucleotides are then added sequentially to the chain, using a coupling reagent. Eventually, the protecting groups are removed, and the synthetic oligonucleotide is then detached from the solid support.

Let us now proceed from the primary DNA structure, the specific base sequences, to the secondary DNA structure: the double helix and the genetic code.

18.8 Secondary DNA Structure; the Double Helix

It has been known since 1938 that DNA molecules have a discrete shape, because x-ray crystallographic studies of DNA threads showed a regular stacking pattern with some periodicity. A key observation by Erwin Chargaff (Columbia University) in 1950 provided an important clue to the structure. Chargaff analyzed the base content of DNA from many different organisms and found that the amounts of A and T are always equal and the amounts of G and C are also equal. For example, human DNA contains about 30% each of A and T and 20% each of G and C. Other DNA sources give different percentages, but the ratios of A to T and of G to C are always unity.

The meaning of these equivalences was not evident until 1953, when Watson and Crick,** working together in Cambridge, England, proposed the double helix model for DNA. They received simultaneous supporting x-ray data for their proposal from Rosalind Franklin and Maurice Wilkins in London. The important features of their model follow:

1. DNA consists of two helical polynucleotide chains coiled around a common axis.
2. The helices are right-handed, and the two strands run in opposite directions with regard to their 3′ and 5′ ends.
3. The purine and pyrimidine bases lie *inside* the helix, in planes perpendicular to the helical axis; the deoxyribose and phosphate groups form the outside of the helix.
4. The two chains are held together by purine–pyrimidine base pairs connected by hydrogen bonds. *Adenine is always paired with thymine, and guanine is always paired with cytosine.*
5. The diameter of the helix is 20 Å. Adjacent base pairs are separated by 3.4 Å and oriented through a helical rotation of 36°. There are therefore 10 base pairs for every turn of the helix (360°), and the structure repeats every 34 Å.
6. There is no restriction on the sequence of bases along a polynucleotide chain. The exact sequence carries the genetic information.

---

*Har Gobind Khorana (Massachusetts Institute of Technology) won the Nobel Prize in medicine in 1968.

**James D. Watson (Harvard University) and Francis H. Crick (Cambridge University), along with Maurice Wilkins, won the Nobel Prize in medicine in 1962.

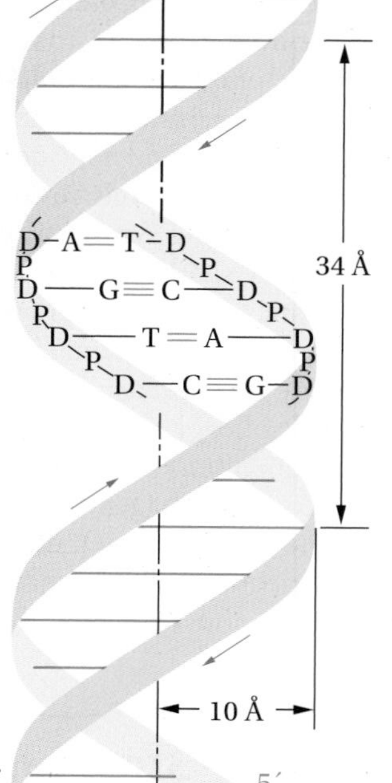

D = deoxyribose
P = phosphate
A = adenine
T = thymine
G = guanine
C = cytosine

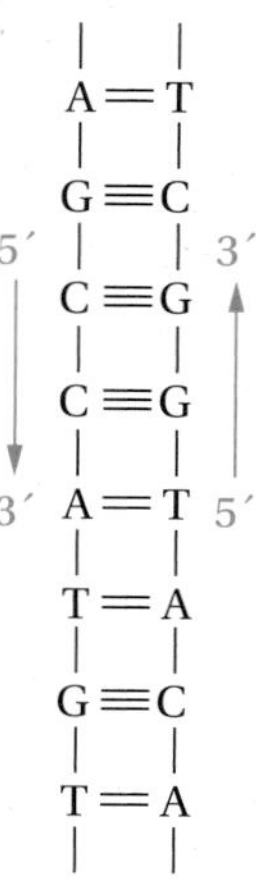

A=T pairs have two hydrogen bonds

G≡C pairs have three hydrogen bonds

■ Figure 18.4
Model and schematic representations of the DNA double helix. The space-filling model at the left shows the base pairs in the helix interior, in planes perpendicular to the main helical axis. The center drawing shows the structure more schematically, including the dimensions of the double helix. At the far right is a schematic method for showing base pairing in the two strands.

Figure 18.4 shows schematic models of the double helix. The key feature of the structure is the complementarity of the base pairing: **A—T** and **G—C**. Only purine–pyrimidine base pairs fit into the helical structure. There is not enough room for two purines and too much room for two pyrimidines, which would be too far apart to form hydrogen bonds. Of the purine–pyrimidine pairs, the hydrogen-bonding possibilities are best for A—T and G—C pairing. The A—T pair is joined by two hydrogen bonds and the G—C pair by three. The geometries of the two pairs are nearly identical (Figure 18.5).

PROBLEM 18.5 Consider the following sequence of bases from one strand of DNA: —AGCCATGT— (written from 5′ to 3′). What will the sequence of bases on the other strand be?

We now know that, although the Watson–Crick model for the double helix is essentially correct, it is oversimplified. Helical conformations of DNA can be classified into three general families, called the A-, B-, and Z-forms. B-DNA, the predominant form, is the regular right-handed helix of Watson and Crick, with the base pairs essentially perpendicular to the helix axis. In the A-form, base pairs may be tilted by as much as 20° to the helix axis, and the sugar rings are puckered differently from the way they are in the B-form. And in the Z-form, we see a 180° rotation of some of the bases about the C—N glycosidic bond, resulting in a *left*-handed helix.

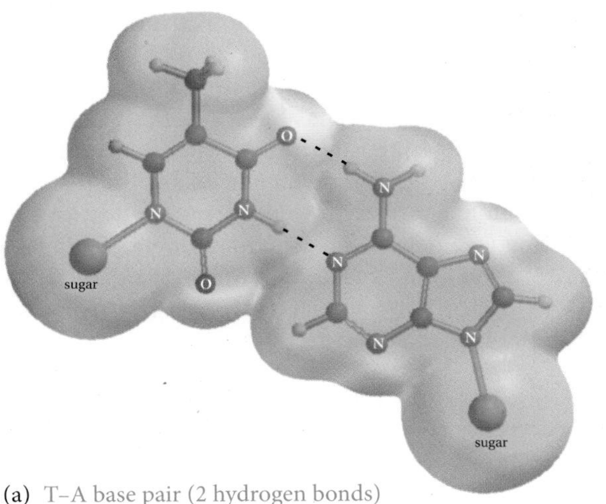

(a) T–A base pair (2 hydrogen bonds)

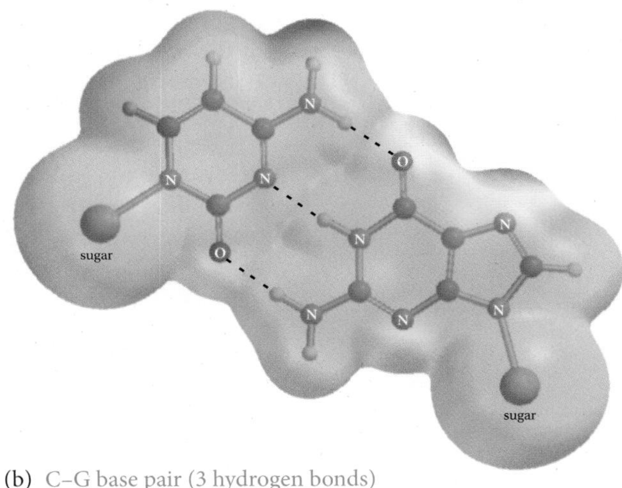

(b) C–G base pair (3 hydrogen bonds)

■ **Figure 18.5**
Electrostatic potential maps illustrating hydrogen bonding between thymine and adenine (a) and cytosine and guanine (b). The oxygens are red, and the nitrogens are blue.

The particular overall conformation adopted by a DNA molecule depends in part on the actual base sequence. For example, synthetic DNAs made of alternating purine–pyrimidine units have different conformations from DNAs made of blocks of purine bases followed by blocks of pyrimidine bases. Also, A—T and G—C base-pairing with different H-bonds from that originally proposed by Watson and Crick has been observed.

These variations in details of DNA structures lead to DNA molecules with bends, hairpin loops, supercoils, single-stranded loops, and even cruciforms in which single intrastrand H-bonded loops are extruded from the double helix. These structural changes add flexibility to the way DNA molecules are able to recognize and interact with other cellular components to perform their functions.*

18.9 DNA Replication

The beauty of the DNA double helix model was that it immediately suggested a molecular basis for transmitting genetic information from one generation to the next. In 1954, Watson and Crick proposed that, as the two strands of a double helix separate, a new complementary strand is synthesized from nucleotides in the cell, using one strand as a template for the other. Figure 18.6 schematically depicts the process.

Though simple in principle, **DNA replication** is quite a complex process in practice. The nucleotides must be present as triphosphates (not monophosphates), an enzyme (DNA-polymerase) adds the nucleotides to a primer chain, other enzymes link DNA chains (DNA-ligase), there are specific places at which replication starts and stops, and so on. Our knowledge of the details of this process has increased considerably since the DNA double helix was proposed nearly five decades ago.

The process by which DNA molecules make copies of themselves is called **DNA replication**.

An advance that has revolutionized the study and analysis of genes is the invention in the mid-1980s of the **polymerase chain reaction (PCR)** technique by Kary Mullis. The PCR technique is a modification of the natural process of DNA replication that makes possible the amplification (production of many copies) of a specific DNA sequence. In brief, DNA is separated into two strands by heating, short complementary

The **polymerase chain reaction (PCR)** is a technique for making many copies of a specific DNA sequence.

*For an excellent, readable article on this subject, see J. K. Barton, *Chemical and Engineering News* **1988**, September 26, pp. 30–42.

oligonucleotide "primers" are combined with the single strands to mark the two ends of the desired sequence, and *Taq* polymerase (a DNA polymerizing enzyme that is stable at high temperatures) is used to link the appropriate nucleotides to the marked sequence.* Thirty to sixty cycles of this process, which has been automated and can be run in one reaction vessel, yield millions of copies of the specific DNA sequence in less than a day. PCR can be used to amplify a DNA sequence from as little material as one DNA molecule! PCR amplification has already found useful applications in such diverse areas as DNA fingerprinting (see page 534), gene mapping, detection of genetic mutations, monitoring of cancer therapy, and studies of molecular evolution.

Before we turn to the role of DNA in protein synthesis, we must first consider another type of nucleic acid, RNA, which plays a key role in this process.

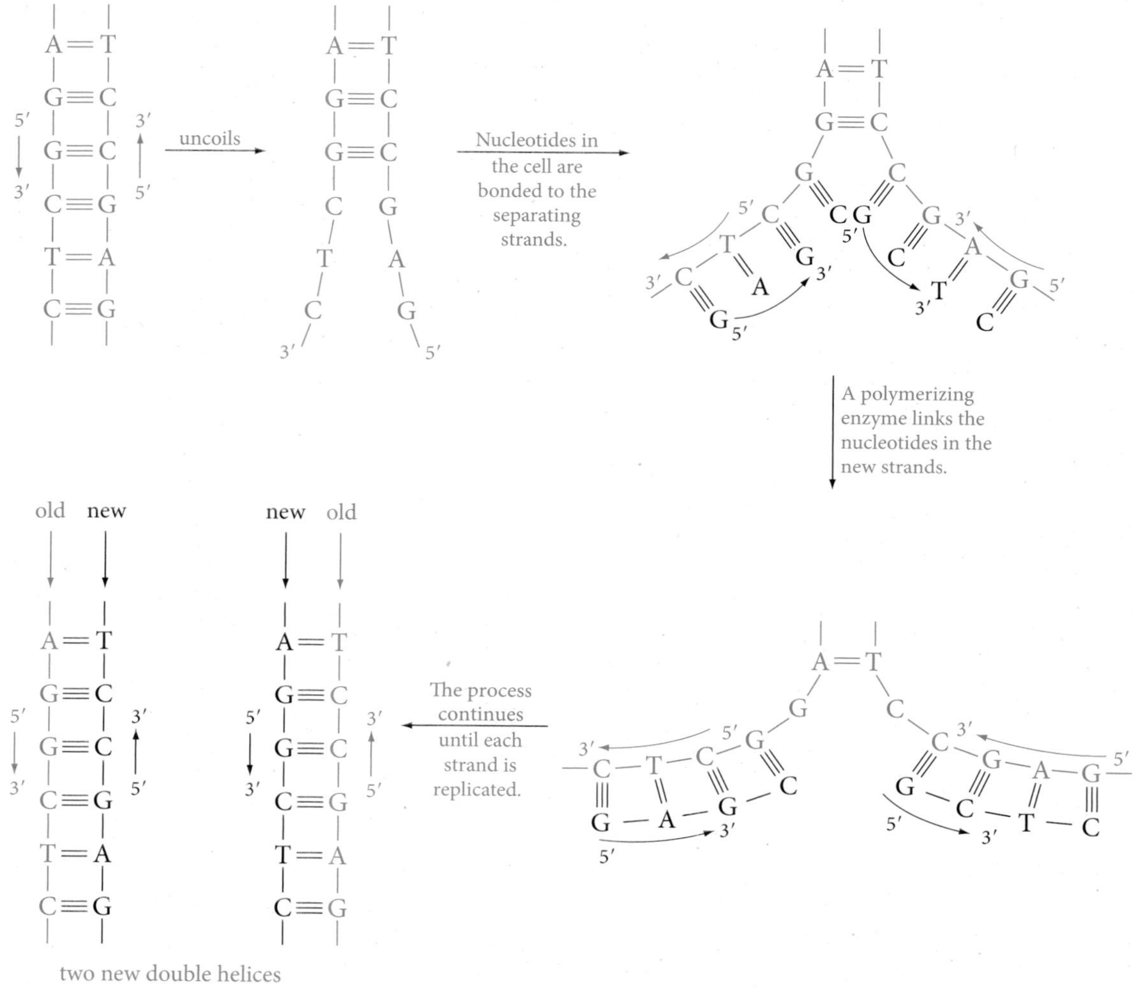

uncoils

Nucleotides in the cell are bonded to the separating strands.

A polymerizing enzyme links the nucleotides in the new strands.

The process continues until each strand is replicated.

old new

new old

two new double helices

■ **Figure 18.6**

Schematic representation of DNA replication. As the double helix uncoils, nucleotides in the cell bond to the separate strands, following the base-pairing rules. A polymerizing enzyme links the nucleotides in the new strands to one another. Both new strands are assembled from the 5′ to the 3′ end.

*For a more detailed description of the PCR technique, see J. D. Watson, M. Gilman, J. Witkowski, and M. Zoller, *Recombinant DNA*, 2nd ed. New York: W. H. Freeman and Co., 1992, Chap. 6. Also see "A Closer Look At . . . the Polymerase Chain Reaction (PCR)" on page 539.

A CLOSER LOOK AT... The Polymerase Chain Reaction (PCR)

Conduct research on the Internet to find information on the polymerase chain reaction and to answer the following questions.

The Polymerase Chain Reaction

1. The polymerase chain reaction (PCR) allows one to make copies of genetic material called DNA.
2. What are the three basic steps of the PCR?
3. Each cycle doubles the number of targeted DNA molecules present in a PCR reactor. If each cycle takes three minutes, how many copies of a DNA molecule can be made in an hour?
4. An enzyme called DNA polymerase is an important reagent in the PCR. Most DNA polymerases are destroyed (denatured) at the temperatures required by the PCR reaction cycle. What is special about *Taq* polymerase, the enzyme used in commercial PCR? From what organism does *Taq* polymerase come?
5. What are two or three examples of practical applications of the PCR technique?
6. How might the invention of the PCR have stimulated the notion that a place such as Jurassic Park could be developed? What evidence is there that the genetic makeup of extinct species could be determined using PCR?

18.10 Ribonucleic Acids; RNA

Ribonucleic acids (RNA) differ from DNA in three important ways: (1) The sugar is D-ribose; (2) uracil replaces thymine as one of the four heterocyclic bases; and (3) most RNA molecules are single stranded, although helical regions may be present by looping of the chain back on itself.

Ribonucleic acids (RNA) contain the sugar D-ribose and the base uracil instead of thymine. Most RNA molecules are single stranded.

The RNA sugar D-ribose differs from the DNA sugar in that it has a hydroxyl group at C-2. Otherwise, the nucleosides and nucleotides of RNA have structures similar to those of DNA.

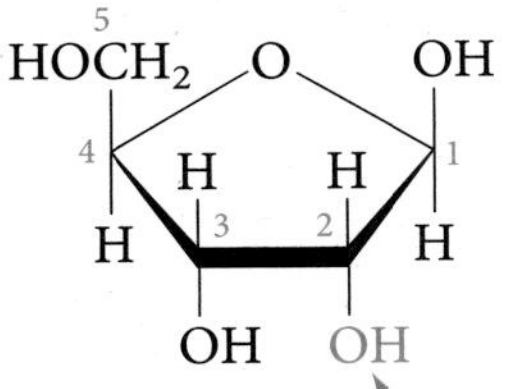

D-ribose

hydroxyl at C-2

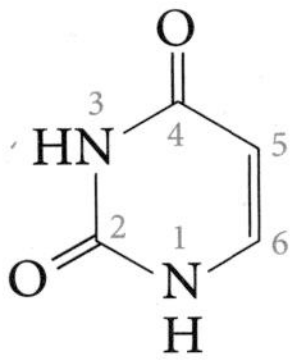

uracil

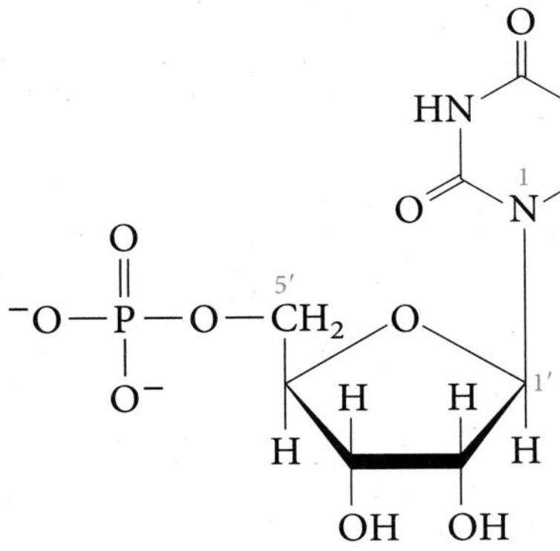

uridine-5′-monophosphate (UMP)

Uracil differs from thymine only in lacking the C-5 methyl group. Like thymine, it forms nucleotides at N-1, and their names are similar to those in Table 18.1. In the abbreviations of these names, letter d (which stands for deoxyribose) is omitted because the sugar is ribose.

PROBLEM 18.6 Draw the full structure of

a. adenosine-5′-monophosphate (AMP)
b. the RNA trinucleotide UCG (written from 5′ to 3′)

Messenger RNA (*m*RNA) transcribes the genetic code from DNA and is the template for protein synthesis.

Cells contain three major types of RNA. **Messenger RNA (mRNA)** is involved in **transcription** of the genetic code and is the template for protein synthesis. There is a specific *m*RNA for every protein synthesized by the cell. The base sequence of *m*RNA is complementary to the base sequence in a single strand of DNA, with U replacing T as the complement of A.

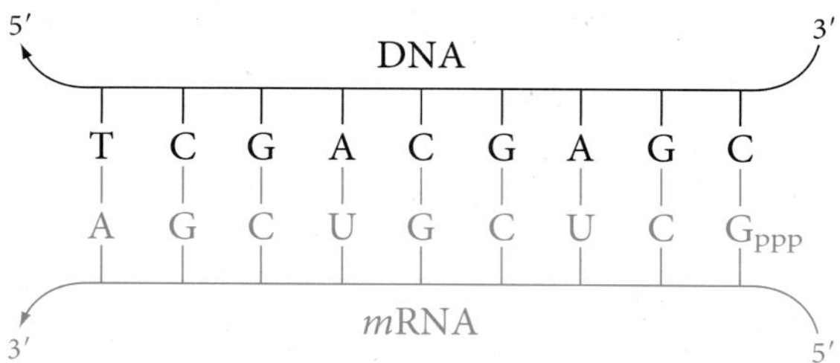

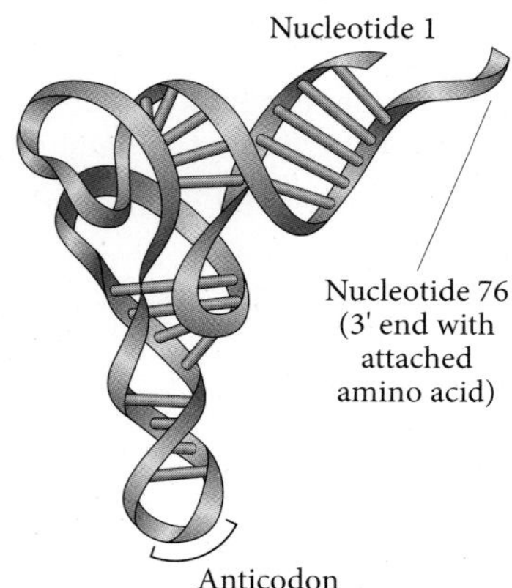

Transcription proceeds in the 3′-to-5′ direction along the DNA template. That is, the *m*RNA chain grows from its own 5′ end. The 5′ terminal nucleotide in *m*RNA is usually present as a triphosphate, not a monophosphate, and is commonly pppG or pppA. An enzyme called RNA-polymerase is essential for transcription. Usually only one strand of DNA is transcribed. It contains base sequences, called *promoter sites,* which initiate transcription. It also contains certain termination sequences, which signal the completion of transcription.

At the 3′ end of *m*RNA, there is usually a special sequence of about 200 successive nucleotide units of the same base, adenine. This sequence plays a role in transporting the *m*RNA from the cell nucleus to the *ribosomes,* the cellular structures where proteins are synthesized.

Transfer RNA (*t*RNA) carries activated amino acids to ribosomes for peptide bond formation. Each *t*RNA has an **anticodon loop** with three base pairs complementary to the three-base code on an *m*RNA for a specific amino acid.

Transfer RNA (tRNA) *carries amino acids in an activated form to the ribosome for peptide bond formation,* in a sequence determined by the *m*RNA template. There is at least one *t*RNA for each of the 20 amino acids. Transfer RNA molecules are relatively small as nucleic acids go, with about 70 to 90 nucleotide units. Each *t*RNA has a three-base sequence, C—C—A, at the 3′ hydroxyl end, where the amino acid is attached as an ester. Each *t*RNA also has an **anticodon loop** quite remote from the amino acid attachment site. This loop contains seven nucleotides, the middle three of which are complementary to the three-base code word on the *m*RNA for that particular amino acid.

Eighty percent of cellular RNA is **ribosomal RNA (*r*RNA)**, the main component of ribosomes.

The third type of RNA is **ribosomal RNA (*r*RNA)**. It comprises about 80% of the total cellular RNA (*t*RNA = 15%, *m*RNA = 5%) and is the main component of the ribosomes. Its molecular weight is large, and each molecule may contain several thousand nucleotide units.

Ribozymes are RNA molecules that function as enzymes.

Until recently, it was thought that all enzymes are proteins. But this dogma of biochemistry was recently overthrown by the discovery that some types of RNA can function as biocatalysts. They can cut, splice, and assemble themselves without any outside help from conventional enzymes. This discovery* of **ribozymes**, as they are called, had a major impact on theories of the origin of life. The question was: Which came first in the primordial soup from which life began, the proteins or the nucleic acids? Proteins could be enzymes and catalyze the reactions needed for life, but they could not store genetic information. The reverse was thought to be true for nucleic acids. But, with the discovery of catalytic activity in certain types of RNA, it now seems almost certain that the earth of 4 billion years ago

*Sidney Altman (Yale University) and Thomas R. Cech (University of Colorado) received the 1989 Nobel Prize in chemistry for this discovery.

was an RNA world, in which RNA molecules carried out all of the processes of life without the help of proteins or DNA—even though the latter now contains the genetic code.

18.11 The Genetic Code and Protein Biosynthesis

It is beyond the scope of this book to give a detailed account of the genetic code, how it was unraveled, and how more than a hundred types of macromolecules must interact to translate that code into the synthesis of a protein. But we can present a few of the main concepts.

The **genetic code** is the relationship between the base sequence in DNA, or its RNA transcript, and the amino acid sequence in a protein. A three-base sequence, called a **codon**, corresponds to *one* amino acid. Because there are 4 bases in RNA (A, G, C, and U), there are $4 \times 4 \times 4 = 64$ possible codons. However, there are only 20 common amino acids in proteins. Each codon corresponds to only one amino acid, but the code is *degenerate;* that is, *several different codons may correspond to the same amino acid.* Of the 64 codons, 3 are codes for "stop" (UAA, UAG, and UGA); they signal that the particular protein synthesis is complete. One codon, AUG, serves double duty. It is the initiator codon, but if it appears again after a chain has been initiated, it codes for the amino acid methionine. The entire code is summarized in Table 18.2.

The **genetic code** is the relationship between the base sequence in DNA and the amino acid sequence in a protein. A three-base sequence located on *m*RNA, called a **codon**, corresponds to *one* amino acid.

How was the genetic code solved? The first successful experiment was performed by Marshall Nirenberg in 1961 (Nobel Prize, 1968). Nirenberg added a synthetic RNA, polyuridine (an RNA in which all of the bases were uracil, U), to a

Table 18.2 ■ The Genetic Code; Translation of the Codons into Amino Acids

First base (5′ end)	Middle base	Third base (3′ end) *U*	*C*	*A*	*G*
U	U	Phe	Phe	Leu	Leu
	C	Ser	Ser	Ser	Ser
	A	Tyr	Tyr	Stop	Stop
	G	Cys	Cys	Stop	Trp
C	U	Leu	Leu	Leu	Leu
	C	Pro	Pro	Pro	Pro
	A	His	His	Gln	Gln
	G	Arg	Arg	Arg	Arg
A	U	Ile	Ile	Ile	Met (start)
	C	Thr	Thr	Thr	Thr
	A	Asn	Asn	Lys	Lys
	G	Ser	Ser	Arg	Arg
G	U	Val	Val	Val	Val
	C	Ala	Ala	Ala	Ala
	A	Asp	Asp	Glu	Glu
	G	Gly	Gly	Gly	Gly

A WORD ABOUT... The Human Genome

Every cell of a living organism generally contains the same DNA. This genetic blueprint is packed in tightly coiled bundles called *chromosomes.* Some of the DNA base sequences in chromosomes provide the code needed for a cell to synthesize a specific protein that performs a biological function. Such a sequence is called a *gene.* Other base sequences found in chromosomes simply separate genes from one another. These sequences of base pairs are called *intergenic DNA.* The entire assembly of genes and intergenic DNA for an organism is called the *cellular genome.*

About 10 years before the turn of the twenty-first century, a program of enormous magnitude was undertaken—sequencing the entire human genome! In 2001, the International Human Genome Sequencing Consortium* and scientists from Celera Genomics** independently published "first draft sequences" of the human genome. These reports provide a blueprint for 85% to 95% of the genome, and many of the gaps have since been filled in.

What has been learned thus far is well beyond the scope of this book, but here are a few interesting numbers. There are approximately 3.2 billion base pairs in the genome. If you were to compare your genome to that of another human, it would be identical 99.9% of the time. Thus, only a small number of base-pair differences are responsible for the great heterogeneity that appears in humans. Finally, the genome is now believed to contain between 26,000 and 40,000 genes!

Applications of human genome research to problems in the fields of medicine and biology are already being reported. For example, information from the genome has been used to find previously unknown enzymes that play roles in the development of asthma, neurological disorders, Alzheimer's disease, and other human maladies. Since the development of enzyme inhibitors can provide access to useful drugs (see "A Word About . . . Prostaglandins, Aspirin, and Pain," page 450 and "A Word About . . . Nucleic Acids and Viruses," pages 544–545), information from the genome project is already providing medicinal chemists with new "targets" for drug development. It is apparent that the sequencing of the human genome has long-range scientific and social consequences that will only become clear in the decades to follow.

See Problems 18.24, 18.25, and 18.29.

*See *Nature* **2001**, *409*, 860.
See *Science* **2001, *291*, 1304.

cell-free protein-synthesizing system containing all of the amino acids. He found a tremendous increase in the incorporation of phenylalanine in the resulting polypeptides. Since UUU is the only codon present in polyuridine, it must be a codon for phenylalanine. Similarly, polyadenosine led to the synthesis of polylysine, and polycytidine led to polyproline. Thus AAA≡Lys, and CCC≡Pro. Later, other synthetic polyribonucleotides with known repeating sequences were found to give other polypeptides with repeating amino acid sequences, and, in this way, the complete code was unraveled.

EXAMPLE 18.3

A polyribonucleotide was prepared from the tetranucleotide UAUC. When it was subjected to peptide-synthesizing conditions, the polypeptide (Tyr—Leu—Ser—Ile)$_n$ was obtained. What are the codons for these four amino acids?

Solution The polyribonucleotide must have the sequence

$$\underline{\text{U A U}}\,\underline{\text{C U A}}\,\underline{\text{U C U}}\,\underline{\text{A U C}}\cdots$$

If we divide the chain into codons, we get

$$\left.\begin{array}{l}\text{UAU—CUA—UCA—AUC}\cdots \\ \text{Tyr — Leu — Ser — Ile}\quad\cdots\end{array}\right\}\text{ this sequence repeats}$$

In this way, the meaning of four codons is disclosed.

PROBLEM 18.7 A polynucleotide made from the dinucleotide UA turned out to be (Tyr—Ile)$_n$. How does this outcome confirm the results of Example 18.3? What new information does this experiment give?

The genetic code is universal for all organisms on earth and has remained invariant through years of evolution. Consider what would happen if the "meaning" of a codon were changed. The result would be a change in the amino acid sequence of most proteins synthesized by that organism. Many of these changes would undoubtedly be disadvantageous. Hence there is a strong natural selection *against* changing the code. It was recently demonstrated, from a statistical analysis of *t*RNAs, that the genetic code cannot be older than 3.8 (±0.6) billion years and thus is not older than, but almost as old as, our planet.*

PROBLEM 18.8 Mutations (caused by radiation, cancer-producing agents, or other means) may replace one base with another or may add or delete a base. What would happen to the protein produced if the sequence UUU were mutated to UCU? If UCU were mutated to UCC? What advantage is there to the genetic code in having redundant codons?

Proteins are biosynthesized using *m*RNA as a template (Figure 18.7). Amino acids, each attached to its own unique *t*RNA, are brought up to the *m*RNA,

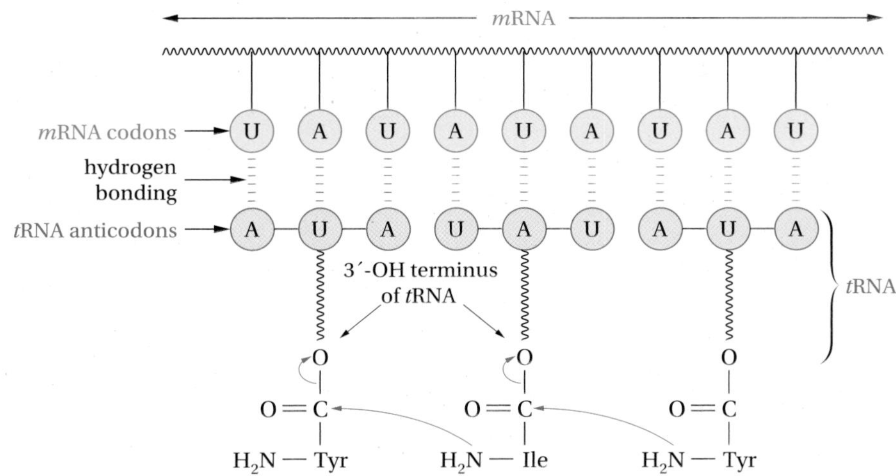

■ **Figure 18.7**
Schematic representation of protein biosynthesis. The amino group of one amino acid (Ile) displaces the carboxyl group of another amino acid (Tyr) from the terminal 3′-hydroxyl group of its *t*RNA (AUA). This continues down the *m*RNA template. The example would give the tripeptide unit Tyr—Ile—Tyr (see Problem 18.7).

*Manfred Eigen (Nobel Prize, 1967) and coworkers, *Science* **1989**, *244*, 673–679.

A WORD ABOUT... Nucleic Acids and Viruses

Viral infections—from influenza and the common cold to the more serious herpes infections and AIDS (acquired immune deficiency syndrome)—account for about 60% of illnesses, contrasted with about 15% for bacterial infections. Since the 1940s, chemists have developed all sorts of highly effective antibiotics (sulfa drugs, penicillins, tetracyclines, and others) that are effective against bacterial infections, but progress with antiviral agents has been slower and more difficult. Why is this so, and what hope do we have for combating viral infections?

The problem is one of selectivity. Any drug must *selectively* kill pathogens in the presence of other living cells. Fortunately, there are sufficient biochemical differences between the metabolisms of bacterial and of mammalian cells to allow selectivity; thus, safe antibiotics have been developed. Viruses present a more difficult problem because, during their replicative cycle, they become physically and functionally incorporated into host cells, and one must find biochemical features that selectively attack the virus without damaging the host.

Viruses are exceedingly small; they consist of a protein coat surrounding an inner core of nucleic acid. The core carries all the genetic information for their reproduction. Broadly, there are two classes of viruses, in which the nucleic acid is either DNA (herpes viruses) or RNA (flu virus and HIV, the human immunodeficiency virus responsible for AIDS). When viruses infect a host cell, they first attach to the membrane of the host cell, then penetrate the membrane and shed their protein coat, and the viral nucleic acid enters the host cell nucleus, where it replicates. The new nucleic acid then leaves the nucleus and combines with structural proteins to form new viruses, which are then expelled from the cell. The normal metabolic machinery of the cell is essentially hijacked and forced into manufacturing the viral components instead of its own. The host cell becomes, in a sense, a virus factory.

Any one of these steps (from attachment to the membrane to final ejection) is a target for interference by an antiviral agent. One trick, for example, might be to design a molecule very much like the DNA nucleosides, but sufficiently different for it to put a roadblock in the DNA synthesis scheme. One example of a successful antiviral agent based on this idea is **acyclovir**, 9-((2-hydroxy-ethoxy)methyl)guanine **(ACV)**, developed by Gertrude B. Elion* and coworkers at the Wellcome Research Laboratories (North Carolina). Notice that ACV mimics one of the natural nucleosides, 2′-deoxyguanosine, differing only (but importantly) in lacking part of the sugar portion of the molecule (shown in color).

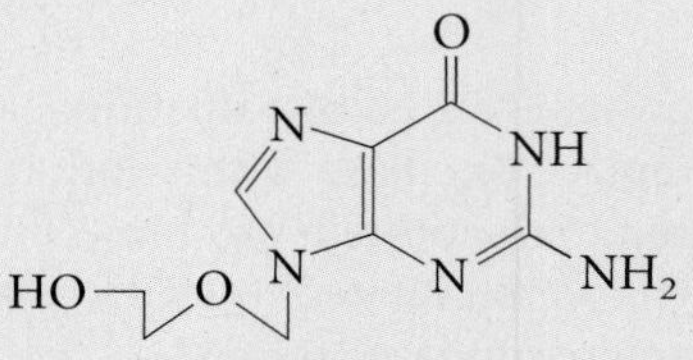

ACV

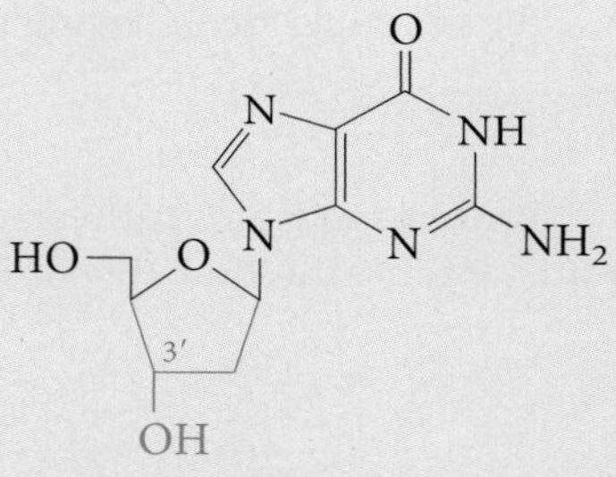

2′-deoxyguanosine

where the anticodon on the *t*RNA matches up, through hydrogen bonding, with the codon on the *m*RNA. Enzymes then link the amino acids, detach them from their *t*RNAs, and detach the *t*RNAs from the *m*RNA so that the process can be repeated.

Coordinated reactions that involve many types of molecules are required for protein biosynthesis. The molecules include *m*RNA, *t*RNA, scores of enzymes, the

ACV is effective against herpes simplex virus (HSV). It has been used clinically since the 1980s and has decreased suffering and saved lives. ACV is especially useful in treating genital herpes. It functions by selectively interfering with at least two enzyme-catalyzed processes essential for viral reproduction. Also, if it is incorporated into a DNA chain, ACV acts as a chain terminator because it lacks the 3′-hydroxyl group essential for attaching the next nucleotide unit.

AIDS, another virally transmitted disease, has evolved into an epidemic of worldwide proportions.** Several nucleoside analogs have been approved as drugs to treat HIV-infected individuals. *Zidovudine* (or *AZT,* 3′-azido-2′-deoxythymine), an analog of the DNA nucleoside 2′-deoxythymidine, has been in use the longest.

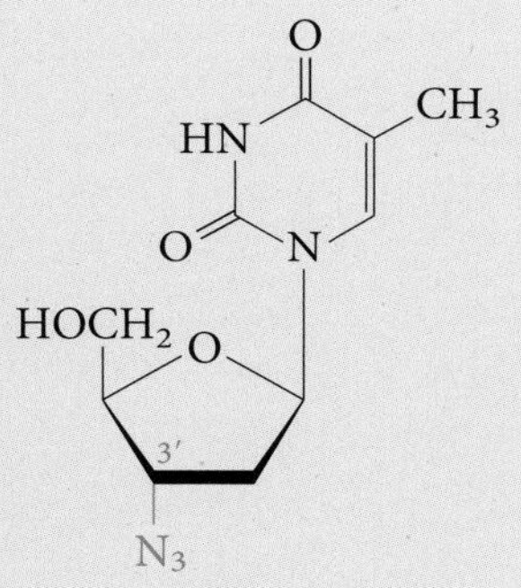

zidovudine (AZT)

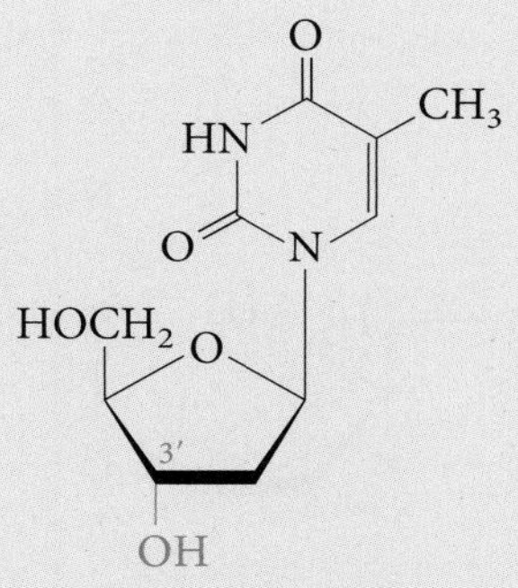

2′-deoxythymidine

Didanosine (*DDI,* 2′,3′-dideoxyinosine) and *zalcitabine* (*DDC,* 2′,3′-dideoxycytidine) are two other nucleoside analogs that are now used to treat the AIDS virus. Numerous other compounds are in clinical trials.

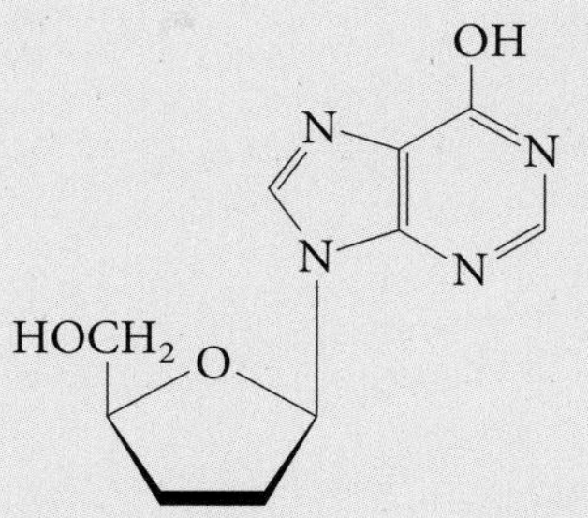

DDI

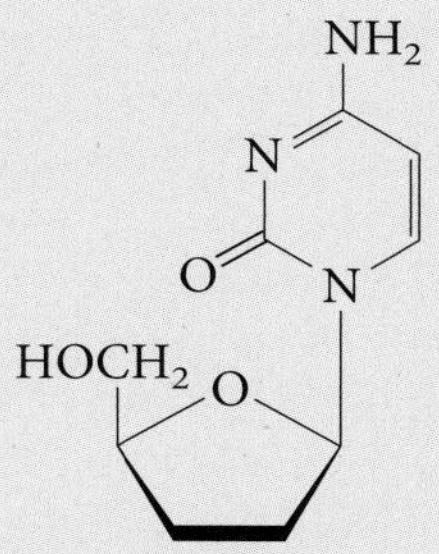

DDC

AZT and other nucleoside analogs interfere with reverse transcriptase, the enzyme HIV uses to transcribe its RNA genome into a DNA copy. The enzyme is fooled into incorporating the drug instead of the natural nucleotide into the growing DNA chain, but the growth then stops because there is no OH group (an azido group or hydrogen atom instead) at the 3′-position. But AZT is hardly an ideal drug, due to its toxicity to bone marrow (where blood cells are produced). Also, the AIDS virus becomes resistant to these drugs, so there is still a great need for new methods of therapy.

See Problems 18.35 and 18.37.

*Gertrude B. Elion, George H. Hitchings, and James W. Black shared the 1988 Nobel Prize in physiology and medicine.

For an overview of recent strategies for the prevention and treatment of AIDS, see *Scientific American* **1998, *279*(1), 81–105.

amino acids, phosphate, and many others. In spite of these requirements, this complex process happens with remarkable speed. It is estimated that a protein containing as many as 150 amino acid units can be assembled biosynthetically in less than a minute! When we compare this with solid-phase peptide synthesis (Sec. 17.12), efficient as it is for a laboratory procedure, we see that as chemists, we still have a long way to go if we are to emulate nature.

18.12 Other Biologically Important Nucleotides

The nucleotide structure is a part not only of nucleic acids, but also of several other biologically active substances. Some of the more important of these are described here.

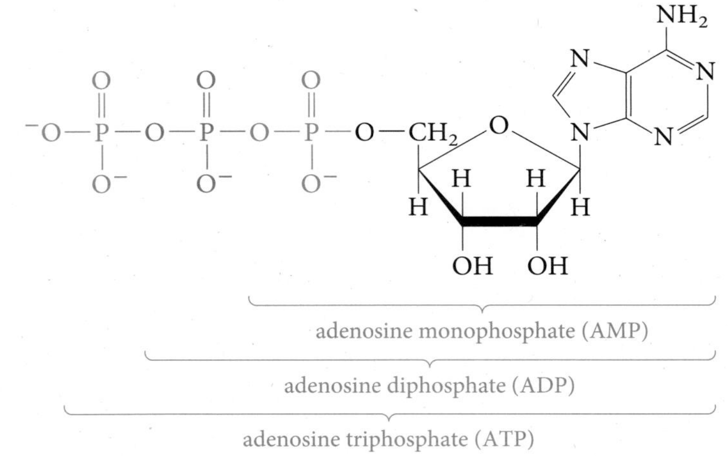

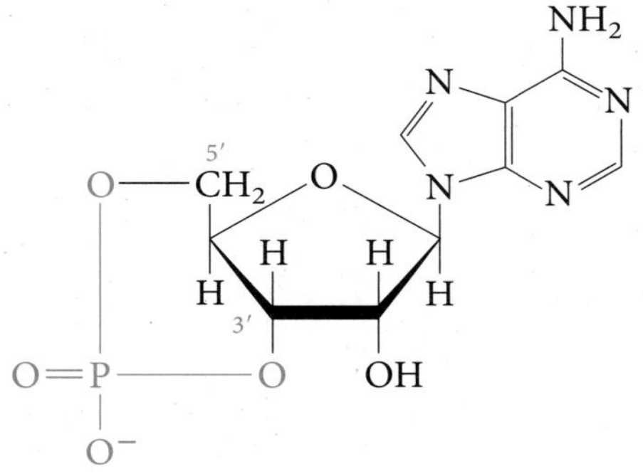

adenosine monophosphate (AMP)

adenosine diphosphate (ADP)

adenosine triphosphate (ATP)

adenosine 3′,5′-cyclic monophosphate (cyclic AMP; cAMP)

The nucleotide **adenosine**, in several phosphate forms, plays a key role in many biological processes.

Adenosine exists in several different phosphate forms. The 5′-*mono*phosphate, *di*phosphate, and *tri*phosphate, as well as the 3′,5′-cyclic monophosphate, are key intermediates in many biological processes.

Adenosine triphosphate (ATP) contains two phosphoric anhydride bonds, and considerable energy is released when ATP is hydrolyzed to adenosine diphosphate (ADP) and further to adenosine monophosphate (AMP). These reactions are used to provide energy for other biological reactions.

Cyclic AMP (cAMP) is a mediator of certain hormonal activity. When a hormone outside a cell interacts with a receptor site on the cell membrane, it may stimulate cAMP synthesis inside the cell. The cAMP in turn acts *within* the cell to regulate some biochemical process. In this way, a hormone does not have to penetrate a cell to exert its effect.

Nicotinamide adenine dinucleotide (NAD) and **flavin adenine dinucleotide (FAD)** are two coenzymes involved in many biological oxidation–reduction reactions.

Four important coenzymes contain nucleotides as part of their structures. We have already mentioned *coenzyme A* (for its structure, see page 312), which contains ADP as part of its structure. It is a biological acyl-transfer agent and plays a key role in fat metabolism. **Nicotinamide adenine dinucleotide (NAD)** is a coenzyme that dehydrogenates alcohols to aldehydes or ketones, or the reverse process: It reduces carbonyl groups to alcohols. It consists of two nucleotides linked by the 5′ hydroxyl group of each ribose unit.

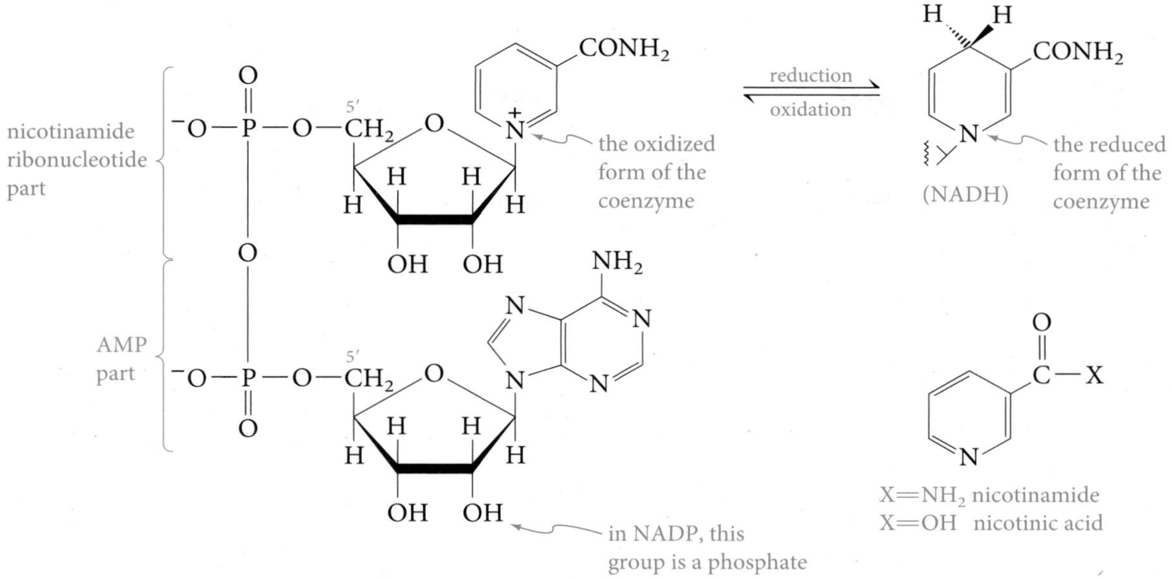

nicotinamide ribonucleotide part

AMP part

the oxidized form of the coenzyme

reduction

oxidation

(NADH)

the reduced form of the coenzyme

in NADP, this group is a phosphate

nicotinamide adenine dinucleotide (NAD)

X=NH_2 nicotinamide
X=OH nicotinic acid

When NADP (a phosphate ester of NAD) oxidizes an alcohol to a carbonyl compound, the pyridine ring in the nicotinamide part of the coenzyme is reduced to a dihydropyridine, giving NADPH. The reverse process occurs when NADPH reduces a carbonyl compound to an alcohol. Nicotinic acid is a B vitamin needed for synthesis of this coenzyme. Its deficiency causes the chronic disease pellagra.

Flavin adenine dinucleotide (FAD) is a yellow coenzyme involved in many biological oxidation–reduction reactions. It consists of a riboflavin part (vitamin B_2) connected to ADP. The reduced form has two hydrogens attached to the riboflavin part.

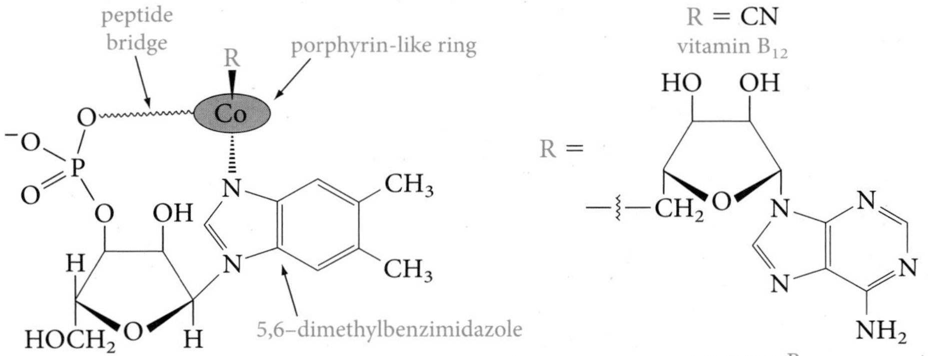

peptide bridge

porphyrin-like ring

5,6-dimethylbenzimidazole

R = CN
vitamin B_{12}

coenzyme B_{12}

Vitamin B_{12} (cyanocobalamine), which is essential for the maturation and development of red blood cells, is an incredibly complex molecule that includes a nucleotide as part of its structure (Figure 18.8). The related **coenzyme B_{12}** contains a second nucleotide unit. Both of these molecules have a central cobalt atom surrounded by a macrocyclic molecule containing four nitrogens, similar to a porphyrin (see "A Word About . . . Porphyrins," page 402). But the cobalt has two additional ligands attached to it, above and below the mean plane of the nitrogen-containing rings. One of these ligands is a ribonucleotide of the unusual base, **5,6-dimethylbenzimidazole**. The other ligand is a cyanide group in the vitamin and a 5-deoxyadenosyl group in the coenzyme. In each case, there is a direct carbon–cobalt bond. The reactions catalyzed by coenzyme B_{12} usually involve replacement of the Co—R group by a Co—H group.

Vitamin B_{12}, essential for the maturation and development of red blood cells, contains a nucleotide. **Coenzyme B_{12}** contains a ribonucleotide with the base **5,6-dimethylbenzimidazole.**

Vitamin B_{12}, which is produced by certain microorganisms, cannot be synthesized by humans and must be ingested. Only minute amounts are required, but pernicious anemia can result from its deficiency.

Vitamin B_{12}, with its remarkable array of functionality and chirality, is one of the most complex molecules ever to have been created in an organic laboratory. Its synthesis was completed in 1973 by Robert Woodward* and Albert Eschenmoser and their students.

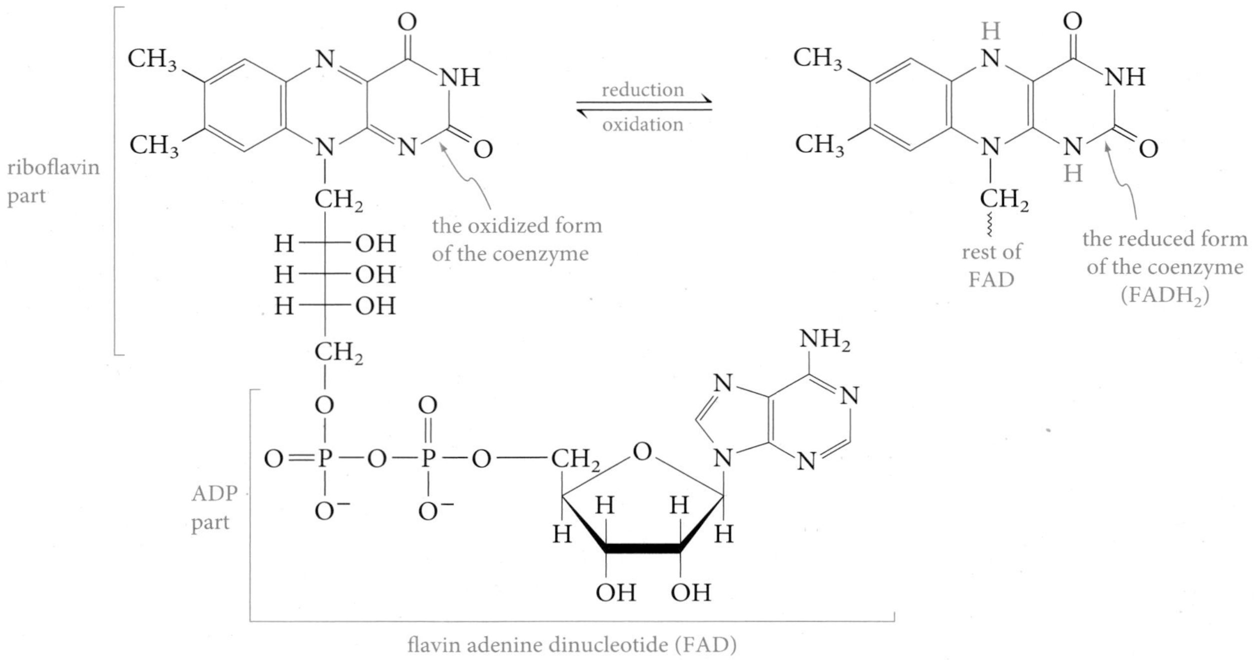

riboflavin part

reduction

oxidation

the oxidized form of the coenzyme

rest of FAD

the reduced form of the coenzyme ($FADH_2$)

ADP part

flavin adenine dinucleotide (FAD)

Figure 18.8
Schematic representation of vitamin B_{12} and coenzyme B_{12}.

*Robert Burns Woodward (1917–1979) received the Nobel Prize in chemistry in 1965 for his many contributions to the "art of organic synthesis." He is regarded by many organic chemists to have been perhaps the greatest practitioner of this art.

REACTION SUMMARY

1. Hydrolysis of Nucleic Acids (Sec. 18.1)

$$\text{DNA} \xrightarrow[\text{enzyme}]{H_2O} \text{nucleotides} \xrightarrow[HO^-]{H_2O} \text{nucleosides} \xrightarrow[H^+]{H_2O} \text{heterocyclic bases} + \text{2-deoxyribose}$$

The behavior of RNA is identical to DNA except ribose is produced instead of 2-deoxyribose.

2. Hydrolysis of Nucleotides (Sec. 18.4)

$$\text{nucleotide} \xrightarrow[HO^-]{H_2O} \text{nucleoside} + {}^-O-P(=O)(O^-)-OH \text{ (inorganic phosphate)}$$

nucleotide nucleoside inorganic phosphate

ADDITIONAL PROBLEMS

OWL Interactive versions of these problems are assignable in OWL.

Nucleosides and Nucleotides: Nomenclature and Structure

18.9 Write the structural formula for an example of each of the following:

a. a pyrimidine base **b.** a purine base **c.** a nucleoside **d.** a nucleotide

18.10 Examine the structures of adenine and guanine (Figure 18.1). Do you expect their rings to be planar or puckered? Explain. What about the pyrimidine bases, cytosine and thymine?

18.11 Draw the structure of each of the following nucleosides:

a. guanosine (from β-D-ribose and guanine)
b. deoxycytidine (from β-2-deoxy-D-ribose and cytosine)
c. deoxyadenosine (from β-2-deoxy-D-ribose and adenine)
d. uridine (from β-D-ribose and uracil)

18.12 Write an equation for the complete hydrolysis of AMP to its component parts.

18.13 Using Table 18.1 as a guide, write the structures of the following nucleotides:

a. dAMP **b.** guanosine-5′-monophosphate **c.** 2′-deoxycytidine-5′-monophosphate

DNA and RNA Structure

18.14 Draw the structures of the following DNA-derived dinucleotides:

a. A—T **b.** G—T **c.** C—A

18.15 Draw the structures of the following RNA-derived dinucleotides:

a. A—U b. G—U c. A—C

18.16 Consider the DNA-derived tetranucleotide G—A—C—T. What products will be obtained when this tetranucleotide is hydrolyzed by each of the following?

a. base b. base, followed by acid

18.17 Draw the structures of the following RNA components:

a. UCC b. CAA c. ACG

18.18 Draw a structure showing the hydrogen bonding between uracil and adenine, and compare it with that for thymine and adenine (page 529).

18.19 A segment of DNA contains the following base sequence:

5′ A—T—G—G—T—C—T—A—G 3′

Draw the sequence of the complementary segment, and label its 3′ and 5′ ends.

18.20 For the DNA segment in Problem 18.19, write the *m*RNA complement, and label its 3′ and 5′ ends.

18.21 Consider the following *m*RNA sequence:

5′ A—G—C—U—C—G—U—A—C 3′

Draw the segment of DNA double helix from which this sequence was derived, using the schematic method shown at the right of Figure 18.4. Be sure to show the 5′ and 3′ ends of each strand.

18.22 Explain how the double-helical structure of DNA is consistent with Chargaff's analyses for the purine and pyrimidine content of DNA samples from various sources.

The Genetic Code

18.23 The codon CAC corresponds to the amino acid histidine (His). How will this codon appear in the DNA strand from which it was transcribed? In the complement of that strand? Be sure to label the 5′ and 3′ directions.

18.24 Provide the names of the amino acids that correspond to the three-base sequences (codons) listed in Problem 18.17.

18.25 Consider Table 18.2. Will any changes occur in the resultant biosynthesized protein by a purine → purine mutation in the third base of a codon? In a pyrimidine → pyrimidine mutation in the third base of a codon? If so, describe the change.

18.26 From Table 18.2, are mutations in the first or second base of a codon more or less serious than mutations at the third base?

18.27 A *m*RNA strand has the sequence

–$^{5'}$CCAUCCGGCAUACCAAAUUACUAAACUAGC$^{3'}$–

What peptide would be produced? (Don't forget the start and stop codons!)

18.28 What would happen if the first U in the sequence in Problem 18.27 were deleted?

18.29 What peptide would be synthesized from the following DNA sequence?

$^{5'}$TTACCGACTGGTCACTCCCAT$^{3'}$

18.30 There are a number of DNA databases available worldwide for comparing the DNA of individuals. The largest database is maintained by the U.S. Federal Bureau of Investigation and is called the Combined DNA Index System (CODIS). The United Kingdom currently has the second largest DNA database for individuals. These databases have grown rapidly as DNA is taken from all felons and, in some states or countries, from anyone who is arrested or is a suspect in a crime. How many records are currently in these two large databases?

Biologically Active Nucleosides and Nucleotides

18.31 What products would you expect to obtain from the complete hydrolysis of NAD? See page 547 for its structure.

18.32 UDP-glucose is an activated form of glucose involved in the synthesis of glycogen. It is a nucleotide in which α-D-glucose is esterified at C-1 by the terminal phosphate of uridine diphosphate (UDP). From this description, draw the structure of UDP-glucose.

18.33 Caffeine, the alkaloid stimulant in coffee and tea, is a purine with the following formula:

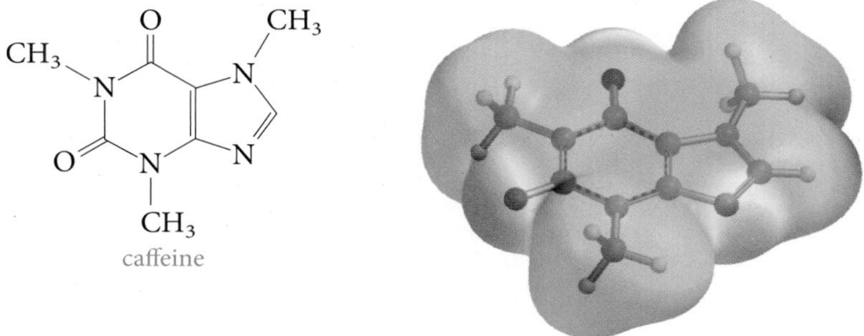

caffeine

Compare its formula with those of adenine and guanine. Do you expect caffeine to form *N*-glycosides with sugars such as 2-deoxy-D-ribose? Use the electrostatic potential map of caffeine to determine which nitrogen is most basic. Why is this in accord with expectations?

18.34 The plant *Theobroma cacao* is the source of chocolate, and theobromine is the primary purine derived from chocolate. Another related alkaloid is theophylline, which is an orally administered asthma drug.

theobromine

theophylline

Note that theobromine, theophylline, and caffeine differ by the location of two methyl groups—caffeine has both methyl groups, while theobromine and theophylline have a single methyl group on only one of the two nitrogens. Such minor changes in the chemical structure can provide significantly different biological responses.

a. What is the most basic nitrogen in theobromine and theophylline?
b. What is the most acidic hydrogen in theobromine and theophylline?
c. Do you expect theobromine and theophylline to form *N*-glycosides with sugars such as 2-deoxy-D-ribose?

18.35 *5-Fluorouracil-2-deoxyriboside* (FUdR) is used in medicine as an antiviral and antitumor agent. From its name, draw its structure. How do you think FUdR performs its function?

18.36 *Psicofuranine* is a nucleoside used in medicine as an antibiotic and antitumor agent. Its structure differs from that of adenosine only in having —CH_2OH attached with α geometry at C-1′. Draw its structure.

18.37 Suggest the origin of the therapeutic activity for *Zidovudine* (AZT), a drug that is used to treat HIV-infected individuals, as discussed in the "A Word About . . . Nucleic Acids and Viruses" on page 545. (*Hint:* Viruses are rapidly replicating their genetic information.) Using AZT as an example, what other compounds could you suggest as potentially viable therapeutics against AIDS?

18.38 The glycosylated nucleoside monosulfate shown below was recently isolated from the venom of the hobo spider, *Tegenaria agrestis.* Which purine base is part of this toxin? What is the stereochemistry of the linkage of the base to the ribose part of the toxin? Which carbon in ribose is sulfated? Is the 6-deoxyhexose portion of the toxin a D- or an L-hexose?

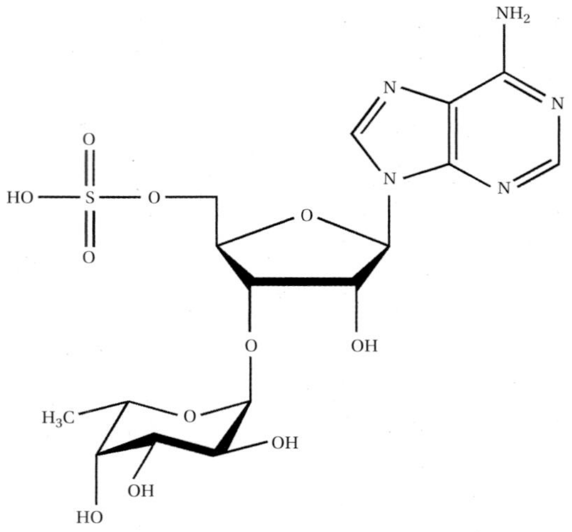

Appendix

Table A — Bond Energies for the Dissociation of Selected Bonds in the Reaction $A{-}X \rightarrow A\cdot + X\cdot$ (in kcal/mol)

I. Single bonds	Bond energies (kcal/mol)								
A—X	***X = H***	***F***	***Cl***	***Br***	***I***	***OH***	***NH_2***	***CH_3***	***CN***
CH_3—X	105	108	84	70	57	92	85	90	122
CH_3CH_2—X	100	108	80	68	53	94	84	88	
$(CH_3)_2CH$—X	96	107	81	68	54	94	84	86	
$(CH_3)_3C$—X	96		82	68	51	93	82	84	
H—X	104	136	103	88	71	119	107	105	124
X—X	104	38	59	46	36			90	
Ph—X	111	126	96	81	65	111	102	101	
$CH_3C(O)$—X	86	119	81	67	50	106	96	81	
$H_2C{=}CH$—X	106								
$HC{\equiv}C$—X	132								

II. Multiple bonds	Bond energies (kcal/mol)
$H_2C{=}CH_2$	163
$HC{\equiv}CH$	230
$H_2C{=}NH$	154
$HC{\equiv}N$	224
$H_2C{=}O$	175
$C{\equiv}O$	257

Table B — Bond Lengths of Selected Bonds (in angstroms, Å)

I. Single bonds				II. Double bonds		III. Triple bonds	
Bond	***Length (Å)***	***Bond***	***Length (Å)***	***Bond***	***Length (Å)***	***Bond***	***Length (Å)***
H—H	0.74	H—C=	1.08	C=C	1.33	C≡C	1.20
H—F	0.92	H—Ph	1.08	C=O	1.21	C≡N	1.16
H—Cl	1.27	H—C≡	1.06			C≡O	1.13
H—Br	1.41	C—C	1.54				
H—I	1.61	C—N	1.47				
H—OH	0.96	C—O	1.43				
H—NH_2	1.01	C—F	1.38				
H—CH_3	1.09	C—Cl	1.77				
F—F	1.42	C—Br	1.94				
Cl—Cl	1.98	C—I	2.21				
Br—Br	2.29						
I—I	2.66						

Table C Typical Acidities of Organic Functional Groups

Name and Example*	pK_a	Conjugate Base
Hydrochloric acid, HCl	−7	Cl^-
Sulfuric acid, H_2SO_4	−3	HSO_4^-
Sulfonic acid	−2–0	
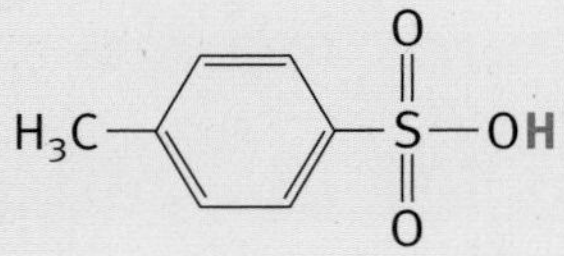 $H_3C-C_6H_4-S(=O)_2-OH$	−1	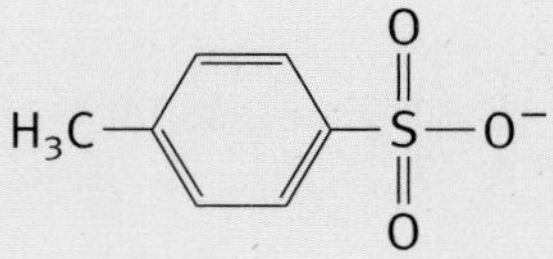 $H_3C-C_6H_4-S(=O)_2-O^-$
Carboxylic acid	3–5	
$CH_3-C(=O)-OH$	4.74	$CH_3-C(=O)-O^-$
Arylammonium ion	4–5	
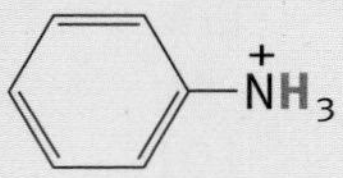 $C_6H_5-\overset{+}{N}H_3$	4.6	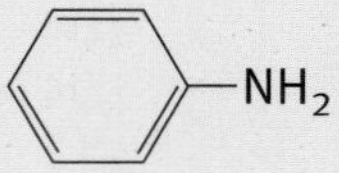 $C_6H_5-NH_2$
Ammonium ion, $\overset{+}{N}H_4$	9.3	NH_3
Phenol	9–10	
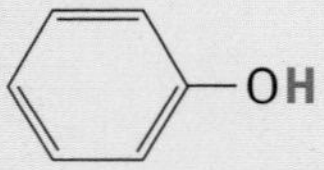 C_6H_5-OH	10	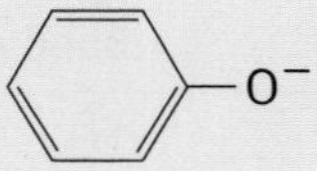 $C_6H_5-O^-$
β-diketone	9–10	
$CH_3-C(=O)-CH_2-C(=O)-CH_3$	9	$CH_3-C(=O)-\bar{C}H-C(=O)-CH_3$
Thiol	8–12	
CH_3CH_2SH	10.6	$CH_3CH_2S^-$
β-ketoester	10–11	
$CH_3-C(=O)-CH_2-C(=O)-OCH_2CH_3$	10.7	$CH_3-C(=O)-\bar{C}H-C(=O)-OCH_2CH_3$
Alkylammonium ion	10–12	
$CH_3CH_2\overset{+}{N}H_3$	10.7	$CH_3CH_2NH_2$
Water, H_2O	15.7	HO^-

Strong Acid ↑ … Weak Acid; Weak Base … Strong Base ↓

Table C Typical Acidities of Organic Functional Groups (continued)

Name and Example*	pK_a	Conjugate Base
Alcohol	15–19	
CH_3CH_2OH	15.9	$CH_3CH_2O^-$
Amide	15–19	
$CH_3-C(=O)-NH_2$	15	$CH_3-C(=O)-\bar{N}H$
Aldehyde, ketone	17–20	
$CH_3-C(=O)-CH_3$	19	$CH_3-C(=O)-\bar{C}H_2$
Ester	23–25	
$CH_3-C(=O)-OCH_2CH_3$	24.5	$\bar{C}H_2-C(=O)-OCH_2CH_3$
Alkyne	23–25	
$H-C\equiv C-H$	24	$H-C\equiv C^-$
Ammonia, NH_3	33	$\bar{N}H_2$
Hydrogen, H_2	35	H^-
Alkylamine	~40	
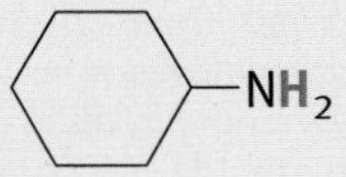 cyclohexyl–NH_2	42	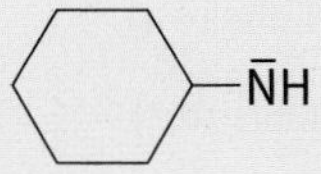 cyclohexyl–$\bar{N}H$
Alkene	~45	
$H_2C=CH_2$	44	$H_2C=\bar{C}H$
Aromatic hydrocarbon	41–43	
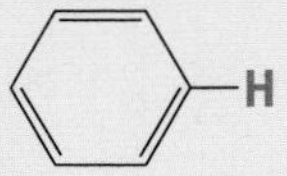 phenyl–H	43	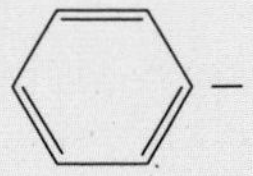 phenyl$^-$
Alkane	50–60	
CH_4	50	$\bar{C}H_3$

Strong Acid ↑ … Weak Acid; Weak Base … Strong Base ↓

*Some inorganic acids are included for comparison.

Index/Glossary

Key terms that appear in **boldface** are followed by their definitions in ***italics***. Page numbers followed by ***n*** refer to notes.